Indian Institute of Metals Series

About the Book Series:

The study of metallurgy and materials science is vital for developing advanced materials for diverse applications. In the last decade, the progress in this field has been rapid and extensive, giving us a new array of materials, with a wide range of applications, and a variety of possibilities for processing and characterizing the materials. In order to make this growing volume of knowledge available, an initiative to publish a series of books in Metallurgy and Materials Science was taken during the Diamond Jubilee year of the Indian Institute of Metals (IIM) in the year 2006. Ten years later the series is now published in partnership with Springer.

This book series publishes different categories of publications: textbooks to satisfy the requirements of students and beginners in the field, monographs on select topics by experts in the field, professional books to cater to the needs of practicing engineers, and proceedings of select international conferences organized by IIM after mandatory peer review. The series publishes across all areas of materials sciences and metallurgy. An eminent panel of international and national experts acts as the advisory body in overseeing the selection of topics, important areas to be covered, and the selection of contributing authors.

More information about this series at https://link.springer.com/bookseries/15453

A. K. Tyagi · Raghumani S. Ningthoujam
Editors

Handbook on Synthesis Strategies for Advanced Materials

Volume-I: Techniques and Fundamentals

Editors
A. K. Tyagi
Chemistry Division
Bhabha Atomic Research Centre
Mumbai, Maharashtra, India

Homi Bhabha National Institute
Mumbai, Maharashtra, India

Raghumani S. Ningthoujam
Chemistry Division
Bhabha Atomic Research Centre
Mumbai, Maharashtra, India

Homi Bhabha National Institute
Mumbai, Maharashtra, India

ISSN 2509-6400 ISSN 2509-6419 (electronic)
Indian Institute of Metals Series
ISBN 978-981-16-1809-3 ISBN 978-981-16-1807-9 (eBook)
https://doi.org/10.1007/978-981-16-1807-9

This Springer imprint is published by the registered company Springer Nature Singapore Pte Ltd.
The registered company address is: 152 Beach Road, #21-01/04 Gateway East, Singapore 189721, Singapore

Series Editor's Preface

The Indian Institute of Metals Series is an institutional partnership series focusing on metallurgy and materials science and engineering.

About the Indian Institute of Metals

The Indian Institute of Metals (IIM) is a premier professional body (since 1947) representing an eminent and dynamic group of metallurgists and materials scientists and engineers from R&D institutions, academia, and industry, mostly from India. It is a registered professional institute with the primary objective of promoting and advancing the study and practice of the science and technology of metals, alloys, and novel materials. The institute is actively engaged in promoting academia–research and institute–industry interactions.

Genesis and History of the Series

The study of metallurgy and materials science and engineering is vital for developing advanced materials for diverse applications. In the last decade, the progress in this field has been rapid and extensive, giving us a new array of materials, with a wide range of applications and a variety of possibilities for processing and characterizing the materials. In order to make this growing volume of knowledge available, an initiative to publish a series of books in metallurgy and materials science and engineering was taken during the Diamond Jubilee year of the Indian Institute of Metals (IIM) in the year 2006. IIM entered into a partnership with Universities Press, Hyderabad, and, as part of the IIM book series, 11 books were published, and a number of these have been co-published by CRC Press, USA. The books were authored by eminent professionals in academia, industry, and R&D with outstanding background in their respective domains, thus generating unique resources of validated expertise of interest in metallurgy. The international character of the authors' and editors has enabled the books to command national and global readership. This book series includes different categories of publications: textbooks to satisfy the requirements of undergraduates and beginners in the field, monographs on selected topics by experts in the field, and proceedings of selected international conferences organized by IIM, after mandatory peer review. An eminent panel of international and national experts constitutes the advisory body in overseeing the selection of topics, important areas to be covered, in the books and the selection of contributing authors.

Current Series Information

To increase the readership and to ensure wide dissemination among global readers, this new chapter of the series has been initiated with Springer in the year 2016. The goal is to continue publishing high-value content on metallurgy and materials science and engineering, focusing on current trends and applications. So far, four important books on state of the art in metallurgy and materials science and engineering have been published and, during this year, three more books are released during IIM-ATM 2021. Readers who are interested in writing books for the Series may contact the Series Editor-in-Chief, Dr. U. Kamachi Mudali, Former President of IIM and Vice Chancellor of VIT Bhopal University at ukmudali1@gmail.com, vc@vitbhopal.ac.in or the Springer Editorial Director, Ms. Swati Meherishi at swati.meherishi@springer.com.

About the Three Volumes of Handbook on Synthesis Strategies for Advanced Materials

The Handbook on "Synthesis Strategies for Advanced Materials" is aimed to provide information on (i) Variety of synthetic methods to prepare advanced materials (stable and metastable hitherto unknown materials, chemically and crystallographically designed materials and assemblies) and their structure, micro-structure, and morphology; and (ii) Functional properties like soft to hard, insulators to superconductors, crystalline to amorphous like glass or polymeric, nano- to thin films to bulk single crystals. Keeping in mind the interests of students and young researchers, and senior faculty members, the basic concepts of synthesis, processing and materials aspects, and their recent developments are covered in three volumes.

The Editors Dr. A.K. Tyagi, Director, Chemistry Group, and Dr. S.R. Ningthoujam, Scientific Officer-F, Chemistry Group, from Bhabha Atomic Research Centre have meticulously edited the three volumes with 20 each chapters for Vols. I and II, and 18 chapters for Vol. III. These chapters have been prepared by the editors as well as well-experienced authors from academia, R&D, and industry. This handbook will be a treasure for those who are interested in learning everything about advanced materials and pursue a career and study in the area of advanced materials. The editors and authors are gratefully acknowledged for their excellent chapters covering wide range of information on the subject matter.

Dr. U. Kamachi Mudali
Editor-in-Chief
Series in Metallurgy and Materials Engineering

Preface

The ever-developing human civilization thrives on materials which may be of technological, health, environmental or geological relevance. The development of materials has, thus, been a constantly evolving process both in nature and by human efforts. Over time immemorial, a continuous evolution of materials for the fulfillment of the needs of healthy living or advanced lifestyle has been witnessed and that makes the twenty-first century a century of materials. This is recognized by the surge in advanced materials in engineering, electronics and communications as well as in healthcare, medicine and societal sectors. The fascinating and ever-growing world of materials extends from soft materials to super-hard materials, insulators to superconductors, extended solids to molecular solids, self-assembled materials, catalysts, materials with tailored thermal expansion, composites and hybrid materials, materials with multi-functionality, ceramics and glasses, metals–alloys–intermetallics, drugs and drugs delivery systems, polymers, biomaterials, nuclear materials, optical materials, fast ionic conductors, soft and hard magnets, etc. Still the quest of humanity for developing better and more efficient materials remains never-ending. The development of materials depends on the ability to synthesize them or to find a more cost- and energy-efficient synthesis methodology or design newer materials with appropriate constituents and functionalities to make them usable. Thus, the synthesis methods play a pivotal role in the materials research. Although the synthesis or synthetic materials chemistry originated just after the Stone Age, the understanding of chemistry and physics of materials with the progress of time only could lead to the discovery of newer materials as well as the targeted materials for desired purposes. This, in turn, resulted in the development of state-of-the-art synthesis procedures. Further, new functional materials are also being designed by the interplay of synthesis methodologies, crystallographic structures, morphologies and dimensionality for desired functional properties. Many a time, thermodynamics and kinetic parameters are controlled to overcome the barriers to achieve the desired materials. Thus, the methodologies for the synthesis of materials became multi-disciplinary which include the approaches from chemists, biologists, physicists, metallurgists and engineers. This has been witnessed as the development of several unconventional synthetic routes that involve parameters such as extremely high temperature, high pressure, radiation, mechanical attrition and unusually reactive intermediates. Some non-traditional synthesis routes have also been developed that follow a gentle chemical reaction favoring an intermediate or alternate pathway to bypass hindrance to reach the targeted material or utilizing the memory of the materials to introduce functionalities. The unconventional synthesis methodologies play important roles in the direction of many new and metastable materials which otherwise were not possible to prepare. Similarly, the multi-functional materials, i.e., the materials which can perform two or more synergistic or antagonistic functionalities, are being achieved by judicious adoption of synthesis methods. In addition, varieties of soft chemical methods have emerged that play important roles in the field of functional materials, in particular medicine and healthcare products, to design materials for desired technological applications. Thus, the material synthesis assumes an unprecedented role in this endeavor and remains a challenge as well as an opportunity to chemists and materials scientists. The synthesis methods and their scopes have been discussed in varieties of monographs as well as compilations and proceedings from time to time. Usage for various synthetic methods for the preparation of newer and exotic materials as well as recent modifications and their potentials as handy information is essentially a need for researchers in today's times and that has been achieved in this present compilation "Handbook on Synthesis Strategies for Advanced Materials," Volumes I, II and III.

This handbook series on "Synthesis Strategies for Advanced Materials" is aimed to provide information on varieties of synthetic methods being adopted by researchers to prepare different kinds of advanced materials covering from the viewpoints of structure, microstructure and morphology of materials, stable and metastable hitherto unknown materials, chemically and crystallographically designed materials and assemblies, as well as from the viewpoints of functional properties like soft to hard, insulators to superconductors, crystalline to amorphous like glass or polymeric, nano- to thin films to bulk single crystal. These have been achieved by adoption, alteration or judicious selection of synthesis methods. Keeping in mind the interests of students and young researchers, and senior faculty members, the basic concepts of synthesis, processing and materials aspects and their recent developments are covered in three volumes, namely Volume I: *Techniques and Fundamentals*, Volume II: *Processing and Functionalization of Materials* and Volume III: *Materials Specific Synthesis Strategies*. Each volume is made independent by taking care of minimal overlap of the topics. Volume I is primarily focused on the principles and procedures of various synthesis methods. The basic principles and scope/limitations of various synthetic methods, like solid-state reaction to gentle molecular aggregation methods and chimie douce, synthesis under high temperature, hot-injection, method, polyol method, metal-organic frameworks, electrochemical method, mechanochemistry, hydro/solvothermal reaction, high-pressure and high-temperature reactions, arc melting, induction heating, melt-quench method, ion exchange process, microwave and visible to gamma radiations, green methods of synthesis, thermolysis, bio-inspired synthesis, etc., are discussed along with the inputs from authors' hands-on experience and expertise. In Volume II, various processing methodologies for the preparation of various types of functional materials or functionalization of materials by chemical, structural or microstructural alterations are presented. This volume covers processing of nanomaterials, porous or sintered materials, composite materials, low dimensional like 1D to 2D materials, thin films, single crystals, template method, self-assembly, biomaterials, inkjet printing, 3D printing, size and shape engineering, etc., in a lucid manner. Volume III is focused on the synthesis aspects of materials like hybrid inorganic–organic, metal oxide frameworks, intermetallics, hydrides, borides, carbides, nitrides, phosphides, silicide, selenides, fluorides, various biomaterials, materials for sensors and detectors, optical materials, carbon-based materials, colloids, noble gas compounds, lithium-based ceramics, materials with unusual oxidation state, organo-selenium and platinum compounds, silicon-based materials and lithium-based ceramics. The evolution and state-of-the-art synthesis methods for practical requirements as well as new concepts with the most recent literatures dealing with their synthesis are presented in this volume. These volumes are expected to serve as handy guides for synthesis and processing of advanced materials of wide range and category.

The editors are immensely thankful to all the authors for their rich contributions toward this book. Although due efforts have been taken to make the book as error-free as possible, some may have crept in as unnoticed. We shall be thankful to the readers for bringing such unintentional errors to our notice. Finally, we sincerely hope that our efforts will be of use to both new and experienced researchers in the field.

Mumbai, India
July 2021

A. K. Tyagi
Raghumani S. Ningthoujam

Contents

About the Editors

Dr. A. K. Tyagi obtained his M.Sc. (Chemistry) degree in 1985 from Meerut University, Meerut, India and joined 29th batch of BARC Training School, Mumbai in the same year. After completing one year orientation course, he joined Chemistry Division, Bhabha Atomic Research Centre (BARC), Mumbai in 1986. Presently, he is Director, Chemistry Group, BARC, Mumbai, and a Senior Professor of Chemistry at Homi Bhabha National Institute (HBNI), Mumbai. His research interests are in the field chemistry of materials, which includes functional materials, nanomaterials, nuclear materials, energy materials, metastable materials, hybrid materials and structure-property correlation. He has published more than 600 papers in journals, several books and has supervised 30 Ph.D. students.

He was awarded Ph.D. by Mumbai University, Mumbai in 1992. He did postdoctoral research at Max-Planck Institute for Solid State Research (MPI-FKF), Stuttgart, Germany during 1995–1996 on a Max-Planck Fellowship. Subsequently, he regularly visited MPI-FKF, Stuttgart as a visiting scientist. In addition, he has also visited Institute of Superior Technology, Portugal; Institute for Chemical Process and Environmental Technology, Ottawa, Canada; Dalhousie University, Halifax, Canada; Moscow State University, Moscow, Russia; Institute for Materials, Nantes, France; University of Malay, Malaysia; National Institute of Materials Science, Tsukuba, Japan; National University of Singapore, Singapore; Royal Institute of Technology, Stockholm, Sweden; Rice University, Houston, USA; Shanghai, China; University of

Valencia, Valencia, Spain; Weizmann Institute of Science, Israel; University of Queensland, Brisbane, Australia; US-Air Force Research Lab, Dayton, USA, Institute for Studies of Nanostructured Materials, Palermo, Italy and iThemba Labs, Cape Town, South Africa.

In recognition of his significant contributions to the field of chemistry of materials, he has been conferred with many prestigious awards, such as Dr. Lakshmi award by the Indian Association of Solid State Chemists and Allied Scientists (2001); Rheometric Scientific-Indian Thermal Analysis Society Award (2002); Gold Medal of Indian Nuclear Society (2003); Materials Research Society of India's Medal (2005); Chemical Research Society of India's Bronze Medal (2006); DAE-Homi Bhabha Science and Technology Award (2006); IANCAS-Dr. Tarun Datta Memorial Award (2007); Rajib Goyal Prize in Chemical Sciences (2007); RD Desai Memorial Award from Indian Chemical Society (2008) and DAE-SRC Outstanding Research Investigator Award (2010); CRSI-Prof. CNR Rao National Prize in Chemical Sciences (2012); ISCB Award for Excellence in Chemical Sciences (2013); MRSI-ICSC Materials Science Senior Award (2014); Coastal Chemical Research Society's Award (2014); Platinum Jubilee Lecture Award in Materials Science from Indian Science Congress Association (2015); Metallurgist of the Year Award (2017), from Ministry of Steel, Government of India; Chemical Research Society of India's Silver Medal (2018); Materials Research Society of India's Prof. C. N. R. Rao Prize in Advanced Materials (2018); JNCASR's National Prize in Solid State and Materials Chemistry (2018) and Acharya PC Ray Memorial Award from Indian Science Congress Association (2020).

He is an elected Fellow of the Indian Academy of Sciences (FASc); National Academy of Sciences, India (FNASc); Maharashtra Academy of Sciences (FMASc); Royal Society of Chemistry, UK (FRSC) and Asia Pacific Academy of Materials.

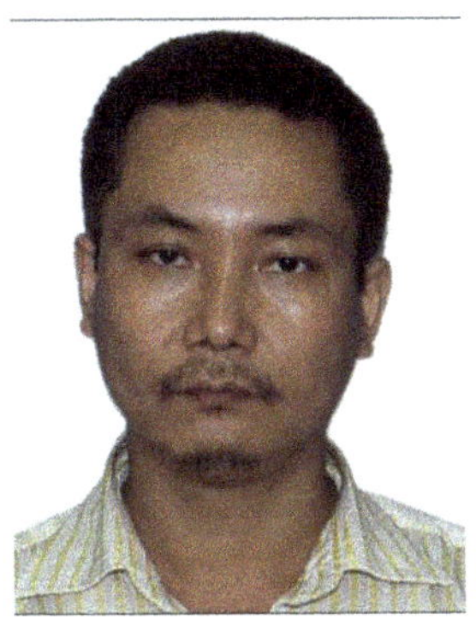

Dr. Raghumani S. Ningthoujam obtained M.Sc. in Chemistry from Manipur University, Imphal, Manipur, India in 1994. He was awarded Ph.D. in Chemistry from IIT Kanpur in area of superconductivity, electron transport and magnetic properties of nanostructured transition metal nitrides in 2004. He joined BARC (Bhabha Atomic Research Centre), Mumbai as Scientific Officer (D) in 2006 after completion of Dr. K. S. Krishnan Research Associate Fellowship. Presently, he is working in area of Luminescent and Magnetic Nanomaterials and their applications in sensors, imaging, diagnosis and therapy. He did Post-Doctoral Fellowship at University of Victoria, Canada in the area of quantum dots. Presently, he is Scientific Officer (F) at Chemistry Division, BARC, Mumbai. He is recognized for Guideship of Ph.D. at Homi Bhabha National Institute (HBNI), Mumbai and Mumbai University, Mumbai. Many research scholars and students finished their projects under his guidance. He has published about 150 Papers in the refereed journals, five review articles and four book chapters. In recognition of his significant contributions to the chemical science, he has been awarded DAE-Scientific & Technical Excellence Award in 2012 and Young Achiever Award, SSPS 2010. He has been elected as a Fellow, The National Academy of Sciences, India (FNASc) in 2016 and Fellow, Maharashtra Academy of Sciences (FMASc) in 2013.

Chapter 1
Solid State Synthesis of Materials

Vinita Grover, Balaji P. Mandal, and A. K. Tyagi

Abstract Solid state synthesis is probably one of the oldest synthetic procedures for materials synthesis. It involves heating the reactants at high temperature for a specified period of time. The temperature and time chosen for the synthesis depend on many factors. A thorough knowledge of all these factors is essential to design a successful solid state synthesis to achieve desired product. The chapter, initially, introduces solid state synthesis in details. A historical perspective of solid state synthesis, its significance and utility has been discussed. Various concepts involved in solid state synthesis such as choice of reactants, equipments employed and reaction conditions have been elaborated in detail. These will give reader an insight into thoughtful planning of synthesis protocol. The concepts have been supported by giving specific examples of the synthesis of compounds belonging to important structural classes. Sometimes, in order to improve the homogeneity of the product or to decrease the reaction time, some modified solid state synthesis techniques are employed. They have also been described briefly. The solid state synthesis was conventionally used to prepare inorganic compounds, but there has been an upsurge in synthesis of solvent-free synthesis of organic compounds termed as solid state organic synthesis which has also been described briefly. The chapter also delves into common characterization techniques which are employed to characterize the synthesized products. It has been realized that metastable compounds show very interesting and technologically significant properties which in turn is the consequence of their structure. The concept behind metastability has been discussed and supported by few examples. Solid state synthesis is one of the synthetic techniques which are easily amenable to scaling up and easy to follow. However, it has certain limitations which have been elucidated to benefit the readers. The chapter covers all the aspects of solid state synthesis to enable even a beginner in this field to embark on the wonderful journey of solid state synthesis.

V. Grover · B. P. Mandal · A. K. Tyagi (✉)
Chemistry Division, Bhabha Atomic Research Centre, Mumbai 400085, India
e-mail: aktyagi@barc.gov.in

V. Grover · B. P. Mandal · A. K. Tyagi
Homi Bhabha National Institute, Mumbai 400094, India

A. K. Tyagi and R. S. Ningthoujam (eds.), *Handbook on Synthesis Strategies for Advanced Materials*, Indian Institute of Metals Series,
https://doi.org/10.1007/978-981-16-1807-9_1

Keywords Solid state reactions · High-temperature heating · Furnaces · Solid solutions · Metastable · Mechanism of solid state reactions

1.1 Introduction

1.1.1 History and Background

> It is the wonder and excitement of finding the unprecedented and unimaginable that makes the research enjoyable, even exhilarating, and worthwhile. John D. Corbett [1]

One of the most important parameters that measure the development index of society is the advancement in materials. In fact, different ages in the history of humankind have been recognized by the kind of materials used in that age whether it is stone-age or iron-age. The growth is indicated by the level of enhancement of the materials used by humankind. In addition to being scientific, synthetic methods and synthesis in general have been compared to an art which involves creativity and innovation of the scientist.

> The unique challenge which chemical synthesis provides for the creative imagination and the skilled hands ensures that it will endure as long as men write books, paint pictures, and fashion things which are beautiful, or practical, or both. Robert Burns Woodward [2]

The research on newer and better materials has thus always been a thrust area of research in all times. The *synthesis* of materials is pivotal to the research on materials. Over the years, various synthesis methods have been developed and pursued. These have been governed by the nature of the end products desired by that synthesis, cost incurred and also the basic scientific curiosity to understand the effect of synthesis protocol followed on the products obtained.

As the journey of solid state synthesis begins, a very basic question arises: Why at all do we do the syntheses? The answers are varied ranging from synthesis necessary for producing materials to satisfy the basic academic quest of creating new structures. However, the underlying answer to this range of replies is that chemists synthesize because there is a need for materials. Materials are needed for energy, pharmaceuticals, electronics, petrochemicals and cosmetics, virtually everywhere. In addition to synthesizing the known materials for meeting up the demand for these industries, there is also a continuous quest for developing new materials for better efficiency, new functionalities and better economics. The role of synthesis in the growing economies worldwide cannot be over-emphasized. Other than the utilitarian purposes, there are also more fundamental questions that the synthesis tends to satisfy. These could be knowing the nature of chemical bonding, the curiosity of developing an altogether new structure as well as developing altogether new morphologies. As long as the civilization is there, there would be both: the need for utilitarian materials and the quest for knowing and developing the unknown, and hence, the material synthesis will always hold a centre stage.

Among all the synthesis routes adopted worldwide by material scientists, solid state synthesis is probably one of the oldest and the simplest preparative methods. In the simplest terms, it involves mixing the components and heating them together at some higher temperature for a particular duration of time to get the desired product. One of the oldest examples of synthesis by heating can be traced back to Mayan civilization concerning the synthesis of Mayan blue. Mayan blue was the blue pigment which was used by the Maya civilization and is still present in mural paintings, sculptures and other objects in Chichen Itza and Yucatan (Mexico). It is an organic–inorganic hybrid of indigo from *Indigofera suffruticos* leaves with clay, which is most probably palygorskite or sepiolite or mixtures of these two (both of these clays consist of hydrated magnesium aluminium silicates) [3–5]. In laboratory, Mayan blue can be synthesized simply by mixing the clay and indigo solution at temperatures ranging from 75 to 150 °C. It is worth mentioning that the composite obtained is not stable but when it is heated at these temperatures for longer durations (several days), it shows remarkable stability which indicates that importance of temperature and time of heating was probably known even at that time.

1.1.2 What Is Solid State Synthesis?

The process of heating two materials together to get a new material is probably the oldest synthesis method known. Often referred to as "heat and beat" method, solid state synthesis has made its place in history of science as one of the most reliable and versatile preparative methods. This is the most common method for synthesizing inorganic solids also called the "ceramic method", wherein synthesis is performed at elevated temperatures and yields thermodynamically stable phases.

In very basic sense, the components required for the desired product are mixed to obtain a homogeneous mixture. The mixture is then pressed into pellets or as such heated at high temperature. Many a time, the synthesis may involve regrinding and repelletizing to obtain the product of homogeneous uniform composition. Then, there are specific conditions for mixing, heating, heating atmospheres, heating conditions, furnaces, crucibles, etc. All these are discussed in detail later in the chapter.

1.1.3 Materials Synthesized by Solid State that Shaped up World of Materials: Rational and Serendipitous

From "Stone" Age to "Silicon" Age, materials are the building blocks of not just science but culture and civilization as well. Like every branch of science, preparative solid state chemistry has had its "stars" which have revolutionized the fields of

Table 1.1 Compounds which revolutionized the field of materials [6]

Solid	Reported in year	Reported by	Application
Yttria-stabilized ZrO_2 [6]	1900	Nernst	Fuel cell, oxygen sensor
Calcia stabilized zirconia [6]	1929	Ruff	Fuel cell, oxygen sensor
Na β-alumina [6]	1926	Stillwell	Solid electrolytes, Na–S battery
$LiNbO_3$ [6]	1937	Sue	Ferroelectrics, nonlinear materials
$BaFe_{12}O_9$ [6]	1938	Adelskod and Schrewelis	Ferrites, memory devices
Amorphous Si [6]	1944	Konig	Solar cells
$BaTiO_3$ [6]	1925	Tammann	Ferroelectrics
$Bi_4Ti_3O_{12}$ [6]	1949	Aurivillius	Ferroelectrics
H_xMoO_3 [6]	1951	Glemser	Proton conductors, electrochromic displays
$K_{0.5}MoS_2$ [6]	1959	Rudorff	Intercalation chemistry, battery cathodes
$K_2[Pt(CN)_4]X_{0.3}$ [6]	1968	Krogmann	1D conductors
MMo_6X_8 (X: S, Se) [6]	1971	Chevrel	High field superconductors
$(CH)_x$ [6]	1977	Chiang and others	Molecular metals
Sialon (Si, $Al)_3(O,N)_4$ [6]	1978	Jack	High-temperature ceramics
Fullerenes [7]	1985	Kroto, Curl and Smalley	Biomedical applications
Ba–La–Cu–O [8]	1986	Bednorz and Muller	High-temperature superconductors
Carbon nanotubes [9]	1991	Sumio Iijima	In composites requiring superior mechanical and electronic behaviour
Graphene [10]	2004	Andre Geim and Kostya Novoselov	First two-dimensional material, thinnest and strongest substance, transparent and electrically conducting

materials. Synthesis of these solids possessing unusual and/or superior properties has opened new arena in research as well as led to their technical application as well in some cases. The table adopted from US Panel Report (Table 1.1) lists a few of them [6]. Some more recent additions have been made to this table to make it more contemporary. However, this list is not exhaustive, and there are many more important materials which have influenced the research on materials in a big way.

Preparative solid state chemistry requires a thorough knowledge (or understanding) of the structure of the desired compound as well as the reactivity patterns of the reactants. This is also termed as rational synthesis of solids. In a rational synthesis, the reactants are chosen taking into consideration the end product. The properties and the structure required in the product are the guiding factors for designing the scheme of the synthetic reaction. The thermodynamics, crystal chemistry and kinetic factors of the reactions play a role and thus are factored before planning the synthesis. This planning involves choice of reactants, choice of heating materials, temperature and duration of heating, atmosphere during the reaction, etc. There are many examples of famous and path-breaking rational synthesis of materials. The synthesis of SIALON is one example wherein parts of Al and O in Al_2O_3 were replaced by Si and N to obtain the specialist refractory ceramics with high strength at ambient and high temperatures, good thermal shock resistance and exceptional resistance to wetting or corrosion. Similarly, the sodium ion fast conductor NASICON with formula $Na_3Zr_2PSi_2O_{12}$ was synthesized from the reactants keeping in the view the coordination polyhedra formed by preferably adopted by the ions in the reactants and the way they join together to form networks [11]. The structure of NASICON is very conducive to three-dimensional sodium ion movement, thus bestowing upon its fast ionic conductivity. Another example of innovative preparative solid state chemistry is evident in synthetic bone material, calcium hydroxyapatite [11]. The properties desired for the bone material were good porosity and 100% connectivity. The two facts that formed the basis of synthesis plan for synthetic bone materials were that certain marine corals ($CaCO_3$) possessed the porosity and connectivity similar to bone materials and also the aragonite form of $CaCO_3$ can be topotactically converted to calcium hydroxyapatite retaining the desired porosity and connectivity by hydrothermal synthesis [11]. Hence, a synthesis of bone materials was accomplished by reacting marine corals with phosphoric acid under hydrothermal conditions. Similarly, the design of synthesis of low expansion or zero thermal expansion solids utilizes the concept that low expansion on heating is favoured by the structure where the framework is formed by corner-sharing polyhedra of highly charged ions, e.g. ZrW_2O_8. These syntheses emphasize that a good knowledge of basics of solid state chemistry can widen the synthesis skills of a chemist.

Despite the fact that there are numerous examples wherein useful materials could be synthesized by careful planning of synthesis route adopted, no branch of science is untouched of *serendipitous* discoveries, and solid state synthesis is no exception to that. There have been a number of important materials that were not planned but were obtained due to some unexpected reactions happening in the system. The synthesis of metal cluster compound, $NaMo_4O_6$, is an example of the same [12]. In an attempt to synthesize sodium zinc molybdate, $NaZn_2Mo_3O_8$ using Na_2MoO_4, ZnO, MoO_2 and Mo at 1100 °C, $NaMo_4O_6$ was obtained which has infinite chains of MoO_6 polyhedra sharing opposite edges. Discovery of phosphorus tungsten bronzes is similarly an example of preparation not proceeding in a planned way. The discovery of graphene is probably one of the biggest examples of serendipity that emerged out of playful experiments of two scientists Geim and Novoselov trying to obtain thinner flakes of graphene with the help of sticky Scotch tape on a graphite block, and this fetched them the Nobel Prize in Physics in 2010 [13].

Nonetheless, as it is said that fortune favours the prepared mind, the serendipity also needs a good scientific intuition to convert the chance discoveries into the "useful" discoveries. The chapter would discuss in detail about the nuances of solid state synthesis which include the concepts behind designing a solid state synthesis route for the desired product along with relevant examples. The basic prerequisite knowledge and tools for solid state synthesis are discussed to equip young researchers to enter into the field of solid synthesis. Various common characterization techniques employed are also briefly discussed. The concept of metastability and the possible ways to achieve metastable products have been addressed. Of late, the solid synthesis has also been employed to synthesize organic compounds. In fact, there is a separate branch known as solid phase supported synthesis for synthesizing peptides. Keeping this in view, a section has been dedicated to solid state synthesis of organic compounds as well. In the end, the chapter discusses the merits and demerits of solid state route vis-a–vis other synthesis procedures.

1.2 Concepts of Solid State Synthesis

As mentioned in the introduction section, solid state synthesis is a simple method to prepare a wide range of materials. However, despite being a simple method, there are many concepts which need to be understood and practised to get a phase pure material with desired properties. This section of the chapter intends to elaborate these concepts or guidelines.

1.2.1 Different Types of Pestle–Mortars, Ball Mills, Grinding Media and Pellet Press

Grinding of the reactants is a very crucial step in solid state synthesis. Any improper grinding may lead to the formation of undesired phase(s) and/or incorporation of impurities in the product. In addition, it is required to increase the surface area of the reactants which, in turn, increases the area of contact of the solid, thereby increasing the rate of product formation in the solid state reaction. By a thorough mixing of the reactants by grinding, the diffusion pathways between the reactants also become shorter which has a strong bearing on the formation of the desired product. The mechanistic aspects of role of mixing of the reactants will be discussed in more detail in subsequent sections. This section just intends to introduce the beginners about various laboratory equipments for mixing and pelletizing the reactants.

The conventional laboratory grinding apparatus is pestle–mortar and ball mill.

Mortar–pestle: These are very basic grinding apparatus that are used for the synthesis at laboratory scale. The mortar–pestle material should be hard, non-reacting, stable and compatible with the reactants to be used. They can be made of various materials like agate (grey or black), granite, porcelain, marble and metal/alloys. Depending upon the usage, a wise choice of mortar–pestle should be made. For example, best mortar–pestle suited for oxide-based preparative solid state chemistry is grey agate, while it is not at all suitable for fluorides. For synthesis of fluorides, mortar–pestle made of alloy Inconel is ideal. Thus, the selection of mortar–pestle needs to be carefully made depending upon the nature of reactants; otherwise, it will lead to the formation of undesired phases or incorporation of impurity from the mortar–pestle material.

Ball mills: This equipment is mostly used for scaling up the synthesis of materials. As the name suggests, it ensures milling of the powder with grinder balls which works on the principle of shock–impact friction, shearing and attrition. The balls actually act as grinders for thoroughly mixing and grinding the reactants and to achieve size reduction. It uses top-down approach for achieving the desired particle size. A ball mill consists of a hollow cylinder made of natural stone rotating about its axis. In this cylinder, the powder mixture and balls are filled. The balls can be made of stainless steel, alumina, zirconia, tungsten carbide, etc. The ball should be hard, abrasion-resistant, non-reacting, stable and compatible with the reactants. The selection of balls should also be done as per the nature of reactants. The balls material should not react with reactants during the milling process. In some cases, the product may also be obtained by ball milling the reactants.

Medium of grinding: The grinding and mixing can be dry or wet. At times, a liquid is added to the powder to make it slurry; this liquid medium increases the efficiency of the grinding, e.g. acetone and ethyl alcohol are commonly used as medium. It is called wet grinding. The grinding medium should be inert and should easily volatilize, so that it can evaporate after the grinding.

Once the reactants powder is thoroughly ground, it needs to be pelletized to increase the contact area and the kinetics of the solid state reaction. Also, to sinter the product, powder needs to be pelletized. It is a well-known fact that better the green density of the pellet, better is the sintered density after appropriate heat treatment. There are different kinds of pellet press machines available for the purpose. Some of the commonly used pellet presses are discussed as follows:

Uniaxial pellet press: It is the conventional hydraulic press machine which can be operated either electrically or manually, wherein the powder is put inside the die and pressure is exerted along one axis, i.e. either from the top or bottom using a plunger. Small pellets can be easily made from this machine but as we go on increasing the thickness, then fracture or scaling or peeling of the pellet is seen due to the residual stress and non-uniform pressure which leads to gradient in the green density.

Biaxial pellet press: In this pellet press machine, the pressure is exerted from both sides, i.e. from top and bottom. Better density is achieved than the uniaxial press machines.

Isostatic press: In this, pressure is exerted equally from all sides and hence better pellet properties, green density is achieved. Usually, polymer moulds are used to fill the powder to get the desired shape of the green pellet.

Cold isostatic press (CIP): Cold isostatic pressing machine is operated via motor up to 250 MPa. It is an excellent tool that helps in obtaining very high-density ceramic pellets, rods, discs and targets. Better compaction is achieved compared to uniaxial and biaxial dry die pressing.

Hot isostatic press (HIP): Hot isostatic pressing machine is operated via motor up to 250 MPa and at high temperature. It is also an excellent tool for preparing high-density ceramic rods, discs and targets. It yields better density as compared to uniaxial dry die pressing.

At times for some powders, the powder characteristics are not suitable for pelletization. In these cases, the sample does not form a good green pellet with desirable density. In order to circumvent this problem and to get better density, pellets binders are used. These binders are organic chemicals that provide better adhesive property to the powder. Some of the binders are glycerol, starch, etc. The required quality of the binder is that it should be non-reactive, should be able to remove at lower temperature without breaking the pellet and should not leave any residual traces on heating. While using the binder, two-step heating is preferred, i.e. initial slow heating at lower temperature to remove the binder and followed by appropriate high-temperature heating. In case if the powder is ultra-fine and sticks to die plunger, then the making of pellet not only becomes difficult but also it degrades the quality of green pellet. In order to address this problem, lubricants are used. The most common lubricant is stearic acid. Dipping the plunger in stearic acid dissolved in acetone before making pellet provides better lubrication and improves the quality of green pellet.

1.2.2 Different Types of Heating Elements of Resistance Furnaces and Thermocouples

Solid state reactions are commonly performed using resistance furnaces, which can be either tubular or muffle type. The principle of resistance furnace is that as a known amount of current is passed through the heating element (electric conductor) having some resistance, heat will be produced which will increase the temperature of heating element and its surroundings. The highest temperature achievable using resistance furnaces depends upon its heating element. Several criteria need to be considered while selecting a heating element, briefly described here. The heating element should have melting point higher than the furnace operating temperature. Thus, it should possess enough resistivity so as to generate the desired temperature. It should be inert to the atmosphere in which solid state reaction is being performed. In particular, it should not get oxidized in oxidizing atmosphere or degrade in the

Table 1.2 Common heating elements used in resistance furnaces

Heating element	Maximum temperature achievable (°C)	Other remarks
Nichrome	About 1200	Cost effective
Kanthol (iron–chrome–aluminium)	About 1200	Most commonly used heating element
Platinum	About 1400	
Platinum–rhodium	About 1700	
Silicon carbide	About 1500	
Molybdenum disilicide	About 1750	
Graphite	About 2200	To be used in vacuum
Molybdenum	About 2500	To be used in vacuum
Tungsten	About 3000	To be used in vacuum

presence of water vapours. It should be able to withstand thermal and mechanical shocks. The common heating elements are compiled in Table 1.2.

Usually, metallic and alloy-based heating elements are used in the form of wire, whereas heating elements like silicon carbide, molybdenum disilicide, etc., are commonly used in the form of horseshoe. Wire-type heating elements are wound on quartz or re-crystallized alumina tubes. Quartz has excellent thermal shock resistance but very poor mechanical shock resistance. On the other hand, alumina does not have a good thermal shock resistance but has good mechanical resistance. These are some simple concepts, if practised, can considerably enhance the life of furnaces.

Different Types of Thermocouples

Thermocouples are also an important hardware used during solid state reaction for measuring temperature. It is made of two dissimilar metal wires, which are welded to create a junction at which temperature is measured by virtue of generation of voltage corresponding to change in temperature. The generated voltage is converted into temperature by using standard reference table for the thermocouple in use. The modern temperature controllers and programmers have in-built provision of converting voltage into temperature which is shown on display or monitor. It is a safe practice to encase thermocouple wires into alumina sleeves or some other suitable sheath to protect them from oxidation and other chemical attacks or atmosphere at higher temperature. In case of highly reactive atmosphere, it is advised to cap thermocouple tip. The ideal position of thermocouple in a furnace is as close to sample as possible. Typical features of some of common thermocouples are given in Table 1.3.

Table 1.3 Common thermocouples, their constituent elements and temperature range

Type	Composition	Temperature range	Remarks
E	Chromel–constantan	−50 to +740 °C	Use in medium-temperature furnaces
J	Iron–constantan	−40 to +750 °C	
K	Chromel–alumel	−200 to +1350 °C	Most commonly used
M	82%Ni/18%Mo–99.2%Ni/0.8%Co (by wt.)	Maximum up to 1400 °C	Commonly used in vacuum furnaces
N	Nicrosil–nisil	−270 and +1300 °C	Excellent oxidation resistance
B	70%Pt/30%Rh–94%Pt/6%Rh (by wt.)	Maximum up to 1800 °C	Commonly used for high-temperature furnaces
R	87%Pt/13%Rh–Pt (by wt.)	Maximum up to 1600 °C	
S	90%Pt/10%Rh–Pt (by wt.)	Maximum up to 1600 °C	

1.2.3 *Selection of Crucible Materials*

The selection of crucible material in solid state reactions is a very crucial step as a wrong selection may lead to incorporation of impurities to the desired materials. In some cases, it may lead to the formation of altogether different undesired products. The incorporation of even small amounts of constituent elements of crucible will adversely affect the properties of targeted product. Hence, the selection of crucible has to be done judiciously, and the same time cost factor needs to be kept in mind. The commonly used crucibles, their incompatibility and some general remarks are given in Table 1.4.

It is amply evident from this table that the selection of crucibles largely depends on the nature of reactants/products and the required temperature at which solid state reaction is to be performed.

1.2.4 *Selection of Reactants and Their Preheat Treatment*

The selection of proper reactants in solid state synthesis is also very significant. The well-defined reactants should always be preferred, e.g. it is important to know the exact amount of water molecules present in the reactant. At times, the number of water molecule given on the bottle label gets altered due to local prevailing conditions such as high humidity. In particular, this kind of problem is commonly observed in hygroscopic materials. In this respect, in general, metal carbonates have an advantage over corresponding metal nitrates which are often hygroscopic.

Table 1.4 Common crucible materials

Crucible materials	Incompatible with	Other remarks
Porcelain	Alkali and alkaline earth metals and strong bases	It should be used maximum up to 600 °C
Quartz	Metal fluorides and alkali metals	It should be used up to about 900 °C
Gold and silver	Metals and metalloids should not be heated in these crucibles	Can be used till 700 °C
Platinum crucibles and boats	Bismuth, lead, cobalt-based compounds. Metals also should not be heated in platinum ware	Can be used till 1400 °C
Alumina	Metal fluorides	Can be used till 1800 °C
Silminite	Metal fluorides	Commonly used for melting glasses. It is economical also
Magnesia, yttria, thoria and YSZ	Metal fluorides	Can be used even above 2000 °C
Tantalum, niobium, tungsten	Tends to form alloys with certain reactive metals	Should be used only in vacuum or inert atmosphere. Can be used even above 2000 °C
$LaPO_4$, YPO_4	Metal fluorides, highly basic materials, alkali metal-based compounds	Can be used for melting reactive metals. Can be used even above 1600 °C
Graphite	Reducible metal oxides	Should be used in vacuum. Can be used even above 2500 °C. Care should be taken to avoid carburization of samples

Enough care needs to be taken while using reactants containing highly electro-positive elements such as La, Cs, Rb, K and Ba. Likewise many metal fluorides, such as CsF, RbF, NH_4HF_2 just to name a few, are also hygroscopic. Thus, it is wise to give a preheat treatment to the reactants to get rid of occluded water, adsorbed CO_2 and hydroxyl groups, etc. The temperature of preheat treatment can be determined by performing thermogravimetry. For example, in order to prepare lanthanum barium copper oxide, the first high Tc oxide superconductor [14], La_2O_3, $BaCO_3$ and CuO can be taken in an appropriate mole ratio so as to get the desired composition with an optimum content of Ba^{2+} in La_2CuO_4 lattice. In general, high Tc oxide superconductors are multi-component mixed oxide systems [15, 16] and it is required to carefully perform solid state reactions among the constituents reactants so as to get the desired single-phasic product. For example, during synthesis of $YBa_2Cu_3O_7$, a solid state reaction between Y_2O_3, $BaCO_3$ and CuO needs to be performed in appropriate mole ratio. Depending upon solid state reaction temperature, several undesired phases such as Y_2BaCuO_5, $Y_2Cu_2O_5$ and $BaCuO_2$ can be encountered, in addition to the desired $YBa_2Cu_3O_7$ phase. It is relatively easy to synthesize simple ternary systems like $LaCrO_3$, for which well-mixed La_2O_3 and Cr_2O_3 need to be heated in 1:1 mol ratio. In solid state reactions, often ageing effect is also observed. The reactivity of solid reactants which are very old is much less than freshly prepared reactants. A metal oxide in situ generated during the reaction by the decomposition of corresponding metal oxalate exhibits much higher reactivity as compared to an old metal oxide. This leads to the formation of the desired product at much lower temperature.

There are some guidelines to store solid reactants and products. It is advised that metal fluorides should not be stored in glassware due to their chemical incompatibility with glasses. Plastic bottles are the best option to store inorganic fluorides. Highly hygroscopic materials are best stored under sealed conditions or glove boxes.

1.2.5 Selection of Temperature, Heating/Cooling Rates, Intermittent Grindings and Atmosphere

Selection of optimum heat treatment is another important step in solid state reactions. Generally, too low temperature leads to incomplete reaction, and too high temperature is often responsible for the formation of undesired phases and reaction with the crucible materials. One should compile the melting temperatures, decomposition temperature, vapour pressure, etc., of the reactants and also nature of crucible material (discussed in details in previous section). The broad guideline while selecting the temperature was given by Tamann's law [17] which states that the diffusion of ions in lattice becomes substantial at 0.5 to $0.7 \times T_m$ (where T_m is melting temperature of reactants). Thus, the optimum temperature of solid state reaction can be about 0.5–0.7 times of melting temperature of at least one of the reactants. Of course, there can be some specific requirements wherein a much higher reaction temperature may be used. There is on old concept called as Hedvall effect [18] which states that the reactivity of solids increases considerably in the vicinity of phase transition. Thus, if one of the reactants has a first-order phase transition, the solid state reaction in the vicinity of phase transition will lead to much enhanced diffusion of ions in the lattice which will have a bearing on the kinetics of reaction.

Likewise, a judicious selection of heating and cooling rates also needs to be made. The optimum heating rate is in the range of 6–10 °C/min. Too high heating rates may compromise with the life of heating elements due to thermal shock. The soaking time (hold time) during solid state reaction also plays a role. Since these are diffusion controlled and mostly sluggish reactions prolong heat treatment with intermittent grindings, it needs to be used for their completion. These aspects will be further elaborated in the next section.

The selection of proper atmosphere during the reaction is also very important for obtaining the desired compounds. Different types of atmospheres such as static air, flowing air, moist air, flowing oxygen, flowing inert atmosphere (argon, nitrogen and helium), reducing atmosphere (H_2 or H_2–Ar mixture) flowing reactive gases (F_2, HF or CO_2, etc.) can be used. The selection of atmosphere depends on the specific requirement. For example, in order to prepare compounds having high or low oxidation state metal ions, flowing oxygen or H_2 or H_2–Ar mixture can be used. The inert atmosphere (argon, nitrogen and helium) is used to protect the product from getting oxidized or hydrolysed. Some specific examples will be discussed in subsequent sections.

1.2.6 Mechanism of Solid State Reactions

As discussed in the previous section, solid state reactions are diffusion-controlled reactions, which can be represented by the following schematic (Fig. 1.1).

The schematic depicts the formation of AB_2O_4 spinel by a solid state reaction of divalent and trivalent metal oxides with general formula AO and B_2O_3. As the reaction between AO and B_2O_3 proceeds, a product layer with the composition AB_2O_4 is formed which grows with time of heat treatment. As the product layer grows up to certain threshold thickness, reactants are not able to diffuse through it for further reaction, and thus, the reaction either stops or becomes very sluggish. It is for this reason most of the solid state reactions follow a parabolic rate law, which explains that the rate of solid state reaction is fast in the beginning and then gradually it becomes slow and finally reaction stops at the stage when the product layer becomes thick enough to not allow diffusion of the reactants. Hence, it is important to employ intermittent grindings which destroy the product layer and ensure that unreacted reactants come into contact of each other. Therefore, a multiple steps solid state reaction with intermittent grindings always gives a better

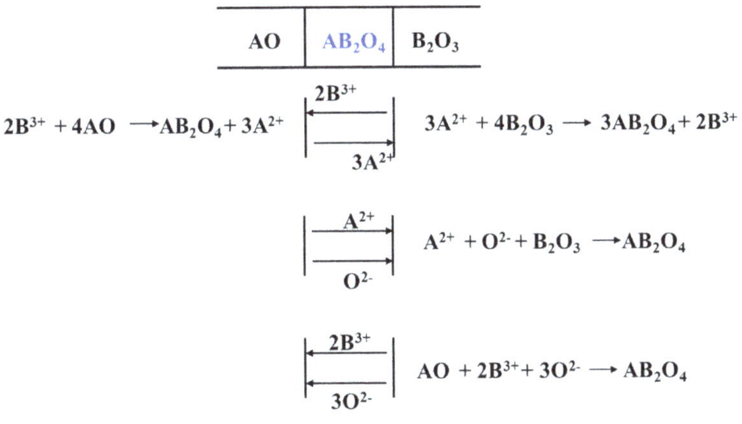

AO | AB_2O_4 | B_2O_3

$2B^{3+}$ / $3A^{2+}$: $2B^{3+} + 4AO \rightarrow AB_2O_4 + 3A^{2+}$; $3A^{2+} + 4B_2O_3 \rightarrow 3AB_2O_4 + 2B^{3+}$

A^{2+} / O^{2-}: $A^{2+} + O^{2-} + B_2O_3 \rightarrow AB_2O_4$

$2B^{3+}$ / $3O^{2-}$: $AO + 2B^{3+} + 3O^{2-} \rightarrow AB_2O_4$

Fig. 1.1 Schematic of mechanism of a typical solid state reaction

control over homogeneity. The kinetics of solid state reactions becomes fast if one of the reactants melts at or below the reaction temperature, due to facile diffusion of ions in molten state.

There are several terms like calcination, sintering and annealing which are used in solid state reaction. Though all these phenomena deal with heat treatment, but they all have different functions, which are described briefly.

Calcination is the process in which reactants are heated to decompose them and to obtain the desired products. By this process, the gaseous components evolved due to heating of reactants escape. A simple example of calcination process is the following solid state reaction:

$$BaCO_3(s) + CuO(s) = BaCuO_2(s) + CO_2(g)$$

During calcination process, an appropriate atmosphere may also be selected. The calcination reactions are mostly performed in the vicinity of thermal decomposition temperature (ca. of metal carbonates or metal oxalates) or near phase transition temperature. Thermodynamically, the calcination temperature is defined as the minimum temperature at which the standard Gibbs free energy for a given solid state reaction becomes zero.

Sintering is another important thermal process which is used for compaction of the pellets by way of reducing porosity during the appropriate heat treatment to achieve desired sintered density. Sintering has a strong bearing on many physical properties such as mechanical strength, electrical and thermal conductivity, dielectric property, thermal expansion behaviour and optical transparency. Sintering is mostly performed just below the melting point of the solid mass of the materials. Sintering process involves the diffusion of atoms or ions across the physical boundaries of the constituent particles, which leads to their fusing into a large and well-packed solid. The sintered density is calculated with respect to theoretical density which is calculated by XRD [19]. There are different types of sintering, e.g. liquid phase-assisted sintering, activated sintering using hot press, microwave-heated sintering and sintering with the help of sintering aids, just to name a few. However, a detailed description of different types of sintering is out of scope of the present chapter. It is possible to prepare sinter-active powders, which will be discussed in the next chapter on combustion synthesis of solids.

Annealing is another heat treatment which is often given to solids. It involves heating of materials at a certain temperature to get rid of bulk point defects, surface defects, line dislocations, residual stress, etc. In case of metals and alloys, annealing process is employed to improve ductility and reduce hardness and also to remove residual stress. It may be noted that line dislocations are often encountered in metals and alloys as compared to ceramics (ionic), where point defects are more predominant. Thus, the annealing process has to be used depending upon the nature of solid sample and the desired functional properties.

There is one more important concept in solid state reactions which need to be discussed. During a solid state reaction, the electro-neutrality condition cannot be violated. It can be seen from the schematic of solid state reaction (Fig. 1.1) that there can be three ways on which ions can diffuse during the reaction in such a way that the electro-neutrality condition can be maintained. This is explained by a typical example of synthesis of AB_2O_4 spinel from the reactants AO and B_2O_3. In order to maintain the electro-neutrality, $3A^{2+}$ ions and $2B^{3+}$ ions have to counter-diffuse in opposite direction or one A^{2+} ions can diffuse along with one O^{2-} ion towards B_2O_3 or $2B^{3+}$ ions along with $3O^{2-}$ ions can diffuse through the reactant AO. These three combinations of diffusion of ions shall maintain the local electro-neutrality. It may be noted that which of these three types of diffusion will be observed depends on the nature of reactants.

1.2.7 Basic Thermodynamics of Solid State Reactions: Enthalpy or Entropy Driven?

Like any other reaction, the free energy of formation of solid state reactions also needs to be negative. The Gibbs free energy is equal to the enthalpy (of a system or process) minus the product of the entropy and the absolute temperature. It can be expressed as follows:

$$\Delta G = \Delta H - T\Delta S$$

In a solid state reaction, i.e. in which there is no gaseous component (e.g. by carbonate decomposition) or liquid component (e.g. melting of reactant), the entropy change is insignificant.

$$A(s) + B(s) = C(s)$$

$$\text{For example :} \quad 2CuO + Y_2O_3 = Y_2Cu_2O_5$$

These types of reactions are enthalpy driven instead of entropy driven. Thus, the $T\Delta S$ factor can be neglected. Now it is evident that in order to have negative ΔG, ΔH has to be negative which is the case for exothermic reactions. Thus, the solid state reactions are always exothermic in nature. However, the solid state reactions in which some gaseous or liquid components are present, and this inference of thermodynamics needs not hold good.

1.2.8 Methods to Introduce Non-stoichiometry

The introduction of an optimum non-stoichiometry is an important way to induce the desired functionality. Several properties such as ionic conductivity, luminescence, dielectric properties, magnetism and order–disorder transitions considerably depend on the presence of optimum degree of non-stoichiometry. It may be added here that no compound is ideal by definition and many important compounds have intrinsic vacancies. Typical examples are TiO_2, ZnO, Fe_2O_3 and SnO_2 which have non-negligible amount of intrinsic anionic vacancies which make them n-type semiconductors. However, it is possible to introduce even a large amount of oxygen vacancies in metal oxides either by an appropriate heat treatment or by aliovalent substitution. If transition metal oxides TiO_2, ZnO, Fe_2O_3 and SnO_2 are heated in reducing atmosphere, considerable oxygen vacancies can be introduced in their lattice. Likewise, it is possible to introduce oxygen excess also by an appropriate heat treatment, in particular in flowing oxygen atmosphere or under high pressure of oxygen. It is rather easy to introduce oxygen vacancies or excess in metal oxides having variable valency metal ions. However, this strategy does not work for metal oxides having single stable oxidation state, e.g. ThO_2, ZrO_2, HfO_2, La_2O_3, Lu_2O_3 just to name a few. In such cases, the substitution of aliovalent ions can be used to introduce oxygen vacancies or excess. Some typical examples are represented as follows [20–24]:

$$(1-x)\mathrm{CeO_2} + x\mathrm{SrCO_3} = \mathrm{Ce}_{1-x}\mathrm{Sr}_x\mathrm{O}_{2-x}$$

$$(1-x)\mathrm{CeO_2} + x\mathrm{GdO_{1.5}} = \mathrm{Ce}_{1-x}\mathrm{Gd}_x\mathrm{O}_{2-x/2}$$

$$(1-x)\mathrm{MO_2} + x\mathrm{YO_{1.5}} = \mathrm{M}_{1-x}\mathrm{Y}_x\mathrm{O}_{2-x/2},$$

$$(\mathrm{M} = \mathrm{Ce\ and\ Th})$$

$$(1-x)\mathrm{BaF_2} + x\mathrm{NdF_3} = \mathrm{Ba}_{1-x}\mathrm{Nd}x\mathrm{F}_{2+x}$$

In these typical examples, the anionic vacancies or excess is created to impart the charge balance consequent to aliovalent substitution.

It may be noted that the isovalent substitution does not lead to the incorporation of anionic vacancy or anion excess. A typical example is discussed as follows [25]:

$$(1-x)\mathrm{ThO_2} + x\mathrm{CeO_2} = \mathrm{Th}_{1-x}\mathrm{Ce}_x\mathrm{O_2}$$

In this case, charge remains balanced as the host cation and substituent cation both are tetravalent. It may also be noted here that it is not possible to introduce anionic vacancies or excess endlessly. Beyond a certain concentration of anionic vacancies, excess lattice energy increases considerably, and it is needed to lower the energy, which can be achieved by several ways such as lowering of symmetry, phase separation and ordering of defects. The classic example is the transformation of fluorite (AO_2) to pyrochlore ($A_2B_2O_7$) by substituting tetravalent A^{4+} cations in AO_2 fluorite by trivalent B^{3+} ions of appropriate size. This transformation can be sequentially depicted as follows:

AO_2 (Fluorite): Parent compound

$A_{1-x}B_xO_{2-x/2}$ (Defect fluorite)

$A_{0.5}B_{0.5}O_{1.75}$ (Extensive defect fluorite)

$A_2B_2O_7$ (Pyrochlore: An ordered structure)

1.2.9 Solid Solutions (Substitutional, Interstitial)

Solid solutions in materials science play an important role in fine-tuning functional properties. Many properties such as thermal expansion, specific heat, dielectric properties, just to name a few, can be fine-tuned by making appropriate solid solutions as by and large they follow rule of addition. Hence, it is very important to understand the concepts of solid solutions, to which this section is devoted to. There are two types of solid solutions: substitutional solid solutions and interstitial solid solutions. Some typical substitutional solid solutions are $Al_{1-x}Ga_xPO_4$ (x = 0.0–1.10) [26], $Zr_{1-x}Y_xO_{2-x/2}$ [27], $Zn_{1-x}Mn_xO$ [28]. The solubility of guest ions in the host lattice in substitutional solid solutions depends on several factors such as their relative ionic radii difference, ionic charge and electro-negativity difference [29]. The heat treatment also has some influence on solubility of guest ion into host lattice. By and large, the substitutional solid solutions are formed if the host and guest atoms are of somewhat similar sizes (difference of less than about 15%). However, in case the host atoms are much larger in size as compared to that of guest atoms, it leads to the formation of interstitial solid solution. The interstitial substitution occurs when atoms or ions occupy interstices (voids) found in the lattice comprising of array of atoms or ions of the host material. Carbon dissolved in iron is a classic example of interstitial solid solution, which leads to a remarkable increase in strength. Figure 1.2 gives the schematics of substitutional and interstitial solid solutions. Several metal carbides have carbon atoms in interstitial positions. Similarly, metal hydrides have hydrogen atoms located on interstitial positions.

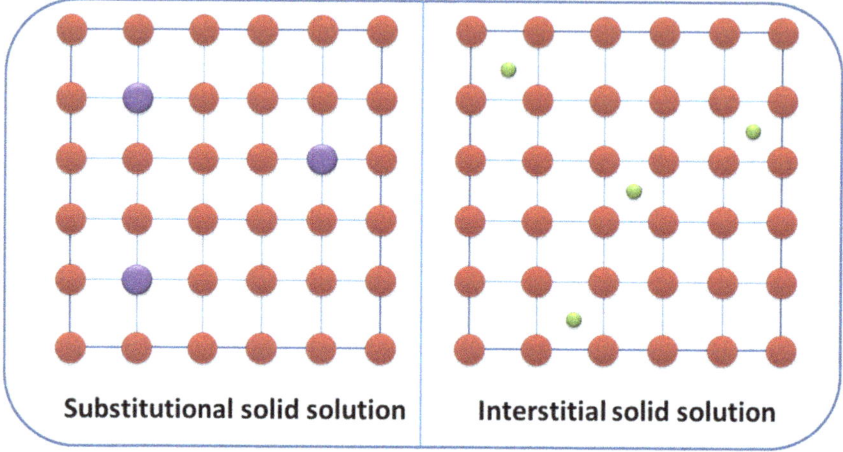

Fig. 1.2 Schematic representation of substitutional and interstitial solid solutions

As mentioned earlier that a number of properties can be fine-tuned by making solid solutions of the appropriate end members, some typical examples are discussed as follows.

1.2.9.1 Tailoring of Magnetic Properties and Band Gap

A series of solid solutions between end members $LaCrO_3$ and $CeCrO_3$, with general composition $La_{1-x}Ce_xCrO_3$ ($0.0 \leq x \leq 1.0$) was synthesized by a two-step synthesis route [30, 31]. The first step consisted of a combustion reaction, and in the second step the combustion synthesized powders were vacuum heat treated in the presence of Zr sponge which functioned as a reducing agent. All $La_{1-x}Ce_xCrO_3$ ($0.0 \leq x \leq 1.0$) compounds in this series exhibited an antiferromagnetic behaviour with a linear increasing trend in Neel temperature (T_N) from 257 to 281.5 K on going from $LaCrO_3$ to $CeCrO_3$. Likewise, the band gap of this series of solid solutions also could be tailored from 3.21 to 3.02 eV on going from $LaCrO_3$ to $CeCrO_3$. In a similar fashion, $Pr_{1-x}Ce_xScO_3$ ($0.0 \leq x \leq 1.0$) solid solutions were also prepared which showed a decreasing trend in the band gap from 4.74 to 2.91 eV [32]. Thus, it may be emphasized that the design of appropriate solid solution is a good approach to fine-tune band gap for various photochemical reaction.

1.2.9.2 Tailoring of Dielectric Properties

Dielectric constant is known to be significantly modified by appropriate substitution. This section is not intended for an elaborate discussion on dielectric properties as there are dedicated books and review articles on this topic. This small section is just intended to briefly emphasize the solid solution approach to tailoring the dielectric behaviour in some recent systems, in particular in lead-free systems. A series of solid solutions with general composition $YIn_{1-x}Fe_xO_3$ ($0.0 \leq x \leq 1.0$) series was prepared [33]. The single-phasic hexagonal nominal compositions, $YIn_{1-x}Fe_xO_3$ ($0.0 \leq x \leq 0.3$), in this series were investigated by impedance analysis. The room temperature dielectric constant was observed to considerably increase from 10 for $YInO_3$ to 1000 for $YIn_{0.7}Fe_{0.3}O_3$. It was further observed that an optimum incorporation of Fe into $YInO_3$ led to tuning of normal dielectric behaviour to relaxor ferroelectric in this series. This particular example highlights the role of solid solution with an optimum composition with the right degree of distortion leading to lead-free relaxor materials. Likewise, in $GdSc_{1-x}In_xO_3$ solid solutions, also the system could be transformed from normal dielectric to classical relaxor ferroelectric behaviour just by a careful compositional and structural fine-tuning [34].

1.2.9.3 Tailoring of Ionic Conductivity

Ionic conductivity is another physical property which can be considerably fine-tuned by solid solution formation strategy. This small section is not a detailed description on ionic conductivity as that is a separate topic. However, it is proposed to briefly discuss some of the important factors affecting the ionic conductivity. The ionic conductivity of solids can be enhanced by introducing an optimum extent of point defects, which in turn can be incorporated by aliovalent substitution. The classic example is yttria-stabilized zirconia [35] in which Zr^{4+} is partially substituted by Y^{3+}, which leads to the stabilization of cubic phase coupled with the incorporation of oxygen vacancies. Another approach is to incorporate a counter cation with lone pair of electrons, which by virtue of higher polarizability will enhance the ionic conductivity. Another approach is to introduce an optimum degree of disorder in the lattice by substitutional solid solution. Mandal et al. [36] have extensively used this approach to enhance the oxide ionic conductivity of pyrochlore materials. A typical example is $Gd_{2-y}Nd_yZr_2O_7$ (y = 0.0, 0.1, 0.4, 0.6, 1.0, 1.4, 1.6 and 2.0) pyrochlores, which were synthesized by solid state route starting from Gd_2O_3, Nd_2O_3 and ZrO_2 in appropriate ratio. All the samples in this series were characterized by XRD and Raman spectroscopy. Their ionic conductivity in the frequency range 100 Hz–15 MHz and temperature range of 622–696 K was measured by AC impedance spectroscopy. The DC conductivity (σ_{DC}) was found to vary as a function of Nd concentration in $Gd_{2-y}Nd_yZr_2O_7$ series. The highest ionic conductivity was observed at the composition $GdNdZr_2O_7$ (y = 1.0), which was attributed to the introduction of an optimum extent of disorder in the lattice which increases the pre-exponential factor, and in turn the ionic conductivity increases considerably. The extent of oxygen ion disorder is determined from X-ray diffraction and Raman spectroscopy. This system amply demonstrates that a significant increase in ionic conductivity can be achieved by suitable doping at the Gd^{3+} site with isovalent rare earth ions like Nd^{3+}. On similar line, Sayed et al. [37] fine-tuned the ionic conductivity in $Sm_{2x}Dy_xZr_2O_7$ system.

1.2.9.4 Tailoring of Thermal Expansion Behaviour

The basic information on thermal expansion behaviour is an important prerequisite for a given material's use under non-ambient temperatures. Thus, it is desired to fine-tune thermal expansion to avoid any adverse effect at higher temperatures. Materials with tailored thermal expansion behaviour exhibit several applications such as in electronic, precision alignment optical instruments, bio-ceramic materials such as artificial bones, dental implants common household items, just to name a few. Thermal expansion can be classified as bulk thermal expansion or lattice thermal expansion. The bulk thermal expansion is determined by a thermo-dilatometer. On the other hand, lattice thermal expansion is measured by X-ray diffractometer with a high-temperature attachment (HT-XRD). In case of dilatometry, rod shape pellets with high bulk density are required. Thermal expansion of solids can be fine-tuned by designing appropriate solid solution. Thermal expansion follows the rule of addition to a great extent. A typical example of tailoring of thermal expansion was depicted in $Al_{1-x}Ga_xPO_4$ low cristobalite (orthorhombic)-type materials [26]. In this system, thermal expansion could be gradually fine-tuned on varying the composition in this series of materials. These materials were extensively studied by high-temperature XRD and DTA. It was observed that low cristobalite (orthorhombic) phase under goes a phase transition to high cristobalite (cubic) phase at higher temperature. This was a displacive phase transition. Interestingly, a considerable decrease in thermal expansion coefficient was shown by high cristobalite-type structures in comparison with the low cristobalite-type structure. Thermal expansion trend could be explained by variation in inter-polyhedral bond angles.

Several technologically important systems like $Th_{1-x}U_xO_2$, $Th_{1-x}Ce_xO_2$ and $U_{1-x}Pu_xO_2$ have been well reported for their thermal expansion behaviour. In these solid solutions, also thermal expansion coefficient varies systematically on going from one end to the other hand of solid solutions. These trends can be correlated with the melting point of the respective end members [38, 39]. In general, higher the melting temperature, lower is the thermal expansion coefficient [36]. It may be emphasized that high-temperature XRD gives information about the intrinsic thermal expansion behaviour of solids. On the other hand, bulk thermal expansion behaviour studied by thermo-dilatometer is affected by microstructure and bulk density and cracks, etc. It may be mentioned that both the techniques have their own merits and demerits for thermal expansion studies. The readers are advised to get more details from literature.

1.3 Different Variations of Solid State Synthesis

As discussed earlier in the chapter, the conventional solid state synthesis needs high temperature, and the heating has to be performed for longer duration due to their diffusion-controlled reaction mechanism (described further in chapter in detail). The high temperatures employed yield, almost always, the thermodynamically stable phases and diminish the possibility of obtaining metastable phases. Metastable phases are important, because many times they possess exotic structures and superior/newer properties. In fact, during solid state synthesis in many systems, certain phases are not observed only because the temperatures employed were higher than their stability zones which precluded their formations. The growth of solid state chemistry, hence, has also been accompanied by various newer methods and variations of solid state synthesis which could yield the metastable products. More importantly, these variants of solid state reactions are also employed in many cases to reduce the reaction time, reaction temperature, to obtain desired powder morphology and many other factors. There is a lucid review by Rao et al. on chemical synthesis of inorganic materials [12a]. Few of them are listed below.

1.3.1 Solid State Metathesis

Conventional solid state reactions not only require high temperature and prolong heating durations but are also concerned with the reactivity of solid phase reactants, which are affected by the formation of diffusion barrier of the product layer. If one can have labile component in the reactant, then the reaction could be carried out at lower temperature with better control over the reaction conditions, with this in view, a conceptual development of new synthesis method named "solid state metathesis" (SSM) was introduced. This method is used for synthesis of various binary or ternary oxides, phosphides, chalogenides, nitrides and silicides. SSM method is amenable for scaling up of synthesis of significant materials as it provides better command over the crystallinity and phase formation.

Solid state metathesis is a simple double exchange reaction in which the anionic group gets exchanged between the reactants. It is most convenient, fast and uncomplicated route to prepare the materials at lower temperature that otherwise needs higher temperature for their synthesis. In addition, sometimes this method also yields compound which are metastable or with different microstructure. The driving force for the SSM reaction is the formation of one highly stable by-product.

The general reaction can be written as follows:

$$AX_{(s)} + BY_{(s)} \rightarrow AY_{(s)} + BX_{(s)/(l)/(g)}$$

The by-product of the reaction can be easily separated by taking the advantage of its physico-chemical properties.

Typical examples are as follows:

$$Na_2WO_4 + CaCl_2 \rightarrow CaWO_4 + 2NaCl \xrightarrow[RT]{\text{Washing, } H_2O} CaWO_4$$

$$Na_2WO_4 + Ca(NO_3)_2 \rightarrow CaWO_4 + 2NaNO_3 \xrightarrow[RT]{\text{Washing, } H_2O} CaWO_4$$

In above case formation of highly stable by-product, i.e. NaCl/$NaNO_3$ ($\Delta H_{f\,(NaCl)}$ = −98 kcal/mol; ($\Delta H_{f(NaNO3)}$ = −111.6 kcal/mol) is the driving force for the reaction. It was observed that when the reactant salt is altered, the crystallite size of the product also changes. The parameters that can be tuned in an SSM reaction are reactant, time and temperature, thus offering better control over the product properties.

Metal chalcogenides can also be prepared by SSM reaction.

$$2MoCl_5 + 5Na_2S \rightarrow 2MoS_2 + 10NaCl + S(\Delta H_{rxn} : -213\,\text{kcal/mol})$$

$$GaI_3 + Na_3As \rightarrow GaAs + 3NaI(\Delta H_{rxn} = -117\,\text{kcal/mol})$$

Some of the examples of high refractory materials that could be prepared by metathesis reaction are given below:

$$4WCl_6 + 6Mg_2Si \xrightarrow{\text{Hot filament}} 3WSi + 12MgCl_2 + W$$

$$6ZrCl_4 + 8Li_3N \rightarrow 6ZrN + 24LiCl + N_2$$

Some reaction of SSM can be carried out easily at room temperature and pressure, while some reactions require temperature and preferred atmosphere. These self-sustaining metathesis reactions are highly exothermic and require only initial ignition or trigger (frictional heating). The heat released in the initial steps is sufficient for the propagation of the further reaction which occurs continuously without any disturbance. In some cases, heating with flux (molten salts, e.g. NaCl) serves the purpose.

The advantages of SSM process are as follows:

(1) SSM reaction are better, fast and self-sustaining, thus require less time and energy.
(2) Flux-mediated reaction can be used for synthesis of refractory materials at temperatures far below melting points.
(3) The temperature and time factors can be easily tuned to obtain better yield and crystallinity of the products.
(4) Nanocrystalline compounds can also be prepared by SSM reaction.
(5) It is possible to prepare some metastable compounds which, otherwise, would exist at high temperature, e.g. cubic/tetragonal ZrO_2.
(6) Compositions with better homogeneity can be easily prepared by this method.

1.3.2 Microwave Solid State Synthesis

The reaction time can be diminished if one of the major reactants is a microwave absorber and is able to couple strongly to the microwave field at room temperature. In that case, under microwave irradiation heating, reactions can be accelerated owing to rapid heating, selective coupling and also enhanced reaction kinetics. An example is the synthesis of industrially important β-SiC which is a high-temperature ceramic [40]. It is conventionally prepared by carbothermal reduction of quartz by coke which is a multi-step highly energy-intensive process. In a microwave reactor, same synthesis can be performed at 1300 °C in 5 min [40], and it yields homogeneous sub-micron-sized particles. There is a dedicated chapter in this book that discusses the nuances of microwave synthesis.

1.3.3 Spark Plasma Sintering (SPS)

This technique also allows faster chemical reactions. In this particular method, the material is internally heated by Joule effect which leads to faster heating rates. The additional simultaneous application of a uniaxial pressure allows contact between the grains and leads to shorter reaction durations and also limits the grains growth. An example is the synthesis of Li_2CoPO_4F (battery material) [41] and Magnelli phases such as V_6O_{11}, wherein spark plasma sintering technique was employed from a mixture of VO_2 and V powders [42].

1.3.4 Solid State Synthesis by Flux Route

In order to reduce the heat, i.e. to carry out synthesis at lower temperatures, one of the routes employed is the synthesis using fluxes which are normally the low melting eutectics. The flux provides a medium for the reaction to occur fast by decreasing the diffusion distance and increasing the mobility of reactants. Generally, alkali chlorides, sulphates, carbonates and hydroxides are used as flux. The melting point of the flux must be lower than the formation temperature of the product. Also, the product must be easily separable from the flux after the reaction is over. In the flux synthesis, the reactants are mixed with flux and heated above the melting point of flux, and after the reaction, the flux is washed out to obtain the products. The flux or the molten salt synthesis has been employed to obtain several technologically important materials like ferrites $BaFe_{12}O_9$ [43] and $Pb(Zr_{1-x}Ti_x)O_3$ [44].

1.3.5 High-Pressure Synthesis

The application of high pressure aids in bringing the reacting moieties in close vicinity, and thus, this synthesis route is used for synthesizing products in which unusually high coordination numbers are desired. This may lead to unusual oxidation states and metastable products as well and the pressure–temperature variables are generally not obtainable at room temperature. The technique has been employed to synthesize interesting class of materials such as "super-hard" materials which include C_3N_4 and materials based on diamond structure, high T_c superconductors, etc. [45]. There is a separate chapter detailing about the nuances of this important branch of high-pressure synthesis, and hence, this topic will not be delved any further here.

1.3.6 Precursor Routes

Many a time, in order to reduce the reaction temperature and time, the precursors to the product are obtained by various soft chemical solvent-based routes as well such as gel combustion, hydrothermal, ultrasonication and spray pyrolysis. These precursors are then fired at elevated temperature (by solid state reaction route) which is much less than what would be required if the synthesis proceeds entirely by the high-temperature route. This leads to reduced reaction temperature as well as the heating duration. An easy example to precursor synthesis is the synthesis of mixed oxides by calcining carbonate solid solutions of calcite structure containing two or more cations in desired ratio [46].

$$x\mathrm{MNO_3} + y\mathrm{M'NO_3} + (1-x-y)\mathrm{CaNO_3} \underset{(\mathrm{NH_4})_2\mathrm{CO_3}}{\longrightarrow} \mathrm{Ca}_{1-x-y}\mathrm{M}_x\mathrm{M}'_y\mathrm{CO_3}$$

$$\mathrm{Ca}_{1-x-y}\mathrm{M}_x\mathrm{M}'_y\mathrm{CO_3} \xrightarrow{\Delta} \mathrm{Ca_2FeCoO_5}(\mathrm{M{:}\ Fe,\ M'}:\ \mathrm{Co};\ x=y=0.25)$$

$$\mathrm{Ca}_{1-x-y}\mathrm{M}_x\mathrm{M}'_y\mathrm{CO_3} \xrightarrow{\Delta} \mathrm{Ca_2Mn_3O_8}(\mathrm{M,\ M'}:\ \mathrm{Mn};\ x=y=0.3)$$

Sometimes, the intercalation or de-intercalation of some compounds is also adopted to obtain desired compounds, e.g. de-intercalation of $LiVS_2$ gives VS_2 which cannot be prepared otherwise. Synthesis of hydrated precursor and then dehydrating have also been employed to obtain desired product [47].

1.4 Specific Classes of Materials Synthesized by Solid State Route

In this section, synthesis of different solid compounds is discussed. Initially, synthesis of different classes of compounds is discussed, and in the later part synthesis of some useful compounds is included.

1.4.1 Synthesis of Fluorite-Based Materials

Fluorite-based materials have chemical formula AB_2 where A cation resides at corner and face-centred position, whereas B ion is located at tetrahedral holes in a FCC lattice. The A cation has eight surrounding B ions which suggest that the ionic radius of A should be such that it can accommodate eight ions. In many cases like zirconia, the Zr^{4+} ion is relatively smaller, therefore it cannot adopt eightfold coordination. Therefore, monoclinic ZrO_2 is doped with relatively bigger cations yields fluorite-type ZrO_2. Around 16 at.% Y^{3+} is required to stabilize monoclinic ZrO_2 in cubic form. This yttria-stabilized zirconia (YSZ) acts as an excellent oxide ion conductor. Similarly, calcia-stabilized zirconia (CSZ) also is an example of oxide ion conductor which is used in oxygen sensors. It should be mentioned here that the anion defects generated also contribute towards stabilization of cubic zirconia. In nuclear industries, yttria-doped thoria (YDT) is used as oxygen sensors. Lightly doped ceria by lanthanides (Ln^{3+}) results in the formation of solid solutions in fluorite structure. [48a, b] This doped ceria, especially Sm-doped ceria and Gd-doped ceria, exhibits superior oxide ion conductivity which find applications in solid oxide fuel cell and oxygen sensors. It has been observed that Sm-doped ceria and Gd-doped ceria exhibit superior ionic conductivity at relatively lower temperature ($\sim$600 °C), therefore, finds suitability as electrolyte in SOFC technology. The constituent oxides such as CeO_2, Gd_2O_3 or Sm_2O_3 are preheated over night to remove the moisture and any other volatile impurity present in the system. These oxides are then mixed and ground well. The mixed powder is then pelletized and heated at higher temperature [49]. The resulted pellets were found to be fluorite in structure.

Another very useful form of fluorite structure is MOX fuels being used in nuclear reactors. MOX fuel refers to mixed oxide fuel for fast breeder nuclear reactors. The nuclear reactors (PWR and PHWR both) generate Pu from U-238. After reprocessing of waste, plutonium oxide is prepared from plutonium nitrate. The conventional fabrication route for synthesis of MOX fuel is mechanical blending of the feed powders: UO_2 and PuO_2, or $(U–Pu)O_2$. As the blended powder is not free flowing and is therefore unsuitable for feeding to a pellet press, the powder is preconditioned by pre-compaction in a slugging press, followed by granulation, the granules being obtained by crushing the slugs. The challenge in this process is to obtain a uniform distribution of the plutonium in the product [47].

Optimizing the ball (or attritor) mill is of paramount importance for achieving uniformity of the plutonium distribution, as well as a good dispersion of the lubricant and of the pore former, if the use of a pore former is required [50].

1.4.2 Synthesis of Pyrochlores

Pyrochlore is a cubic structure with space group is Fd-3m, (Z = 8, a ≈ 10 Å), and the formula is ideally $^{VIII}A_2^{VI}B_2^{IV}X_6^{IV}Y$ (coordination numbers are indicated as superscript), where A is a mostly trivalent rare earth ion and B is mostly transition metal at their suitable oxidation state [51]. In simpler way, pyrochlore structure ($A_2B_2O_7$) is superstructure form of fluorite (MX_2)-type structure, where A and B cations are at two crystallographically distinct sites and one-eighth of the oxygen atoms are absent. The phase stability of pyrochlore structure is basically decided by the ionic radius ratio of A and B cations. In the structure $A_2B_2O_7$ where the sizes of A and B cations are closer enough (r_A/r_B close to 1.4), the cations A and B have the tendency to swap their sites which leads to the formation of disordered fluorites instead of ordered pyrochlores. However, if the radius ratio is higher, then degree of disorder is less. For instance, $Er_2Zr_2O_7$ having $r_A/r_B = 1.39$ prefers to crystallize in disordered fluorite form, whereas $Er_2Ti_2O_7$ with $r_A/r_B = 1.66$ adopts ordered pyrochlore form [52]. On the other hand, if the radius ratio is more than the stability of pyrochlore, then it prefers to stabilize in layered-type monoclinic structure (e.g. $Nd_2Ti_2O_7$ etc.). Interestingly, the $A_2B_2O_7$ structure with lower radius ratio has tremendous radiation stability because the cations can exchange their sites to dissipate the extra energy due to radiation. Therefore, $Y_2Zr_2O_7$ is found to demonstrate better radiation resistance since the r_Y/r_{Zr} ratio is 1.41 which is at the borderline of pyrochlore to defect fluorite phase.

In these types of materials, AR grade rare earth oxide and ZrO_2 are heated at 900 °C for overnight to remove moisture and other volatile impurities, if any. Stoichiometric amounts of the reactants were taken to synthesize the compositions corresponding to $RE_2Zr_2O_7$. The starting reagents were mixed thoroughly and then heated at three different temperatures with intermittent grindings. The ground mixtures were pelletized and heated at 1200 °C for 36 h, followed by heating at 1300 °C for 36 h and at 1400 °C for 48 h. In the synthesis of defect fluorite and pyrochlore, heating rate and cooling rates are important. The slow cooling rate favours ordering of cations and anions which leads to the formation of pyrochlore. By suitably tuning the degree of ordering, ionic conductivity can be varied. It has been shown that in case of Nd-doped $Gd_2Zr_2O_7$ pyrochlore, when the ratio of Gd/Nd = 1, then oxide ion conductivity becomes the highest [36]. Fast cooling or quenching leads to the formation of defect fluorite structure in case of borderline pyrochlore. Very interestingly, if $Gd_2Zr_2O_7$ is quenched from high temperature, it crystallizes in defect fluorite structure. Among the other rare earth-based zirconate pyrochlore, synthesis of $Ce_2Zr_2O_7$ is very interesting. Initially, oxygen-rich $Ce_2Zr_2O_8$ forms by reaction between nano-CeO_2 and ZrO_2. The extra oxygen can be de-intercalated by heating in reducing atmosphere. Mandal et al. have been reported that $Ce_2Zr_2O_8$, an oxygen-rich pyrochlore transforms to stoichiometric pyrochlore $Ce_2Zr_2O_7$ at 1000 °C in reducing atmosphere. However, $Ce_2Zr_2O_7$ can be again transformed to anion-rich pyrochlore by gentle oxidation [53].

Another interesting case is the formation of perovskite-type $Eu_2Ti_2O_7$. $EuTiO_3$ is could be synthesized by heating well-mixed Eu_2O_3 and TiO_2 in 1:2 mol ratio at 1000 °C for 24 h in Ar + H_2 atmosphere. When $EuTiO_3$ is heated in open air, it initially transforms to an amorphous perovskite-related intermediate and remains like that till 650 °C. Upon further heating till 750 °C, layered perovskite polymorph of $Eu_2Ti_2O_7$ begins to crystallize, and it remains stable up to 800 °C. However, further heating leads to the formation of pyrochlore polymorph [54].

Synthesis of vanadate pyrochlores is quite interesting. Since the radius ratio of vanadium pyrochlores of Tm^{3+}, Yb^{3+} and Lu^{3+} is well within the limit, therefore, these could be prepared by solid state route by reacting Ln_2O_3 (Ln = Tm, Yb, Lu) with V_2O_3 and V_2O_5 in sealed silica tube at 1250 °C for 72 h [55]. The other vanadate pyrochlores (Tb, Dy, Ho, Er, Y) are difficult to prepare even though radius ratio is still within limit as V^{4+} disproportionate to V^{3+} and V^{5+}.

The compounds manganate pyrochlore ($Y_2Mn_2O_7$) could be synthesized by high oxygen partial pressure and high temperature (1100 °C); however, $Y_2Cr_2O_7$ could not be prepared in single phase [56]. $Ln_2Nb_2O_7$ pyrochlores could not be prepared in pure form by solid state synthesis. These synthesis methods always lead to the formation of $LnNbO_4$ as an impurity. Molybdate pyrochlores are also very interesting as all these pyrochlore ranging from Ln = Sm to Lu could be prepared by reaction with MoO_2 in sealed silica tube as Mo has strong tendency to oxidize to MoO_3. Ruthenate pyrochlores could be prepared by standard solid state reaction [57, 58].

Among the pyrochlores with 5D elements at B site, hafnate, osmate, iridate and platinate pyrochlores are reported in literatures [58]. The rare earth hafnates (La–Dy) are easy to prepare as the starting materials can be taken into stoichiometric ratio and heated at higher temperature. In most of the cases, three-step heating protocols are preferred as this gives rise to phase pure materials. Hafnates with heavier rare earth ions crystallize in defect fluorite structure as radius ratio does not permit it to crystallize in ordered pyrochlore structure. The osmate pyrochlores ($A_2Os_2O_7$, A = Ln, Y, Bi) are prepared at high temperature and ambient pressure. However, the Tl analogue ($Tl_2Os_2O_7$) synthesis needs high oxygen pressure [59].

The platinate pyrochlores ($A_2Pt_2O_7$, A = Pr–Lu, In, Y) could also be prepared at high temperature and high pressure. Other than transition elements at B site, Si and Ge have also been tried by several researchers for the synthesis of $A_2B_2O_7$ pyrochlores (A = Lu, Y, Sc, Tl). High pressure and high temperature are essential for phase pure products. However, tin-based pyrochlores do not need high pressure. These pyrochlores ($A_2Sn_2O_7$, A = La–Lu and Y) could be prepared at high temperature and at ambient pressure [58].

1.4.3 Synthesis of Perovskite-Based Materials

Solid state sintering method is the most simple and versatile method for synthesis of perovskite-type materials also. Simply, the starting oxides or carbonates are mixed and heated at higher temperature. The heating temperature depends on melting temperature or decomposition temperature of constituent oxides or carbonates. The heating time and heating and/or cooling rate are very important for synthesis of technologically important ceramics. Mostly, the soaking time to carry out the reaction ranges from 8 to 24 h to allow the diffusion of cations through the crystalline grains to form perovskite structures. Like synthesis of other solid materials, the cations diffuse from the bulk to interface and form the product. Intermittent grinding is essential to break the chunks and bring the new surface for reaction. The proceeding of solid state reaction is depicted in Fig. 1.3.

The physical and chemical properties of the perovskite materials are strongly influenced by either the preparation or sintering conditions. This excellent class of materials exhibits various properties due to flexibility in its compositions and its tolerance towards ionic defects. Such diversity in compositions enables perovskite-related structures to exhibit a wide variety of physical properties.

$LaGaO_3$ co-doped with Sr^{2+} and Mg^{2+} (LSGM) is one of the high performing oxygen ion conducting electrolytes. Also, various authors have worked with Sr-doped $LaMnO_3$ as cathode. Attempts have been made to improve the electrical conductivity of these oxides through various doping at both the cationic sites with the aim to increase the number of oxygen vacancies. These perovskite materials have been prepared by mixing and heating the constituent oxides [60, 61].

Solid state synthesis of Bi-based perovskite ($BiFeO_3$, $BiMnO_3$ etc.) compounds is really challenging due to high volatility of Bi_2O_3 which is one of the reactants. This phenomenon results Bi-deficient phase which has severe impact on the properties of the desired compounds. $BiFeO_3$ is a well-known multiferroic material due to non-centrosymmetric nature of its structure. The preferential evaporation of Bi at higher temperature leads to the formation of sillenite-type $Bi_{25}FeO_{39}$ and the mullite-type $Bi_2Fe_4O_9$ impurity phases. These secondary phases are intermediate phases for formation of the ferrite; however, their elimination through a simple solid state reaction route has proven very difficult. The Bi-rich phase, i.e. $Bi_{25}FeO_{39}$ can be removed by leaching followed by drying at relatively higher temperature. In

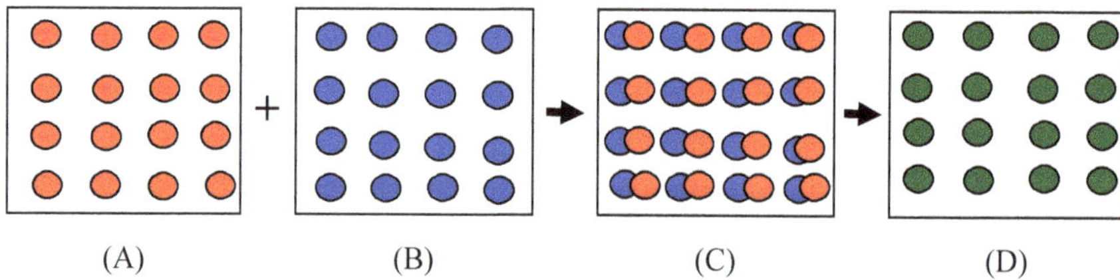

Fig. 1.3 Reactants (**A** and **B**) diffuse from bulk to the interface (**C**) and then form product (**D**)

other words, synthesis of $BiFeO_3$ should be done using about 10% excess Bi_2O_3 with Fe_2O_3. The mixture needs to be calcined at 650–675 °C for 1.5 h. This temperature is sufficient enough for reaction between Bi_2O_3 and Fe_2O_3. At lower temperature (<650 °C), the reactants do not react completely. However, in the narrow range of temperature, i.e. 650–675 °C, the reaction is completed. Since some excess Bi_2O_3 is taken at the beginning, therefore, Bi-rich phase ($Bi_{25}FeO_{39}$) forms simultaneously with $BiFeO_3$. This Bi-rich phase can be easily removed by leaching with dilute nitric acid (0.3–0.5 M). If the mixture is heated at higher temperature (>700 °C), the $BiFeO_3$ gets decomposed into Bi-rich and Bi-deficient phases following the reaction

$$49BiFeO_3 = 12Bi_2Fe_4O_9 + Bi_{25}FeO_{39}$$

It has been reported earlier that three-phase coexisting system is thermodynamically stable during the solid state reaction [62]. The coexistence of different distinct phases is explained by the diffusion of Bi ions into Fe_2O_3 phase. In more precise term, the nuclei of the $Bi_2Fe_4O_9$ phase are formed in the Fe_2O_3 core, whereas the outer shell remains as Bi-rich $Bi_{25}FeO_{39}$ phase. With increase in temperature, Bi ions of Bi-rich phase slowly diffused to Bi-lean phase and leads to the formation of phase pure $BiFeO_3$. Simultaneously, the core $Bi_2Fe_4O_9$ crystallizes with increase in temperature, and this by-product tends to block the formation of $BiFeO_3$. Therefore, higher calcination temperature always leads to the formation of $Bi_2Fe_4O_9$ which cannot be removed by leaching with nitric acid. In conclusion, it can be stated that while synthesizing $BiFeO_3$ by solid state reaction, it is recommended to heat well-mixed Bi_2O_3 and Fe_2O_3 at 650–675 °C for 1.5 h followed by leaching with dilute HNO_3. The residue can be made free from any acid just by washing with copious amount of water. The wet powder needs to be calcined again to get well crystalline product [63].

Other Bi-based analogues like $BiMnO_3$ and $BiCrO_3$ are prepared using high pressure and high temperature.

Double Perovskites

The double perovskites ($A_2B'B''O_6$) like Sr_2FeMoO_6, Sr_2CoBiO_6, etc., are so named because the unit cell is twice that of perovskite. It has the same architecture of 12 coordinated A sites and six coordinated B sites, but two cations are ordered on the B site. The B′ and B″ atoms are ordered in a 3D chessboard-type fashion. There are some Os-based double perovskites which exhibit wide range of properties. Ba_2MOsO_6 (M = Ca^{2+}, Li^+, Na^+), Sr_2MOsO_6 (M = Li^+, Na^+, Mg^{2+}, Ca^{2+}), A_2NiOsO_6 (A = Ca, Sr), Ln_2LiOsO_6 (Ln = La, Pr, Nd, Sm), and Ln_2NaOsO_6 (Ln = La, Pr, Nd) have been synthesized by solid state method. Some of them show remarkable magnetic and electronic properties such as the Os-containing double perovskite oxide Sr_2CrOsO_6, which shows an outstanding Curie temperature (725 K) [64]. The magnetic properties may reflect its potential uses in various applications and materials science. Similarly, polycrystalline compound Ca_2InOsO_6 has been synthesized by means of a solid state reaction. Stoichiometric amounts of

CaO_2, Os, In and $KClO_4$ and powders were mixed using an agate mortar in a glove box; the mixture was then put into a platinum capsule which was sealed from both the sides. The sealed capsule was heated at 1400 °C for 1 h in 6 GPa pressure [65]. However, lots of other double perovskites are there which do not need high pressure.

1.4.4 *Layered Perovskites*

Perovskites may be structured in layers. The ABO_3 units are separated by thin slabs of another materials or ions. Different forms of intrusions are reported in literature.

1.4.4.1 Ruddlesden–Popper (R-P) Phase

In R-P phases, the intruding layer occurs between every one (n = 1) or two (n = 2) layers of the ABO_3 lattice. The general formula of R-P phase is $A_{n+1}B_nX_{3n+1}$ (A_{n-1} $A'B_nX_{3n+1}$) where A and A′ represent alkali, alkaline earth or rare earth metal, whereas B is transition metal. In R-P phase, the A cation has 12-fold coordination and is located in the perovskite layer. The A′ cations are surrounded by nine anions and are positioned at the perovskites boundary with an intermediate layer [66]. B cation is usually located inside the octahedra or pyramid or squares. The first R-P phase was reported in 1958. These R-P phases are prepared by heating the constituent oxides or carbonates in stoichiometric ratio.

1.4.4.2 Aurivillius Phase

The Aurivillius family $\left(A'_2O_2\right)\left(A_{n-1}B_nX_{3n+1}\right)$ of oxides is well known for having ferroelectric properties at high temperatures. The intruding layer is $[Bi_2O_2]^{2+}$ ion, residing between ABO_3 layers, leading to an overall chemical formula of $[Bi_2O_2]$-$A_{(n-1)}B_2O_7$. It is expected that some compounds of this family may replace the existing lead zirconate titanate (PZT) in near future as the present candidate contains lead which is not a favourable choice due to the presence of lead ion in the compound. Lead-free $SrBi_2Ta_2O_9$ and $Bi_4Ti_3O_{12}$ are prominent member with high ferroelectricity. However, high ferroelectric fatigue in $Bi_4Ti_3O_{12}$ and lower remnant polarization in $SrBi_2Ta_2O_9$ are serious concerns for practical application.

The compounds with Aurivillius structure are formed by stacking of two different blocks of $\left(A'_2O_2\right)^{2+}$ and $(A_{n-1}B_nX_{3n+1})^{2-}$ along the c-axis in the crystal structure. The $\left(A'_2O_2\right)$ slab is of red-PbO type and A′ is mostly Bi(III) cation or other cations with lone pair of electrons (e.g. Te(IV), Pb(II), Sb(III)). The $(A_{n-1}B_nX_{3n+1})$ block follows pseudo-perovskite structure, where A and B have coordination number 12 and 9, respectively. The anion is mostly oxygen; however, in some cases it may be partially substituted by F.

The compound like Bi_2WO_6 exhibits high ferroelectricity, whereas the Mo analogue, i.e. Bi_2MoO_6 demonstrates low-temperature co-fired ceramics (LTCC) properties. However, some Aurivillius-type compounds also exhibit high oxide ionic conductivity. Some Aurivillius phase can be synthesized from D-J phases also by metathesis reaction. Sivakumar et al. reported the transformation of the Dion–Jacobson phases, $KLaNb_2O_7$ and $RbBiNb_2O_7$ to the Aurivillius phase, $(PbBiO_2)LaNb_2O_7$ and $(PbBiO_2)BiNb_2O_7$ in a metathesis reaction with the Sillen phase, $PbBiO_2Cl$ [67].

1.4.4.3 Dion–Jacobson (D-J) Phase

In this kind of phases, the intruding layer is composed of alkali metal (M) in every ABO_3 layers, which leads to overall formula as $M + A_{(n-1)}B_nO_{(3n+1)}$. D-J phase like $KLaNb_2O_7$ and $RbBiNb_2O_7$ has been prepared by reacting stoichiometric quantities of KNO_3, La2O_3, Nb2O_5 at 1100 °C for 2 days with intermediate grinding. Slight excess KNO_3 (25 mol%) needs to be added before reaction to compensate K loss due to volatilization. $RbBiNb_2O_7$ was prepared by reacting stoichiometric quantities of $RbNO_3$, Bi_2O_3 and Nb_2O_5 at 1100 °C for 2 days with intermediate grinding. In this case, also slight excess $RbNO_3$ (20 mol%) needs to be added to take care Rb loss due to volatilization.

1.4.4.4 Brownmillerite Structure

The brownmillerite structure, general formula, $A_2B_2O_5$, is a perovskite-related structure with a oxygen vacancy per one formula unit. The oxygen vacancies are fully ordered in brownmillerite structure where BO_6 octahedra and BO_4 tetrahedra are alternatively arranged. $Bi_2In_2O_5$ is one of the famous electrolytes which has been explored in recent past. In brownmillerite form, the oxide ion conductivity is not very high due to ordering of oxygen ions. However, it transforms to cubic perovskite, which is a disordered variant, upon heating. The high temperature phase exhibits very high oxide ion conductivity (0.1 S/cm at 900 °C). The high-temperature cubic phase could be stabilized at room temperature by suitable substitution of A and B cations. Yao et al. [68] have shown that by doping Ga^{3+} at In^{3+} site of $Ba_2In_2O_5$, the disordered can be introduced at relatively low temperature. In case of $Ca_2Fe_2O_5$ compounds, it has been observed that transition temperature can be reduced by 15 °C by doping with relatively smaller cation like Al^{3+} at Fe site. This brownmillerite compound is G-type antiferromagnetic in nature.

1.4.5 Spinel Structure

Another interesting class of compounds is spinel. Its chemical formula is AB_2X_4. It consists of eight FCC cells made by oxygen ions in the configuration $2 \times 2 \times 2$. Depending upon the coordination of A and B cations, the structure can be normal or inverse spinel. When all the A cations occupy the tetrahedral sites and B cations occupy octahedral sites, then the structure is called normal spinels. All the aluminates like $MgAl_2O_4$, $FeAl_2O_4$, etc., and some ferrites like $ZnFe_2O_4$, $CdFe_2O_4$ are examples of normal spinel. In case of inverse spinel, the tetrahedral site is occupied by B cation and the A cations reside at octahedral site of the structure. Most of the ferrites like Fe_3O_4, $CoFe_2O_4$, etc., adopt inverse spinel structure. In addition, these spinels might adopt mixed characters, i.e. partial inversion, also.

$LiMn_2O_4$ an important cathode material for Li ion battery adopts spinel structure. The Li ion from the structure can be intercalated and de-intercalated electrochemically without appreciable structure distortion. $Li_4Ti_5O_{12}$ anode also adopts spinel structure where some Li ion resides at Ti site. This structure can accommodate extra lithium in it, and it can be turned into $Li_7Ti_5O_{12}$ (rock salt structure) upon heavy lithiation. Both of these materials can be successfully prepared by solid state reaction method. Since lithium, a volatile element is being used; therefore, some excess lithium is needed in the beginning of synthesis.

Degree of mixing of A and B cations depends on synthesis condition. The site occupancy by iron can be traced by Mossbauer spectroscopy. Other spectroscopic tools like IR, EXAFS, etc., are also used to determine the coordination number of A and B cations.

1.4.6 Hexaferrite Synthesis

In addition to cubic spinel ferrites (AFe_2O_4), there are other types of ferrites which are hexagonal in structure. These types of ferrites are called hexaferrites. These hexaferrites are technologically and commercially important material. These materials are found to be useful in the field of magnetic recording, data storage, etc.

The synthesis of hexaferrites is really tricky, and its formation mechanism is not very clear even after more than five decades of research. Standard solid state reaction method is used to prepare the hexagonal ferrites. Initially, iron oxide and barium carbonate powders are mixed and heated at higher temperature to obtain the desired phase. High temperature and longer reaction time lead to the formation of product with coarse grain size. However, if reactants are in nano regime, then reaction temperature as well as time can be reduced drastically.

1.4.7 Tungsten Bronze

Tungsten bronzes were discovered by Wohler way back in 1824. Since these compounds have metallic properties, therefore, the name has been applied. These materials do not have any relation with the bronze alloy of copper and tin. These materials are neither alloy nor intermetallic. There are several tungsten bronze types of materials. It was observed that alkali metal tungstate becomes golden in colour if heated in hydrogen atmosphere. These are non-stoichiometric compounds of general formula $M'WO_3$ where M is metal ion mostly an alkali, and x is <l.

It has been observed that not only tungsten present in the structure, Mo, V, etc., can also be present in the structure and display similar kind of properties. This class of compounds exhibits various properties like (a) possess excellent electronic conductivity, (b) it is chemically inert, (c) intensely coloured and (d) sequence of solid phases occur with variation of x in $M'_xM''_yO_z$ The colour of the compositions depends on the extent of alkali metal ions present in the composition.

The structure of tungsten bronze, $NaWO_3$ ($x = 1$), adopts cubic perovskite structure. Lower value of x signifies that $(1 - x)$ is the vacancy present in the structure. On the other hand, WO_3 is like distorted ReO_3-type structure. In this structure, tungsten atoms are slightly off-centre in adjacent unit cells such that the W-W distances are alternately long and short. As the sodium concentration decreases, the cubic structure also lowers its symmetry, and it passes through two tetragonal (type I and II) phases to distorted monoclinic WO_3 structure.

The other bronze structures with Mo and V as the central cation have much more complex structure.

The synthesis of Na_xWO_3 is very interesting. The finely grounded mixtures of Na_2WO_4, WO_3 and W are mixed thoroughly and heated in vacuum at 850 °C. This leads to the formation of Na_xWO_3. In this way, composition with maximum $x = 0.8$ could be achieved [69, 70].

1.4.8 High Tc Oxides

High-temperature superconductors (sometime abbreviated HTSC) are materials that behave as superconductors at relatively high temperatures. Bednorz and Müller [71] were awarded the 1987 Nobel Prize in Physics for their work in high Tc superconductor $La_{2-x}Ba_xCuO_4$.

In case of "ordinary" or metallic superconductors, usually the transition temperatures lie below 30 K; therefore, these materials need to be cooled by liquid helium to realize superconductivity. On the other hand, high Tc materials exhibiting Tc more than 138 K should be cooled to superconductive state by liquid nitrogen (77 K) [72]. Until 2008, only certain cuprate compounds were reported to demonstrate high Tc properties. Some iron-based compounds are also known to show superconductivity at relatively high temperatures [73].

The first real high Tc superconductor was reported to be $YBa_2Cu_3O_{7-x}$. [74]. Its structure is closely related to distorted oxygen-deficient multi-layered perovskite structure. The most important characteristic of the crystal structure of oxide superconductors is alternating multi-layer of CuO_2 planes in the structure. It has been observed that with increase in number of layers of CuO_2, transition temperature increases.

The simplest method to prepare high Tc superconductors is a solid state reaction involving mixing, grinding, calcining and sintering. The required amounts of precursor powders, mostly oxides, carbonates or any other salts, are mixed thoroughly in a mortar–pestle or ball mill. These mixed powders are sintered in the temperature range from 800 to 950 °C for few hours with intermittent grinding. Subsequently, the powders are pelletized and sintered at 950 °C. The sintering parameters like temperature, soaking time, atmosphere and heating/cooling rate have important role in synthesis of high T_c superconducting materials. As for example, $YBa_2Cu_3O_{7-x}$ compound is prepared by calcination and sintering of a homogeneous mixture of Y_2O_3, $BaCO_3$ and CuO in stoichiometric ratio. Initial calcination is done at relatively lower temperature (900–950 °C) in ambient atmosphere, whereas sintering is done at 950 °C in the presence of oxygen atmosphere. The oxygen stoichiometry in $YBa_2Cu_3O_{7-x}$ compound is very important to get superconductivity. The formed product is $YBa_2Cu_3O_6$ which transforms to $YBa_2Cu_3O_{7-x}$, on slow cooling in oxygen atmosphere. The latter compound is superconducting in nature. Several other superconductors could also be prepared by solid oxide synthesis method.

1.4.9 GMR Materials

Magnetoresistance of any material is defined by relative change in resistance upon application of magnetic field. It is defined as MR (%) = $(\rho_H - \rho_O) \times 100/\rho_O$ where ρ_H and ρ_O are resistivity in the presence and absence of magnetic field. When the change is very high, then it is called giant magnetoresistance (GMR). It was initially observed in multi-layer metallic system. Later on, it has been observed in oxides also. When the La site of $LaMnO_3$ substituted by alkaline earth metal (like Ca or Sr), the material turns into metallic state. In fact, upon substitution of divalent cation in place of La leads to the formation of Mn^{4+} in the structure and around 30% Mn^{4+} is optimum to observe GMR properties. Other than manganites, some alkaline earth-doped lanthanide cobaltites ($La_{1-x}A_xCoO_3$) also exhibit GMR properties [75]. Among pyrochlores, $Tl_2Mn_2O_7$ exhibits GMR properties. This oxide has only Mn^{4+} (d^3), still it shows GMR properties [76].

These materials are prepared by standard solid state method by using the oxides of constituent oxide and heating them at high temperature.

1.4.10 Energy Storage Materials

Lot of materials are being investigated in as energy storage materials. Among energy materials, electrodes of batteries and supercapacitors are most important. Initially, the electrodes were synthesized by solid state method only. Subsequently, these materials were synthesized by soft chemical routes to improve the energy storage capabilities.

Prof. J. B. Goodenough first identified Li_xCO_2 compound as an excellent electrode material to be used in rechargeable Li ion battery. Later, Sony successfully commercializes this lithium ion battery using this cathode material. The synthesis of this compound was first done by solid state route only. $LiFePO_4$, $LiMn_2O_4$, $Li_4Ti_5O_{12}$, etc., are other electrode materials for the use in lithium ion battery. There is lithium ion channel in these compounds which make them suitable lithium ion conductor. In addition, the electronic conductivity of these compounds is reasonable which is prerequisite for a material to be electrode material.

1.4.11 Negative Thermal Expansion Materials

Research on negative thermal expansion materials remained very vibrant research topic for some time. Among the ceramics, ZrW_2O_8 is the very famous example that exhibits negative thermal expansion behaviour in the temperature range 0.3–1050 K. Above 1050 K, the compound decomposes [77]. In addition to this example, there are other members of the AM_2O_8 (A = Zr or Hf, M = Mo or W) family of materials and ZrV_2O_7 which also show negative thermal expansion behaviour. $A_2(MO_4)_3$ family also exhibits controllable negative thermal expansion. During synthesis of ZrW_2O_8 material, ZrO_2 and WO_3 are mixed in 1:2 ratio, and then the mixed powders were grounded well. The mixed powder needs to be heated at 1150–1200 °C for 6 h. This leads to the formation of phase pure ZrW_2O_8. Other materials could also be prepared by similar method [78, 79].

1.5 Solid State Organic Synthesis

The awareness about reducing the environmental pollution associated with using solvents for organic synthesis has led the scientists to develop and employ successfully the methodologies which use no solvents. The solvents are used for obvious reasons, and they provide a media for interaction of reactants and facilitate faster reactions. Sometimes, the solvents influence reactions by other means as well such as forming intermediates with the reactants which ultimately yield the products; however, they are always released at the end of the reaction. Various solvents that are employed include water, ammonia, hydrocarbons, ethers, alcohols, halogenated hydrocarbons, etc. The solvent-free routes to synthesize organic compounds are relatively benign, environment friendly and are termed as solid state synthesis. Also, in many cases, the reactions in solid state proceed more efficiently and selectively in comparison with their solution counterparts since molecules in solids are arranged in more regular and arranged manner [80].

There are two types of solid state organic synthesis as follows:

1. Solid phase organic synthesis without any solvent
2. Organic synthesis using a solid support.

1.5.1 Solid Phase Organic Synthesis Without Any Solvent

These synthetic processes follow the solid state path, wherein the reactants are mixed together in mortar and pestle (or by other means) and kept at room temperature or subjected to elevated temperatures (in suitable apparatus) and/or radiations such as UV, microwave, etc., to carry out the reaction. The elimination of solvents from the synthetic procedures also reduces the handling costs and greatly improves the simplicity of the procedure in terms of reaction work up, etc., these factors become more important when the syntheses are scaled up to industrial level. The progress of solid state reactions can be followed by measurement of IR, UV and circular dichroism (CD) spectra. The formation of the products also depends on the particle size sometimes. It has been showed that the smaller the particle size, faster is the reaction rate. For example, the Baeyer-Villiger oxidation of benzophenone derivatives with *m*-chloroperbenzoic acid in the solid state proceeded ~10 times faster when the particle diameter of both components was halved to 50 μm from 100 μm. However, in some of the cases, the size of the particle is not related to reaction rate and reaction can take place smoothly by simple mixing of reactants without grinding.

One of the earliest examples of solid state synthesis in dry state could possibly be the rearrangement of allyl phenyl ether to *o*-allyl phenol (Claisen rearrangement) [81]. Another classic example of solid state organic synthesis is the formation of urea in 1828 by Wohler [82], which formally led to the birth of synthetic organic chemistry.

$$NH_4NCO \longrightarrow [solid]\Delta NH_2CONH_2$$

Inclusion complexes can easily be formed by solid state reaction and can occur readily by mixing and grinding powdered host and guest. For example, a mixture of powdered 1,1,6,6-tetraphenyl hexa-2,4-diyne-1,6-diol and equi-molar amount of benzophenone was found to be identical to their 1:1 inclusion complex which is normally prepared by recrystallization of the two components from solution, and this shows that complex formation occur very rapidly [83]. The organic solid state reactions even occur enantioselectively. Some of the oxidations of ketones with *m*-

chloroperbenzoic acid (Baeyer-Villiger oxidation) occur faster in solid state than in $CHCl_3$, and even the yields are better. For example, the following reaction occurs in 24 h, and in solid state the yield is 85%, whereas in $CHCl_3$ the yield is observed to be 13% [84].

$$PhCOPh \rightarrow PhCOOPh$$

Reduction of ketones with $NaBH_4$ can also be made to occur solid state with impressive yields. The solid state reaction of diketones with hydrazine hydrate to obtain pyrazole derivatives by a solvent-free route has been reported by Wang and Qin [85]. Some other reactions that can be/are carried out in the absence of solvents are Michael addition (addition of a nucleophile to carbon double bond having a strong electron withdrawing group), Aldol condensation [86], Tishchenko reaction, etc. Solvent-free synthesis of various azines, pyrazoles and pyridazinones can be carried out with high yields (>97%) by grinding solid hydrazine with dicarbonyl compounds. The syntheses did not require any catalysts or additives to promote the reactions [87]. Pericyclic reactions such as Diels–Alder reactions can also occur in solid state. Herein, a diene and the dienophile can combine to yield a number of [4 + 2] additions. Thus, a variety of solid state reactions can be carried out under solvent-free conditions. For an extensive review of solid state organic synthesis, a lucid review by Tanaka et al. [80] can be referred.

1.5.2 Organic Synthesis Using a Solid Support

There is another methodology of solid state synthesis of organic moieties known as solid phase synthesis, wherein one of the reactant molecules is attached to a solid support through a linker. The product is formed on the solid support as the reaction proceeds, and then the product is separated by cleaving the linker. Hence, the idea of the solid phase supported synthesis is simple: the substrate is attached to a polymer support, the desired chemical reaction is carried out, and the excess reagents are washed off. This cycle of reaction wash is repeated until the synthesis is ready and then the product is cleaved off from the support. This methodology was originally developed for peptide synthesis but later was extended to general organic synthesis. It was termed as “solid phase peptide synthesis” by Merrifield in 1963 who used a polymer as the insoluble support to synthesize a tetrapeptide [88]. Merrifield received Noble Prize in 1984 for “his development of methodology for chemical synthesis on a solid matrix”. The advances in solid phase peptide synthesis were also motivated by the fact that the separation of the product is relatively easier since it remained attached to the solid support and the excess reagents/by-products can be removed by rinsing. This was soon extended to general organic synthesis. The major attraction for adopting this method was the ease of chemistry. The reaction involves adding the reagent, filtering and washing. Even the purification of the product is not needed at each stage, and only the final product needs to be purified. Also, since it is relatively easier to separate out the reagent, excess amounts of reagents may be added to drive the reaction faster. This method also allows for the selectively converting one functional group in a poly-functional molecule by binding it covalently it to the support. Many organic and inorganic supports have been utilized for the purpose. The basic criterion is that it should be amenable to chemical modification so as to generate group that can be utilized to bind the substrate molecule. In addition to chemical criteria, there are various physical properties as well that the material proposed to be used as support should possess. More is the number of chemically modified sites, more are the sites where substrates can attach. Hence, the surface area, porosity, pore volume, etc., are the properties that are important for a material to qualify at solid support. The support materials should show good swelling behaviour in the solvent. Better swelling behaviour is desired so that the resin is permeable to solvent and reagents providing better access to molecule to "attach' to the resin. It should be stable under thermal and chemical conditions of synthesis. It should also be cheap, easily available and compatible with reagents and solvents. There are two types of supports that exist (a) that simply act as the carrier that provide the support for reaction and the product is cleaved after formation (passive supports) and (b) that act as the catalyst also and catalyses the reaction (active supports). Different types of polymers, organic and inorganic materials have been employed as solid supports. Cross-linked polystyrene/DVB copolymers are the most common polymer that is employed for the purpose for they are cheap and possess excellent chemical stability [89]. Other than this, polyacrylamide, poly methyl methacrylate and poly hydroxyl methyl methacrylate have also been used as solid supports. PEG grafted onto cross-linked polystyrene is also employed. It has lower stability than polystyrene resins but much better solvent spectrum. In order to improve general resin performance, the resin based on bifunctional styrene derivatized PEG chains is employed to cross-link polystyrene. Polymer resins are cross-linked to impart enhanced mechanical strength and better diffusion and swelling behaviour to resin. Cross-linking introduces some permanent sites of entanglements that maintains structural integrity; otherwise, the polymer chains would dissolve under favourable thermodynamic conditions. Better swelling behaviour is desired so that the resin is permeable to solvent and reagents providing better access to molecule to "attach" to the resin.

The inorganic matrices that have been used as support include silica, zeolites and modified glasses. These have hydroxyl groups on their surface that may bind with substrate. Naturally occurring polymers such as cellulose and chitin are also the preferred choices. The efficacy of polymer support depends upon degree of cross-linking of polymer resin, the solvation characteristics of the resins and the functional groups available on the resin that would bind to the substrate.

Other than the support materials, choosing a proper linker is also essential to solid phase synthesis. The linker covalently attaches substrate to the solid support and hence provides a control for attachment and cleavage. The stability of the linker affects the scope of the reaction. The method to cleave the linker depends on the type of the linker moiety. The historical resins such as the one used by Merrifield have acid labile linkers. There are varied types of linkers which can be cleaved using a nucleophile or photolysis, etc.

The simplicity, ease of operation and effectiveness of solid supported synthesis led to its extensions to other areas of organic chemistry, e.g. synthesis of depsipeptides [90], polyamides [91], oligonucleotides [92–95], oligosaccharides [96–98], peptide nucleic acids [99, 100] and to solid phase organic synthesis (SPOS).

Since 1990, a lot of interest has been directed towards synthesizing small organic molecules on solid support. These syntheses can be classified based on the nature of bond formation taking place. These could be C–C bond formation reactions, C–N bond formation reactions, pericyclic reactions such as Diels–Alder reactions and electrophillic aromatic substitutions. Numerous examples [101] of cyclization reactions have been shown to be more efficient in the solid phase than in solution where oligomerizations compete readily. This is generally attributed to site isolation, a feature that also permits the solid support to act as a protecting group for one of the two identical functional groups in the same molecule [102, 103].

The solid supported synthesis involving anchoring of the substrate to an insoluble matrix has several advantages. Some of them are listed as follows. Simplified work up since excess reagents as well as soluble by-products can be removed easily by washing and filtration [104]. The ease of separation allows for using larger excess of reagents which enables the reaction to proceed to completion and increases yield [104]. The extraction and/or chromatographic separation of the intermediate products can be avoided. Usually, the resin (or support) need not be dried, and this speeds up the synthesis procedure. This synthesis route is particularly amenable for industrial processes where automation of repetitive synthesis steps in a sealed environment brings about speed, better yields and reproducibility as compared to the manual work.

Click Chemistry in Solid Phase

Click chemistry is an approach that makes use of efficient and reliable reactions, such as Cu(I)-catalysed azide–alkyne cycloaddition (CuAAC) in order to bind two molecular building blocks and synthesize cyclic structures. CuAAC has wide applications in medicinal chemistry as well as other fields of chemistry. It has become a tool for universal modification of biomolecules such as DNA, proteins as well as for conjugate preparation and fluorescent labelling. Few of its advantages are selectivity, insensitivity to pH, shorter duration, and it is quantitative. The concept of click chemistry was established by Kolb et al. [105]. The advantage of using solid phase syntheses in case of reactions proceeding by click chemistry is particularly in context of product purification. In general, to obtain effective coupling, the syntheses are conducted with excesses of conjugating group and in the solid phase approach. The excess reagents, along with undesired side products, can then simply be washed away [106].

1.6 Metastable Materials

There are several definitions of metastable state. As per the common physics definition, metastable state is in a dynamical system with respect to the state of least energy. Another definition is that the metastable state corresponds to the local free energy minima. According to thermodynamics, metastability corresponds to non-equilibrium state. Metastable state tends to return to the lowest energy state. However, metastable materials are not unstable materials. They are kinetically stable materials and may exist for a prolonged period also. A classic example is diamond which is metastable but has been existing for a long period. The metastable state can be schematically represented in Figs. 1.4 and 1.5.

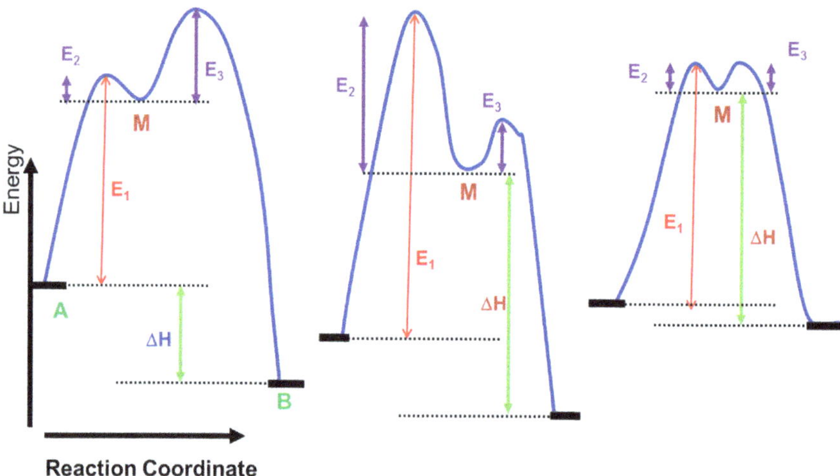

Fig. 1.4 Schematic representation of metastable state in terms of energy and reaction coordinates. The metastable state is denoted by M in these schematics

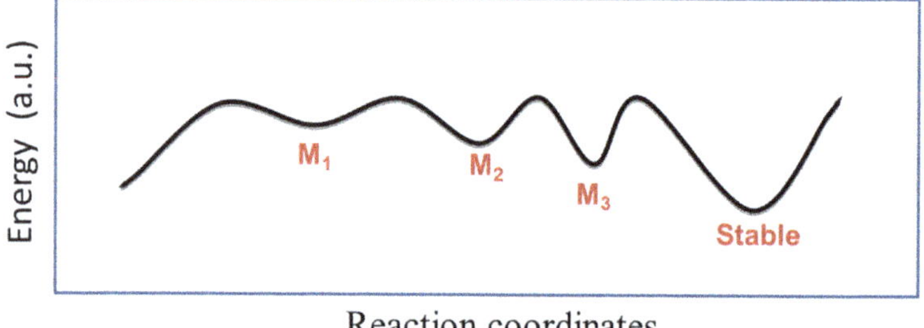

Fig. 1.5 Another way of representing metastable state. Three metastable states are denoted by M_1, M_2 and M_3, respectively, which correspond to local free energy minima. The last minimum is global energy minimum, which corresponds to the stable state in this imaginary system

1.6.1 Typical Examples of Metastable Materials

Some typical examples of metastable materials are given in the following table.

(a) Glasses (Natural: obsidian glasses, basaltic glasses)
(b) Man-made (sodium borosilicate glasses, metallic glasses)
(c) Diamond, cubic BN
(d) Steel (martensite), superalloys (Ni-based super alloy Ni–Al–Mo)
(e) High entropy alloys
(f) New forms of carbon (fullerenes, CNT, graphene)
(g) Minerals like gibbsite, silica (cristobalite, stishovite), opal, topaz aluminium silicates, several other gem stones
(h) Perovskite-type silicates, zeolites
(i) Many ceramics (t, c-ZrO_2, γ-Al_2O_3, ZrW_2O_8, anatase TiO_2, c-ZnO)
(j) Noble gas compounds
(k) Cement is also metastable
(l) Many drugs are also metastable.

1.6.2 Origin for Metastability

There can be several factors which can impart metastability to a given system. Some of them are elaborated as follows (Table 1.5).

1.6.3 Synthesis of Metastable Materials

Many conventional synthesis methods such as solid state synthesis do not yield metastable materials. This is because conventional solid state synthesis mostly used under equilibrium conditions, and hence, it results in stable materials. However,

Table 1.5 Some of the factors leading to metastability

Factor	Examples	References
Unusually high packing	$ThGeO_4$ (Scheelite)	[107]
Unusually high packing	α-$VO(PO_3)_2$	[108]
Unusually high oxidation state	$HoCrO_4$, $CaCrO_4$, etc.	[109]
Unusually low oxidation state	$CeCrO_3$ and $La_{1-x}Ce_xCrO_3$	[30, 32]
Unusually high coordination number	$LiScF_4$	[110]
Unusual order	Pyrochlore-type $Ce_2Zr_2O_8$	[111]
Unusual disorder	C-type $REInO_3$	[112]

heating in solid state followed by quenching can be used to prepare many metastable materials. Heating of reactants or stable products under high pressure and/or high temperature is another strategy to prepared metastable materials. Intercalation or de-intercalation is yet another simple method to get metastable materials. Several metastable materials which are almost impossible to prepare by other known methods can be easily prepared by an ion exchange method. There are several soft chemical methods which often lead to metastable materials. Several of these methods will be discussed in subsequent chapters, and readers are advised to refer to these chapters.

1.7 Common Characterization Techniques

1.7.1 X-ray Diffraction

Once a solid sample is formed, it needs to be characterized by suitable techniques to confirm phase purity, crystal structure, crystallite size, etc. In this context, XRD is the first technique of choice to characterize the solid sample. In this technique, a monochromatic X-ray beam falls on sample, and it gets diffracted following Bragg's law. Subsequently, a suitable detector detects the diffracted beam. The data is processed to obtain an X-ray diffractogram which contains a lot of information like phase purity, crystal structure, crystallite size of the sample, etc. These diffractograms could be read by suitable software.

Principle of X-ray Diffraction

X-rays are electrically neutral, electromagnetic radiations with their wavelength (λ) range from approximately 0.004–100 nm. In terms of energy, X-rays lie between ultraviolet (UV) and gamma radiations. Even though it is roughly stated that energy of X-ray is lower than gamma rays, it may not be true always. The difference lies in source of radiation, gamma rays are originated from nucleus of atom, whereas X-rays are extra-nuclear phenomena. Perhaps, discovery of X-ray is the most famous event in scientific arena in last decade of nineteenth century. X-rays were invented by Wilhelm Conrad Rontgen in Germany in 1895. He is also named it as "X-rays" keeping similarity with unknown quantity.

Characteristic X-rays are produced from the inner (core) level electronic transitions in an atom. The electromagnetic radiations are also produced when a charged particle undergoes deceleration. This radiation is also called as white radiation or Bremsstrahlung radiation. The moving charged particle loses its kinetic energy and converts to radiation by this way it follows conservation of energy.

The X-rays (both white radiation and characteristic radiation) get scattered when they fall on solid material. The regular arrangement of atoms in solid materials acts as grating for the incident X-rays. When the coherent scattering of X-rays interferes with each other constructively, then it gives rise to bright fringes, whereas when these X-rays interact destructively, they generate dark fringes. The scattered rays interfere constructively only at a certain angle when the path difference between the two rays is an integral multiple of incident wavelength. This phenomenon is called as X-ray diffraction. Bragg law of diffraction is given by the following equation.

$$n\lambda = 2d \sin \theta$$

where λ is wavelength of X-rays, θ is the incident angle, d is inter-planar spacing, and n is order of diffraction. Bragg diffraction was proposed by Lawrence Bragg and his father William Henry Bragg in 1913. Lawrence Bragg modelled the crystal as a set of discrete parallel planes composed of atoms and are separated by a constant parameter "d". Incident X-ray radiation would produce a Bragg peak if the diffracted beams from various planes of the sample interfered constructively.

Atoms, ions and molecules are periodically arranged in a crystal lattice. These parallel arrangements of atoms in the crystal lattice are equivalent to the parallel lines of the diffraction grating. Therefore, the inter-planar spacing of the atoms in crystal could be easily measured from the position of bright fringes in the diffraction pattern.

X-rays being electromagnetic radiations interact with the electrons of the atom present in the crystals, which vibrate with the same frequency as that of the X-rays. Then, these oscillating electrons act as source and re-emit radiation of the same frequency. Thus, the incident radiation appears to be scattered by the atoms.

Experimental technique

X-ray diffraction experimental set-up requires an X-ray source, sample and a detector to pick up the diffracted X-rays.

The X-rays are generally produced by bombarding high-energy electrons on a metal target in a sealed X-ray tube. During this process, the target element gets heated, and hence, it requires a continuous cooling. The X-rays are produced in all the directions and hence, slightly divergent X-ray beams emitted in a particular direction with respect to the exciting electron beam, are allowed to pass out through beryllium (Be) window. The background and β-radiation are filtered using β filters. The β filters essentially absorb the radiation other than characteristic radiation. The X-ray beam passes through the soller and divergence slits and then fall on the sample, which is mostly kept on a glass slide. X-rays scattered (diffracted) from the sample pass though the soller, receiving slits successively and then fall on a monochromator before reaching to detector. The monochromator filters out the unwanted wavelengths as well as fluorescent radiation arising from the sample. The details of the X-ray production and the typical X-ray spectra are explained in several Refs. [113, 114].

Data Collection and Analysis

The output of the diffraction measurement is obtained as plot of intensity of diffracted X-rays beam versus twice of incident angle. X-ray peaks are obtained at those angles where it follows Bragg's law. The data collection protocols often depend on the specific purpose of the data collections. In general, a short time scan in the 2θ range of 10–70° is required for the characterization of a well crystalline inorganic solids. However, the scan time should be optimized to get intense peaks. The obtained diffraction patterns are compared with standard Joint Committee on Powder Diffraction Standards (JCPDS, 1974) data available for reported crystalline compounds. Different parameters like cell parameters, thermal parameters and stress within the lattice can be calculated using different softwares like FULPROF, GSAS, etc.

1.7.2 Thermal Techniques

Thermogravimetric analysis (TGA) and differential thermal analysis (DTA) are also essential techniques for characterization of solids. The variation of mass as a function of temperature could be detected by thermogravimetric analysis. The weight changes can be monitored during heating as well as cooling or as a function of time (isothermal) in a specified atmosphere. In differential thermal analysis (DTA), difference in the temperatures between the sample under study and an inert reference material is recorded as a function of temperature, as both are simultaneously heated or cooled at a predetermined rate. If there is no heat effect in the sample in the temperature range of interest, the DTA scan will show a steady base line with a slight shift sometimes. On the other hand, if there is any physical or chemical change taking place in the sample accompanied by a significant heat effect, the temperature of the sample will either lag behind or lead with respect to that of the reference, and the difference in the temperature is recorded as a peak. The area under the peak represents the total heat change. The peak obtained for endothermic and exothermic reactions is in opposite directions.

The sample and reference material (here alumina) are kept in two similar platinum crucibles. The sample starts weight loss with increase in temperature. The weight loss is monitored by a thermo-balance, whereas two thermocouples are used to measure the difference in temperatures between sample and reference material and to measure the temperature of the sample. The simultaneous TG/DTA technique was used to determine (i) the thermal stability and decomposition behaviour of the precursor (ii) nature of the combustion reaction and (iii) the minimum calcination temperature.

Table 1.6 General comparison of solid state and solution reactions

Solid state reactions	Solution reactions
Slow	Fast
Mostly need higher temperatures	Room temperature or slightly above RT
Diffusion controlled	Diffusion is not very important
No well-defined rate laws	Well-defined rate laws
Sample history influence	No such influence
Defects enhance the reactivity	No such concept exists
Progress of reaction is difficult to monitor	Easy to monitor
Larger particles are obtained	Fine particles are obtained

1.8 Merits and Demerits of Solid State Method

Several of the subsequent chapters will be devoted to the solution-based synthesis methods. In view of this, a brief comparison of solid state method with solution-based synthesis method is given in Table 1.6. More details will be discussed in the respective chapters.

1.9 Summary and Future Scope

Solid state synthesis of materials is one of the oldest and widest methods used for synthesis of a wide range of materials. This chapter was intended to introduce various concepts of solid state synthesis method to the beginners in the field of materials sciences. Attempts were made to write a comprehensive article on various aspects of solid state synthesis. The readers are advised to refer the books [115–118] which deal extensively with experimental solid state chemistry. These will also apprise the young readers with new directions and trends in solid state chemistry. Several aspects such as scaling of product and precise cost comparison could not be discussed in this chapter as they were somewhat out of scope. However, still we feel that this chapter will inculcate deep interest among the young students in the field of materials.

References

1. Corbett John D (2000) Inorg Chem 39:5178–5191
2. Woodward RB (1963) Art and science in the synthesis of organic compounds: retrospect and prospect. In: Pointers and pathways in research. CIBA of India, Bombay
3. Giustetto R, Sreenivasan K, Bonino F, Ricchiardi G, Bordiga S, Chierotti MR, Gobetto R (2011) J Phys Chem C 115:16764–16776

4. Dejoie C, Martinetto P, Dooryhee E, Strobel P, Blanc S, Bordat P, Brown R, Porcher F, del Rio MS, Anne M, Appl ACS (2010) Mater Interfaces 2:2308–2316
5. (a) Del Rio MS, Domenech A, Domenech-Carbo MT, Pascual MLVA, Suarez M, Garcia-Romero E The maya blue pigment. In: Developments in palygorskite-sepiolite research: a new outlook on these nanomaterials, 1st ed.; (b) Galan E, Singer A (eds) (2011), vol 3. Elsevier, Oxford, p 520, (c) Lei t ão, I nês MV, Sér gio Seixas de Melo J (2013) Maya blue, an ancient guest—host pigment: synthesis and models. J Chem Edu
6. Warren JL, Geballe TH (1981) Mater Sci Engg 50:149
7. Bednorz JG, Müller KA (1986) Z Physik B—Condensed Matter 64:189
8. Kroto HW, Heath JR, Obrien SC, Curl RF, Smalley RE (1985) Nature 318:162–163
9. Iijima S (1991) Nature 354:56–58
10. Novoselov KS, Geim AK, Morozov SV, Jiang D, Zhang Y, Dubonos SV, Grigorieva IV, Firsov AA (2004) Science 306:666–669
11. J. Gopalakrishnan (1984) Proc Indian Acad Soc (Chem Sei) 93:421–432
12. (a) Rao CNR (1993) Mater Sci Eng, BI8 (1993) 1–21; (b) Torardi CC, McCarley RE (1979) J Am Chem Soc 101:3963
13. https://www.graphene.manchester.ac.uk/learn/discovery-of-graphene/
14. Bednorz JG, Müller KA (1986) Z Phys B 64:189–193
15. Wu MK, Ashburn JR, Torng CJ, Horst PH, Meng RL, Gao L, Huang ZJ, Wang YQ, Chu CW (1987) Phys Rev Lett 58:908–910
16. Chu CW (2012) Cuprates—superconductors with a Tc up to 164 K. In: Rogalla H, Kes PH (eds) 100 years of superconductivity. CRC Press/Taylor & Francis Group, Boca Raton, pp 244–254. ISBN: 9781439849484
17. Tammann G (1929) Lehrbuch der Metallkunde, 4th edn. VerlagVoss, Berlin
18. Hedwall JA (1938) Reaktionsfahigkeit Festen Stoffe. J. A. Barth, Leipzig
19. Tyagi AK, Achary SN (2009) In: Tyagi AK, Roy M, Kulshreshta SK, Banerjee S (eds) Advanced techniques for materials characterization. Trans Tech Publication, Switzerland, pp 1–36. ISBN: 13-978-0-87849-379-1
20. Chavan SV, Patwe SJ, Tyagi AK (2003) J Alloys Comp 360:189–192
21. Grover V, Tyagi AK (2004) Mater Res Bull 39:859–866
22. Chavan SV, Mathews MD, Tyagi AK (2004) J Amer Ceram Soc 87:1977–1980
23. Mathews MD, Ambekar BR, Tyagi AK (2005) J Nucl Mater 341:19–24
24. Grover V, Achary SN, Patwe SJ, Tyagi AK (2003) Mater Res Bull 38:1101–1111
25. Mathews MD, Ambekar BR, Tyagi AK (2000) J Nucl Mater 280:246–249
26. Achary SN, Jayakumar OD, Tyagi AK, Kulshresththa SK (2003) J Solid State Chem 176:37–46
27. Masatomo Y, Katsuya O, Haruo A, Masato K, Masahiro Y (1994) J Appl Phys 74:7603–7605
28. Jayakumar OD, Tyagi AK (2011) J Mater Chem 21:12246–12250
29. Hume-Rothery W, Powell HM (1935) Z Krist 91:23
30. Shukla R, Manjanna J, Bera AK, Yusuf SM, Tyagi AK (2009) Inorg Chem 48:11691–11696
31. Shukla R, Bera A, Yusuf SM, Deshpande SK, Tyagi AK, Hermes W, Eul M, Pöttgen R (2009) J Phys Chem C 113:12663–12668
32. Sayed FN, Shukla R, Tyagi AK (2015) Dalton Trans 44:16929–16936
33. Shukla R, Sayed FN, Grover V, Deshpande SK, Guleria A, Tyagi AK (2014) Inorg Chem 53:10101–10111
34. Grover V, Shukla R, Jain D, Deshpande SK, Arya A, Pillai CGS, Tyagi AK (2012) Chem Mater 24:2186–2196
35. Yanagida H, Koumoto K, Miyayama M (1996) The chemistry of ceramics. John Wiley & Sons, ISBN: 0-471-956279
36. Mandal BP, Deshpande SK, Tyagi AK (2008) J Mater Res 23:911–916
37. Sayed FN, Grover V, Bhattacharyya K, Jain D, Arya A, Pillai CGS, Tyagi AK (2011) Inorg Chem 50:2354–2365

38. Tyagi AK, Mathews MD, Ambekar BR, Ramachandran R (2004) Thermochim Acta 421:69–71
39. Tyagi AK, Ambekar BR, Mathews MD (2002) J Alloys Comp 337:277–281
40. Satapathy LN, Ramesh PD, Agrawal D, Roy R (2005) Mater Res Bull 40:1871–1882
41. Dumont-Botto E, Bourbon C, Patoux S, Rozier P, Dolle M (2011) J Power Sources 196:2274–2278
42. Joos M, Cerretti G, Veremchuk I, Hofmann P, Frerichs H, Anjum DH, Reich T, Lieberwirth I, Panthöfer M, Zeier WG, Tremel W (2018) Inorg Chem 57:1259–1268
43. Uitert LGV, O'Bryan HM, Lines ME, Guggenheim HJ, Zydig G (1977) Mater Res Bull 12:261
44. Arendt RH, Rosolowski JH, Szmasek JW (1979) Mater Res Bull 14:703
45. Ma HA, Jia XP, Chen LX, Zhu PW, Guo WL, Guo XB, Wang YD, Li SQ, Zhou GT, Zhang G (2002) J Phys Condensed Matter 14:44
46. Vidyasagar K, Gopalakrishnan J, Rao CNR (1984) Inorg Chem 23:1206–1210
47. Ganapathi L, Ramanan A, Gopalakrishnan J, Rao CNR (1986) J Chem Soc Chem Commun 62
48. (a) Mandal BP, Grover V, Roy M, Tyagi AK (2007) J Am Ceram Soc 90:2961–2965; (b) Mandal BP, Grover V, Tyagi AK (2006) Mater Sci Eng A 430:120–124
49. Balazs GB, Glass RS (1995) Solid State Ionics 76:155–162; Mandal BP, Grover V, Tyagi AK (2006) Mater Sci Eng A 430:120–124
50. https://www-pub.iaea.org/MTCD/Publications/PDF/TRS415_web.pdf
51. Mandal BP, Banerji A, Sathe V, Deb SK, Tyagi AK (2007) J Solid State Chem 180:2643–2648
52. Mandal BP, Garg N, Sharma SM, Tyagi AK (2006) J Solid State Chem 179:1990–1994
53. Mandal BP, Shukla R, Achary SN, Tyagi AK (2010) Inorg Chem 49:10415–10421
54. Henderson NL, Baek J, Shiv Halasyamani P, Schaak RE (2007) Chem Mater 19:1883–1885
55. Knoke GT, Niazi A, Hill JM, Johnston DC (2007) Phys Rev B 76:
56. Fujinika H, Kinomura N, Koizumi M, Miamoto Y, Kume S (1979) Mater Res Bull 14:1133
57. Hubert PH (1974) Bull Soc Chim. France 2385
58. Subramanian MA, Aravamudan G, Subba Rao GV (1980) Mat Res Bull 15:1401
59. Lazarev VB, Shaplygin IS (1978) Res Bull 13:229
60. Zinkevich M, Geupel S, Aldinger F, Durygin A, Saxena SK, Yang M, Liu Z (2006) J Phys Chem Solids 67:1901–1907
61. Zhang Q, Saito F (2000) J Alloys Compd 297:99–103
62. Selbach SM, Einarsrud MA, Grande T (2009) Chem Mater 2:169
63. Han H, Lee JH, Jang HM (1916) Inorg Chem 56(2017):11911–11916
64. Krockenberger Y, Mogare K, Reehuis M, Tovar M, Jansen M, Vaitheeswaran G et al (2007) Phys Rev B 75:
65. Feng HL, Sathish CI, Li J, Wang X, Yamaura K (2013) Phys Procedia 45:117–120
66. Ruddlesden SN, Popper P (1958) Acta Crystallogr 11:54–55
67. Sivakumar T, Gopalakrishnan J (2005) Mater Res Bull 40:39–45
68. (2000) Solid State Ionics 132:189–198
69. Dickens PG, Whittingham MS (1968) Quart Rev Chem Soc 22:30
70. Conroy LE, Yokokowa T (1965) Inorg Chem 4:994
71. Bednorz JG, Müller KA (1986) Zeitschrift für Physik B 64:189–193
72. Antipov EV, Putilin SN, Kopnin EM, Capponi JJ, Chaillout C, Loureiro SM, Marezio M, Santoro A (1994) Physica C 235–240:21–24
73. Leggett A (2006) Nat Phys 2:134–136
74. Wu MK, Ashburn JR, Torng CJ, Hor PH, Meng RL, Gao L, Huang ZJ, Wang YQ, Chu CW (1987) Phys Rev Lett 58(9):908–910
75. Ju HL, Kwon C, Li Q, Greene RL, Venkatesan T (1994) Appl Phys Lett 65:2108
76. Sushko YuV, Kubo Y, Shimakawa Y, Manako T (1996) Czech J Phys 46:2003

77. Mary TA, Evans JSO, Vogt T, Sleight AW (1996) Science 272(5258):90–92
78. Evans JSO, Mary TA, Vogt T, Subramanian MA, Sleight AW (1996) Chem Mater 8:2809–2823
79. Tyagi AK, Achary SN, Mathews MD (2002) J Alloy Comp 339:207–210
80. Tanaka K, Toda F (2000) Chem Rev 100:1025–1074
81. Ahluwalia VK et al (2004) New trends in green chemistry. Anamaya Publishers, New Delhi, India
82. Wohler F (1828) Poggendorff's Ann Phys 12:253–256
83. Toda F, Tanaka K, Sekikawa A (1987) Chem Soc Chem Commun:279–280
84. Toda F (1995) Acc Chem Res 28:480–486
85. Wang ZX, Qin HL (2004) Green Chem 6:90–92
86. Raston CL, Scott JL (2000) Green Chem 2:49–52
87. Byeongno L, Philjun K, Kyu Hyung L, Jaeheung C, Wonwoo N, Won Koo L, Hwi Hu H (2013) Tetrahedron Lett 54:1384–1388
88. Merrifield RB (1963) J Am Chem Soc 85:2149–2154
89. Labadie JW (1998) Curr Opin Chem Biol 2:346–352
90. Gisin BH, Merrifield RB, Tosteson DC (1969) J Am Chem Soc 91:2691–2695
91. Kusch P (1966) Angew Chem Int Ed Engl 5:610
92. Letsinger RL, Mahadevan VJ (1966) Am Chem Soc 88:5319–5324
93. Froehler BC, Ng PG, Matteucci MD (1986) Nucleic Acids Res 14:5399–5407
94. Garegg PJ, Lindh I, Regberg T, Stawinski J, Strömberg R (1986) Tetrahedron Lett 27:4051–4054
95. Beaucage SL, Caruthers MH (1981) Tetrahedron Lett 22:1859–1862
96. Frechet JM, Schuerch CJ (1971) Am Chem Soc 93:492–496
97. Danishefsky SJ, McClure KF, Randolph JT, Ruggeri RB (1993) Science 260:1307–1309
98. Plante OJ, Palmacci ER, Seeberger PH (2001) Science 291:1523–1527
99. Nielsen PE, Egholm M, Berg RH, Buchardt O (1991) Science 254:1497–1500
100. Thomson SA, Josey JA, Cadilla R, Gaul MD, Hassman CF, Luzzio MJ, Pipe AJ, Reed KL, Ricca DJ, Wiethe RW, Noble SA (1995) Tetrahedron 51:6179–6194
101. Leznoff CC (1974) Chem Soc Rev 3:65–85
102. Crowley JI, Rapoport H (1976) Acc Chem Res 9:135–14412
103. Mazur S, Jayalalekshmy P (1979) J Am Chem Soc 101:677–681
104. Blackburn C (1998) Biopolymers (Peptide Science) 47:311–351
105. Kolb HC, Finn MG, Sharpless KB (2001) Angew Chem Int Ed Engl 40:2004–2021
106. Gillis EP, Burke MD (2008) J Am Chem Soc 129:6716–6717
107. Errandonea D, Kumar RS, Gracia L, Beltran A, Achary SN, Tyagi AK (2009) Phys Rev B 80:094101 (7 pp)
108. Achary SN, Patwe SJ, Tyagi AK (2008) J Alloys Comp 461:474–480
109. Shukla R, Bedekar V, Yusuf SM, Srinivasu R, Vinu A, Tyagi AK (2009) J Nanosci Nanotech 9:501–505
110. Tyagi AK, Balog P, Weber J, Köhler J (2005) J Solid State Chem 178:2620–2625
111. Achary SN, Sali SK, Kulkarni NK, Krishna PSR, Shinde AB, Tyagi AK (2009) Chem Mater 21:5848–5859
112. Shukla R, Grover V, Srinivasu K, Paul B, Roy A, Tyagi AK (2018) Dalton Trans 47:6787–6799
113. Knoll GF Radiation detection and measurements. John Wiley & Sons, Inc, ISBN 0-471-07338-5
114. Jensen T, Aljundi T, Gray JN, Wallingford R (1996) A model of X-ray film response. In: Thompson DO, Chimenti DE (eds) Review of progress in quantitative nondestructive evaluation, vol 15A. Springer, Boston, MA, p 441
115. Rao CNR (1994) Chemical approaches to the synthesis of inorganic materials. John Wiley, New York

116. Rao CNR, Gopalakrishnan J (1986) New directions in solid state chemistry. Cambridge University Press; Paper back 1989, Russian translation 1990, Chinese Translation 1990, Second Edition (1997)
117. Rao CNR, Honig JM (eds) (1981) Preparation and characterization of materials. Academic Press, New York
118. West AR Solid state chemistry and its applications. John Wiley and Sons, ISBN: 978-1-119-94294-8

Chapter 2
Combustion Synthesis: A Versatile Method for Functional Materials

Rakesh Shukla and A. K. Tyagi

Abstract Among several synthesis methods, combustion technique is an efficient method capable of producing material in shorter duration of time at lower temperature. Combustion method of synthesis has now been considered as an advanced synthesis protocol that can be used for synthesis of many pure and doped functional materials. This method has made an impact in the field of material science as it can be used for development of many stable and metastable materials like metals, alloys, ceramic, etc. This chapter discusses the basic of this synthesis technique to the recent advanced development in this field. Synthesis of materials by combustion will be elaborated followed by applications of few functional materials (catalytic, electrical, magnetic, and optical properties).

Keywords Self-sustained combustion · Combustion synthesis · Gel combustion · Sol–gel combustion · Glycine–nitrate precursor · Auto-ignition · Pechini method

2.1 Introduction

The world of science has observed an incredible rise in the field of nanoscience and nanotechnology in last two decades. Materials are at the heart of any technology, some technologies desire highly sintered form of the material while some need the highly porous form, some processes require the size of the product in the micron form, while some in the nanoform, some processes require highly stable form of the material while some require the material in metastable form. Thus, the demand of materials (new or old) in the customized form is a challenge which has been catered well by the scientist and engineer throughout the globe.

R. Shukla · A. K. Tyagi (✉)
Chemistry Division, Bhabha Atomic Research Centre, Mumbai 400085, India
e-mail: aktyagi@barc.gov.in

R. Shukla · A. K. Tyagi
Homi Bhabha National Institute, Mumbai 400094, India

A. K. Tyagi and R. S. Ningthoujam (eds.), *Handbook on Synthesis Strategies for Advanced Materials*, Indian Institute of Metals Series,
https://doi.org/10.1007/978-981-16-1807-9_2

In the process of designing these materials, not only the conventional materials were reinvestigated for their properties but also new and better materials have been invented, that can offer better properties and have potential to replace the conventional materials. Among the various technologies, combustion synthesis is fast and novel protocol that is used to produce new materials with distinctive properties.

2.2 History

The process of combustion has its genesis in ancient times as well, e.g., the process of combustion was used to synthesize "kajal" a high surface area carbon black, for eye application. Moreover, there are evidence of presence of nanomaterials in the ayurvedic medicinal preparation called "Bhasmas" which were used for the treatment of various diseases. This process was called "Bhasmikaran" which means controlled combustion [1]. Similarly glimpses of nanotechnology could be seen in ancient times, e.g., Dichroic Lycurgus Cup is an example of fourth-century nanoscience. This Roman glass cup was designed with colloidal gold and silver that show a different color depending on whether or not light was passing through it [2]. Nanotechnology in seventeenth century could be seen in the "Damascus sword". The process of smelting for preparing the wootz steel material utilized wood and plant product as fuel during the carbothermic combustion process which resulted in the carbon nanotube formation. The analysis revealed that the key to its properties is nanotechnology, unintentionally used by blacksmiths centuries before the birth of modern science. This sword achieved its sharpness, strength, and flexibility due to incorporation of carbon nanotubes in the matrix [3]. Thus, it could be seen that the field of combustion and nanotechnology was already existing albeit without any proper scientific name.

2.3 Concepts of Combustion Synthesis

Before understanding the nuances of combustion synthesis, it is important to understand the concept of combustion. Combustion is an effect of interaction between a fuel and an oxidant in which enormous amount of heat is generated in a short time. It can be defined as an exothermic redox reaction in which the fuel gets oxidized and oxidant gets reduced and enthalpy of the reaction is observed as heat/flame. The process of combustion reactions is generally transient, characterized by very high temperatures and large evolution of gases.

In combustion synthesis, a redox reaction occurs that liberates heat and this exothermicity of the reaction is utilized to get the desired product. The temperature of the reaction ranges from as low as 400 K to as high as 4000 K during the combustion reaction [4, 5]. Temperature observed is a function of the quantity and quality of the fuel. The heat release during the reaction falls around 10^{12}–10^{14} W/m^3 [5]. Various metals, intermetallics, alloys, cermets, oxides, composites, etc., can be prepared by combustion process. With special attention, some carbides, nitrides, and borides can also be prepared by this process [6–8]. Combustion synthesis is considered to be simple, economic, rapid, reproducible, scalable, and efficient process for the synthesis of materials.

Various types of combustion process based on the type of combustion are briefly discussed below.

2.3.1 Self-Propagating High-Temperature Synthesis (SHS) [9]

This type of combustion occurs with ignition (auto or through any process) and then it propagates through the medium without supply of any extra heat. The reaction is highly exothermic. Due to the lower rate of heat of dissipation compared to heat generation, proper combustion flame propagation is observed.

2.3.2 Volume Combustion Synthesis (VCS) [9]

This type of combustion occurs with continuous heating, supply of heat is continuously required until the reaction is complete. Due to the higher rate of heat dissipation compared to heat generation, a propagating flame is generally not observed.

Another way of classifying the combustion process is as follows [10]:

(a) Gas phase combustion: Synthesis that includes flame/plasma reactor synthesis of nanoparticles such as CNT, aerosol-assisted combustion synthesis of oxides from salts.
(b) Solution phase combustion: One of the most versatile methods used to prepare oxide nanomaterials.
(c) Solid phase combustion: All metallothermic reduction, solid-state pyrolysis are classified under solid phase combustion.

2.4 Classical Combustion Reaction

2.4.1 *Metallothermic Combustion or Thermite Reduction [11]*

This method is generally used to synthesize metals/intermetallics/alloys by taking advantage of the enthalpy of oxidation of highly electropositive elements such as Mg, Al, and C.

$$\text{e.g. } Fe_2O_3 + 3C \rightarrow 2Fe + 3CO + \text{heat}$$

$$Fe_2O_3 + Al \rightarrow 2Fe + Al_2O_3 + \text{heat}$$

The relative reducing capability of the element can be understood by the Ellingham diagram that shows the temperature dependence of the stability for compounds. It is a plot of ΔG ($RTlnP_{O2}$) versus temperature. The equilibrium temperature between the metal, metal oxide, and oxygen is the parameters related in this diagram. The lower the position of a metal's line in the Ellingham diagram, the greater is the stability of its oxide.

The thermite combustion reaction is named based on the element used for reduction, e.g., carbothermic, aluminothermic, magnesiothermic reaction/reduction.

This chapter will be focus only on solution combustion reaction as other types of combustion processes are out of scope of this chapter.

2.4.2 *Sol–Gel Combustion*

Gel combustion is basically an exothermic reduction–oxidation process that occurs among the oxidants and fuel [12, 13]. Oxidants are usually the metal nitrates/oxynitrates, which supply the required oxygen for the combustion reaction. Fuels are organic moiety capable of complexing the ions and getting combusted with an oxidant at low ignition temperature. The compositional homogeneity throughout the constituents should be maintained by the fuel. The most commonly used fuels are urea, glycine, hydrazine, citric acid, hexamethylene tetraamine (HMTA), etc. The powder properties like crystallite size, surface area, etc., can be tailored by changing the fuel and also its oxidant-to-fuel ratio. To get the maximum benefit of heat of combustion (ΔH) and to complete the combustion process in one step, the quantity of fuel is determined by the principle of propellant chemistry [14]. The combustion process comprises of two steps: (i) preparation of fuel-oxidant precursor or gel formation (formation of homogenous transparent precursor) and (ii) gel combustion (auto-ignition of the precursor gel to give the product).

2.4.2.1 Preparation of Fuel-Oxidant Precursor/Gel Formation

This stage includes dissolution of the corresponding nitrate/oxynitrate/acetate salts of the metals (in stoichiometric molar ratio) in an aqueous media or dilute nitric acid to obtain a clear solution. An appropriate fuel is added as per the calculated amount to this mixed metal–nitrate clear solution. On thermal dehydration of this solution on a hot plate (at about 80–120 °C) a viscous transparent semi-solid (gel) is obtained. A pictorial representation of the gel formation step is given below (Fig. 2.1).

Entire process of thermal drying of the solution to obtain the gel is a crucial step, as any excess traces of water left behind will provide a sluggish combustion thereby declining the phase purity and powder quality. To avoid any precipitation in the solution, the nature of the fuel, amount of fuel, and pH of the solution are optimized so as to get a clear transparent gel precursor.

2.4.2.2 Combustion of the Fuel-Oxidant Precursor or Auto-Ignition

The gel precursor obtained is then subsequently heated at elevated temperature ($\sim$300 °C) that initiates the combustion of gel. The combustion is either a self-propagating combustion or volume/bulk combustion. The exothermic decomposition of the gel is observed in the form of fire, that swiftly propagates and covers the entire gel in a single step. This process does not require further external heating and therefore, it is also termed as "auto-ignition". In case of volume or bulk combustion, the continuous combustion happens with large evolution of gases but

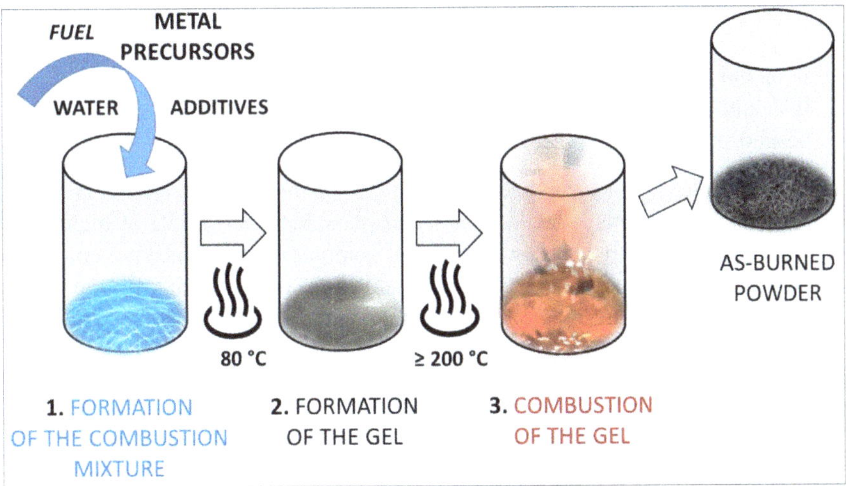

Fig. 2.1 Main steps in combustion process. Reproduced from Ref. [9b], with the permission of Elsevier publishers

continuous heating is required. In the auto-ignition process, the flame continues only for around couple of seconds. During this small period of combustion, the product formation takes place. Once the auto-ignition process is complete, a large voluminous powder as a combustion product is obtained. The resultant combustion product obtained could be the required phase, or partially decomposed precursor with carbonaceous residue, due to the nature and amount of fuel used in the process. Throughout the combustion process, significant amounts of gaseous products are released. These gases disperse the heat energy and fragment the solid residual product, thereby providing fine powdered products.

Fuel in combustion plays a very important role, nature of the fuel and its amount in the reaction decides the exothermicity of the reaction and guides the process toward attaining the powder of the desired product in terms of phase and powder properties. Therefore, it is important to discuss about the oxidant and fuel in the gel-combustion reaction.

2.5 Oxidant in the Gel Combustion

An oxidant in gel-combustion reaction is defined as a moiety/salt of the desired element (of which the compound is to be prepared) that provides oxygen for the fuel in the combustion reaction.

2.5.1 Desired Characteristic of Oxidants

1. It should be soluble in water or dilute nitric acid.
2. It should not allow precipitation of other cations.
3. It should provide the oxygen for the combustion reaction.
4. It should give out gaseous by-products so that at the end of the reaction only the desired product is obtained.

Metal nitrates, oxynitrates, and acetates fulfill all the requirement criteria and hence are selected as oxidants. Metal sulfates and phosphates are stable at higher temperature and do not provide their oxygen for combustion and hence not considered as oxidant for gel-combustion reaction. Metal halides do not qualify for the oxidant as per the definition and requirement of the oxidant in the gel combustion.

2.6 Fuel in the Gel Combustion

A fuel in a gel-combustion reaction is defined as an organic substance/moiety that binds with oxidant and forms a complex with low ignition temperature. It is consumed in a reaction to produce energy and gaseous products. The primary and most important properties of the fuel in the gel-combustion process are capability of maintaining the homogeneity among all the components and must have combustion with the all the reactants (oxidants) simultaneously at a lower ignition temperature.

2.6.1 Desired Characteristic of Fuel

1. It should maintain compositional homogeneity among constituents throughout the pH range of combustion.
2. It should form a clear transparent gel.
3. It should undergo proper combustion with the constituents.
4. It should have low ignition temperature.
5. It should be non-hygroscopic, economic, and readily available.

A large variety of organic materials and some inorganic molecules qualify as a fuel for the combustion reaction. However, choice of fuel should be done on the basis of cation used and desired powder properties of the product. Some examples of fuels are, e.g., citric acid, ascorbic acid, glycine, urea, hydrazine, hexamethylene tetra amine (HMTA), other-α-amino acids, dimethyl urea, EDTA, etc. Some of the commonly used fuels will be discussed in the next section.

2.6.1.1 Glycine (NH_2CH_2COOH)

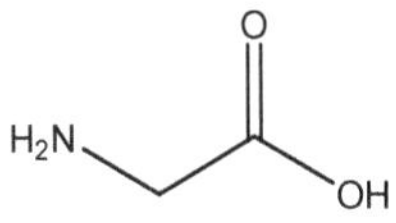

O

H_2N OH

Glycine

It is one of the best, most commonly used and economic α-amino acids for gel-combustion reaction. It not only acts as a good complexing/coordinating agent for several metal ions but also stabilizes them by giving a high-quality transparent and homogenous gel on heating. Glycine has a carboxylic acid (–COOH) group at one end and an amino (–NH_2) group at the other end [15]. This imparts a zwitterionic character to it which is effectively utilized to bind the metal ions of varying ionic potentials to form complexes and thereby preventing the selective precipitation. The complexes of glycine maintain the compositional homogeneity among all the reactant constituents. Combustion with glycine gives out a lot of gases, mainly

H_2O, CO_2, N_2 and NO_x that make the material highly porous. The reactions are spontaneous with high exothermicity resulting with harder agglomerates due to partial local sintering.

2.6.1.2 Citric Acid ($C_6H_8O_7$)

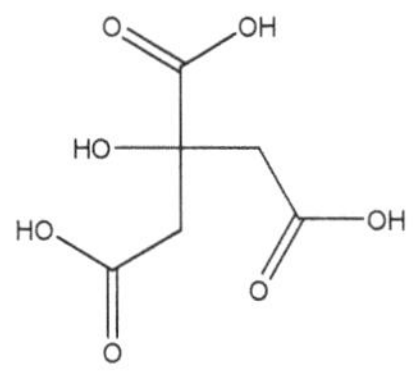

Citric acid

Citric acid (CA) is also among the most used economic fuel that can be used for gel-combustion reaction. It forms citrate complexes with metal ions that have low ignition temperature (~250 °C). CA has three carboxyl group (–COOH) and one hydroxyl group (–OH) for coordinating the metal ions to maintain the compositional homogeneity through gel formation [16]. Citric acid has no nitrogen atoms and is high carbon containing fuel. Combustion of citric acid gel gives out lots of gases, mainly CO_2 that makes the material highly porous. The reactions are sluggish in nature with less exothermicity resulting with softer agglomerates and better sinter-active powders.

2.6.1.3 Ascorbic Acid ($C_6H_8O_6$) [17] and Tartaric Acid ($C_4H_6O_6$) [18]

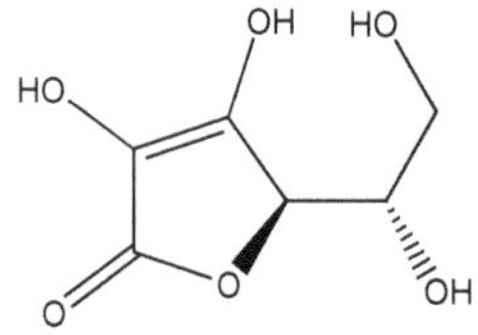

Ascorbic acid

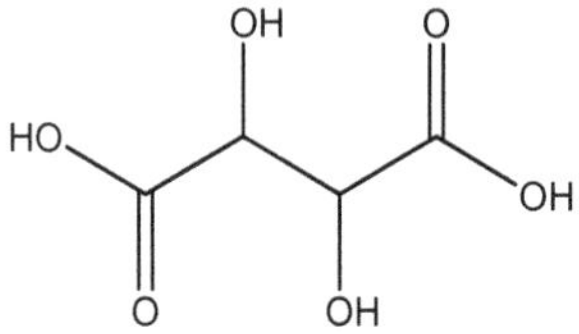

Tartaric acid

Ascorbic acid and tartaric acid are used for metal ion combustion reaction when less exothermicity is required. They are fuels with high carbon and no nitrogen content, especially used for sluggish combustion reaction. The complexation is weak with both the fuels and precipitation can occur early with multiple cation oxidants. The products formed are charred due to the high carbon of the fuel and in turn requires longer duration of calcination.

2.6.1.4 Hydrazine Hydrate ($N_2H_4 \cdot H_2O$) [19]

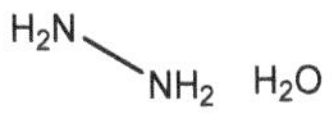

Hydrazine monohydrate

It is an inorganic liquid with density ~ 1.03 g/cc at 20 °C. It is one of the strongest known reducing agent fuels. Hydrazine is specially used when a control of oxidation state is required during the combustion process. It is a carbonless fuel that gives very high temperature during the auto-ignition process. Hydrazine vapor forms explosive mixtures with air (concentration is >4.7% by volume). This fuel is also used in a mixture as rocket propellant.

2.6.1.5 Hexamethylene Tetramine ($(CH_2)_6N_4$) (HMTA) [20]

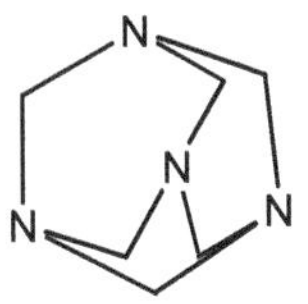

Hexamethylenetetramine

HMTA is an organic heterocyclic tertiary amine that binds very well with the metal ion. The combustion with HMTA is highly exothermic and explosive. The reaction has high enthalpy of combustion because of which hexamethylene tetramine is used as one of the base components to produce research developed explosives (RDX).

2.6.1.6 Aspartic Acid ($C_4H_7NO_4$) [21] and Glutamic Acid ($C_5H_9NO_4$) [22]

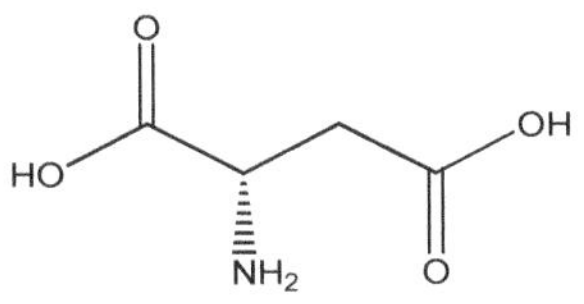

Aspartic acid

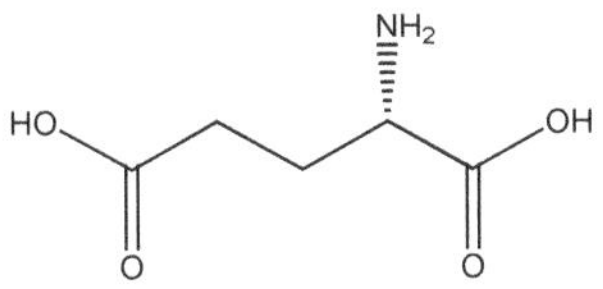

Glutamic acid

Both of the compounds are high carbon fuel "**acidic-α-amino acid**". Aspartic acid has total four carbon, while glutamic acid has five carbon atoms in the molecule. They contain two carboxylic (–COOH) groups and one amino (–NH_2) group. These acids show better coordination with the metal ions through gel formation as they form polychelate complex with the cations but sometime selective partial precipitation is observed due to competing stability of the complex. These fuels are undergoing combustion at low temperatures (~250 °C). The combustion process with these acids as fuel show controlled sluggish combustion which yield product with better powder properties.

2.6.1.7 Arginine ($C_6H_{14}N_4O_2$) [23]

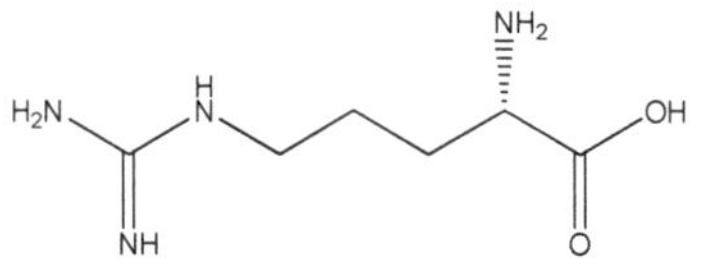

Arginine

Arginine is a strong basic α-amino acid containing two amino ($-NH_2$) groups and one carboxylic (–COOH) group. They show better ability for complex formation with hard metal ions giving a consistent gel precursor that gives combustion at lower temperature in basic pH range.

2.6.1.8 Tryptophan ($C_{10}H_{10}N_2O_2$) [24]

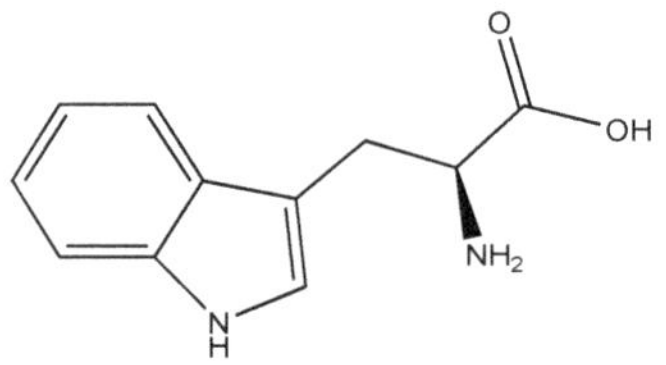

Tryptophan

Tryptophan is a weakly basic α-amino acid containing one amino group and a heterocyclic moiety containing nitrogen as the heteroatom. Tryptophan is capable of complexing with the soft metal ions. Sometimes partial precipitation is the problem with this fuel.

2.6.1.9 Valine ($C_5H_{11}NO_2$) [25] and Phenyl Alanine ($C_9H_{11}NO_2$) [25]

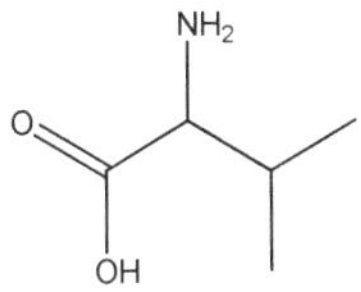

Valine

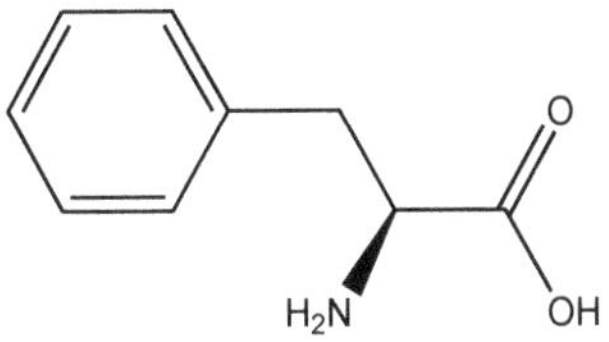

Phenylalanine

They are well-known neutral α-amino acids with one amino and one carboxyl group. Their combustion reaction is similar to the glycine reaction but a number of moles of gases evolved are more.

2.6.1.10 Urea (CH_4N_2O) and Dimethyl Urea ($C_3H_8N_2O$) [25]

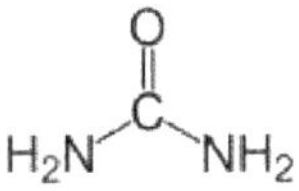

Urea

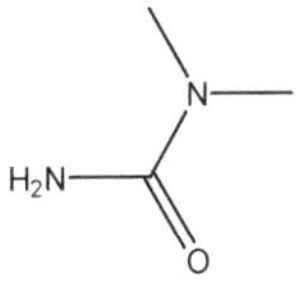

Dimethyl urea

Urea and dimethyl urea are good chelating agents and hence unique set of fuels. They undergo violent combustion with very high exothermicity. They are also used as combustion fuel in some of the combustion reaction where nitrogen doping is required, e.g., nitrogen-doped titania [26].

2.6.1.11 Ethylene Diamine Tetra Acetic Acid (EDTA) ($C_{10}H_{16}N_2O_8$) [27]

EDTA is a well-known complexing agent and a unique fuel with two amino and four carboxylic acid groups. In the case where precipitation with other fuel is problem, then EDTA as fuel is a better option due to its better complexing ability attributed to multi-denticity of this ligand.

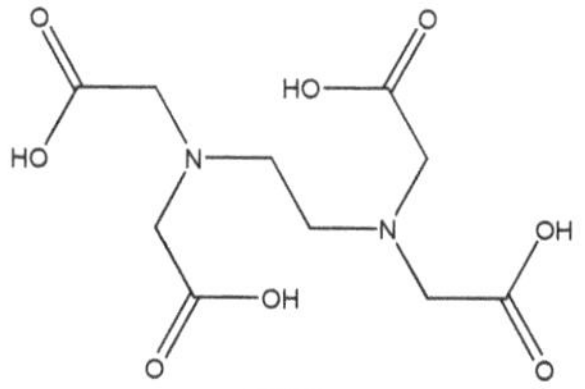

EDTA

Various other fuels have also been explored as reported in the literature just to name a few, oxalic acid, ethylene glycol, carbohydrates (starch, sucrose, glucose, etc.), waste-derived fuels, polymeric fuels, microstructure guiding fuels and shape directing fuels. Sometimes use of single fuel does not serve the purpose of gelation/ complexation of the mixed metal ions which calls for the use of mixed fuel. In case of mixed fuel, two or more fuels in different molar ratio are added so as to get the desired phase and desired microstructure. Therefore, in order to get the better control over the combustion and the product's powder properties, mixed fuels are taken. Some examples of combination of fuels are EDTA-citrate, sucrose-PEG/EG, CA-PVA, CA-succinic acid, etc. [28–31]. A very interesting review has been published by Deganello and Tyagi which carries in detail information about the fuel utilization in combustion reaction [32].

2.7 Role of Fuel

Gel combustion involves reaction/complex formation between fuel and oxidant, and nitrates/oxynitrates/acetates fulfill the requirement of oxidant by providing the oxygen for combustion of the fuel. A suitable fuel in required amount is dissolved with the mixed metal nitrate solution. As mentioned earlier, the fuel serves the dual purpose of not only providing the heat of reaction during the redox combustion process, but also binding the reacting metal ions so as to bring them in close proximity to one another. The fuel should have better stability of complexes, as it should bind all the metal ion so as to avoid precipitation. However, more stability results in high temperature of ignition, hence optimization of better stability and low ignition temperature is required for the combustion reaction.

The fuel selected should be able to form a uniform transparent gel on heating. The gel thus formed maintains an intimate blending between fuel and an oxidant, which is necessary and important step for the combustion process. The decomposition of the gel should preferably occur in single step, uniform and rapid to ensure no side products are formed. As discussed earlier, the thermal dehydration process should be complete otherwise a sluggish combustion and inferior quality product powder is obtained.

2.8 Selection of Fuel

Selection of fuel is a very important step in a combustion reaction. As fuel forms complex with oxidant and finally forms a gel, quality of final product, to some extent, depends on this step. The stability of the complexes, the nature of the gel formed, pH of the precursor gel, the auto-ignition temperature, all depend upon the selection of proper fuel. The formation of desired phase and powder properties of the material thus depend lot on the fuel selection.

The criteria for selection of fuel varies from on case to case. However, some guidelines that can help in selecting the appropriate fuel are given below:

1. Temperature requirement for the formation of product,
2. Thermodynamic and thermal stability of the product,
3. Stability of the gel obtained for auto-ignition.

To obtain the desired powder quality, one should optimize the size of the batch. The flame temperature, quantity of gaseous by-product, and reaction kinetic of the combustion reaction are also points of consideration.

For example, if there are two competing phases (high- and low-temperature form) for a compound and isolation of each phase is required, then one should select citric acid in fuel-deficient ratio and glycine in stoichiometric ratio. Citric acid will provide low heat of combustion, therefore low temperature phase could be easily obtained, while glycine has high heat of combustion hence high-temperature modification can be obtained.

For mixed-cation oxidants, the stability of their complexes should be similar otherwise partial precipitation of one the cations would occur before ignition resulting in biphasic mixture instead of single phasic compound.

2.9 Amount of Fuel

Amount of fuel plays a very important role in gel-combustion method. As per the rule of propellant chemistry, to get a stoichiometric redox reaction between a fuel and an oxidant, the ratio (**Φ**) of net oxidizing valency of oxidants to net reducing valency of fuel (O/F) should be one [16]. An example of calculation of these valencies is elucidated below.

For synthesis of nanocrystalline Eu_2O_3.

The oxidizing valencies calculated for the oxidant, viz. $Eu(NO_3)_3{\cdot}6H_2O$ is 15−. The reducing valency of the fuel, viz. glycine is 9+. The oxidizing and the reducing valencies are calculated as follows:

(i) $Eu(NO_3)_3{\cdot}6H_2O$: (oxidizing valency of Eu = 3+, N = 0) (reducing valency of O = 2−).
Thus, the net oxidizing valency = 15−

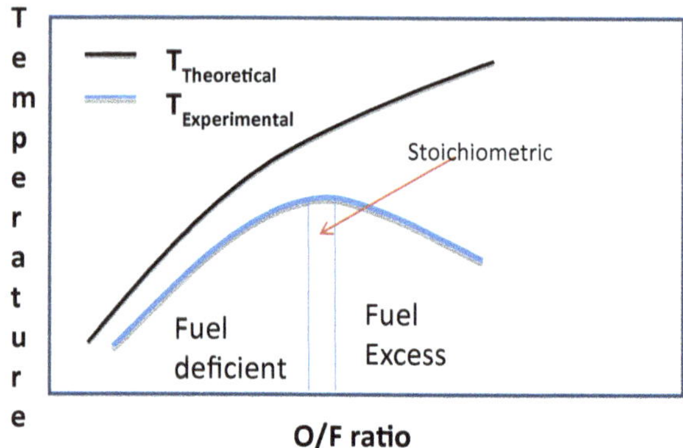

Fig. 2.2 Theoretical and the experimental trend of temperature at increasing Φ values. Reproduced from Ref. [32]: Prog Cryst Growth & Character Mater 64(2018) 23–61 with the permission of Elsevier Publishers

(ii) NH_2CH_2COOH: (oxidizing valency of N = 0, H = 1+, C = 4+).
Thus, the net reducing valency = 9+.

The valencies of C and N are taken as 4+ and 0, as these species are lost as CO_2 and N_2, respectively, during the combustion process [12]. The ratio of O/F valency then turns out to be 1:1.66 is called as (Φ), and this is the stoichiometric ratio. The O/F ratio lesser than 1.66 is termed as the fuel-deficient ratio and O/F ratio greater than 1.66 is called as fuel-excess ratio. This O/F ratio is an important factor as it governs the flame temperature and heat of the combustion process. Theoretical and the experimental trend of temperature at increasing Φ values is shown in Fig. 2.2.

2.10 Selection of Fuel to Oxidant Ratio

Combustion can be performed in various oxidant-to-fuel ratio, i.e., from fuel-deficient ratio to fuel-excess ratio. The nature of combustion and obtained powder would vary in properties for different oxidant-to-fuel ratio.

The nature of combustion can be categorized into following categories.

2.10.1 Extreme Fuel-Deficient Gel-Combustion Reaction

When the amount of fuel used is less than the 40% of the stoichiometric ratio, the combustion is considered as "Extremely fuel-deficient gel-combustion reaction". This combustion gives flameless smoldered reaction with evolution of large volume of gases like H_2O, CO_2, NO_x, etc. This combustion is performed when the thermal stability of the product is low. As this reaction is bulk/volume combustion, continuous external heating is required for completion of the reaction.

2.10.2 Fuel-Deficient Gel Combustion

When the amount of fuel used is more than the 40% but less than stoichiometric ratio, the combustion is considered as "Fuel-deficient gel-combustion reaction". Fuel-deficient combustion gives sluggish combustion with large evolution of gases but yields powder that are soft agglomerates with better sinter activity. In this case also, external heating is continuously required for completion of reaction. This combustion gives generally the best powder characteristics.

2.10.3 Stoichiometric Gel Combustion

The combustion with the amount of fuel equivalent to the ratio of oxidizing-to-reducing valency is the stoichiometric gel-combustion reaction. Stoichiometric gel-combustion reaction gives self-propagating auto-ignition process in violent manner. The reaction occurs with proper flame that exists for 3–5 s. Exothermicity of the reaction is very high and high temperature up to 1800 °C can be observed. Hard agglomerates are formed in stoichiometric gel combustion due to local partial sintering of the nanocrystalline powder.

2.10.4 Slight Fuel Excess Gel-Combustion Reaction

When the amount of fuel is taken slightly excess than the stoichiometric ratio but less than ratio 1:1.15, then the combustion is termed as slight fuel excess gel-combustion reaction. The combustion happens similar to the stoichiometric gel-combustion process. In order to increase the exothermicity, sometimes an extra oxidant, i.e., ammonium nitrate in calculated amount is added. The reaction proceeds in violent manner and hence proper care and safety is required for this reaction.

2.10.5 Extreme-Fuel Excess Gel-Combustion Reaction

When the fuel is taken in relatively higher amount than the stoichiometric ratio, i.e., more than 1:1.2, then the combustion is termed as extreme fuel excess gel-combustion reaction. These reactions undergo decomposition of gel without flame. As sufficient oxidant is not available, the fuel decomposition occurs with less oxygen resulting in charring of the products. The product obtained are with very high carbon content and need longer duration calcination to get rid of the carbonaceous impurity.

Exothermicity of the reaction is governed by fuel-to-oxidant ratio. The fuel-deficient combustion yields highly porous and soft aggregates due to large evolution of gaseous product, whereas stoichiometric combustion results in compound with hard agglomerates, so the powder properties are different and varies as a function of nature of fuel, fuel content and in turn exothermicity of the reaction. If one needs a material for catalyst, then soft agglomerate forming fuel-deficient combustion should be preferred, but if one needs material for column chromatographic application then stoichiometric ratio is the best as it will provide hard agglomerate powders providing additional mechanical strength to the column packing material and avoiding choking of the column.

2.11 Selection of Reaction Vessels

Selection of reaction vessels should be such that it withstands the entire combustion process, should be non-corrosive, and should not add any impurity to the powder. The best suitable vessels for laboratory purpose are high-quality borosilicate glass. For large-scale industrial purposes, the reactor chamber should be low cost, durable, and inert, and hence, stainless steel is one of the preferred choices for combustion.

2.12 Role of PH

The pH level of the solution must be maintained crucially in slight acidic range in order to prevent any precipitation of any of the metal hydroxide or other salts. Partial or complete precipitation of reactants affects the composition of the product. In fuels with amino acids, due to the presence of basic and acidic groups in the same molecule, they are capable of forming zwitter/dual-natured ions in aqueous solutions with a basic part, ammonium ion (NH_3^+) and an acidic part, carboxyl anion (COO^-). Thus, the overall acidic or basic behavior and its complexation behavior are dependent on the pH or hydrogen-ion concentration of the solution, and for each fuel, there is a specific pH value, at which it behaves as neutral. This specific pH value is termed as isoelectric point. Therefore, every fuel has its own isoelectric point and its prior knowledge helps in the selection of optimum pH for the combustion reaction.

Gelation conditions also play a major role in combustion synthesis. Gelation is formation of 3D complex network of fuel and the oxidant. Gelation allows intimate blending of atoms/ions in the solution. The rate of water evaporation also plays a very important role in the formation of gel. Slow evaporation of water is always beneficial as it provides a better quality gel which in turn gives homogenous product powder. The parameters that can be optimized to get the desired product are time for complexation of sol–gel, pH of sol–gel, aging of gel, water content of sol–gel, temperature of gelation, etc.

2.13 Modified Gel-Combustion Method

Over the year, scientists have applied their innovation and modified the conventional gel-combustion method to achieve better material with desired properties. Some of the modified synthesis methods are discussed below:

(a) ***External template-assisted combustion method***: In this method, an external template either of organic origin such as cellulose, waste-derived biofuels, or inorganic origin such as silica, carbon, alumina, other salts are added to get a desired environment/impregnation or impregnated material for the application [32]. Various nanocomposites can be easily prepared by this method.

(b) ***Non-oxidizer combustion process***: This modified combustion process uses non-oxidizer compounds instead of nitrate mixtures. Best example are lithium silicates, namely Li_2SiO_3, $Li_2Si_2O_5$, and Li_4SiO_4, which were obtained by this method depending on the reaction conditions. Li_4SiO_4 compound with small impurity was synthesized by Cruz et al. [33] using this modified combustion method.

(c) ***Emulsion combustion synthesis***: This method developed by Takatori et al. utilizes formation of emulsion and then combustion is carried out by introducing it into the flame. After combustion, the resulting powder is collected. The emulsion combustion method is a highly scalable production method for synthesis of nanopowders with special powder properties. Metal ions dissolved in the aqueous droplets forms an emulsion with the oil and surfactant and are swiftly oxidized by the ignition of the encapsulating flammable fluid. The process provides thin-walled homogenously distributed hollow particles. Aluminum oxide, zirconium oxide, zirconia–ceria solid solutions, and barium titanate were synthesized by the ECM process [34].

(d) ***Biowaste fuel combustion synthesis***: Naturally existing materials or biowaste such as fruit peel/seed/pulp extracts are used for the combustion as fuel. Azizi et al. have discussed the synthesis of biomorphic ZnO for biomedical applications using such fuel [35].

(e) ***Combination or two-step method***: This method is combination of two methods of which one is combustion technique. The combustion method is used to prepare material in nanoform. The nanopowder is then treated to obtain the desired product, e.g., for synthesis of lower valent compounds, vacuum reduction in presence of zirconium is carried out. To synthesize $CeScO_3$ initially a gel combustion is carried out to obtain the precursor nanopowder, which is fluorite-type solid solution of ceria and scandia ($Ce_{0.5}Sc_{0.5}O_{1.75}$) [36]. The calcined powder after pelletization was wrapped in tantalum (Ta) or platinum (Pt) foil. The wrapped foil is then transferred to a quartz tube having thin pelletized zirconium sponge. The assembly is online vacuum sealed at 10^{-6} mbar pressure. The ampoule is heated to 1100 °C for 24 h to get phase pure $CeScO_3$. Zirconium metal at high temperature has got very high affinity for oxygen. The reaction of formation of zirconium oxide is irreversible. Hence, during this heating process, in the presence of zirconium powder, the residual

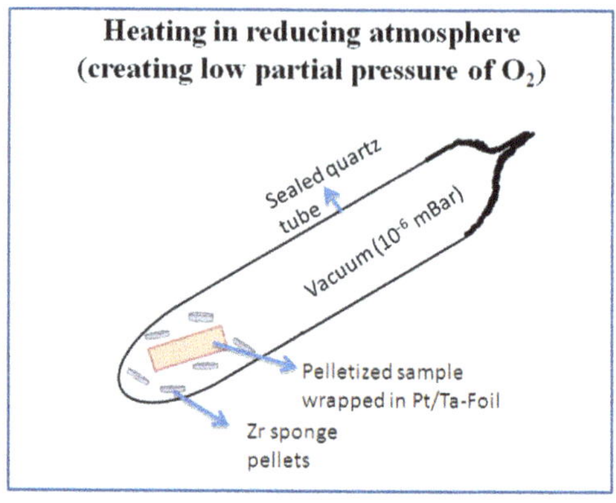

Fig. 2.3 Schematic of assembly to achieve low partial pressure of oxygen

oxygen in the sealed ampoule is taken up by zirconium metal and there is an additional reduction of partial pressure of O_2, which creates a reducing atmosphere and facilitates reduction leading to stabilization of lower valent ion in the compound. The schematic representation is shown in Fig. 2.3.

Some important features to be noted

1. While carrying out gel combustion with barium or strontium nitrate solution with glycine as fuel, excess of nitric acid leads to precipitation of the alkali metal near the gel-formation step and hence small amount of water is added continuously to prevent the precipitation. A common example in this regard: during synthesis of $BaZrO_3$, under too much acidic condition, Ba^{2+} complex precipitates out. Under basic condition, zirconium hydroxide precipitates out, hence optimization of pH is very much required during the synthesis of these kinds of compounds.
2. Extra care of maintaining pH has to be taken when the combustion is performed with zirconyl ion as precipitation is sensitive to not only pH but also to the water content in gel. Frequently few drop of basic water (i.e., ammoniated water 0.1% NH_3 in water) is added during evaporation till the gel formation.

2.14 Precautions and Limitations

The auto-ignition process of gel combustion is prompt and vigorous with a large amount of heat and gas release for very short duration. Combustion process is known for its simplicity, but a few precautionary measures are necessary:

(i) Proper information about the fuel and its behavior should be compiled. If the data is not available, a small TG-DTA experiment should be carried out to understand the ignition temperature and heat release. Initially small batch reactions are useful to understand the behavior of the fuel and the oxidants.

(ii) The combustion experiments should be carried out in fume hood. For a large batch of combustion, an arrangement for proper and safe ventilation should be done.
(iii) Combustion process should be carried out in a wide mouthed apparatus, with large volumes to avoid any pressure build up.
(iv) Combustion-synthesized powders are highly reactive as their surface area is large and surface energy is high. Proper conditions of calcinations and storage are required to avoid any deterioration in powder properties.

Thus, it can be said that with proper knowledge of reactants, combustion process, and safety guidelines, combustion reaction can be easily carried out to synthesize innumerable products that can be used for various applications. In next section, few examples of material synthesized by gel-combustion route will be discussed.

2.15 Typical Examples of Materials Synthesized by Gel-Combustion Route

Various products synthesized via combustion route have been adopted in various technologies and some of them will be discussed briefly. The combustion technique has been employed by the various researchers to get the desired products.

1. Synthesis of metals [37], alloys [38], simple oxides [39, 40], doped oxides [41, 42], mixed or complex oxides [43, 44], cermets [45], to evaluate the structure and functionalities, e.g., recently precursors of high entropy alloys in oxide forms were prepared by combustion method using citric acid as fuel by Niu et al. Single phasic high entropy alloys could be easily prepared by GC followed by heat treatment method compared to mechanical alloying method [37].
2. Various compounds with fluorite [46], rutile [47], perovskite [48], spinel [49], garnet [50], pyrochlore [51], hexagonal perovskite related structure [52], bixbyite structure [52], corundum [53], etc., structure have been synthesized and explored.
3. The combustion-synthesized product has been used in production of function materials for various application such as conventional ceramics, electroceramics, bioceramics, magnetic oxides, porous materials for sorbents, catalysts such as catalyst for CO oxidation, catalyst for selective oxidation and reduction, catalyst for Suzuki coupling, and optical materials. Some examples are discussed below for each application.

Simple oxide-based conventional ceramics form the major part of the ceramic materials. Alumina (Al_2O_3) is one of the widely used ceramic [54]. Studies indicate that solution combustion offers a flexible process for preparation of nanocrystalline Al_2O_3 and ZrO_2 powders which can be used for making compact ceramics with desired properties. [55–60] Stabilized zirconia using Y_2O_3, MgO, CaO substitutions has found significant utilization in industries. Also, CeO_2, magnetite (Fe_3O_4) zinc oxide (ZnO), etc., are in great demand [56–60]. Nanocrystalline ceramic oxides synthesized using combustion method show better characteristics in terms of crystallite size, surface area, agglomeration, sintered density, etc. Functional properties and demand of such nanoceramic oxides have seen significant growth in recent few decades [55].

Electroceramics are the materials that have found applications in electrical and electronic field. Many oxides possess dielectric/ferroelectric/semiconductor properties. Several electroceramics synthesized by gel combustion have been explored for, e.g., barium titanate ($BaTiO_3$) [61], strontium titanate ($SrTiO_3$) [62], $Bi_4Ti_3O_{12}$ [63], etc. Several examples of electroceramic material synthesized by gel-combustion method have been discussed in by Varma et al. [55]. The fields of solid oxide fuel cells (SOFC), solid oxide electrolyzer cells SOEC, direct methanol fuel cells (DMFC) have seen tremendous advancement by utilizing this modern method of synthesis, e.g., Liu et al. have synthesized $Ba_{0.5}Sr_{0.5}Co_{0.8}Fe_{0.2}O_{3-\delta}$ and $La_{0.9}Sr_{0.1}Ga_{0.8}Mg_{0.2}O_{3-\delta}$ composite cathode for solid oxide fuel cell with desired grain size, microstructure, better chemical compatibility, and electrochemical performance [64]. The properties such as specific resistance (ρ) and the activation energies (ΔE) of this material were found to be superior for the combustion-synthesized material compared to the conventional material. Solution combustion-synthesized wüstite-based material was explored for the water-splitting activity and production of hydrogen [65]. The regeneration factor was better in the solution combustion-synthesized sample compared to the solid-state-synthesized samples of the same composition which was attributed to the higher concentration of structural defects formed in the combustion-synthesized sample. Jayalakshmi et al. reported synthesis of nanosized ZnO/carbon composite with better specific capacitance, compared to bulk ZnO powder and utilization of it for supercapacitor application [66].

Kam et al. investigated glycine–nitrate combustion-synthesized titanium-incorporated nickel manganese cobalt oxide (NMC) cathode materials [67]. Electrochemical properties evaluation of this material was carried out, and better cycling performance of the battery was attributed to the synthesis method and Ti substitution which results in decreased cell volume change during deintercalation of lithium.

A lucid review by Xu et al. discusses the important battery materials synthesized by solid-state and solution combustion method. V_2O_5 nanowire clusters prepared by gel-combustion method using biofuel, cassava starch, exhibited better rate capability and cycling ability (188 mAh g^{-1} and capacity retention $\sim$90%) [68, 69]. The combustion synthesis method aides control of the morphology and particle size of the synthesized products.

Of late ceramics are increasingly used for biomedical applications. Bhatkar et al. [70] discussed the combustion synthesis of akermanite ($Ca_2MgSi_2O_7$and $Sr_2MgSi_2O_7$), which are known to have better in vitro and in vivo bioactivities. Srinivas et al. [71] discussed the use of combustion-synthesized yttria-stabilized zirconia as defect bone filler. Yttrium tetragonal zirconia polycrystals (Y-TZP), an all-ceramic system, is used for crowns and fixed partial dentures [72].

Ferrites are the most commonly known magnetic oxides, and they show different structures with various properties. Magnetic oxides have several applications such as in recording media, cooling devices, sensing materials, environmental remediation, catalysis, and other biomedical utilization [73–75]. Solution combustion synthesis of different iron oxides, i.e., α- and γ-Fe_2O_3 and Fe_3O_4, has been reported by Deshpande et al. [76]. They showed that by utilizing complex fuels and complex oxidizers and other reaction parameters, a better control of the product composition and properties could be obtained. Other magnetic oxides have been reported in the informative review by Varma et al. [55].

Highly porous magnesium oxide is synthesized via combustion method using $Mg(NO_3)_2$ and ethylene glycol [77]. The as-prepared porous magnesium oxide showed large BET surface area of 203.8 m^2/g after calcination. The porous magnesium oxide exhibited superior adsorption removal properties with maximum adsorption capacity of ~1088 mg/g in removing Congo red a toxic dye [77]. Venkatesham et al. discuss the synthesis of high surface area zinc oxide by gel combustion and its utilization as an efficient adsorbent to remove lead (II) (Pb^{2+}) from aqueous solution. Adsorption characteristic was studied by varying the effect of lead concentration, adsorbent dosage, and pH [78]. Luu et al. have applied combustion method to synthesize different crystallite sizes of Ca-doped $LaCoO_3$ nanoparticles using polyvinyl alcohol and metal nitrate by varying the reaction conditions. Total oxidation of CO with this nanocatalyst was observed well below 250 °C. The enhanced activity was attributed to the synthesis protocol, grain size, nanocrystalline phase, and morphology of the nanoparticles [79]. Nanocrystalline $CeZrO_{4-\delta}$ was synthesized via gel-combustion method using glycine as a fuel. The material was used as a substrate for photodeposition of palladium. This photodeposited palladium on nanocrystalline $CeZrO_{4-\delta}$ support was evaluated for CO oxidation, as a function of particle size and pretreatment. Based on the studies, it was inferred that Pd with average crystallite size of 7 nm exhibited excellent catalytic activity for CO oxidation. Detailed analysis revealed that Pd deposition occurs preferentially at oxygen-vacancy sites and subsequent metal-support interaction influences the catalysis [80]. Pawar et al. have used $BaFe_2O_4$ synthesized via citrate gel-combustion method for selective oxidation of styrene to benzaldehyde [81]. Phase pure Pd-doped $CeZrO_{4-\delta}$ and undoped $CeZrO_{4-\delta}$ were prepared via gel-combustion method. Photodeposition of 1 wt% Pd on reduced $CeZrO_{4-\delta}$ support displays significantly higher catalytic activity than other catalysts for Suzuki cross-coupling reaction. Pd on redox $CeZrO_{4-\delta}$ fresh catalysts and reduced form of 5% Pd on $CeZrO_{4-\delta}$ catalyst exhibited 100% conversion in 2.5 and 1 h [82]. Costa et al. have prepared highly sintered ceramics of laser host material Nd: YAG by gel combustion at almost half the temperature (850 °C) of the conventional synthesis method (1600 °C) [50]. An efficient red emitter phosphor Cr^{3+} doped $MgAl_2O_4$ was synthesized by the combustion method using urea as fuel. The luminescent properties were also compared with the structural parameters [83].

2.16 Comparison of Solid-State Reaction and Combustion Reaction

Solid-state reactions	Combustion reactions
As solid-state reactions are diffusion controlled, they are kinetically sluggish and mostly need high-temperature furnace	Combustion reactions are carried out in solution; hence, they are kinetically rapid and can be carried out on a simple hot plate
Micron-size particles are obtained	Nanoparticles are obtained
For reaction heating, grinding and pelletizing at each heating step required	One-time calcination required to get rid of carbonaceous impurity
Thermodynamically stable product is formed	Metastable product can be formed and also could be isolated
Complete product formation requires days or weeks	Product is formed within few minutes to hours

Each synthesis method has its own merits and demerits over the other. However, it is up to us, how to converge toward the best synthesis method for a typical reaction and achieve the best desired properties.

2.17 Merits and Demerits of Gel-Combustion Process

Merits

1. Gel-combustion synthesis is a versatile process giving monophasic homogenous powders at lower temperatures.
2. This technique provides an atomistic level of blending among the reacting constituents.
3. This preparatory method does not require any elaborate experimental setup and can be employed in the scale up of materials with high production rate.
4. The combustion process utilizes cheap raw materials and hence is economical.
5. The process is capable of stabilizing metastable phases.
6. The powder products obtained using gel-combustion technique consists of fine crystals of high purity.
7. As these combustion-synthesized powders are loosely agglomerated, they can be sintered to high densities at lower temperature in a short time.
8. The desired powder characteristics such as particle size and surface area can be achieved by altering the process parameters such as nature and amount of the fuel.

Demerits

1. As the combustion-synthesized powders are voluminous in nature, their handling and transferring require special care.
2. The powders obtained are found to be agglomerates of fine crystals due to the exothermicity of the combustion reaction.
3. These powders usually exhibit high-dimensional shrinkage during the sintering process.

2.18 Sinterability and Nanopowders

Sinterability is one of the important mechanical properties of material. The densification of a material depends upon large number of factors such as

(i) nature of starting materials, (ii) preparation method, and (iii) heat treatment conditions such as atmosphere and rate of heating.

Sintering is driven by the reduction of surface energy of the constituent particles, which by coming close together result in lowering of free energy of the system. The main purpose of sintering is to create a solid dense body of the desired size and shape, with as little porosity as possible. Ideally the material should achieve near theoretical density with uniform grain size and homogenous microstructure. Densification takes place by transport of matter through diffusion of atoms, ions, or other charged species. Sintering is a thermally activated process that follows an Arrhenius-type behavior [84]. The sintering behavior of any material relies on various parameters like temperature, time, composition, particle size, and surface area. As a result, material transport and sintering occur faster at higher temperatures. Finer the particle size, faster is the sintering process, as less diffusion path has to be followed. It has been observed that nanoparticles obtained after combustion get sintered at comparatively much lower temperatures, than those in solid-state reactions.

There is a strong correlation between the sintered density and powder properties for nanocrystalline powders. Combustion method provides nanocrystalline powders with very high surface area and high surface-to-volume ratio. Due to their high surface energies resulting from the unsaturated bonds on their surfaces, these nanoparticles tend to react among themselves resulting in large secondary particles termed as agglomerates. The nature and the extent of agglomeration of the nanopowders is a crucial property which has a direct impact on their sinterability. Depending upon the nature of bonding among the nanoparticles, there are two types of agglomerates, viz. soft and hard. The term "agglomerate" is generally used for "soft" agglomerates in which the particles are attached together by weak van der Waals forces. "Hard" agglomerates often referred as aggregates are formed due to the partial local sintering among the nanoparticles. The crystallites within hard agglomerates are more strongly bonded together with predefined pores. Such powders are not required for the fabrication of nanoceramics as they will result in

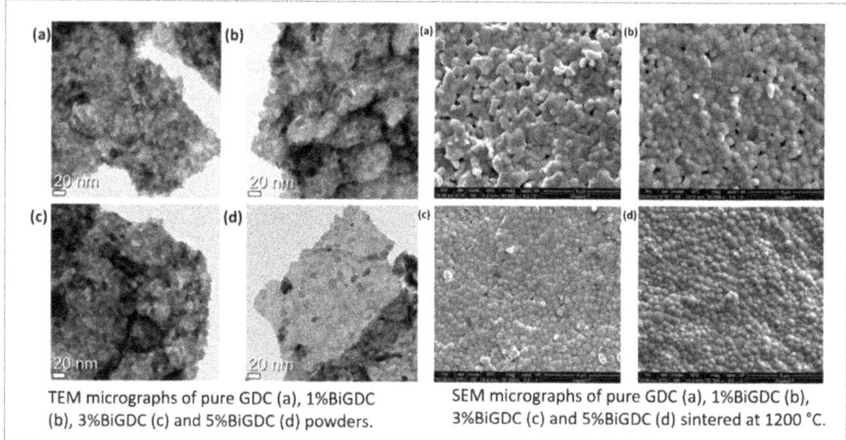

Fig. 2.4 TEM and SEM micrograph showing powder morphology and sintered microstructure of GDC, 1% Bi co-doped GDC, 3% Bi co-doped GDC, and 5% Bi co-doped GDC. Reproduced from Ref. [85]: Ceramics International, 44(2018) 3800–3809 with the permission of Elsevier Publishers

non-uniform dispersion of grain densities. It is challenging to synthesize nanopowders with minimum agglomeration and uniform particles, e.g., Accardo et al. [85] have synthesized phase pure gadolinium-doped ceria (GDC) and bismuth co-doped GDC (BiGDC) by sol–gel combustion method [83]. The synthesis method and Bi co-doping not only considerably reduce the sintering temperature by 300 °C but also high densification ~99.7% theoretical density is achieved. This material showed superior performance as a solid electrolyte in intermediate-temperature solid oxide fuel cells. The TEM and SEM micrograph showing powder morphology and sintered microstructure of GDC and Bi doped GDC are shown as Fig. 2.4.

2.19 Conclusions and Future Scope

Among all the available synthesis methods, combustion technique has the capability of delivering ultra-fine nanopowders in shorter duration of time with a lower calcination temperature. The nanocrystalline powder obtained via combustion process shows better powder properties such as finer crystals, large surface area, and higher sinterability. The better characteristic of the material obtained allows it to be useful for electrical, optical, catalytic, magnetic, and other high-end application. One of the important aspects regarding the combustion process is the tunability of the processing conditions to obtain metastable phases. With varieties of modification coming up in this process, this technique has opened up a new direction of synthesis which in future will cater the needs of the scientist to design the material for technological demands.

References

1. Sharma R, Prajapati PK (2016) Nanotechnology in medicine: leads from ayurveda. J Pharm Bioallied Sci 8:80–81
2. Freestone I, Meeks N, Sax M, Higgitt C (2007) The lycurgus cup—a roman nanotechnology. Gold Bull 40:270–277
3. Srinivasan C (2007) Do Damascus swords reveal India's mastery of nanotechnology? Cur Sci 92:279–280
4. Munir ZA, (1998) Field-effects in self-propagating solid-state reactions. Z. Physik Chemie 207:39–57
5. Martirosyan KS, Luss D (2006) Combustion synthesis of ceramic composites from lunar soil simulant. Lunar and planetary science XXXVII
6. Li Z, Zhou W, Su X, Luo F, Zhu D, Liu P (2008) Preparation and characterization of aluminum-doped silicon carbide by combustion synthesis. J Am Ceram Soc 91:2607–2610
7. Yeh CL, Chuang HC (2004) Combustion characteristics of SHS process of titanium nitride with TiN dilution. Ceram Int 30:705–714
8. Yeh CL, Chen WH (2006) Preparation of niobium borides NbB and NbB_2 by self-propagating combustion synthesis. J Alloys Comp 420:111–116
9. (a) Tyagi AK, Purohit RD, Chavan SV, Bedekar V (2011) Nanocrystalline oxide ceramics through solution combustion. Encycl Nanosci Nanotechnol 17:89–117; (b) Deganello F (2017) Nanomaterials for environmental and energy applications prepared by solution combustion based-methodologies: role of the fuel. Mater Tod: Proc 4:5507–5516
10. Lackner M (2010) Combustion synthesis: novel routes to novel materials. Bentham Science Publishers
11. Wang LL, Munir ZA, Maximov YM (1993) Thermite reactions: their utilization in the synthesis and processing of materials. J Mater Sc 28:3693–3708
12. Patil KC, Aruna ST, Ekambaram S (1997) Combustion synthesis. Curr Opin Solid State Mater Sci 2:158–165
13. Aruna ST, Mukasyan AS (2008) Combustion synthesis and nanomaterials. Curr Opin Solid State Mater Sci 12:44–50
14. Jain SR, Adiga KC, Verneker VRP (1981) A new approach to thermochemical calculations of condensed fuel-oxidizer mixtures. Combust Flame 40:71–79
15. Chick LA, Pederson LR, Maupin GD, Bates JL, Thomas LE, Exarhos GJ (1990) Glycine-nitrate combustion synthesis of oxide ceramic powders. Mater Lett 10:6–12
16. Purohit RD, Tyagi AK (2002) Auto-ignition synthesis of nanocrystalline $BaTi_4O_9$ powder. J Mater Chem 2:312–316
17. Tsay J, Fang T (1999) Effects of molar ratio of citric acid to cations and of pH value on the formation and thermal-decomposition behavior of barium titanium citrate. J Amer Ceram Soc 82:1409–1415
18. Kareiva A, Karppinen M, Niinisto L (1994) Sol–gel synthesis of superconducting $YBa_2Cu_4O_8$ using acetate and tartrate precursors. J Mater Chem 4:1267–1270
19. Yu X, An X (2009) Enhanced magnetic and optical properties of pure and (Mn, Sr) doped $BiFeO_3$ nanocrystals. Solid Stat Comm 149:711–714
20. Garcia R, Hirata GA, McKittrick J (2001) New combustion synthesis technique for the production of $(In_xGa_{1-x})_2O_3$ powders: hydrazine/metal nitrate method. Mater Res 16:1059–1065
21. Prabhu YT, Rao KV, Sai Kumar VS, Siva Kumari B (2013) Synthesis of ZnO nanoparticles by a novel surfactant assisted amine combustion method. Adv Nanopart 2:45–50
22. Merino MCG, Lascalea GE, Sánchez LM, Vázquez PG, Cabanillas ED, Lamas DG (2010) Nanostructured aluminium oxide powders obtained by aspartic acid–nitrate gel-combustion routes. J Alloys Comp 495:578–582
23. Krishnamurthy N, Gupta CK (2015) Extractive metallurgy of rare earths, 2nd edn. CRC Press, pp 615–616

24. Muresan L, Cadis A, Perhaita I, Ponta O, Silipas D (2015) Thermal behavior of precursors for synthesis of $Y_2Si_2O_5$:Ce phosphor via gel combustion. J Therm Anal Calor 119:1565–1576
25. Mukherjee ST, Bedekar V, Patra A, Sastry PU, Tyagi AK (2008) Study of agglomeration behavior of combustion-synthesized nano-crystalline ceria using new fuels. J Alloys Comp 466:493–497
26. Du X, Du HL, Shi X, Wang J, He JJ (2015) Photocatalytic activity of solar-light-active N-doped TiO_2 by sol-gel combustion method. Mater Sci Forum 809–810:800–806
27. Merino MCG, Nasisi LDT, Montoya WM, Aguilera JNU, Emilia Fernandez de Rapp M, Lascalea GE, Vázquez PG (2015) Combustion syntheses of Co_3O_4 powders using different fuels. Proc Mater Sci 8:526–534
28. Patra H, Rout SK, Pratihar SK, Bhattacharya S (2011) Effect of process parameters on combined EDTA–citrate synthesis of $Ba_{0.5}Sr_{0.5}Co_{0.8}Fe_{0.2}O_{3-\delta}$ perovskite. Powd Technol 209:98–104
29. Deganello F, Liotta LF, Marcı G, Fabbri E, Traversa E (2013) Strontium and iron-doped barium cobaltite prepared by solution combustion synthesis: exploring a mixed-fuel approach for tailored intermediate temperature solid oxide fuel cell cathode materials. Mater Renew Sustain Energy 2:8, 14 pp
30. Subramania A, Angayarkanni N, Vasudevan T (2007) ffect of PVA with various combustion fuels in sol–gel thermolysis process for the synthesis of $LiMn_2O_4$ nanoparticles for Li-ion batteries Mater Chem Phys 102:19–23
31. Lertpanyapornchai B, Yokoi T, Ngamcharussrivichai C (2016) Citric acid as complexing agent in synthesis of mesoporous strontium titanate via neutral-templated self-assembly sol-gel combustion method. Micropor Mesopor Mater 226:505–509
32. Deganello F, Tyagi AK (2018) Solution combustion synthesis, energy and environment: best parameters for better materials. Prog Cryst Growth Character Mater 64:23–61
33. Cruz D, Bulbulian S (2005) Synthesis of Li_4SiO_4 by a modified combustion method. J Am Ceram Soc 88:1720–1724
34. Takatori K, Tani T, Watanabe N, Kamiya N (1999) Preparation and characterization of nano-structured ceramic powders synthesized by emulsion combustion method. J Nanoparticle Res 1:197–204
35. Azizi S, Mohamad R, Mahdavi Shahri M (2017) Green microwave-assisted combustion synthesis of zinc oxide nanoparticles with citrullus colocynthis (L.) schrad: characterization and biomedical applications. Molecules 22:301
36. Shukla R, Arya A, Tyagi AK (2010) Interconversion of perovskite and fluorite structures in Ce−Sc−O system. Inorg Chem 49:1152–1157
37. Manukyan KV, Cross A, Roslyakov S, Rouvimov S, Rogachev AS, Wolf EE, Mukasyan AS (2013) Solution combustion synthesis of nano-crystalline metallic materials: mechanistic studies. J Phys Chem C 117:24417–24427
38. Niu B, Zhang F, Ping H, Li N, Zhou J, Lei L, Xie J, Zhang J, Wang W, Fu Z (2017) Sol-gel autocombustion synthesis of nanocrystalline high-entropy alloys. Sc Reports 7:3421
39. Hwang CC, Huang TH, Tsai JS, Lin CS, Peng CH (2006) Combustion synthesis of nanocrystalline ceria (CeO_2) powders by a dry route. Mater Sc Eng: B 132:229–238
40. Purohit RD, Saha S, Tyagi AK (2006) Combustion synthesis of nanocrystalline ZrO_2 powder: XRD, Raman spectroscopy and TEM studies. Mater Sci Eng B 130:57–60
41. Ningthoujam RS, Shukla R, Vatsa RK, Duppel V, Kienle L, Tyagi AK (2009) $Gd_2O_3:Eu^{3+}$ particles prepared by glycine-nitrate combustion: phase, concentration, annealing, and luminescence studies. J Appl Phys 105:084304
42. Singh SK, Kumar K, Rai SB (2009) Multifunctional Er^{3+}–Yb^{3+} codoped Gd_2O_3 nanocrystallinephosphor synthesized through optimized combustion route. App Phys B: Lasers Optics 94:165–173
43. Pradhan GK, Martha S, Parida KM, (2012) Synthesis of multifunctional nanostructured zinc–iron mixed oxide photocatalyst by a simple solution-combustion technique. Appl ACS Mater Interfaces 4:707–713

44. Singh KA, Pathak LC, Roy SK (2007) Effect of citric acid on the synthesis of nano-crystalline yttria stabilized zirconia powders by nitrate-citrate process. Ceram Int 33:1463–1468
45. Marinsek M, Zupan K, Maèek J (2002) Ni–YSZ cermet anodes prepared by citrate/nitrate combustion synthesis. J Power Sources 106:178–188
46. Biswas M, Ojha PK, Durga Prasad C, Gokhale NM, Sharma SC (2012) Synthesis of fluorite-type nanopowders by citrate-nitrate auto-combustion process: a systematic approach. Mater Sci App 3:110–115
47. Deorsola FA, Vallauri D (2008) Synthesis of TiO_2 nanoparticles through the Gel Combustion process. J Mater Sci 43:3274–3278
48. Chakroborty A, Das Sharma A, Maiti B, Maiti HS (2002) Preparation of low-temperature sinterable $BaCe_{0.8}Sm_{0.2}O_3$ powder by autoignition technique. Mater Lett 57:862–867
49. Hwang BJ, Santhanam R, Liu DG (2001) Effect of various synthetic parameters on purity of $LiMn_2O_4$ spinel synthesized by a sol–gel method at low temperature. J Power Sources 101:86–89
50. Costa AL, Esposito L, Medri V, Bellosi A (2007) Synthesis of Nd-YAG material by citrate–nitrate sol–gel combustion route. Adv Eng Mater 9:307–312
51. Mandal BP, Shukla R, Achary SN, Tyagi AK (2010) Crucial role of the reaction conditions in isolating several metastable phases in a Gd−Ce−Zr−O system. Inorg Chem 49:10415–10421
52. Shukla R, Sayed FN, Grover V, Deshpande SK, Guleria A, Tyagi AK (2014) Quest for lead free relaxors in $YIn_{1-x}Fe_xO_3$ ($0.0 \leq x \leq 1.0$) system: role of synthesis and structure. Inorg Chem 53:10101–10111
53. Peng T, Liu X, Dai K, Xiao J, Song H (2006) Effect of acidity on the glycine–nitrate combustion synthesis of nano crystalline alumina powder. Mater Res Bull 41:1638–1645
54. Zhuravlev VD, Bamburov VG, Beketov AR, Perelyaeva LA, Baklanova, Sivtsova OV, Vasil'ev VG, Vladimirova EV, Shevchenko VG, Grigorov IG (2013) Solution combustion synthesis of α-Al_2O_3 using urea. Ceramics Int 39:1379–1384
55. Varma A, Mukasyan AS, Rogachev AS, Manukyan KV (2016) Solution combustion synthesis of nanoscale materials. Chem Rev 116:14493–14586
56. Manicone PF, Iommetti PR, Raffaelli L (2007) An overview of zirconia ceramics: Basic properties and clinical applications. J Dent 35:819–826
57. Kelly JR, Denry I (2008) Stabilized zirconia as a structural ceramic: an overview. Dent Mater 24:289–298
58. Jiang J, Hu X, Shen W, Ni C, Hertz JL (2013) Improved ionic conductivity in strained yttria-stabilized zirconia thin films. Appl Phys Lett 102:143901
59. Scherrer B, Schlupp MVF, Stender D, Martynczuk J, Grolig JG, Ma H, Kocher P, Lippert T, Prestat M, Gauckler LJ (2013) On proton conductivity in porous and dense yttria stabilized zirconia at low temperature. Adv Funct Mater 23:1957–1964
60. Chao CC, Park JS, Tian X, Shim JH, Gur TM, Prinz FB (2013) Enhanced oxygen exchange on surface-engineered yttria-stabilized zirconia. ACS Nano 7:2186–2191
61. Zhong Z, Gallagher P (1995) Combustion synthesis and characterization of $BaTiO_3$. J Mater Res 10:945–952
62. Klaytae T, Panthong P, Thountom S (2013) Preparation of nanocrystalline $SrTiO_3$ powder by sol–gel combustion method. Ceram Int 39:S405–S408
63. Krengvirat W, Sreekantan S, Mohd Noor AF, Chinwanitcharoen C, Muto H, Matsuda A (2012) Influences of pH on the structure, morphology and dielectric properties of bismuth titanate ceramics produced by a low-temperature self-combustion synthesis without an additional fuel agent. Ceram Int 38:3001–3009
64. Liu B, Zhang Y, Zhang L (2008) Characteristics of $Ba_{0.5}Sr_{0.5}Co_{0.8}Fe_{0.2}O_{3-\delta}$–$La_{0.9}Sr_{0.1}Ga_{0.8}Mg_{0.2}O_{3-\delta}$ composite cathode for solid oxide fuel cell. J Pow Sourc 175:189–195
65. Agrafiotis C, Roeb M, Konstandopoulos AG, Nalbandian L, Zaspalis VT, Sattler C, Stobbe P, Steele AM (2005) Solar water splitting for hydrogen production with monolithic reactors. Solar Energy 75 409–421

66. Jayalakshmi M, Palaniappan M, Balasubramanian K (2008) Single step solution combustion synthesis of ZnO/carbon composite and its electrochemical characterization for supercapacitor application. Int J Electrochem Sci 3:96–103
67. Kam KC, Mehta A, Heron JT, Doeff MM (2012) Electrochemical and physical properties of Ti-substituted layered nickel manganese cobalt oxide (NMC) cathode materials. J Electrochem Soc 159:A1383–A1392
68. Xu J, Lin F, Doeff MM, Tong W (2017) A review of Ni-based layered oxides for rechargeable Li-ion batteries. J Mater Chem A 5:874–901
69. Ramasami AK, Reddy MV, Nithyadharseni P, Chowdari BVR, Balakrishna GR (2017) Gel-combustion synthesized vanadium pentoxide nanowire clusters for rechargeable lithium batteries. J Alloys Comp 695:850–858
70. Bhatkar VB, Bhatkar NV (2011) Combustion synthesis and photoluminescence study of silicate biomaterials. Bull Mater Sci 34:1281–1284
71. Srinivas M, Buvaneswari G (2006) A study of in vitro drug release from zirconia ceramics. Trends Biomater Artif Organs 20:24–30
72. Pilathadka S, Vahalová D, Vosáhlo T (2007) The Zirconia: a new dental ceramic material. An overview. Prague Med Rep 108:5–12
73. Pullar RC (2012) Hexagonal ferrites: A review of the synthesis, properties and applications of hexaferrite ceramics. Prog Mater Sci 57:1191–1334
74. Shukla R, Ningthoujam RS, Umare SS, Sharma SJ, Kurian S, Vatsa RK, Tyagi AK, Gajbhiye NS (2008) Decrease of superparamagnetic fraction at room temperature in ultrafine $CoFe_2O_4$ particles by Ag doping. Hyperfine Interac 184:217–225
75. Cullity BD (1972) Introduction to magnetic materials. Addison-Wesley, London
76. Deshpande K, Mukasyan A, Varma A (2004) Direct synthesis of iron oxide nanopowders by the combustion approach: reaction mechanism and properties. Chem Mater 16:4896–4904
77. Li S (2019) Combustion synthesis of porous MgO and its adsorption properties. Int J Ind Chem 10:89–96
78. Venkatesham V, Madhu GM, Satyanarayana SV, Preetham HS (2013) Adsorption of lead on gel combustion derived nano ZnO. Procedia Eng 51:308–313
79. Luu TH, Nguyen XD, Huyen Phan TM, Schulze S, Hietschold M (2015) Synthesis and microstructure of $La_{1-x}Ca_xCoO_3$ nanoparticles and their catalytic activity for CO oxidation. Adv Nat Sci Nanosci Nanotechnol 6:025016
80. Burange AS, Reddy KP, Gopinath CS, Shukla R, Tyagi AK (2018) Role of palladium crystallite size on CO oxidation over $CeZrO_{4-\delta}$ supported Pd catalysts. Molecul Catal 455:1–5
81. Pawar RY, Pardeshi SK (2018) Selective oxidation of styrene to benzaldehyde using soft $BaFe_2O_4$ synthesized by citrate gel combustion method. Arab J Chem 11:282–290
82. Burange AS, Shukla R, Tyagi AK, Gopinath CS (2016) Palladium supported on fluorite structured redox $CeZrO_{4-\delta}$ for heterogeneous suzuki coupling in water: a green Ppotocol. Chem Select 1:2673–2681
83. Singh V, Chakradhar RPS, Rao JL, Kim DK (2009) Combustion synthesized $MgAl_2O_4$:Cr phosphors-An EPR and optical study. J Lumin 129:130–134
84. Othmer K (1993) Encyclopedia of chemical technology, vol 5. John Wiley, New York
85. Accardo G, Frattini D, Ham HC, Han JH, Yoon SP (2018) Improved microstructure and sintering temperature of bismuth nano-doped GDC powders synthesized by direct sol-gel combustion. Ceram Int 44:3800–3809

Chapter 3
Microwave-Assisted Synthesis of Inorganic Nanomaterials

Dimple P. Dutta

Abstract Microwave-assisted synthesis has seen rapid development in the last two decades as it is being heralded as a green synthesis technique. It is based on the effective absorption of microwave energy by the reactants/solvents in a chemical reaction which results in efficient heat transfer through dielectric heating. The reactions occur at a must faster time scale compared to conventional thermal heating and helps in the reduction of carbon footprint. In this chapter, the effect of microwaves in chemical reaction and the advantages of microwave heating compared to thermal methods has been discussed in detail. The different components in domestic and laboratory microwave reactors have been reviewed. The microwave-assisted synthesis of various class of compounds, particularly inorganic nanomaterials, that has been recently reported in literature, has been explored. The precautions to be observed while planning a microwave synthesis reaction has been explained. The chapter ends with a concise report on future outlook and prospects of microwave-assisted synthesis technique.

Keywords Microwave energy · Synthesis · Nanomaterials · Dielectric heating · Microwave reactor

3.1 Introduction

Microwaves are electromagnetic waves having frequency in the range of 0.3–300 GHz and are located between the infrared and radiofrequency region of the electromagnetic spectrum. The application of microwaves in mobile communication, satellite broadcasting, and in RADAR systems is well known. Besides these, they are also being used in industrial, biomedical, chemical, and in scientific

D. P. Dutta (✉)
Chemistry Division, Bhabha Atomic Research Centre, Mumbai 400085, India
e-mail: dimpled@barc.gov.in

D. P. Dutta
Homi Bhabha National Institute, Mumbai 400094, India

A. K. Tyagi and R. S. Ningthoujam (eds.), *Handbook on Synthesis Strategies for Advanced Materials*, Indian Institute of Metals Series,
https://doi.org/10.1007/978-981-16-1807-9_3

research applications. The energy associated with the microwave radiation varies from 1.24×10^{-6} to 1.24×10^{-3} eV which is insufficient to break chemical bonds or even trigger Brownian motion [1]. However, in 1945, Percy LeBaron Spencer, an American engineer at the Raytheon Co., accidentally chanced upon the heating effect of microwaves. Spencer was engaged in developing magnetron for generating microwave radio signals with application in combat radars. He noticed that a bar of chocolate melted in his pocket when the magnetron was switched on [2]. In his quest to find an explanation to this observation, he discovered the heating effect of microwaves. The first commercial microwave oven was built by Spencer in 1947, and it weighed around 300 kg and was approximately 5.5 ft tall. With passage of time, the magnetrons used to generate microwaves became cheaper and microwave ovens gradually found application in household cooking and heating of food. The utility of microwaves in inorganic synthesis in liquid phase was first explored by Komarneni and Roy in 1985 [3]. In 1986 Robert Gedye, George Majetich and Raymond Giguere published papers relating to application of microwave radiation in organic synthesis [4, 5]. This triggered an interest in the scientific community regarding microwave heating technology, and it has emerged to be a popular choice as heat source for conducting chemical synthesis at a much-reduced time compared to that possible using conventional heating techniques [6]. In order to avoid interference with cable communication and mobile phone frequencies, all domestic microwave ovens and dedicated microwave reactors for chemical synthesis operate at a frequency of 2.45 GHz. Earlier, most of the chemical reactions were performed in household microwave ovens which had poor control over the experimental parameters and hence there was a reproducibility issue with the reported experiments. However, in recent times, microwave-assisted synthesis has gained prominence with the availability of microwave reactors with special features like built-in magnetic stirrers and temperature and pressure sensors. Consequently, its application in material sciences, nanotechnology, organic/peptide synthesis, polymer science, and biochemistry has gained immense prominence [7].

Organic synthesis using microwave-assisted technique is being extensively explored by drug companies for drug screening and discovery. Since huge number of compounds are synthesized for screening in the pharmaceutical industry, the fast kinetics and increased variety of reaction offered by microwave-assisted organic synthesis reduce the time for the production [8]. A large number of review articles are available in literature which focus on both the process of microwave-assisted organic synthesis as well as its related applications [8–10]. The microwave-assisted preparation of inorganic nanomaterials in solution phase is currently a fast-growing area of research. Review articles on metal, carbon, inorganic and colloidal nanostructures, as well as metal oxide nanoparticles supported on carbon nanotubes, polymer nanocomposites, and nanoporous materials are available in literature [7, 11–18]. In this chapter, the focus is on microwave-assisted synthesis of inorganic nanostructures with special emphasis on materials synthesized in the past decade. A background on effect of microwaves in chemical reaction has been given along with the differences in microwave and conventional heating. The effect of solvents and open or closed reaction environment on microwave synthesis has been discussed. Microwave-assisted synthesis of metals, semimetals, alloys, metal oxides, metalchalcogenides, inorganic biomaterials, inorganic–inorganic nanocomposites, and inorganic–organic nanocomposites, in various solvents including water, polyols, ionic liquids, etc., have been reviewed in detail. Finally, the challenges and future prospects of microwave-assisted synthesis has been deliberated upon to analyze the potential of this technique at industrial scale.

3.1.1 *Effect of Microwaves in Chemical Reaction*

A 2.45 GHz microwave oven/reactor does not produce enough energy to break typical chemical bonds. The energy is less than that of Brownian motion and can only affect molecular rotations. Microwaves are non-ionizing and only thermally activate the system. Therefore, it does not change the molecular structure of the compounds being heated. Microwave chemistry is based on efficient heating of materials and is different from photochemical processes where chemical reactions occur by direct absorption of electromagnetic radiation. The heating effect of microwaves (MW) is triggered by mechanism of dipolar polarization and ionic conduction depending on the materials used for the reaction. When polar solvent or reagents are involved, the dipoles in the sample strive to align themselves in the direction of the applied electric field under influence of microwaves. Since the electric field oscillates, the molecular dipoles attempt to realign themselves accordingly along the alternating electric field, and this leads to loss in energy through molecular friction, rotation, and dielectric loss. The heating effect is related to the ability of these molecular dipoles to realign themselves with the applied field frequency. Heating effect is disrupted if the dipoles align too quickly which happens with low frequency radiation or if unable to reorient themselves which occurs when the frequency of MW radiation is high. In ionic conduction mechanism, as the name suggests, the ions formed in the solution collide with the neighboring atoms or molecules in the presence of the oscillating electric field and energy is lost in the form of heat. The heating effect through ionic conduction is stronger compared to that obtained via dipolar rotation mechanism [7]. This is evident when equal volume of distilled water and tap water is heated under identical radiation power and fixed time in a MW reactor. The final temperature is higher in the tap water sample containing dissolved ions as in this case both dipolar polarization and conductive mechanisms contribute to the heating effect. The microwave heating effect of any material is expressed by its loss tangent/loss factor tan $\delta = \delta''/\delta'$ where δ'' is the conversion efficiency of the electromagnetic radiation into heat (dielectric loss) and δ' is the polarizability of molecules in the electric field (dielectric constant) [19]. The angle δ represents the phase lag between the polarization of the material and the applied electric field and helps in determining the efficiency of the microwave heating process. The parameters which affect loss tangent are temperature and frequency, and they also play a role in the penetration depth of MW radiation in any material. Penetration depth refers to the point in the material which retains 37% of the initial irradiation power [20].

It has been observed that microwave heating accelerates the rate of chemical reaction which reduces the reaction time considerably. Also, certain reactions which do not occur under conventional heating methods can be performed under identical condition via microwave heating. Two different models have been proposed for this microwave heating induced increase in reaction rate. The first model is based on thermal effects leading to generation of superheated hot spots in the reaction system. For any reaction to proceed, the reactants need to acquire the activation energy to get activated to the transition state (TS) (Fig. 3.1). Thus, activation energy (E_{TS}) is absorbed by the reactants from the reaction environment and ultimately leads to the formation of products with lower energy E_P. Application of microwave irradiation does not alter the activation energy but provides a great momentum to complete the reaction more quickly and efficiently as compared to conventional heating. This is due to the fact that MW transfers energy in 10^{-9} s in each cycle whereas the kinetic molecular relaxation occurs within a time period of 10^{-5} s. Since energy transfer rate is faster compared to molecular relaxation, it leads to non-equilibrium condition and higher instantaneous temperature or hot spots within the reaction medium which impacts the reaction kinetics and accelerates the reaction rate [21].

The second model is based on non-thermal effects which include possible excitement of rotational or vibrational transitions, varied activation energy, and increased collision efficiency by mutual orientation of polar molecules [22]. However, critical analysis of the research literature available in support of non-thermal mechanism indicates that in majority of cases the observed rate enhancements actually originate from a purely thermal/kinetic effect [20]. The rate accelerations observed in MW-assisted synthesis is a result of the microwave dielectric heating mechanism which cannot be achieved by conventional heating.

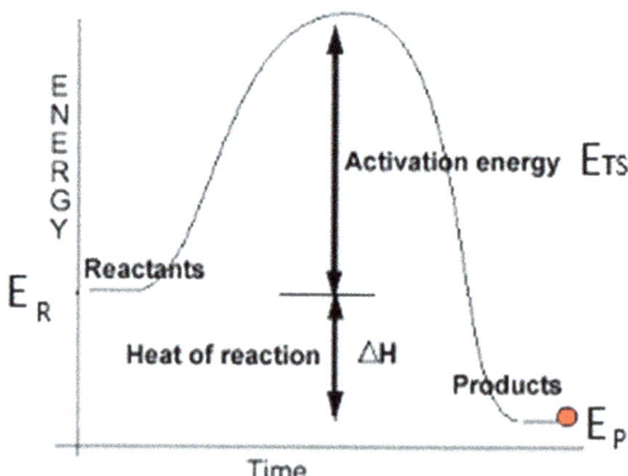

Fig. 3.1 Kinetic energy profile of an exothermic reaction

3.1.2 Microwave Heating Vis-à-Vis Conventional Heating

Microwave heating via electromagnetic waves leads to faster kinetics and shorter reaction time compared to conventional heating. The heating mechanism used in microwave synthesis is via dielectric polarization and ionic conduction. Conventional heating involves the use of electric furnace or oil bath which transfers heat to wall of the reaction vessel and then passes it on to the reactants via convection or conduction. These lead to the development of a thermal gradient in the reaction media which results in non-uniform reaction sites [23]. In contrast, in microwave heating, direct heat transfer to target material is possible without heating the entire furnace or oil bath, which conserves time and energy. In conventional heating, generally, the maximum temperature of reaction is limited by the boiling point of the reaction medium whereas in microwave, the temperature can be raised higher than the boiling point, i.e., superheating may take place. As a result, microwave heating has opened the possibility of realizing fast chemical synthesis in very short time periods compared to conventional heating methods, leading to a relatively low cost, energy saving, and high efficiency technique for materials production.

3.1.3 Effect of Solvents in Microwave Synthesis

The intensity of microwave heating in a sample depends on its electrophysical properties, geometric size, frequency and intensity of the applied field, and penetration depth of the electromagnetic waves into the sample. The type of solvent used in microwave synthesis plays a major role in the overall process. Solvents having a high loss tangent (tan δ) at the standard operating frequency of 2.45 GHz of a MW reactor is required for adequate absorption of energy and efficient heating [24]. The loss tangent values of a few solvents are given in Table 3.1. For use in MW synthesis, solvents can be classified as high- (tan $\delta > 0.5$), medium- ($0.1 \leq$ tan $\delta \leq 0.5$), and low-absorbing (tan $\delta < 0.1$). Consequently, water and alcohol can be considered as good medium for microwave-assisted synthesis of inorganic nanostructures. Ethylene glycol with its high boiling point (198 °C) and reducing properties along with high loss tangent value finds application in open vessel microwave synthesis.

Solvents with high loss factor at the incident radiation frequency will heat at a faster rate from core to surface. The loss tangent value of a solvent is inversely proportional to the microwave penetration depth (D_p). When microwaves hit at a 90° angle on the surface of the solvent, its intensity decreases gradually due to dissipation inside the volume of the solvent. D_p denotes the depth at which the microwave power density is reduced to 1/e of its initial value. The penetration depths (D_p) is given by the following equation in case of low loss dielectric medium:

Table 3.1 Loss tangent (tan δ) of various solvents at 2.45 GHz, 20 °C

Solvent	tan δ	Solvent	tan δ
Ethylene glycol	1.350	Dimethylformamide (DMF)	0.161
Ethanol	0.941	1,2-dichloroethane	0.127
Dimethyl sulfoxide (DMSO)	0.825	Water	0.123
2-propanol	0.799	Chlorobenzene	0.101
Methanol	0.659	Acetone	0.054
1,2-dichlorobenzene	0.280	Tetrahydrofuran (THF)	0.047
N-Methyl-2-pyrrolidone (NMP)	0.275	Toluene	0.040
Acetic acid	0.174	Hexane	0.020

$$D_p = \left(\frac{\lambda_0}{2\pi}\right)\left(\frac{\sqrt{\varepsilon'}}{\varepsilon''}\right) \quad (3.1)$$

Here, λ_0 is the wavelength of the microwave radiation used which is 0.122 m for a 2.45 GHz MW reactor. Thus, penetration depth is dependent on the frequency of microwave used. In case of water, which is the most common solvent for majority of microwave reactions, the values of D_p are ~1.8 and 0.34 cm at 22 °C for the 2.45 and 5.8 GHz microwaves, respectively [25].

Polar solvents can couple with the microwave energy, which can lead to a rapid rise in temperature and increase in the reaction rate. The dielectric parameter of the solvent and the penetration depth depend strongly on temperature and hence fluctuates during heating. In case of ethanol, the loss tangent value decreases from 0.941 to 0.080 with rise in temperature from 20 to 200 °C [7]. The microwave heating mechanism in organic solvents is predominantly dipolar polarization which decreases due to reduced viscosity and molecular friction in the heated solvents. In contrast to it, ionic liquids like [bmim][PF_6] (1-Butyl-3-methylimidazolium hexafluorophosphate) display an increase in loss tangent from 0.185 to 3.592 with increase in temperature to 200 °C. In ionic liquids, the microwave heating occurs through ionic conduction mechanism and hence its ability to absorb microwave increases with temperature. Since ionic conduction mechanism leads to more efficient microwave heating compared to dipolar polarization, ionic liquids are considered to be better microwave absorbers in both low- and high-temperature microwave reactions.

MW synthesis can also be carried out in non-polar alkanes or alkenes which are almost MW transparent by addition of small quantity of good MW absorbers in the medium. This kind of mixed solvent medium can lead to greater control in chemical composition, structure, size, and morphology of the product by variation of the type of solvents and their volumes. For example, addition of small amount of 1,3-dialkylimidazolium iodide in hexane, the temperature reached ~217 °C after microwave heating at 200 W for 10 s compared to only ~46 °C achieved in the absence of the ionic liquid [26].

3.1.4 Microwave-Assisted Hydro/Solvothermal Synthesis

Inorganic nanostructures in liquid phase can be synthesized via microwave technique in open reaction system under atmospheric pressure using high boiling solvents having moderate to high loss tangent values. This kind of open reaction systems are easy to design and are relatively safe and cheap method of production. As mentioned earlier, high boiling ionic liquids, ethylene glycol, and glycerol are the solvents of choice for such open system reactions. Refluxing is also a choice for performing microwave-assisted reactions at the boiling point of a solvent in an open reaction system. However, if a low boiling point solvent like methanol is used, even by refluxing method, the temperature zone of chemical reaction cannot exceed beyond 65 °C, and this restricts the scope of the chemical reaction. The temperature of the microwave reaction can be increased beyond the boiling point of the solvent used, if the reaction is performed under pressurized condition in a closed system. The boiling point of ethanol can be increased to 164 °C under 12 atm pressure which can increase the rate of microwave-assisted synthesis by thousand-fold compared to that achievable in the open system [19]. Whenever autogenous pressure is allowed to develop over the solvent level in a closed reaction system via microwave heating, the process is termed as microwave-assisted solvothermal technique. If the solvent used is water, the process is termed as microwave-assisted hydrothermal method. By this method, solvents can be brought to temperatures well above their boiling points by increasing autogenous pressures resulting from heating. Microwave-assisted hydrothermal/solvothermal techniques employ autoclaves made from high-strength polymeric materials that are transparent to microwaves, but the reaction systems are rapidly heated to a temperature much higher than its boiling point which is in turn governed by the autogenous pressure [27]. Due to drastically reduced reaction time in combined microwave-assisted hydro/solvothermal synthesis, this technique can be regarded a fast chemistry method.

This is a truly low-temperature, energy saving, and environment-friendly method for the preparation of nanophase materials of different morphology, because the reactions take place in a shut isolated system condition. The nanophase materials can be produced in either a batch or continuous process.

The pioneering work of comparing the differences between hydrothermal syntheses performed with conventional means of heating and also in special autoclaves with microwaves was started in 1990 by Komarneni et al. [28]. Since then various functional materials like oxides (binary and ternary), oxyhydroxides, zeolites, etc., have been produced by microwave-assisted hydrothermal synthesis [29]. Besides microwave power and time, other reaction parameters such as the solvent, pH of the solution, temperature and pressure also affect microwave synthesis. Therefore, more advanced synthesis technology is required to control pressure and temperature in the subcritical region of water. It should be noted that microwave-assisted hydrothermal synthesis is an effective method for the production of nanoparticles, because the particle size can be controlled by tuning the mechanisms of nucleation and growth kinetics through the appropriate choice of the synthesis parameters.

Usually, in MW reactors, there is auto-power cut if the reaction pressures rise higher than 20 bars. The maximum obtainable concentration is dependent on the properties of the reactants and the solvent(s) used. A unimolecular reaction is independent of concentration and can be performed in a very dilute solution. On the other hand, bi- or tri-molecular reactions are essentially concentration dependent with higher concentration offering faster rate of reaction.

3.2 Components of a Microwave Reactor

The design of any MW reactor is based on its application. The main focus is to achieve maximum heating efficiency which is essential for economy and longevity of the equipment and for obtaining repeatable and reliable results. On entering a reactor chamber, microwaves are reflected by the walls of the cavity. Eventually, a 3D standing wave pattern is generated within the cavity and it is called mode. Microwave reactors operate either in the monomode or the multimode model. The monomode reactor is based on the standing wave model which is produced by interference of microwaves having different directions but same amplitude. This results in generation of nodes and antinodes with minimum and maximum MW intensity, respectively. The sample vessel is placed in the antinodal position of the standing electromagnetic waves to maximize the effect of microwave-assisted synthesis (Fig. 3.2). This system leads to uniform heating which results in higher reproducibility. Higher field strengths can be generated using monomode MW reactors which also increases the rate of heating. The monomode MW reactors are most suitable for chemical synthesis experiments.

The common domestic microwave ovens operate in the multimode manner for higher dispersion of the radiation which ultimately increases the area of effective heating [8]. However, this technique also leads to non-uniform heating within the sample with regions of hot and cold zones which is unsuitable for heating of small amounts of sample. This lack of temperature uniformity results in inhomogeneity

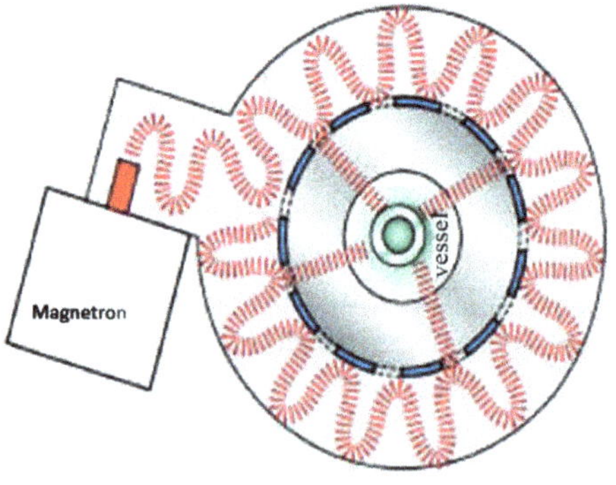

Fig. 3.2 Placement of sample vessel in monomode MW reactor

and non-reproducibility of experimental results and hence chemical reactions done in such apparatus are usually non-reliable. A MW reactor has four crucial components: power source, waveguide, oven cavity, and reaction vessel.

3.2.1 Power Source

The main component of every MW reactor is a high-voltage system generating microwaves, commonly known as the magnetron power supply. The magnetron was invented in 1921 and the current improved versions of it generate megawatts of power with 1–40 GHz frequencies and lifetime of $\sim$5000 h. It is a thermionic diode having directly heated cylindrical cathode and hollow circular anode (possessing small cavities) with a magnetic field imposed in the space between them. The magnetic field is generated using strong permanent magnets, which leads to a magnetic field parallel with the axis of the cathode. The direction of the magnetic field is such that the electron proceeds to the anode in a curve rather than a direct path. In presence of strong magnetic field, the electrons cannot reach the anode but form a rotating space charge. The resonant cavities of the anode accelerate/decelerate the electrons, and this leads to electron bunches which move around the cathode at microwave frequencies. For most MW reactors, the magnetrons emit microwaves of 2.45 GHz with bandwidth of a few MHz.

3.2.2 Waveguide

The waveguide is a rectangular channel made of reflective metal tubes which transmits the microwaves generated by the magnetron to the MW chamber. The reflective walls prevent leakage of MW radiation and increase the efficiency of the MW reactor [30]. For generation of standing waves, there is a maximum wavelength, λ_{max}, which can be transported by the waveguide. For a given direction perpendicular to the axis, the optimum inner tube size is $\lambda_{max}/2$, which is 6.1 cm in case of a 2.45 GHz MW reactor. The microwaves leaving the waveguide are polarized since only one inner dimension is greater than the cut-off length.

3.2.3 Oven Cavity

The microwave cavity is the internal space of the MW reactor in which the reaction vessel is placed for irradiation. The cavity has walls with reflective coating to prevent the leakage of microwaves which increases the efficiency of the reactor. If the microwaves are not absorbed, they can get reflected back into the waveguide and damage the magnetron. Hence, it is essential to run a MW reactor only in the loaded condition. In domestic microwave oven, the dimension of the rectangular cavity is several times larger than the wavelength of the MW radiation. The oven works in multimode fashion and the precise conditions in it depend on the position and nature of material that is placed in the cavity and also the evolution of the material's dielectric properties during heating. Hence, chemical reactions performed in domestic microwave ovens have serious reproducibility issues. The commercial instruments available for laboratory scale as well as industrial synthesis generally work in single mode model. In this design, the field is well defined in space and the material to be heated can be placed at a point of known and relatively even intensity. Important reaction parameter like temperature, pressure, power, and time can be controlled automatically, and these instruments have ramp up, hold, and ramp down features with accessories for stirring and pumping [31]. Certain commercial instruments can operate both in single as well as multimode technique as per need [32]. A point to note is that all these MW reactors are good for solution-based synthesis, and the temperature and pressure of reaction achieved are rarely >300 °C and 20 atm, respectively. For solid-state microwave synthesis, the solid precursors are mixed thoroughly and pressed to form a pellet, which is then deposited in porcelain/alumina/SiC crucible and placed in multimode domestic microwave oven. In an ideal situation, some of the precursors will be good MW absorbers and will facilitate MW heating of the reagents leading to product formation in a short time span. When none of the precursors are good MW absorbers, a susceptor material having a high dielectric loss tangent (e.g., graphite, amorphous carbon, SiC, CuO) can be added to it. The susceptor material can be in direct contact with the precursors or the reaction vessel can be kept in a container of the susceptor material. However, sometimes direct contact with susceptor can lead to contamination of the final product and hence during scale up of a process, it is generally avoided. In rare cases where susceptor cannot be used, MW-induced plasma (MIP) is the alternative solution [33]. The plasma transfers energy between the MW radiation and the reactants and can heat up solids which are not MW absorbers. Use of the MIP also avoids contamination of the product by the susceptor and issues of unwanted side reactions. For certain reactions, the plasma may also serve as a source of reactive species (e.g., N_2 or NH_3 plasma for nitriding solids) [34].

3.2.4 Reaction Vessel

The reaction vessel used for MW-assisted synthesis should not be MW absorbers as that will lead to reduction in efficiency of MW heating of the reactants. The reaction vessels are generally made of material that is virtually transparent to microwaves at the operating frequency. Commonly used materials are quartz, borosilicate glass, or polytetrafluoroethylene. For microwave-assisted hydro/solvothermal synthesis, the vessel should be robust to endure the pressure and temperature changes associated with the reaction and here Teflon is the material of choice. Metallic containers are avoided as it gets over heated and cause high-voltage arcing due to preferential absorption and reflection of microwaves.

3.3 Synthesis of Nanostructures Using Microwave

The nucleation and growth of inorganic nanostructures are highly sensitive to the reaction conditions and hence it benefits a great deal from the efficient and controlled heating provided by MW irradiation. The most common solvent for synthesis of inorganic nanostructures is water. The abundance of water makes it a cheap and good solvent of choice as it is non-toxic, noncorrosive, non-flammable and has relatively low vapor pressure as compared to other organic solvents [35]. Being a polar solvent, water is a good microwave absorber. The loss tangent of any solvent is dependent on the relaxation time of its molecule and in case of water, the relaxation time at 20 °C is 9×10^{-12} s which corresponds to a relaxation frequency of 18 GHz. Hence, at 18 GHz, water will have the most efficient conversion of microwave energy to thermal energy. However, even though the commercially available MW reactors operate at 2.45 GHz, water is still capable of acting as an effective medium for microwave dielectric heating because its loss tangent is sufficiently large. Polyalcohols with high boiling points like ethylene glycol, glycerol, 1,3-propanediol have high loss tangents and hence they are also excellent solvents for microwave-assisted synthesis of inorganic nanostructures. Due to their propensity to form hydrogen bonds, these solvents are highly viscous and hence have longer relaxation times [36]. Their high boiling point makes them good solvent for open vessel microwave synthesis. Many other solvents such as ionic liquids, alcohols, propylene carbonate, ethylene diamine, DMF, trioctylphosphine (TOP), DMSO oleylamine, toluene, dichlorobenzene, and low molecular weight polyethylene glycol (PEG) have also been studied for their application in microwave-assisted preparation of inorganic nanostructures [37–58].

3.3.1 Metals, Non-metals, and Alloys

Metal nanoparticles have myriad applications in diverse areas such as molecular diagnostics, electronics, catalysis, drug delivery or sensing. MW-assisted synthesis is a fast and environment-friendly technique for obtaining these technologically important metal nanoparticles. In aqueous medium, a soluble metal salt is used as the metal source and the metal nanostructure is formed by the reduction of metallic ions with a reducing reagent under microwave irradiation in the presence/absence of additives/surfactants. Commonly studied metal nanostructures include Au, Ag, Pt, Pd, and Cu [12]. Among these, Au nanostructures have been most intensively investigated due to their vast applications in catalysis and biomedicine. For synthesis of Au nanoparticles, $HAuCl_4$ is generally the precursor material of choice which is reduced from Au(III) to Au^0 state in presence of various reductants. Hydrazine hydrate, α-D-glucose, sucrose, maltose, citric acid or sodium citrate, $NaBH_4$, amino acids, and poly(vinylpyrrolidone) (PVP) have been used as reductants [59–67]. The addition of surfactants like PVP, sodium dodecyl sulfate (SDS), cetyltrimethylammonium bromide (CTAB), Triton X-and pluronic acid prevents the agglomeration of the Au nanoparticles by providing a coating on their surface. It has been established that the molar ratio of surfactant to metal ion as well as the molar ratio of reactants play a major role in determining the final morphology of Au nanostructures [68]. In recent years, several environment-friendly microwave-assisted approaches have been initiated for the preparation of Au nanostructures by using biocompatible natural materials, for example, bovine serumalbumin (BSA), human serum albumin, chicken egg white lysozyme, beet juice, and grape pomace/black seed extract [20]. Similarly, Ag nanostructures have been synthesized from $AgNO_3$ using microwave-assisted technique in presence of a variety of biocompatible natural materials, such as beet juice, red wine or grape pomace extract, guava/sugar cane leaf extract, starch/ascorbic acid, amino acids, glutathione, sucrose, and glucose [20]. In case of microwave-assisted synthesis of Pt, Pd, Cu, Te, Rh nanostructures, the precursors used were H_2PtCl_6, $PdCl_2$, $CuCl_2$, Na_2TeO_3, $RhCl_3$, respectively, which were reduced using various reductants in presence/absence of surfactants. Contrary to synthesis in aqueous solution, in the polyol process, the polyol can act as both the solvent and the reductant and hence additional reducing reagent is not required. When ethylene glycol is used as reducing agent as well as surfactant, the following reactions occur under microwave irradiation which lead to the formation of metal nanostructures [69].

$$HOCH_2 - CH_2OH \rightarrow CH_3CHO + H_2O \tag{3.2}$$

$$CH_3CHO + M^{n+} \rightarrow M + CH_3COCOCH_3 + H_2O \tag{3.3}$$

Pt, Pd, Au, Ag, Ru, Rh, Cu, Fe, Co, Ni, Bi, nanostructures have been synthesized using the microwave-assisted polyol method [20]. Also, microwave-assisted synthesis in presence of ionic liquids was found to be suitable for the rapid synthesis of Au, Co, Mn, Cr, Mo, W, Re, Ru, Os, Rh, Ir, nanostructures. The ionic character and high polarizability of ionic liquids make them highly susceptible to microwave irradiation and hence they are excellent microwave absorbers resulting in faster formation of products. Similarly, metal nanostructures of various morphologies have been synthesized using a variety of other organic solvents [20].

Carbon is the most studied non-metal nanostructure that has been synthesized using the microwave-assisted technique. A variety of microwave synthesized carbon dots, nanoparticles, nanospheres, and graphene sheets are reported in literature [70–83]. Microwaving carbohydrate (dextrin, glycerol, glycol, glucose, sucrose, etc.) in presence of an inorganic ion (phosphate, sulfate, etc.) for a few minutes produces carbon dots which exhibit increased absorption, higher quantum yield, and longer fluorescence lifetime than those prepared by the conventional reflux heating. Carbon dots were also obtained via 2 min MW irradiation of ionic liquid *N*-octylpyridinium hexafluorophosphate which acts both as solvent and carbon source [84]. Graphene sheets can be produced by either exfoliating expanded graphite in aqueous ammonia solution by exposing it to MW irradiation or by reduction of graphite oxide in presence of a reductant (ascorbic acid, hydrazine hydrate, etc.) and MW irradiation.

Bimetallic alloys like Fe-Ru, Bi-Rh, $FeNi_3$, $Co_{0.8}Ni_{0.2}$ have been synthesized via microwave-assisted polyol method by reducing a mixture of the corresponding salts in presence of a suitable surfactant [85–88]. Microwave-assisted preparation of Fe–Pt and Fe–Pd nanoparticles have been reported using $Na_2Fe(CO)_4$ and platinum/palladium acetylacetonate $Pt(acac)_2/Pd(acac)_2$ in nonadecane or octyl ether. The solvent, surfactant, and reaction system (open/closed) played a vital role in the final crystal form of the alloy obtained [89].

3.3.2 *Metal Oxides*

Metal oxide nanostructures find wide application in many fields due to their interesting properties and high stability. In a typical reaction of microwave-assisted synthesis of metal oxide nanostructures, a water-soluble metal salt is used as the metal source, an alkaline reagent is used to create an alkaline environment, and an additive or a surfactant is often added to control the morphology and size of the product. Using this method, various oxide materials of metals like Zn, Mg, Pt, Sn, Fe, Co, Ti, Zr, W, Mo, Nd, Gd, V, etc., have been synthesized [20, 90]. Synthesis of $NiWO_4$, $CaWO_4$, $SrWO_4$, $BaWO_4$, $FeWO_4$, $CoWO_4$, and $MnWO_4$ nanoparticles have been accomplished via MW irradiation of corresponding metal nitrate and $Na_2WO_4{\cdot}2H_2O$ solution in deionized water at 110 or 150 °C for 10 or 20 min. Addition of PVP/CTAB altered the morphology leading to formation of tungstate nanosheets and nanobelts [91–93]. Various spinel ($ZnFe_2O_4$, $ZnAl_2O_4$, $NiFe_2O_4$, $MnFe_2O_4$, $CoFe_2O_4$) nanoparticles have been synthesized within 30 min via microwave-assisted hydrothermal technique using corresponding metal nitrate salts. Perovskites like $BaTiO_3$, $BiFeO_3$, $PbTiO_3$, $SrTiO_3$ can be synthesized via MW irradiation of corresponding metal nitrate solution in alkaline condition [94–97]. $BiVO_4$, $CeVO_4$, YVO_4, $LaVO_4$ synthesis using MW irradiation is possible in closed reaction condition with corresponding metal nitrate and $NaVO_3/NH_4VO_3$ as precursor material [98–102]. Similarly, various metal oxide, vanadate, tungstate, molybdate as well as spinel and perovskite compounds can be synthesized using microwave-assisted polyol method. In this case, a metal salt is used as the metal source, and an alkali and/or an additive is added in a polyol solvent [20]. In the microwave-assisted ionic liquid synthesis of metal oxide nanostructures, a metal compound is used as the metal source, an alkaline reagent provides an alkaline environment, and an additive or a surfactant is sometimes added to control the morphology and size of the product. The time period of MW irradiation as well as the type of ionic liquid used plays a major role in the final morphology of the product [20, 103].

3.3.3 Metal Chalcogenides

Metal chalcogenides (sulphides, selenides, and tellurides) are important functional materials with varied applications ranging from nanoelectronics, nanotribology, heterogeneous catalysis, and also as source materials for energy applications. Microwave-assisted rapid synthesis of metal chalcogenide nanostructures is particularly advantageous in terms of very short processing time. In the microwave-assisted synthesis of metal chalcogenide nanostructures in aqueous solution, a water-soluble metal salt is usually used as the metal source, a chalcogenide source is used to provide X^{2-}(X = S, Se, Te) ions, and addition of additive or surfactant helps to control the morphology and particle size of the product. Binary metal sulfides like CdS, ZnS, PbS, HgS, CuS, Cu_9S_8, Ag_2S, Bi_2S_3, as well as ternary metal sulfides of the type $AgInS_2$, $AgIn_5S_8$, $CuInS_2$, $CuInSe_2$, $CdIn_2S_4$, and Cu_2ZnSnS_4 have been synthesized via MW irradiation using Na_2S, CS_2, $Na_2S_2O_3$, NH_2CSNH_2, $NH_2NHCSNH_2$, CH_3CSNH_2, and 3-mercaptopropanoic acid as the sulfur source [104–120]. For example, PEG-coated ZnS nanoparticles have been synthesized via MW-assisted technique using $Zn(NO_3)_2$, CH_3CSNH_2 and PEG as precursor materials. MW irradiation leads to fast decomposition of thioacetamide

$$CH_3CSNH_2 + H_2O \rightarrow CH_3CONH_2 + H_2S \tag{3.4}$$

$$H_2S \rightarrow 2H^+ + S^{2-} \tag{3.5}$$

The Zn^{2+} ions in aqueous solution react with the sulphide ions to form ZnS which gets coated with PEG. The various literature reports indicate the remarkable synthetic flexibility of MW-assisted technique over the traditional methods. A larger variety of precursors with poor solubility can be used, more efficient reaction as well as induction of temporally short spatially homogeneous nucleation due to volumetric heating and accelerated growth [7]. Compared to metal sulfides, metal selenide syntheses via MW irradiation are generally less reported as selenide sources are expensive and rare and their synthesis is a little more challenging [20]. The rate acceleration for reactions in MW technique is more appealing in case of selenides and consequently there are reports on synthesis of binary selenides like CdSe, PbSe, CuSe, ZnSe using corresponding metal sulfates/acetates with $NaSeSO_3$ as the Se source and $NaBH_4$ as the reductant [121, 122]. Ternary $CuInSe_2$ synthesis using MW irradiation has also been reported [20]. The other commonly used Se precursors are Na_2Se, NaHSe, H_2Se, etc. Among all the metal chalcogenides, synthesis of metal tellurides is the most challenging. Metal tellurides are important functional semiconductor materials having application in optical infrared windows and solar cells. The traditional aqueous solution route to the preparation of CdTe nanocrystals is usually time consuming, and this can be avoided using MW technique which reduces the reaction time considerably. This method results inquicksynthesis of a series of CdTe nanocrystals emitting at the green to near-infrared spectral window (505–733 nm) at moderate temperatures. The commonly used source of tellurium includes NaHTe and Na_2TeO_3.

Apart from aqueous solution, MW-assisted polyol synthesis is also quite popular for the synthesis of metal chalcogenides. The process is similar to synthesis in aqueous solution but here instead of water, a polyol (ethylene glycol (EG), tetraethylene glycol (TEG), 1,3-propanediol, triethylene glycol, glycerol, etc.) is used as solvent. This technique has been used to synthesize various sulphides/selenides like PbS, CdS, CoS, CuS, ZnS, SnS, SnS_2, FeS_2, Cu_3BiS_3, Bi_2Se_3, $MoSe_2$, $CuInSe_2$, etc. [20]. Binary tellurides like Bi_2Te_3, Sb_2Te_3, Ni_2Te_3, PbTe, Cu_7Te_5 have been obtained via MW irradiation only when EG has been used as the reducing agent. In polyol method, Se/Te powder can be used directly as the chalcogenide source.

MW-assisted synthesis of metal chalcogenides is extremely fast when ionic liquids are used as solvent. The ionic liquid plays a major role in the final morphology of the product. For example, in MW-assisted synthesis of Bi_2Te_3 using Bi $(NO_3)_3{\cdot}5H_2O$, Te powder as precursors in alkaline medium (KOH) with EG as reductant, addition of 1-butyl-3-methylimidazolium bromide changes the morphology of the product. As the concentration of ionic liquid increases, instead of mixture of nanorods and nanoflakes of Bi_2Te_3, hexagonal nanoplates are obtained [123]. Also, ethylene diamine, DMF, DMSO, diethanolamine, oleylamine have been used as solvents for synthesis of metal chalcogenides [20]. DMSO acts both as solvent and as source of sulfur ions.

3.3.4 *Inorganic Biomaterials*

Biomaterials are biocompatible inorganic or organic materials that can be implanted in the human body to substitute and repair failing tissue. Hydroxyapatite (HAp) and calcium phosphates (CaP) are important inorganic biomaterials and have wide applications in the biomedical fields. Hydroxyapatite (HAp) is a calcium phosphate ($Ca_{10}(PO_4)_6(OH)_2$) with hexagonal structure and a stoichiometric Ca/P ratio of 1.67, which is identical to bone apatite. Calcium phosphate (CaP) biomaterials ($Ca_3(PO_4)_2$) are of special interest because they are bioactive and can form functional interfaces with neighboring bone. CaP is commonly used in orthopedic and dental surgery as bone void fillers or as a coating material on metallic implants [124]. Microwave processing of HAp particles was initially employed for sintering of HAp ceramics to produce a dense material with improved physical and mechanical properties [125]. Currently this technique is used to synthesize HAp nanoparticles in a less energy consuming and more reproducible manner. The fast homogenous nucleation results in microwave synthesis of HAp nanoparticles in less than 30 min and the product has smaller size, higher purity, and narrower size distribution. A typical reaction uses calcium nitrate and H_3PO_4 or NaH_2PO_4 as precursor materials. The MW irradiation is done under alkaline condition in presence of surfactants like CTAB or EDTA. The power of the MW irradiation affects both the particle size and the particle shape. It has been established that applied microwave power and mole ratio of Ca/P are among the important factors affecting the characteristics of HAp. According to the results, low microwave power of 450 W and Ca/P ratio of 1.57 leads to mixed calcium phosphate compounds including $Ca(OH)_2$, $CaHPO_4$ and HAp, whereas running the reaction at 550 W power and Ca/P 1.67 results in monophasic HAp [126]. The shape of HAp crystals can be easily controlled by adjusting the stability of Ca–surfactant complex and the hindrance effect of OH^- ions on the crystallite facets [127]. For inorganic phosphate ion sources, supersaturation can lead to rapid nucleation and a disordered growth and the morphology and size of the particles get difficult to control. Currently, MW-assisted synthesis of HAp is being done using biocompatible phosphorus containing biomolecules as organic phosphorus sources (adenosine 5′-triphosphate (ATP), fructose 1,6-bisphosphate, creatine phosphate, and pyridoxal-5′-phosphate). The advantage of this technique is that the phosphorus source exists in the form of phosphate groups in an organic biomolecule and no PO_4^{3-} ions exist in the precursor solution, thereby avoiding the fast nucleation and disordered growth of the calcium phosphates [20]. Microwave-assisted hydrothermal synthesis of HAp nanowires has been reported using $CaCl_2 \cdot 2H_2O$ and ATP in aqueous solution without any surfactant rendering it to be an environment-friendly method [128].

CaP-based materials consist of different phases like octacalcium phosphate ($Ca_8(HPO_4)_2(PO_4)_4 \cdot 5H_2O$), α-tricalcium phosphate (α-$Ca_3(PO_4)_2$), β-tricalcium phosphate (β-$Ca_3(PO_4)_2$), amorphous calcium phosphate ($Ca_xH_y(PO_4)_z \cdot nH_2O$, $n = 3$–4.5, 15–20% H_2O), hydroxyapatite ($Ca_{10}(PO_4)_6(OH)_2$), fluorapatite ($Ca_{10}(PO_4)_6F_2$), etc., and they differ in their structure and properties. However, controlling the phase of CaP, especially tricalcium phosphate (TCP), is very challenging under mild conditions, particularly when using one preparation protocol for all CaP phases. The crystal phase and morphology of synthetic CaP depend critically on the method of synthesis. Using MW-assisted technique, various CaP, from low-temperature to high-temperature phases, have been successfully fabricated without any additional surfactants. As a result, the synthesis process turned out to be less expensive and simple. The nucleation rate of CaP is determined by the concentration of calcium and phosphate ions in solution which is dependent on the calcium acetate monohydrate (precursor) solubility in different reaction solvents. The calcium and phosphate ions, from calcium acetate monohydrate and phosphoric acid, chemically interact and either forms precipitates or form the intermediates for additional CaP phases. Calcium acetate monohydrate is sparingly soluble in methanol and almost insoluble in ethanol, hence, the Ca^{2+} ion concentration in ethanol at room temperature would be lower than in methanol. Hence, in methanol, amorphous calcium phosphate, with Ca/P molar ratio of 1.2–2.2, was obtained. The reaction time also affected the size and morphology of the crystals; larger crystals were obtained with longer reaction times. As the reaction time was extended, the needle-shaped β-TCP particles transformed into regular elliptical shapes in methanol.

3.3.5 *Miscellaneous Compounds*

Apart from metal, non-metals, alloys, metal oxides, and chalcogenides, various other metal compounds can be synthesized using the microwave-assisted method. Nanostructured metal hydroxides can be synthesized by MW irradiation of an aqueous solution of metal salt and alkali, with/without the presence of surfactant. $Mg(OH)_2$, $Ni(OH)_2$, $Co(OH)_2$, $In(OH)_3$, $Nd(OH)_3$, $Pr(OH)_3$ nanoparticles have been synthesized using various additives like HMTA, urea, L-cysteine, etc. The microwave-assisted method has also been adopted for the rapid synthesis of nanostructured metal fluorides in various solvents like water, polyols, ionic liquids, etc. Commonly NaF, $NaBF_4$, KBF_4, and CF_3COOH are used as the fluorine source of which CF_3COOH is a toxic compound. Fluorine containing ionic liquids such as $[BMIM][BF_4]$ can decompose thermally and hydrolyze under the microwave heating conditions to release F^- ions. Hence, the ionic liquids containing the anion of $[BF_4]^-$ can be used as an environment-friendly fluorine source for the rapid microwave-assisted synthesis of nanostructured metal fluorides. Consequently, metal fluorides like FeF_2, MgF_2, CoF_2, ZnF_2, LaF_3, YF_3, SrF_2 and REF_3 (RE = rare earths) have been synthesized via microwave irradiation of the corresponding metal salt in presence of various tetrafluoroborate ionic liquids which acts both as solvent as well as fluorine source [20, 129]. The ionic liquid used strongly influences the structure, morphology, and luminescent properties of the obtained metal fluorides. Nanostructures ranging from spherical nanoparticles to nanorods with aggregates shaped in the form of bundles, columns, etc., have been formed. This method is fast and facile and most reactions complete within minutes compared to hours taken in case of conventional thermal synthesis. Nanostructured metal phosphates can also be synthesized by the microwave-assisted method. Generally, H_3PO_4, Na/K/$NH_4H_2PO_4$ or $Na_3PO_4 \cdot 12H_2O$ is used as the phosphate anion source. The reaction involves microwaving the metal salt and phosphate source in slightly acidic condition. A variety of metal phosphates like $Mg_3(PO_4)_2$, $CePO_4$, $LnPO_4$, $BiPO_4$, $Ti_2P_2O_7$, $LiFePO_4$ has been successfully synthesized in aqueous solution using the above method. $LaPO_4$ and iron hydroxyl phosphate have also been synthesized via one step microwave-solvothermal ionic liquid method.

Microwave-assisted synthesis of various metal bromides, sulfates, carbonates, silicates, molybdates, and tungstates have also been reported [20]. Nanoparticles of lanthanide orthovanadates $CeVO_4$, $PrVO_4$, and $NdVO_4$ with sizes in the nanometer range have been synthesized by microwave irradiation using water as a solvent. $BaCO_3/SrCO_3$ nanoparticles have been synthesized using $Sr(NO_3)_2$ or $Ba(NO_3)_2$ and Na_2CO_3 in EG under 20 min of cyclic microwave irradiation [130]. Nanostructures of oxyhalides like Zn(OH)F have been synthesized under microwave irradiation in the presence of ionic liquids like 1,2,3-trimethylimidazolium tetrafluoroborate or [BMIM][BF_4] [131]. BiOBr nanostructures have been synthesized using the microwave-assisted polyol method with EG as solvent and CTAB acting as as both bromide source and soft template.

3.3.6 *Inorganic–Inorganic Nanocomposites*

A multiphase solid material with at least one of the phases with dimension of less than 100 nm qualifies as nanocomposite material [132]. Generally, it comprises of a bulk matrix interspersed with a nanodimensional phase which differs in properties due to its structural and chemical differences. The properties of a nanocomposite material differ markedly from its constituents and they are mostly multifunctional in nature which is uncommon in traditional materials. Microwave-assisted methods are especially advantageous for producing nanocomposites in terms of simplicity, rapidness, low cost, and energy saving. The microwave-assisted technique has been used to synthesize a variety of inorganic/inorganic nanocomposites from aqueous solution. The nanocomposites synthesized include Pt/carbon nanotubes, metal/graphene, Au/graphene oxide, carbon nanotubes/graphene oxide, Pd/Au, Pt/Au, Pt/C, Ag/C, PtNi/C, PtCo/C, Se/C, Au/SiO2, Au/ZnO, Ag/SnO_2, Ag/TiO_2, Cu/iron oxide, C/SnO_2, C/ZnO, C/Fe_3O_4, C/$LiMPO_4$ (M = Mn, Fe, and Co), carbon nanotubes/ZnO, carbon nanotubes/MnO_2, carbon nanotubes/RuO_2, graphene/SnO_2, graphene/ZnO, graphene/iron oxide, graphene/Co_3O_4, graphene/carbon nanotubes/SnO_2, carbon nanotubes/ZnS, carbon nanotubes/ZnxCd_{1-x}S, graphene/ZnS, graphene/CdS, graphene/TiO_2, $LiFePO_4$/C/graphene, α-ZrO_2/Al_2O_3, Bi_2O_3/Bi_2CrO_6, Fe_2O_3/ZnO, Co_3O_4/CoO, ZnO/zinc aluminum hydroxide, ZnO/CdS, ZnO/ZnS, ZnO/ZnSe, Pb_3O_4/CdS, CdO/CdS, SiO_2/CdS, Cd$(OH)_x$/CdSe, ZnS/ZnS:Ag, CdS/CdSe, ZnS/CdSe, CdS/CdTe, CdSe/CdTe, CdSeS/ZnS, CdSeTe/ZnS/SiO_2, Mn:ZnSe/ZnS, CdSe/CdS/ZnS, CdTe/CdS/ZnS. Ag/AgCl, and nickel boride/MgO as summarized in a recent review article [20]. Reduction of $HAuCl_4$ with Pt/Pd nanopowder in presence of MW irradiation leads to the formation of bimetallic Au–Pt/Pd nanocomposites. It has been demonstrated that the microwave-assisted method is suitable for the preparation of carbon-supported nanostructures in aqueous solution. Carbon supported PtNi nanoparticles have been synthesized by the reduction of K_2PtCl_4 and $NiCl_2 \cdot 6H_2O$ with hydrazine using PVP as a stabilizer and in the presence of carbon by microwave heating at 100 W for a short time of 10 min. Microwave-assisted urea-mediated homogeneous hydrolysis of $SnCl_4$ in aqueous solution yields graphite/SnO_2 nanocomposite. Carbon-coated $LiMPO_4$ (M = Mn, Fe, and Co) nanocomposite which finds application as cathode materials in lithium ion batteries can be synthesized via MW irradiation ofan aqueous solution of LiOH, H_3PO_4, glucose, and the corresponding metal sulfate. This method offers advantages of being rapid, cost-effective, and energy saving compared to traditional solid-state synthesis. Metal nanoparticles dispersed on carbon nanotube (CNT) which is easily synthesized via MW irradiation of metal salts in presence of reductants and preformed CNTs for a short period of time. For metal oxide dispersed on CNT, an alkaline solution of metal salt needs to be irradiated with MW. Metal sulfide dispersed CNT can be synthesized via MW irradiation of metal salt and thioacetamide in presence of preformed CNT. Pd nanoparticles/graphene is synthesized using tannic acid as a reducing agent in aqueous solution by the microwave-assisted method. For most of the metal nanoparticles dispersed on graphene, simultaneous reduction of graphite oxide and a variety of metal salts is carried out by the reductant, thus resulting in the dispersion of metallic and bimetallic nanoparticles supported on graphene sheets. The loading amount and size of metal nanoparticles on graphene sheets could be controlled by the ratio of raw materials and microwave irradiation time. Various metal/metal oxide and metal oxide/metal oxide nanocomposites have been synthesized by rapid microwave-assisted approaches in aqueous solution. Microwave irradiation for 3 min (800 W) of an aqueous solution containing ZnO nanorods and $HAuCl_4$ results in the formation of Au nanoparticles dispersed on the surface of ZnO nanorods. Nanocomposite with smaller crystallite size and higher crystallinity could be obtained by the microwave-assisted method. Metal oxide/metal sulphide nanocomposites can be synthesized via surface sulfidation of the metal oxide. Here, metal oxide nanorods can be synthesized by conventional methods and used as template. A layer of metal sulphide gets deposited on the metal oxide nanorods with thickness and particle size being dependent on the concentration of the sulfur source and the time period of MW irradiation. Metal chalcogenide/metal chalcogenide nanocomposites have also been prepared by the microwave-assisted method in aqueous solution. In microwave synthesis of CdSe/CdS nanocomposites, CdSe seeds were first obtained by the reaction between NaHSe and Cd^{2+} ions, and then CdSe/CdS quantum dots were rapidly produced under microwave irradiation in the presence of 3-mercaptopropionic acid as a sulfide source. Ag/AgCl nanocomposite has been synthesized by MW irradiation of $AgNO_3$, NaCl, and beet juice in aqueous medium which is a fast and environment-friendly process.

Microwave-assisted polyol method is also a popular choice for the synthesis of various metal inorganic/inorganic nanocomposites. Au/Ag core–shell nanostructures can be synthesized through a two-step microwave-polyol process with EG as solvent. Absence/presence of Cl^- ions lead to formation of predominantly spherical Ag nanoparticles and AgCl, respectively. A small amount of Cl^- anions ($\sim$0.3 mM) was found to be a key factor for the preferential formation of Au/Ag core–shell nanostructures with well-defined shapes. Ag core/Ni shell nanoparticles can be synthesized by reduction of a mixture of Ag and Ni salts in EG in the presence of NaOH and PVP by microwave heating. Bi/Bi_2O_3 composite has been synthesized via a microwave-assisted solvothermal route in which $BiCl_3$ was used as the bismuth source, glucose as a reductant, and EG as the solvent. Pt–Fe nanoparticles/carbon nanotubes nanocomposite was prepared by the microwave-polyol method where the ratio of Fe/Pt and the attached density of PtFe nanoparticles on the surface of carbon nanotubes could be varied by adjusting the experimental parameters. The microwave-assisted polyol synthesis of nanocomposites containing metal oxide and carbon has also been reported. Fe_3O_4/CNT was synthesized by in situ decomposition of iron(III) acetylacetonate in polyol under microwave irradiation. Compared to metal oxides, synthesis of metal telluride composites using polyol process is much less explored. Sb-doped $PbTe/Ag_2Te$ core–shell composite has been synthesized by microwaving $SbCl_3$, $AgNO_3$, Pb $(CH_3COO)_2{\cdot}3H_2O$, Na_2TeO_3, NaOH, and $NaBH_4$ in EG solvent [133].

The microwave-assisted ionic liquid method has also been adopted for rapid synthesis of nanocomposites. Carbon-coated Cu/Ni/Pt nanoparticles can be synthesized by MW irradiation of metal salts with a reductant in ionic liquid as solvent within a short time span of 5 min [134]. Graphene/Ru as well as graphene/Rh core/shell nanocomposites have been synthesized by microwave decomposition of the corresponding metal carbonyl precursors under argon atmosphere in a suspension of graphene in an ionic liquid [BMIM][BF_4]. Alcohols have been proven to be suitable solvents for the microwave-assisted preparation of a variety of nanocomposites. $Fe/Fe_{3-x-y}Mn_xZn_yO_4$ nanocomposites have been prepared by microwave-solvothermal treatment of alcoholic solutions of chloride precursors and sodium ethoxide in a short period of time. Iron oxide supported on porous silica or MCM-41 as well as non-porous biomaterials like chitosan and cellulose has also been synthesized using MW technique. Here, $FeCl_2{\cdot}4H_2O$ acts as the iron precursor and ethanol as the solvent and reaction is carried in presence of the porous or non-porous support material.

3.3.7 Inorganic–Organic Nanocomposites

Inorganic–organic nanocomposites are another class of hybrid materials offering a wide range of possibilities in terms of processing and chemical and physical properties [135]. In these nanocomposites, inorganic nanoparticles with high surface area and unsaturated surface bonds are interspersed in an organic polymer which enhances its properties. In the conventional method of synthesis, the polymer and inorganic nanoparticles are physically mixed together. The inhomogeneity in the product leads to severe aggregation of the inorganic nanoparticles resulting in nanocomposites with uneven property. The high pressure and temperature requirement in certain materials makes it a cost intensive process. The microwave-assisted method has also been used for synthesis of a variety of metal/polymer nanocomposites. However, here the polymerization occurs via microwave-assisted crosslinking and then the metal nanoparticle/polymer composite is obtained by reacting the respective metal salt with the polymer. Metallic and bimetallic PVA crosslinked nanocomposites of Pt–In, Pt–Fe, Pt–Pd, Ag–Pt, Cu–Pd, and Pd–Fe with various shapes, such as nanospheres, nanodendrites, and nanocubes, have been obtained using this technique [136]. Transition metal/CMC (carboxymethyl cellulose) nanocomposites can be synthesized by reacting respective metal salts with the sodium salt of CMC in aqueous media. However, for noble metal/CMC nanocomposites, microwave irradiation at 100 °C is needed. This is a green synthesis method which occurs in the absence of toxic reductants [137].

Metal oxide/polymer nanocomposites can be synthesized using the microwave-assisted preparation method. The monomer and metal oxide nanoparticles are added in reaction vessel and the in situ polymerization process is hastened via MW irradiation. Calcium phosphate/polyacrylamide nanocomposites with nanostructures of the inorganic biomaterial calcium phosphate homogeneously dispersed in the polymer matrix can be synthesized in aqueous solution by the single-step microwave-assisted method using calcium salt, phosphate, and acrylamide monomer [138].

Reports on one-step synthesis of inorganic/organic nanocomposites are relatively scarce [139, 140]. Reverse microemulsion technique has been used for synthesis of monodisperse Fe_3O_4@PANI (Polyaniline) nanocomposites at ambient temperature [141]. Au@polymer nanostructures can be synthesized via a spontaneous redox reaction between a $HAuCl_4$ precursor and the amphiphilic block copolymer poly(styrene-alt-maleic acid) (PSMA) [142]. A novel single-step microwave-assisted polyol method has been developed for the rapid synthesis of inorganic/organic nanocomposites. MW irradiation of corresponding metal salt and acrylamide monomer in EG gives polyacrylamide (PAM)/metal (M = Pt, Ag, Cu) nanocomposites. Here, EG acts as a good MW absorbing solvent and reductant. Thus, complicated work-up procedures for removal of the additives can be avoided. Use of sulfur powder in the reaction powder can result in the formation of metal sulphide/PAM nanocomposites [140].

3.4 Safety Precautions While Using Microwaves

There are a few safety features which needs to be considered while using microwaves for chemical synthesis. Microwaves having frequency of 2.45 GHz and 915 MHz are absorbed by human tissue and have a penetration depth of 2 cm and 3 cm, respectively. This can increase the temperature of blood and tissue and lead to serious side effects. Constant exposure of DNA to microwaves has been reported to lead to its complete degeneration [143]. Hence, all microwave reactors are designed to safeguard the operators from exposure to microwaves. The maximum microwave energy exposure limit has been limited to 5 mW cm^{-2} at a distance of 5 cm from the MW reactor. Care should be taken to ensure that there is no leakage of microwave radiation from the reactor and a damaged reactor should never be used. During any microwave synthesis, apart from the reactants, solvents, catalysts, and supports used in the reaction, the reaction vessel is also subjected to the effect of microwaves. Common materials used for making reactor vessels include quartz, porcelain, borosilicate glass and Teflon, all of which have low loss tangent values. Hence, their heating up rate is very slow and they have minimum contribution in heating during the MW reaction. However, SiC is a material with high loss tangent value and SiC reaction vessels can be used for inefficient heat absorbing reactants. Local overheating of reactant material should be avoided at all cost as it can lead to fire hazard and explosion. In case the reactants ignite, the microwave oven door should be kept closed and the power connection needs to be shut off. A good knowledge about the heat capacity of reactants, solvents, catalysts, and reaction vessel can help design a safer MW reaction method. Radioactive and hazardous chemicals should not be used in MW reactors. When reactions are done in open vessels, adequate precautions should be taken to avoid spillage due to boiling of the solvent. This can be achieved by connecting the reactor vessel with an upright condenser outside of the microwave cavity. The small amounts of solvent vapors leaking out of joints can be expelled by internal fans which are generally in-built in MW reactors. The entire set up should be set in a fume hood or connected to a powerful exhaust system to prevent build-up of flammable/toxic solvent vapors. It should be noted that the reaction temperatures in a microwave-heated and a conventionally heated reflux experiment are similar (i.e., boiling point of the used solvent). Consequently, similar results are obtained for both heating modes since the reaction progress primarily depends on the reaction temperature. In closed reaction systems, MW reactors should be equipped with automatic power cut off sensors to prevent overheating or excessive pressure build-up in the reaction vessel. Metal objects should not be placed in MW reactors as they reflect MW which can cause arching in oven cavity leading to accidents. Any materials which are being used as joints/seals should be placed in MW reactor in advance and irradiated for a short time with low power. If the material gets heated, then it should not be used as joints/seals in MW reaction.

3.5 Conclusions and Future Prospects

From the above reports, it can be inferred that the advantages of MW-assisted synthesis are rapid uniform volumetric heating, higher chemical reaction rate and selectivity, shorter reaction time, higher product yield, and green synthesis as compared to the conventional heating methods. This method leads to the formation of inorganic nanostructures in shorter time which results in low cost of production, energy savings, and higher reaction efficiency. Various solvents with different MW absorbing power can be used for synthesis which can alter the chemical composition, size and morphology of the nanoparticles which have vast applications in the field of materials science. Hence, microwave heating technology is gradually gaining popularity for performing chemical reactions. Though considerable research has been done on this topic for the last two decades, the earlier reports are unreliable since most of the reactions were done using domestic microwave ovens where the conditions of the experiment were not standardized. The repeatability of such experiments and reproducibility of the results is thus dubious for these reports. Currently, the MW reactors have precise temperature and pressure monitoring by various probes and sensors, built-in magnetic stirring, power control, cooling system, and software operation which has increased the popularity of this technique. Another lacuna is the lack of in-depth analysis of the reaction mechanism which leads to shorter reaction time in MW-assisted reactions compared to conventional heating methods. The non-thermal effects of microwave-assisted reactions is controversial and there are no proper scientific experiments to support this theory. Therefore, more carefully designed comparative experiments need to be planned to elucidate this controversial effect. Also, most of reactions have been implemented in the laboratory scale which needs to be scaled up for application in industry. However, the penetration depth of MW in solvents having large dielectric constant is short which prohibits the scale up to a large batch reactor. In this respect, development of a flow reactor under precise control of experimental parameter will be necessary for the industrial application of MW-assisted syntheses. Efforts have to be made to tap the commercial potential of this technique for its greater general applicability.

References

1. Kerr JA (1992) CRC handbook of chemistry and physics, 76th edn. CRC Press, Boca Raton, Ann Arbor, London, Tokyo, p 51
2. Klinowski J, Almeida Paz FA, Silva P, Rocha J (2011) Microwave-assisted synthesis of metal–organic frameworks. Dalton Trans 40:321–330
3. Komarneni S, Roy R (1985) Titania gel spheres by a new sol-gel process. Mater Lett 3:165
4. Gedye R, Smith F, Westaway K, Ali H, Baldisera L, Laberge L, Rousell J (1986) The use of microwave ovens for rapid organic synthesis. Tetrahedron Lett 27:279
5. Giguere RJ, Bray TL, Duncan SM, Majetich G (1986) Application of commercial microwave ovens to organic synthesis. Tetrahedron Lett 27:4945
6. Rao KJ, Vaidhyanathan B, Ganguli M, Ramakrishnan PA (1999) Synthesis of inorganic solids using microwaves. Chem Mater 11:882
7. Baghbanzadeh M, Carbone L, Davide Cozzoli P, Oliver Kappe C (2011) Microwave-assisted synthesis of colloidal inorganic nanocrystals. Angew Chem Int Ed 50:11312–11359
8. Lidstrom P, Tierney J, Wathey B, Westman J (2001) Microwave assisted organic synthesis—a review. Tetrahedron 57:9225–9283
9. Ravichandran S, Karthikeyan E (2011) Microwave synthesis - A potential tool for greenchemistry. Int J Chem Tech Res 3:466–470
10. Surati MA, Jauhari SA, Desai KR (2012) A brief review: Microwave assisted organic reaction. Arch Appl Sci Res 4:645–661
11. Nadagouda MN, Speth TF, Varma RS (2011) Microwave-assisted green synthesis of silver nanostructures. Acc Chem Res 44:469
12. Tsuji M, Hashimoto M, Nishizawa Y, Kubokawa M, Tsuji T (2005) Microwave-assisted synthesis of metallic nanostructures in solution. Chem Eur J 11:440
13. Zhang XY, Liu Z (2012) Recent advances in microwave initiated synthesis of nanocarbon materials. Nanoscale 4:707
14. Park SE, Chang JS, Hwang YK, Kim DS, Jhung SH, Hwang JS (2004) Supramolecular interactions and morphology control in microwave synthesis of nanoporous materials. Catal Surv Asia 8:91
15. Motshekga SC, Pillai SK, Ray SS, Jalama K, Krause RWM (2012) Recent trends in the microwave-assisted synthesis of metal oxide nanoparticles supported on carbon nanotubes and their applications. J Nanomater 2012:691503.
16. Bogdal D, Prociak A, Michalowski S (2011) Synthesis of polymer nanocomposites under microwave irradiation. Curr Org Chem 15:178

17. Bilecka I, Niederberger M (2010) Microwave chemistry for inorganic nanomaterials synthesis. Nanoscale 2:1358
18. Tompsett GA, Conner WC, Yngvesson KS (2006) Microwave synthesis of nanoporous materials. Chem Phys Chem 7:296
19. Gabriel C, Gabriel S, Grant EH, Halstead BSJ, Mingos DMP (1998) Dielectric parameters relevant to microwave dielectric heating. Chem Soc Rev 27:213
20. Zhu Y-J, Chen F (2014) Microwave-assisted preparation of inorganic nanostructures in liquid phase. Chem Rev 114:6462–6555
21. Anastas PT, Warner JC (1998) Green chemistry, theory and practice. Oxford University Press, Oxford
22. de la Hoz A, Díaz-Ortiz Á, Moreno A (2005) Microwaves in organic synthesis. Thermal and non-thermal microwave effects. Chem Soc Rev 34:164
23. Schanche J-S (2003) Microwave synthesis solution from personal chemistry. Mol Diversity 7:293
24. Leadbeater NE (2011) Microwave heating as a tool for sustainable chemistry. CRC, Boca Raton, FL
25. Horikoshi S, Sakai F, Kajitani M, Abe M, Serpone N (2009) Microwave frequency effects on the photoactivity of TiO_2: dielectric properties and the degradation of 4-chlorophenol, bisphenol A and methylene blue. Chem Phys Lett 470:304
26. Leadbeater NE, Torenius HM (2002) A study of the ionic liquid mediated microwave heating of organic solvents. J Org Chem 67:3145
27. Rajamathi M, Seshadri R (2002) Oxide and chalcogenide nanoparticles from hydrothermal/ solvothermal reactions. Curr Opin Solid State Mater Sci 6:337
28. Komarneni S, Roy R, Li QH (1992) Microwave-hydrothermal synthesis of ceramic powders. Mat Res Bull 27:1393–1405
29. Prado-Gonjal J, Schmidt R, Morán E (2013) Perovskite: crystallography, chemistry and catalytic performance. In: Zhang J, Li H (eds) Novascience Publishers, Hauppauge (USA)
30. Somani R, Prabhakar S, Dandekar R, Gide P, Pal T, Kadam V (2010) Optimization of microwave assisted synthesis of some Schiff's bases. Int J Chem Tech Res 2:172–179
31. Ferguson JD (2003) Controlled microwave heating in modern organic synthesis. Mol Diversity 7:281
32. Frecentese F, Fiorino F, Perissutti E, Severino B, Magli E, Esposito A, De Angelis F, Massarelli P, Nencini C, Viti B, Santagada V, Caliendo G (2010) Efficient microwave combinatorial synthesis of novel indolic arylpiperazine derivatives as serotoninergic ligands. Eur J Med Chem 45:752
33. Douthwaite RE (2007) Microwave-induced plasma-promoted materials synthesis. Dalton Trans 10:1002
34. Kitchen HJ, Vallance SR, Kennedy JL, Tapia-Ruiz N, Carassiti L, Harrison A, Gavin Whittaker A, Drysdale TD, Kingman SW, Gregory DH (2014) Modern microwave methods in solid-state inorganic materials chemistry: From fundamentals to manufacturing. Chem Rev 114:1170–1206
35. Polshettiwar V, Varma RS (2010) Green chemistry by nano-catalysis. Green Chem 12:743
36. Cintas P, Tagliapietra S, Calcio Gaudino E, Palmisano G, Cravotto G (2014) Glycerol: a solvent and a building block of choice for microwave and ultrasound irradiation procedures. Green Chem 16:1056–1065
37. Vollmer C, Thomann R, Janiak C (2012) Organic carbonates as stabilizing solvents for transition-metal nanoparticles. Dalton Trans 41:9722
38. He J, Zhao XN, Zhu JJ, Wang JJ (2002) Preparation of CdS nanowires by the decomposition of the complex in the presence of microwave irradiation. Cryst Growth 240:389
39. Liu XY, Tian BZ, Yu CZ, Tu B, Zhao DY (2004) Microwave-assisted solvothermal synthesis of radial ZnS nanoribbons. Chem Lett 33:522
40. Palchik O, Kerner R, Zhu Z, Gedanken A (2000) Preparation of $Cu_{2-x}Te$ and HgTe by using microwave heating. J Solid State Chem 154:530

41. Wu CC, Shiau CY, Ayele DW, Su WN, Cheng MY, Chiu CY, Hwang B (2010) Rapid microwave-enhanced solvothermal process for synthesis of $CuInSe_2$ particles and its morphologic manipulation. J Chem Mater 22:4185
42. Liu JS, Cao JM, Li ZQ, Ji GB, Zheng MB (2007) A simple microwave-assisted decomposing route for synthesis of ZnO nanorods in the presence of PEG400. Mater Lett 61:4409
43. Grisaru H, Pol VG, Gedanken A, Nowik I (2004) Preparation and characterization of Cu_2SnSe_4 nanoparticles using a microwave-assisted polyol method. Eur J Inorg Chem 1859
44. Wada Y, Kuramoto H, Anand J, Kitamura T, Sakata T, Mori H, Yanagida S (2001) Microwave-assisted size control of CdS nanocrystallites. J Mater Chem 11:1936
45. Atkins TM, Louie AY, Kauzlarich SM (2012) An efficient microwave-assisted synthesis method for the production of water soluble amine-terminated Si nanoparticles. Nanotechnology 23
46. He R, Qian XF, Yin J, Zhu ZK (2003) Preparation of Bi_2S_3 nanowhiskers and their morphologies. J Cryst Growth 252:505
47. Karan S, Majumder M, Mallik B (2012) Controlled surface trap state photoluminescence from CdS QDs impregnated in poly(methyl methacrylate). Photochem Photobiol Sci 11:1220
48. Cao XB, Zhao C, Lan XM, Gao GJ, Qian WH, Guo Y (2007) Microwave-enhanced synthesis of Cu_3Se_2 nanoplates and assembly of photovoltaic CdTe−Cu_3Se_2 clusters. J Phys Chem C 111:6658
49. Hu XL, Yu JC (2008) High-yield synthesis of nickel and nickel phosphide nanowires via microwave-assisted processes. Chem Mater 20:6743
50. Ferrer E, Nater S, Rivera D, Colon JM, Zayas F, Gonzalez M, Castro ME (2012) Turning "on" and "off" nucleation and growth: microwave assisted synthesis of CdS clusters and nanoparticles. Mater Res Bull 47:3835
51. Wang QB, Seo DK (2006) Synthesis of deep-red-emitting CdSe quantum dots and general non-inverse-square behavior of quantum confinement in CdSe quantum dots. Chem Mater 18:5764
52. Pein A, Baghbanzadeh M, Rath T, Haas W, Maier E, Amenitsch H, Hofer F, Kappe CO, Trimmel G (2011) Investigation of the formation of $CuInS_2$ nanoparticles by the oleylamine route: comparison of microwave-assisted and conventional syntheses. Inorg Chem 50:193
53. Yang X, Xu J, Xi LJ, Yao YL, Yang QD, Chung CY, Lee C-S (2012) Microwave-assisted synthesis of Cu_2ZnSnS_4 nanocrystals as a novel anode material for lithium ion battery. J Nanopart Res 14:931
54. Yamauchi T, Tsukahara Y, Yamada K, Sakata T, Wada Y (2011) Nucleation and growth of magnetic Ni−Co (Core−Shell) nanoparticles in a one-pot reaction under microwave irradiation. Chem Mater 23:75
55. Muthuswamy E, Iskandar AS, Amador MM, Kauzlarich SM (2013) Facile synthesis of germanium nanoparticles with size control: microwave versus conventional heating. Chem Mater 25:1416
56. Saraswat PK, Free ML (2013) Unidirectional growth of Methyl 2-amino-5-bromobenzoate crystal by Sankaranarayanan–Ramasamy method and its characterization. J Cryst Growth 372:87
57. Borja-Arco E, Jiménez Sandoval O, Escalante-García J, Sandoval-González A, Sebastian P (2011) Microwave assisted synthesis of ruthenium electrocatalysts for oxygen reduction reaction in the presence and absence of aqueous methanol. J Int J Hydrogen Energy 36:103
58. Cano M, Benito A, Maser WK, Urriolabeitia EP (2011) Reduced graphene oxide: firm support for catalytically active palladium nanoparticles and game changer in selective hydrogenation reactions. Carbon 49:652
59. Shang L, Yang LX, Stockmar F, Popescu R, Trouillet V, Bruns M, Gerthsen D, Nienhaus GU (2012) Microwave-assisted rapid synthesis of luminescent gold nanoclusters for sensing Hg^{2+} in living cells using fluorescence imaging. Nanoscale 4:4155

60. Pal A, Shah S, Devi S (2007) Synthesis of Au, Ag and Au–Ag alloy nanoparticles in aqueous polymer solution. Colloids Surf A 302:51
61. Kundu S, Wang K, Liang H (2009) Size-selective synthesis and catalytic application of polyelectrolyte encapsulated gold nanoparticles using microwave irradiation. J Phys Chem C 113:5157
62. Uppal MA, Kafizas A, Ewing MB, Parkin IP (2010) The effect of initiation method on the size, monodispersity and shape of gold nanoparticles formed by the Turkevich method. New J Chem 34:2906
63. Fang Y, Ren YP, Jiang M (2011) Co-effect of soft template and microwave irradiation on morphological control of gold nanobranches. Coll Polym Sci 289:1769
64. Arshi N, Ahmed F, Kumar S, Anwar MS, Lu JQ, Koo BH, Lee CG (2011) Microwave assisted synthesis of gold nanoparticles and their antibacterial activity against Escherichia coli (E. coli.). Curr Appl Phys 11:S360
65. Tsuji M (2017) Microwave-assisted synthesis of metallic nanomaterials in liquid phase. Chem Select 2:805
66. Mallikarjuna NN, Varma RS (2007) Microwave-assisted shape-controlled bulk synthesis of noble nanocrystals and their catalytic properties. Cryst Growth Des 7:686
67. Aswathy B, Suji S, Avadhani GS, Aswathy R, Suganthi S, Sony G (2011) Microwave assisted one pot synthesis of biocompatible gold nanoparticles in Triton X-100 aqueous micellar medium using tryptophan as reducing agent. J Mol Liq 162:155
68. Kundu S, Peng LH, Liang H (2008) A new route to obtain high-yield multiple-shaped gold nanoparticles in aqueous solution using microwave irradiation. Inorg Chem 47:6344
69. Fievet F, Lagier JP, Blin B, Beaudoin B, Figlarz M (1989) Homogeneous and heterogeneous nucleations in the polyol process for the preparation of micron and submicron size metal particles. Solid State Ionics 32(33):198
70. Mitra S, Chandra S, Patra P, Pramanik P, Goswami A (2011) Novel fluorescent matrix embedded carbon quantum dots for the production of stable gold and silver hydrosols. J Mater Chem 21:17638
71. Puvvada N, Kumar BNP, Konar S, Kalita H, Mandal M, Pathak A (2012) Synthesis of biocompatible multicolor luminescent carbon dots for bioimaging applications. Sci Technol Adv Mater 13
72. Wang QL, Zheng HZ, Long YJ, Zhang LY, Gao M, Bai WJ (2011) Microwave–hydrothermal synthesis of fluorescent carbon dots from graphite oxide. Carbon 49:3134
73. Jiang J, He Y, Li SY, Cui H (2012) Amino acids as the source for producing carbon nanodots: microwave assisted one-step synthesis, intrinsic photoluminescence property and intense chemiluminescence enhancement. Chem Commun 48:9634
74. Jaiswal A, Ghosh SS, Chattopadhyay A (2012) One step synthesis of C-dots by microwave mediated caramelization of poly(ethylene glycol). Chem Commun 48:407
75. Wang Q, Liu X, Zhang LC, Lv Y (2012) Microwave-assisted synthesis of carbon nanodots through an eggshell membrane and their fluorescent application. Analyst 137:5392
76. Song YC, Shi W, Chen W, Li XH, Ma HM (2012) Fluorescent carbon nanodots conjugated with folic acid for distinguishing folate-receptor-positive cancer cells from normal cells. J Mater Chem 22:12568
77. Chandra S, Das P, Bag S, Laha D, Pramanik P (2011) Synthesis, functionalization and bioimaging applications of highly fluorescent carbonnanoparticles. Nanoscale 3:1533
78. Zhu H, Wang XL, Li YL, Wang ZJ, Yang F, Yang XR (2009) Microwave synthesis of fluorescent carbon nanoparticles with electrochemiluminescence properties. Chem Commun 5118
79. Cui RJ, Liu C, Shen JM, Gao D, Zhu JJ, Chen HY (2008) Gold nanoparticle–colloidal carbon nanosphere hybrid material: preparation, characterization, and application for an amplified electrochemical immunoassay. Adv Funct Mater 18:2197
80. Zhang M, Liu S, Yin XM, Du ZF, Hao QY, Lei DN, Li QH, Wang TH (2011) Fast synthesis of graphene sheets with good thermal stability by microwave irradiation. Chem Asian J 6:1151

81. Choi BG, Park H, Yang MH, Jung YM, Lee SY, Hong WH, Park TJ (2010) Microwave-assisted synthesis of highly water-soluble graphene towards electrical DNA sensor. Nanoscale 2:2692
82. Long J, Fang M, Chen GH (2011) Microwave-assisted rapid synthesis of water-soluble graphene. J Mater Chem 21:10421
83. Vadahanambi S, Jung J-H, Oh I-K (2011) Microwave syntheses of graphene and graphene decorated with metal nanoparticles. Carbon 49:4449
84. Safavi A, Sedaghati F, Shahbaazi H, Farjami E (2012) Facile approach to the synthesis of carbon nanodots and their peroxidase mimetic function in azodyes degradation. RSC Adv 2:7367
85. Köhler D, Heise M, Baranov AI, Luo Y, Geiger D, Ruck M, Armbrüster M (2012) Synthesis of BiRh nanoplates with superior catalytic performance in the semihydrogenation of acetylene. Chem Mater 24:1639
86. Jia JC, Yu JC, Wang YXJ, Chan KM, (2010) Magnetic nanochains of $FeNi_3$ prepared by a template-free microwave-hydrothermal method. Appl ACS Mater Interface 2:2579
87. Guo XH, Li Y, Liu QY, Shen WJ (2012) Microwave-assisted polyol-synthesis of CoNi nanomaterials. Chin J Catal 33:645
88. Du JQ, Zhang Y, Tian T, Yan SC, Wang HT (2009) Microwave irradiation assisted rapid synthesis of Fe–Ru bimetallic nanoparticles and their catalytic properties in water-gas shift reaction. Mater Res Bull 44:1347
89. Nguyen HL, Howard LEM, Giblin SR, Tanner BK, Terry I, Hughes AK, Ross M, Serres AI, Bürckstümmer H, Evans JSO (2005) Synthesis of monodispersed fcc and fct FePt/FePd nanoparticles by microwave irradiation. J Mater Chem 15:5136
90. Dutta DP, Tyagi AK (2011) White light emission from microwave synthesized spin coated Gd_2O_3:Dy:Tb nano phosphors. Proc Natl Acad Sci India Sec A 81:53–58
91. Luo ZJ, Li HM, Xia JX, Zhu WS, Guo JX, Zhang BB (2007) Microwave-assisted synthesis of barium tungstate nanosheets and nanobelts by using polymer PVP micelle as templates. Mater Lett 61:1845
92. Thongtem T, Kaowphong S, Thongtem S (2008) Characterization of $MeWO_4$ (Me= Ba, Sr and Ca) nanocrystallines prepared by sonochemical method. Appl Surf Sci 254:7765
93. Hu B, Wu LH, Liu SJ, Yao HB, Shi HY, Li GP, Yu SH (2010) Microwave-assisted synthesis of silver indium tungsten oxide mesocrystals and their selective photocatalytic properties. Chem Commun 46:2277
94. Wang LY, Han YY, Jia G, Zhang CM, Liu YJ, Liu L, Wang CZ, Cao XG, Yin KW (2011) Electrochemical impedance spectroscopy of zinc oxide nanoparticles after deposition on screen printed electrode. J Nanosci Nanotechnol 11:5207
95. Jhung SH, Lee JH, Yoon JW, Hwang YK, Hwang JS, Park SE, Chang JS (2004) Effects of a donor concentration on the structure of Nb-doped nano-sized $BaTiO_3$ powders prepared by microwave-hydrothermal synthesis methods. Mater Lett 58:3161
96. Sulaeman U, Yin S, Sato T (2010) Solvothermal synthesis of designed nonstoichiometric strontium titanate for efficient visible-light photocatalysis. Appl Phys Lett 97
97. Dutta DP, Mandal BP, Abdelhamid E, Naik R, Tyagi AK (2015) Enhanced magneto-dielectric coupling in multiferroic Fe and Gd codoped $PbTiO_3$ nanorods synthesized via microwave assisted technique. Dalton Trans 44:11388–11398
98. Zhang HM, Liu JB, Wang H, Zhang WX, Yan H (2008) Adsorption-controlled growth of $BiVO_4$ by molecular-beam epitaxy. J Nanopart Res 10:767
99. Zhu ZF, Du J, Li JQ, Zhang YL, Liu DG (2012) An EDTA-assisted hydrothermal synthesis of $BiVO_4$ hollow microspheres and their evolution into nanocages. Ceram Int 38:4827
100. Wang H, Meng YQ, Yan H (2004) Rapid synthesis of nanocrystalline $CeVO_4$ by microwave irradiation. Inorg Chem Commun 7:553
101. Xu HY, Wang H, Meng YQ, Yan H (2004) Rapid synthesis of size-controllable YVO_4 nanoparticles by microwave irradiation. Solid State Commun 130:465

102. Wang QM, Zhang ZY, Zheng YH, Cai WS, Yu YF (2012) Multiple irradiation triggered the formation of luminescent LaVO4: Ln^{3+}nanorods and in cellulose gels. Cryst Eng Comm 14:4786
103. Wang WW, Zhu YJ (2004) Shape-controlled synthesis of zinc oxide by microwave heating using an imidazolium salt. Inorg Chem Commun 7:1003
104. He XL, Demchenko IN, Stolte WC, van Buuren A, Liang H (2012) Synthesis and transformation of Zn-doped PbS quantum dots. J Phys Chem C 116:22001
105. Mu CF, Yao QZ, Qu XF, Zhou GT, Li ML, Fu SQ (2010) Controlled synthesis of various hierarchical nanostructures of copper sulfide by a facile microwave irradiation method. Colloids Surf A 371:14
106. Liao XH, Chen NY, Xu S, Yang SB, Zhu JJ (2003) A microwave assisted heating method for the preparation of copper sulfide nanorods. J Cryst Growth 252:593
107. Han ZH, Yang Q, Shi J, Lu GQ, Lewis SW (2008) Well-dispersed cadmium sulfide prepared in the presence of laponite by microwave irradiation. Solid State Sci 10:563
108. Kundu S, Lee H, Liang H (2009) Synthesis and application of DNA− CdS nanowires within a minute using microwave irradiation. Inorg Chem 48:121
109. Shao MW, Xu F, Peng YY, Wu J, Li Q, Zhang SY, Qian YT (2002) Microwave-templated synthesis of CdS nanotubes in aqueous solution at room temperature. New J Chem 26:1440
110. Hu Y, Liu Y, Qian HS, Li ZQ, Chen JF (2010) Coating colloidal carbon spheres with CdS nanoparticles: microwave-assisted synthesis and enhanced photocatalytic activity. Langmuir 26:18570
111. Zhao Y, Hong JM, Zhu JJ (2004) Microwave-assisted self-assembled ZnS nanoballs. J Cryst Growth 270:438
112. Mehta SK, Khushboo UA (2011) Highly sensitive hydrazine chemical sensor based on mono-dispersed rapidly synthesized PEG-coated ZnS nanoparticles. Talanta 85:2411
113. Xing RM, Liu SH, Tian SF (2011) Microwave-assisted hydrothermal synthesis of biocompatible silver sulfide nanoworms. J Nanoparticle Res 13:4847
114. Thongtem T, Pilapong C, Kavinchan J, Phuruangrat A, Thongtem S (2010) Microwave-assisted hydrothermal synthesis of Bi_2S_3 nanorods in flower-shaped bundles. J Alloys Compd 500:195
115. Shao MW, Kong LF, Li Q, Yu WC, Qian YT (2003) Microwave-assisted synthesis of tube-like HgS nanoparticles in aqueous solution under ambient condition. Inorg Chem Commun 6:737
116. Zhang WJ, Li DZ, Chen ZX, Sun M, Li WJ, Lin Q, Fu XZ (2011) Microwave hydrothermal synthesis of $AgInS_2$ with visible light photocatalytic activity. Mater Res Bull 46:975
117. Zhang WJ, Li DZ, Sun M, Shao Y, Chen ZX, Xiao GC, Fu XZ (2010) Microwave hydrothermal synthesis and photocatalytic activity of $AgIn_5S_8$ for the degradation of dye. J Solid State Chem 183:2466
118. Bensebaa F, Durand C, Aouadou A, Scoles L, Du X, Wang D, Le Page Y (2010) A new green synthesis method of $CuInS_2$ and $CuInSe_2$ nanoparticles and their integration into thin films. J Nanopart Res 12:1897
119. Apte SK, Garaje SN, Bolade RD, Ambekar JD, Kulkarni MV, Naik SD, Gosavi SW, Baeg JO, Kale BB (2010) Hierarchical nanostructures of $CdIn_2S_4$ via hydrothermal and microwave methods: efficient solar-light-driven photocatalysts. J Mater Chem 20:6095
120. Shin SW, Han JH, Park CY, Moholkar AV, Lee JY, Kim JH (2012) Quaternary Cu_2ZnSnS_4 nanocrystals: facile and low cost synthesis by microwave-assisted solution method. J Alloys Compd 516:96
121. Zhu JJ, Palchik O, Chen SG, Gedanken A (2000) Microwave assisted preparation of CdSe, PbSe, and $Cu_{2-x}Se$ Nanoparticles. J Phys Chem B 104:7344
122. Qian HF, Qiu X, Li L, Ren JC (2006) Microwave-assisted aqueous synthesis: a rapid approach to prepare highly luminescent ZnSe (S) alloyed quantum dots. J Phys Chem B (110):9034

123. Ji GB, Shi Y, Pan LJ, Zheng YD (2011) Effect of ionic liquid amount ($C_8H_{15}BrN_2$) on the morphology of Bi_2Te_3 nanoplates synthesized via a microwave-assisted heating approach. J Alloys Compd 509:6015
124. Lee JS, Murphy WL (2013) Functionalizing calcium phosphate biomaterials with antibacterial silver particles. Adv Mater 25:1173–1179
125. Fang Y, Agrawal DK, Roy DM, Roy R (1994) Microwave sintering of hydroxyapatite ceramics. J Mater Res 9:180–187
126. Han JK, Song HY, Saito F, Lee BT (2006) Synthesis of high purity nano-sized hydroxyapatite powder by microwave-hydrothermal method. Mater Chem Phys 99:235–239
127. Sadat-Shojai M, Khorasani M-T, Dinpanah-Khoshdargi E, Jamshidi A (2013) Synthesis methods for nanosized hydroxyapatite with diverse structures. Acta Biomater 9:7591–7621
128. Qi C, Tang QL, Zhu YJ, Zhao XY, Chen F (2012) Microwave-assisted hydrothermal rapid synthesis of hydroxyapatite nanowires using adenosine 5'-triphosphate disodium salt as phosphorus source. Mater Lett 85:71
129. Jacob DS, Bitton L, Grinblat J, Felner I, Koltypin Y, Gedanken A (2006) Are ionic liquids really a boon for the synthesis of inorganic materials? A general method for the fabrication of nanosized metal fluorides. Chem Mater 18:3162
130. Tipcompor N, Thongtem T, Phuruangrat A, Thongtem S (2012) Characterization of $SrCO_3$ and $BaCO_3$ nanoparticles synthesized by cyclic microwave radiation. Mater Lett 87:153
131. Zhu LJ, Zheng YT, Hao TY, Shi XX, Chen YT, Ou-Yang J (2009) Synthesis of hierarchical ZnO nanobelts via Zn (OH) F intermediate using ionic liquid-assistant microwave irradiation method. Mater Lett 63:2405
132. Ajayan PM, Schadler LS, Braun PV (2003) Nanocomposite Science and Technology. Wiley-VCH, Weinheim
133. Dong GH, Zhu JY (2012) One-step microwave-solvothermal rapid synthesis of Sb doped PbTe/Ag_2Te core/shell composite nanocubes. Chem Eng J 193:227
134. Liu ZM, Sun ZY, Han BX, Zhang JL, Huang J, Du JM, Miao SD (2006) Microwave-assisted synthesis of Pt nanocrystals and deposition on carbon nanotubes in ionic liquids. J Nanosci Nanotechnol 6:175
135. Sanchez C, Gó mez-Romero P (2004) Functional hybrid materials. Wiley VCH, Weinheim
136. Nadagouda MN, Varma RS (2007) Preparation of novel metallic and bimetallic cross-linked poly (vinyl alcohol) nanocomposites under microwave irradiation. Macromol Rapid Commun 28:465
137. Nadagouda MN, Varma RS (2007) Synthesis of thermally stable carboxymethyl cellulose/ metal biodegradable nanocomposites for potential biological applications. Biomacromolecules 8:2762
138. Tang QL, Wang KW, Zhu YJ, Chen F (2009) Single-step rapid microwave-assisted synthesis of polyacrylamide–calcium phosphate nanocomposites in aqueous solution. Mater Lett 63:1332
139. Zhu JF, Zhu YJ (2006) Microwave-assisted one-step synthesis of polyacrylamide− metal (M= Ag, Pt, Cu) nanocomposites in ethylene glycol. J Phys Chem B 110:8593
140. Zhu JF, Zhu YJ, Ma MG, Yang LX, Gao L (2007) Simultaneous and rapid microwave synthesis of polyacrylamide−metal sulfide (Ag_2S, Cu_2S, HgS) Nanocomposites. J Phys Chem C 111:3920
141. Kumar S, Jain S (2014) One-step synthesis of superparamagnetic Fe_3O_4@ PANI nanocomposites. J Chem. Article ID 837682
142. Liu T-M, Yu J, Chang CA, Chiou A, Chiang HK, Chuang Y-C, Wu C-H, Hsu C-H, Chen P-A, Huang C-C (2014) Enhanced Raman sensitivity and magnetic separation for urolithiasis detection using phosphonic acid-terminated Fe_3O_4 nanoclusters. Sci Rep 4:5593
143. Charde MS, Shukla A, Bukhariya V, Chakole RD (2012) A review on: a significance of microwave assist technique in green chemistry. Int J Phytopharm 2:39–50

Chapter 4
Sonochemical Synthesis of Inorganic Nanomaterials

Dimple P. Dutta

Abstract Development of novel synthesis method for nanomaterials having desired size, morphology and composition has been the cornerstone of nanotechnology. The application of high-intensity ultrasound for synthesis of nanostructured materials has been studied extensively in the last three decades. Sonochemistry occurs mostly under ambient conditions without application of external high pressure or temperature and hence proves to be advantageous compared to other conventional synthesis techniques. It is based on the acoustic cavitation phenomenon which occurs when ultrasonic waves move through a reaction medium. In sonochemical synthesis, the extreme high pressure (1000 atm), high temperature (≥ 5000 K), high cooling rates ($\sim 10^{10}$ Ks^{-1}) and enhanced mass transport, generated transiently during the cavitation process, lead to accelerated chemical reaction rates and conditions which are not achievable normally. In ultrasonic spray pyrolysis, nebulization of the precursor solution by ultrasonic waves leads to formation of mist which acts as isolated microreactors in which the chemical reaction can occur. In this chapter, the principles of sonochemistry and the effect of various parameters on sonochemical reaction have been discussed. The design of various ultrasonicators used in the laboratory has been reviewed. The sonochemical as well as ultrasound spray pyrolysis synthesis of various classes of compounds, particularly inorganic nanomaterials, that has been recently reported in literature, has been explored. The chapter ends with a concise report on future outlook and prospects of sonochemical synthesis technique.

Keywords Sonochemistry · Synthesis · Nanomaterials · Ultrasonic spray pyrolysis · Ultrasonicator

D. P. Dutta (✉)
Chemistry Division, Bhabha Atomic Research Centre, Mumbai 400085, India
e-mail: dimpled@barc.gov.in

D. P. Dutta
Homi Bhabha National Institute, Mumbai 400094, India

A. K. Tyagi and R. S. Ningthoujam (eds.), *Handbook on Synthesis Strategies for Advanced Materials*, Indian Institute of Metals Series,
https://doi.org/10.1007/978-981-16-1807-9_4

4.1 Introduction

Chemical synthesis using ultrasonic waves has gained prominence in the last century following the pioneering work of Richard and Loomis in [1] when they reported their observations on the effect of ultrasound on chemical reactions. Sound waves are longitudinal in nature and propagate through a medium via alternate areas of compression and rarefaction. Ultrasonic waves encompass sound energy with frequencies higher than 20 kHz. This high-frequency sound is inaudible to the human ear. The ultrasonic range spans between 20 kHz and 5 MHz, and the energy associated with it is not even sufficient to excite molecular rotations [2]. High-frequency (2–5 MHz) ultrasonic waves with low energy are used to study the medium on which it is applied and find application in medical diagnostics and imaging which will not be discussed in this chapter [3]. Here we will concentrate on low-frequency (20 kHz to 2 MHz) ultrasonic waves with high energy which produces certain chemical effects in the reaction medium. There is no direct interaction between ultrasonic waves and the chemical species leading to any bond breaking and bond making. However, it was observed that application of ultrasound speeds up certain chemical reactions and the reasons for this have been explored intensively.

4.1.1 Principle of Sonochemistry

In a liquid, the molecules are oscillating about their mean position. When ultrasonic waves propagate through a liquid medium, it generates a dynamic tensile stress in the molecules which alters the average distance between them. In case of compression, the molecules are subjected to higher positive pressure whereas when rarefaction occurs, the pressure felt is negative. In case there is sufficient negative pressure, the average distance between the molecules is increased beyond the critical distance needed to uphold the integrity of the medium and leads to creation of voids or cavitation bubbles. In aqueous medium, where the critical distance is considered to be $\sim 10^{-8}$ cm, a tensile pressure of almost 1000 atm is needed to create a vapor-filled cavitation bubble [4]. However, the presence of dissolved gas/gas bubbles and particulate matter in the medium acts as soft spots for generation of cavitation bubbles under much lower pressure. The violent implosion of these cavitation bubbles results in generation of transient (10^{-6} s) high temperature (~ 5000 K) and pressure (~ 1000 atm) conditions, leading to shock waves, which brings about the various effects observed in a reaction with application of ultrasound [5]. It is to be noted that the process of cavitation bubble collapse is nonlinear in nature and is heavily dependent on the local environment and hence varies considerably in homogenous and heterogenous systems.

4.1.2 Effect of Ultrasound on Chemical Reaction

The acoustic cavitation occurs within a frequency range of 10–10^6 Hz. However, most of the ultrasonic instruments used in the laboratory operate within 20–40 kHz. There are two dominant effects of ultrasound: physical and chemical. The physical effects are manifested when the instrument operates at lower range of frequencies, whereas the chemical effects are more pronounced at higher frequency range. The shock waves created by violent collapse of the cavitation bubbles generate mechanical stress in the reaction medium. In case of a heterogenous reaction, the cavitation bubbles do not have a homogenous surrounding and hence undergo asymmetrical growth. During implosion, such a distorted bubble forms microjets which impinge with high force on the nearby available surface and this leads to surface modification via pitting and erosion. The high pressure and velocity of the shock wave can lead to enhanced mass transport via rapid mixing of reactants, alteration in particle size and morphology due to increased collisions and hence play a very significant role in the nature of product obtained from the reaction. Application of ultrasound has been reported to produce particle agglomeration in case of malleable materials, size reduction for brittle materials and exfoliation in materials with lamellar structure [6]. The chemical effect of ultrasound has been studied extensively in case of water which is one of the most commonly used reaction medium. The passage of ultrasonic waves leads to sonolysis of water generating $H^{\bullet}$ and $OH^{\bullet}$ radicals. These radicals can combine and form H_2 and H_2O_2 which act as strong reductants and oxidants, respectively, in various sonochemical reactions conducted in aqueous solution. Non-aqueous organic solvents also form radicals through cavitational collapse provided it has low vapor pressure. Volatile organic solvents have high vapor pressure which lessens the impact of bubble collapse. Solvents like hydrocarbon form H_2 and CH_4, acetonitrile forms N_2, H_2 and CH_4, carbon tetrachloride forms Cl_2, toluene forms H_2, CH_4, benzene, xylenes, etc. through combination of free radicals which plays a major role in the overall reaction.

Another major effect of applying ultrasound to aqueous and organic solvent is the production of light, and this phenomenon is termed sonoluminescence. Sonoluminescence was observed during sonication of water by the scientist duo Frenzel and Schultes [7]. The origin of sonoluminescence is attributed to acoustic cavitation, and in water it exhibits a peak at 310 nm with a broad continuum extending well into the visible range. The former is due to the generation of excited OH• radical, and the latter is attributed to formation of other chemical species. On a similar note, sonoluminescence spectrum of hydrocarbons is dominated by emission from the excited diatomic C_2 unit which is similar to that observed in flame chemistry. The intensity of emission increases with decrease in vapor pressure of the organic solvent confirming its dependence on the cavitational collapse. Sonoluminescence is used as spectroscopic probe to detect the temperature and pressure conditions in a reaction when it is being ultrasonicated. The above observations and many subsequent detailed researches have confirmed that sonoluminescence is a form of chemiluminescence [8–12].

4.1.3 Effect of Various Parameters on Sonochemical Synthesis

Various parameters affect the process of cavitation, viz. frequency and power of the ultrasonic waves, viscosity and temperature of the reaction medium, type of atmosphere and reactor geometry used during synthesis, etc. [13]. Increase in frequency of ultrasonic waves leads to decreased production of cavitation bubbles in the medium. Increasing frequency means shorter compression and rarefaction cycles, and hence, there is less time for the cavitation bubbles to grow to an optimum size which can facilitate their collapse. Also, the power requirement increases ten times with increase in ultrasonic frequency from 10 to 400 kHz. Hence, most of the ultrasonic reactors in laboratories work in an optimum frequency range of 20–50 kHz. The acoustic intensity expressed in W cm^{-2} is the acoustic power at the transducer tip. When acoustic intensity increases, there is an increase in the size of the cavitation bubbles during the rarefaction cycle. The collapse of larger bubbles generates stronger shock waves and increases the rate of reaction. However, the bubble collapse time is also dependent on the size of the bubble, and it increases with size which leads to insufficient cavitation within a fixed period of time. Hence, an optimum power (50–200 W cm^{-2}) is necessary to maximize the rate of sonochemical reaction.

The viscosity of the solvent used affects the rate of acoustic cavitation. For generation of the empty/vapor-filled voids, the negative pressure during the rarefaction cycle should be more than the cohesive force in the solvent. Hence, the cavitation process becomes more difficult in a viscous medium. However, solvents with low surface tension having high vapor pressure are also not desirable since they lead to less intense cavitation collapse by providing a cushioning effect in the vapor-filled bubble. The sonochemical reaction rate is also affected by the temperature of the reaction medium. Higher temperature leads to vaporization of dissolved gas in the medium and also the solvent itself which decreases the efficiency of the bubble collapse. Also, higher temperature decreases the surface tension of the medium and reduces the heat generated during the bubble collapse which results in lower transient temperature arising from the implosion. An optimum external pressure is also required for effective cavitation since high pressure prevents bubble formation as acoustic field has to overcome both surface tension and ambient pressure to generate the cavitation bubbles. However, cavitation can be produced at very high ultrasonic intensity when the external pressure is increased since it leads to a decrease in the bubble collapse time.

The presence of dissolved gas in the medium facilitates sonochemical reaction. It is to be noted that the maximum temperature (T_{max}) during the adiabatic collapse of a cavitation bubble depends on the ratio of specific heats of the gas mixture (γ), the experimental temperature (T_0) and bubble pressure at maximum size (P) and is given by:

$$T_{\max} = T_0 \left[\frac{P_m(\gamma - 1)}{P} \right] \tag{4.1}$$

Hence, monoatomic gases with high γ leads to better cavitation effect compared to diatomic ones. Gases with low thermal conductivity and high solubility also increase the effectiveness of a sonochemical reaction by dissipating less heat from the cavitation site and facilitating the generation of a greater number of cavitation nuclei, respectively. Large volume of gas is undesirable since it lowers cavitation threshold and also lessens the intensity of cavitation bubble implosion.

The reactor design also affects a sonochemical reaction. Reactions done using ultrasonic baths lead to indirect sonication since the reaction vessel is generally immersed in the bath containing a coupling fluid. The ultrasonic power experienced by the reaction is actually much less than that supplied and there are reproducibility issues since the placement of the bath affects the amount of power reaching the reaction vessel. The reaction vessel geometry particularly the bottom design affects the wave pattern even when it is placed at the same position in the bath. The increase in the temperature of the coupling fluid during sonication makes it difficult to maintain isothermal condition which affects cavitation. When ultrasonic probes are used, the sonication is direct but erosion and pitting of the probe tip affects cavitation and also leads to impurity generation in the system.

4.2 Design of Ultrasonic Reactors

There are three main components in any ultrasonic reactor: ultrasonic power generator, ultrasonic transducer and the reaction vessel. The ultrasonic power generator provides electrical energy to the ultrasonic transducer. It converts the energy from the power source to the required voltage and frequency needed for operation of the ultrasonic reactor. The generator can vary the quantum of power supplied to the transducer and also alter its frequency sweep rate as per the requirement. The transducer is a device that can convert energy from one form to other. The generator pumps in a particular frequency, the transducer resonates and amplifies it, and this leads to the generation of alternate regions of compression and rarefaction in the reaction medium. The transducers do this by either piezoelectric method or magnetostriction. In piezoelectric transducer, piezo-ceramic materials (quartz, lead zirconate titanate) are used which can change shape with application of electrical energy. When AC current passes through, the continuous alternating voltage signal leads to resonance vibration in the metal housing around the horn tip attached to the transducer. The magnetostrictive effect is manifested when iron rich material is used which expand/contract with application of magnetic field. The AC current produces an alternating magnetic field and consequent expansion and contraction of the material leads to resonant vibration generated in the transducer. The piezo-electric transducers are favored as they consume less power. This is due to the fact that in this case, there is a single-step energy conversion from electrical to mechanical. However, electrical energy converts to magnetic energy followed by conversion to mechanical energy in magnetostrictive transducers and lot of energy gets dissipated as heat resulting in loss of efficiency. The vessel used for the sonochemical reaction depends on the type of ultrasonic reactor used. The most common ones are ultrasonic baths, ultrasonication probes and batch flow and continuous flow ultrasonicators.

4.2.1 Ultrasonication Bath

Ultrasonic baths are mostly used for cleaning and have transducers attached to the bottom of the bath containing the coupling fluid. These are the most common ultrasonication instrument used in the laboratories as it is relatively inexpensive. If reactions are carried out using ultrasonic bath, the vessel is placed in the coupling fluid and indirect sonication occurs. The bath can also be filled with the reaction solvent, but it might react with the materials and cause corrosion in the lining resulting in impurities in the final product. The process is inefficient as most of the ultrasonic intensity is dissipated before reaching the reaction vessel, and it has poor reproducibility. The warming up of the coupling fluid during sonication also affects the reaction as adiabatic conditions are difficult to maintain. The optimum power, time of reaction and placement of reaction vessel have to be optimized for every bath for maximizing efficiency of the process. The geometry of the reaction vessel should also remain unaltered since the wave pattern generated is influenced by it.

4.2.2 Ultrasonication Probe

Ultrasonic probes called horns are being used mostly in laboratories for carrying out sonochemical reaction. The horn tips are dipped directly into the reaction medium, and it transfers large amount of power to it in a regulated manner by varying the amplitude delivered to the transducer. The tips can undergo pitting and corrosion over time which can add impurities in the product. Regular examination of the tips is needed to enhance the efficiency of the sonication process. The ultrasonic intensity area is dependent on the power delivered to the transducer by the ultrasonic generator. As the power delivered increases, the ultrasonic intensity increases at the center of the reactor and dissipates in the radial direction. The pressure intensities were also found to be highest at the tip and then decrease exponentially with the distance from the tip of the horn. The sonochemical reaction done with ultrasonic horns generally produces more yield compared to that done using ultrasonic baths.

4.2.3 Batch Flow and Continuous Flow Ultrasonicators

For industrial sonochemical processes, both batch and continuous flow reactors are required. These large-sized vessels have multiple transducers attached to its sides and bottom. The reactor has an inbuilt stirring mechanism, an external jacket for isothermal control and reaction ports which allow for operation in the batch, semi-batch or continuous mode. The transducers do not act in tandem and are protected from atmospheric disturbances by caps. These reactors aim to increase the reproductivity and efficiency of sonochemical process compared to ultrasonic baths and transducers.

4.3 Synthesis of Nanostructures Using High Intensity Ultrasound

The applications of ultrasound are manifold. It is used for exfoliation and delamination of brittle and layered solids, for cleaning of surfaces, and also helps in aggregation of metal powders and ceramics which is the result of the enhanced collisions and immense transient heat generated during the cavitation bubble collapse. Sonochemistry finds wide application in organic synthesis as it helps in achieving faster reaction rates at lower temperatures, increases the yield of the desired product and can also lead to different product compared to mechanical agitation. There are review articles available in literature which extensively covers organic synthesis via use of sonochemistry [14]. In this chapter, the focus is on sonochemical synthesis of inorganic nanostructures and the following subsections will highlight the recent results where sonochemistry has been used to synthesize nanoforms of metals, alloys, non-metals, metal oxides, metal chalcogenides and metal carbides. Nanomaterials have sizes ranging between 1 and 100 nm at least in one dimension, and the excess surface energy alters its properties considerably compared to its bulk counterpart. This leads to their important applications in the fields of optoelectronics, electronics, energy storage, catalysis, environment science, biomedical science, etc. [15]. Hence, efficient synthesis of nanomaterials forms an integral part of modern research. Surface deposition of nanomaterials facilitated by ultrasound is another interesting aspect, and this has also been discussed. Sonochemical polymerization and depolymerization reactions form an integral part of polymer mechanochemistry, and some salient features of this topic have been covered.

4.3.1 Metals, Non-metals and Alloys

The commonly used method for synthesis of metal nanoparticles is reduction of its salt with chemical reagents like sodium borohydride, hydrogen, etc. However, with passage of ultrasound, metal nanoparticles can be obtained without addition of chemical reducing agents and in a much shorter time. The sonolysis of aqueous/ ethanolic solution of metal salts generates H• as well as other secondary radicals which act as reducing agents which leads to the formation of metal nanoparticles. The presence of alcohols and other surfactants inhibits the growth of nanoparticles since they can act as capping agents which results in their smaller size. Ag nanoclusters have been synthesized by application of ultrasound to polymethylacrylic acid (PMMA)-added $AgNO_3$ aqueous solution [16]. Core–shell metallic nanoparticles have been synthesized by sonochemical reduction of two different metallic salts. This is done by astute selection of metals having a difference in their reduction potential. Au/Ag and Pt/Ru core–shell nanoparticles have been synthesized using this technique [17, 18]. The use of ultrasound in synthesis of nanoparticles was demonstrated by Suslick et al. by sonochemical-assisted decomposition of volatile iron pentacarbonyl in non-aqueous solvent which led to the formation of iron nanoparticles [19, 20]. Oleic acid and polyvinyl-pyrrolidone (PVP) have been used as capping agents which restrict the agglomeration of the amorphous iron metal formed due to the very high cooling rate of the collapsed cavitation bubble. The non-aqueous solvents generate vapors essential to control the intensity of the cavitational collapse and controls the reaction. Similarly, other volatile organometallic compounds like $Ni(CO)_5$, $Co(CO)_3NO$ have been used as precursors for sonochemical synthesis of amorphous Ni and Co, respectively. Passage of ultrasound through a mixture of $Fe(CO)_5$ and $Co(CO)_3NO$ in various stoichiometric ratio produces Fe–Co alloy having different compositions [20]. The metal nanoparticles synthesized using this technique are highly reactive and can react with other reactants yielding a host of products as shown in Fig. 4.1. For example, ultrasound-assisted decomposition of $Mo(CO)_6$ in 1,2,3,5-tetramethylbenzene under Ar cover leads to the formation of amorphous Mo which can react in situ with sulfur powder and produce MoS_2 which has spherical morphology and is different from the plate-like morphology seen in case of MoS_2 synthesized using conventional technique [21].

The addition of polymers and surfactants in the reaction mixture acts as capping as well as structure directing agent and can lead to non-spherical morphologies like nanorods, nanobelts, etc. There are reports on synthesis of Au nanobelt in presence of a-D-glucose and Au nanodecahedra in presence of PVP [22, 23]. Trigonal selenium nanowires have been obtained by acoustic cavitation of amorphous Se colloids (synthesized via reduction of selenious acid using hydrazine). The smaller particles agglomerate due to the localized intense heat generated during the cavitational bubble collapse [24].

Among non-metals, the application of sonochemistry in synthesis of carbon nanostructures has been studied in detail. The conventional synthesis methods for carbon nanotubes (CNT), nanoscrolls, etc. require application of high temperature,

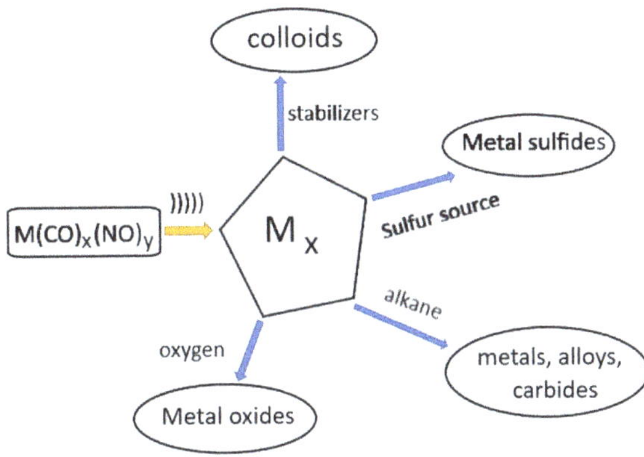

Fig. 4.1 Volatile organometallic compounds leads to formation of metal nanoparticles and various other products. *Note* M_x is metal cluster and all products are in nanoform

catalysts, arc discharge conditions and vacuum compatible systems. However, using ultrasound, carbon nanostructures can be synthesized under ambient conditions and in the absence of any catalyst. Multiwalled CNTs have been synthesized by sonicating $CHCl_3$, CH_2Cl_2 and CH_3I in presence of Si nanowires. The ultrasound leads to decomposition of the solvents and their subsequent reaction with Si surface which leads to growth of CNTs [25]. Similarly, sonication of ferrocene and silica powder in *p*-xylene leads to formation of single-walled CNTs [26]. With CNTs finding widespread application in electronic devices and sensors, their limited solubility in water and organic solvents pose a big problem as it hinders the fabrication process. Sonochemistry finds immense application in increasing the dispersibility and solubility of CNTs. *p*-doped CNTs have been synthesized by sonicating carbon nanotubes in *o*-dichlorobenzene. Addition of sodiumdodecylbenzene sulfonate (SDS) leads to water dispersible CNTs. Carbon sonogels have been synthesized using resorcinol and formaldehyde as precursors. Ultrasonic waves are passed through the mixture till gelation point and pyrolysis after appropriate aging of the gel yields highly porous carbon sonogel [27]. The time of gelation is greatly reduced on application of ultrasound compared to the conventional process which has been attributed to the formation of free radical species in the former case. The physical effect of ultrasound is also evident in its utility in separating out the sp^2 hybridized carbon sheets of graphite. This is needed to introduce guest ions/molecules in between the sheets or completely exfoliate them to produce graphene. The graphene sheets can form nanoscrolls on application of ultrasound which finds application as hydrogen storage materials [28, 29].

4.3.2 *Metal Oxides*

The synthesis of nanostructured metal oxides is one of the most useful applications of the sonochemical method. In a typical reaction, the metal salt solution in water is sonicated under air at normal atmospheric pressure and room temperature. The pH

Fig. 4.2 **a** TEM image of $BiFeO_3$. **b** SAED pattern of $BiFeO_3$ (Ref. [47])

of the solution is made slightly alkaline by addition of ammonia or urea. Ultrasound waves increase the rate of hydrolysis, and the shock waves generated from the cavitation bubble collapse sometimes lead to unusual morphology in the metal oxide nanoparticles. Compared to conventional synthesis technique, metal oxide nanoparticles obtained via sonochemical synthesis have increased surface area, better size distribution, enhanced phase purity and the added advantage of shorter reaction time. There are numerous reports available in literature on sonochemical synthesis of metal oxide nanoparticles, viz. TiO_2, CeO_2, In_2O_3, Gd_2O_3, CoO, CdO, ZrO_2, spinel compounds like $CoCr_2O_4$, $ZnAl_2O_4$, $ZnGa_2O_4$, $FeAl_2O_4$, perovskites like $BiFeO_3$, $BaTiO_3$, $PbTiO_3$, $SrTiO_3$, $ZnTiO_3$, and also mixed metal oxides like $Bi_2Fe_4O_9$, $Bi_4Fe_3O_{12}$, $ZnSb_2O_6$, $CdSb_2O_6$ and $BaSb_2O_6$ [30–53]. Morphological modification of the product is possible through judicious selection of additives in the reaction. The addition of surfactants like tetra-ethylene glycol (TEG) in the aqueous metal salt solution mixture have resulted in the formation of nanorods as reported in case of $BiFeO_3$ synthesis by ultrasonication as shown in Fig. 4.2 [47]. Nanocubes of $In(OH)_3$ have been synthesized sonochemically in presence of CTAB as surface directing agent [53]. The increase in sonication time from 0.5 to 1 h

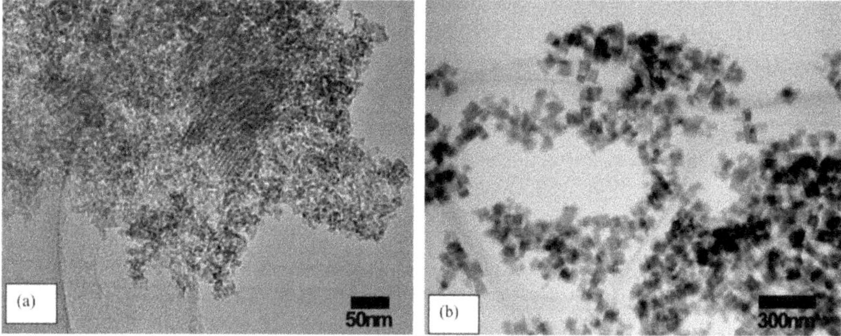

Fig. 4.3 TEM micrograph of $In(OH)_3$ particles obtained after **a** 0.5 h and **b** 1 h sonochemical treatment. Reprinted Ref. [53] with copyrights permission from American Chemical Society

changed the morphology of the $In(OH)_3$ nanostructure (Fig. 4.3). Hollow haematite as well as Co_3O_4 nanospheres has been synthesized sonochemically using carbon/CNT as the sacrificial template and corresponding Fe and Co precursors [54, 55]. Sol gel synthesis is another popular method for synthesis of metal oxide nanoparticles. In sol gel synthesis, the initial step is hydrolysis which is followed by the condensation process leading to the formation of the gel precursor after aging. In this process, cross-linked polymer networks in liquid form which in turn gets converted to the solid phase. The solid product is obtained from the gel either by applying heat or by addition of solvent. The hydrolysis rate is enhanced on application of ultrasound in sol gel process, and this leads to better size uniformity and phase purity in the metal oxide nanoparticles formed. This has been amply demonstrated in the synthesis of oxides of Ti, Zn, Ce, Mo, V and In using sonochemically assisted sol gel synthesis. Mesoporous TiO_2 and TiO_2 hollow nanospheres have been synthesized by 6 h/3 h sonication of $[Ti(OPr^i)_4]$ precursor in presence of de-, do- and octadecylamine surfactants or block copolymer, respectively [56, 57]. The reduced time of gelation has been attributed to the enhanced mass transport facilitated by the passage of ultrasonic waves. The increased surface area of these TiO_2 nanostructures results in their better catalytic performance compared to TiO_2 synthesized using conventional techniques.

4.3.3 Metal Chalcogenides

The use of ultrasound in metal chalcogenide synthesis was first reported by Suslick's group [21]. Ultrasonication of molybdenum hexacarbonyl in 1,2,3,5-tetramethylbenzene under Ar gas cover in presence of sulfur powder yielded MoS_2. The higher surface area and presence of defects in the sonochemically synthesized MoS_2 considerably enhance their catalytic activity for hydrodesulfurization of thiophene. However, the most reported method for sonochemical synthesis of chalcogenides involves ultrasonication of an aqueous metal salt solution in presence of chalcogen donors like thiourea, thioacetamide, selenourea, etc. The S/Se• radicals generated through sonolysis form H_2S or H_2Se, which reacts with the metal ions and lead to faster kinetics. Nanostructures with various morphologies (rods, cubes, spheres, etc.) have been reported by using surfactants during the sonochemical synthesis which act as structure directing agents [58–61]. Hollow nanostructures of metal chalcogenides act as better catalysts and photonic materials, and hence, a lot of emphasis has been laid in its synthesis. The conventional method uses silica/polymer templates as sacrificial agents which complicates the synthesis procedure and makes it expensive. Hollow CdSe nanosphere has been synthesized via one-step ultrasonication of aqueous solution of $CdCl_2$ and $NaSeSO_3$ in presence of ammonia. The $Cd(OH)_2$ formed acts as a sacrificial template for the formation of CdSe hollow nanospheres [62]. ZnO nanospheres can act as templates in sonochemical synthesis of ZnS/ZnSe hollow nanospheres in presence of appropriate sulfur/selenium sources [63]. The S^{2-}/Se^{2-} generated gets adsorbed on the ZnO surface and forms ZnS/ZnSe which act as nucleation sites and leads to complete dissolution of the ZnO structure leaving hollow nanospheres of the Zn chalcogenides. Various other hollow metal chalcogenide nanostructures have been synthesized using the same technique [64, 65]. Also, there are reports on synthesis of oxide/chalcogenide core–shell nanostructures (SnO_2/CdS, ZnO/CdS) using similar sonochemical synthesis method [66]. A two-step sonochemical synthesis of highly luminescent CdSe/ZnS core–shell nanoparticles has also been reported.

4.3.4 Metal Carbides

Conventional synthesis of refractory materials like Mo_2C and W_2C requires very high temperature heating of the corresponding metal and carbon. Hence, synthesis of nanostructures of carbides through conventional synthesis is not possible since high temperature leads to agglomeration of the product. Sonochemical synthesis proves invaluable in the production of such carbide nanostructures since acoustic cavitation leads to transient hot spot generation but it does not affect the temperature of the reaction in the macroscale. Sonication of organometallic $Mo(CO)_6$ and W $(CO)_6$ in presence of hexadecane leads to formation of metal oxycarbides which when heated in a reducing CH_4/H_2 atmosphere yields the corresponding metal carbide nanoparticles. The nanostructures are porous which lends them a high surface area which results in their enhanced catalytic activity for dehydrogenation reactions [67].

4.3.5 Surface Deposition

Apart from the chemical effects of ultrasound for nanoparticle synthesis, the physical effects of ultrasound are also useful in surface deposition of nanoparticles on various substrates. The generation of high-speed microjets and intense shock waves is a result of the cavitation process, and this is manifested as the physical effect of ultrasound. There are reports on sonochemical deposition of noble metal particles on silica, carbon, polymer substrates without the need of surface tailoring [68–70]. Using ultrasound, the coating obtained is very uniform. Uniform deposition of silica on Fe_3O_4 nanoparticles as well as indium tin oxide has been reported via passage of ultrasound in mixture of Fe_3O_4 nanoparticles/indium tin oxide and tetraethyl orthosilicate [71, 72]. The thickness of the silica coating can be adjusted by altering the time of sonication. Sonicating a dispersion of $Mo(CO)_6$ and ZSM-5 in hexadecane under inert gas cover yields Mo_2C coating over ZSM-5 resulting in an egg shell configuration which acts as an efficient catalyst for aromatization of methane [73]. De-intercalation of layered compounds is also considerably facilitated by application of ultrasound as it weakens the attractive forces between the layers. Formation of the reducing agent KC_8 is done easily by sonication of graphite in toluene in presence of potassium metal [74]. Layered graphite oxide has reduced van der Waals attraction between layers compared to graphite and is easily separated by application of ultrasound waves. On reduction with appropriate reductants, it leads to the formation of graphene sheets. Zn and Sn aqueous salt solution on being subjected to ultrasound forms low-melting Zn and Sn, and the enhanced mass transfer makes these metal clusters impinge on one another with sufficient intensity to induce melting at the point of contact [75]. However, this phenomenon is observed only in case of metals having melting point less than 3000 k and an optimum size which is neither too small nor large. The size determines the velocity with which the particles impinge on each other and must have enough mechanical energy to induce agglomeration. Sonochemically synthesized FePt nanoparticles have been deposited on silica nanoparticles by application of ultrasound. Consequent treatment with HF dissolved the silica and leaves behind FePt hollow nanospheres. Polystyrene and polymethylmethacrylate nanospheres are also used as sacrificial templated for synthesis of hollow nanostructures. Uniform deposition of metal/metal oxide/metal chalcogenide nanoparticles is achieved on the polymer nanospheres via sonochemical method. The polymer core is then removed by thermal treatment or washing with suitable solvent leaving behind hollow nanostructures of desired material.

4.3.6 Inorganic–Polymer Nanocomposites

The power of the physical effects of ultrasound is also manifested in its mechanochemical action on covalent bond which leads to the shortening of polymer chain length. High-molecular-weight polymers undergo radical formation and bond breakage on passage of ultrasound. The breakage can occur at specified points in the polymer chain if easily sliceable groups are attached to it. The groups easily cleaved by ultrasonication include azo, peroxide, and strained rings. Apart from shortening of polymer chain length, ultrasonication also leads to polymer formation from monomers via radical polymerization process. This method leads to polymerization in a much shorter time and at much lower reaction temperatures compared to the conventional process. Ultrasonication is also used for synthesis of polymer nanocomposites with inorganic materials like CNT, oxides, clay, etc. CNT dispersion in solvents is a problem as it has a high van der Waals force of attraction which leads to agglomeration. Ultrasonication of CNT in organic solvents improves their dispersibility. Polymer CNT nanocomposites are synthesized by sonication of solvent-dispersed CNT in polymer matrix. For example, multiwalled CNT/PMMA nanocomposites have been synthesized via a simple sonochemical method. The CNTs are treated with acid and then sonicated in liquid MMA in presence of 2,2-azoisobutyronitrile (AIBN) which serves as an initiator for the radical polymerization process. The radicals are generated on CNT and aid in the in situ polymerization of MMA to PMMA and in the process, which leads to disentanglement of the CNTs and their better dispensability in the polymer matrix [76]. The nanocomposites when dissolved in chloroform and cast on Teflon plates yield free standing films with minimum aggregation of CNTs. The quality of films obtained is much superior to that obtained without in situ polymerization via sonication. Multiwalled CNT/polystyrene (PS) nanocomposites have also been synthesized via sonochemically induced radical generation from monomer decomposition in absence of any other initiator [77]. The CNTs are found to be well dispersed in the PS matrix. However, when simple solution mixing process is used for synthesis of these nanocomposites, and the CNTs are found to be in an agglomerated state. This proves that both chemical and physical effects of sonication are beneficial for synthesis of such CNT/polymer nanocomposites. Similarly, sonochemically assisted synthesis of carbon black/polyvinyl alcohol nanocomposites has been reported, and the carbon black shows better stability and dispersion in the composite matrix compared to that in water or PVA solution [78].

Polymer/clay forms another class of nanocomposites which finds application as better filler materials due to their enhanced mechanical and thermal properties. In order to synthesize these nanocomposites, it is essential to have chemical compatibility between the nanoclay and polymer material, and the former should be distributed homogenously in the latter. Ultrasonication has been established as a facile technique for synthesis of such nanocomposites. The dispersion of clay in polymer matrix via sonication yields materials with steady shear viscosity compared to the conventional chemical and mechanical methods. PMMA/MMT and poly(glycidyl methacrylate)/MMT nanocomposites have been synthesized via *in situ* polymerization by sonication of a chloroform solution of the precursors [79]. These nanocomposites find application in optoelectronic devices. Similarly, poly (glycidyl methacrylate)/MMT nanocomposites have also been synthesized via in situ polymerization using ultrasonication which reduced the time of synthesis and yielded nanocomposites with better dispersion of the MMT clay compared to solution blending method [80]. The superior dispersion has been attributed to the generation of powerful microconvection in the solution via acoustic cavitation.

4.4 Ultrasonic Spray Pyrolysis

Spray pyrolysis is a process in which aerosols are thermally decomposed under a gas flow. This technique is used for film deposition and also for synthesis of submicron as well as nanosized particles. When the aerosols are generated using an ultrasonic nebulizer, the process is known as ultrasonic spray pyrolysis (USP). The ultrasonic generation of aerosols is an energy-efficient process which leads to the formation of aerosols with medium velocity, and this in turn increases the yield of product. Being a continuous flow process, it is easily scalable to industrial production level and is a cost-effective technique. In sonochemistry, low-frequency (20 kHz) and high-intensity ultrasonic waves are used, whereas in USP, the frequency used is generally 2 MHz. The ultrasound nebulizers produce micron-sized droplets which act as microreactors and get heated thermally leading to decomposition of the precursors and eventually yield the product. The Lang equation aptly describes the dependence of the diameter of the droplet (D) on the applied frequency:

$$D_{\text{droplet}} = 0.34\left(\frac{8\pi\gamma}{\rho f^2}\right)^{1/3} \tag{4.2}$$

Here, γ denotes the surface tension, ρ is the density of the solution, and f is the applied ultrasonic frequency. It is evident that for synthesis of nanomaterials, the ultrasonic frequency should be high (≥ 1 MHz).

The schematics of an USP apparatus are shown below (Fig. 4.4). The carrier gas carries the ultrasonically nebulized precursor mist into the heated furnace where it undergoes decomposition to form the product which is generally deposited on a Si/glass substrate placed in the cooler region of the furnace. The sequence of events occurs in the following order: Ultrasonic nebulization leads to droplet formation, solvent in the drop gets evaporated under thermal treatment leading to supersaturation, solutes diffuse and precipitate, the precipitated solids decompose under heat, the gases trapped in decomposed product escape, and the material gets somewhat

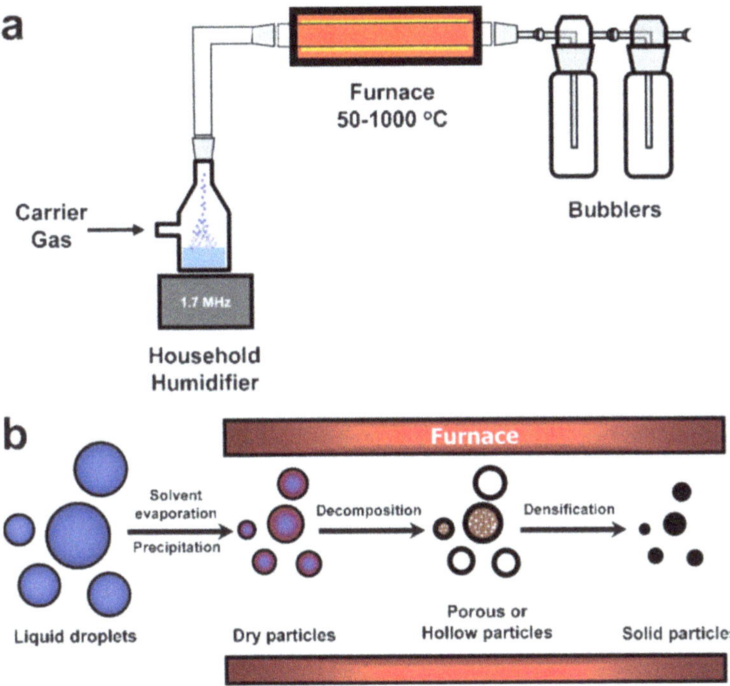

Fig. 4.4 **a** Schematics and **b** reaction procedure for USP apparatus. Reprinted from Ref. [82] with copyrights permission from Wiley-VCH, New York/ACS

sintered. The morphology of the final product depends on the rate of evaporation of the solvent and the extent of solubility of the precursors in the solvent. USP finds application in production of fine micron-sized powders of metals, alloys by thermal decomposition of metal nitrates under inert gas cover. This procedure has been used extensively for synthesis of Ag, Au, Pd, Ag-Pd particles [2]. For the synthesis of Cu, Ni, Co particles, a reducing atmosphere is desirable and hence the thermal treatment is generally carried out under Ar/H_2 atmosphere. An even better method is to heat the metal nitrates/acetates in presence of alcohol. The latter decomposes to produce H_2 along with carbon monoxide and methane, which can provide the reducing atmosphere in situ, and thus simplifies the synthesis process [81]. Synthesis of metal oxides and chalcogenides using USP is very common, and the procedure involves ultrasonic nebulization of metal acetate/nitrate/chloride solution under air or in presence of a chalcogen source, respectively [82]. However, the major advantage of USP is evident in synthesis of composite powders. The conventional method of synthesis of crystalline composites is liquid precipitation technique which needs high-temperature annealing leading to agglomeration, hydrothermal method which can be used for synthesis of only a few specific composites and solid-state synthesis which produces powders with minimum control of size and morphology. In USP, the reactions occur in isolated chemical microreactors due to the ultrasonic nebulization and form crystalline composites with minimum agglomeration.

The last decade has seen phenomenal amount of research being devoted to synthesis of nanostructured composites using USP. The metal/metal oxide/metal chalcogenide composites are formed with sacrificial templates which can be later removed to leave behind porous nanostructures. The use of silica as template for nanostructure synthesis is now well exploited, and MoS_2/SiO_2 nanocomposite has been synthesized via decomposition of $(NH_4)_2MoS_4$ in presence of silica colloid [83]. Porous MoS_2 with controlled morphology is obtained by dissolving the composite in HF which removes the silica template. The nature and morphology of the silica colloid play a major role in controlling the level of porosity in MoS_2 nanostructures. The extent of deposition of the desired material on the silica colloids also affects the morphology. This is evident in USP synthesis of $ZnS:Ni^{2+}$ samples which produced hollow microspheres at low temperatures and nanoparticles at high temperature after removal of the silica colloid template [84]. At high temperature, the extent of ZnS deposition on silica is larger than that of the silica colloid and hence on removal of the template, the hollow structure is not supported and leads to the formation of nanoparticles. Using sucrose and silica colloid as template, porous carbon nanospheres have been synthesized via USP. Manipulation of pore size was done by varying the sucrose to silica colloid ratio during the synthesis, and these materials find application in H_2 storage [85]. Macroporous SiO_2/TiO_2 has been synthesized via USP using polymer colloids as template [86, 87]. Colloidal dispersion of silica/titania and polystyrene is subjected to ultrasonic nebulization, and sequential thermal treatment forms SiO_2/TiO_2/polystyrene composite at low temperature and macroporous SiO_2/TiO_2 spheres at high temperature which pyrolyzes the polystyrene. Since polymers are expensive, a better approach has been developed to reduce the cost of synthesis. Here monomers are used in the USP process, and the temperature is gradually increased during ultrasonic nebulization so that polymer composites are formed and further increase in temperature leads to pyrolysis of the in situ generated polymer, leaving behind macroporous material [88].

One of the major problems in synthesis of nanomaterials using USP is the tendency of the numerous nuclei generated forming multiple nanocrystallites to agglomerate. This leads to formation of micron-sized material. To circumvent this problem, metal salt (Li/Na/K chloride or nitrate) or their mixtures are added to the precursor solution while performing USP. The salts act as a solvent to the precipitated nanocrystallites and trap them preventing their agglomeration. The salt can be removed from the final product by repeated washing. Using this molten salt-modified USP technique, various metal, alloy, metal oxide and metal chalcogenide nanoparticles have been synthesized [2]. The amount of metal salt added influences the morphology of the final product. Use of less amount of metal salt leads to the formation of porous nanomaterials. The salts are mostly inexpensive, non-toxic and reusable, which renders the process more attractive. Porous carbon spheres and nanocages can also be synthesized via USP of alkali metal salts of chloro/dichloroacetates [89]. This single-step method is much preferred compared to the complicated multistep conventional procedure for porous carbon synthesis.

USP has also matured as a method of choice for synthesis of semiconducting metal chalcogenide quantum dots. In this case, the precursors are mixed in high-boiling organic solvents diluted with low-boiling organic solvent to reduce their viscosity. After ultrasonic nebulization, the low-boiling solvent evaporates at the low temperature heating zone leaving a concentrated dispersion of precursor in the high boiling solvent. On passage through the high-temperature zone, the precursors decompose to yield the desired semiconductor quantum dots (binary and ternary CdS/Se/Te) [90].

4.5 Conclusions and Future Prospects

The application of high-intensity ultrasound in synthesis of nanomaterials is possible due to its physical and chemical effects on the reaction mixture, which is brought on by the phenomenon of acoustic cavitation. On varying the reaction conditions and the composition of the precursor, a wide variety of nanomaterials with controlled size and morphology can be synthesized. However, industrial application of this technique is rather limited since scaling up of the synthesis requires better technical support. Though laboratory-scale ultrasonicators are now quite common and comparatively inexpensive, large-scale ultrasonicators are relatively scarce in supply. The energy efficiency of the process needs to be improved since the coupling of ultrasound to generate chemically fruitful acoustic cavitation is not as easy as conversion of electrical energy to ultrasonic waves. In this regard, much work needs to be done to increase the extent of chemically fruitful acoustic cavitation in the reaction mixture. In ultrasonic spray pyrolysis technique, the ultrasonic waves nebulize the precursor solutions producing micron-sized droplets that act as microreactors and leads to product formation. Though the last decade has seen explosive development of the USP method for formation of various stable nanomaterials, the major emphasis has been on nanocomposites and semiconductor materials. The report on metal oxide synthesis using USP is relatively scarce since their formation requires longer period of thermal treatment, which is not available because of the short residence time of the microreactor droplets in furnace. Lack of proper precursors for synthesis of metal oxides nanoparticles is another bottleneck which needs attention. Hence, emphasis should be focussed on development of novel precursor materials and better control of various reaction parameters to get products with desired size, morphology and composition.

References

1. Richards WT, Loomis AL (1927) The chemical effects of high frequency sound waves I. A preliminary survey. J Am Chem Soc 49:3086
2. Bang JH, Suslick KS (2010) Applications of ultrasound to the synthesis of nanostructured materials. Adv Mater 22:1039
3. Carovac A, Smajlovic F, Junuzovic D (2011) Application of ultrasound in medicine. Acta Inform Med 19(3):168
4. Lorimer JP, Mason TJ (1987) Sonochemistry. Part 1—the physical aspects. Chem Soc Rev 16:239
5. Suslick KS (1990) Protein microencapsulation of nonaqueous liquids. Science 247:1439
6. Xu H, Zeiger BW, Suslick KS (2013) Sonochemical synthesis of nanomaterials. Chem Soc Rev 42:2555
7. Frenzel H, Schultes H (1934) Sonoluminescence from metal carbonyls. Z Phys Chem 27b:421
8. Flannigan DJ, Suslick KS (2005) Plasma formation and temperature measurement during single-bubble cavitation. Nature 434:52
9. Eddingsaas NC, Suslick KS (2007) Evidence for a plasma core during multibubble sonoluminescence in sulfuric acid. J Am Chem Soc 129:3838
10. Didenko YT, Suslick KS (2002) The energy efficiency of formation of photons, radicals and ions during single-bubble cavitation. Nature 418:394
11. Flannigan DJ, Hopkins SD, Camara CG, Putterman SJ, Suslick KS (2006) Measurement of pressure and density inside a single sonoluminescing bubble. Phys Rev Lett 96:204301
12. Flannigan DJ, Suslick KS (2005) Plasma line emission during single-bubble cavitation. Phys Rev Lett 95:044301
13. Mason TJ (2000) Sonochemistry. Chemistry Primers, Oxford
14. Cravotto G, Cintas P (2006) Power ultrasound in organic synthesis: moving cavitational chemistry from academia to innovative and large-scale applications. Chem Soc Rev 35:180
15. Yin Y, Talapin D (2013) The chemistry of functional nanomaterials. Chem Soc Rev 42:2484
16. Xu HX, Suslick KS (2010) Sonochemical synthesis of highly fluorescent Ag nanoclusters. ACS Nano 4:3209
17. Anandan S, Grieser F, Ashokkumar M (2008) Sonochemical synthesis of Au− Ag core− shell bimetallic nanoparticles. J Phys Chem C 112:15102
18. Vinodgopal K, He Y, Ashokkumar M, Grieser F (2006) Sonochemically prepared platinum−ruthenium bimetallic nanoparticles. J Phys Chem B 110:3849

19. Suslick KS, Choe SB, Cichowlas AA, Grinstaff MW (1991) Sonochemical synthesis of amorphous iron. Nature 353:414
20. Suslick KS, Fang M, Hyeon T (1996) Sonochemical synthesis of iron colloids. J Am Chem Soc 118:11960
21. Mdleleni MM, Hyeon T, Suslick KS (1998) Sonochemical synthesis of nanostructured molybdenum sulfide. J Am Chem Soc 120:6189
22. Sánchez-Iglesias A, Pastoriza-Santos I, Pérez-Juste J, Rodríguez-González B, García de Abajo FJ, Liz-Marzán LM (2006) Synthesis and optical properties of gold nanodecahedra with size control. Adv Mater 18:2529
23. Zhang J, Du J, Han B, Liu Z, Jiang T, Zhang Z (2006) Sonochemical formation of single-crystalline gold nanobelts. Angew Chem Int Ed 45:1116
24. Gates B, Mayers B, Grossman A, Xia Y (2002) A sonochemical approach to the synthesis of crystalline selenium nanowires in solutions and on solid supports. Adv Mater 14:1749
25. Jeong SH, Ko JH, Park JB, Park WJ (2004) A sonochemical route to single-walled carbon nanotubes under ambient conditions. J Am Chem Soc 126:15982
26. Li CP, Teo BK, Sun XH, Wong NB, Lee ST (2005) Hydrocarbon and carbon nanostructures produced by sonochemical reactions of organic solvents on hydrogen-passivated silicon nanowires under ambient conditions. Chem Mater 17:5780
27. Tonanon N, Siyasukh A, Wareenin Y, Charinpanitkul T, Tanthapanichakoon W, Nishihara H, Mukai SR, Tamon H (2005) Improvement of mesoporosity of carbon cryogels by ultrasonic irradiation. Carbon 43:525
28. Viculis LM, Mack JJ, Kaner RB (2003) A chemical route to carbon nanoscrolls. (Brevia). Science 299:1361
29. Savoskin MV, Mochalin VN, Yaroshenko AP, Lazareva NI, Konstantinova TE, Barsukov IV, Prokofiev LG (2007) Carbon nanoscrolls produced from acceptor-type graphite intercalation compounds. Carbon 45:2797
30. Fulekar J, Dutta DP, Pathak B, Fulekar MH (2018) Novel microbial and root mediated green synthesis of TiO_2 nanoparticles and its application in wastewater remediation. J Chem Tech Biotech 93:736
31. Dutta DP, Roy M, Maiti N, Tyagi AK (2016) Phase evolution in sonochemically synthesized Fe_{3+} doped $BaTiO_3$ nanocrystallites: structural, magnetic and ferroelectric characterisation. Phys Chem Chem Phys 18:9758
32. Dutta DP, Tyagi AK (2016) Weak room temperature ferromagnetism and ferroelectric behavior in sonochemically synthesized bismuth and iron codoped $SrTiO_3$ nanoparticles. Mater Lett 164:368
33. Dutta DP, Tyagi AK (2016) Facile sonochemical synthesis of Ag modified $Bi_4Ti_3O_{12}$ nanoparticles with enhanced photocatalytic activity under visible light. Mater Res Bull 74:397
34. Dutta DP, Mandal BP, Abdelhamid E, Naik R, Tyagi AK (2015) Enhanced magneto-dielectric coupling in multiferroic Fe and Gd codoped $PbTiO_3$ nanorods synthesized via microwave assisted technique. Dalton Trans 44:11388
35. Dutta DP, Singh A, Tyagi AK (2014). Ag doped and Ag dispersed nano $ZnTiO_3$: improved photocatalytic organic pollutant degradation under solar irradiation and antibacterial activity. J Environ Chem Eng 2:2177
36. Dutta DP, Mandal BP, Mukadam MD, Yusuf SM, Tyagi AK (2014) Improved magnetic and ferroelectric properties of Sc and Ti codoped multiferroic nano $BiFeO_3$ prepared via sonochemical synthesis. Dalton Trans 43:7838
37. Dutta DP, Ballal A, Nuwad J, Tyagi AK (2014) Optical properties of sonochemically synthesized rare earth ions doped $BaTiO_3$ nanophosphors: probable candidate for white light emission. J Lumines 148:230
38. Dutta DP, Ballal A, Singh A, Fulekar MH, Tyagi AK (2013) Multifunctionality of rare earth doped nano $ZnSb_2O_6$, $CdSb_2O_6$ and $BaSb_2O_6$: photocatalytic properties and white light emission. Dalton Trans 42:16887
39. Dutta DP, Tyagi AK (2013) Enhanced multiferroic properties in scandium doped $Bi_2Fe_4O_9$. AIP Conf Proc 182:1512

40. Dutta DP, Mandal BP, Naik R, Lawes G, Tyagi AK (2013) Magnetic, ferroelectric and magnetocapacitive properties of sonochemically synthesized Sc doped $BiFeO_3$ nanoparticles. J Phys Chem C 117:2382
41. Dutta DP, Roy M, Tyagi AK (2012) Dual function of rare earth doped nano Bi_2O_3: white light emission and photocatalytic properties. Dalton Trans 41:10238
42. Dutta DP, Sudakar C, Mocherla PSV, Mandal BP, Jayakumar OD, Tyagi AK (2012) Enhanced magnetic and ferroelectric properties in scandium doped nano $Bi_2Fe_4O_9$. Mater Chem Phys 135:998
43. Mukherjee S, Dutta DP, Manoj N, Tyagi AK (2012) Sonochemically synthesized rare earth double-doped zirconia nanoparticles: probable candidate for white light emission. J Nanoparticle Res 14:814 (10 pp)
44. Dutta DP, Manoj N, Tyagi AK (2011) White light emission from sonochemically synthesized rare earth doped ceria nanophosphors. J Lumines 131:1807
45. Dutta DP, Tyagi AK (2011) White light emission from microwave synthesized spin coated Gd_2O_3:Dy:Tb nano phosphors. Proc Natl Acad Sci India Sec A (Special Issue: Emerging Energy technologies) 81:53
46. Bedekar V, Dutta DP, Tyagi AK (2010) Sonochemical synthesis of doped CeF_3 nanoparticles exhibiting room temperature ferromagnetism and white light emission. J Nanosci Nanotech 10:8234
47. Dutta DP, Jayakumar OD, Tyagi AK, Girija KG, Pillai CGS, Sharma G (2010) Effect of doping on the morphology and multiferroic properties of $BiFeO_3$ nanorods. Nanoscale 2:1149
48. Dutta DP, Tyagi AK, Sharma G (2010) Optical properties of sonochemically synthesized dysprosium doped CdO nanoparticles. J Indian Chem Soc 87:23
49. Dutta DP, Ghildiyal R, Tyagi AK (2009) Luminescent properties of doped zinc aluminate and zinc gallate white light emitting nanophosphors prepared via sonochemical method. J Phys Chem C 113:16954
50. Dutta DP, Manjanna J, Tyagi AK (2009) Magnetic properties of sonochemically synthesized $CoCr_2O_4$ nanoparticles. J Appl Phys 106
51. Dutta DP, Sharma G, Manna PK, Tyagi AK, Yusuf SM (2008) Room Temperature Ferromagnetism in CoO nanoparticles obtained from sonochemically synthesized precursors. Nanotechnology 19:245609
52. Bedekar V, Dutta DP, Mohapatra M, Godbole SV, Ghildiyal R, Tyagi AK (2009) Rare earth doped gadolinia based phosphors for potential multicolor and white light emitting deep UV LEDs. Nanotechnology 20:125707 (9 pp)
53. Dutta DP, Sudarsan V, Srinivasu P, Vinu A, Tyagi AK (2008) Indium oxide and europium/ dysprosium doped indium oxide nanoparticles: Sonochemical synthesis, characterization and photoluminescence studies. J Phys Chem C 112(17):6781–6785
54. Bang JH, Suslick KS (2007) Sonochemical synthesis of nanosized hollow hematite. J Phys Chem C 129:2242
55. Du N, Zhang H, Chen BD, Wu JB, Ma XY, Liu ZH, Zhang YQ, Yang DR, Huang XH, Tu JP (2007) Porous Co_3O_4 nanotubes derived from $Co_4(CO)_{12}$ clusters on carbon nanotube templates: a highly efficient material for Li-Battery applications. Adv Mater 19:4505
56. Wang Y, Tang X, Yin L, Huang W, Hacohen YR, Gedanken A (2000) Sonochemical synthesis of mesoporous titanium oxide with wormhole-like framework structures. Adv Mater 12:1183
57. Zhang L, Yu JC (2003) A sonochemical approach to hierarchical porous titania spheres with enhanced photocatalytic activity. Chem Commun 2078
58. Zhou SM, Feng YS, Zhang LD (2003) Sonochemical synthesis of large-scale single crystal CdS nanorods. Mater Lett 57:2936
59. Zhou SM, Feng YS, Zhang LD (2003) Sonochemical synthesis of large-scale single-crystal PbS nanorods. J Mater Res 18:1188
60. Zhu JJ, Wang H, Xu S, Chen HY (2002) Sonochemical method for the preparation of monodisperse spherical and rectangular lead selenide nanoparticles. Langmuir 18:3306

61. Qiu X, Burda C, Fu R, Pu L, Chen H, Zhu J (2004) Heterostructured Bi_2Se_3 nanowires with periodic phase boundaries. J Am Chem Soc 126:16276
62. Zhu JJ, Xu S, Wang H, Zhu JM, Chen H-Y (2003) Sonochemical synthesis of CdSe hollow spherical assemblies via an in-situ template route. Adv Mater 15:156
63. Geng J, Liu B, Xu L, Hu FN, Zhu JJ (2007) Facile route to Zn-based II– VI semiconductor spheres, hollow spheres, and core/shell nanocrystals and their optical properties. Langmuir 23:10286
64. Wang SF, Gu F, Lu MK (2006) Sonochemical synthesis of hollow PbS nanospheres. Langmuir 22:398
65. Zhou H, Fan T, Zhang D, Guo Q, Ogawa H (2007) Novel bacteria-templated sonochemical route for the in situ one-step synthesis of ZnS hollow nanostructures. Chem Mater 19:2144
66. Gao T, Wang T (2004) Sonochemical synthesis of SnO_2 nanobelt/CdS nanoparticle core/shell heterostructures. Chem Commun 2558
67. Hyeon T, Fang M, Suslick KS (1996) Sonochemical synthesis of iron colloids. J Phys Chem Soc 118:5492
68. Pol VG, Gedanken A, Calderon-Moreno J (2003) Deposition of gold nanoparticles on silica spheres: a sonochemical approach. Chem Mater 15:1111
69. Pol VG, Motiei M, Gedanken A, Calderon-Moreno J, Mastai Y (2003) Sonochemical deposition of air-stable iron nanoparticles on monodispersed carbon spherules. Chem Mater 15:1378
70. Pol VG, Grisaru H, Gedanken A (2005) Coating noble metal nanocrystals (Ag, Au, Pd, and Pt) on polystyrene spheres via ultrasound irradiation. Langmuir 21:3635
71. Morel A-L, Nikitenko SI, Gionnet K, Wattiaux A, Lai-Kee-Him J, Labrugere C, Chevalier B, Deleris G, Petibois C, Brisson A, Simonoff M (2008) Sonochemical approach to the synthesis of Fe_3O_4@SiO_2 core–shell nanoparticles with tunable properties. ACS Nano 2:847
72. Chen Q, Boothroyd C, Tan GH, Sutanto N, Soutar AM, Zeng ZT (2008) Silica coating of nanoparticles by the sonogel process. Langmuir 24:650
73. Dantsin G, Suslick KS (2000) Sonochemical preparation of a nanostructured bifunctional catalyst. J Phys Chem Soc 122:5214
74. Jones JE, Cheshire MC, Casadonte DJ, Phifer CC (2004) Facile sonochemical synthesis of graphite intercalation compounds. Org Lett 6:1915
75. Prozorov T, Prozorov R, Suslick KS (2004) High velocity interparticle collisions driven by ultrasound. J Phys Chem Soc 126:13890
76. Park SJ, Cho MS, Lim ST, Choi HJ, Jhon MS (2003) Synthesis and dispersion characteristics of multi-walled carbon nanotube composites with poly (methyl methacrylate) prepared by in-situ bulk polymerization. Macromol Rapid Commun 24:1070
77. Jia ZJ, Wang ZY, Xu CL, Liang J, Wei BQ, Wu DH, Zhu SW (1999) Study on poly (methyl methacrylate)/carbon nanotube composites. Mater Sci Eng A 271:395
78. Li QY, Wu GZ, Ma YL, Wu CF (2007) Grafting modification of carbon black by trapping macroradicals formed by sonochemical degradation. Carbon 45:2411
79. Prado BR, Bartoli JR (2018) Synthesis and characterization of PMMA and organic modified montmorilonites nanocomposites via in situ polymerization assisted by sonication. Appl Clay Sci 160:132
80. Cherifi Z, Boukoussa B, Zaoui A, Belbachir M, Meghabar R (2018) Structural, morphological and thermal properties of nanocomposites poly (GMA)/clay prepared by ultrasound and in-situ polymerization. Ultrason Sonochem 48:188
81. Vivekchand SRC, Gundiah G, Govindaraj A, Rao CNR (2004) A new method for the preparation of metal nanowires by the nebulized spray pyrolysis of precursors. Adv Mater 16:1842
82. Kodas TT, Hampden-Smith M (1999) Aerosol processing of materials. Wiley-VCH, New York
83. Skrabalak SE, Suslick KS (2005) Porous MoS_2 synthesized by ultrasonic spray pyrolysis. J Phys Chem Soc 127:9990
84. Bang JH, Helmich RJ, Suslick KS (2008) Adv Mater 20:2599

85. Hu Q, Lu Y, Meisner GP (2008) Preparation of nanoporous carbon particles and their cryogenic hydrogen storage capacities. J Phys Chem C 112:1516
86. Iskandar F, Mikrajuddin A, Okuyama K (2002) Controllability of pore size and porosity on self-organized porous silica particles. Nano Lett 2:389
87. Iskandar F, Nandiyanto ABD, Yun KM, Okuyama HJCJK, Biswas P (2007) Enhanced photocatalytic performance of brookite TiO_2 macroporous particles prepared by spray drying with colloidal templating. Adv Mater 19:1408
88. Suh WH, Suslick KS (2005) Magnetic and porous nanospheres from ultrasonic spray pyrolysis. J Phys Chem Soc 127:12007
89. Skrabalak SE, Suslick KS (2006) Porous carbon powders prepared by ultrasonic spray pyrolysis. J Am Chem Soc 128:12642
90. Didenko YT, Suslick KS (2005) Chemical aerosol flow synthesis of semiconductor nanoparticles. J Am Chem Soc 127:12196

Chapter 5
Hydrothermal Method for Synthesis of Materials

V. S. Tripathi

Abstract Hydrothermal method of synthesis has emerged as the primary choice for synthesizing several strategic materials. The application of this method has diversified in the last few decades into several advanced fields of material science with the progress in the understanding of the process and with the evolution of better instrumentation. The journey of hydrothermal synthesis started with the preparation of minerals particularly quartz crystals using the temperature gradient method with the emphasis on quality of the product in terms of purity, defects, and size. This method gained immense prominence with the emergence of mesoporous zeolites which act as excellent catalyst for the cracking of petroleum. Hydrothermal synthesis is the most suitable route for the preparation of zeolites and other related mesoporous structures with engineered pores. Detailed studies have been carried out to understand the growth mechanism in the quest of designing the framework with required porosity. Hydrothermal synthesis has emerged as the preferred route for the synthesis of metal oxide nanoparticles. Enhanced dehydration and overall kinetics of the process due to increased temperature result in the formation of the desired product. Designing the hydrothermal synthesis process to tailor the morphology of the product at nanoscale has led to the development of several interesting semiconducting nanoparticles and nano-structured arrays. Variations in the technique like using microwave-assisted hydrothermal method or continuous hydrothermal flow synthesis have helped in further improving the quality of product. The above aspects related to the hydrothermal method of synthesis have been described in this chapter.

Keywords Hydrothermal synthesis · Framework · Dehydration · Mesoporous · Nanoparticles

V. S. Tripathi (✉)
Radiation & Photochemistry Division, Bhabha Atomic Research Centre,
Mumbai 400085, India
e-mail: vst_apcd@barc.gov.in

A. K. Tyagi and R. S. Ningthoujam (eds.), *Handbook on Synthesis Strategies for Advanced Materials*, Indian Institute of Metals Series,
https://doi.org/10.1007/978-981-16-1807-9_5

5.1 Introduction

Hydrothermal method of synthesis involves performing a heterogeneous reaction in an aqueous medium at temperatures above 100 °C and pressure above 1 bar. It is one of the important methods of synthesis and it has been used in both the arena of research for materials development and in industries for bulk synthesis. The term "hydrothermal" has geological origin as the related studies were initially used to understand the geochemical formation of various minerals. The first reported hydrothermal synthesis was that of the quartz microcrystals in a Papin's digester using silicic acid by German Geologist Karl Emil von Schafhäutl in 1845. Papin digester had a pressure relief mechanism for safety purpose. The method was subsequently adopted with some modifications in the design of the vessel by several researchers for the synthesis of crystals. A thick-walled glass tube was used by Bunsen in 1848 for the synthesis of $BaCO_3$ and $SrCO_3$ from an ammoniacal solution. Glass vessel provided an additional advantage of visual observation. In 1851, de Sénarmont used sealed glass ampoules as reaction vessels which in turn were placed inside the gun barrel partly filled with water. These gun barrels were closed by welding before heating. de Sénarmont had extensively used this technique to synthesize a variety of materials ranging from sulfides to sulfates, fluorides, and carbonates among others. This arrangement served as the preliminary basis for the future designs of autoclaves which were used for hydrothermal synthesis for a substantial duration. The further advancement was made by the advent of Morey's vessel in 1914. This vessel was a noble metal lined autoclave which could be used for entire pH range. The vessel used a flat plunger system for sealing the inventory as shown in Fig. 5.1. This design is finally the one which is continued

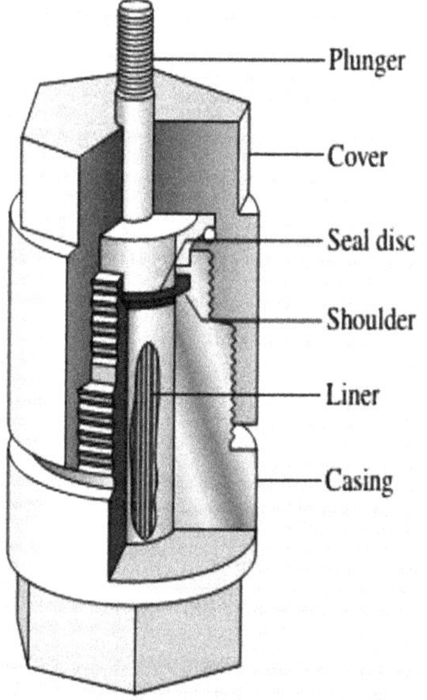

Fig. 5.1 Morey vessel. [3] Reproduced with permission from Ref. Copyright 2015 Elsevier

to be used as autoclaves employed for hydrothermal synthesis, albeit with some modifications. Bridgman changed the material of construction of the seal which helped in achieving higher pressure [1] while Tuttle made the sealing arrangement conical instead of flat and thereby removing the requirement of the noble metal liner [2]. The design of autoclaves used in the laboratory has further evolved and the common variants are shown in Fig. 5.2. The interest in the field of hydrothermal synthesis has diversified from the material scientists who were interested in growing large crystals to those who were interested in the synthesis of microporous materials and subsequently to the researchers who are interested in developing semiconducting metal nanoparticles. The methodology of hydrothermal synthesis has undergone continuous modification as per the required applications. The initial application in the area of crystal growth required systems with a temperature gradient. This is achieved by keeping the upper portion of the autoclave at a lower temperature than the lower one. The saturated solution of the hot zone crystallizes in the cold zone. The movement of the solution is due to convection. In certain systems where retrograde solubility is seen and also with the change in solvent, there is a reversal in the process, viz., crystallization occurs in the hot zone. Isothermal methods are used for the synthesis applications like deposition of oxide layer on metallic substrates, synthesis of microporous materials like zeolites, etc. Recent advancements have introduced novel techniques like microwave-assisted hydrothermal synthesis which have significantly improved the quality of product. Microwave-assisted hydrothermal synthesis requires much shorter time to reach the reaction temperature which limits the time for particle growth of the product. Thus, the particle size of the synthesized product is much smaller as compared to the product formed in the conventional hydrothermal synthesis. Microwave-assisted hydrothermal synthesis also ensures uniform heating in the entire medium as compared to the convective heat transfer in the conventional hydrothermal synthesis. This results in the better quality of the crystal with lesser defects. Conventional hydrothermal method is a batch synthesis method with limitations of miniscule yield per batch and inter-batch variations in the quality of product due to

Fig. 5.2 Common variant of the autoclaves which is currently used in hydrothermal synthesis. [4] Reproduced with permission from Ref. Copyright 2007 Elsevier

Table 5.1 Various phases of growth in the field of hydrothermal synthesis

Time duration	Key objectives	Apparatus	References
1845–1950	Synthesis of minerals (particularly quartz with large crystal size)	Sealed capsules, Morey vessel	[5, 6]
1950–1980	High-temperature high-pressure phase diagram	Autoclave with Tuttle or Bridgman modification	[7–9]
1980–1990	Microporous materials, binary and ternary oxides	Teflon lined, advanced material-based autoclaves	[10–13]
Beyond 1990	Size and morphology controlled synthesis	Microwave hydrothermal	[14–17]

variations in the exact experimental conditions. Continuous hydrothermal synthesis technique using a high-pressure injection pump overcomes the issues related to batch processes. A brief summary of advancement in the field of hydrothermal synthesis is given in Table 5.1.

5.1.1 Role of Water as Medium

The thermophysical properties of water vary with the varying hydrothermal conditions due to the variation in the extent of hydrogen bonding and its strength with the varying temperature and pressure conditions. The dissociation constant increases with the increase in temperature and pressure. At a temperature of 1000 °C, water is completely dissociated into H_3O^+ and OH^- and its properties become similar to that of ionic liquids. Viscosity of water decreases with the increase in temperature, while the increase in pressure leads to the increase in the viscosity of water. The dielectric constant of water also decreases with the increase in temperature. This leads to the reduction in the polar nature of water. As a result, the minerals which were completely dissolved under ambient condition tend to associate and precipitate out as crystals at high temperature.

The fill volume of the reaction vessel determines its pressure at the working temperature. The fill volume must be limited to the level which allows for the accommodation of the expansion of fluid. In case of water, increase in volume is relatively slower up to 200 °C (15% increase at 200 °C), but it starts increasing rapidly beyond 300 °C. The increase in volume of water with varying temperature is enlisted in Table 5.2. The PT diagram of water with varying fill volume is shown in Fig. 5.3. As evident from Fig. 5.3, the fluid gas interface is curved upward for a fill volume of 32% and above and it is curved downwards for the fill volume below 32%. Such a fill volume would favor dehydration process. Most of the hydrothermal synthesis involves higher fill volume.

Table 5.2 Liquid volumes and vapor pressures for water in a closed vessel at elevated temperatures[a]

Temperature (°C)	Density of the liquid (kg/m^3)	Vapor pressure (bar)	Volume multiplier (Sp.V$_T$/Sp.V$_{25\,°C}$)	% volume increase
25	996.79	–	1.00	0
100	958.04	0	1.04	4
200	864.46	14.55	1.15	15
250	796.94	38.75	1.25	25
282	745.04	65.36	1.34	34
300	711.93	84.81	1.40	40
321	664.67	113.76	1.50	50
349	576.20	162.03	1.73	73
363	508.52	191.67	1.96	96
371	434.10	211.67	2.30	130
372	416.06	215.12	2.40	140
373	390.69	217.87	2.55	155
374	318.46	219.94	3.13	213

(Sp.V: specific volume)
[a]Data from J. H. Keenan, F. G. Keyes, P. G. Hill and J. G. Moore "Steam Tables," John Wiley & Sons, Inc., New York (1969)

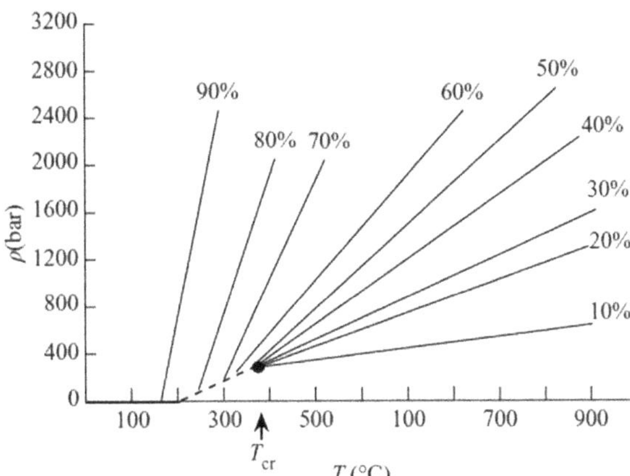

Fig. 5.3 Variation in pressure with changing fill volume of the reaction vessel as a function of temperature. [18] Reproduced with permission from Ref. Copyright 2001 Elsevier

5.2 Synthesis of Different Types of Materials

The industrial scale application of the hydrothermal synthesis started with the preparation of quartz crystals for oscillators which requires stringent control on the purity, defects, and size. Quartz crystals for this purpose are synthesized by the temperature gradient method [19]. Crystal growth by hydrothermal route can be understood as a type of chemical transport reaction and for the sparingly soluble species; a suitable mineralizer is added to increase its solubility. The solubilized species is transported to another part of the autoclave where it can crystallize. Crystallization is achieved by using a seed which is usually a substrate of the required crystal itself. Kinetics of crystal growth depends on several parameters in the hydrothermal method and it is difficult to control precisely. This results in variation in the quality of the product obtained in different batches. The crystal growth rate is lower in the hydrothermal method as compared to that in the melt method. This leads to the formation of crystals with lesser defects. Also, by this method, the product is obtained at a relatively lower temperature as compared to that in the melt technique resulting in lower thermal strain and defects in the final product. The control parameters for crystal growth by hydrothermal method are temperature, pressure, temperature gradient, and mineralizer concentration. Supersaturation, which represents the extent of higher concentration as compared to the saturation concentration of the species in the solution, controls the nucleation and growth rate [20]. Supersaturation leads to nucleation wherein clusters of crystals nucleate out of the solution phase. These crystals further grow in size and the process is irreversible [21]. The crystal growth process is either sequential or concurrent, wherein the growth units are incorporated into the existing crystal entities. The growth units have same chemical composition as the crystal entities but their structure may or may not be the same. Various steps of crystal growths are: transportation of crystal growth units through the solution, growth units getting attached to the surface, migration of growth units on the surface, and finally their attachment to the growth sites. Crystallization phase diagram is used to identify the supersaturation region for a given species. The crystallization diagram is generally represented as a temperature-composition relation or as a composition-composition relation as per the required information. There are three distinct zones of the supersaturation region, viz., metastable, nucleation, and precipitation zones. There is a time delay in nucleation in the metastable zone while in the nucleation zone, crystalline aggregates are formed at the nucleation centers. In the precipitation zone, the excess concentration precipitates out in form of amorphous aggregates [22].

Hydrothermal Synthesis of Zeolites

Aluminosilicates with engineered pore sizes are an important class of material which are popularly known as zeolites. Zeolites have immense application in the industrially important arenas of catalysis, sorption and ion exchange. The structure of zeolite is porous at molecular level and has a regular array of open structure with a typical pore diameter in the range of 3–15 Å, thus, resulting in a molecular sieve. These are size and shape selective material and hence are very important in synthesis. Zeolites are used for catalytic cracking of petroleum at a mass scale. Zeolites are most commonly prepared by hydrothermal route [23]. The synthesis is carried out either in alkaline medium in a temperature range of 80–200 °C or in a lower pH medium with addition of fluoride as a mineralizer to the system. At lower pH, larger crystal sizes are obtained due to lower nucleation rates. Materials consisting of other elements with similar structure have also been synthesized [24] and these are called zeotypes. Lower pHs are more suitable for zeotypes synthesis since the heteroatoms would remain in solution at lower pHs. Zeotype-like titanosilicate is commercially used as an oxidation catalyst. The significant advance in the field of zeolite synthesis could be achieved by the use of alkylammoniumcations instead of simple cations [25]. This alteration helped in increasing the silicon to aluminum ratio in the zeolites. This was indicated by the low nitrogen content of the product and by the difficulty of accounting the positive charge of the product framework in terms of bulky quaternary ammonium cations. Also, this alteration led to the template supported synthesis, wherein the zeolite framework formed around the organic templates. The templates are subsequently removed from the structure by calcination. Zeolite β with high silica in the range of $5 < Si/Al < 100$ was reported in 1967 by Wadlinger [26] which had hitherto the highest known silica content in any material. With increasing silica ratio, the crystal growth rate decreases. Thus, the hydrothermal synthesis requires longer duration and higher temperatures with increasing silica ratio. The zeolites with high silica ratio are mostly hydrophobic in nature and this behavior can be tailored by varying the Si/Al ratio. There is a smooth variation in the properties of the zeolite with the variation in the concentration of aluminum. The synthesis process usually involved using a mixture of tetrapropylammonium and sodium cations in the hydrothermal synthesis with a temperature range of 125–175 °C and duration of around one week. The mechanism of crystal growth had been initially studied with the help of XRD of the product formed at different intervals of time at a given temperature of synthesis [27]. An S-shaped growth kinetics was observed indicating the initial induction period followed by a rapid growth. The crystal growth was attributed to a series of depolymerization-polymerization reactions occurring in the solid phase. Later studies for elucidating the growth mechanism have established two limiting cases [28]. At one extreme, crystal growth is governed by the solution mediated mechanism, wherein the precursor gel dissolves to yield soluble species which are responsible for the crystal growth. On the other hand, in situ rearrangement of the precursor gel yielding the final product in a pure solid-state process is the other extreme. The subsequent works have demonstrated the synthesis of high silica-containing crystalline zeolites like ZSM-5 without using an organic template [29]. Also, with the Na-ZSM-5 synthesis in absence of any organic template, the similarity of the mechanism of formation for the high silica-containing varieties as compared to that for the high aluminous varieties was established [30]. However, there is a wider range of composition for the different inorganics to be synthesized when organic template is used. Application of fluoride ion as a mineralizing agent in place of hydroxide ion has also been demonstrated [31] and its advantages in the preparation of large crystals, substituted hetrostructures, and novel materials [32] have been established. Organic polymeric templates like polyelectrolytes have also been evaluated as a template. However, the covering of the entire polymeric chain by the framework crystals and final removal of the polymeric template are some of the challenges posed by this alteration [33]. Crystal growth rate as a function of temperature has been evaluated by Zhdanov as shown in Fig. 5.4 [34].

A direct correlation between the linear crystal growth rate and temperature was observed. Also, the linear growth rate is constant for a significant duration at all temperatures. Using the constant linear growth rate and the product size distribution profile, Zhdanov separated the contributions of nucleation and growth in the overall

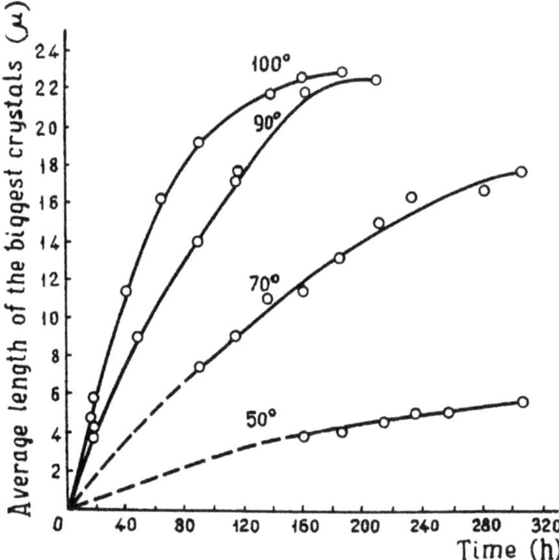

Fig. 5.4 Crystal growth rate of zeolite A for same precursor composition and varying temperature. [34] Reproduced with permission from Ref. Copyright 1974 American Chemical Society

growth rate profile. Detailed studies on similar lines by including incubation time, seeding effect, and other such parameters led to the development of the overall scheme of zeolite crystallization. This scheme includes the formation of elementary aluminosilicate blocks and crystal nuclei by condensation reaction. The subsequent crystal growth leads to the final product.

A significant advancement in the field of microporous materials was achieved when aluminophosphate synthesis was reported by Wilson et al. [35]. The hydrothermal synthesis required acidic or neutral medium instead of the alkaline medium used for the zeolite synthesis. Subsequent studies demonstrated the synthesis of several new materials such as substituted aluminophosphates (SAPO) and metal substituted aluminophosphates establishing the feasibility of hetero-substitution of the zeolite framework. The studies have continued in the direction of evolving new materials with a better open ring diameter for sorption of bigger molecules. The trend of this evolution is shown in Fig. 5.5.

In further improvisations in the synthesis protocols, ammonium and tetra alkyl quaternary ammonium systems were used to obtain large crystals for the systems where alkali metal ions were hindering the crystal growth [37]. Also, these systems do not require the final step of processing in form of ion exchange treatment and calcination itself provides the final product. Studies have also been carried out in nonaqueous medium like ethylene glycol [38], with an aim of preparing new zeolites but it did not yield the desired result. However, larger crystals could be grown out of such altered media [39]. Introducing hetero-elements by hydrothermal treatment of an existing zeolite has also been evaluated [40]. This indicates that the zeolite lattice is amenable to further substitution. Synthesis of titanium silicate [41] and its remarkable chemical activity has made it a very useful oxidation catalyst.

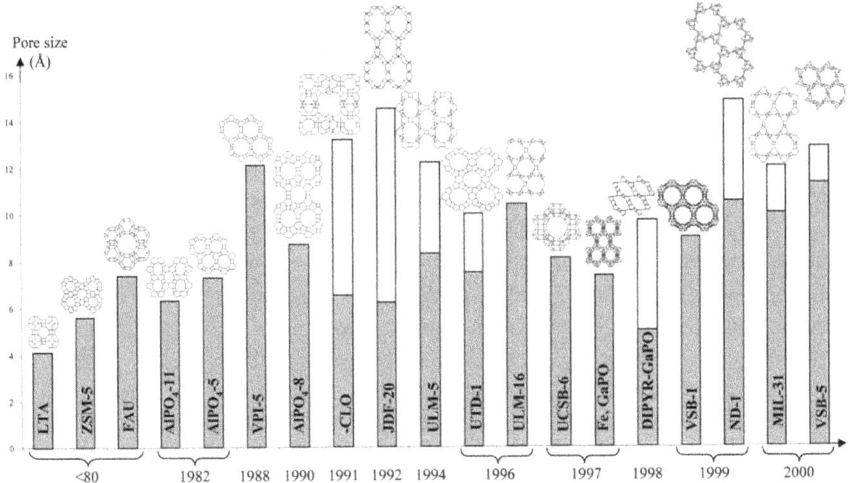

Fig. 5.5 Evolution of pore dimension with advancements in synthetic methodologies. [36] Reproduced with permission from Ref. Copyright 2001 American Chemical Society

Use of microwave as a rapid heating source has been found to be beneficial for the product in terms of its crystallinity [42]. Also, there is a selective crystallization of the desired product since there is a marked difference in the nucleation and growth rates of the competing phases at higher synthesis temperatures [43]. Synthesis of some of the zeolites has been carried out at much higher pressures (kilobar) as well [44]. In few cases like bikitaite, synthesis could be achieved at 2 kbar pressure but not at 1 kbar pressure. The field of zeolites and mesoporous material synthesis by hydrothermal route is an active one, wherein several types of new materials with interesting framework structures keep on emerging.

5.3 Hydrothermal Synthesis of Metal Oxide Nanoparticles

In the past few decades, metal oxide nanoparticles have caught the eyes of many researchers. Metals can form a variety of oxides which possesses varying structural properties along with electronic properties that results in insulator, semiconductor, or metallic behavior of the oxides. The properties of the metal oxides can be altered by controlling their size. Particularly, when the size is below 100 nm, the contribution of the surface atoms to the properties of the material increases significantly. For these reasons, metal oxide nanoparticles are used in various applications like microelectronic devices, solar cells, fuel cells, catalysts, sensors, targeted drug delivery system, cancer therapy, and many more. Almost all of these applications require predetermined particle size. Thus, the synthesis of these oxides with controlled particle size is a very important parameter. Some of the commercially useful oxides which have unique physical, chemical, and catalytic properties are SiO_2, Al_2O_3, TiO_2, V_2O_5, ZnO, etc. There are different synthetic approaches for preparing metal oxides nanoparticles, for example, decomposition, precipitation, chemical vapor deposition, electrospinning, sol–gel techniques, reverse micelles, hydrothermal process, and so on. Various applications of nanomaterials indicate that these have a strong potential for use in industries, but the synthesis of these materials in bulk is still a challenge. Hydrothermal route is a good option for the synthesis of these materials at industrial scale. For the synthesis of nanomaterials, hydrothermal route has certain advantages over other methods of synthesis since the solvated ions have higher mobility in the solvent medium and hence have better mixing characteristics [45]. This route yield materials of fine powders, controlled stoichiometry, controlled particle size, controlled morphology, less defects, uniformity, higher crystallinity, and higher purity among others. A schematic of the product quality of the nanoparticles obtained by hydrothermal route vis-a-vis other routes is shown in Fig. 5.6.

In recent past, several oxide nanoparticles with controlled morphology and controlled size have been synthesized. The morphology, particle size, and other physical and chemical properties of the oxide nanoparticles are determined by several parameters: additives in the given reaction mixture, pH of the solvent, reaction temperature, pressure, duration of reaction, and many more. Supercritical water or supercritical fluids have significantly different dielectric constant and solvent density, this also affects the reaction kinetics, and finally leads to smaller particle size. In order to optimize the experimental parameters, computational aid in terms of stability field diagrams is evolved for a system. These diagrams are evolved by considering all experimental parameters in a comprehensive manner.

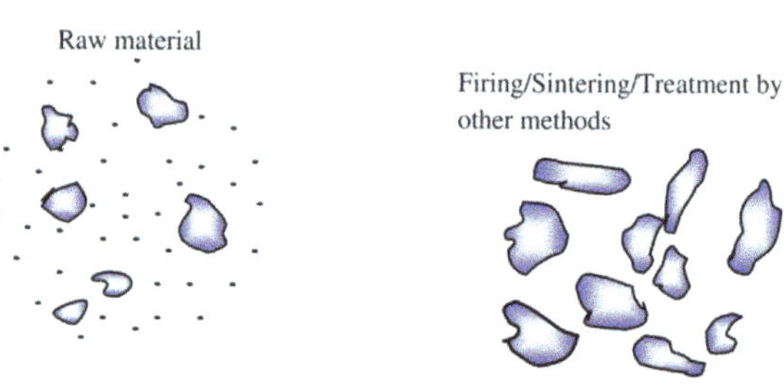

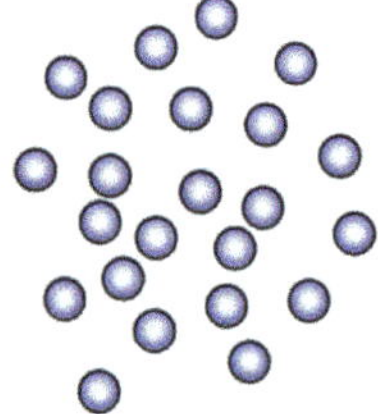

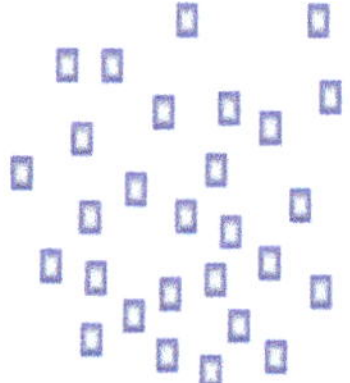

Fig. 5.6 Comparison of the products obtained by hydrothermal route with other routes. [4] Reproduced with permission from Ref. Copyright 2007 Elsevier

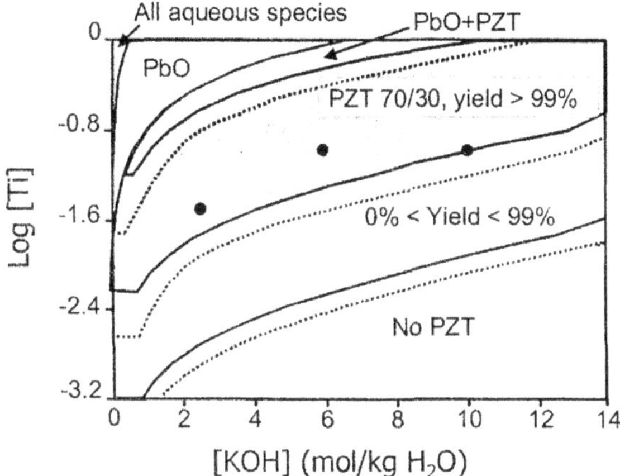

Fig. 5.7 Calculated stability field diagram for PZT at 180 °C. [46] Reproduced with permission from Ref. Copyright 2002 Elsevier

The parameters included in the calculation are activity coefficient of all species present in the reaction mixture, ionic strength, pH, pressure, and temperature. An example of the stability field diagram for lead zirconium titanate (PZT) is shown in Fig. 5.7.

One of the important parameters which control the morphology and particle size of nanomaterials is additives. Jiao et al. [47] have demonstrated how the morphology of zirconia can be changed by changing the additives. Phase pure tetragonal (t) and monoclinic (m) zirconia have been prepared selectively via hydrothermal route by maintaining all other experimental conditions identical while changing the additives. When tri-ethanolamine or glycerol is used as additive, phase pure tetragonal zirconia was obtained since this additive acts as a chelating agent favoring the formation of t-ZrO_2. However, using alkyl halides as additives resulted in the formation of m-ZrO_2. In addition, the additives also affect the particle size of the products. The t-ZrO_2 was formed with smaller particle size and sphere-like morphology, whereas the m-ZrO_2 exhibited larger particle size with spindle-like morphology. Reaction temperature and reaction time also play an important role on the morphology of nanomaterials. Sun et al. [48] reported the formation of CeO_2 nanorods where these factors affect the morphology of the product. They performed the experiments at three different temperatures: 120, 160, and 180 °C while maintaining other experimental conditions unmodified. Below 120 °C, relatively smaller fraction of nanorods was formed, whereas at 180 °C, aggregation occurred. The highest amount of product was obtained at 160 °C. Similarly, varying the reaction time has established that with shorter reaction time, smaller size of nanorods with lesser yield would be obtained. Increasing reaction time increased both size and yield but beyond 72 h there is no further increase in the yield. Other than these two factors, the amount of cerium precursor and the use of surfactants are also important parameters which can alter the quality of the product. pH of the reaction mixture also plays an important role in determining the size and morphology of the product. Byrappa et al. [49] reported the synthesis of TiO_2

nanoparticles where they had demonstrated how the pH affects the formation of the particles. At low pH (1 or 2), rutile phase of TiO_2 nanoparticle has been formed. Upon further lowering the pH, the product contained a small amount of anatase phase as well. With the increase in the pH of the medium (higher than 2), the composition of the anatase phase kept increasing. And when the pH finally exceeded 12, an amorphous phase was obtained. Not only the phase but the particle size is also controlled by the pH of the medium. Wang et al. [50] reported the synthesis of Dy_2O_3 nanorods where the pH was maintained between 7 and 8 by using 10% KOH solution. Cote et al. [51] reported the synthesis of $CoFe_2O_4$ nanoparticles, demonstrating that it is necessary to control the pH to get the desired phase within 100 nm particle size distribution. Using supercritical water as reacting medium can result in the formation of smaller-sized oxide nanoparticles. Adschiri and co-workers [52] have demonstrated that very fine oxide nanoparticles can be formed from the nitrate precursors. In the first step, the nitrates hydrolyze in the aqueous medium resulting in the formation of metal hydroxides. In the second step, these hydroxides readily dehydrate before their significant growth, which leads to smaller-sized metal oxide product. The overall reaction can be represented as:

$$M(NO_3)_x + xH_2O \rightarrow M(OH)_x + xHNO_3 \quad (5.1)$$

$$M(OH)_x \rightarrow MO_{x/2} + 1/2xH_2O \quad (5.2)$$

The rate of synthesis is drastically enhanced in the supercritical water medium as a result of rapid dehydration. In this medium, the dehydration is not governed by the process of diffusion through a solid particle which is the rate determining step for liquid medium. Also, rapid mass transfer is facilitated in the supercritical medium due to its high diffusivity and low viscosity and this also contributes to the increased rate of synthesis. Also, the high temperature increases the reaction rate. Due to these advantages several nanoparticles have been prepared by this route like Co_3O_4, α-Fe_2O_3, CeO_2, NiO, Fe_3O_4, ZrO_2, etc. Sorescu et al. [53] reported nanocrytallinerhombohedral In_2O_3 by hydrothermal route by maintaining a temperature of 200 °C for 4 h. To get final product, post-annealing of the nanoparticles at 500 °C was carried out. Wu et al. [54] have synthesized nanowire arrays of Co-doped magnetite taking cobalt chloride, ferrous chloride, and sodium hydroxide as precursors. These nanowires have formed a single magnetic domain and were regarded as small-wire like magnet. TiO_2 nanoparticles are an important class of materials owing to their high photocatalytic activity. More than 1000 publications are there which describe the synthesis of TiO_2 nanoparticles with desired particle size and morphology. Generally, the TiO_2 nanoparticles are synthesized in small Morey type autoclave at low temperature (<150 °C) and pressure (<100 bars). It gives desire particle size with desired morphology. However, there are other hydrothermal methods of synthesizing TiO_2 nanoparticles. Qian et al. [55] presented a two-step process to prepare ultrafine powder of TiO_2. First step leads to oxidation of titanium in the presence of aqueous solution of H_2O_2 and ammonia which results in the formation of a gel which is the hydrated titania. Second step is the hydrothermal treatment of this gel which gives rise to TiO_2 nanoparticles. Chen et al. [56] reported TiO_2 powder that is prepared by oxidation hydrothermal combination method. They have also showed the effect of additives on particle shape and crystalline structure.

5.4 Hydrothermal Synthesis of Semiconducting Nanoparticles

Semiconducting nanomaterials is one of the emerging areas being pursued by the material scientists due to their unique electric, magnetic, catalytic, and optical properties. Because of their unique properties, these materials have a wide application in solar cells, light emitting diodes, biosensors, laser technology, waveguides, catalysts, and so on. Recently, several nanomaterials have been reported which are semiconductors, wherein this property is exhibited only in the nano-form and not in the bulk form of the same material. Two major factors which govern the properties of nanomaterials are large surface area and quantum confinement effect. Though the field of semiconductor nanomaterials is still budding at the research stage, there are many variations of the semiconductor nanomaterials that have shown good results. Generally, II-VI, III-V, and IV-VI are formed as semiconductor nanomaterials like CdS, CdSe, GaN, GaP, InP, ZnS, ZnO, etc. The synthesis of these nano-semiconductors can be carried out in several ways like sol–gel technique, co-precipitation, and hydrothermal or solvothermal method. One of the key factors which control the properties of nanomaterials is the quantum confinement effect that depends on size, morphology, and crystal structure of the nanoparticle which is better controlled in the hydrothermal method as compared to other methods. In addition, this technique is environmentally and economically viable. By varying the parameters like temperature, pressure, additives, etc., the morphology and particle size of the product can be controlled. There are several strategies which are used for the synthesis of these particles like organic additives-free or template-free synthesis, organic additives-assisted synthesis, and template-assisted synthesis among others. In addition, microwave-assisted and magnetic field-assisted synthesis methods are also known.

5.4.1 Direct Hydrothermal Synthesis Methods

In the organic additives and template-free synthesis, there are least chances of contamination of the product. The crystal growth can either be kinetically controlled or thermodynamically controlled. Mainly, the morphology is controlled by Oswald ripening process, where smaller particle gets smaller and larger particles becomes larger. Yang and Zeng [57] reported the synthesis of hollow anatase TiO_2

nanospheres where localized Oswald ripening mechanism has been followed. There are also two other Oswald ripening mechanism, "symmetric Oswald ripening" and "asymmetric Oswald ripening". For example, homogeneous ZnO core–shell nanospheres are synthesized by symmetric Oswald ripening, whereas Co_3O_4 semi-hollow nanospheres are synthesized by asymmetric Oswald ripening [58]. There is another mechanism, known as oriented attachment (OA) mechanism where the self-organization of particles results in the formation of nanocrystals [59]. For example, ZnS nanoparticles, TiO_2 nanosheets, and $Ag_2V_4O_{11}$ nanobelts are formed in this way [60–63]. The organic additives and template-free hydrothermal synthesis can be undertaken in several ways like recrystallization of metastable precursors, reshaping bulk materials, indirect-supply reaction source, and decomposition of single-source precursor. For recrystallization, a metastable precursor is initially obtained by co-precipitation reaction via high speed stirring; the precursor may be composed of nanocrystals or amorphous colloids. The precursors are subsequently treated under high pressure and temperature in hydrothermal conditions which results in the formation of semiconductor nanomaterials. Several metal oxide semiconducting nanomaterials have been formed using this process. Zhu et al.[64] reported ZTO (Zn_2SnO_4) nanorods using this route by taking $ZnCl_2$, $SnCl_4$, and hydrazine as raw materials. Other reported multi-metal oxide nanomaterials are $TbMn_2O_5$ nanorods, $CuFe_2O_4$ nanocrystals, Bi_2WO_6 flower-like nanostructures assembled with nanoplates, and so on. Another way of preparing semiconductor nanoparticles is a top-down approach, wherein larger-sized and irregular-shaped materials are initially prepared and these are subsequently reduced to smaller size with desired morphology by suitable treatment. Kasuga et al. [65] have first reported the synthesis of TiO_2 nanotubes by rutile TiO_2 powder. Menzel et al. [66] have synthesized TiO_2 nanorods, hollow nanotubes from the raw materials, like rutile TiO_2, anatase TiO_2, respectively. Xu et al. [67] have reported the synthesis of $Tb(OH)_3$, $Dy(OH)_3$, $Ho(OH)_3$, and $Y(OH)_3$ using this method. In another methodology, the nuclei of the required product are generated using a solution of inorganic salts in autoclave while maintaining a particular temperature and pressure during the hydrothermal treatment. The nuclei further grow out to the desired product by the indirect-supply reaction process, wherein the source of reactant could be the ions produced in the redox reaction or the dissolved oxygen present in the medium. The advantage of this method is the formation of uniform product with narrow size distribution. Ohgi et al. [68] have reported SnO_2-based nanostructures using this method by taking SnF_2as precursor. α-Fe_2O_3 nanorods and nanosphere, CeO_2 nanorods, SnO_2-based nanoplates, and many more have been prepared by this method. Another useful method of synthesizing semiconductor nanoparticles is through the decomposition of a single precursor. Single source of precursor has to be taken in this method and with proper conditions hydrothermal synthesis can be carried out to get the desired product. This process is relatively cleaner and can avoid expensive or toxic materials. Many sulfide nanostructures such as FeS_2 cubes, Sb_2S, and Bi_2S_3 nanorods are synthesized by this method.

5.4.2 Organic Additive-Assisted Synthesis

This method yields more controlled morphology and uniformity of nanostructures as compared to the previous one. There are five basic strategies according to the nature of organic additives, viz., the surfactant-assisted route: where the surfactant forms micelles in the aqueous medium and hydrophobic groups form the core of aggregates which give various morphologies depending on the nature of surfactants. The morphology can be controlled by three parameters: absorbing agent, soft template, and etching agent. VO_x nanotubes and PbTe nanotubes by CTAB surfactants, $PbZr_{0.52}Ti_{0.48}O_3$ (PZT) nanorods by polyvinyl alcohol surfactants are the examples which follow this route. Another method is the biomolecule-assisted route. Due to the excellent self-assembly property of biomolecules, many semiconductor nanostructures have been prepared by this route. Self-assembly, reducing property, and presence of anions (such as S^{2-}, OH^-) of biomolecules play important role on the growth mechanism of nanostructures. N–TiO_2 nanocrystals, CeO_2 hollow nanospheres, and many more are synthesized by this method. Ionic liquid-assisted route has its own advantages. Ionic liquids have numerous advantages over organic solvents like negligible vapor pressure, low toxicity, broad electrochemical potential window, good thermal stability, and high ionic conductivity among others. These ionic liquids can also act as capping agents or soft templates for the synthesis of nanomaterials. Some examples like CuO nanoplates, NiO microspheres, Fe_2O_3 nanorods, etc., are synthesized by this method. Organic acid-assisted route is also a useful way of preparing semiconductor nanoparticles because of the presence of the carboxylic group in acids. Metal–organic acid complexes are formed which help in synthesizing semiconducting nanomaterials. Some typical examples are Fe_2O_3 nanotubes and nanoflakes, In_2O_3nanospheres, CaF_2 nanoflowers. Alcoholic solvent-assisted route is another feasible option. Alcohols have "OH" group which is amenable to form hydrogen bonds. Also, alcoholic solvents are used to control shape and they can also be a reducing agent or hydrolyzation inhibitor [69–71]. Ethanol, ethylene glycol, glycerol, etc., are used as alcohol source. Some typical examples are ZnO flower-like structures, Bi_2WO_6 clew-like nanostructures, and Lu_2O_3 nanoflowers.

5.4.3 Template-Assisted Synthesis

The main advantage of this approach is the templates which can regulate morphology and size of the nanomaterials. Template-assisted synthesis can be based on providing an additive to the reaction mixture which has to be finally removed. Carbon spheres, multiwalled carbon nanotubes, anodized aluminum oxide, etc., are used as templates. These are introduced before the hydrothermal treatment and then are finally removed after the reaction by calcination or dissolution. Zhang et al. [72] have reported the synthesis of cage-like Fe_2O_3 hollow spheres by this method.

Another variation of this method is the addition of such a template which would be removed by itself at the end of the synthesis. Here, the template which is added initially reacts with other reagents to synthesize the required product in the desired morphology during the hydrothermal treatment. Hence, there is no further requirement of any treatment to remove the template after the reaction is over, which is the main advantage of this process. Qian et al. [73] reported the synthesis of semiconductor $CdIn_2S_4$ nanorods by this method. There are ways in which the template would be added as self-additive, and finally, it would be self-removed from the product. Here, the intermediates formed during the progress of reaction themselves act as the template for providing the required morphology of the product. The challenge of this route is to control the morphology of the product. Fan et al. [74] have reported the synthesis of CoTe nanotubes by this hydrothermal route.

Semiconducting nano-structured arrays have potential application in several area owing to their excellent charge transfer properties. Controlled fabrication of arrays can be carried out in a cost-effective manner by hydrothermal route. There are several routes to synthesize nano-structured arrays. A metal substrate can act as both the substrate as well as source of metal ions for the nanoparticle deposition. In this method, the metal foil is initially dissolved to form complexes or metal ions. These complexes or metal ions are subsequently used for the synthesis of the required semiconductor nano-structured arrays on the metal foil substrate. Yang et al. [75] have reported the synthesis of ZnO nano-rod array with zinc foil in presence of 1% ammonia. Another methodology for synthesizing nano-structured array is to initially deposit the semiconductor nanocrystals on the substrate which would act as a seed for further growth of the array during hydrothermal treatment. This method provides the flexibility of using any substrate of choice. Hsu et al. [76] have shown that the conductivity of the ZnO nano-rod arrays can be tailored by varying the seed layer composition.

5.5 Microwave-Assisted Hydrothermal Synthesis

Microwaves are the electromagnetic waves with the frequency range of 300 MHz to 300 GHz. When incident on some of the materials, microwaves can pass through these without any attenuation and such materials are known as transparent to microwaves. Most of the polymer materials and nonmetallic materials fall in this class. There is another type of materials which are completely opaque to microwaves and these would reflect the microwaves. Metals and alloys fall in this class of materials. However, there is a third type of materials as well. These materials/media would partly absorb the microwave energy and this energy would manifest in terms of heat. Dipole polarization and ionic conduction in the medium are responsible for the heating effect [77]. Microwaves coupling with the electric dipoles of the medium and dielectric and magnetic loss heating of the medium are responsible for the rapid energy transfer to the medium. Thus, the overall synthesis time is drastically reduced in the microwave-assisted hydrothermal synthesis. Also, there is lower power consumption in this method as compared to the conventional hydrothermal synthesis. Microwave-assisted hydrothermal synthesis results in faster kinetics of crystallization [78]. Microwave heating is more efficient in the polar solvents and the efficiency of the interaction of the microwave with the medium is expressed in terms of loss tangent, tan δ:

$$\tan\delta = \varepsilon''/\varepsilon' \quad (5.3)$$

where ε'' is the loss factor which is a measure of efficiency of conversion of microwave energy into heat and ε' is the dielectric constant of the medium. The solvents with higher tan δ values would more efficiently absorb the microwave energy. 2.45 GHz is the most common frequency used in the commercial microwave apparatus, whereas the dielectric constant of a medium would vary with the temperature. Thus, the loss factor for a medium will depend on the temperature. Loss tangents for some common solvents are given in Table 5.3. Komarmeni et al. [79] reported the synthesis of TiO_2, ZrO_2, Fe_2O_3, and binary oxides $KNbO_3$ and $BaTiO_3$ using microwave-assisted hydrothermal while establishing the effect of other experimental parameters like concentration, temperature, and duration on the crystallization rate. Wilson et al. [80] have established better crystallinity for TiO_2 prepared through microwave-assisted hydrothermal route as compared to that through conventional hydrothermal route. Microwave-assisted hydrothermal synthesis has been used widely to prepare a vast variety of materials ranging from mixed oxides including perovskites [81], bioceramics [82], thin films [83], vanadates [84], and garnets [85] among others. These studies established that a better crystallinity is achieved in a much shorter duration by using this route. The effect of microwave interaction on morphology of the product is also established. This alters several properties of the material like their photocatalytic activity.

Table 5.3 Loss tangent for common solvents at 2.45 GHz at 20 °C [86–89]

High (tan δ > 0.5)		Medium (0.1 < tan δ < 0.5)		Low (tan δ < 0.1)	
Solvent	tan δ	Solvent	tan δ	Solvent	tan δ
Ethylene glycol	1.17	2-Butanol	0.45	Chloroform	0.091
Ethanol	0.94	1-Hexanol	0.34	Ethyl acetate	0.059
2-Propanol	0.80	Acetic acid	0.17	Acetone	0.054
Formic acid	0.72	DMF	0.16	Toluene	0.040
Methanol	0.66	Water	0.12	Hexane	0.020

5.6 Continuous Hydrothermal Flow Synthesis

In the continuous hydrothermal flow synthesis route, a pressurized stream of preheated solvent is mixed with a solution of metal precursor which is initially kept under ambient conditions. The flow condition of this mixture is engineered to obtain the required metal oxide nanoparticle as the product. The nature of the products obtained through this route is unique. Metastable phases can be obtained by this route which cannot be synthesized by conventional methods. This is possible since the residence time of the precursors in the high-temperature zone can be controlled by varying the flow rate of the high-pressure pump which is used to circulate the preheated solvent. The advantage of having a control over the flow of stream can be effectively used to control the nucleation and thus the particle size. Provision for injection at high temperature also provides the flexibility of post-treatment of the product like adding a surfactant coat to the product. The quality of the product synthesized through this route would be consistent across the batches. Thus, this method efficiently overcomes the limitations of the conventional hydrothermal batch reactors (autoclaves). The static autoclaves have limitations in terms of batch operations which have lower throughput, no control over nucleation, and growth process and hence have variation in the quality of the product of different batches [90].

For the production of nanoparticles, the use of superheated water as a solvent is a better choice in this method. The saturation concentration of the metal ion precursor would be lower in the superheated water owing to the reduction in the dielectric constant of water with increasing temperature. This would result in the formation of a strongly supersaturated solution of the metal ion upon mixing which in turn would increase the rate of nucleation of particles, and thus, the resulting particles would be smaller in size [22]. Also, the number of water molecules in the first hydration shell of the metal ion reduces with increasing temperature [91]. At the same time, metal water reaction becomes more feasible at higher temperatures. Thus, mixing of metal ion solutions at ambient condition with superheated water would result in its rapid hydrolysis and dehydration to oxides. The anions would undergo decomposition and in certain cases the degradation products may play an important role in altering the final product. Schematic of a simple continuous hydrothermal flow system is shown in Fig. 5.8.

Considerable efforts have been spent into designing the system to carry out effective mixing. This includes evaluating different tee geometries, a 90° tee, a 50° tee, and a swirling tee [93]. Swirling tee has yielded a smaller particle sized $LiFePO_4$. Other designs like vertical counter-current flow mixers have also been studied [94]. A plus joint type mixer with the supercritical fluid entry from bottom arm and metal solution flowing through the side arms gives a rapid mixing and collection of the product from the top arm without any blocking [95]. These coaxial confined jet mixers are found to be scalable to plant level with no variation in the quality of the products.

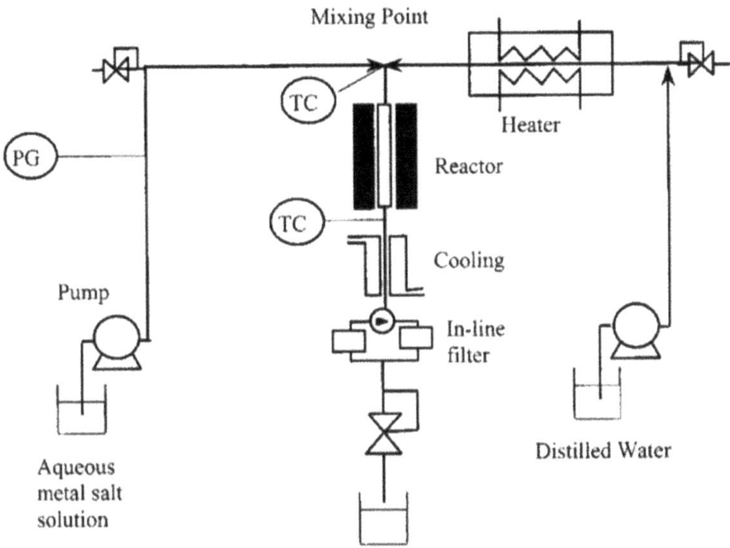

Fig. 5.8 Diagram of a continuous hydrothermal flow system. [92] Reproduced with permission from Ref. Copyright 2000 American Chemical Society [92]

This is a fast-evolving methodology which is improving with the experimental inputs of the users. There is a scope of improvement in the synthesis of binary or ternary compounds as occasional incomplete mass transfer may hinder the formation of desired heterogeneous product. Supplementing such experimental approaches with computational fluid dynamics has also helped in better understanding of the hydrothermal flow method of synthesis.

5.7 Conclusions

Hydrothermal method of synthesis, for the synthesis of advanced materials, has evolved significantly over the years. Coupling the conventional method with other techniques like the microwave-assisted hydrothermal or using a continuous flow setup has further enhanced the capabilities of this method. Use of alternate solvents has also drastically changed the required pressure and temperature conditions for the synthesis of several advanced materials. Refining the experimental conditions with computational aids has also improved the scope of synthesis process significantly. The method has gained vast grounds in synthesizing advanced materials, particularly with nano-dimensions. In the area of optical application advanced materials like nanophosphors, UV attenuating nanoparticles, refractive index modifiers, nonlinear optics, thermochromic nanoparticles are been developed at a rapid pace by this method. In the field of healthcare applications advancements like therapeutic nanoparticles, magnetic particles for health care, antimicrobials, bioceramics, nanocoatings could evolve because of the developments in the synthesis methodologies. Nanophosphors for imaging applications, electronics applications including semiconductor particles, transparent conducting particles based on indium tin oxide, gas sensors, dielectrics, piezoelectrics, multiferroics, and magnetic particles have also been developed recently. Catalyst applications including photocatalysts, heterogeneous catalysts, materials for energy harvesting and energy storage materials, and several other important areas have seen significant advancements due to the application of hydrothermal methods of synthesis. It has become an indispensible method in the synthesis of advanced materials and it would be useful in addressing the emerging advanced material requirements.

References

1. Rabenau A (1985) Angew Chem Int Ed Engl 24:1026
2. Roy R, Tuttle OF (1956) Phys Chem Earth 1:138
3. Byrappa K, Keerthiraj N, Byrappa SM (2015) Hydrothermal growth of crystals—design and processing in handbook of crystal growth. Elsevier
4. Byrappa K, Adschiri T (2007) Progr Cryst Growth Character Mater 53:117
5. Morey GW, Niggli P (1913) J Am Chem Soc 35:1086
6. Barrer RM (1946) Nature 157:734
7. Walker AC (1953) J Am Ceram Soc 36:250
8. Helgeson HC, Garrels RM, Mackenzie FT (1969) Geochim Cosmochim Acta 33:455
9. Laudise RA, Nielsen JW (1961) Solid State Phys 12:149
10. Dove PM, Crerar DA (1990) Geochim Cosmochim Acta 54:955
11. Tani E, Yoshimura M, Somiya S (1983) J Am Ceram Soc 66:11
12. Dawson W (1988) J Am Ceramic Soc Bull 67:1673
13. Ohmoto H, Lasaga AC (1982) Geochim Cosmochim Acta 46:1727
14. Li W-J, Shi E-W, Zhong W-Z, Yin Z-W (1999) J Crystal Growth 203:186
15. Yaghi OM, Li H (1995) J Am Chem Soc 117:10401
16. Liu B, Aydil ES (2009) J Am Chem Soc 131:3985
17. Jiang J, Zhao K, Xiao X, Zhang L (2012) J Am Chem Soc 134:4473
18. O'Hare D (2001) Hydrothermal synthesis in encyclopedia of materials: science and technology. Elsevier
19. Buisson X, Arnaud R (1994) J de Physique Colloque 4(C2):25
20. Andelman T, Tan MC, Riman RE (2010) Mater Res Innov 14:9
21. Rupp B (2015) Acta Cryst F71:247
22. Kashchiev D (1982) J Chem Phys 76:5098
23. Flanigen EM (2001) Stud Surf Sci Catal 137:11
24. Kresge CT, Leonowicz ME, Roth WJ, Vartuli JC, Beck JS (1992) Nature 359:710
25. Barrer RM, Denny PJ (1961) J Chem Soc 201:971
26. Wadlinger RL, Kerr GT, Rosinski EJ (1997) U.S. Patent 3,308,069
27. Flanigen EM, Breck DW (1960) 137th Meet, ACS, Div Inorg Chem 33-M
28. Guth JL, Caullet P (1986) J Chim Phys 83:155
29. Grose RW, Flanigen EM (1981) U.S. Patent 4 257 885
30. Lowe BM, Nee JRD, Casci JL (1994) Zeolites 14:610

31. Flanigen EM, Patton RL (1978) U.S. Patent 4,073,865
32. Kessler H (1989) Stud Surf Sci Catal 52:17
33. Daniels RH, Kerr GT, Rollmann LD (1978) J Am Chem Soc 100:3097
34. Zhdanov SP (1974) Some problems of zeolite crystallization in molecular sieve zeolites-I. Am Chem Soc
35. Wilson ST, Lok BM, Messina CA, Cannan TR, Flanigen EM (1982) J Am Chem Soc 104:1146
36. Férey G (2001) Chem Mater 13:3084
37. Mueller U, Unger KK (1988) Zeolites 8:154
38. Bibby DM, Dale MP (1985) Nature 317:157
39. Kuperman A, Nadimi S, Oliver S, Ozin GA, Garces JA, Olken MM (1993) Nature 365:239
40. Liu X, Thomas JM (1985) J Chem Soc Chem Commun 1544
41. Taramasso M, Perego G, Notari B (1983) U.S. Patent 4,410,501
42. Thompson RW (1998) Molecular sieves. Springer, Berlin
43. Zhao JP, Cundy CS, Dwyer J (1997) Stud Surf Sci Catal 105:181
44. Ghobarkar H, Schaf O, Guth U (2001) High-Pressure Res 20:45
45. Yoshimura M, Byrappa K (2007) J Mater Sci 43:2085
46. Riman RE, Suchanek WL, Lencka MM (2002) Ann Chim Sci Mat 27:15
47. Jiao X, Chen D, Xiao L (2003) J Cryst Growth 258:158
48. Sun C, Li H, Zhang H, Wang Z, Chen L (2005) Nanotechnology 16:1454
49. Byrappa K, LokanathaRai KM, Yoshimura M (2000) Env Tech 21:1085
50. Wang G, Wang Z, Zhang Y, Fei G, Zhang L (2004) Nanotechnology 15:1307
51. Cote LJ, Teja AS, Wilkinson AP, Zhang ZJ (2002) J Mater Res 17:2410
52. Adschiri T, Kanaszawa K, Arai K (1992) J Am Ceram Soc 75:1019
53. Sorescu M, Diamandescu L, TarabasanuMihaila D, Teodorescu VS (2004) J Mater Sci 39:675
54. Wu M, Xiong Y, Jia Y, Ye J, Zhang K, Chen Q (2005) Appl Phys A 81:1355
55. Qian Y, Chen Q, Chen Z, Fan C, Zhou G (1993) J Mater Chem 3:203
56. Chen Q, Qian Y, Chen Z, Zhou G, Zhang Y (1995) Mater Lett 22:77
57. Yang HG, Zeng HC (2004) J Phys Chem B 108:3492
58. Liu B, Zeng HC (2005) Small 1:566
59. Banfield JF, Welch SA, Zhang HZ, Ebert TT, Penn RL (2000) Science 289:751
60. Shi WD, Wang C, Wang HS, Zhang HJ (2006) Cryst Growth Des 6:915
61. Yang XH, Li Z, Sun CH, Yang HG, Li CZ (2011) Chem Mater 23:3486
62. Zhang J, Lin Z, Lan Y, Ren GQ, Chen DG, Huang F, Hong MC (2006) J Am Chem Soc 128:12981
63. Shen GZ, Chen D (2006) J Am Chem Soc 128:11762
64. Zhu H, Yang D, Yu G, Zhang H, Jin D, Yao K (2006) J Phys Chem B 110:7631
65. Kasuga T, Hiramatsu M, Hoson A, Sekino T, Niihara K (1998) Langmuir 14:3160
66. Menzel R, Peiró AM, Durrant JR, Shaffer MSP (2006) Chem Mater 18:6059
67. Xu AW, Fang YP, You LP, Liu HQ (2003) J Am Chem Soc 125:1494
68. Ohgi H, Maeda T, Hosono E, Fujihara S, Imai H (2005) Cryst Growth Des 5:1079
69. Ashoka S, Nagaraju G, Tharamani CN, Chandrappa GT (2009) Mater Lett 63:873
70. Li SZ, Zhang H, Ji YJ, Yang DR (2004) Nanotechnology 15:1428
71. Srinivasan R, Chavillon B, Doussier-Brochard C, Cario L, Paris M, Gautron E, Deniard P, Odobel F, Jobic S (2008) J Mater Chem 18:5647
72. Zhang LZ, Ai ZH, Jia FL, Liu L, Hu XL, Yu JC (2006) Chem Eur J 12:4185
73. Qian HS, Yu SH, Gong JY, Luo LB, Wen LL (2005) Cryst Growth Des 5:935
74. Fan H, Zhang YG, Zhang MF, Wang XY, Qian YT (2008) Cryst Growth Des 8:2838
75. Yang HQ, Song YZ, Li L, Ma JH, Chen DC, Mai SL, Zhao H (2008) Cryst Growth Des 8:1039
76. Hsu YF, Xi YY, Tam KH, Djurišić AB, Luo JM, Ling CC, Cheung CK, Ng AMC, Chan WK, Deng X, Beling CD, Fung S, Cheah KW, Fong PWK, Surya CC (2008) Adv Funct Mater 18:1020
77. Sun J, Wang W, Yue Q (2016) Materials 9:231

78. Komarneni S (2003) Curr Sci 85:1730
79. Komarneni S, Roy R, Li Q (1992) Mater Res Bull 27:1393
80. Wilson GJ, Will GD, Frost RL, Montgomery SA (2002) J Mater Chem 12:1787
81. Liu S-F, Abothu IR, Komarneni S (1999) Mater Lett 38:344
82. Yu H-P, Zhu Y-J, Lu B-Q (2018) Ceram Int 44:12352
83. Zhou Z, Bowland CC, Patterson BA, Malakooti MH, Sodano HA, Appl ACS (2016) Mater Interfaces 8:21446
84. Prado-Gonjal J, Molero-Sánchez B, Ávila-Brande D, Morán E, Pérez-Flores JC, Kuhn A, García-Alvarado FJ (2013) Power Sources 232:173
85. Cho YS, Burdick VL, Amarakoon VR (1997) J Am Ceram Soc 80:1605
86. Hayes BL (2002) Microwave synthesis: chemistry at the speed of light. CEM Publishing, North Carolina
87. Gabriel C, Gabriel S, Grant EH, Halstead BSJ, Mingos DMP (1998) Chem Soc Rev 27:213
88. Nemmaniwar BG, Kalyankar NV, Kadam PL (2013) Orbital Electron J Chem 5:1
89. Kappe CO (2013) Chem Soc Rev 42:4977
90. Arita T, Moriya K-I, Minami K, Naka T, Adschiri T (2010) Chem Lett 39:961
91. Simonet V, Calzavara Y, Hazemann JL, Argoud R, Geaymond O, Raoux D (2002) J Chem Phys 117:2771
92. Adschiri T, Hakuta Y, Arai K (2000) Ind Eng Chem Res 39:4901
93. Hong S-A, Kim SJ, Chung KY, Chun M-S, Lee BG, Kim JJ (2013) Supercrit Fluids 73:70
94. Lester E, Huddle T (2014) Worldwide patent WO2014111703-A2
95. Gruar RI, Tighe CJ, Darr JA (2013) Ind Eng Chem Res 52:5270

Chapter 6
Synthesis of Materials Under High Pressure

S. N. Achary and A. K. Tyagi

Abstract Preparative chemistry under high pressure becomes an interesting and challenging process for the preparation of novel functional materials where the common crystallographic and thermodynamics constraints can be easily deviated. Thus, the reaction of materials under high pressure and temperature can lead to newer products which otherwise cannot be obtained by the conventional high temperature reactions. Also, such procedures can stabilize the unusual coordination number, valence states, densely packed metastable compounds. In this chapter, a brief introduction to the chemical reactions under high pressure and high temperature, and effect of pressure and pressure/temperature on the chemical equilibrium are presented. Subsequently, the evolution of high pressure synthesis experimental setups and different types of common setups used in laboratory are discussed. Finally, some of the examples of materials synthesized under high pressure conditions are presented.

Keywords High pressure · High pressure-high temperature · Inorganic materials · Unusual compounds · Metastable materials · Synthesis

6.1 Introduction

In general, synthesis of material can be described as transforming some materials known as reactants or precursors to another material, known as product, having different composition, structure, and/or properties than the original materials. The reactants lose their identity by breaking the bonds between the different atoms

S. N. Achary (✉) · A. K. Tyagi
Chemistry Division, Bhabha Atomic Research Centre, Mumbai 400085, India
e-mail: sachary@barc.gov.in

A. K. Tyagi
e-mail: aktyagi@barc.gov.in

S. N. Achary · A. K. Tyagi
Homi Bhabha National Institute, Mumbai 400094, India

A. K. Tyagi and R. S. Ningthoujam (eds.), *Handbook on Synthesis Strategies for Advanced Materials*, Indian Institute of Metals Series,
https://doi.org/10.1007/978-981-16-1807-9_6

within them and diffuse in the matrix to form another set of bonds leading to formation of products. The reactants are usually subjected to some thermodynamic perturbations, like temperature or chemical environment to overcome activation energy in the path of the product formation [1–4]. However, it may happen that the products have almost similar bond lengths and coordination around the ions in the first coordination sphere as the reactants, but possess different structure or properties. Also, the products often have lower free energy than the reactants. However, there may be cases where the free energy of the products may be more than the reactants but can exist due to kinetic or thermodynamic barrier to revert back. Such materials are known as metastable materials. Thus, in all chemical reactions, the chemical equilibrium and free energy of reactants and products are essentially considered for designing an appropriate synthesis procedure. Since the involvement of breaking and formation of chemical bonds is the key feature of synthetic chemical reaction, temperature has been exploited as an external thermodynamic parameter for synthesis of material. In addition, the understanding of the effect of pressure, temperature, and compositions on chemical equilibrium provides information for the suitable or most desired thermodynamics parameters for designing a synthesis procedure for any material. Additionally, the tailored preparation strategies, like designing alternate reactants, mediums of reactions, secondary active or passive materials are also used to facilitate the product formation. Several of such synthesis methods have been covered in the present monograph also. Of late, synthesis and behavior of materials under pressure have become a widely investigated research area due to its scope to prepare non-conventional materials as well as to mimic minerals and understanding the formation of materials under deep geological conditions. In this chapter, the syntheses of materials utilizing pressure as an external thermodynamic parameter are presented.

The use of pressure for chemical and physical processes knowingly or unknowingly has been exploited by the mankind since long. However, the effective use of pressure and requirement of high or low pressure came to picture of materials synthesis after the development and understanding of thermodynamics and kinetic of chemical reactions [1, 2]. Thus, for a given reaction scheme, temperature and pressure are considered as the most important thermodynamic parameters for synthetic chemists, where the chemical equilibrium and hence the direction of chemical reaction can be controlled. These aspects have been extensively explored to alter the chemical equilibrium and to prepare the desired products. A classic example of controlled pressure and temperature synthesis is Haber's process for synthesis of ammonia from N_2 and H_2 which has been exploited for industrial synthesis of ammonia.

As mentioned earlier, the driving force for any chemical reaction is governed by the free energy of the component reactants and products, as that governs the stability of the materials under given thermodynamic conditions. It can be mentioned here that the pressure and temperature oppositely alter the structural and chemical properties like pressure decreases the inter-atomic distances while temperature increases it. In both cases, the inter-atomic distance deviates from their equilibrium values and hence increases the energy of the system from equilibrium lowest energy state as shown in Fig. 6.1a. Thus, the reactivity of material increases under both the

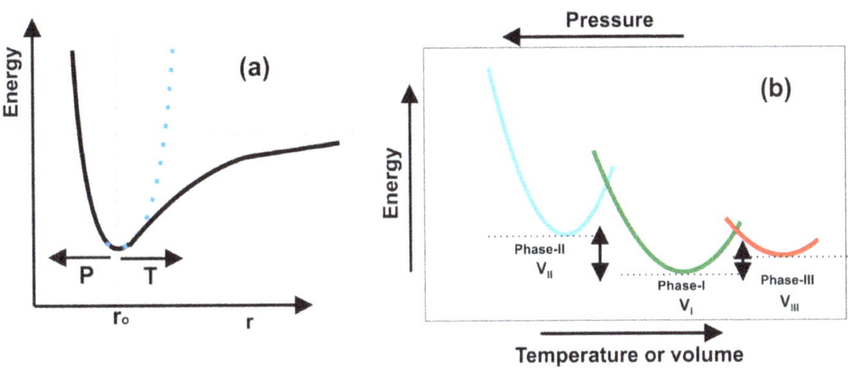

Fig. 6.1 **a** Variation of potential energy with interatomic distance, and **b** variations of free energy of different phases with temperature, pressure, and volume. The minima of each parabola represent lowest energy state of the respective phase and the corresponding volume represents the equilibrium volume. The intersection points of the parabola represent temperature, pressure and volume at which both phases have the same energy and attribute to the transition temperature or pressure. Based on free energy the phase I with V_I is the stable phase which can transform at high temperature to phase III by increasing the volume to V_{III}, while on cooling it can transform to phase II by decreasing the volume to $V_{II.}$ Thus, with increasing temperature, the transition sequence of the material can be as Phase II → Phase-I → Phase III, and with lowering the temperature or increasing the pressure the transition sequence can be Phase III → Phase-I → Phase II

cases. Also, the system can lower the net energy either by phase transition, decomposition or reaction with other components forming a new compound. The lowest possible energy under each condition suggests the reaction pathway as shown in Fig. 6.1b. As the system tend to exist in lower free energy condition, the reaction proceeds in forward or backward direction depending on the relative values of free energy. This concept also holds for the phase transition which may be melting, boiling, decomposition, or solid-solid phase transitions.

Since pressure has a significant effect on volume, the free energy considered for high pressure synthesis as well as pressure-induced phase transition becomes dependent on volume of the sample. Thus, the Gibb's free energy (G) is defined by two terms like $G = H - TS$, where the first term represents combined internal energy which is related to its net energy arising from cohesive interactions and distances between the atoms or ions in it and energy due to pressure and volume, while the second term arises from degree of disorderness or entropy in it and temperature. Thus, the change in Gibb's free energy can be represented as

$$\Delta G = \Delta H - T\Delta S$$

Thus, depending on the sign of the entropy change, the second term ($-T\Delta S$) of free energy can decrease or increase with increasing temperature. In most cases, the entropy increases with increasing temperature, and hence the ΔS has a positive sign.

However, in the case of solid materials (reactants and products), the entropy change is negligible or zero. Thus, the enthalpy (ΔH) which is contributed by net pressure and volume of the system increases with increasing temperature, and that governs reaction of the solids. Hence, a situation where the enthalpy (ΔH) decreases with temperature, favors the reaction. Thus, the solid reactions are usually an enthalpy-driven process where the lowering of enthalpy by the formation of products occurs, and hence mostly they are exothermic processes. The details of conventional solid-state reactions have not been covered in this chapter. However, it can be mentioned here that in the presence of pressure, the free energy becomes significantly dependent on the volume change in the material as the mechanical stress on the materials can effectively influence their volume. Thus, the free energy dependency of the volume can be expressed as

$$\Delta G = P\mathrm{d}V - S\mathrm{d}T$$

Again, if the reactants and products are in solid states, the change in entropy becomes insignificant and thus the lowering of free energy is achieved by the decrease in volume. Thus, the equilibrium of chemical reaction carried out under pressure shifts more towards the lower volume, i.e. lower molar volume side. Hence, the pressure-induced syntheses are usually driven towards the higher density products. The schematics of synthesis of materials under simultaneous pressure and temperature can be represented as in Fig. 6.2.

As mentioned earlier both temperature and pressure have destabilization effects on the materials which increase their reactivity, but the mode of destabilizations due to these parameters are different, one due to expansion while other due to contraction. Also, the effect due to pressure on inter-atomic distance and in turn on the molar volume of solids and free energy is more prominent than that due to temperature, viz. the change in free energy of an element with pressure is about ~ 10 kJ/mol/GPa, much larger compared to 0.1 kJ/mol/K [5]. Another important difference can be attributed to the change in the inter-atomic distances in solid, viz. the increase in inter-atomic distances can be around 10% of original in solid state in case of temperature while this can be much larger like 20–30% in the case of pressure [6]. Thus, on application pressure, materials transform to denser states while on application of temperature they become less dense. Also, the usability of pressure for a reaction can vary in a wide range. For example, the pressure in intergalactic space is $\sim 10^{-32}$ bar, while that at the centre of neutron stars is $\sim 10^{31}$ bar [7]. Of course, there can be anomalous behaviors in the effect of pressure and temperature. This anomalous behavior is related to the nature of bonding, physical states, and crystal and/or electronic structure of the material. Besides, these structural changes, they also cause noticeable effects on the non-structural parameters, like pores in the samples, better physical contacts between reactants.

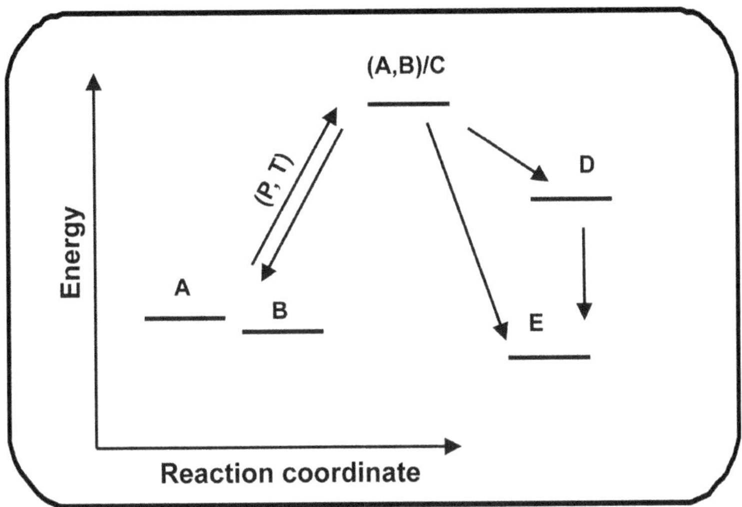

Fig. 6.2 Typical reaction schemes in HP-HT synthesis. A and B are reactants in their stable energy states. Under application of pressure and increasing temperature, both A and B transform to their higher energy states. In case of no reaction, their energy states will be dependent on the nature of A and B and they revert back to their original energy states on decrease the pressure and temperature to ambient conditions. Sintering effect may be observed in both A and B, as the case of pressure-assisted sintering like hot isostatic pressure sintering. If A and B react together under the influence of pressure and temperature, a new compound C is formed and it remains in high-energy state. If C is stable product, on releasing pressure and temperature it remains as C with normal energy stable or metastable state or transforms to a new compound D or E as metastable or stable compound, respectively. In general, all the HP-HT synthesized materials are metastable and can be transformed into components or some other stable products at some temperature or time

In general, pressure can bring crystallites of the reactants to maximum possible proximity and hence the diffusion of ions among the reactants becomes easier. But it may also hinder the diffusion of ions within a solid due to the shrinkages of channels responsible for ion migration. Also, the bond-strengths between pair of atoms increases and hence can also hinder breaking of bonds for a chemical reaction. Thus, application of temperature is essentially required to overcome such effects or a careful optimization of pressure is required for enhancing the chemical reactivity. Though this is valid in general, exceptions are also known. Under pressure, several solid–solid/liquid/gas, liquid/gas, liquid/liquid or gas/gas reactions can occur without application of temperature or can occur at ambient or even at lower temperature also. For example, the metal oxidation can occur faster under high pressure of oxygen, several hydrides can also be prepared by interacting with hydrogen gas at higher pressure at ambient temperature. Such reactions may be enhanced at higher temperature or in certain optimized temperature.

6.2 Brief Historical Picture on High Pressure Effects on Physical and Chemical Processes

The physical effect of pressure on materials has been exploited in a number of technologies, viz. the usage of siphons and double cylinder pumps for transfer of liquids, effect of steam jet for driving wheels, uses of explosives to generate high speed to firearms, etc. are to name a few. However, the chemical and physico-chemical effects of pressure came to use after the development of thermodynamics of materials and development of setups for artificial generation of pressure, and developments of newer materials to withstand the stress due to pressure and/or temperature. The applications of pressure were used extensively in chemical industry for synthesis of materials [8–10]. Also, the wide scope of such high pressure and high temperature synthesis methods has been discussed in several literatures [11–15]. Several examples of such high pressure synthesis which have been exploited extensively in industry to increase the quality and yield in shorter time have been explained in literatures [8–10] and some of the typical examples are mentioned below.

a. Haber-Bosch synthesis of ammonia from N_2 and H_2: The reaction is carried out under the pressure of about 200 atm and temperature 400 °C in presence of Fe as catalyst.
b. Fischer-Topsch synthesis of hydrocarbon from CO and H_2: The reaction is carried out using a compressed gas mixture of CO and H_2 at around 15 bars over catalysts like Fe, Co, or other noble metals at around 250 °C.
c. Production of polyethylene: Ethylene gas polymerize to solid polyethylene at 1–3 kbar pressure and 80–300 °C in presence of small amount of O_2.
d. Hydrogenation or dehydrogenation of fatty acids: Different degrees of saturation in the oils are controlled by heating them with gaseous hydrogen at certain pressure, milibar to several bars, like 2–3 bars.
e. Synthesis of artificial diamond: Heating carbon or graphite at temperature around 2000–3000 °C and pressure around 100 kbar.
f. Forging of metal and alloys: Probably this was known to the mankind since the metal ages. High compressive stress is applied on a metal or alloy either in cold or hot conditions to introduce line or plane defects or alter the crystallization process to enhance the mechanical strength.
g. Food processing and preservation: The application of pressure up to 100 MPa increases the shelf life of several food and drug by killing the microbes responsible for degradation. This procedure becomes a competitive market over other pasteurization methods and known as pascalization or bridgminaization.

Besides these industrially exploited high pressure processes, number of discoveries violating the conventional "thermodynamically stable" concepts are known under high pressure and high pressure and high temperature chemistry [11–15]. Details of preparation methods adopted of some materials are explained in subsequent sections of this chapter.

6.3 General Expected Features of Materials Prepared Under HP-HT Conditions

The syntheses under high pressure and phases formed under high pressure are the cumulative effects of several factors and some of the important factors are as below.

- Bond lengths shorten and interatomic repulsion becomes dominant.
- Force constant increases, rigid polyhedra are formed.
- Leads to denser structure by reduction of empty/free volume in the structure.
- Alters the electronic configuration, like delocalization of charges, separation of charges, and alteration of electronic arrangement in orbitals.
- Coordination number increases and introduces newer interaction between atoms or ions.
- Decreases in vibration parameters (U).
- Tends to form closely packed structure.
- Leads to geometrical frustration and charge frustration states.
- Decomposes the sample under study.

Due to such structurally important factors happening in the materials, synthesis under pressure can make possible to achieve novel metastable and unusual materials which otherwise would not have existed under normal thermodynamic conditions. In addition, discovery of minerals with such unusual structural properties formed under geological conditions provides the opportunity for high pressure (HP) and high temperature (HT) methods for synthesis of wide varieties of materials [11–16]. The synthesis of artificial diamond is a breakthrough initiative for the development of high pressure setups for preparation. Several of such unusual phenomena observed in high pressure and temperature synthesized materials are mentioned below.

- Enhanced melting of H_2, melting temperature can increase up to 600 °C and so.
- Metallization of H_2, O_2, like molecules.
- Polymerization of molecules like CO_2, N_2, CO, O_2.
- Synthesis of super hard materials, like c-BN, OsB_2 etc.
- Perovskite type $MgSiO_3$, $MgGeO_3$ and pyrochlore type $Y_2Ge_2O_7$, $Y_2Mn_2O_7$, $Dy_2V_2O_7$ like materials with off radius ratio cations materials.
- Compounds with high oxidation states, like $SrFe^{4+}O_3$ (under 340 bar O_2 pressure) and $Cs_2Cu^{4+}F_6$ (4 kbar F_2 gas pressure).
- Compounds with low oxidation state, like $CaTiF_3$, $KTiF_3$, etc.
- Compounds with octahedrally coordinated Si^{4+}, P^{5+} and V^{5+} like smaller cations. Stishovite type SiO_2, $CaCl_2$ type $AlPO_4$ etc., fluorite type MgF_2, TiO_2, GeO_2, and Ti^{4+} and Sc^{3+} with coordination number higher than 6 are examples of high pressure and high temperature synthesized materials.
- Increasing super conducting transition temperature T_c. At around 150 GPa, H_2S becomes superconducting with $T_c \sim 203$ K, LaH_{10} is predicted to have

superconducting $T_c \sim$ 274–286 K at 210 GPa and experimentally exhibits $T_c \sim$ 280 K at 200 GPa.
- New stoichiometric compounds like $BiAlO_3$, Bi_2FeCoO_6, etc. with polar structure.

The observations of such unusual properties elude scientists and technologists to explore feasibility of HP and HP-HT experiments, and to develop experimental facilities for synthesis. Such experimental investigations not only aim to discover newer materials but also to understand the behavior of material under such conditions. These studies also lead to pressure-temperature-composition phase diagram which are essential to understand interiors of earth and other planets.

In this chapter, an overview of synthesis of material under high pressure is presented. Briefly, the setups used for generation of pressure as well as both pressure and temperature together for reactions are explained. Optimum temperature and pressure for a probable reaction, as well as possible precursors for syntheses of desired materials are explained briefly. Some of the materials synthesized under high pressure and their difference from the normal compositions or would not have existed under normal pressure are discussed. Though the hydrothermal or solvothermal methods are also cases of high pressure synthesis, they are not discussed as they form a method on their own. The syntheses under high pressure of gaseous environments are briefly touched upon as they can also generate similar results as the high pressure solid-state reactions.

6.4 Experimental Methods and Instrumentations

6.4.1 Pressure

A usual understanding is that the pressure arises from the external force applied on unit surface area of a material. Since the atmosphere exists over any material, always a material experiences a pressure depending height of atmosphere over it. One "atmosphere" (atm) pressure is considered as a force exerted on unit area by the column of atmosphere at sea level. All the higher or lower pressure situations are defined in comparison to this "atmosphere" unit. Pressure due to atmosphere is equivalent to the force exerted by column of mercury of 760 mm on an area of 1 cm^2. As we move up from the sea level the pressure decreases due to decreasing height of atmosphere over it, while it increases as we move down from the sea level. This can be simply written as force exerted by column (mass of the column $\times$ acceleration due to gravity; i.e., ρgh; where ρ = density of air, h = height of the air column, g = acceleration due to gravity). Thus, the decreasing height decreases the pressure, and the atmospheric pressure drops at a rate of 0.1 atm per 1 km rise above the sea level. Similarly, it increases below the sea level, and reaches to maximum pressure at the centre of earth (364 GPa). A general standardized unit used for comparison of pressure is a standard unit "bar" which is

Table 6.1 Commonly used units for pressure and their inter-relations

Unit	Definition	Symbol	Equivalence
Pascal (SI unit)	1 N force on area of 1 m^2	Pa	1 N/m^2
Atmosphere	Pressure due to atmosphere on the sea level pressure, and is equivalent to force exerted by 760 mm Hg column on 1 cm^2 area	atm	1.01325×10^5 Pa
Bar	Pressure due to force exerted by 750 mm Hg column on 1 cm^2 area	bar	10^5 Pa
Pounds per square inch	Pressure due to one-pound weight on a square inch area	psi	6.90×10^3 Pa
Torr	Pressure due to 1 mm of Hg column	Torr	133.3 Pa
mtorr	10^{-3} Torr	mTorr	133.3×10^{-3} Pa
Giga Pascal	10 kbar	GPa	10^9 Pa
1 mbar	10^{-3} bar	mbar	10^2 Pa
1 Mbar	10^6 bar	Mbar	10^{11} Pa

equal to the force exerted by a mercury column of 750 mm height on an area of 1 cm^2. Standard SI unit of pressure is 1 N/m^2 which is named as 1 Pa and 1 bar = 10^5 Pa. Different pressure unit conventionally used for quantifying pressure and their relations are given in Table 6.1.

6.4.2 *Pressure Generation*

Generation of pressure for experiments or by nature is based on two principle process like increasing the force on fixed area or decreasing the area for a fixed force. Both the procedures are used for generating pressure on the medium. In a closed compartment, pressure can be generated by gases either by increasing kinetic energy or number density or both of the gases. Such procedures are generally used in hydrothermal process or high pressure gas-filled sealed tubes. Such processes are usually termed as 'bomb' as the hot molecules of gases are confined in a container which can cause an explosive burst. The pressure in such cases is dependent on temperature, where the increasing kinetic energy exerts more force on the sample. In all the cases, the maximum pressure that can be retained depends on the burst strength of container materials.

Syntheses under higher gas pressure are carried out in autoclaves or sealed tubes where the reactive gas (to be used as reactants) or non-reactive gas like inert gases, are filled continuously from external source up to the maximum permissible limits. Also boiling of suitable solvent and decomposition of appropriate solids inside a close compartment are used to generate pressure. At a particular temperature, the liquid and gas equilibrium govern the pressure inside it. In a gas-filled system, a metallic tube of high mechanical strength is filled with desired reactive gases like ammonia, O_2, F_2, H_2, etc. or non-reactive gas like argon, N_2, He, etc. though an inlet valve and then closed. Both sorts of assemblies are heated in a furnace where simultaneously temperature and pressure are imparted on the reactants. Pressure from gas at a particular temperature is measured using the typical pressure-volume-temperature (PVT) relations for the gases. The pressure and temperature limits are dependent on the strength of the container, boiling points of the liquid inside it. Similar to water used in hydrothermal process, liquids, like alcohol, ketones, ethers, mild organic and inorganic acids, etc. and liquid CO_2, N_2, NH_3, SO_2, CS_2, CCl_4, etc. are used in the reactor vessel for pressure generation. Several materials which can decompose producing gaseous species, like $KMnO_4$, $KClO_3$, $KClO_4$, MnO_2, $NaNO_3$, etc. are also used to autogenously generate pressure and highly oxidizing environment. The liquid or gas is selected on the basis of their inertness towards the reactor material in the operation time, temperature, and pressure range. In order to avoid reactivity often protective liners from inert materials, like Teflon, titanium and noble metals, like platinum, silver, etc. are used in the pressure vessels. A schematic of pressure vessel used for high temperature and high pressure reactions is shown in Fig. 6.3.

However, these high pressure tubular reactors can be used for pressure range from bar to kbar at maximum, which can be related to reactivity of gas at high temperature and pressure, and mechanical strength of the reactor vessel. Several

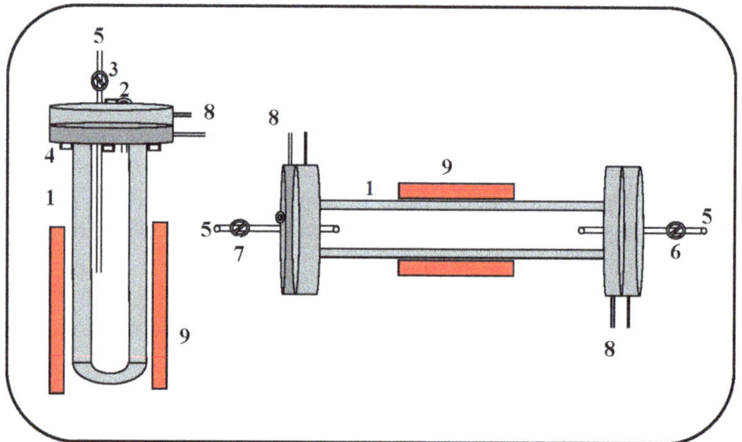

Fig. 6.3 Schematics of setup used for synthesis of materials in high gas pressure (left: vertical and right horizontal reactors). (1) Reactor; (2) flanges with release valve, (3 and 6) gas inlet regulator, (4) screw to tight and leak proof the recator and flange, (7) gas out let regulator, (5) gas flow tube, (8) water circulations tube to cool flanges, (9) heater or furnace. The dimensional is not be scaled as the specific experiments have their specific design and dimensions. The drawings are based on laboratory autoclaves and controlled atmosphere horizontal or vertical furnaces. The operation temperature and pressure are primarily governed by materials and such designs are widely used as vacuum furnace, controlled atmosphere, and over pressure gas flow furnaces

types of reactor materials and design modifications have been adopted to increase the pressure sustainability of such reactors. For lower gas pressure (10–100 bars) and moderate temperature (200–300 °C), brass, steel, and platinum, gold, silver like metal tubes can be used. Platinum can sustain higher temperature and compressive pressure, but less tensile stress. Ti and Ti alloys, Fe and Fe alloys are generally used for high gas or vapor pressure synthesis. But they are also restricted within bars to kbar, usually <10 kbar, ranges of pressure. Tungsten and its alloys, and WC are used for higher pressure reactors. Also, to increase pressure concentric tubes are used as collars to the pressure tube [17].

Alternatively, the mechanical pressure is applied to a sample in confined space for high pressure generations. The confined space is usually created in a cylindrical material with cylindrical hole at the centre [18]. Two cylindrical pistons are inserted into the hole from top and bottom keeping the sample in between them, and the setup is commonly known as piston-cylinder (Fig. 6.4). This basic design of the piston cylinder has also been modified by reducing the cross-section as well as using supported collar of hardened materials, like tungsten carbide to increase the burst pressure strength and hence to increase the pressure limits. Such modification with reducing the cross-sections area leads to the uses of anvils for high pressure generation, and pressure in the range GPa are generally generated by anvils. Hardened steel and more generally WC are used as anvils in high pressure synthesis. Extremely high pressure synthesis relies only on the diamond anvil cell (DAC). Some of the common anvils used for high pressure experiments are explained below.

The concepts and design of anvils have evolved after the pioneering work of Prof. P. W. Bridgman on piston cylinder and belt apparatus in Geophysical laboratory, USA [17, 20, 21]. In brief, the anvils consist of a pair blocks having protruded flat tips with small surface area on each. A schematic of Bridgman anvil is shown in Fig. 6.5. The sample is kept in between the tips and force is applied from back side of each tips. The space in between the two anvil tips is called pressure chamber. In order to increase the pressure, the collars of different types of alloys are used around the pressure chamber and anvils support. By such a suitable apparatus Bridgman could generate pressure of the order of 10 GPa with and without temperature which he had extensively used to investigate compressibility and synthesis of materials as well as for studying the electrical and thermoelectric properties of a wide variety of materials [17, 20, 21]. For the pioneering work on generation of high pressure and synthesis of mew materials under pressure, Prof. P. W. Bridgman won the Nobel Prize in physics in the year 1946.

Further, the high pressure Bridgman anvil apparatus has seen several improvements and modifications both in design as well as in construction materials for anvils. These inventions strengthened the high pressure syntheses and studies of materials under high pressure. The second boost on high pressure synthesis can be attributed to the synthesis of artificial diamond by Bundy et al. [22] and cubic boron nitride by Wentorf et al. [23]. The evolved geometries of pressurization procedures like belt apparatus and torroid anvil apparatus etc. are routinely used for generation of pressure and temperature, and to carry out synthesis of newer materials by

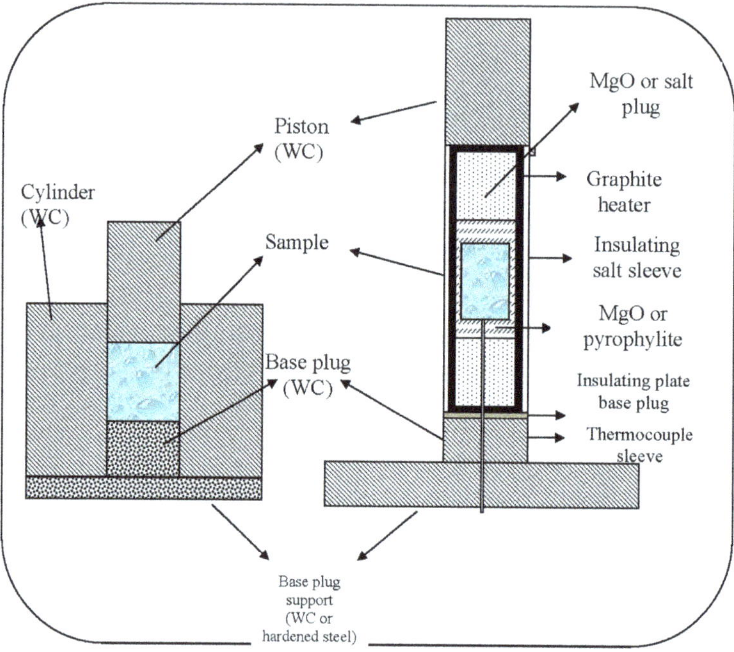

Fig. 6.4 Schematic sketch of a piston cinder apparatus for HP and HP-HT synthesis of materials. Sample is a mixture of reactants with appropriate stoichiometry kept inside a non-reacting MgO or pyrophyllite cup with caps and placed inside a graphite cup which acts as heater also. Pressure is applied through the piston to the sample chamber and temperature is generated with high current and low voltage power supply. Appropriate thermal and electrical insulations are used to protect the piston. The volume of the sample is decided by the diameter of the cylinder and hence can be used for synthesis of larger volume of sample. The pressure can be of several GPa and temperature up to 2000 °C can be attained in such setup. Also, the time for reaction can be of several hours. The dimensions in the figure are not to be scaled and they are dependent on the actual design for pressure and temperature and nature of experiments. The schematic is adapted from Ref. [19], H. T. Hall, Some high pressure, high temperature apparatus design considerations: equipment for use at 100,000 atmospheres and 3000 °C, Rev. Sci. Instrum. 29(1958):267–275, with the permission of AIP publishing

simultaneous application of both high pressure and high temperature. Typical supported high pressure anvil commonly used for high pressure and temperature experiments is the belt apparatus (Fig. 6.6) [24]. The key feature of the geometry is concerned to distribute the load and to generate largest pressure at specific volume.

In all such high pressure setups used have common instrumental features such as anvil, pressurizing or pressure transmitting medium, pressure chamber, spacer, or gaskets. Anvils are a pair of moveable heads made from super hard materials like hardened steel, WC, diamond, etc. aligned along a single axis such a way to render almost no shear stress on any of them when pressurized. The anvils approach each other and squeeze the pressure transmitting medium in between them. The gaskets hold the pressure transmitting mediums as well as prevent the anvils to touch each

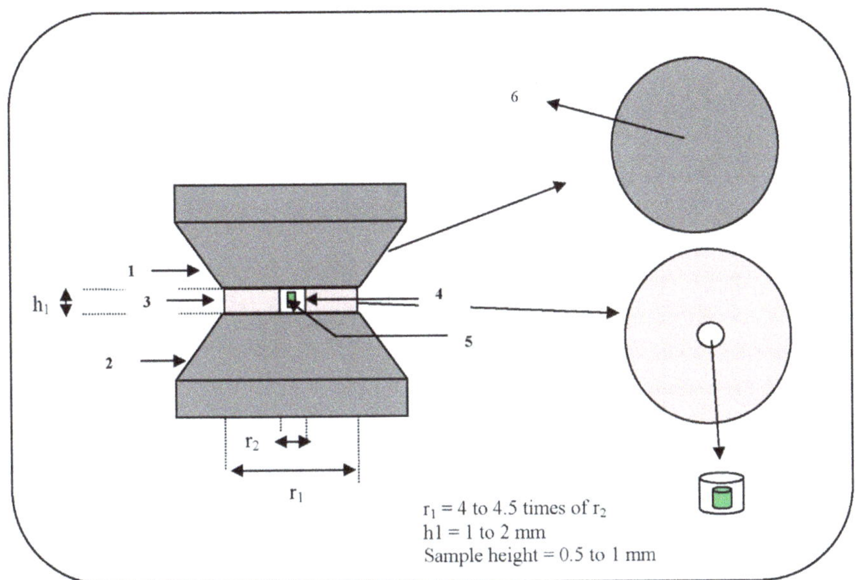

Fig. 6.5 Typical sketch of a Bridgman anvil. Left: side view, right: cross-sectional views of anvil and assembly. (1 and 2) Upper and lower anvils made from hardened steel, (6) tungsten anvil, (3) pyrophyllite gasket, (4) talc sample holder, (5) sample under consideration. Sample is pressed in a small cylindrical disk of height about 0.5–11 mm and placed in inside a thin pyrophyllite gasket. Two pyrophyllite disks are placed above and below of the sample. Also, certain cases, the powdered samples are placed inside a pyrophyllite or BN cups and the cups are covered with a lid of the same materials. The sample assembly is inserted into the pyrophyllite gasket. The outer diameter of the gasket is same as the diameter of the anvils. Force is applied from top and a bottom support ram of the anvils by hydraulic mechanism. Pressure is increased and held for different times. For high temperature generations, the sample assembly cup is inserted into a graphite or $LaCrO_3$ tube which acts as resistive heater. By applying current to the heater temperature is raised. Such assemblies are also used to measure electrical properties by using probe wires on the top and bottom of the sample cylinder. Figure and cell assembly are drawn with inputs from Dr. S. N. Vaidya, High Pressure Physics Division, BARC, Mumbai, India. For more details on anvil and assembly refer to the references, P. W. Bridgman, Recent works in the field of high pressure, Rev. Mod. Phys. 18(1946):1–95; S. N. Vaidya, D. K. Joshi, and C. Karunakaran, Anvil apparatus for the measurement of electrical resistances up to 80 kbar, Indian J. Technol. 14.12(1976):679–680; S. N. Achary, G. D. Mukherjee, A. K. Tyagi, and S. N. Vaidya, Preparation, thermal expansion, high pressure and high temperature behavior of $Al_2(WO_4)_3$, J. Mater. Sci. 37(2002):2501–2509

other and house the sample under study. Thus, the pressure is essentially exerted on the pressurizing medium. Ideally, the pressure transmitting medium behaves like a fluid and transmits the pressure uniformly, i.e., hydrostatically to the sample. Based on these concepts a number of setups, like piston cylinder, torroid anvil, diamond anvil (DAC) and large volume press (LVP) with multiple anvils, etc. have been developed for generation of pressure and they are extensively used for synthesis as well as for studying the high pressure behavior of materials in present days (Figs. 6.4, 6.6, 6.7, 6.8, 6.9). High temperature in such press can be generated using

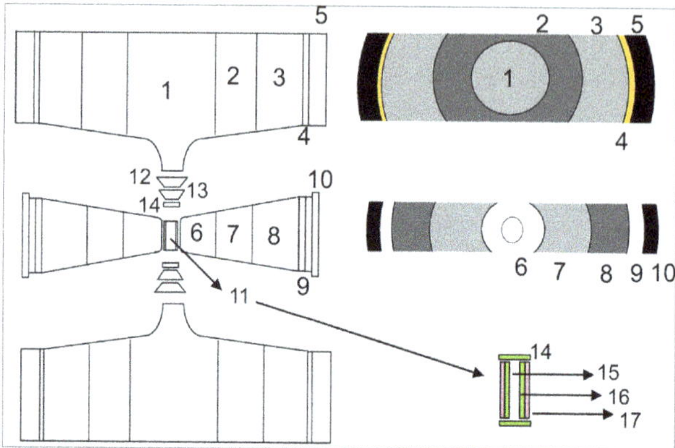

Fig. 6.6 Schematics of a belt apparatus used for high pressure and high temperature experiments. The system is modified version of supported Bridgman anvil and is usable for pressure of about 10 GPa and temperature 2000 °C for prolonged period. (1) Carboloy piston which push the sample from top and bottom and current is applied through these pistons to the sample assembly (11) though the steel rings (12, 13, and 14). The pistons are supported by binding rings (2 and 3) made from hardened steel, (12 and 13) are steel rings and pyrophyllite disks. (6) Carboloy ring for pressure chamber and (7 and 8) are supporting binding rings made from hardened steel. (9) Cooling water circulation ring, (4 and 10) are soft steel counter rings to prevent flying the objects in binding ring failure, (5) is copper counter pressure ring. The sample assembly is made from Ni cylinder (16) with nickel disks (14) on top and bottoms surrounded by a pyrophyllite cylinder (17) to provide thermal and electrical insulation. (15) Sample under experiment. Adapted from Ref. [24], H. T. Hall, Ultrahigh pressure, high temperature apparatus: the "Belt", Rev. Sci. Instrum. 31(1960):125–131 with the permission of AIP Publishing

different types of heater or heating modes. Some of such anvils are explained later in this section.

6.4.3 Pressure Transducer

Pressure transmitter or transducer plays a crucial role in the high pressure experiments. The pressure from the anvil is transferred uniformly and hydrostatically to samples through this medium. Soft solids, liquids as well as condensed gases are used for transmitting the pressure. Ideally, the pressure transmitting medium (PTM) should behave like fluid with zero shear strain and zero differential stress in the studied pressure range. Since the properties of materials varies with pressure as well as the liquid and gases can also solidify and hence exhibit directional properties, all the pressure transducer transmit pressure an hydrostatically beyond certain pressure and hence render inhomogeneous pressure on the sample. Other important considerations of pressure transmitter are: their reactivity with the anvil material, gasket, sample, heater, etc. as well as their own pressure and temperature stabilities. Thus, different pressure transmitting medium for different types of

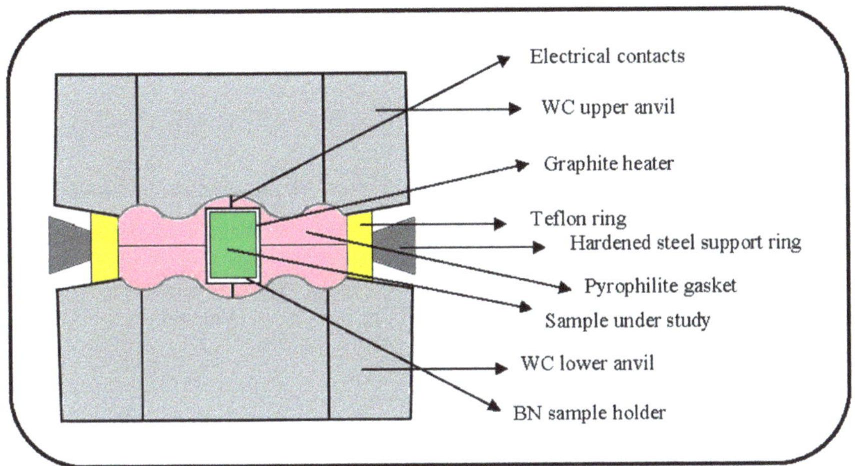

Fig. 6.7 Schematics of a torroid anvil used for high pressure and high temperature experiments. The sample of diameter of 4–5 mm and height 6–8 mm is located at the center of the graphite furnace which is heated by high current low voltage power supply. For metallic samples, electrical insulation is provided by sleeves made of NaCl. Pressure is generated at the sample by applying force from lower or lower anvil. The dimensions in the figure are not to be scaled. Figure is drawn with inputs from Dr. S. N. Vaidya, High Pressure Physics Division, BARC, Mumbai, India. For a typical design refer: S. N. Vaidya, D. K. Joshi, and C. Karunakaran, Anvil apparatus for the measurement of electrical resistances up to 80 kbar, Indian J. Technol. 14.12(1976):679–680, C. Karunakaran and S. N. Vaidya, Toroid-anvil high pressure high temperature AC conductivity measurements on fast-ion conductors, High-temp. High-press. 26(1994):393–400

experiments are used. Solid pressure transmitting medium like talc, pyrophyllite, AgCl, CsCl, NaCl, MgO, $CdBr_2$, CdI_2, $CoCl_2$, $CuBr_2$, HgI_2, Ag_2S, metallic lead, indium, sodium, hard paraffin, polyethylene, etc. are commonly used as pressure transmitting media in synthesis by LVP. Talc and pyrophyllite can become rigid at a moderate pressure, viz. around 5 GPa, and thus lead to large an hydrostatic pressure around the sample. Thus, they have applicability only in a limited range pressure. But still, talc and pyrophyllite are the very commonly used pressure transducer for high pressure synthesis experiments due to their chemical inertness and ease machinability. AgCl and CsCl are another widely used PTM due to their higher chemical stability and low melting point and high flowability. Other pressure transmitting mediums like lithium hydride, carbon, hexagonal boron nitride, and Teflon. For very high pressure and temperature, viz. >10 GPa, porous or semi-sintered MgO, BeO, cubic ZrO_2, Fe_2O_3, Al_2O_3, Mg_2SiO_3, etc. are also used as PTM [4, 8, 11, 25].

Lighter alcohols, like methanol, ethanol, water, low viscosity mineral oils like silicone oils, hydrocarbons like neo or isopentene, terpenols, fluorocarbons, several polymers are also used as pressure transmitting medium depending on temperature, pressure, and nature of the sample. Also, gases like, N_2, O_2, He, Ne, Ar, Kr, Xe, H_2, D_2, etc. are also used for pressure transmission. Practically for pressure around 9–

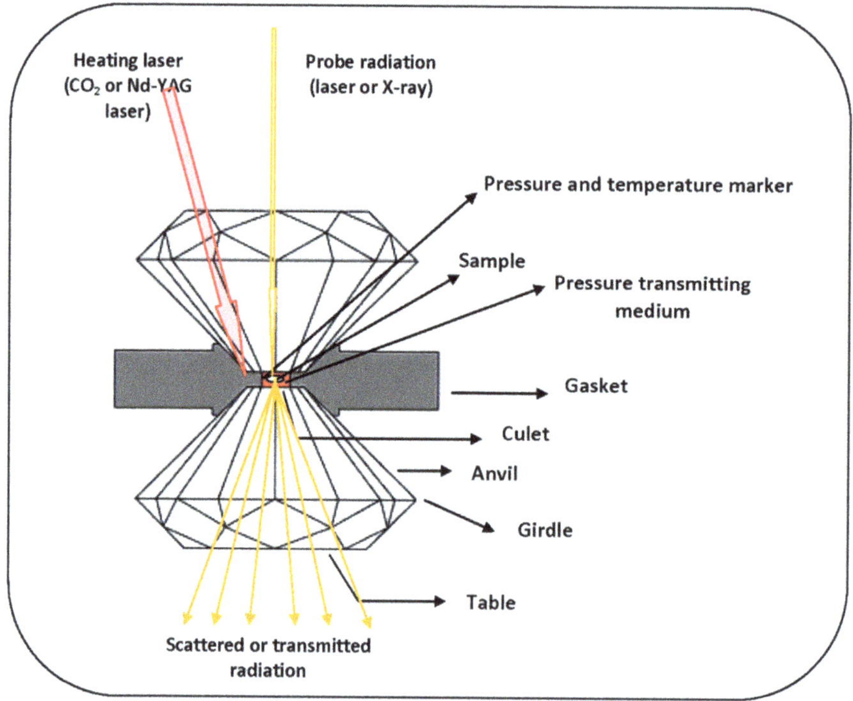

Fig. 6.8 Schematics of a diamond anvil cell for high pressure and high temperature experiments as well as in situ probing of the reaction or transition by X-ray or laser. Temperature and pressure are measured from the output marker. The diamonds anvils were supported on WC supports. For more details on design, assembly and sample environments in diamond anvil cells refer to Refs. [24–28]

10 GPa, methanol-ethanol (4:1) mixture is generally used while methanol-methanol-water (16:3:1) mixture is used for hydrostatic pressure up to around 12 GPa. Since retention and conferment of gases in between the gasket and anvils is difficult, they are used by condensing them or using highly compressed gases. Helium can behave like fluid even at higher pressure, like 20 GPa. For high pressure and temperature reactions NaCl, MgO like isotopic solids are used. The usability and hydrostatic limits for a number pressure transmitting medium have been explained in literature [4, 25].

6.4.4 Generation of Temperature Under Pressure

The desired temperature for reaction in high pressure can be generated by only in limited procedures. Only in few instances the sample container or pressure cell can be directly heated. However, the temperature and pressure in such instances are

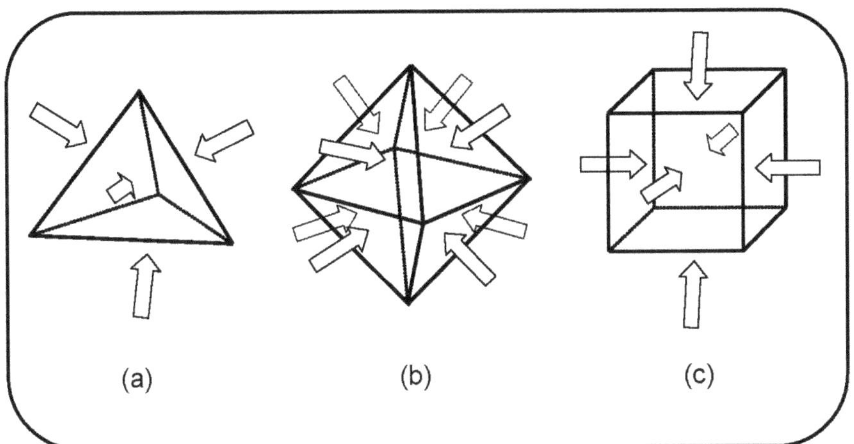

Fig. 6.9 Schematics of pressure generation in multi-anvil press (**a**, **b**, **c** are tetrahedra, octahedral and cubic press). The figure is based on the pressurizing anvil positions and does not represent the actual multi-anvil apparatus and sample environment. The polyhedra represent the pressure chamber which houses the sample capsule along with heater and thermocouples. Force is applied along all the triangular faces of tetrahedra or octahedral, or squares of the cube. The anvils faces are exactly similar area as that of the medium. For details refer to Refs. [8, 24, 33–35]

limited and the limitations mostly arise from the strength of the container materials. For example, high gas pressure, hydrothermal and solvothermal experimental methods use direct heating of pressurized sample container. In all these cases the pressure is also limited at high temperature as the system may explode due to softening materials by temperature or limited strength of materials at high temperature. Thus, they always need a safety consideration either to release the excess pressure or to generate pressure as per the consideration of strength of materials. For cryogenic or moderately above ambient temperature, viz. 400–500 °C, several types of diamond anvil cell can also be heated externally. Higher temperature under pressure is commonly generated by a resistive heater placed near the sample under study.

Usually graphite, $LaCrO_3$, SiC and $MoSi_2$ like resistive heating elements are used as collar to the samples and placed inside the pressure chamber. Heater is usually operated in high current and low voltage mode, and so low restive heaters are generally used. The temperature can also be generated by tungsten or molybdenum coil or foil around the sample. As high as 2000 °C under pressure can be generated by such heaters.

In the high pressure and high temperature synthesis, appropriate reactants are placed inside a cylindrical cup and caped. The sample container can be a graphite, boron nitride, platinum, alumina depending on the requirements and compatibilities. The sample container is again placed inside a cylindrical heater or in inside the pressure transducer where tiny heater rods are inserted. The electrical contacts to the heaters are provided either by the anvil itself or by appropriate connecting wires inside the pressure transducer. All these procedures for generation of heat are possible in large volume high pressure apparatus. In case of diamond anvil cell, heating the diamond by external source can be carried out for low to moderate temperatures. But for high temperature such procedures are not practically feasible and hence extremely high temperature is generated by heating the sample by laser beam [26–28]. Continuous beam of Nd–YAG (Nd^{3+} ion doped $Y_3Al_5O_{12}$) laser is most commonly used to heat the sample inside the diamond anvil. High temperature around 5000 °C in the diamond anvil can be generated by such laser heating. Extremely high pressure and temperature in the range of TPa and 10^5–10^6 K can be generated by pulsed high power laser, inertial conferment in laser ignition facilities and nuclear explosion devices. Dynamic pressure and temperature are generated by hitting target with a high-speed projectile. The use high speed ballistic can cause large dynamic pressure, like 300–400 GPa even higher but only for momentarily. Such conditions, though rarely used for synthesis purposes, but metastable phases of metals like Zr, Ti, W, etc. have been obtained by such dynamic shock pressure.

Successively, the high pressure apparatus with one-dimensional press is evolved to multidimensional large volume press (LVP), like tetrahedral, octahedral, and cubic where the force is applied on the regular faces of such polyhedra (Fig. 6.9). The Walker-type large volume multi-anvil press has evolved as a modern synthesis setup for preparation of unconventional and novel materials. Such large volume pressure can generate pressure even up to about 15 GPa and temperature up to 2500 °C depending on the heater assembly. Further development of high pressure generation is done by diamond anvils. The usage of diamond, the hardest material known, opens a practically simpler mode for generation of high pressure and nowadays used in almost all laboratories for high pressure studies. Pressure at the sample is achieved by squeezing it in between parallel faces of two diamonds and the pressure depends on the areas of the faces of diamond anvils. Further the uses of beveled diamond and nano-diamond increase pressure limit to even up to 150 GPa. The nano-twinned diamond can achieve Vickers hardness up to 200 GPa which significantly higher than the natural synthesis diamond [29–32]. However, the amount of sample that can be prepared by such diamond anvil cells is very small and mostly suitable for in situ structural and spectroscopic or microscopic characterizations. Yet the DAC is an extremely useful apparatus for synthesis of sample under extreme pressure and temperature. For practical materials synthesis procedure, most commonly the belt apparatus, torroid and multi-anvils, and diamond anvil are routinely used worldwide.

The high pressure synthesis procedures can be directly carried out at ambient temperature, using such anvils directly. In such cases, the reaction of solids or stable products rarely occurs. However, the solid-gas or solid-liquid reactions can occur in such cases. Gas inclusion and intercalated compounds, host-guest compounds, and adducts can occur in such conditions. Materials exhibiting pressure-induced irreversible phase transition can also be a mode for synthesis of materials. Some of such syntheses are explained in subsequent synthesis. However, most of the reactions and very commonly reactions of solids as well as several irreversible phase transitions need temperature along with pressure. Thus, high pressure and high temperature conditions are usually adopted for synthesis of materials. Some of the well-established synthesis procedures are tabulated earlier in this article. Since the pressure is generated in a small volume, generation of temperature under pressure is in general different than conventional methods. For details of these anvils refer to Refs. [4, 8, 33–35].

6.4.5 Pressure and Temperature Measurements

Similar to the generation of pressure and temperature for high HP-HT synthesis, their measurements in an experiment are often tedious due to the cumbersome setup and influence of both temperature and pressure on the measured signals [36]. In piston-cylinder and anvils, pressure is usually calibrated with dead weight gauge with respect to the applied stress. A number of pressure gauges, like strain gauges, Bourdon gauge, Macleod gauge, and Hg column are used in lower pressure limits. However, there are several other measurements modes like variation of electrical resistances, piezoelectric materials, luminescent materials, etc. used to measure the pressure. The variation of resistance of a foil or a coil, usually manganin, an alloy of Cu, Mn, and Ni as $Cu_{85}Mn_{12}Ni_2$, is used for both pressure and temperature measurements. The variation of electrical resistance with pressure or temperature is used as calibration curve to read pressure or temperature. The electrical properties of manganin under pressure have been calibrated in a large number of reports since the Bridgmann himself calibrated up to 3 GPa [36–38]. Manganin has a positive pressure coefficient of resistance and shows appreciably larger and linear variation of resistance with pressure [28, 39, 40]. The variation of resistance of manganin with pressure has been calibrated by Fujioko et al. [39] up to 220 kbar, and linear pressure dependency up to 180 kbar has been reported. The typical pressure coefficient of manganin under static pressure is given below [39]

$$\frac{\Delta R}{R_0} = \frac{(R_{\mathrm{p}} - R_0)}{R_0} = 2.322 \pm 0.008 \times 10^{-3} \times P\ (\mathrm{kbar})$$

Additionally, variations of resistances of zeranin ($Cu_{90.7}Mn_7Sn_{2.3}$) [40] and Fe [37] are also used for pressure measurements. The discontinuity in resistance of Fe is also used as a fixed-point pressure calibration.

The change in the resistances at the phase transition of certain standard materials (viz. BiI–Bi-II, Bi-II–Bi-III, Yb-I–Yb-II, etc.) are also commonly used to calibrate the pressure [20, 21, 41–43]. The structural transitions associated with volume discontinuity like B1-B2 transition of NaCl is also used as fixed-point pressure celebrant [44–46]. Most of these discontinuities are also structural transitions and occur with discontinuity in molar volume. Thus, they are used to calibrate pressure by X-ray diffraction experiments also. Some of the fixed-point pressure calibrants and their properties are summarized in Table 6.2.

Table 6.2 Fixed point pressure calibrate for HP-HT setups [20, 21, 41–46]

P (GPa)	Transitions	Comments	$-\Delta V/V$ (%)	Resistance discontinuity
2.55	Bi-I to Bi-II	*R*-3*m* to C2/*m*	4.7	Y
7.7	Bi III to Bi IV	$P2_1/m$ to $P2_1/n$	1	Y
0.68	Ti II to Ti III	α to ω	15	Y
5.5	Ba I to Ba-II	bcc to hcp	1.9	Y
12.3	Ba II to Ba III	hcp-tetragonal	0.95	Y
9.4	Sn I to Sn III	bct-bct (β to γ)	1.1	Y
45	Sn III to Sn IV	Bct to bcc	7.6	Y
26.8	NaCl (B1-B2)	NaCl-CsCl	4.8	Y

Unlike pressure, the temperature can be measured directly by using suitable thermocouple or measurement resistances of metallic or semiconducting wires. Since pressure affects the electrical properties of such materials, the temperature coefficient of resistance or *emf* needs a pressure calibration. Manganin wire is also used as a temperature measurement mode. The temperature coefficient of resistance of manganin is close to 1×10^{-6} per degree centigrade [47, 48]. The temperature is also quite often measured from the thermo-power supplied to the heater. This procedure generally requires calibration from independent temperature measurement. However, such measurements are usable only at limited temperature range, i.e., around 1000 °C.

In large volume apparatus as well as in piston-cylinder or toroid anvils, thermocouples can be used to measure the temperature. W-Re (Type C: 95%W/5%Re–74%W/26%Re; Type D: 97%W/3%Re–75%W/25%Re; Type G (100%W–74%W/26%Re and Pt-Rd (Type R: 87%Pt/13%Rh–100%Pt, Type S: 90%Pt/10%Rh–100%Pt, (all in wt %) thermocouples are inserted into the pressure chamber suitably and temperature is measured from the generated EMF. Both R and S type thermocouples can be used to measure temperature up to 1600 °C, while the C, D, and G type can be used to measure temperature up to 2329 °C. The measurement range by the C, D, and G type thermocouples can be extended to about 3000 °C in inert atmosphere. Since the pressure assembly is mainly non-reactive and non-oxidizing, the W-Re thermocouples are routinely used for temperature measurements up to 3000 °C. However, over this limit, indirect modes like phase transition, melting or optical probes are used to measure temperature.

At higher pressure and temperature, fluorescence markers are commonly used to read pressure and temperature [49–51]. Fluorescence characteristics of some standard markers are given in Table 6.3. Ruby (Al_2O_3:Cr^{3+}), SrFCl:Eu^{2+} and SrFBr:Eu^{2+}, SrB_4O_7:Sm^{2+} etc. are commonly used fluorescence marker for high pressure or high temperature experiments carried out in diamond anvil cell. The shift in emission wavelength with pressure or temperature as well as for both is calibrated to read the pressure and temperature. Ruby has been extensively used for pressure measurements in a transparent high pressure medium [52, 53]. Micron sized grains of synthetic ruby crystals is loaded along with the sample inside a pressure

Table 6.3 Typical parameters of luminescent pressure and temperature sensors

	λ (nm)	$d\lambda/dP$ (nm/GPa)	$d\lambda/dT$ (nm/10^3 K)	Ref.
Cr^{3+}-Al_2O_3 (R1)	694.2	+0.365	+6.8	i
Cr^{3+}-$YAlO_3$	722.8	+0.70	+7.6	i
Eu^{3+}-YAG	590.6	+0.197	−0.5	ii
Sm^{2+}-SrB_4O_7	685.4	+0.255	−0.1	iii
Sm^{2+}-SrFCl	690.3	+1.10	−2.3	iv

(i) J. D. Barnett, S. Block, and G. J. Piermars, Rev. Sci. Instrum., 40, 1(1973)
(ii) Y. Chi, S. Liu, W. Shen, L. Wang, and G. Zou, Physica, 139 & 140B, 555(1986)
(iii) A. Lacam and C. Chateau, J. Appl. Phys., 66,366(1989)
(iv) Y. R. Shen, T. Gregorian & W. B. Holzapfel, High Pressure Research 7, 73(1991)

transmitting medium and excited by a probe laser, usually a He–Ne or similar visible light laser. The ruby emission, R1 and R2 lines at standard pressure (1 atm) and temperature (300 K) are located at 694.25 and 692.86 nm, respectively and they are due to the 2E excited states to $4A^2$ ground state transitions [54]. The shift in emission wavelength and the spacing ($\Delta\lambda$) between them are sensitive to pressure and temperature. The pressure dependency of the shift of R1 line of ruby is about 0.365 nm/GPa up to about 20 GPa and then becomes nonlinear [53, 55]. Typical ruby luminescence lines at two pressures are shown in Fig. 6.10.

Experimentally, the calibration curves for the shift in wavelength for R1 line has been refined and widely accepted relation for Ruby emission is given below [52, 55–58]

$$P\ (\mathrm{GPa}) = 1870\frac{\Delta\lambda}{\lambda_0}\left(1 + 6.0\frac{\Delta\lambda}{\lambda_0}\right)$$

$$P\ (\mathrm{GPa}) = \frac{A}{B}\left(\left(\frac{\lambda_\mathrm{p}}{\lambda_0}\right)^B - 1\right)\ (\text{up to 150 GPa})$$

where λ_0 is wavelength of R1 line at 298 K (694.24 nm) and $\Delta\lambda = \lambda_\mathrm{p} - \lambda_0$, λ_p = wavelength of R1 at pressure P. A- and B are empirical constants determined as $A = 1876$, $B = 10.71$, respectively [52, 58].

Also, *PV* and *PVT* equation of states (EOS) of several standard materials, like Pt, Au, Cu, Ag, Al, W, etc. are used for measuring pressure and temperature. At higher pressure, EOSs of NaCl, KCl, MgO etc. are used for pressure and temperature measurements. These methods are very commonly used in the diffraction studies, where simultaneously the structural changes in the pressure marker is observed along with the samples. The change in unit cell parameter (volume) of marker is compared with the established equation state of the same marker to determine pressure. The standard bulk modulus and pressure derivative of some standard materials are summarized in Table 6.4. Commonly used third-order Birch-Murnaghan (BM-III) equation of state for measurement of pressure is given below.

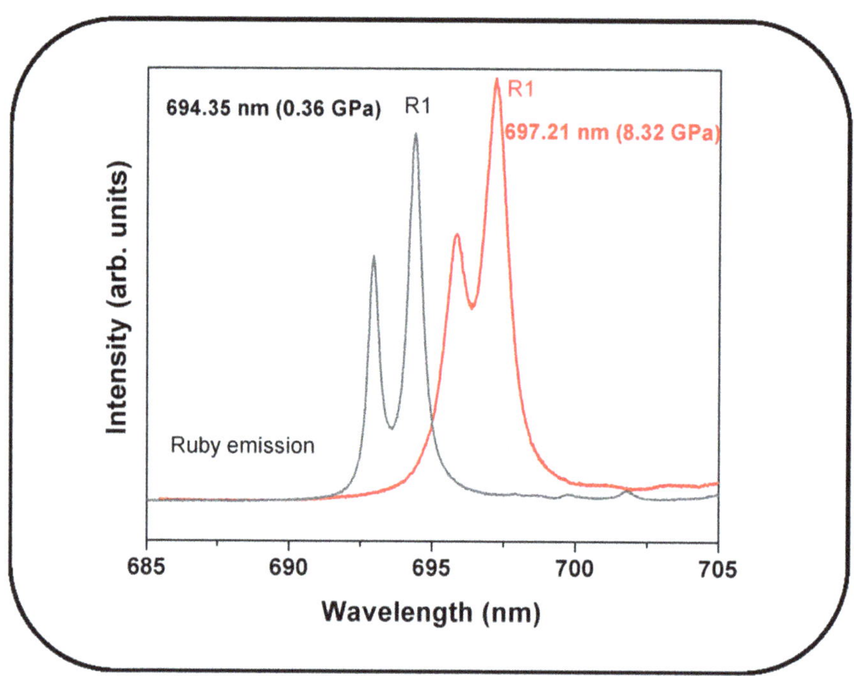

Fig. 6.10 Emission lines of ruby at two different pressures. Spectra were recorded using a diamond-anvil cell and methanol–ethanol–water (16:3:1) mixture as pressure-transmitting medium. Emission spectra were recorded on a Horiba Jobin Yvon LabRAM Raman spectrometer using 632.8 nm line of the HeNe laser as excitation source. R1 is more intense than R2. With increasing pressure emission wavelength sifts to higher side and rate of shift is approximately 0.365 nm/GPa under hydrostatic conditions up to 20 GPa, while the shifts are around 0.0073 nm/K up 600 K. The line shape remains more or less similar with pressure while broadening as well as $\Delta\lambda$ (the separation of R1 and R2, i.e., Δ(R1 − R2)) varies with temperature. The separation of R1 and R2 also depends on the concentration of Cr^{3+}. The separation becomes unreliable at high temperature due to merging of the R1 and R2 lines to a single asymmetrically broadened line. Life time of emission has also been proposed as a mode for measurements of pressure and temperature. Besides at higher pressure, the phase transition of corundum to Rh_2O_3 type structural transition though does not alter the Cr^{3+} environment in the structure, they are reflected in the separation of R1 and R2 lines [54]. For more details on ruby scale and emission mechanisms of Cr^{3+} to Ref. [54] and references cited therein

$$P = \frac{3B_0}{2}\left[\left(\frac{V_0}{V}\right)^{7/3} - \left(\frac{V_0}{V}\right)^{5/3}\right] \times \left(1 + \frac{3}{4}(B_0' - 4)\left(\left(\frac{V_0}{V}\right)^{2/3} - 1\right)\right)$$

where B_0 = Bulk modulus, V_0 = Volume at pressure, $P = 0$, B_0' = Pressure derivative of bulk modulus.

However, in the synthesis apparatus, the equation of states is not easy to use as the sample is surrounded by the anvil and pressure transmitting media. Thus,

Table 6.4 EOS of pressure markers

Standard	B_0 (GPa)	B'_0	Ref.
Ag	118	3.8	a
Au	167	5.5	a
Cu	133.5	5.3	b
Al	72.8	4.6	b
Fe	158	5.8	a
Pt	275.3	5.2	b
W	312	3.83	a
Re	372	4.05	a
bcc-Bi	54.74	4.905	c
C	441.5	3.98	b
NaCl-B1	25.18	4.80	d
NaCl-B2	31.21	4.84	d
MgO	160.3	4.2	b
Ne	1.097	9.23	e
α-Al_2O_3	253	4.30	f

(a) P. C. Sahu and N. V. Chandrasekhar, Resonance 12(2007):10
(b) T. S. Sokolova et al., Russian Geology and Geophysics 54(2013):181
(c) Y. Akahama, H. Kawamura, and A. K. Singh, J. Appl. Phys. 92(2002):5892
(d) P. I. Dorogokupets and A. Dewaele, High Pressure Research, 27(2007):431
(e) S. M. Dorfman, V. B. Prakapenka, Y. Meng, and T. S. Duffy, J. Geophys. Res. 117(2012): B08210
(f) K. Syassen, High Pressure Research, 28(2008):75

usually thermocouples, bourdon or strain/piezo gauges are used for measuring the temperature and pressure. In addition, it is necessary to calibrate the pressure with the fixed-point celebrants prior to experiments.

6.5 Synthesis Under High Pressure and/or High Temperature

6.5.1 Synthesis of Artificial Diamond

High pressure and high temperature synthesis are developed from the origin at artificial diamond synthesis which dates back to late nineteenth century. The early experiment on chemical synthesis of diamond from graphite in iron tube by high temperature generated by electric discharge indicated the requirement of pressure for graphite to diamond transformation. Successive theoretical and experimental developments in understanding the graphite to diamond transformation and carbon atom to diamond transformation become industrial methods for production of synthetic diamonds. The continuous research and development in this aim are now becoming one of the most profitable businesses in the world. The vapor phase of carbon can be condensed to diamond which has been exploited in vapor deposition growth of diamond. In both cases, the diamond is formed as a metastable compound. In this section, the HP-HT methods for synthesis of diamonds are only briefed while the later approach is beyond the scope of this chapter.

The pressure-temperature (PT) phase diagram of graphite-diamond transformation is shown in Fig. 6.11 [59]. From the phase boundary of graphite and diamond in the phase diagram, it can be suggested that with increasing temperature the pressure required for this transition needs to be higher. The synthesis of large-size gem-quality diamond is relied on the PT phase boundary of diamond and graphite. The initial successful synthesis of diamond has been carried out in belt apparatus with pressure around 10 GPa and temperature around 2000 K [60]. The graphite either in a rod or powder form is mounted inside a cup made of Pt, Mo, W, Ta or Fe, Co., Ni or $LaCrO_3$ and the assembly is placed inside a pressure transmitting gasket like MgO, BeO, and ZrO_2. The system is first pressurized and then temperature is increased while the sample is under stress. The mentioned metals act as heater for the assembly as well as catalysis for the formation of diamond. The time, temperature, and pressure are optimized to obtain large crystals of diamond. Successively catalysts, like Fe, Co, Ni, etc. have been employed to accelerate the

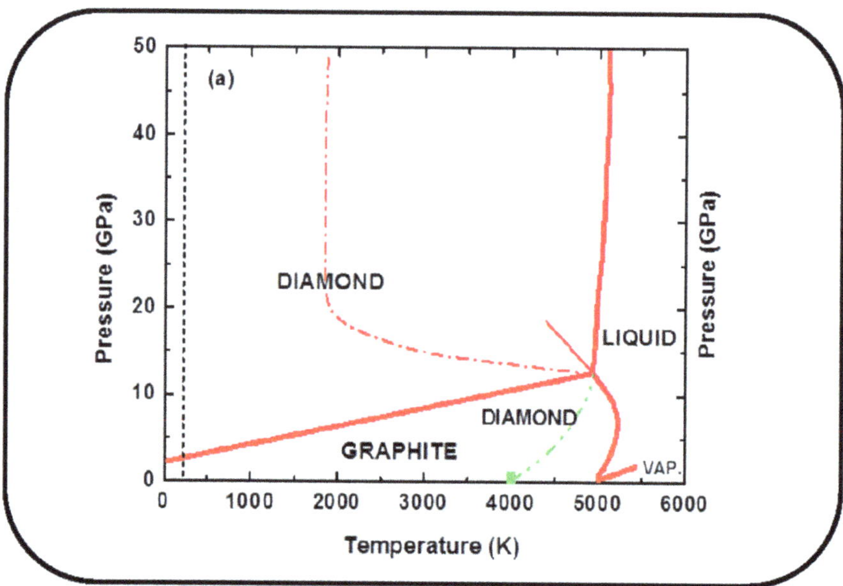

Fig. 6.11 Pressure-temperature phase diagram of carbon [59]. Graphite can transform to diamond by solid to solid transformation under pressure and temperature or molten carbon crystallizes to diamond under pressure. The catalytic process is shown as green line. Reproduced with permission from J. Narayan and A. Bhaumik, Novel phase of carbon, ferromagnetism, and conversion into diamond, J. Appl. Phys. 118(2015):215303. Copyright 2015, AIP Publishing

graphite to diamond conversion. Hu et al. [61] have prepared B and N co-doped diamond crystal using B and N elements along with carbon in a gradient temperature and pressure (viz. 5.3–5.8 GPa and 1300–1550 °C) condition and Fe as catalyst. Presently, varieties of multianvils are being used to prepare diamond from graphite in industrial scale. This represents the first technological and commercial success of high pressure and high temperature synthesis techniques.

Nanocrystalline diamond has been prepared under high pressure and temperature using $C_3H_5N_3O$ (cyanoacetic acid hydrazide) as an organic additive in NiMnCo-C system by Guo et al. [62]. The synthesis is typically carried out at 5.5–6.2 GPa and 1280–1320 °C in a large volume cubic anvil apparatus. Graphite powder mixed with $C_3H_5N_3O$ and pressed into a cylindrical pellet is used as precursors and NiMnCo–C alloy mixed with $C_3H_5N_3O$ is used as catalysts. Single crystal of highly pure diamond is used as a seed crystal to grow larger crystal. This assembly is placed inside a ZrO_2 +MgO insulating materials and then encapsulated by a sleeve made from NaCl and ZrO_2 mixture. Further the sleeve is surrounded by a graphite heater and housed inside a pyrophyllite and dolomite composite block. This sample assembly is placed inside a cubic anvil for high pressure and temperature treatment. Pressure, temperature, and time as well as composition of initial carbon sources precursor mixture and catalysts are adjusted to prepared single crystalline diamond. The typical images of synthesized diamonds are reproduced in Fig. 6.12. However,

Fig. 6.12 Optical images of the diamond crystals synthesized from the NiMnCo–C system with $C_3H_5N_3O$ additive (wt% of additive is shown in the figure) by pressure and high temperature conditions. Well defined crystals with high transparency are formed with o additives while darker crystals are formed with additives [62]. Reproduced with permission from L. Guo, H. Ma, L. Chen, N. Chen, X. Miao, Y. Wang, S. Fang, Z. Yang, C. Fang, and X. Jia, Cryst. Eng. Comm. 20 (2018):5457–5464. Copyright (2018) The Royal Society of Chemistry

the impurities like N, O, and H are inherently incorporated in the diamond crystals and they govern their color and transparencies. Sintered nanocrystalline diamond aggregates prepared by such methods can be used as anvils for ultrahigh pressure studies [63]. Gotou et al. [63] have reported pressures and temperatures up to about 30 GPa and 1700 K can be easily achieved using such sintered nanocrystalline diamonds.

6.5.2 Synthesis of Superhard Materials

Similar high pressure and high temperature conditions are also adopted for preparation of super hard materials. Cubic BN is prepared from hexagonal BN under high pressure and temperature conditions, typically between pressure 5–10 GPa and temperature 1500–300 K [64]. Demazeau has reviewed wide varieties of conditions and fluxes, like metallic Li, Mg, Ca, Sb, Sn, Pb, fluorides, like NaF, $LiBF_4$, NH_4F, $(NH_4)_2SiF_6$, nitrides Li_3N, Mg_3N_2, Ca_3N_2, etc. for assisting the transformation of h-BN to c-BN under pressure and temperature [64].

Chen et al. [65] have prepared B_6O by heating a mixture of B, B_2O_3, and B_4C at temperatures of 1500–1900 K under pressure. The authors have prepared B_6O and B_6O-B_4C composites using the pressure in the range of 3–4.8 GPa inside a cubic press [65]. Single phase B_6O has been prepared by reacting B with B_2O_3 under high pressure and temperature. The mechanistic aspects of this synthesis indicate that B dissolves in B_2O_3 melt forming randomly distributed B_6O units which subsequently grows to larger crystallites with time [66]. The B_6O formed under high pressure and temperature has a rhombohedral (*R*-3*m*) structure which is formed by eight icosahedra at the vertices of a rhombohedral lattice and 2 oxygen atoms at the interstices aligned along $\langle 111 \rangle$ direction. Each icosahedron of B_{12} is formed by 12 B atoms and thus the net composition of the sub-oxide formed is B_6O. Due to shorter inter-atomic distances, they form strong covalent bond and exhibit superior hardness and wear-resistant and chemical inertness [66]. Typical synthesis is carried by multianvil using sintered octahedral of MgO as pressure transmitting medium. Pellets of homogenous mixture of B_2O_3 and amorphous boron is placed inside an h-BN capsule and placed insides graphite or $LaCrO_3$ heater and that is inserted into the octahedral MgO pressure transmitting medium. Pressure, time and temperature have been optimized to grow larger crystallite of B_6O. Crystallites of about 30 μM could be obtained at pressure 4 GPa, Temperature 1700 °C and time 2 h using the initial mixture of B and B_2O_3 as (16:3) [66]. Typical SEM images and crystal structure of B_6O are shown in Fig. 6.13. The B_6O crystals prepared at 5.5 GPa and 2100 °C for 60 min shows hardness equivalent to c-BN [67].

Nitrides of platinum like PtN_x has been prepared by heating Pt to 2000 K at 40–50 GPa pressure in a medium of N_2 in a laser-heated diamond anvil cells. The bulk modulus of PtN_x can reach upto 392 GPa depending on N content [69]. The N atoms are filled inside the voids of Pt lattice and form strong covalent bond with platinum [69]. Similarly, the platinum carbides (PtC) phase with high bulk modulus (301 GPa) has been prepared by laser-heated diamond anvil cell at 75 GPa. The

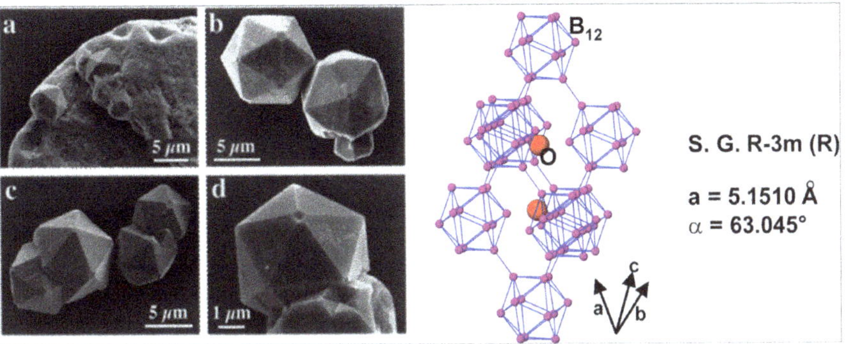

Fig. 6.13 SEM images of icosahedral B_6O crystals [66] and crystal structure of B_6O [68]. Reproduced with permission from H. Hubert, L. A. J. Garvie, B. Devouard, P. R. Buseck, W. T. Petuskey, and P. F. McMillan, Chem. Mater. 10(1998):1530–1537. Copyright (1998) American Chemical Society. Crystal structure of B_6O drawn from the data reported in Ref. [68]

high pressure phase of PtC has a cubic rock salt type structure and remains as a metastable phase on decompression to ambient condition [70]. The typical experiment is carried out by loading homogenous mixture of Pt and C powder inside a diamond anvil cell and pressured to around 70 GPa and Ar gas is used as pressure transmitting medium. Subsequently, the temperature is raised to about 2600 °C by either Nd-YLF ($YLiF_4$) or Nd–YAG ($Y_3Al_5O_{12}$) laser. The formation of PtC at these conditions is observed from the in situ XRD studies and also can be retained at ambient conditions by quenching the sample to ambient temperature. The metastable nature is confirmed by subsequent heating under pressure where it decomposes to metallic Pt and diamond above 47 GPa.

6.5.3 Compounds with Atoms of Inert Gas and Molecular Gas

The advancements of high pressure technology have been extensively exploited on molecular gases or inert gases to discover new compounds and new phases with unexpected physical or chemical properties. In general, gases are highly compressible and transforms to liquid or solid under compression and also new bonding is introduced in such systems [15]. Such new bonding leads to newer phases of the system. In case of molecular gases, instead of high temperature, often the low temperature is effective to bring the structural changes. Several of such molecular gases have been studied under high pressure and temperature in diamond anvil cell with external heating and cooling arrangements. The assembly containing the gas is cooled by cryostats or heated by resistive heaters. Also, laser heating is also employed to generate temperature in the pressure chamber. The high thermal conductivity of diamond can easily transfer the heat to the samples under study. In situ XRD and Raman spectroscopy are generally used to study the formation and structure of the phases. As typical examples, few cases on molecular gases are discussed here.

6.5.4 N_2 Molecules Under Pressure and Temperature

N_2 gas forms a solid at low temperature and high pressure, and shows series of structural transition with temperature or pressure. However, none of the high pressure polymorphs of N_2 can be stabilized at ambient conditions upon releasing the pressure. Yet the crystallization and phase transition of N_2 widen the understanding the effect pressure on breaking the strong interatomic bonds as well as delocalization of electrons. The pressure-temperature phase diagram reported for N_2 indicates the formation of a cubic (Pa3) type structure by quadrupole-quadrupole interaction of N_2 molecules at low temperature and moderate pressure [71]. With increasing pressure, viz. 11 GPa, this quadrupolar interactions become less dominating and hence structures with orientation disordered are formed [72, 73]. Also, it has been predicted formation of non-molecular metallic or insulating nitrogen crystal at pressure around 100 GPa [74, 75] and even at still higher pressure like 200 GPa, it suggests for structure and properties similar to other V group elements, like P and As [75–77].

6.5.5 O_2 Molecules Under Pressure and Temperature

The effect of high pressure and temperature on oxygen molecule is closely similar to that of N_2, but the difference in electronic configurations leads to ferromagnetic oxygen crystal. At normal pressure O_2 forms a monoclinic solid (α-O_2, space group C2/m) at low temperature while under pressure of about 5.5 GPa at room temperature O_2 forms a rhombohedral (β-O_2, space group R-3m) solid. On further increasing pressure to 9.6 GPa at ambient temperature, the β-O_2 transforms to an orthorhombic structure (α'-O_2, space group Fmmm) [78, 79]. Also, formation of several other phases of O_2 at higher temperature and pressure has been reported in literature [15]. Metallic solid O_2 has been prepared in DAC at around 95 GPa [80]. Further superconducting solid oxygen (ζ-O_2) with T_c 0.6 K has been prepared around 100 GPa [81]. The complex and novel phases of oxygen discovered experimentally by in situ high temperature and pressure studies summarized by Goncharov et al. [82] is depicted in Fig. 6.14.

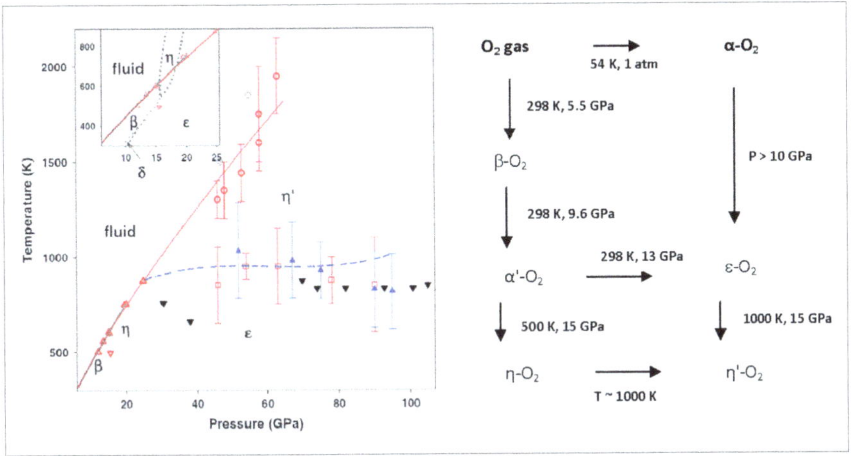

Fig. 6.14 Different phases of O_2 discovered by high pressure and temperature studies on molecular oxygen in laser-heated diamond anvil cell [82]. Reproduced with permission from A. F. Goncharov, N. Subramanian, T. R. Ravindran, M. Somayazulu, V. B. Prakapenka, and R. J. Hemley, Polymorphism of dense, hot oxygen, J. Chem. Phys. 135(2011):084512. Copyright (2011) American Institute of Physics publishing. Some of the metastable phases of O_2 identified by in situ XRD or Raman under HP and HT conditions are summarized in right hand side

6.5.6 CO_2 Molecule Under Pressure and Temperature

Novel phases of CO_2 have also been prepared by high pressure and temperature experiments in diamond anvil cell [83]. Despite CO_2 has a simpler molecular unit, it transforms to diversified crystalline and amorphous structures under pressure [83]. CO_2 can be solidified by cooling below 194 K at ambient pressure (1 atm) which has cubic structure (CO_2-I, Space Group Pa-3) and is well known as dry ice. This phase is a normal molecular solid and the structure is formed with linear CO_2 molecules and is stable below 11 GPa [83]. The CO_2-III phase can be prepared by applying pressure above 11 GPa and temperature from 293 to 450 K [84], which has an orthorhombic structure (Cmca). A new phase of CO_2 (CO_2-II) having density about 9.6% higher than CO_2-III phase has been prepared at pressure 28 GPa and 680 K [85]. The linear structure of CO_2 molecules are retained in this phase, but the alignment of the molecules occurs in such a way that a pseudo-six fold coordination can be assumed around the C atom. Park et al. [86] have prepared a new polymorph of CO_2 (CO_2-IV) by heating solid CO_2-III to temperatures between 500 and 750 K at pressures between 11 and 20 GPa. The structure of phase-IV is characterized with the bent CO_2 molecules and is closely related to α-cristobalite (SiO_2) and α-PbO_2. A disordered quartz-like phase of CO_2 (CO_2-V phase) having polymeric CO_2 units has been prepared from CO_2-III phase by heating to 1800 K under pressure of 40 GPa in laser-heated DAC [87].

Similarly, a number of new polymorphs of molecular materials, like H_2, ice, CO, etc. and also of several organic molecules, like CH_4, C_2H_2, etc. have also been synthesized by application of pressure and temperature [15].

6.5.7 Synthesis of Structures with Unusual Coordination

As mentioned earlier, the pressure increases the density and coordination number of atoms in a structure. The increasing compactness of the structures brings the molecules closer together and thus new bonds are created, rearrangements of bonds occur, and also the nature of bonds is altered. Thus, high pressure and high temperature synthesis becomes a routine for synthesis of compounds with unusual coordination number. In some cases, the new compounds are retained as stable form even on releasing the pressure and temperature while in some cases they form only under the pressurized and/or high temperature conditions. For examples, the polymorphs of O_2, N_2 and CO_2 are stable under the pressurized conditions only while several complex and multi-elements compounds are stable even at ambient conditions. The stishovite type SiO_2 and GeO_2 can be prepared by heating under pressure above 20 GPa [88, 89]. Some of such unusual compounds are discussed in this section.

6.5.7.1 Perovskites

It is well known that the structure of ABX_3 type perovskites is formed only with selected combination of cations, one suitable to form octahedral coordination while the other suitable to form 12 coordinated polyhedra. Thus, for ideal perovskite structure, the tolerance factor defined as: $t = \frac{r_A + r_X}{\sqrt{2} \times (r_B + r_X)}$, where r_A and r_B are ionic radii of cations and r_X is ionic radius of anion governs the formation boundary. For the ideal perovskite structure $t = 1.0$, but as it deviates from 1, the structure is retained with some distortion. Since the structure essentially needs cations with ionic radii suitable to form a framework of BX_6 octahedra, and thus there exists a wider choice for *A* cations. Thus, the application pressure is often useful to prepare perovskites with smaller ionic radii cations, where the increased coordination number of cations helps to stabilize the perovskite structures. Due to such effect of pressure, a number silicates also exist as perovskites with unusual SiO_6 octahedra in deep geological conditions [90]. In addition, perovskite-type materials with cations in higher and unstable oxidation states can also be prepared under such conditions. The synthesis of such materials can also be carried out under high gas pressure and temperature or using a strong oxidant like $KClO_3$ in a confined reaction medium. Some of the perovskite-type compounds prepared under high pressure and temperature conditions are explained in this section.

Silicates of Ca and Mg exist as an appreciable fraction of earth crust, about 7% by volume. These are formed by the reaction of the Ca^{2+} and Mg^{2+} ions with silica. Under ambient pressure solid-state reaction of CaO or MgO with SiO_2 leads to pyroxene type $MgSiO_3$ or $CaSiO_3$, where the structure is formed by the SiO_4 tetrahedral units. The pyroxene type $MgSiO_3$ and $CaSiO_3$ show pressure-induced structural transition and then they undergo amorphization or decomposition. In all such cases the tetrahedral SiO_4 configuration is retained. However, the reaction carried out under pressure leads to the perovskite-type $MgSiO_3$ or $CaSiO_3$ with SiO_6 octahedral units.

The observation of perovskite-type $MgSiO_3$ and $CaSiO_3$ in deep mantle has motivated large number P-V-T studies to understand their formation and stabilities. Above 10 GPa pressure and temperature above 1200 K in a laser-heated DAC, $CaSiO_3$ forms a perovskite structure which has a pseudo-tetragonal (c/a $\sim$0.007) structure [91]. Similarly, the ambient pyroxene phase of $MgSiO_3$ can be converted to perovskite phase by the application of HP and HT [92]. For the synthesis of perovskite-type $MgSiO_3$, Yagi et al. [92] have pressurized pyroxene type $MgSiO_3$ in diamond anvil cell to about 27 GPa and then heated to about 1000 °C by using infrared laser-like Nd–YAG. In general the formation pressure of orthorhombic perovskites ($Z = 4$) increases with decreasing the unit cell volume, but the temperature for reaction is always higher than 1000 °C. The formation pressures of several perovskites have been summarized by Yagi et al. [92] as in Fig. 6.15.

In addition to the normal ABO_3 perovskites, a large number of cation-ordered double perovskites have also been synthesized by high pressure and high temperature conditions. Leinenweber et al. [93] have prepared cation ordered Ca_2TiSiO_6 perovskite with ordered SiO_6 and TiO_6 octahedra. The high pressure and high temperature conditions for the formation of several perovskites with cation ordering, vacant cation sites, and with layered structures are reviewed by Rodger et al. [94].

In recent years, wide varieties of perovskites type materials have been prepared in LVP presses, like belt apparatus and multi-anvil press. Belik [95] has reviewed the synthesis of several bismuth-based perovskite-type materials with transition metal ions exhibiting multiferroic and magnetodielectric properties. A number of $BiMO_3$, with wide varieties of cation as M ion have been prepared by high pressure and temperature process. Typical synthesis conditions for some of the $BiMO_3$ materials are summarized in Table 6.5 and structure $BiAlO_3$ prepared under such conditions is shown in Fig. 6.16.

Similarly, the use of high oxygen pressure in the solid-state reactions is also useful to prepare perovskite-type compounds with cations in unusually higher oxidation states. Perovskite type $RNiO_3$ or $RCoO_3$, with R = Sm, Eu, Gd, Dy, Ho, Y have been prepared by under high oxygen pressure about 200 bar or under hydrostatic pressure in anvils using $KClO_4$ as oxygen source [97, 98]. Demazeau and Darracq have prepared $La_{1-x}Sr_xCuO_3$ with Cu^{3+} and Cu^{4+} using high oxygen pressure at high temperature [99].

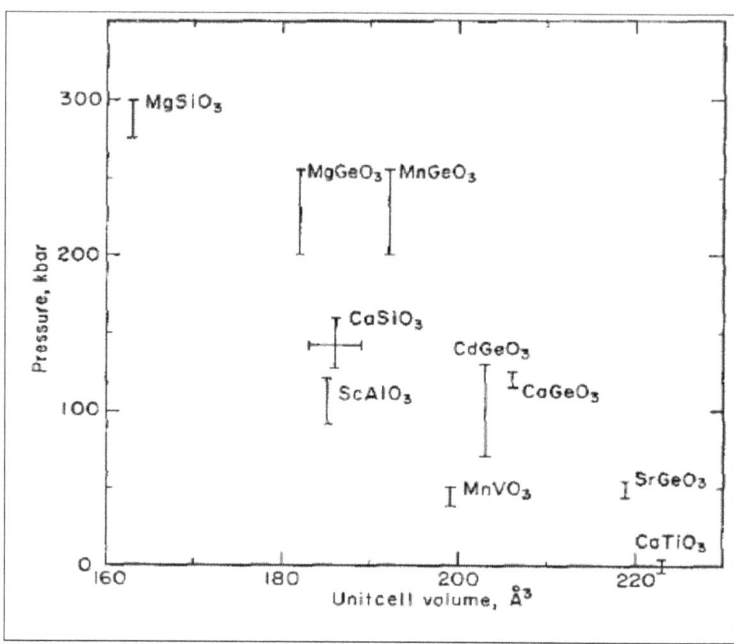

Fig. 6.15 Formation pressure of orthorhombic (Pnma) perovskite-type ABO_3 compounds [92]. Reproduced with permission from T. Yagi, H.-K. Mao and P. M. Bell, Structure and crystal chemistry of perovskite-type $MgSiO_3$, Phys. Chem. Minerals 3(1978):97–110. Copyright (1978) Elsevier. Inc

6.5.7.2 Pyrochlore Type Materials

As the case of the perovskites, the $A_2B_2O_7$ compounds with pyrochlore-type structures are also formed only with suitable cations. Subramanian et al. [100] have proposed the tolerance limits for the pyrochlore structure from the radius ratio of A and B type cations (r_A/r_B). For a normal cubic pyrochlore structure, the tolerance limit is predicted as: $1.46 \leq r_A/r_B \leq 1.80$. The cation pairs having r_A/r_B lower than 1.46 forms cation disordered fluorite structure due to their indiscriminate behavior for forming eight coordinated polyhedra while for the pair having r_A/r_B higher than 1.90 tend to form perovskite or other structures depending on the preference for ideal coordination polyhedra of A and B cations. For example, for $La_2Ti_2O_7$, has perovskite type structure due to preference of La to form larger coordination polyhedra than eight required for pyrochlore structures. The $Ln_2Si_2O_7$ (Ln = Lanthanides) form pyrosilicates type structures due to preference of Si^{4+} for tetrahedral coordination. However, the upper limit of tolerance can be extended up to 2.3 by adopting high pressure and high temperature synthesis procedures. This can be attributed to the favorable conditions introduced by pressure to increases the coordination number of the smaller B site cations [101, 102].

Table 6.5 Typical preparation conditions for some Bi-based perovskites [95]

Sample	Reactants	Temp. (K)	Pressure (GPa)	Time (min)	Symmetry	Ref.
$BiAlO_3$	$Bi_2O_3 + Al_2O_3$ pretreated at 1023 K for 8 h	1273	6	40	R3c, $Z = 6$	i
$BiGaO_3$	$Bi_2O_3 + Ga_2O_3$ pretreated at 1023 K for 8 h	1473	6	15 min and then 3.2 GPa at RT	Cm, $Z = 2$	ii
$BiScO_3$	$Bi_2O_3 + Sc_2O_3$	1413	6	40	C2/c, $Z = 8$	iii
$BiInO_3$	$Bi_2O_3 + In_2O_3$	1273	6	80	Pna21, $Z = 4$	iv
$BiMnO_3$	$Bi_2O_3 + Mn_2O_3$	1383	6	70 s	C2/c, $Z = 8$	v
$BiCrO_3$	$Bi_2O_3 + Cr_2O_3$	1413	6	40	C2/c, $Z = 8$	iii
$BiCoO_3$	Bi_2O_3, Co_3O_4, and $KClO_3$ (9:6:1 molar ratio)	1243	6	100 min	P4mm, $Z = 1$	vi
$BiNiO_3$	Bi_2O_3 and Ni dissolved in HNO_3 and dried and heated 750 °C for 12 h (Precursor) Precursor + $KClO_4$ (4:1 molar ratio)	1273	6	30 min	$P - 1$, $Z = 4$	vii

Data compiled with permission from A. A. Belik, Polar and nonpolar phases of $BiMO_3$: a review, J. Solid State Chem. 195(2012):32–40. Copyright (2012), Elsevier. Inc

(i) A. A. Belik, T. Wuernisha, T. Kamiyama, K. Mori, M. Maie, T. Nagai, Y. Matsui and E. Takayama-Muromachi, Chem. Mater. 18(2006):133

(ii) H. Yusa, A. A. Belik, E. Takayama-Muromachi, N. Hirao, Y. Ohishi, Phys. Rev. B, 80 (2009):214103

(iii) A. A. Belik, S. Iikubo, K. Kodama, N. Igawa, S.-i. Shamoto, M. Maie, T. Nagai, Y. Matsui, S. Yu. Stefanovich, B. I. Lazoryak and E. Takayama-Muromachi, J. Am. Chem. Soc., 128(2006):706

(iv) A. A. Belik, S. Yu. Stefanovich, B. I. Lazoryak, E. Takayama-Muromachi, Chem. Mater., 187 (2006):1964

(v) A. A. Belik, S. Iikubo, T. Yokosawa, K. Kodama, N. Igawa, S. Shamoto, M. Azuma, M. Takano, K. Kimoto, Y. Matsui and E. Takayama-Muromachi, J. Am. Chem. Soc., 129(2007):971

(vi) A. A. Belik, S. Iikubo, K. Kodama, N. Igawa, S.-i. Shamoto, S. Niitaka, M. Azuma, Y. Shimakawa, M. Takano, F. Izumiand, E. Takayama-Muromachi, Chem. Mater. 18(2006):798

(vii) S. Ishiwata, M. Azuma, M. Takano, E. Nishibori, M. Takata, M. Sakata and K. Kato, J. Mater. Chem., 12(2002):3733

A number cubic pyrochlore with such unusual combination of cations has been successfully prepared by using high pressure and high temperature techniques. Sleight et al. have explored the synthesis of pyrochlore type materials with unusually smaller cations, like Si^{4+}, Ge^{4+}, Pb^{4+}, Pt^{4+} and Pd^{4+} etc. as B cation [102]. Shannon and Sleight have proposed requirement of high pressure and temperature conditions the synthesis $A_2B_2O_7$ type cubic pyrochlores with several unusual cations [103]. The stability field for $A_2B_2O_7$ pyrochlore proposed by Shannon and Sleight is depicted in Fig. 6.17 [103].

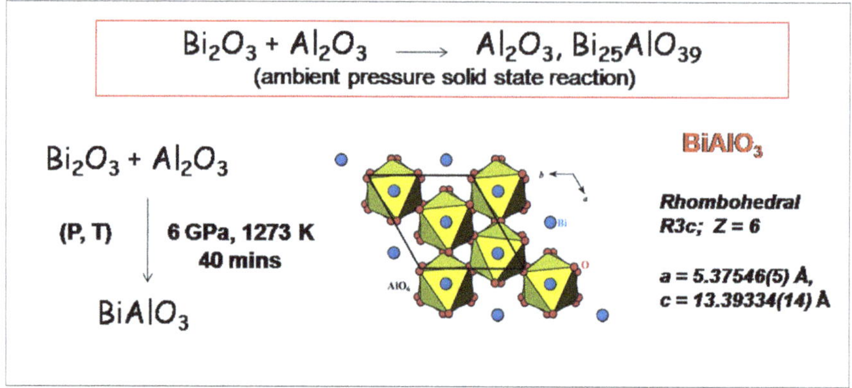

Fig. 6.16 Preparation details and structure of perovskite type $BiAlO_3$ [96]. Adapted with permission from "A. A. Belik, T. Wuernisha, T. Kamiyama, K. Mori, M. Maie, T. Nagai, Y. Matsui and E. Takayama-Muromachi, High-pressure synthesis, crystal structures, and properties of perovskite-like $BiAlO_3$ and pyroxene-like $BiGaO_3$, Chem. Mater. 200618(2006):133–139". Copyright (2006) American Chemical Society

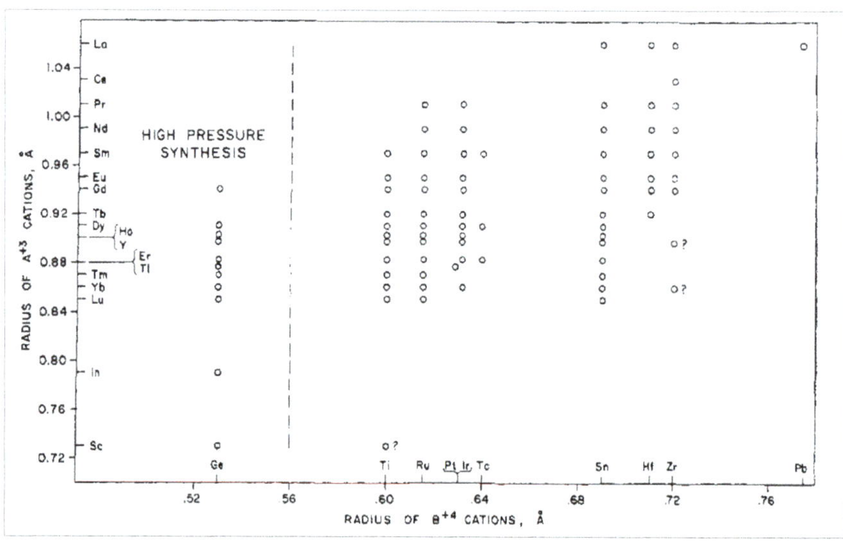

Fig. 6.17 Stability field of cubic $A_2B_2O_7$ pyrochlore as proposed by Shanon and Sleight [103]. Reproduced with permission from R. D. Shannon and A. W. Sleight, Synthesis of new high pressure pyrochlore phases, Inorganic Chem. 7(1968):1649–1651. Copyright (1968) American Chemical Society

Silicates like $Sc_2Si_2O_7$ and $In_2Si_2O_7$ with pyrochlore type structure have been prepared by Reid et al. [104] using high pressure and high temperature techniques. Reid et al. [104] have used thortveitite type $Sc_2Si_2O_7$ or $In_2Si_2O_7$ prepared by high temperature solid-state reaction as starting material. Finely powdered thortveitite type silicates were pressed to 10 GPa and then heated at 1000 °C for 5 min. Pyrochlore type manganates, like $Y_2Mn_2O_7$, $Ho_2Mn_2O_7$, and $Yb_2Mn_2O_7$ have been prepared by heating R_2O_3 (R–Y, Ho, Yb) and MnO_2 in a sealed gold tube at 500 °C in a hydrostatic pressure of 3 kbar [105]. The stability of Mn^{4+} is enhanced by the pressure and hence retained the pyrochlore structure instead of forming the $RMnO_3$ type hexagonal perovskite. Cubic pyrochlore type cubic $R_2Pt_2O_7$ (R = Er, Yb) have been prepared from R_2O_3 and PtO_2 mixture by heating at 1000 °C under a pressure of 4 GPa in a multi-anvil press [106]. Cubic $Cd_2Ru_2O_7$ and $Cd_{2-x}Ca_xRu_2O_7$ pyrochlores with Ru^{5+} have been prepared by heating stoichiometric mixture of CdO, CaO, and RuO_2 in a sealed gold capsule at 4 GPa and 1000 °C for 30 min in a multi-anvil set up [107]. The reactions are carried out in presence of 20–30 wt% of $KClO_4$ to provide excess oxygen for oxidizing Ru^{4+} to Ru^{5+}.

6.5.8 Metastable ABX_4 Type Compounds

The ABX_4 type compounds are known to exist in diversified structures depending on the nature and ionic radii of A and B cation and X anions. The cation ordered compounds can form zircon, scheelite, fergusonite, wolframite, monazite, barite, and their related structures as well as varieties of silica, rutile, and fluorite related structures [108]. Usually, in all such structures, the coordination numbers of cations are four or more than four. The cation ordering is related to the difference in the nature of the A and B cations, like electronic configuration, ionic radii and oxidation states, and usually they form different coordination polyhedra and at least symmetrically different crystallographic sites. In ABX_4 type compounds, with B as B^{3+}, Si^{4+}, Ge^{4+}, P^{+5}, V^{5+}, Cr^{6+}, S^{6+}, Se^{6+}, W^{6+}, Mo^{6+}, Te^{6+}, Tc^{6+}, Re^{6+} etc., a tetrahedrally coordinated polyhedra is formed around them due to the smaller ionic radii. However, in ABX_4 type compounds, the Mo^{6+}, W^{6+}, Te^{6+}, Re^{6+} etc. as V site can have octahedral coordination, and still larger ions as B site cation can form even eight coordinated polyhedra. The coordination polyhedra around the A cation varies from 4 to 12 depending upon ionic radii of the cations. Due to wider choices of cations and amenability to accommodate wide varieties of oxidation states and ionic radii, the ABX_4 type materials are widespread in nature, and also have been prepared in laboratory. As an example, the rare-earth phosphates or vanadates are formed with tetrahedrally coordinated P^{5+} or V^{5+} ions while the coordination polyhedra around the rare-earth ion are usually 8 + 1 or 8. With eight coordinated lanthanide ions also the structures are formed as zircon or scheelite which are related to the connections of AO_8 or BO_4 polyhedra. The rare-earth phosphates or vanadates, in general, exists either in zircon (xenotime) or monazite type structures, viz. larger cations like La, Ce, Pr, Nd, etc, monazite is preferentially formed while rest forms zircon or xenotime structures. Since all the structures in ABX_4 type compounds are related with respect to the coordination polyhedra around the cations and their packing, pressure plays a crucial role in synthesizing these compounds and hence pressure has been also extensively used to prepare different metastable polymorphs of such compounds. In this section, several examples of synthesis of some metastable ABX_4 type compounds are discussed. Also, only oxide ion as X site anion is considered. A brief discussion of ABX_4 type fluorides are presented in another chapter.

From the ionic radius of V^{5+}, the stable coordination polyhedron around it is tetrahedra. However, on application of pressure the coordination polyhedra around V^{5+} can increase reversibly or irreversibly to octahedra. Similarly, higher coordination polyhedra around P^{5+}, Si^{4+} has also been observed under pressure. However, the stabilities are less in such cases, and lowering the pressure the original tetrahedral configuration reverts back. The pressure-induced irreversible transformations of VO_4 to VO_6 units have been observed in $InVO_4$ and $FeVO_4$ [109, 110]. The $FeVO_4$ and $InVO_4$ with wolframite type structure can be prepared by application of pressure itself at ambient temperature the pressure. Typical preparation of $InVO_4$ with octahedrally coordinated V^{5+} from the normal $InVO_4$ is explained in Fig. 6.18 [109].

Similar pressure-induced irreversible natures of zircon to scheelite or zircon to monazite transition under pressure have also been employed to prepare metastable scheelite and monazite type phases of a number of rare-earth vanadates [111–119]. All these transitions do not require temperature and hence can be obtained by applying pressure in diamond anvil, Bridgman or belt apparatus. All the vanadates of rare-earth earth ions except La^{3+} prepared by solid-state reactions have zircon type tetragonal structure at ambient conditions. But the scheelite type rare-earth vanadates for all the rare-earth earth ions can be prepared by compressing them to certain pressure. The transition pressure is dependent on the ionic radii of the rare-earth ions, and smaller the ionic radii larger the pressure required for transition. Also, the vanadates with larger rare-earth ions transform preferentially to monazite type structure than the zircon structures. The preparation procedure for the monazite type $PrVO_4$ and scheelite type $SmVO_4$ is shown in Fig. 6.19 [111].

Similarly, $LiScF_4$ composition does not form a single-phase compound due to smaller ionic radius of Sc^{3+} ions and preferentially form Li_3ScF_6 and ScF_3 mixture [116]. However, under high pressure and temperature conditions $LiScF_4$ form a scheelite type structure with ScF_8 bisdisphenoid and LiF_4 tetrahedra. More details on the synthesis of LiScF4 and preparation of newer phases of $NaYF_4$ are explained in the chapter "Synthesis of inorganic fluorides" by Achary et al. of this book.

6.5.9 *Other Miscellaneous Metastable Phases*

The complex oxides with relatively higher density and closely packed structure can be carried out by using both pressure and temperature. The reconstruction of the

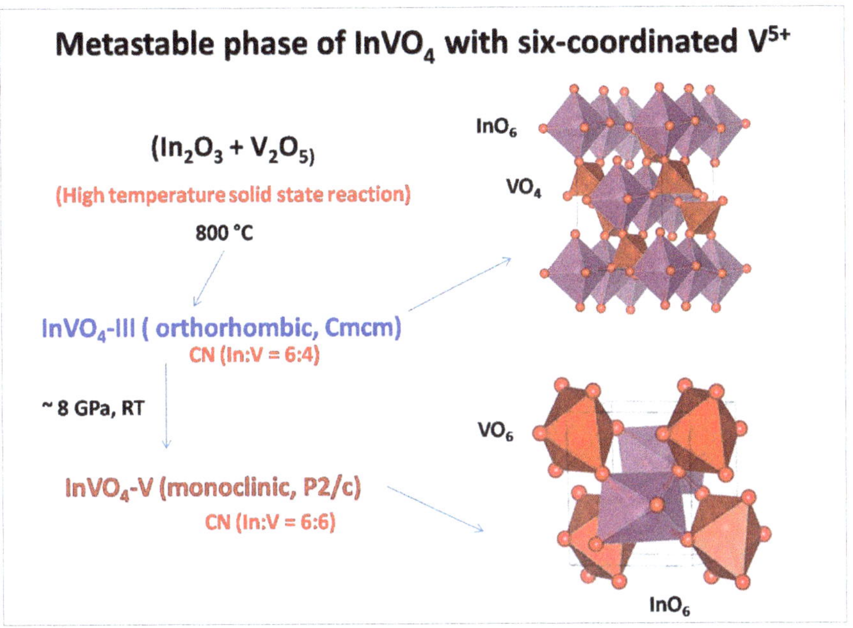

Fig. 6.18 High-pressure metastable modification of $InVO_4$. The stable orthorhombic phase transforms to a metastable wolframite type phase around 6 GPa. The structural details of $InVO_4$ orthorhombic phase are: $a = 5.75194(5)$ Å, $b = 8.52148(8)$ Å, $c = 6.58426(6)$ Å; $V = 322.728(5)$ Å^3, $Z = 4$; InO_6 octahedra: In-O2 = 2.200(4) × 2; In-O1 = 2.162(3) × 4; VO_4 tetrahedra: V-O1 = 1.785(5) × 2; V-O2 = 1.593(5) × 2 and the structural details of monoclinic phase at 8.2 GPa are: $a = 4.714(5)$ Å, $b = 5.459(6)$ Å, $c = 4.903(5)$ Å, $\beta = 93.8(3)$ Å; $V = 125.89(2)$ Å^3, $Z = 2$; InO_6 octahedra: In-O1 = 2.027(6) Å (× 2); In-O2 = 2.140(5) (× 2); In-O2 = 2.210(6) (× 2); VO_6 octahedra: V-O1 = 1.673(5) Å (× 2), V-O2 = 1.886(5) Å (× 2), V-O2 = 2.217(6) Å (× 2). In both structures the InO_6 octahedra share two opposite edges of forming linear chains while the VO_4 tetrahedra links these chains forming a three-dimensional structure in orthorhombic $InVO_4$. Under pressure, as the lattice is compressed, the tetrahedral VO_4 comes closer together and they are connected each other forming two additional bonds. Thus, a linear chain of VO_6 octahedra is formed which link the InO_6 chains as in wolframite structure. This high pressure phase partially transforms to ambient phase on decompression. The retention of HP phase under ambient conditions is due to the first order reconstructive transition [109]. Adapted with permission from D. Errandonea, O. Gomis, B. García-Domene, J. Pellicer-Porres, V. Katari, S. N. Achary, A. K. Tyagi, C. Popescu, New polymorph of $InVO_4$: a high pressure structure with six-coordinated vanadium, Inorg. Chem., 52(2013):12790–12798. Copyright (2013) American Chemical Society

crystal structures in the transition often favor to stabilize the high pressure polymorphs by lowering the pressure and temperature slowly to ambient conditions or by quenching the temperature to ambient conditions. A typical example of preparation of monoclinic $HfMo_2O_8$ by high pressure and thigh temperature [119–121] is explained here. Under ambient pressure, the high temperature solid state reaction of HfO_2 and MoO_3 leads to the formation a trigonal phase of $HfMo_2O_8$. Finely powdered sample of trigonal $HfMo_2O_8$ is filled inside a BN cup and placed inside a

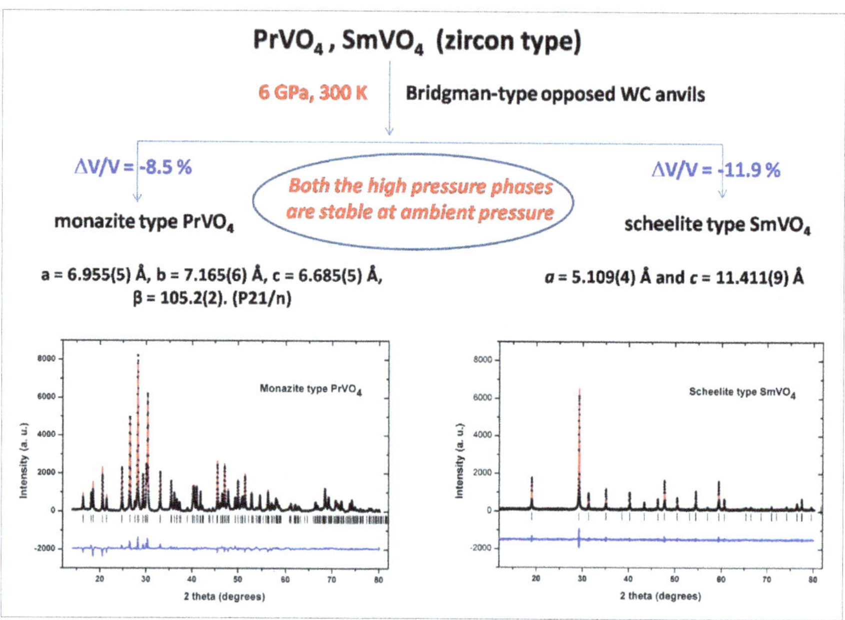

Fig. 6.19 Preparation procedure of metastable monazite type $PrVO_4$ and scheelite type $SmVO_4$. The preparation is carried out in a Bridgman anvil using the zircon type samples as starting materials inside a pyrophyllite gasket. However, materials transforming at higher pressure, the experiments are always carried out in DAC. Also, it can be mentioned here that the transition can also change with the degree of anhydrostaticity and that may also lead to formation of orthorhombic $BaSO_4$ type structures. Most of the time, the phases observed at very high pressure are not retained on releasing pressure [111, 112, 115]. The Rietveld refinement plots of XRD patterns of monazite type $PrVO_4$ and scheelite type $SmVO_4$ recorded at ambient condition using CuK_α radiation is shown. The reaction scheme and Rietveld refinements of XRD data are based on the data reported in Ref. [111]

graphite heater. The assembly is pressurized in a torroid anvil cell. Torroidal pyrophyllite gaskets were used to transmit pressure to the sample. Pressure is increased slowly to 2.15 GPa and then the temperature is increased to 560 °C by applying current to the graphite heater and held for 15 min, and then quenched by putting off the power to the heater. The product is characterized by a monoclinic lattice (Sp. Gr. C2/c) with unit cell parameters $a = 11.4138(6)$, $b = 7.9105(4)$, $c = 7.4395(3)$ Å and $\beta = 122.35(0)$, $V = 567.45(5)$ Å^3, $Z = 4$. Typical crystal structures of trigonal and monoclinic polymorphs of $HfMo_2O_8$ are shown in Fig. 6.20. Compared to the ambient pressure trigonal phase, the prepared HP phase is about 20% more denser, and is a metastable phase. Also, the MoO_4 tetrahedral units transforms to MoO_5 units in the transition. On heating to 700 °C this phase reverts back to the ambient pressure phase [119, 121].

Also, a complete rearrangement of structures without any change in coordination number can also happen in high pressure and high temperature synthesized

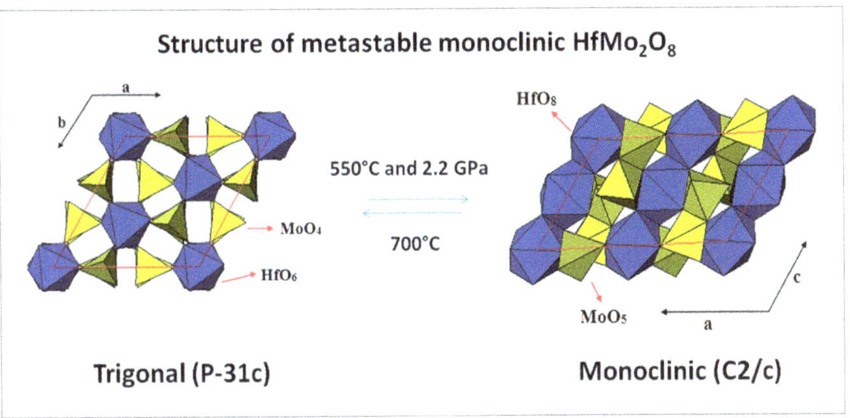

Fig. 6.20 Crystal structures of ambient pressure trigonal and monoclinic high pressure polymorph of $HfMo_2O_8$. The trigonal phase is formed by the regular HfO_6 octahedral and MoO_4 tetrahedral units. HfO_6 units share all the corner oxygen atoms with six MoO_4 units while the MoO_4 units share three of their corner oxygen atoms. This arrangement leads to a sheet $HfMo_2O_8$ and they are stacked along the *c*-axis of hexagonal unit cell of trigonal lattice. Such arrangements lead to a loosely packed sheet-type structure for trigonal $HfMo_2O_8$. Under pressure, this structure shows anisotropic compression and transitions due to rotation of polyhedra which reverts back to original phase on releasing pressure. However, on application of temperature under pressure, the structural rearrangements occur by formation of new bonds and thus transforms to a quenchable denser polymorph, monoclinic $HfMo_2O_8$. The sheet stricture transforms to close-packed stricture by increasing the coordination number of Hf to 8 and Mo to 5. Two of MoO_5 units are also linked together by sharing one edge and thus forming stable Mo_2O_{10} dimer units. The monoclinic $HfMo_2O_8$ is a metastable that can be transformed to stable trigonal phase by heating to overcome the thermodynamic activation barrier [119–121]. The crystal structures are redrawn using the data reported in Ref. [119]

materials. The reorientation or rearrangement of polyhedral and distortions in lattices form various kinds of high pressure metastable phases [122–125]. The distortion-induced transitions are usually stable only under pressure and revert to stable phases on releasing the pressure and often occur without any appreciable volume discontinuity. In certain cases, such transitions can also occur with a reconstruction and they can be stabilized at ambient conditions. A case of preparation of tetragonal $NbOPO_4$ phase by high pressure and high temperature reactions of the ambient monoclinic $NbOPO_4$ is depicted in Fig. 6.21.

6.6 Summary

In summary, this chapter presents a brief overview on the high pressure and high temperature method for preparation of newer and metastable compounds. The effect of pressure on the reaction has been explained briefly. The general features for the

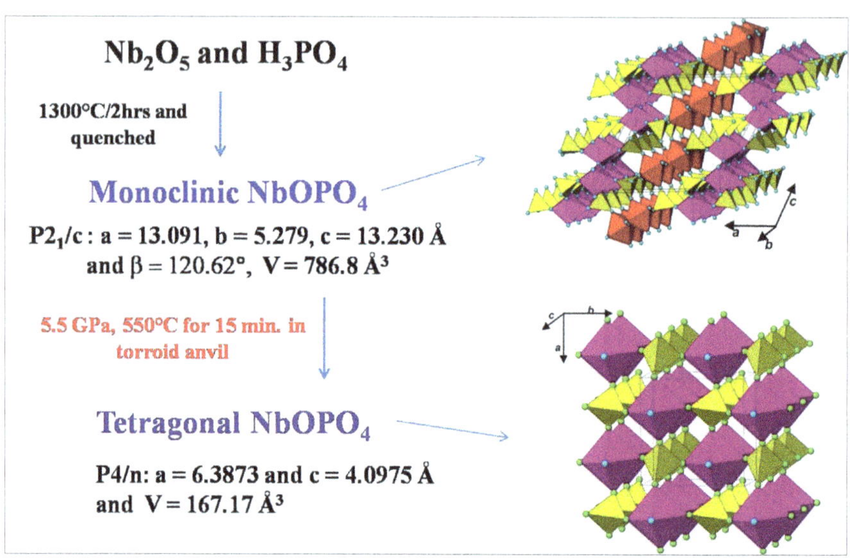

Fig. 6.21 Preparation of tetragonal $NbOPO_4$ from monoclinic $NbOPO_4$. Both the structures have octahedral NbO_6 and PO_4 units, but the higher symmetric tetragonal structure is formed by reorientation of the NbO_6 octahedra. In monoclinic $NbOPO_4$, the octahedral NbO_6 units are connected by sharing adjacent edges and they transformed to opposite edge connected octahedral NbO_6 octahedral units. This occurs with an appreciable volume reduction leading to about 18% increase in density [124]. The crystal structures and reaction scheme are based on data reported in Ref. [124]

high pressure phases are explained. Also, typical experimental setups used for high pressure and high temperature reactions are explained. The scopes of getting unusually coordinated compounds and hitherto unknown compounds which can be prepared by such method are explained. Several case studies ranging from molecular solids to dense inorganic materials are explained. In this chapter, the physical or chemical properties of the high pressure phases are not discussed to limit the chapter within the scope of synthesis. The rapid development of machinery, devices, and experimental methodologies is making the high pressure and high synthesis method is one of the most common preparation techniques.

References

1. Cheetham AK, Day P (1992) Solid state chemistry: volume II: compounds. Clarendon Press. ISBN-10: 0198551665
2. Cheetham AK, Day P (1987) Solid state chemistry: volume 1: techniques. Oxford University Press, Oxford. ISBN: 0198552866
3. Rao CNR (1995) Chemical approaches to the synthesis of inorganic materials. Wiley International. ISBN: 978-0-470-23431-0

4. Xu R, Xu Y (2017) Modern inorganic synthetic chemistry. Elsevier, Amsterdam
5. Badding JV, Parker LJ, Nesting DC (1995) J Solid State Chem 117:229
6. Hazen RM, Finger LW, Mariathasan JWE (1985) J Phys Chem Solids 46:253
7. Herbst E (1995) Annu Rev Phys Chem 46:27
8. Bertucco A, Vetter G (2001) High pressure process technology: fundamentals and applications. Elsevier Science B.V. Amsterdam, The Netherlands
9. Degarmo EP, Black JT, Kohser RA (2011) Materials and processes in manufacturing, 11th edn. Wiley. ISBN: 978-0-470-92467-9
10. Smelt JPPM (1998) Trends Food Sci Technol 9:152
11. Manaa MR (2005) Chemistry under extreme conditions. Elsevier Science, New York
12. Grochala W, Hoffmann R, Feng J, Ashcroft NW (2007) Angew Chem Int Ed 46:3620
13. McMillan PF (2006) Chem Soc Rev 35:855
14. Schettino V, Bini R (2007) Chem Soc Rev 36:869
15. Hemley RJ (2000) Ann Rev Phys Chem 51:763
16. Sorb YA, Bundy FP, DeVries RC (2016) Diamond: high-pressure synthesis, reference module in materials science and materials engineering. Elsevier Inc. https://doi.org/10.1016/b978-0-12-803581-8.03772-3
17. Bridgman PW (1946) Rev Mod Phys 18:1
18. Boyd FR, England JL (1960) J Geophys Res 65:741
19. Hall HT (1958) Rev Sci Instrum 29:267–275
20. Bridgman PW (1941) Proc Amer Acad Arts Sci 74:21
21. Bridgman PW (1935) Phys Rev 48:893
22. Bundy FP, Hall HT, Strong HM, Wentorf RH (1955) Nature 176:51
23. Wentorf Jr RH (1957) J Chem Phys 26:956
24. Hall HT (1960) Rev Sci Instrum 31:125
25. Sharma SM, Garg N (2017) Material studies at high pressure, in materials under extreme conditions [Tyagi AK, Banerjee S (eds)]. Elsevier Inc
26. Jayaraman A (1983) Rev Mod Phys 55:65
27. Sahu PC, Chandra Shekar NV (2007) Resonance 12:10
28. Subramanian N, Chandra Shekar NV, Sanjay Kumar NR, Sahu PC (2006) Curr Sci 91:175
29. Dubrovinsky L, Dubrovinskaia N, Prakapenka VB, Abakumov AM (2012) Nat Comm 3:1163(1–7)
30. Sumiya H, Harano K (2012) Diamond and related materials. Diam Relat Mater 24:44
31. Dubrovinskaia N, Dub S, Dubrovinsky L (2006) Nano Lett 6:824
32. Nakamoto Y, Sakata M, Sumiya H, Shimizu K, Irifune T, Matsuoka T, Ohishi Y (2011) Rev Sci Instrum 82:066104(1–2)
33. Irifune T, Isobe F, Shinmei T (2014) Phys Earth Planet Inter 228:255
34. Liebermann RC (2011) High Press Res 31:493
35. Warxrnr D, Canpnurnn MA, Hrrcn CM (1990) Am Mineral 75:1020
36. Wioeniewski R (1999) Pressure sensors. In: Webster JG (ed) Wiley encyclopedia of electrical and electronics engineering. Wiley. https://doi.org/10.1002/047134608X.W3999
37. Fuller PJA, Price JH (1962) Nature 103:263
38. Bridgman PW (1950) Proc Roy Soc A 203:1
39. Fujioka N, Mishima O, Endo S, Kawai N (1978) J Appl Phys 49:4830
40. Rein C (1993) Meas Sci Technol 4:1194
41. Giardini AA, Samara GA (1965) J Phys Chem Solid 26:1523
42. Yu Tonkov E, Ponyatovsky EG (2000) Phase transformations of elements under high pressure. CRC press LLC, Florida, Boca Raton
43. McQueenand RG, Marsh SP (1960) J Appl Phys 31:1253
44. Fei Y, Angel R, Frank M, Mibe K, Shen G, Prakapenka V (2007) PNAS USA 104:9182
45. Decker DL (1965) J Appl Phys 36:157
46. Brown JM (1999) J Appl Phys 86:5801
47. Zeto RJ, Vanfleet HB (1969) J Appl Phys 40:2227
48. Beavitt AR (1969) J Phys D Appl Phys 2:1675

49. Shen YR (1991) High Press Res 7:73
50. Datchi F, Dewaele A, Loubeyre P, Letoullec R, Le Godec Y, Canny B (2007) High Press Res 27:447
51. Trots DM, Kurnosov A, Ballaran TB, Zhuravlev STK, Prakapenka V, Berkowski M, Frost DJ (2013) J Geophys Res 118:5805
52. Chijioke AD, Nellis WJ, Soldatov A, Silvera IF (2005) J Appl Phys 98:114905
53. Bauer JD, Bayarjargal L, Winkler B (2012) High Press Res 32:155
54. Syassen K (2008) High Press Res 28:75
55. Datchi F, LeToullec R, Loubeyre P (1997) J Appl Phys 81:3333
56. Dewaele A, Torrent M, Loubeyre P, Mezouar M (2008) Phys Rev B 78:104102
57. Mao HK, Xu J, Bell PM (1986) J Geophys Res 91:4673
58. Ragan DD, Gustavsen R, Schiferl D (1992) J Appl Phys 72:5539
59. Narayan J, Bhaumik A (2015) J Appl Phys 118:215303
60. Bundy FP, Bassett WA, Weathers MS, Hemley RJ, Mao HU, Goncharov AF (1996) Carbon 34:141
61. Hu M, Bi N, Li S, Su T, Hu Q, Ma H, Jia X (2017) Cryst Eng Comm 19:4571
62. Guo L, Ma H, Chen L, Chen N, Miao X, Wang Y, Fang S, Yang Z, Fang C, Jia X (2018) Cryst Eng Comm 20:5457
63. Gotou H, Yagi T, Frost DJ, Rubie DC (2006) Rev Sci Instrum 77:035113
64. Demazeau G (1993) Diamond Relat Mat 2:197
65. Chen C, He D, Kou Z, Peng F, Yao L, Yu R, Bi Y (2007) Adv Mater 19:4288
66. Hubert H, Garvie LAJ, Devouard B, Buseck PR, Petuskey WT, McMillan PF (1998) Chem Mater 10:1530
67. He D, Zhao Y, Daemen L, Qian J, Shen TD, Zerda TW (2002) Appl Phys Lett 81:643
68. Bolmgren H, Lundstroem T, Okada S (1991) AIP Conf Proc 231:197
69. Gregoryanz E, Sanloup C, Somayazulu M, Badro J, Fiquet G, Mao HK, Hemley RJ (2004) Nat Mater 3:294
70. Ono S, Kikegawa T, Ohishi Y (2005) Solid State Commun 133:55
71. Manzhelii VG, Freiman YA (1997) Physics of cryocrystals. College Park, MD
72. Hanfland M, Lorenzen M, Wassilew-Reul C, Zontone F (1998) Rev High Press Sci Technol 7:787
73. Bini R, Jordan M, Ulivi L, Jodl HJ (1998) J Chem Phys 108:6849
74. McMahan AK, LeSar R (1985) Phys Rev Lett 54:1929
75. Martin RM, Needs R (1986) Phys Rev B 34:5082
76. Scheerboom MIM, Schouten JA (1993) Phys Rev Lett 71:2252
77. Mailhiot C, Yang LH, McMahan AK (1992) Phys Rev B 46:14419
78. Yagi T, Hlrsch KR, Holzapfel WB (1883) J Phys Chem 87:2272
79. Schiferl D, Cromer DT, Schwalbe LA, Mills RL (1983) Acta Cryst B 39:153
80. Desgreniers S, Vohra YK, Ruoff AL (1990) J Phys Chem 94:1117
81. Shimizu K, Suhara K, Ikumo M, Eremets MI, Amaya K (1998) Nature 393:767
82. Goncharov AF, Subramanian N, Ravindran TR, Somayazulu M, Prakapenka VB, Hemley RJ (2011) J Chem Phys 135:084512
83. Dziubek KF, Ende M, Scelta D, Bini R, Mezouar M, Garbarino G, Miletich R (2018) Nat Comm 9:3148
84. Simon A, Peters K (1980) Acta Crystallogr B 35:2750
85. Yoo CS, Kohlmann H, Cynn H, Nicol MF, Iota V, LeBihan T (2002) Phys Rev B 65:104103
86. Park J-H, Yoo CS, Iota V, Cynn H, Nicol MF, Le T (2003) Phys Rev B 68:
87. Iota V, Yoo CS, Cynn H (1998) Science 283:1510
88. Prakapenka VP, Shen G, Dubrovinsky LS, Rivers ML, Sutton SR (2004) J Phys Chem Solid 65:1537
89. Ono S, Hirose K, Murakami M, Isshikic M (2002) Earth Planet Sci Lett 197:187
90. Hirose K, Sinmyo R, Hernlund J (2017) Science 358:734
91. Shim S-H, Duffy TS, Shen G (2000) J Geophys Res B 105:25955
92. Yagi T, Mao H-K, Bell PM (1978) Phys Chem Miner 3:97

93. Leinenweber K, Parise J (1997) Am Miner 82:475
94. Rodgers JA, Williams AJ, Attfield JP (2006) Z Naturforsch 61b:1515
95. Belik AA (2012) J Solid State Chem 195:32
96. Belik AA, Wuernisha T, Kamiyama T, Mori K, Maie M, Nagai T, Matsui Y, Takayama-Muromachi E (2006) Chem Mater 18:133
97. Alonso JA, Martínez-Lope MJ, Casais MT, García-Muñoz JL, Fernández-Díaz MT, Aranda MAG (2001) Phys Rev B 64:941021
98. Alonso JA, Martínez-Lope MJ, de la Calle C, Pomjakushin V (2006) J Mater Chem 16:1555
99. Demazeau G, Darracq S (1994) High Press Res 12:323
100. Subramanian MA, Aravamudan G, Subba Rao GV (1983) Prog Solid State Chem 15:55
101. Zhou H, Wiebe CR (2019) Inorganics 7:49
102. Sleight AW (1968) Mater Res Bull 3:699
103. Shannon RD, Sleight AW (1968) Inorg Chem 7:1649
104. Reid AF, Li C, Ringwood AE (1977) J Solid State Chem 20:119
105. Greedan JE, Raju NP, Maignan A, Simon Ch, Pedersen JS, Niraimathi AM, Gmelin E, Subramanian MA (1996) Phys Rev B 54:7189
106. Cai YQ, Cui Q, Li X, Dun ZL, Ma J, dela Cruz C, Jiao YY, Liao J, Sun PJ, Li YQ, Zhou JS, Goodenough JB, Zhou HD, Cheng J-G (2016) Phys Rev B 93:014443
107. Jiao YY, Sun JP, Shahi P, Cui Q, Yu XH, Uwatoko Y, Wang BS, Alonso JA, Weng HM, Cheng J-G (2018) Phys Rev B 98:075118
108. Errandonea D, Manjón FJ (2008) Prog Mater Sci 53:711
109. Errandonea D, Gomis O, García-Domene B, Pellicer-Porres J, Katari V, Achary SN, Tyagi AK, Popescu C (2013) Inorg Chem 52:12790
110. López-Moreno S, Errandonea D, Pellicer-Porres J, Martínez-García D, Patwe SJ, Achary SN, Tyagi AK, Rodríguez-Hernández P, Muñoz A, Popescu C (2018) Inorg Chem 57:7860
111. Errandonea D, Achary SN, Pellicer-Porres J, Tyagi AK (2013) Inorg Chem 52:5464
112. Errandonea D, Kumar RS, Achary SN, Tyagi AK (2011) Phys Rev B 84:224121
113. Errandonea D, Lacomba-Perales R, Ruiz-Fuertes J, Segura A, Achary SN, Tyagi AK (2009) Phys Rev B 79:184104
114. Errandonea D, Manjón FJ, Muñoz A, Rodríguez-Hernández P, Panchal V, Achary SN, Tyagi AK (2013) J Alloy Compd 577:327
115. Mittal R, Garg AB, Vijayakumar V, Achary SN, Tyagi AK, Godwal BK, Busetto E, Lausi A, Chaplot SL (2008) J Phys Condens Matter 20:075223
116. Tyagi AK, Kohler J, Balog P, Weber J (2005) J Solid State Chem 178:2620
117. Pellicer-Porres J, Vázquez-Socorro D, López-Moreno S, Muñoz A, Rodríguez-Hernández P, Martínez-García D, Achary SN, Rettie AJE, Mullins CB (2018) Phys Rev B 98:214109
118. Popescu C, Garg AB, Errandonea D, Sans JA, Rodriguez-Hernández P, Radescu S, Muñoz A, Achary SN, Tyagi AK (2016) J Phys Condens Matter 28:035402
119. Achary SN, Mukherjee GD, Tyagi AK, Godwal BK (2002) Phys Rev B 66:1841061
120. Achary SN, Mukherjee GD, Tyagi AK, Godwal BK (2003) Powder Diffr 18:147
121. Tyagi AK, Achary SN (2010) Diffraction and thermal expansion of solids. In: Chaplot SL, Mittal R, Choudhury N (eds) Ch-6 in thermodynamic properties of solids. Wiley-VCH Verlag GmbH & Co, Weinheim
122. Achary SN, Errandonea D, Santamaria-Perez D, Gomis O, Patwe SJ, Manjón FJ, Hernandez PR, Muñoz A, Tyagi AK (2015) Inorg Chem 54:6594
123. Mishra KK, Achary SN, Chandra S, Ravindran TR, Pandey KK, Tyagi AK, Sharma SM (2016) Inorg Chem 55:227
124. Mukherjee GD, Vijaykumar V, Karandikar AS, Godwal BK, Achary SN, Tyagi AK, Lausi A, Busetto E (2005) J Solid State Chem 178:8
125. Bevara S, Achary SN, Garg N, Chitnis A, Sastry PU, Shinde AB, Krishna PSR, Tyagi AK (2018) Inorg Chem 57:2157

Chapter 7
Synthesis of Metallic Materials by Arc Melting Technique

Dheeraj Jain, V. Sudarsan, and A. K. Tyagi

Abstract Arc melting technique, which is widely used for preparation of alloys, intermetallics, and metal-based composites, is discussed in this chapter. Basic principle of electrical arc generation and its utilization for melting of materials has been outlined. Design and operation of laboratory-scale arc melting furnace as well as melt-casting furnace have been presented. Operational principle of graphite arc furnace and consumable electrode-based industrial arc furnace is also mentioned. Advantages and limitations of arc melting technique are summarized with few examples towards the conclusion. Contents of this chapter would be useful to students and researchers engaged in development of metallic materials.

Keywords Arc-melting technique · Metals · Alloys · Intermetallics

7.1 Introduction

Metals form the majority of the Periodic Table as more than 95 elements exhibit metallic behavior among the 118 elements known till date. In chemically bonded form with non-metals/metalloids, they form wide variety of materials. While pure metals too have several applications, they are omnipresent in the form of alloys, inter-metallic compounds and composites. The latter types of materials have superior physico-chemical properties as compared to their constituents. For example, aluminum alloys are used as beverage cans. Use of stainless steel is all-pervading in our lives. Titanium-based alloys are used in a wide range of applications from bio-medical implants to aircraft construction. From dental implants to construction of large ships, from electrical solder to giant turbines, from locomotive tracks to precision components of wrist watches, alloys and inter-

D. Jain (✉) · V. Sudarsan · A. K. Tyagi
Chemistry Division, Bhabha Atomic Research Centre, Trombay, Mumbai 400085, India
e-mail: jaind@barc.gov.in

D. Jain · V. Sudarsan · A. K. Tyagi
Homi Bhabha National Institute, Mumbai 400094, India

A. K. Tyagi and R. S. Ningthoujam (eds.), *Handbook on Synthesis Strategies for Advanced Materials*, Indian Institute of Metals Series,
https://doi.org/10.1007/978-981-16-1807-9_7

metallics are used almost everywhere. Interestingly, in nature, very few metals such as noble metals are found in elemental form while others exist as compounds with non-metals/metalloids in the form of inorganic minerals or ores. Extraction of pure metals from their ores forms the subject area of extractive metallurgy. Pure metal is combined with other metals/non-metals to prepare alloys and compounds. For this, constituents need to be uniformly mixed; preferably at atomic scale, so that compositionally homogeneous products are obtained. Alloys/intermetallics can be prepared by various methods. For example, metal powders can be mixed in stoichiometric amounts; compacted in the form of compressed ingots (pellets) and heated at high temperatures under inert atmosphere to obtain desired product by solid-state reaction. Since solid-state reactions are seldom complete in one heating, multiple cycles of powdering–mixing–compacting–heating are usually needed to obtain phase-pure product. Also, unlike ceramics that are brittle, it is not easy to powder metals/alloys/intermetallics due to their ductility and malleability. In order to avoid oxidation, metallic powders should also be handled under inert atmosphere. These requirements make the solid-state powder processing route less popular for preparation of metallic compounds. Another preparative method is based on reduction of a mixture of metal oxide powders, usually carried out under flowing hydrogen (H_2 + Ar) atmosphere.

$$MO_{x(s)} + M'O_{y(s)} + (x+y)H_{2(g)} \rightarrow M_{(s)} + M'_{(s)} + (x+y)H_2O_{(g)}$$

or

$$MO_{x(s)} + M'O_{y(s)} + (x+y)H_{2(g)} \rightarrow MM'_{(s)} + (x+y)H_2O_{(g)}$$

Reduced metal particles/grains may in-situ react to from desired alloy/intermetallic compound. This method has limitations in terms of presence of residual oxides due to incomplete reduction as well as issues related to handling of reactive metallic powder as product(s). Also, not all metal oxides can be reduced to corresponding metals by hydrogen gas, which can be understood from their Ellingham diagrams. An upcoming method to prepare metallic compounds in powdered form is based on hydriding-dehydriding approach. In this method, alloy/intermetallic compound is taken in the form of bulk ingot and hydrogenated to obtain metallic hydrides. This is usually accomplished at relatively lower temperatures (<800 K) for a large number of metallic systems. When these hydrides are heated, hydrogen (H_2) gas is released and powdered solid residue (alloy/intermetallic) is obtained. Repeated hydriding-dehydriding cycles pulverize the starting ingot into very fine (micron to sub-micron sized) powders. If starting metallic ingot is pure and hydrogen used is free from reactive impurities (moisture, oxygen, nitrogen, CO/CO_2, halogen, etc.), high purity metallic powders can be prepared by this method. Use of high purity alloy/intermetallic ingots is essential for this. Aqueous/non-aqueous solution-based chemical reduction methods are less popular to obtain metallic compounds since multiple process steps are involved and stringent control of chemical environment (solvent) is required to avoid non-metallic by-products.

Methods discussed above involve handling of metallic powders. Also, products obtained by these methods are usually powders or sintered monoliths, which need to be processed further to fabricate the end-product in desired shape and size. For majority of applications, dense components (close to theoretical density) are required. These can either be melt-casted or machined from melt-solidified dense ingots. Melting and casting techniques are therefore important for development of metallic materials. Methods that are used for preparation of alloys/intermetallics/metal-ceramic composites are discussed below.

A straightforward method to prepare high-density alloys and intermetallic compounds is to melt the constituents in stoichiometric ratio to obtain a homogeneous molten mass. Cooling the molten mass leads to solidification and formation of dense ingot in the shape of molten matrix. Alternatively, molten mass (liquid) can be cast into pre-designed moulds to obtain melt-cast ingot. Constituents can be melted in several ways, which include direct heating in flame or resistive furnace, electric arc melting, induction melting, electron beam melting, laser-induced melting, etc. It is pertinent to mention here that metal-ceramic composites can also be prepared by direct melting approach provided that at least one of the constituents melts and others (solid(s)/liquid(s)) are uniformly distributed in the molten phase.

For melting by direct heating in flame or resistive furnace, it is usually required to heat the constituents up to the melting temperature of the highest melting constituent. Flame heating is mostly employed for non-oxidizing noble-metal alloys (Au, Ag, Pt, Cu, etc.). Furnace heating is time-consuming and energy extensive method. If constituents have widely varying melting temperatures, the ones having lower melting temperatures would remain molten till the highest melting constituent is liquefied. This may lead to vaporization loses as well as reaction of low melting constituents with crucible/pot used for melting. Also, except for noble metal alloys, inert atmosphere should be maintained in the furnace during entire process. This method is commonly used either for materials having lower melting temperatures (Na, K, Li, In, Sn, Zn, Pb, Ga, Mg, Al, Bi, Sb, Cd, etc.) or for alloys in which constituents have lower vapor pressure in the molten state. For alloys/intermetallics that are composed of constituent elements with widely varying melting temperatures, direct heating in furnace is not suitable.

Induction melting technique is another useful method for preparation of metallic materials, especially for melting of refractory metal-based alloys as well as low conductive oxides and glasses. In this method, heat is generated in the charge (material to be melted) by eddy currents when it is subjected to electromagnetic induction, usually applied via suitable induction coil. Typical system consists of an electromagnet, an oscillator that passes high-frequency AC (alternating current) through the electromagnet. Due to oscillating magnetic field passing through the material, eddy currents are generated, which upon flowing through the material heat it up by resistive Joule heating and thus melting it. For magnetic materials (ferromagnetic/ferromagnetic), additional heating may result from magnetic hysteresis losses. Since heat is generated within the material, rapid melting is possible.

For electrically conducting materials, this method has another advantage of contactless melting (levitation melting), which minimizes contamination from process (melting) pot or induction coil itself. Induction melting furnaces are useful for both lab-scale (few grams) as well as industrial-scale preparation of alloys. Two principle variants of industrial induction furnaces are induction crucible furnace and induction channel furnace, which are used for melting and holding-pouring purposes, respectively. For melting of very high purity alloys as well as oxide-based glasses, cold crucible induction furnaces are used.

In laser-induced melting, short pulses (nanosecond to femtoseconds) from high-intensity lasers are used for localized melting. Laser pulse is absorbed by small volume region of material and intense heat generated is sufficient to melt/evaporate the sample in the small volume region [1–3]. This method is useful for metallic materials, ceramics, refractories, etc. and requires sophisticated equipment and safety requirements for operation of high-power lasers. Large-scale melting of bulk materials by this method requires further developments. One of the interesting applications of laser-induced melting is to study the melting behavior of nanocrystalline refractory oxides such as ThO_2 [4].

Electron beam melting is another useful technique for processing very high purity metals, alloys, and intermetallics [5]. In this technique, electrons generated from a suitable source (filament/electron gun) are accelerated with high potential (several tens of kV). Accelerated electron beam is focused using electro-magnetic lenses and allowed to fall on the target material. The intense electron beam is scanned over the target material (scan rates = 10^2–10^4 mm/s) to cause pre-heating up to $\sim 0.8\ T_m$ using high beam currents (≥ 30 mA) and melting ($T > T_m$) at relatively lower beam currents (5–10 mA) to obtain molten material. Here, 'T_m' is the melting temperature of metal/alloy/intermetallic compound. Electron beam melting systems require high vacuum ($P \leq 10^{-4}$ Torr) for operation. This technique is useful for mass processing of large components with very fine microstructures and also used for purification of metallic materials. However, unlike laser-induced melting, it can only be used for electrically conducting materials and sophisticated equipments that require extensive maintenance are involved.

The most common and widely used method for preparation of metallic materials is arc melting technique. This method is discussed in detail in this chapter. The discussion is intended to make readers aware about the physical basis of this method, overall description of the process/method, advantages, limitations, and application areas.

7.2 Arc Melting Method

The initial requirement for development of metallic alloys/intermetallics/composites is to prepare them in bulk quantities with minimum contamination/impurities and uniform distribution of constituents. Methods that enable rapid melting of constituents are useful for this purpose. Arc melting method fulfils above requirements and is used for preparation of metallic materials from lab-scale to industrial scale. Equipment used for arc melting of metallic materials is commonly known as Arc melting furnace or Electric arc furnace. As the name suggests, material to be melted is heated by direct application of electric arc between oppositely charged electrodes. Therefore, material (or at least one of the major constituents) needs to be electrically conducting in both solid and molten state so that arc is sustained during the melting process by passage of electrical current through the material. These furnaces can be designed in small-sized laboratory arc melting furnace for preparing high purity alloys/intermetallics in few grams-scale quantities. Also, very large-scale industrial arc furnaces are used to melt several tons of alloys. Steel-making furnaces are typical example of industrial arc furnaces. Arc can be generated by application of direct current and alternating electrical current. Corresponding arc furnaces are known as DC-arc furnace (or direct arc furnace) and AC-arc furnace, respectively. If the material to be melted itself is taken as one of the electrodes, which is consumed as the arc furnace is operated and molten alloy is formed, it is known as consumable electrode electric arc furnace. On the other hand, if the electrode material is not a constituent of the alloy and is only used to generate arc, the furnace is known as arc furnace with non-consumable electrodes. Several industrial furnaces, particularly used for purification of alloys/ intermetallic are used in the consumable electrode configuration. Large-scale steel-making arc furnaces (also known as three-phase graphite arc furnaces) are also used in consumable electrode configuration where a fraction of carbon required for steal making is obtained from consumable graphite electrode(s). Most of the laboratory arc furnaces, which are used for synthesis of wide variety of metallic materials, are used in non-consumable (inert) electrode configuration. A high-temperature refractory conductor such as tungsten-ThO_2 composite (also known as thoriated tungsten) is used as electrode in non-consumable electrode laboratory arc furnaces. It is important to note that temperature in industrial arc furnaces can reach as high as 2,100 K, while laboratory arc furnaces can reach even up to 3,300 K.

7.2.1 Physics of Arc Generation

Electrical arc is an important application of plasma physics. Arc is defined in terms of current and voltage drop across two electrodes. It is generally accepted that arc is formed when currents through an electrical discharge exceed values between 100 mA and 1 A between two smooth electrode surfaces (without any sharp tip, edge, or points) and the voltage drop remains in the range between few volts to few tens of volts [6]. Upper limit of current is usually not specified for arc and is quite large in the order of thousands of amperes. At currents lower than the threshold range as mentioned above, two other types of discharges namely, glow discharge (current ~100 μA to 100 mA; voltage ~few hundred volts) and dark discharge (current <1 μA; voltage ~ 1–2 kV) are observed. Transition from glow discharge to arc discharge occurs with a sudden increase in voltage with increasing current followed by rapid drop of voltage and higher currents flowing through the arc (up to ~ 1 kA or higher). Presence of sharp edges or points over the electrode surfaces may give rise to unwanted events such as corona discharge or streamer discharge [6]. Depending upon the gas pressure present in the vicinity of electrodes, electric arc is classified as high-pressure arc and low-pressure arc. While there are no sharp boundaries between these two types, it is generally accepted that arcs generated below 1 mm Hg gas pressure are low-pressure arcs while those generated above 100 mm Hg gas pressure are high-pressure arcs. Region between 1 and 100 mm Hg is accepted as transition region. For low-pressure arcs, ion temperature (T^+) and gas temperature (T_g) are comparable ($T^+ \approx T_g$ ~ few hundred Kelvin) while electron temperature (T^-) is significantly higher (10^4–10^5 K). On the other hand, all three temperatures are reasonably close to each other in high-pressure arc region ($T^+ \approx T^- \approx T_g \sim 10^4$–$10^5$ K). Each of these temperatures is defined in terms of kinetic energy of the component (3/2 kT = ½ mv^2), where 'k' = Boltzmann's constant, 'm' is mass of particles and 'v' is root mean square velocity. Arc melting furnaces operated at ambient pressure or above are high-pressure arc-type systems and the arc plasma is considered in a state of 'local thermodynamic equilibrium'. An arc generated between two electrodes is typically defined as consisting of three regions. These are (i) the cathodic region, (ii) arc-region (or positive region), and (iii) the anodic region. These regions are schematically are shown in Fig. 7.1 along with the variation of voltage in each region as a function of inter-electrode separation.

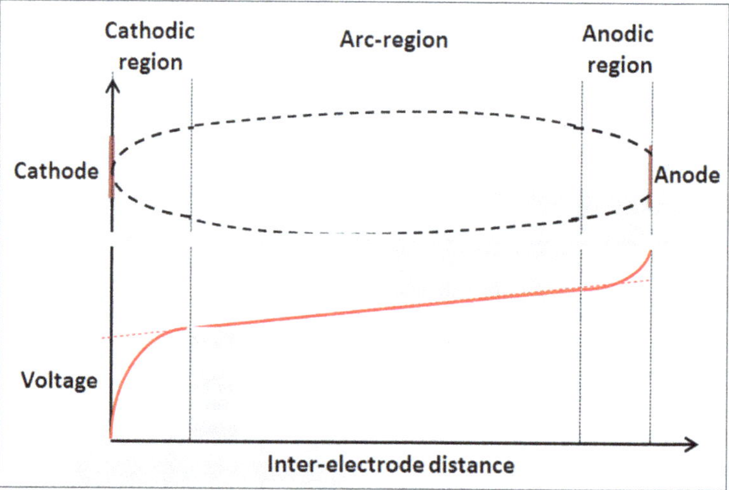

Fig. 7.1 Different regions of electric arc and corresponding voltage variation with inter-electrode separation

The cathodic region is close to cathode surface, in which voltage rapidly and non-linearly increases with increasing inter-electrode distance. Shape of arc in the cathodic region is conical with the tip of cone emerging from cathode. The arc-region is the region where arc is stable and attains near cylindrical shape. It is characterized by slow and linear increase in voltage with inter-electrode distance. The anodic region is confined towards the anode surface, where the shape of arc again becomes constrained (conical) with the arc confining at the anode surface. Voltage again increases non-linearly but less rapidly as compared to the cathodic region, in the anodic region. It is important to remember that in typical arc melting furnaces, the inter-electrode separation is not constant and one of the electrodes (usually cathode) is moved. Regions shown in Fig. 7.1 therefore are representative of a typical electrode configuration.

7.2.2 Utilization of Electric Arc as a Source of Heat

Electrical arc acts as region of electrical conduction and can be utilized as a source of light as well as heat. Fluorescent lamps filled with different gases at low pressures (halogen, neon, sodium vapors, etc.) and inner surface coated with suitable phosphors are examples of utilization of low-pressure arcs for lighting applications. Heat generated in the arc can be used for (i) processing of metals by arc-welding, (ii) preparation of metallic alloys and intermetallics by arc melting, (iii) vaporization of metallic materials used in the form of electrodes, (iv) radiative heating and (v) convection heat transfer by gas circulation to the region where heat is needed. In case of arc-based heating applications, classical equation of Joule heating does not hold good in general and several secondary effects are involved during heating by arcs. Total energy balance must therefore be considered while assessing various heating effects induced by electrical arcs. For example, even though total power of an arc (P_{arc}) is equal to the product of arc current (I_{arc}) and voltage drop (V_{arc}), it is not true for the three different regions of arc separately. For example, heat released in the cathodic region (Fig. 7.1) includes the effect of (i) kinetic energy released by positive ions, which are accelerated through cathodic voltage drop to hit the cathode surface, and (ii) energy corresponding to recombination of positive ions with electrons to become neutral atoms/molecules. Electrons that are removed from cathode surface have a cooling effect that corresponds to product of electron current and work function (in voltage units). Similarly, electrons that hit anode have an opposite (i.e., heating) effect. For arc-systems with gas circulation (such as plasma torch), ionized gas carries away heat corresponding to gas temperature as well as latent heat of plasma formation. In most applications involving arc-melting of various metals/alloys, the plasma recombines and the latent heat of plasma formation is utilized. Speed of gas circulation in arc melting system also affects the overall heat transfer. It should also be kept in mind that a sizable power (heat) is also dissipated in the form of radiation during arc melting process, unless it is reflected back to the region where it is absorbed to generate equivalent heat.

Therefore, while electric arcs are excellent source of heat energy, which can be utilized for synthesis and processing of a variety of metallic materials, estimation of heat energy actually utilized for a given purpose is usually not possible from current and voltage specifications of the arc-setup (such as electric arc furnace).

7.3 Examples of Arc Melting Furnaces

7.3.1 *Laboratory DC Arc Melting Furnace*

A laboratory-scale electric arc furnace; also known as DC arc melting furnace is one of the best tools to prepare high purity metallic compounds. It is a non-consumable electrode type arc furnace, which is widely utilized in research labs for preparation of alloys, solid solutions, intermetallic compounds, metal-based composites, etc. Fig. 7.2 shows a typical single sample laboratory arc melting setup. Inset (right) shows the copper hearth (crucible) used for melting the samples. A typical as-melted alloy in the shape of a button is also shown in the inset (left).

Reactants (metals, metalloids, alloys) are taken in pre-defined stoichiometry and can be quickly heated to very high temperatures (up to $\sim$3300 K) over a span of few seconds. As a result, constituents undergo rapid melting and a homogeneous molten mass of reactants is formed. Rapid melting is achieved by electric arc generated between a cathode (usually made of thoriated tungsten alloy in the form of rod) and a metallic anode (usually made of copper). Copper is chosen since it has very high thermal conductivity (>400 W m^{-1} K^{-1} at 298 K) and high melting temperature (1,358 K). Copper can easily be machined to fabricate the components for arc furnace and it is also highly cost-effective as compared to other metallic options (e.g., silver; >430 W m^{-1} K^{-1} at 298 K), which have higher thermal conductivity than copper. Anode may also serves as hearth (crucible/container), in which melting is carried out. Hearth is continuously cooled with water (inlet temperature $\sim$298 K) so that the hearth surface temperature remains near ambient and melting is carried out without significant contamination with copper. Using arc-furnaces as shown in Fig. 7.2, metallic materials can be routinely prepared with copper contamination from furnace not exceeding few tens of ppm (parts per million).

Arc is generated by applying low voltage–high current ($\sim$30–60 V, $\sim$300–600 A) DC power between cathode (thoriated tungsten) and anode (copper). Power supplies used for TIG (tungsten inert gas) welding are suitable for such arc melting furnaces. DC power is usually applied through multi-strand water-cooled copper cables running from the output terminal of DC power supply to the electrodes. These electrodes are separated from each other by electrically insulating, optically transparent, and high temperature compatible (up to 1,200 K) cylindrical quartz tube, which also enables viewing the sample during arc melting. Other designs of laboratory arc melting furnaces have metallic dome-type enclosure (instead of quartz cylinder shown in Fig. 7.2) with optically transparent view ports for viewing

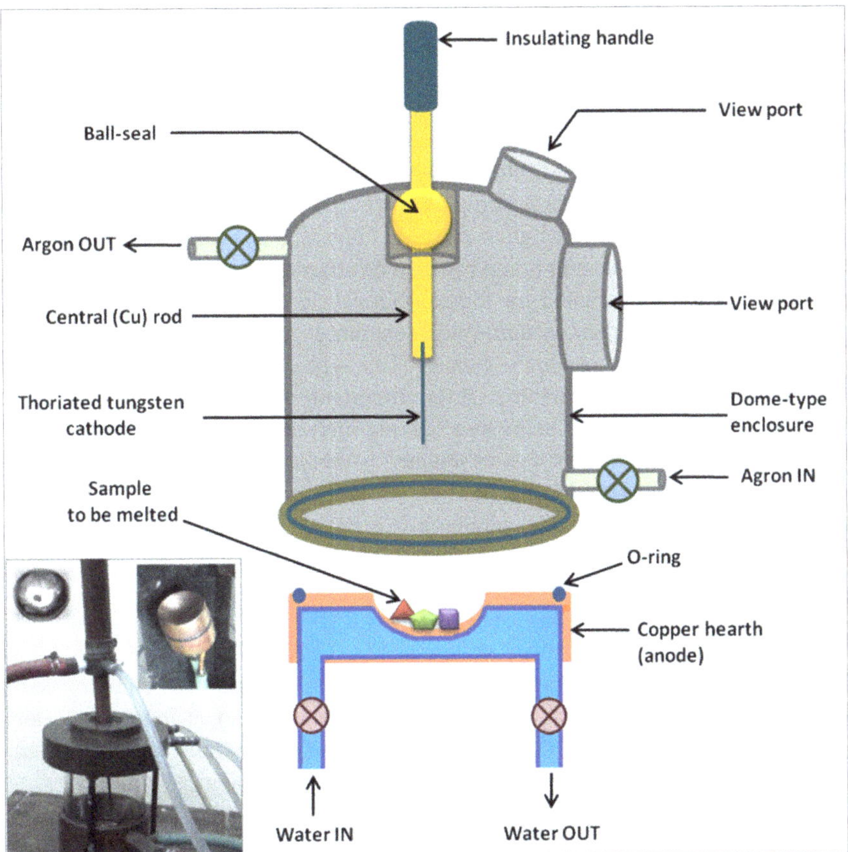

Fig. 7.2 Schematic design of a single sample DC arc melting furnace having dome-type enclosure; inset (left bottom) shows photograph of a laboratory arc melting furnace where dome-type enclosure is replaced with quartz cylinder. Copper hearth (right top frame in inset) and as-melted alloy button (left top frame in inset) can also be seen

the sample as well as guiding the electrode movement while carrying out melting operation. Quartz tube is fixed to water-cooled top and bottom plates (usually made of brass/copper) in a leak-proof manner using neoprene/viton O-rings. A crucial component of arc melting furnace is ball-seal, which is a sphere-shaped, leak-tight metallic seal attached to the top metallic plate (or top of metallic dome-type enclosure). Thoriated tungsten cathode is connected to a metal rod (copper) that passes through this ball-seal. Other end of the metal rod is covered with insulating material (wood/bakelite/polymer) for moving the electrode by hand during furnace operation. The ball-seal arrangement allows (i) linear (vertical) motion of cathode, which enables arc generation when two electrodes come in close proximity and (ii) radial (angular) motion of cathode so that arc can be moved over the sample for uniform melting. Sample to be melted is kept in cleaned copper hearth, which is then fixed into the bottom plate (brass/copper) with O-ring seal. In case of arc-furnaces with metallic dome-type enclosure, copper hearth is usually fixed at the base-plate and only requires to be covered by metallic enclosure in a leak-tight manner. Cathode is kept away from the anode by moving the central rod upwards through the ball-seal. Once sample is loaded, the furnace is flushed with inert gas (high purity argon; 100–200 ml/min) with gas inlet near the copper hearth and outlet from the port available in top plate (or at the top side of metallic dome-type enclosure). This ensures that lighter air moves upwards and sample is covered with heavier argon gas. A vacuum pump can also be attached to gas ports so that furnace atmosphere can be evacuated for forced removal of air followed by purging with inert gas. Repeating this vacuum-purge operation multiple times improves the inertness of atmosphere inside the furnace. This type of setup is commonly referred as vacuum arc melting furnace. It is important to mention here that while vacuum-purge arrangement helps in achieving inert atmosphere inside the furnace within few minutes, the same can be attained without vacuum-purge by continuous flushing of argon for 20–30 min. It is essential to use very high purity argon (or helium; though used rarely) and adequate flow rate ($\sim$100–200 ml/min) to prepare high purity alloys. Commercially available high purity gases (such as 99.999% argon) can be further purified to remove trace impurities (O_2, N_2, CO, H_2O, hydrocarbons, etc.) by passing over moisture traps (silica gel, anhydrous $CaCl_2$, etc.), followed by heated getter materials (Zr, Cu, Ti, U, etc.) to obtain inert atmospheres with oxygen partial pressures (P_{O_2}) less than 10^{-15} atmospheres. Purging by oxygen, nitrogen or air is not recommended, even for melting noble metal-based materials. The reason is that while oxygen/air/nitrogen do not contaminate noble metals, it may cause surface oxidation of copper hearth.

Once inert atmosphere is maintained inside the furnace, water cooling is enabled to cool furnace hearth, electrodes, top and bottom plates (or metallic enclosure), and DC power cables. It is essential to attain inert atmosphere prior to water-cooling so that moisture content inside the furnace can be minimized. After water flow is enabled, power supply is activated and cathode tip is carefully lowered to approach the anode surface. Operator must wear welding view glasses (or face shield with welding view port) during system operation to avoid exposure of eyes and skin to ultra-violet (UV) radiation that is emitted during arcing. At close proximity of electrodes (few millimetres), effective resistance (due to inert gas medium) between the electrodes is small enough to allow large electric current. A large current passing through gas medium generate spark, which is instantaneously followed by sustained arc at high DC power (>10 kVA). This can be understood in terms of current flowing through an ionized column of gas. While positive ions approach towards cathode and get neutralized electrons move away towards anode. Dissipation of large energy (kinetic energy of ions, electrons, and recombination energy) during arcing produces intense heat, which can be utilized to melt a material, if it allows passage of electrical current. Similar phenomenon is responsible for arc welding (TIG welding) of metallic materials.

For uniform melting of sample, cathode is slowly moved over the sample surface (without touching the sample) with minimum vertical movement. In less than a minute, sample melting is completed (up to tens of grams scale) and a molten mass contained in water-cooled copper hearth can be seen through the viewing glass (or view ports). Arc is uniformly moved over the molten sample to ensure homogenous mixing of constituents. Once melting is complete, DC power is put off to stop the arc. Molten sample quickly solidifies in the form of button (or ingot) as shown in inset of Fig. 7.2 (left top panel). Cooled sample (button/ingot) is inverted (usually with the help of cathode tip) and melting is repeated to obtain uniform distribution of constituents in as-melted alloy. An oxygen scavenger (zirconium metal) is usually melted prior to melting any sample so that oxide layer over the surface of hearth (copper cup) is removed during zirconium melting and sample with minimum contamination is obtained. Melting zirconium prior to sample is quite useful for arc melting furnaces that have multiple hearths so that one hearth is used for zirconium melting while the other(s) are charged with sample(s). This enables removal of residual oxygen from the furnace environment during zirconium melting and minimizes sample contamination during melting process. For single hearth setups as shown in Fig. 7.2, zirconium melting prior to sample may be avoided if anode surface is thoroughly polished with diamond impregnated cloth and cleaned by ultrasonication. With common experimental care, very high purity metallic materials with no surface contamination/oxidation can be successfully melted up to several grams scale.

7.3.2 Melt-Casting by Arc Melting Method

This is another useful application of arc melting furnaces, wherein hearth design can be modified to facilitate melt-casting of molten materials into desired shapes (rod, bar, disc, etc.). Figure 7.3 shows a typical casting hearth designed to make metallic materials in rod shape. In the first stage, alloy/intermetallic compound is formed by conventional arc melting. Melt-casting is done under partial suction (through the modified hearth) so that as soon as alloy re-melts, it acquires the shape of hearth pit in the form of rod, bar, disc, etc. This technique is very useful for fabrication of alloy/intermetallic samples for various characterization/applications.

7.3.3 Graphite Arc Furnace

Figure 7.4 shows a schematic diagram of three-electrode industrial graphite arc furnace, which is usually employed for large-scale (several tons) alloy melting. Steel-making arc furnaces are of this design. These furnaces operate on alternating current (AC) power. It essentially consists of a large vessel made of refractory ceramics that are capable to hold the molten alloy (such as steel) at high

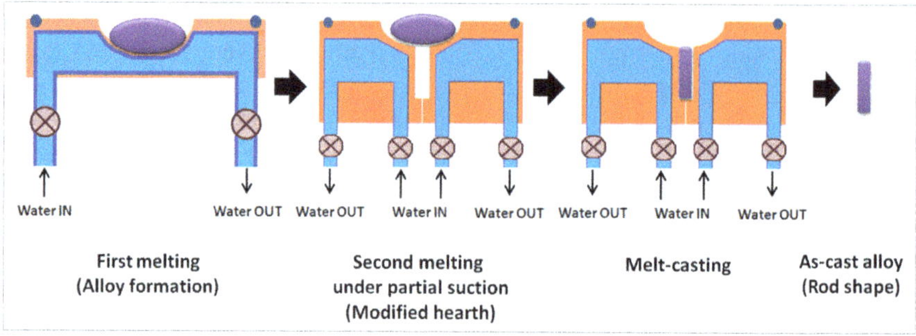

Fig. 7.3 Schematic of melt-casting process from metallic ingot to rod-shaped material by laboratory arc melting furnace

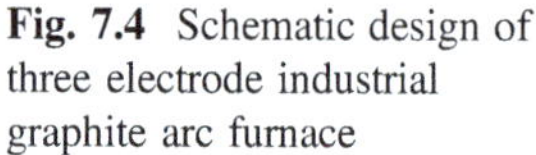

Fig. 7.4 Schematic design of three electrode industrial graphite arc furnace

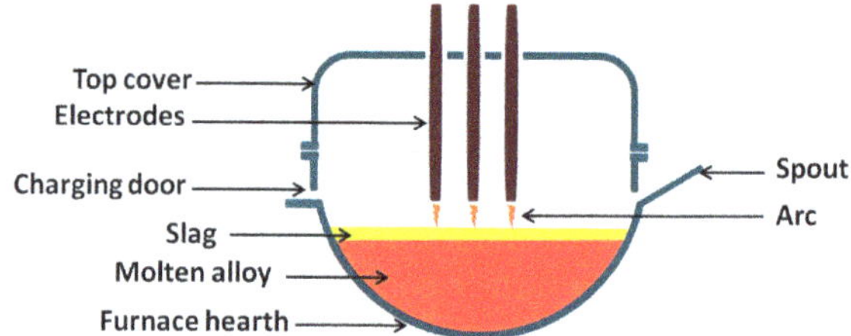

temperatures. This vessel is known as furnace hearth and is lined with resin-bonded magnesia (MgO)-carbon bricks. When furnace is heated, resin burns out and residual carbon makes a network to bind the refractory magnesia grains so that furnace lining is neither wetted by molten slag nor eroded due to chemical interaction of molten alloy/slag with the lining.

Hearth is designed in such a manner that its movement (swinging/tilting) is feasible when the alloy is in molten state so that both alloy as well as slag can be removed independently. Usually, slag is lighter than alloy and floats over the molten alloy. By tilting, slag can be preferentially removed (known as de-slagging). The hearth is covered by cylindrical dome and three graphite electrodes are held vertical from the top of the furnace through three independent holes in the top cover of the dome. Electrodes are clamped to AC power supply using water-cooled electrode holders and AC power is provided through water-cooled cables. Electrodes used in these furnaces are of consumable type and are partially utilized during the alloy melting process. Three electrodes are clamped in such a manner that vertical movement of each electrode can be independently regulated from the remotely operated control panel. These furnaces require careful power control (lower during the beginning of melting and successively increases as the melting progresses). Dome walls are also water cooled. These furnaces are very useful for direct melting of steel scrap.

7.3.4 *Consumable Electrode Vacuum Arc Re-melting Method*

This method is another variant of electric arc melting technique by which metallic materials with high purity and better compositional homogeneity are prepared. As the name suggests, electrodes are consumable in this method and are made of the metallic material itself, which is to be melted. Usually air melted/vacuum induction melted ingots are taken as consumable electrodes. The system consists of water-cooled furnace chamber, melting hearth (made of high purity copper with water jacket), consumable electrode, stinger rod, and its drive mechanism. Under high vacuum or inert gas atmosphere, a DC arc is generated between the electrode and starting material (consumable electrode). The high arc current between the electrodes is sufficient to melt the consumable electrode and molten mass is collected in to form of liquid pool in water-cooled copper hearth and solidifies as product ingot. Product ingot has excellent compositional homogeneity and chemical purity as compared to starting consumable electrode. These types of furnaces are used for melting refractory metals (Nb, Ta, etc.) and their alloys.

7.4 Advantages of Arc Melting Method

Arc melting method is widely used for preparation of metallic materials, both at laboratory scale for research and development purpose, and industrial scale for commercial production of metallic materials. Following are the salient advantages of this technique.

- Arc melting furnaces can be rapidly started/stopped. Therefore, it can be used to quickly produce materials as per demand unlike resistive heating technique.
- For development of new materials, rapid melting/freezing attributes of arc melting method result in minimum sample contamination from impurities present in the cover gas (argon) or crucible material itself.
- Arc melting setup is simple and requires minimum maintenance in terms of routine electrical safety measures and checking of leak-proof water and gas tubes.
- This method is suitable for preparation of metal-ceramic composites where metallic phase forms a continuous network in the molten phase.
- Melting can be carried out in oxidizing or inert atmospheres and at low pressures (partial evacuation) provided that the arc is sustained through the short duration of operation.
- Since arc melted samples are fast cooled to ambient temperatures, high temperature phases may get quenched leading to formation of metastable materials. Materials prepared by arc melting can therefore be used to study

physico-chemical and structural properties of high temperature metastable phases.

- Method is suitable for preparation of small quantity/size samples with negligible loss during melting (due to evaporation or sticking with sample crucible), which is an advantage for preparation of precious/rare metal alloys (based on Pt, Au, Rh, Ir, Pd, Ag, etc.).

7.5 Limitations of DC Arc Melting Method

Following are some of the limitations/challenges associated with arc melting technique:

- For preparation of new materials using laboratory arc melting furnaces, excellent compositional homogeneity is obtained for small quantity/size samples. As the sample size becomes large, melting furnace requires modifications in terms of size/shape of hearth, cathode design, DC power supply rating, etc. For large size samples, if arc is not able to spread through the melting material uniformly and partial cooling occurs at different locations, it results in compositional inhomogeneity. Such samples can be efficiently prepared by other methods such as vacuum induction melting.
- For multi-component alloys, where difference in melting temperature of constituent elements is large, arc melting may cause deviation from intended stoichiometry due to preferential vaporization or loss of low melting (or highly vaporizing) constituents. This can, however, be minimized to some extent if constituent having higher melting temperature is exposed to arc before the constituent that has lower melting temperature and the two constituents make an intermediate compound or solid solution with crystal structure of higher melting constituent.
- Arc melting is not suitable for preparation of ceramics since they are usually electrically insulating so that arc does not sustain through ceramic phase. For electrically conducting ceramics, higher melting temperatures could also make melting difficult.
- Care must be taken to ensure that neither the cathode touches the anode/sample nor it is unevenly moved over the melting mass. The former leads to termination of arc and contamination of sample while the latter results into inhomogeneous alloy. However, these limitations can be overcome with experimental care.
- Industrial arc furnaces require very dynamic quality of the electrical power in terms of furnace load. Stable power supply is therefore a necessary requirement as random power fluctuations such as flicker noise, harmonic distortion may adversely affect the operation and hamper production cycles. Power supply, therefore, requires frequent technical checkups/maintenance.

7.6 Examples (Alloy Preparation by Laboratory Arc Melting Technique)

7.6.1 *Preparation of Ti_2CrV Alloy*

Ti_2CrV-based alloys are being developed for solid-state hydrogen storage applications since they offer up to 4.3 wt% storage capacity and attractive hydriding-dehydriding behavior for the purpose hydrogen-fuelled vehicular application. These alloys are routinely prepared in authors' lab using DC arc-melting method. Constituent metals are taken in the form of rods/sheets/ingots and thoroughly cleaned to remove surface oxide layer by surface polishing (using file/abrasive tools) or mild chemical etching (using dilute acidic medium). Cleaned metals are taken in stoichiometric ratio and placed together in copper hearth. Alloy is melted in arc melting furnace (total furnace volume ~750 ml) under flowing argon atmosphere. Purity of alloy is limited by purity of (i) constituent metals and (ii) argon gas. It is essential to thoroughly clean the inner walls of arc melting furnace (copper hearth, quartz view ports) as well as tungsten electrode to minimize alloy contamination. Argon gas should be purified by passing over moisture traps (silica gel, molecular sieve, fused calcium chloride) and oxygen/nitrogen trap (heated copper/titanium sponge/depleted uranium metal) before flowing into the leak-tight furnace to obtain alloys with minimum oxidation during melting. Argon flow rates may be maintained between 300 and 500 ml/min during purge (~20 min prior to melting) as well as during melting. Compositional homogeneity of alloys is maximized by repeated melting of alloy ingot up to four or five times. It is to be ensured that sharp electrode always remains away from the molten alloy during the melting process such that arc is sustained and no contamination from electrode is incorporated. Alloy ingots up to 20–30 g scale (per hour) can be routinely prepared using similar laboratory-scale arc melting furnaces.

7.6.2 *Preparation of Metallic Alloy Nuclear Fuels*

Development of thorium and uranium-based metallic alloys as prospective nuclear fuels is another area of research in authors' lab. Development of these alloy fuels requires comprehensive basic research to evaluate physico-chemical properties of these alloys for which high-quality alloy samples are needed. These are prepared by arc melting technique. While the experimental procedure remains the same as described for Ti_2CrV alloys in preceding sub-section, additional care must be taken for handling of actinide metals, which are highly reactive to oxygen, air, and moisture and are weakly radioactive also. Most of the metal treatment (metal cleaning, cutting, and weighing) prior to arc melting is carried out inside inert atmosphere glove box and metals are transported under inert atmosphere (in argon-filled vials or immersed in dry hexane). High purity of purge gas (argon) is crucial for preparation of these nuclear alloys (Th–U, Th–Zr, U–Zr, Th–U–Zr). For this, after moisture removal trap, argon is passed over uranium metal turnings heated at ~600 K to remove traces of reactive impurities (N_2, O_2, H_2O, CO, CO_2, etc.). Oxygen partial pressure of argon gas purified by this method can be brought down to as low as $\sim 10^{-17}$ atmospheres, which can be measured by high temperature EMF cell (operating at 1,073 K) using gadolinia doped ceria (or yttria-stabilized zirconia) oxide ion-conducting electrolyte. Alloys weighing 15–20 g with <0.01 wt% oxidation during arc melting are routinely prepared in this manner.

7.7 Conclusions

Arc melting is one of the most popular techniques for preparation of metallic materials at both, lab-scale as well as industrial scale. For development of new materials, it is perhaps the most preferred technique. In authors' lab, arc melting technique is being extensively utilized for preparation of wide variety of alloys. Some examples include (i) thorium and uranium-based alloys (Th–U, U–Zr, U–Mo binaries; Th–U–Zr ternary alloys, etc.), which are potential metallic nuclear fuels [7] and (ii) transition metal-based alloys/intermetallics (Ti–V–Cr alloys, Zr–Ti alloys, Mg-based alloys, etc.), which are being developed for solid-state hydrogen storage application [8, 9]. Brief experimental procedures and important precautions that are needed for preparation of these alloys are discussed in previous section. The technique has been found extremely useful for lab-scale preparation of highly air-sensitive nuclear fuel alloys, which are susceptible to oxidation upon exposure to trace level of air/moisture. Under carefully controlled inert atmosphere conditions (high purity argon), such alloys are rapidly prepared on routine basis.

References

1. Musumeci P, Moody JT, Scoby CM, Gutierrez MS, Westfall M (2010) Laser-induced melting of a single crystal gold sample by time-resolved ultrafast relativistic electron diffraction. Appl Phys Lett 97:063502
2. Liu PL, Yen R, Bloembergen N (1979) Picosecond laser-induced melting and resolidification morphology on Si. Appl Phys Lett 34:864–866
3. Koubassov V, Laprise JF, Theberge F, Forster E, Sauerbrey R, Muller B, Glatzel U, Chin SL (2004) Ultrafast laser-induced melting of glass. Appl Phys A 79:499–505
4. Cappia F, Hudry D, Courtois E, Janßen A, Luzzi L, Konings RJM, Manara D (2014) High-temperature and melting behaviour of nanocrystalline refractory compounds: an experimental approach applied to thorium dioxide. Mater Res Express 1:025034
5. Murr LE, Gaytan SM, Ramirez DA, Martinez E, Hernandez J, Amato KN, Shindo PW, Medina FR, Wicker RB (2012) Metal fabrication by additive manufacturing using laser and electron beam melting technologies. J Mater Sci Technol 28(1):1–14
6. Hoyaux M (1968) Arc physics. Springer

7. Jain D (2020) Preparation, characterization and thermophysical investigations on thorium-based nuclear fuels and related systems. PhD thesis, Homi Bhabha National Institute, Mumbai
8. Ruz P, Banerjee S, Halder R, Kumar A, Sudarsan V (2017) Thermodynamic, kinetic and microstructural evolution of $Ti_{0.43}Zr_{0.07}Cr_{0.25}V_{0.25}$ alloy upon hydrogenation. Int J Hydrogen Energy 42:11482–11492
9. Ruz P, Sudarsan V (2015) An investigation of hydriding performance of $Zr_{2-x}Ti_xNi$ (x = 0.0, 0.3, 0.7, 1.0) alloys. J Alloy Compd 627:123–131

Chapter 8
Synthesis of Materials by Induction Heating

Ratikanta Mishra

Abstract Induction heating is a convenient, efficient, and rapid method of heating conducting material under control atmosphere. In this technique, a conducting body is subjected to a rapidly changing magnetic field producing induced current that causes heating effect. The method is widely used for synthesis of alloys, intermetallic phases, and high melting ceramics. Induction heating also finds applications in material processing like annealing, hardening, tempering, brazing, etc. In the present chapter, the principles of induction hearing, description of different components of induction heater, its operation, and some applications of induction heating in material synthesis & processing will be discussed.

Keywords Induction heating · Eddy current · Synthesis of alloys · Welding · Selective heating · Brazing

8.1 Introduction

Synthesis of materials by thermal heating is one of the most commonly adopted techniques. Heating by resistive furnace, arc melting furnace, induction furnace, laser irradiation method, and plasma sputtering method are examples of frequently employed heating methods for the synthesis and processing of materials. Each of these heating methods has certain advantages as well disadvantages. The choice of the heating technique depends on the nature and amount of material to be heated, temperature of synthesis, duration of heating cycle and requirement of gaseous environment over the sample, etc. Induction heating is one of the most efficient methods for the large-scale synthesis and processing of air-sensitive metallic samples. In addition to the synthesis of metallic phases, it is extensively used for

---

R. Mishra (✉)
Chemistry Division, Bhabha Atomic Research Centre, Mumbai 400008, India
e-mail: mishrar@barc.gov.in

Homi Bhabha National Institute, Anushaktinagar, Mumbai 400094, India

A. K. Tyagi and R. S. Ningthoujam (eds.), *Handbook on Synthesis Strategies for Advanced Materials*, Indian Institute of Metals Series,
https://doi.org/10.1007/978-981-16-1807-9_8

material processing such as hardening, annealing, tempering and brazing, etc. In induction method, an electrically conducting object is heated by the means of electromagnetic induction, where rapidly changing magnetic field penetrates the conducting object, producing electric current in the conductor. The flow of current in the conducting object causes heat effect due to resistive heating.

8.2 Principle of Induction Heating

Inductive sample heating is based on the principle of electromagnetic induction. When a metallic object is subjected to a changing magnetic field, the magnetic flux passing through the sample produces induced current. According to Faraday's law of induction, the induced electromotive force (emf) E is proportional to the rate of change of the magnet flux.

$$E = -\frac{n\mathrm{d}\varnothing_B}{\mathrm{d}t} \tag{8.1}$$

where, n is number of turns. The expression for the magnetic flux can be given by the relation:

$$\Phi_B = \mu_0 I_c n \pi r_0^2 \tag{8.2}$$

where, μ_0 is magnetic permeability of free space, I_c is the alternating current passing through the primary coil, n is number of turns in the coil and πr_0^2 (r_0 is the radius of the coil) is the area of single turn of the primary coil. The induced emf causes flow of current which is known as eddy current. Figure 8.1 depicts the generation of eddy current by high-frequency ac current.

Heating power 'P' due to eddy currents in conducting material placed in changing magnetic field can be expressed as $P = E^2/R$, where R is the resistance of the conducting material 'load'. Thus, the magnitude of the eddy current in a loop is proportional to the strength of the magnetic field, the area of the loop, the rate of change of flux, and inversely proportional to the resistivity of the material. In order to have high and varying current in the work-coil, an oscillatory circuit formed by inductor and capacitor in series or parallel connection mode is connected to the working coil.

8.2.1 *Factors Affecting Induction Heating*

The nature of magnetic flux produced by the alternating current passing through a coil largely depends on the coil geometry. The coil should be designed in such a manner that it should fit the heating element perfectly resulting in a rapid and

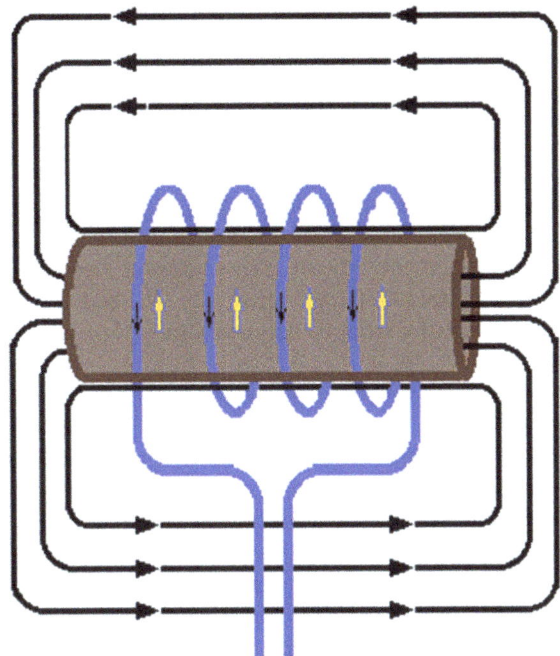

Fig. 8.1 Generation of eddy current by high-frequency ac current in the induction heater

uniform heating [1, 2]. Higher the rate of change of magnetic flux higher is the heating rates. The rate of change of magnetic flux, in turn, depends on the frequency (f) of alternating current. The frequency (f) of the alternating current related to the inductance (L) and capacitance (C) of the circuit by the relation:

$$f = \frac{1}{2\pi\sqrt{LC}} \tag{8.3}$$

For a given coil with fixed value of inductance (L), frequency can be varied by changing the capacitance of the circuit. Higher heating rates can sometimes lead to inhomogeneous heating in the sample [1]. Coupling distance of the heating element from the coil is yet another design parameter that can drastically affect the heating process. By reducing the distance of the heating coil, it is possible to achieve efficient and homogeneous heating with higher heating rate. However, this can be sometimes associated with local overheating [3, 4].

The heating by electromagnetic induction process can be divided into three physical processes such as: (a) transfer of energy from the inductor to the sample to be heated through electromagnetic fields, (b) conversion of the electric energy into heat by Joule effect ($P = I^2R$) and (c) transmission of the heat by thermal conduction. The rate of heating of the object called "susceptor" not only depends on the frequency and the intensity of the induced current, but also on the mass, specific heat, magnetic permeability, and resistance of the material to the flow of current. In the heating process, there is no direct physical contact between the susceptor and the inductor (usually a coil) carrying high-frequency ac current. In induction process, only a limited part of the susceptor is heated due to skin effect. The skin effect is the phenomenon, in which an ac electric current redistributes within the conductor with largest current density at the surface of the conductor. The current density decreases exponentially with the depth of the conductor. The origin of the skin effect is the flow of eddy current in opposite direction formed by changing magnetic field as shown in Fig. 8.1. The eddy current '*I*' flowing at the depth '*d*' from the surface can be expressed as

$$I = I_s e^{-d/\tau} \tag{8.4}$$

where, 'I_s' is the eddy current flowing at the surface of the conductor and τ is the skin depth. From the above equation, skin depth can be defined as the depth of the conductor at which the eddy current decreases by a factor $1/e$ of the value at the surface or in other words skin depth is the depth from the surface of the conductor surface where 87% of power is developed. The depth of induced electric current flowing through the susceptor is related to frequency of the alternating current '*f*', conductivity of the material 'σ' and magnetic permeability of the material 'μ', by the relation

$$\tau = \frac{1}{\sqrt{\pi f \sigma \mu}} \tag{8.5}$$

The dependence of skin depth on frequency of alternating current for copper and heating steel (AISI-316) is given in Table 8.1 [5]. The data indicate that as the frequency of the AC current increase 10^2–10^6 Hz the skin depth for steel (AISI-316) decreases progressively from 43.32 to 0.43 cm. Similarly, skin depth of copper decreases from 6.68 to 0.67 cm as the ac frequency increases from 10^2 to 10^6 Hz.

Table 8.1 Comparison of skin depth as a function of frequency for copper and steel (AISI-316) material

Frequency (Hz)	Copper ($\mu_r \sim 1$)	Steel (AISI-316) ($\mu_r \sim 1$)
1,000,000	0.067	0.43
200,000	0.15	0.97
100,000	0.21	1.37
10,000	0.67	4.33
1,000	2.11	13.70
100	6.68	43.32

8.3 Construction of Induction Heater

An induction furnace basically consists of a power source with adequate power output and frequency range, an induction coil, a chiller unit, a heating element called susceptor and a temperature measurement device, etc. The sketch diagram induction heating process is presented in Fig. 8.2.

The sketch diagram of the induction heating set-up used for small-scale laboratory synthesis of alloys and intermetallic phases is shown in Fig. 8.3.

8.3.1 Power Supply Unit

The power supply unit provides stable and desired radio frequency (RF) alternating current to the induction coil, which acts as a primary circuit of a transformer and generates alternating magnetic field. The metallic load "susceptor" acts as a secondary coil of the transformer, producing eddy current. The electronics of induction furnace is designed for continuous operation of the furnace with long duration at a set power. The modern induction furnace has an Insulated Gate Bipolar Transistor (IGBT) power supply for fine control of temperature. It has parallel compensated power generator with a frequency ranging from 0 to 500 kHz and a variable power ranging from zero to several kW. The maximum power of the induction unit depends on the heater capacity required to achieve the objective of experiment. The designed power considers the size and shape of the heating chamber, various modes of heat losses viz., radiation, conduction and convection, heat capacity of load, heat of fusion, heat of vaporization, and heat of reaction. A typical equivalent circuit representing induction heating system is given in Fig. 8.4.

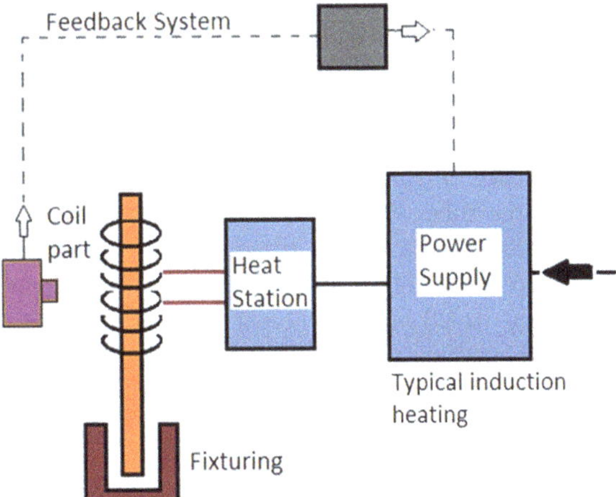

Fig. 8.2 Sketch diagram of induction heating unit

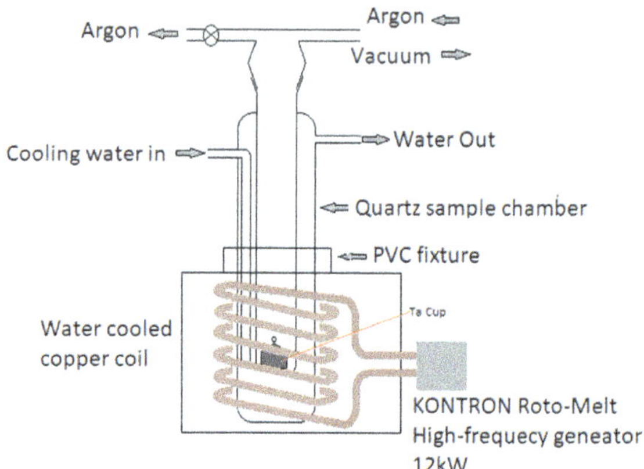

Fig. 8.3 Sketch diagram of an induction heating system for laboratory uses

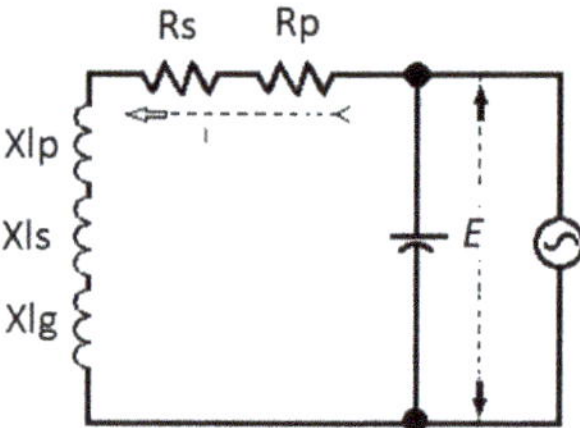

Fig. 8.4 Typical equivalent circuit for an induction heating load

Where, E is voltage, R_p is primary coil resistance, R_s is the secondary coil viz., load resistance, $X_{Ip,}$ X_{IS} and X_{IG} primary coil, secondary coil, and air gap reactance, respectively,

8.3.2 *Induction Coil*

The induction coil is made up of a conducting flexible tube with sufficient mechanical strength. Coil geometry plays important role in induction setup in fixing coupling distance, coil current, and electrical power. Proper design of the induction coil is therefore important in achieving desired temperature and power efficiency. The induced emf and the power rating of the furnace depend on the number of turns in and the tube diameter of the coil. Generally, water-cooled multi-turn helical copper induction coils of different sizes and shapes are used for various applications. For laboratory use flexible water-cooled copper coils with tubes diameters in the range 10–20 mm and height 100–200 mm are employed. Single-turn coils are used for producing a narrow band of hot zone. This type of coil is important for purification of metals by zone refinement process. For heating one side of a material a Pancake type coil is employed. Induction coils of different shapes are shown in Fig. 8.5.

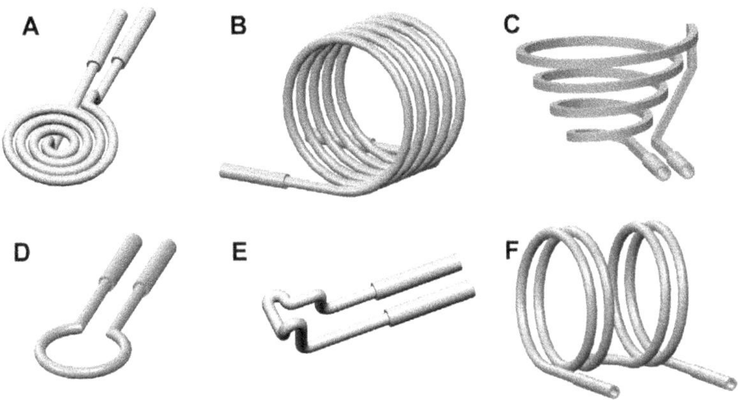

Fig. 8.5 Induction coils of different shape (reproduced with permission from Elsevier, Ref. [6])

8.3.3 *Heating Element*

All types of metallic samples can be directly heated by subjecting them to varying magnetic field employing induction method. Non-metallic samples, which are bad conductors of electricity, cannot be heated directly using induction coil. However, such types of materials can be heated by placing them inside a metallic container. The container metal should have high melting point and low vapor pressure and should be chemically compatible with the sample at high temperatures.

The container materials for induction can be made from high melting conducting materials like tantalum, tungsten, platinum, graphite, etc. The size and shape of the container/load should be such that the entire body of the container is well within the hot zone of induction heating. This ensures uniform temperature within the sample. For heating, a volatile sample container vessel must be designed in such a manner that vapor does not escape during heating. For heating volatile samples by induction method, it is always better to seal the sample in a high melting compatible metallic vessel such as tantalum or tungsten under inert atmosphere. Pöttgen et al. have synthesized a large number of intermetallic phases containing volatile elements like Mg, Sb using induction method by sealing them in tantalum tubes [7–9]. By such process not only the nominal composition of the sample is preserved, but also areal oxidation due to reaction with atmospheric oxygen can be prevented. In induction heating method, it is also possible to fine control of atmosphere over the sample like many other techniques. With proper design, it is possible to heat the materials under ultra-high vacuum employing vacuum pumps or under control atmosphere by passing inert gas over the sample. The oxygen partial pressure around the sample can be controlled by evacuation and purging of the high purity inert gas like argon or helium. The residual oxygen in ppm level present in the inert gas can be purified by passing the carrier gas over oxygen getter such as titanium sponge maintained at 900 °C. Further one can control the oxygen partial pressure by placing non-volatile oxygen getters such as Ti and Zr sponge near the vicinity of sample. In order to arrest the change in composition of the sample particularly for the volatile materials, sample is usually sealed in a tantalum tube. For low temperature sample preparation, sealed quartz ampoules can be used. In such cases, care must be taken to ensure that the sample components do not react with container of the sealed tube. The proper mixing of the sample during heating process is possible by levitation technique, where the sample holder can be rotated to a certain extent.

8.3.4 Chiller Unit

Induction heating system requires a common cooling facility for cooling power source and the inductor coil and the surrounding container heated through radiation heating. The chiller unit contains de-ionized water that circulates through power supply and the coil for cooling purposes. Chiller consists of a non-corrosive SS-304c tank for storing cooling medium, a non-corrosive pump, chilling unit with CFC-free refrigerant, and a leak tight top lid. The cooling medium equilibrated at a desired temperature is circulated by the pump. The flow velocity of the circulating fluid is maintained such a way that the outlet temperature of the fluid is close to the room temperature.

8.3.5 Temperature Measurement

The temperature of the load/susceptor can be measured using thermocouples, radiation detection system such as optical pyrometer or IR detector, eddy current detection system and by measuring voltage, current, phase angle, frequency of the working coil. Among these, optical pyrometer is the most commonly used for measuring temperature in the range 700–2,500 °C. In the low temperature measurement range, either IR detector or a shielded thermocouple can be employed. The output of the pyrometer or IR detector or the thermocouple is fed to the power supply unit for controlling the power source at any given time. A PID-based temperature controller/programmer with feedback of the thermocouple and pyrometer output is used to control the temperature of the load. The precision of the temperature control depends on the actual temperature of the body. The variation of measured temperature increases with increase in temperature. For example, if the temperature variation is ±1 °C at 1,000 °C, it can be ±10 °C at 2,000 °C. All induction heating systems have inbuilt safety features such as inter locks to vessel temperature, coolant temperature, and coolant flow from the power source and safeguards for the induction equipment, circuit breaker, semiconductor fuses, door interlock, etc.

8.4 Advantages of Induction Heating

Induction heating process has many distinct advantages over conventional heating which makes it a preferred technique in industrial application. Induction heating is primarily a more energy-efficient heating process compared to conventional resistive heating using electrical energy or fossil fuel. In the furnace heating the heat produced by the heating elements dissipates its energy in all the 4π-directions. In resistance heating only a fraction of the total energy is utilized for heating the material of interest, and rest of the energy is lost to the surroundings. However, in induction heating the sample is directly heated by flow of eddy current produced by induction process. In this case, the only pathway for loss of energy is through radiation, conduction, convection, and hysteresis loss. These heat loss paths can be effectively reduced by properly shielding the heated object, minimizing the conduction and convection heat loss through proper geometrical design of the system. Since induction method is a localized heating process, it is possible to heat the required portion of material selectively by proper experimental design.

The second most important aspect of induction heating is that it reduces the heating time. One can achieve a temperature of the order 2,000 °C in few minutes which takes several hours to reach in conventional heating processes. In induction process, as the desired sample temperature is achieved fast, the material loss due to vaporization of volatile components during heating reduces considerably. Similarly, the areal oxidation of the metals or alloys also gets reduced. Thus, induction process is the most preferred heating process for large-scale production of metals and alloys.

Induction process can be applied to magnetic particles to produce heating effect. With low AC frequency (of the order few kHz), the magnetic particles can generate temperature slightly higher than the body temperature. This magnetic heating finds applications in the area of hyperthermia-based cancer treatment effect to kill cancer cells selectively [10].

8.5 Applications of Inductive Heating

Induction heating is a convenient, efficient, and a rapid method of heating a conducting material under control atmosphere. The method is widely used for synthesis of alloys, intermetallic phases, and high melting ceramics. Induction heating technique finds enormous applications in heavy industries for melting large number of ferrous and nonferrous alloys. The use of induction plasma method for the synthesis of ceramics, alloys and metals, chlorides, fluorides, nitrates, oxalates, etc. in their nanoform have been demonstrated in a recent US patent [11]. Besides solid materials, induction heating process can also be employed for heating of conducting liquid and gaseous substances. In semiconductor industries, induction heating process is used to process high purity silicon. For contactless heating process, vacuum furnaces make use of induction process for making special alloys. The technique finds applications in the welding of metals as well as some plastics doped with ferromagnetic ceramics. Induction stoves used in the kitchen works on the inductive heating principle, for brazing carbide to shaft, for tamper-resistant cap sealing on bottles and pharmaceuticals and for plastic injection modeling machine to improve energy efficiency for injection.

8.5.1 Synthesis of Alloys and Intermetallic Phases

Induction heating is the most suitable method for synthesis of alloys and intermetallic phases. The technique has been adopted by large number of researchers for synthesis of alloys and intermetallic phases [12–15]. Pöttgen et al. have synthesized large numbers of binary and ternary intermetallic phases and studied their crystal structure and magnetic properties. Details of the synthesis, characterization, and properties of intermetallic phases have been published in recent books [16, 17]. Gutmen et al. [18] have reported synthesis of zirconia-yttria and alumina-based ceramic employing induction method. Huang et al. [19] synthesized Al_2O_3–AlB_{12}–Al composite by laser induction complex heating method and measured its mechanical properties. Zhu et al. [20] have reported combustion synthesis of NiAl alloy employing induction method. Effect of nickel boride additive on simultaneous densification and phase decomposition of TiB_2–WB_2 solid solutions by pressure-less sintering using induction heating has been reported in Ref. [21]. Sosnowchik et al. [22] demonstrated a simple and rapid method for synthesis carbon nanotubes at room temperature employing induction technique employing nickel as susceptor.

8.5.2 Induction Process in Material Processing

The method is also employed for material process treatments like hard brazing, tin soldering, thermal hardening, annealing, tempering, etc. The technique has been employed for targeted surface heating, melting, soldering process. Polymers and composites are not susceptible to induction heating. For heating of such materials by induction process special arrangements like introducing metallic or graphitic carbon fibre susceptors are required. When the working piece holding the sample is metallic, addition of extra susceptor is not necessary. Some examples of induction heating of polymers and composites are described as follows.

Fig. 8.6 Continuous induction welding robot developed by IVW GmbH (reproduced with permission from Elsevier, Ref. [6])

8.5.2.1 Welding of Thermoplastic Composite

Binding two composite materials or composite material with a metal can be possible by welding them together using induction technique [23–25]. Welding of two polymers is done by heating two closely contact materials together above their melting points under control atmosphere. In this welding process, a conducting susceptor is placed surrounding the contact joint. One can also use metallic mesh, graphitic carbon fibres or ferromagnetic materials as induction susceptor to melt the joint. A Continuous induction welding robot developed by IVW GmbH is shown in Fig. 8.6.

8.5.3 Thermoset Curing

Besides melting and welding, induction method is also frequently employed for curing materials at desired temperatures. Tay et al. [26] carried out detailed comparison of curing by induction process with the other conventional methods and observed that for the same extent of curing there is a drastic reduction in the curing time in induction heating i.e. 15 min as compared to 1 h in oven heating method. Similar observations were also recorded by Wetzel et al. in the curing of resins with less degradation. Miller et al. cured epoxy resin employing $FeCo/(FeCo)_3O_4$ nanoparticles as induction susceptor and reported that smaller size particles imparted larger heating rates. So far, the thermoset curing by induction process is being applied in small scale.

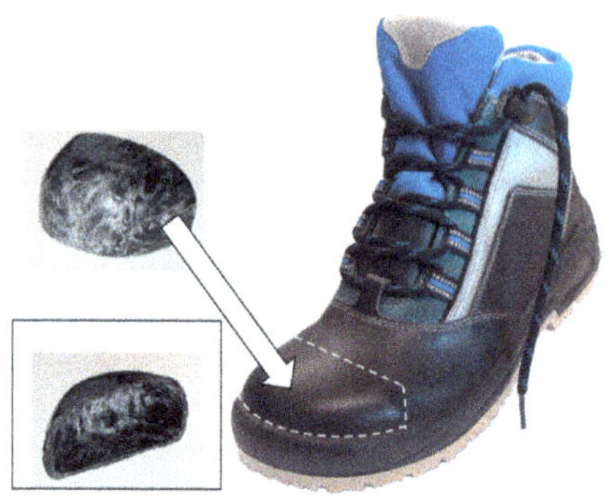

Fig. 8.7 Thermoformed toe caps made from inductively heated PP/HDPE sheets (reproduced with permission from Elsevier)

8.5.4 Selective Heating

Induction heating is very handy for selective heating of material in micro-scale. In such case, the geometry of induction coil is very crucial in controlling the temperature of the selected portion of the material due to limited induction coupling [1, 2, 4, 27]. For non-conducting material position and shape of the susceptor decides the heating zone. Selective heating of polymer-reinforced composite materials using dispersed ferromagnetic susceptor have been reported by many authors [3, 6, 28–30].

Toe cap of a shoe (PP/HDPE) thermoformed by selectively heated induction is shown in Fig. 8.7 [6]. Local thermoforming of component of a sewing machine by inductive heating method has been reported by Lahr et al. [31].

8.6 Summary

Induction heating is a versatile method for the preparation of metals, alloys, and intermetallic phases. In this chapter, principle of induction heating, construction of induction heater, and general applications in the synthesis and processing of metallic samples have been described. The uses of induction heating for selective heating, welding, and thermoset curing have been brought out.

References

1. Zinn S, Semiatin L (1988) Elements of induction heating—design, control, and applications. ASM International, Metals Park
2. Davies EJ (1990) Conduction and induction heating. Peter Peregrinus Ltd., London
3. Bayerl T (2012) Application of particulate susceptor for the inductive heating of polymer–polymer composites. IVW GmbH, Kaiserslautern

4. Moser L (2012) Experimental analysis and modelling of susceptorless induction welding of high performance thermoplastic polymer composites. IVW GmbH, Kaisersalutern
5. www.ceia-power.com
6. Bayenl T, Duhovic M, Mitscharg P, Bhattachayya D (2014) Composite: part A 57:27–40
7. Fickenscher T, Pöttgen R (2001) Synthesis and crystal structure of REAgMg (RE = La, Ce, Nd, Eu, Gd, Tb, Ho, Tm, and Yb). J Solid State Chem 161:67–72
8. Mishra R, Hoffmann R-D, Pöttgen R (2001) New magnesium compounds RE2Cu2Mg (RE = Y, La-Nd, Sm, Gd-Tm, Lu) with Mo2FeB2 type structure. Z Naturforsch 56b:239–244
9. Mishra R, Pöttgen R, Hoffmann R-D, Kaczorowski D, Piotrowski H, Mayer P, Rosenhahn C, Mosel BD (2001) Ternary rare earth (RE) gold compounds REAuCd and RE2Au2Cd. Z Anorg Allg Chem 627:1283–1291
10. Joshi R, Perala RS, Srivastava M, Singh BP, Ningthoujam RS (2019) Heat generation from magnetic fluids under alternating current magnetic field or induction coil for hyperthermia-based cancer therapy: basic principle. Radition Cancer Res 10:156–164
11. United States Patent No. US8013, 269B2 dated 2.9.2011
12. Jeitschko W (1970) The crystal structure of TiFeSi and related compounds. Acta Crystallogr B 26:815–822
13. Pöttgen R, Albering JH, Kaczorowski D, Jeitschko W (1993) Structure refinements of actinoid transition metal stannides $AnT_{2-x}Sn_{2-y}$ (An = Th, U; T = Co, Ni, Cu) with defect $CaBe_2Ge_2$ type structure. J Alloy Compd 196:111–115
14. Chauhan S, Verma V, Prakash U, Tewari PC, Khanduja D (2017) Studies on induction hardening of powder-metallurgy-processed Fe–Cr/Mo alloys. Int J F Miner Metall Mater 24:918–925
15. Kandavel K, Panneerselvam T, Karthikeyan P (2015) Optimization of deformation and densification properties of the sintered plain carbon steel. Mater Manuf Process 30:1240–1249
16. Pöttgen R, Johrendt D, Gruyter D (2014) Intermetallics. Berlin
17. Pöttgen R, Johrendt D, Gruyter D (2019) Intermetallics, 2nd edn. Berlin
18. Gutman VI, Reznikov AG, Lyubalin MD (1992) Induction melting for preliminary synthesis of oxide ceramics in the ZrO2-Y2O3-Al2O3 system. Refractories 33:416–422. https://doi.org/10.1007/BF0128338
19. Huang KJ, Yan L, Kou HM, Xie CS (2010) Synthesis of Al2O3/AlB12/Al composite ceramic powders by a new laser-induction complex heating method and a study of their mechanical properties. Appl Mech Mater 29–32
20. Zhu X, Zhang T, Morris V, Marchant D (2010) Combustion synthesis of $NiAl/Al_2O_3$ composites by induction heating. Intermetallics 18:1197–1204
21. Shibuya M, Kawata M, Ohyanagi M, Munir ZA (2003) Titanium diboride-tungsten diboride solid solutions formed by induction-field-activated combustion synthesis. J Am Ceram Soc 86:706e10
22. Sosnowchik BD, Lin L (2006) Rapid synthesis of carbon nanotubes via inductive heating. Appl Phy Lett 89:193112
23. Yarlagadda S, Fink BK, Gillespie JW Jr (1998) Resistive susceptor design for uniform heating during induction bonding of composites. J Therm Compos Mater 11:321–337
24. Moser L, Mitschang P (2010) In the shifting magnetic field. Kunstst Int 100(7):26–28
25. Wacker M, Moser L, Schlarb AK, Tradt HR (2008) Flexible 3D joining process for complex fibre composite components. Join Plast 2(4):266–271
26. Tay TE, Fink BK, McKnight SH, Yarlagadda S, Gillespie JW Jr (1999) Accelerated curing of adhesives in bonded joints by induction heating. J Compos Mater 33:1643–1664
27. Sosnowchik D, Lin L (2006) Rapid synthesis of carbon nanotubes via inductive heating. Appl Phys Lett 89:193112
28. Yang HA, Lin CW, Peng CY, Fang W (2006) On the selective magnetic induction heating of micron scale structures. J Micromech Microeng 16:1314–1320
29. Svoboda J (1987) Magnetic methods for the treatment of minerals. Elsevier, Amsterdam

30. Dutz S, Hergt R, Mürbe J, Müller R, Zeisberger M, Andrä W (2007) Hysteresis losses of magnetic nanoparticle powders in the single domain size range. J Magn Magn Mater 308:305–312
31. Lahr R (2007) Partielles thermoformen endlosfassverstärkter thermoplaste. Kaiserslautern, IVW GmbH

Chapter 9
Synthesis Strategy for Functional Glasses and Glass-Ceramics

Mohsin Jafar and V. Sudarsan

Abstract Glass and glass-ceramics play vital roles in today's technology development. Although glasses are known to mankind from pre-historic period, its wide applicability as a candidate for various technological applications was realized during the past century. Advancement in communication technology was mainly possible with fabrication of high quality glass fibers. Glass-ceramics are derived from glasses and possess both the advantages of glasses and ceramics. In this chapter, initially, the origin of glasses and glass-ceramics are discussed followed by their classification depending up on the nature of glass forming constituents. Synthesis routes for preparation of different types of glasses are discussed in detail. Thermodynamics and kinetics of glass formation and its conversion to glass-ceramics are explained towards end of the chapter. Structure property correlation is an important aspect in any studies involving material development and this aspect is explained in detail in this chapter. Techniques like solid-state NMR, XPS, EXAFS which give mainly information on short-range order in glasses have been explained with examples. Importance of thermal techniques for understanding glass and ceramic science has been brought out in this chapter. Finally, the chapter ends with certain representative applications of glasses and glass-ceramics.

Keywords Glass · Glass-ceramics · Synthesis · Glass transition · Crystallization · XRD · FTIR · MAS · NMR · DSC

M. Jafar · V. Sudarsan (✉)
Chemistry Division, Bhabha Atomic Research Centre, Mumbai 400085, India
e-mail: vsudar@barc.gov.in

M. Jafar · V. Sudarsan
Homi Bhabha National Institute, Mumbai 400094, India

A. K. Tyagi and R. S. Ningthoujam (eds.), *Handbook on Synthesis Strategies for Advanced Materials*, Indian Institute of Metals Series,
https://doi.org/10.1007/978-981-16-1807-9_9

9.1 Introduction

9.1.1 Origin of Glass and Glass-Ceramics

Glasses can be considered as one of the oldest materials used by mankind over several centuries. It is amorphous in nature and is generally formed by mixing several solid substances, heating to melting stage followed by super-cooling to an extent to obtain a single solid-like structure. Despite its appearance as solid form, it is not termed as a solid and is considered as a semi-solid, fluid, or super-cooled liquid with very high viscosity. In a broader sense, glass is defined as a non-crystalline or amorphous solid which exhibits glass transition phenomenon upon heating. Glasses find wide technological applications due to their ease of synthesis and tailor-made properties.

Natural glasses mostly include obsidian (volcanic glass), tektites (meteorite impacts), fulgurites (formed when lightning strikes sand), metamicts (amorphization by radiation from neighboring radioactive atoms), etc. Ancient people used obsidian glasses as ornaments, valuables, and weapons. Scarcity of natural glasses paved the way for the development of man-made glasses. Archaeological excavations and studies revealed that first man-made glasses date to around 3500 BC in Egypt and Eastern Mesopotamia [1]. In medieval ages, around 1300 AD, glass making belonged to Venetian craftsmen who were considered to be expert in this field. George Ravenscroft of England made an important advancement in this field in 1674 when he used the technique of adding lead oxide to molten glass to increase its lifetime. The Industrial Revolution in England paved the way for increasing automation in glass making and converted this art into an industry. The process of making flat glass in the most convenient and reproducible way is credited to Émile Fourcault of Belgium. The method known as Fourcault process was commercialized in 1914. However, modern day glass-making process took shape from Sir Alastair Pilkington's float glass manufacturing process in 1959 by which more than 90% of glasses are manufactured till date.

Glass-ceramics, on the other hand, are a comparatively newer class of materials with history dates to 1957 AD. Glass-ceramic consists of an amorphous phase (matrix) and one or more embedded crystalline phases produced as a result of controlled crystallization of the matrix. Prominence of glass-ceramic materials in today's world is highly related to the demands of new technological advancements. An ideal glass-ceramic material should have fabrication advantage of glass coupled with special or required properties of ceramics. Glass-ceramic materials usually contain crystalline phase to the tune of 30–90% with ceramic properties such as opacity, non-brittleness, and high-temperature stability.

Glass-ceramics were discovered accidently by Stanley Donald Stookey [2]. He was working on the development and improvement of properties of Fotoform glasses. However, one night in 1954 he mistakenly heated a piece of experimental Fotoform glass to 900 °C instead of 600 °C. The next morning, he found out that the furnace was overheated. To his astonishment, he noticed an opaque solid which bounced back when accidentally dropped on floor. It was concluded that this may be attributed to presence of microscopic crystals within the matrix. His knowledge of chemistry of crystallization led him to tailor-make the experimental conditions and this resulted in the first glass-ceramic FotoCeram [2]. This accidental discovery led to the development of Corning Ware in the year 1957. Corning Ware (a variety of cookware), the first glass-ceramic material, is considered to be one of the major discoveries by Corning Glass Works and one of Stookey's million dollar inventions. Till date, major developments in the field of glass-ceramics are being carried out for the design of new types of cookwares and induction oven cooktops. A famous quote by the creator of glass-ceramic goes as:

> Why did such an important discovery occur so late in the history of glass, and why was an accident necessary to bring it about?—Donald Stookey, 1977

9.1.2 General Properties of Glass and Glass-Ceramics

Glass-ceramics are primarily synthesized by the method of regulated crystallization of glass melt. However, the controlling step involves heating rate, heating temperature, cooling rate, etc. Variation in each or all of these parameters leads to the generation of a wide variety of physical and chemical properties of glass-ceramics as compared to glass. Thus, it can be said that even we start with same chemical composition to obtain a uniform glass matrix or a glass-ceramic material, and the properties of these two compositions can be entirely different depending upon the process control. Hence, an understanding of some of the general properties of glassy material and its evolution under different experimental conditions is essential for developing materials for different applications.

Some of the general properties of glassy materials are follows:

(a) Most of the glasses used in our daily life are transparent or translucent and allow light to fully or partially transmit through it, depending upon our requirements. Lack of grain boundaries in glasses is responsible for this.
(b) Thermal expansion coefficient (TEC) of most of the glassy materials is of the order of 10^{-6} K^{-1}.
(c) Glass is a brittle material. This particular property is attributed to the presence of surface flaws or distorted structural units.
(d) Mechanical strength of glass is very high due to the absence of any form of porosity within its structure as it is a super-cooled liquid.
(e) Non-metallic glasses are mostly semiconductors or insulators at ambient temperature. At elevated temperatures, ionic movement within the glass matrix becomes energetically feasible leading to ionic conduction.
(f) The strong bonding involved in glass network makes them chemically durable. Absence of grain boundaries, the primary zones of etching in glassy network,

renders their stability toward aqueous or acidic media. However, the presence of alkali or alkaline earth metal ions contributes to leachability of glass.

(g) Glasses can be made entirely recyclable to be reused for indefinite times.

(h) Glasses do not have fixed melting or boiling point. They tend to undergo state change over a range of temperature.

(i) Glasses possess the property of glass transition temperature (T_g) beyond which the glassy or brittle nature is replaced by a viscous or rubbery material. Value of T_g is always lower than the melting point of glass.

Glass-ceramics can circumvent some of the shortcomings of glasses for technological applications. Extensive studies have been carried out in this field in order to fine-tune material properties to satisfy user requirements. A brief description of some of the unique properties of synthesized glass-ceramics is given below.

(a) The thermo-physical property such as thermal expansion coefficient (TEC) can be fine-tuned by modifying the processing condition involved in glass to glass-ceramic conversion [3]. It is now possible to design glass-ceramic materials with very low or even negative thermal expansion coefficient [4, 5]. Glass-ceramic materials also show exotic optical properties. Rare earth elements like Nd-doped glass-ceramics are potential candidates for laser application [6, 7]. Glass-ceramics possess almost zero porosity with lesser extent of surface flaws. Hence, the mechanical strength of glass-ceramic materials is relatively high. Defects present in glass-ceramics make glass-ceramics slightly less brittle. Depending upon the process of fabrication and composition, glass-ceramics can show either higher resistivity or higher conductivity as per requirements. Further, glass-ceramics show less amount of leachability as compared to glassy materials.

In addition to these properties, the glass component of glass-ceramic exhibits glass transition temperature (T_g), whereas the ceramic component will exhibit sharp melting point (T_m). It may however be noted that the T_g and/or T_m may shift to higher or lower temperatures depending upon the chemical environment in these materials as compared to base glass and ceramics. Thus, from the above discussion, it is clear that glass-ceramics are those materials which cover both the advantages of glass and ceramic materials. This eventually led to development of materials with improved functional properties. This aspect is discussed in subsequent part of this chapter.

9.1.3 Functional Glasses and Glass-Ceramics

The word "functional" means "having a specific activity, task, or usefulness". In general, functional glasses and glass-ceramics mean materials having some useful and characteristic properties enabling them suitable for definite technological applications. Functional glasses and glass-ceramics can be classified into four types

Table 9.1 Classification of functional glasses and glass-ceramics

Functional glass and glass-ceramics			
Chemical composition	Fabrication	Property	Application
• Oxide-based (phosphate, silicate, borate, etc.) • Non-oxide-based (chalcogenide, halide, metallic, etc.)	• Thermal evaporation • Sputtering • CVD • Melt-quenching • Glow discharge decomposition • Radiation-induced damage	• Optical • Photonic • Electrical • Thermal • Electronic • Mechanical • Chemical	• Energy • Display • Health care • Nuclear • Defense • Space

based on (a) chemical composition, (b) fabrication, (c) property, and (d) application as can be seen from Table 9.1.

(a) **Chemical composition based**: Chemical composition-based glasses and glass-ceramics can be classified into two groups, namely the oxide-based and the non-oxide-based. Oxide-based glasses and glass-ceramics are those materials whose network is formed from oxides such as phosphorous pentoxide (P_2O_5), silica (SiO_2), boron trioxide (B_2O_3), and germanium oxide (GeO_2). On the other hand, non-oxide-based glasses and glass-ceramics mostly hover around chalcogen elements (sulfur (S), selenium (Se), tellurium (Te), arsenic (As), and antimony (Sb). Other elements such as germanium (Ge), gallium (Ga), inorganic halides such as fluorine (F), chlorine (Cl), bromine (Br), iodine (I), and metals or alloys are involved in the formation of non-oxide-based glasses. This will be discussed in detail in Sect. 9.2.

(b) **Fabrication based**: Functional glasses or glass-ceramics can be classified based on their methods of preparation or fabrication. Some important fabrication methods for glass and glass-ceramics are thermal evaporation, sputtering, chemical vapor deposition (CVD), melt-quenching, glow discharge decomposition, radiation-induced damage, etc. These techniques will be discussed in detail in Sect. 9.3.

(c) **Property based**: Different properties associated with materials are basis for the origin of a class of materials known as "functional materials." Glasses and glass-ceramics are classified into six broad categories based on optical, photonic, electrical, thermal, electronic, mechanical, and chemical properties.

(d) **Application based**: Although this is not a criterion for classification of materials, it is closely related to the properties of materials. Many times, properties of materials only decide the application in which the material can be employed. The application or technology includes area such as energy, environment, health care, defense, and space. These points will be dealt in detail in applications part, i.e., Sect. 9.9.

9.2 Different Types of Functional Glasses and Glass-Ceramics

As mentioned earlier, functional glasses and glass-ceramics can be classified under different criteria as shown in Table 9.1. In the following section, different types of functional glasses and glass-ceramics are discussed on the basis of their composition.

9.2.1 *Oxide-Based Glasses*

Glasses formed by bridging of oxygen atoms (O) with other suitable atoms (X) are commonly referred as oxide glasses. The choice of "X" for oxide glass formation is generally guided by rules proposed by a famous Norwegian-American physicist, Wiliam Houlder Zachariasen, in the year 1932. The rules are summarized below

(i) An O atom should be connected to not more than 2 X atoms.
(ii) There should not be more than 3 to 4 O atoms connected with one X atom.
(iii) The polyhedra formed by the coordination of X–O bond should be only corner shared. Edge sharing and face sharing polyhedra do not result in glass formation.
(iv) Three corners of polyhedra should be shared in order to sustain a 3-D glassy network.

In addition to the above rules, other terminologies like network modifier and intermediates are used to understand and explain glass formation. Network formers are those atoms (X) that can form highly crosslinked 3-D network of covalent chemical bonds. Some common network formers are oxide of silicon (Si), boron (B), phosphorous (P), and germanium (Ge). Network modifiers are mostly metals such as calcium (Ca), lead (Pb), sodium (Na), potassium (K), etc., which prefers to remain in cationic form inside glassy network. The favored cationic forms of these elements alter glass network, whereby a bridging X–O–X bond breaks resulting in the generation of non-bridging O species. There are certain elements such as aluminum (Al), titanium (It), beryllium (Be), and lead (in certain cases) which can act both as network former or network modifier. Hence, these elements are termed as intermediates as their role varies with variation in glass composition. Nomenclature of oxide glasses is made based on their constituent X–O structural units.

(a) **Phosphate glasses**: Glassy materials based on the network former PO_4^{3-} (X = P) are broadly called as phosphate glasses. The phosphate glass network consists of one terminal P = O and three bridging P–O–P bonds. The most common starting material for preparation of this type of glass is P_2O_5. Highly moisture sensitive nature of P_2O_5 renders these glasses to be of less technological importance. Hence, suitable network modifiers or intermediates are always required during preparation of phosphate glasses for making them technologically relevant. In general, these glasses possess very high thermal expansion coefficient and melt formation occurs at relatively lower temperatures.

(b) **Silicate glass**: These are the most versatile glassy materials having SiO_4^{4-} (X = Si) as their constituent building blocks and the starting material for preparation of such glasses is SiO_2. This type of glass network mostly consists of four bridging Si–O–Si bonds. The most common types of such glasses include alkali and alkaline earth silicate glasses where introduction of these elements breaks the bridging Si–O–Si bonds resulting in the generation of a negative charge at oxygen ($Si–O^-$) which is balanced by suitable number of mono or divalent cations. Replacement of rigid bridging bonds with less rigid non-bridging linkages results in increase of thermal expansion coefficient and decrease in micro-hardness of these glasses. Higher viscosity of melt and rapid volatilization of alkali metals are problems encountered during preparation of silicate glasses with very low alkali metal concentration.

(c) **Borate glass**: The most common starting material for single component borate glasses is boron oxide (B_2O_3) or boric acid (H_3BO_3), and the main structural building blocks in different borate glasses are BO_3 and BO_4 units along with boroxol ring. Pure borate glasses are difficult to prepare because of affinity toward moisture and this can be avoided by preparing glasses at very low atmospheric pressure of around 1 mm of Hg. Borate glasses are mostly found to be low melting and hence find use as solder glasses.

(d) **Borosilicate glass**: The starting materials for the preparation of these glasses are boron oxide (B_2O_3) or boric acid (H_3BO_3) and SiO_2. Typical network follows with four bridging Si–O–Si bonds and three bridging B–O–B bonds. However, depending upon the ratio of these two starting materials, an average network of B–O–Si is obtained with coordination number between 3 and 4. In order to maintain proper network of four bridging bonds, some network modifiers such as Na_2O, BaO, etc., are added during glass formation. Generally, it is considered that at low modifier concentration, Na_2O reacts with BO_3 to form BO_4 structural units, whereas at higher modifier concentration, Na^+ is equally shared by BO_4- and SiO_4 network. At a given alkali content, addition of B_2O_3/H_3BO_3 to silicate glasses markedly reduces the thermal expansion coefficient and enhances the chemical durability specially to attack by acids.

(e) **Aluminate glass**: Alumina (Al_2O_3) being a refractory material, single component alumina glass is difficult to prepare. However, with addition of suitable alkali or alkaline earth metal oxides binary and ternary glasses with Al_2O_3 can be formed. Addition of silica or germanium oxide (GeO_2) transforms Al_2O_3-based glasses into technologically relevant material which will be discussed in Sect. 9.9.

(f) **Aluminosilicate glass**: These glasses mainly consist of around 20 mol% of alumina (Al_2O_3) with small amounts of calcium oxide (CaO), magnesium oxide (MgO) along with major content of silica (SiO_2). Aluminum mainly exists as tetrahedrally and octahedrally coordinated species in these glasses. Further, in these glasses, Si atom is tetrahedrally coordinated with oxygen having different number of Al as the next-nearest neighbors. These glasses can withstand very high temperatures and thermal shock.
(g) **Germanate glass**: Pure germanium oxide (GeO_2) has the ability to readily form glassy phase. However, for practical applications, GeO_2 along with SiO_2 are added to calcium aluminate glasses. The main building block of these glasses is GeO_4 tetrahedra resulting in the formation of an extended network. Introduction of alkali metal ions in low concentration results in transformation of GeO_4 tetrahedra to GeO_6 octahedra. Under higher concentration, these modifiers result in the formation of non-bridging oxygen atoms attached with Ge in the glass structure.
(h) **Tellurite glasses**: Tellurite glasses were first prepared and studied by Stanworth in 1952. Single component TeO_2 (pure TeO_2) forms glass when melted in gold crucible. The most probable reason may be reaction of alumina crucible with the starting material in the former case. Nevertheless, addition of alkali or alkaline earth metal oxides, PbO, WO_3, etc., is found to facilitate tellurite glass formation.
(i) **Vanadate glasses**: Pure V_2O_5 melts around 660 °C and forms glass only when it is cooled rapidly. Glass formation with V_2O_5 is quite common upon addition of a number of oxides like P_2O_5, TeO_2, B_2O_3, and GeO_2. Such glasses are potential candidates for different applications.

9.2.2 *Non-oxide-Based Glasses*

These are the glasses which are mostly formed by bridging of suitable atoms other than oxygen with metal or nonmetals. The major non-oxide-based glasses are discussed in brief as follows:

(a) **Chalcogenide glasses**: These are glasses consisting of one or more chalcogens such as sulfur (S), selenium (Se), and tellurium (Te) apart from oxygen and polonium (Po). Chalcogenide materials are important due to the following reasons:

1. They can be prepared in the amorphous form in a variety of ways.
2. Glass formation occurs over a wide range of composition and the physical properties vary in a continuous fashion.
3. The bandgaps of chalcogenide glasses are of the order of 1–3 eV, and hence, these materials show semiconducting behavior.

Table 9.2 Representative chalcogen-based glasses

System	Examples
1. Pure chalcogen	S, Se, Te, S_xSe_{1-x}
2. Pnictogen (V–VI)	As_2S_3, P_2Se
3. Tetragen-chalcogen (IV–VI)	$SiSe_2$, GeS_2
4. III–VI	B_2S_3, In_xSe_{1-x}
5. Metal chalcogenide	MoS_3, WS_3, Ag_2S–GeS_2
6. Halogen-chalcogenide	As–Se–I, Ge–S–Br, Te–Cl

4. Many of these glasses are candidates for technological applications which include infrared transmission, switching devices in computer memories, etc. Representative chalcogenide glasses are given in Table 9.2.

(b) **Halide glasses**: These are glasses consisting of metal halides such as beryllium fluoride (BeF_2), zinc chloride ($ZnCl_2$), zirconium fluoride (ZrF_4), hafnium fluoride (HfF_4), etc. Other glass forming halide systems are mostly based on binary or ternary phases such as AgCl, AgBr, AgI, and $PbBr_2$, $PbCl_2$–$BaCl_2$, $SnCl_3$–PbI_2, etc. Ionic radius of Be^{2+} is quite close to that of Si^{4+} and that of F^- and O^{2-} are almost identical. This is the main reason for similar resemblance of corner connected tetrahedral network of vitreous BeF_2 and vitreous SiO_2. Glasses based on $ZnCl_2$ are also formed by corner shared tetrahedra of Zn–Cl bonds quite similar to that existing in BeF_2 glass. However, the water solubility of zinc chloride glasses renders them ineffective for important technological applications. On the other hand, glasses based on ZrF_4 and HfF_4 are found to be highly ionic in nature. The basic building blocks in these glasses are not well defined and studies suggest that coordination number between Zr^{4+}/Hf^{4+} and F^- ions varies from 6 to 8. This is the main reason for the ionic nature of these glasses.

(c) **Organic glasses**: Organic glasses mostly consist of severely entangled C–C chains which prohibits crystallinity upon rapid cooling. The network chains in these glasses are found to be crosslinked and are similar to that existing in chalcogenide glasses. Extent of cross-linking dictates the viscosity of melt and the glass transition temperature of such glasses. Their general properties are mostly similar to those of chain-based inorganic glasses.

(d) **Metallic glass**: Metallic glasses can be obtained by rapid cooling of alloy (consisting of only metals or both metals and metalloids) melt. Nucleation and growth of crystal become an important competing factor to glass formation in these materials. Metallic glasses can be prepared by preventing nucleation under very fast cooling of the order of 10^4 Ks^{-1}. Such cooling rates help to generate glass in film or ribbon forms. Some common metallic glasses are $Pd_{80}Si_{20}$, $Ni_{80}P_{20}$, $Fe_{40}Ni_{40}P_{14}B_6$, etc. Structural model of these materials can be explained by assuming random packing of nanosized spheres with smaller spheres occupying the interstitial spaces. Addition of metalloid atoms hinders reorganization of the atoms during cooling, thereby preventing unwanted crystalline domains within the glass network.

Most of the above-mentioned types of oxide-based functional glasses can form glass-ceramics by suitable tuning of synthesis procedure. Modulation of heating and cooling rates, heating at an intermediate temperature during processing, etc., can lead to glass-ceramics with tailor-made properties in these types of materials. Preparation of metallic glass-ceramics requires minimal nucleation at distant points within the melt to prevent precipitation of larger crystals. Greater extent of crystallization in metallic glass leads to a composite system of an alloy and glass which may not be compatible with each other, thereby rendering unsuitable for applications.

9.3 Different Routes for Synthesis of Glass

Some of the common methods for preparation of glasses are as follows:

9.3.1 Thermal Evaporation

This technique is one of the most extensively used methods for preparation of amorphous or glassy thin films [8]. Material to be amorphized is kept in powder form inside an evacuated chamber at around $\sim 10^{-6}$ torr pressure. A substrate is also kept inside the chamber over the sample to be evaporated. The assembly is then heated by Joule heating (or resistive heating) for low melting materials or using electron beam for refractory materials. The powdered sample becomes mobile at higher temperature and percolates to the substrate. Low temperature of the substrate allows the mobile atoms to freeze on reaching the contact surface resulting in the formation of a glassy or amorphous phase. Schematic diagram of this preparative method is given in Fig. 9.1. Amorphous chalcogenide semiconductors like Si, Ge, GaAs, etc., are prepared by this method. Tunability of properties of the prepared films can be achieved by optimization of various process parameters such as substrate temperature, substrate nature, separation and orientation of base, deposition rate and gas pressure inside the chamber. A major disadvantage of this process is preferential evaporation of low melting component which depletes the source leading to compositional in-homogeneities in the prepared material. Moreover, the volatility of components sometimes results in change of compositional identity in the vapor phase compared to the feed source. A common example in this regard is the presence of As_4S_4 molecules in vapor phase despite the starting reactant being stoichiometric As_2S_3 in solid phase.

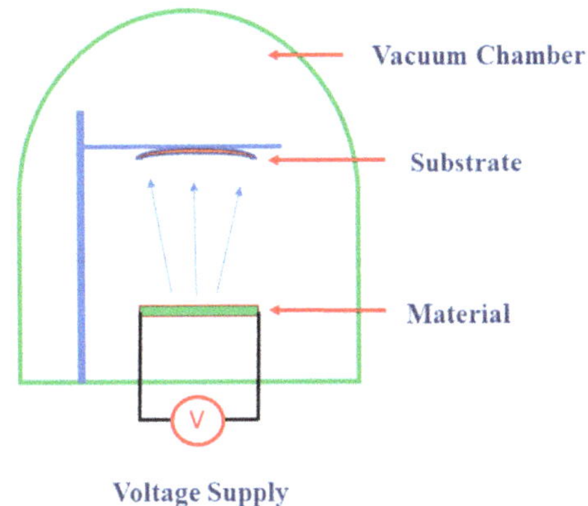

Fig. 9.1 Schematic diagram of setup for thermal evaporation method

9.3.2 *Sputtering*

Sputtering method comprises of bombardment of a target material by high energy ions obtained from plasma generated at low pressure. As a consequence, atoms or cluster of atoms are removed from target and subsequently gets deposited over a substrate in the form of a thin film. The mechanism of synthesis by this process involves the generation of plasma by suitable radio frequency which is struck in between the target material and substrate. During each negative half cycle, the positive ions are attracted toward the target. Since the mobility of electrons is much higher compared to that of corresponding cations generated, accumulation of negative bias takes place. This negative bias attracts the positively charged ions from the plasma as a result of which the positive charged ions are deposited over the substrate. Amorphous thin films of Si, Ge, SiO_2, etc., are prepared by this method [9–11]. The major advantage of sputtering technique is that relatively more homogenous films of uniform thickness can be prepared by this method as compared to thermal evaporation method. The homogeneity of produced films can be attributed to the sputtering rates which do not have a wide variation among different species. Almost similar sputtering rates preserve the compositional homogeneity of the starting material as compared to thermal evaporation methods. A schematic diagram of sputtering set up is shown in Fig. 9.2.

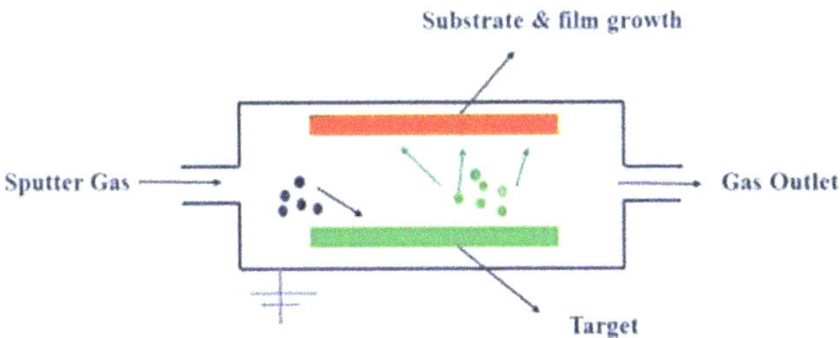

Fig. 9.2 Schematic diagram of sputtering setup

The only disadvantage of the process lies in optimization of the process control parameters such as ratio of partial pressure of reactive to inert gas, bias voltage applied to the target material, and radio frequency power applied to the target.

9.3.3 Glow Discharge Decomposition

Glow discharge decomposition is quite similar to sputtering technique discussed above. The only difference lies in the fact that a chemical reaction is initiated in the gas phase by creating a radio frequency-induced glow discharge of the reactant gas instead of plasma ejecting material in the sputtering process. Glow discharge results in deposition of solid over a suitable substrate kept inside the chamber. The discharge can be generated not only in a mixture of carrier and reactant gas but also with varied combination of reactant gases along with carrier gas for forming films of different functional materials such as SiO_2, Si_3N_4, etc. [12, 13]. The major disadvantage of this method is optimization of process control parameters for preparation of tailor-made materials. Schematic diagram of glow discharge decomposition setup is shown in Fig. 9.3.

Chemical vapor deposition: Chemical vapor deposition technique is quite similar to glow discharge method of preparation. In this technique, the decomposition of the reactant gas is carried out by applying thermal energy (pyrolytic process). As a consequence, temperatures of the order of 10^3 K are commonly used and this is one of the major disadvantages of the process. Some of the materials with different functionalities prepared by this process include amorphous hydrogenated Si, B, and P-doped amorphous Si, etc. [14]. Schematic diagram of chemical vapor deposition (CVD) setup is shown in Fig. 9.4.

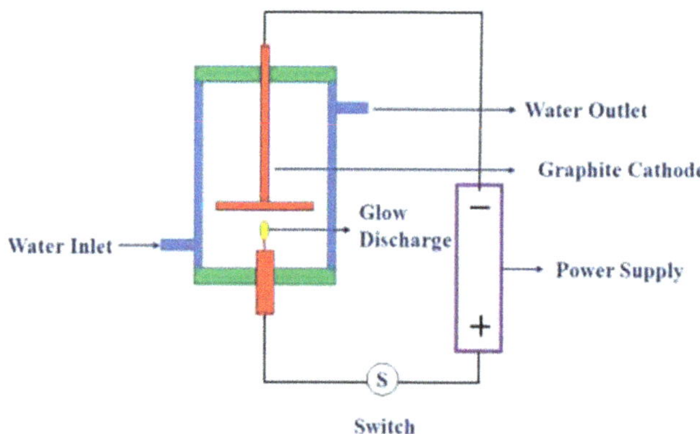

Fig. 9.3 Schematic diagram of glow discharge decomposition setup

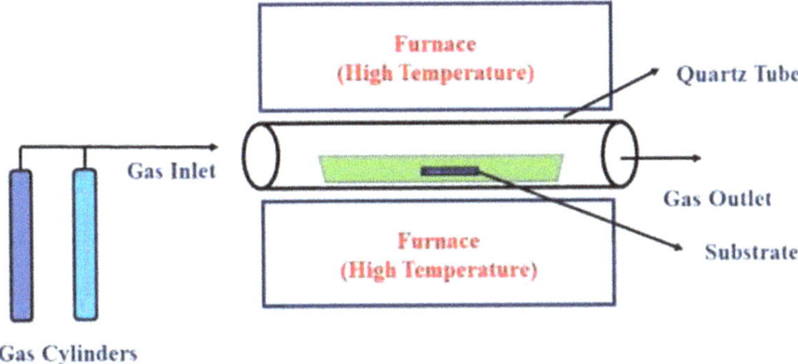

Fig. 9.4 Schematic diagram of chemical vapor deposition (CVD) setup

9.3.4 Melt-Quench Technique

Melt-quench technique is considered the oldest method for the preparation of glassy and amorphous materials. This method involves calcination and melting of glass precursors (mostly oxides, nitrates, acetates, etc.) in compatible or non-reactive crucibles such as platinum, rhodium, alumina, zirconia, sillimanite, followed by pouring of the melt at room temperature or under liquid nitrogen (Fig. 9.5). Depending upon the degree of quenching required, the melting containers are made up of good thermal conducting materials such as copper and graphite. Quenching of the melt is required to avoid nucleation and crystal growth which can lead to phase segregation form the glass matrix. Rate of cooling of the glass melt is an important criterion for glass production by this method. For oxide glasses having glass formers such as B_2O_3, P_2O_5, and SiO_2, a cooling rate of 1 Ks^{-1} is found to be

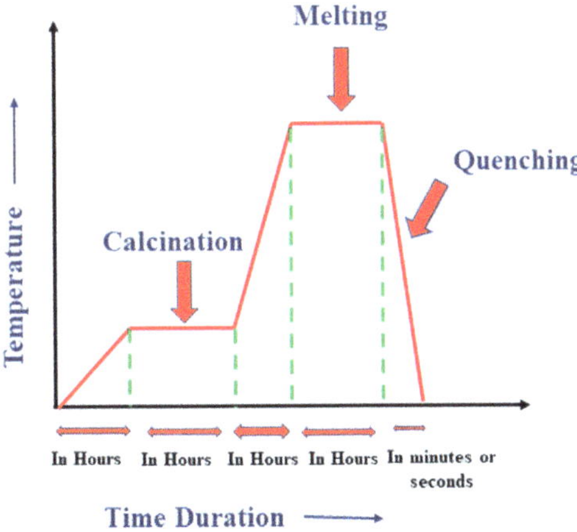

Fig. 9.5 Schematic temperature and time profile of melt-quench method of glass preparation (diagram not to scale)

sufficient for getting good quality glasses, whereas for metallic glasses, cooling rate must be over the range of 10^6 Ks^{-1}.

Most of the oxide glasses are prepared by melt-quench method because of simplicity and cost-effectiveness of the process and non-reactive nature of the product under atmospheric conditions. Only minor disadvantage of this process is that for industrial scale preparation of glass, continuous or intermittent stirring is necessary to avoid inhomogeneity of the viscous melt. Covering the reaction crucible will help in preventing evaporation of high volatile components, thereby avoiding significant changes in the composition of final prepared glass. In order to prepare chalcogenide and chalco-halide glass by this technique, melting and pouring of glass melt is required to be carried out under inert atmosphere or in vacuum sealed crucibles [15]. Moreover, the furnace has to be designed with rocking mechanism for proper mixing of components in [15] sealed crucibles for better homogeneity of the melt. It is preferable to have starting precursor materials as oxides since nitrates or acetates will release NOx, acetic acid vapors, respectively, which may damage or even burst the crucible.

9.3.5 Sol-Gel Method

In general, sol is considered as a colloidal suspension of solid in liquid, whereas gel is considered to be a colloidal suspension of liquid in solid. Thus, it is quite evident from the definition itself that gel can be produced from a viscous colloidal solution by facilitating polymerization of the solid particles. The homogenous and amorphous gel is then heated to remove volatile constituents to obtain an amorphous solid. This is then subjected to final sintering process at suitable temperatures to obtain a dense solid material. The oxide particles obtained by this method show fine dispersion and are found to be in nanoscale dimensions. The major advantage of sol-gel processing is the formation of materials containing refractory-based components at relatively lower temperatures. Suitable processing and purification of precursor materials by various methods before sol-gel processing results in formation of products with very high level of purity [16]. Composition and properties of the products obtained can vary depending upon conditions such as temperature, solvent concentration, reaction time, and sintering temperature. Moreover, the process of drying and sintering of gel are time consuming. Availability of suitable metal precursors also limits the applications of this method for preparation of novel materials for catering to the demands of newer technologies.

9.3.6 Electrolytic Deposition

This is one of the most important methods for the preparation of glassy oxide materials. Using metal surface as anode, it can be oxidized to an amorphous oxide layer in an electrolytic cell having suitable electrolytes. On passing a DC voltage through an electrolytic cell, cations migrate toward the cathode and O^{2-} ions migrate toward the metal anode. These anions then react with the metal to yield an amorphous glassy oxide layer over the anode surface which can grow up to $\sim 10^3$Å thickness. This method is widely used to prepare glassy oxide films of Al, Zr, Ti, Nb, and so on [17]. The major disadvantage of this method is the difficulty in preparing non-oxide-based glasses.

9.3.7 *Radiation Bombardment*

Bombarding by high energy particles such as neutron, alpha, or heavy charged particles (HCP) on crystalline samples leads to formation of amorphous phase. When a particle with very high kinetic energy is impinged over any substrate, enormous amount of local energy deposition takes place followed by rise in local temperature of the order of $\sim 10^3$ K for a very short span of time (typically for $10^{-10} - 10^{-11}$ s). This sudden spurt in temperature results in melting followed by very fast cooling of the substrate. Quartz and cristobalite (allotropes of crystalline silica) are progressively amorphized by radiation resulting in the generation of vitreous silica [18]. Major advantage of this method is incorporation of radioactive species in the material which has immense potential to be used in nuclear medicine and other research applications. Limited experimental facility setup available for carrying out radiation damage studies is the major disadvantage of this method.

The methods discussed above are considered to be the most widely used methods for preparation of glassy or amorphous materials. In addition to these methods, high pressure shock waves, slow mechanical grinding, explosive compaction, hydrogen inclusion, etc., are also used for glass preparation. It may be noted that these methods are specific for certain systems and cannot be extrapolated or optimized for even related systems, thereby lacking a wider applicability in glass manufacturing industry. Till date, melt-quench technique is considered as the best and the most widely used method for glass preparation both in industries as well as in research laboratories worldwide.

9.4 Different Routes for Synthesis of Glass-Ceramics

In spite of the fact that the presence of crystalline phase is undesirable during glass formation, controlled crystallization of desired phases in glass matrix is primary requirement for formation of functional glass-ceramic material. Two most important words in the above sentence are "controlled crystallization" and "desired phase" which are the main guiding factors for preparation of glass-ceramics.

Functional glass-ceramics can be prepared by any of the glass preparation methods followed by modification in heat treatment. The major difference lies in the fact that the one-step process involved in glass preparation has to be converted to a minimum two-step process for glass-ceramic preparation. The most common method is the preparations of glass by a standard glass manufacturing process like melt-quench method. The glass material is then shaped, cooled, and reheated (sometimes repeatedly) above its glass transition temperature to prepare a desired glass-ceramic material. During heat treatments, occasionally nucleating agents such as noble metals, TiO_2, and Fe_2O_3 are added to aid the formation of desired crystals of suitable dimensions required for improving functionality. Synthesis of glass-ceramic materials by this method leads to particle crystallization inside the glass matrix, thereby resulting in a homogenous glass-ceramic material.

The second technique used for preparation of glass-ceramics is induction of controlled internal crystallization during cooling cycle of a glass forming process. This technique involves heating the precursor materials to melting followed by slow cooling above the glass transition temperature followed by quenching below it. This method also involves addition of nucleating agents (sometimes in larger amount) as the reaction kinetics is quite fast. This method generally leads to, relatively coarse glass-ceramics as there exists relatively less control over the whole process. Moreover, the crystalline phase may not be homogeneously distributed within the glass matrix and this leads to reduction in quality of the material for different applications.

The third most common process involves separate preparation of desired glass and ceramic formulations followed by concurrent sinter-crystallization of both the materials to obtain glass-ceramic formulations. The major advantage of this process lies in the fact that one need not be concerned for the formation of "desired ceramic phase" as it is prepared separately. Other advantages are absence of any nucleating agents as the glass and ceramic interface behaves as nucleating sites, leading to faster ceramic particle growth in the matrix. Major disadvantage of the process is residual porosity within the matrix which needs to be eliminated by hot-pressing techniques, thereby making the process costly as compared to others.

9.5 Thermodynamic and Kinetic Aspects of Glass Synthesis

As mentioned earlier in the introduction section, glass is considered as a super-cooled liquid, semi-solid, or fluid in spite of its appearance as a solid form. Thus, technically, it can be said that glass is a liquid having very slow or negligible crystallization kinetics. Crystalline state of any material always has lower energy than liquid state, and hence, the possibility of crystal formation is a thermodynamically favorable process. Crystals are generally obtained on cooling of a liquid melt, whereas the presence of crystals poses a hindrance to glass formation. Hence, the glass fabrication always requires bypassing of stable thermodynamic path. On the other hand, it is also known that glassy materials can exist for centuries without undergoing substantial damage. It is generally accepted that both thermodynamics and kinetics go hand in hand during glass formation. The concept of chemical or thermodynamic stability of glassy materials can be understood with the help of a potential energy diagram (Fig. 9.6). Red balls in Fig. 9.6 can be assumed as a reference glass material. At position A, if the ball is pushed, it will roll down the slope as shown above and settle at position B. Similar will be case for position C and E. In all these three positions (peak), the ball will roll down the slope and settle in a position with lower potential energy (valley). These peak positions (A, C, and E) are thermodynamically called unstable state and any given perturbation leads to change in the energy which is lowered by performing some reactions. Hence, it can be concluded that any material present at elevated potential energy state will be highly reactive. If the ball is at position D, considerable perturbation is required to roll the ball upwards to cross the barrier C or E before it can finally rest. Position D has the lowest potential energy in the above energy profile diagram, and hence, it is difficult for the material being at this position to undergo chemical changes. Position D is termed as the stable state for the reference material.

As the stable state of any liquid upon cooling is the crystalline solid state, glass, which is a super-cooled liquid cannot be called as thermodynamically stable product or thermodynamically controlled product. Assuming a scenario that the ball is rolled from A with very high perturbation will lead to it resting finally at position D. This is the case for all thermodynamically controlled chemical reactions. However, if the reaction is quenched at rapid rate such that the ball cannot rise through barrier C, it will limit its movement up to position B. This position B is called metastable state where the ball is found to be temporarily stable unless the perturbation given makes it crosspoint C. Limited perturbation given at position B can keep the ball chemically unreactive for a long time. This is the condition where a liquid does not convert into crystalline solid but remains stable for longer duration. Any solid obtained at position B will not possess the properties of pure solid (position D) or pure liquid (position A, C and E). It will possess some mixed properties of pure solid and pure liquid, and it can correctly be termed as super-cooled liquid.

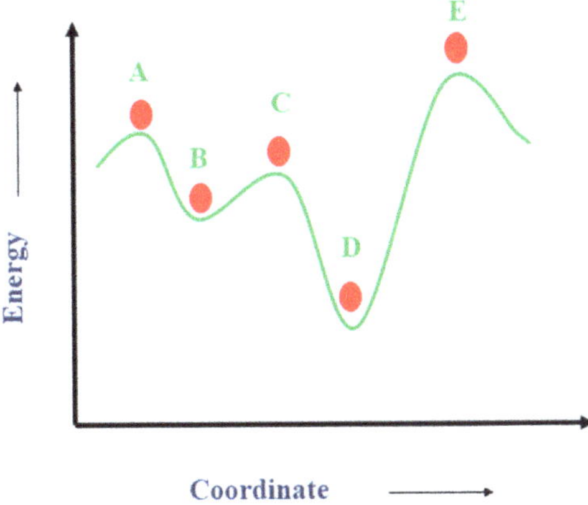

Fig. 9.6 Schematic of a potential energy diagram for understanding the concepts of stability, instability, and metastability

Thermodynamically, this is termed as the glassy state where the material remains in a metastable equilibrium with its surroundings without undergoing further degradation (devitrification) into the crystalline solid state.

The major question here arises regarding the way to restrict chemical changes to position B. Restriction at this position implies zero or negligible crystal formation in the glass matrix. Formation of crystal in any melt occurs via two major steps namely nucleation and crystal growth. Nucleation occurs when solute molecules dispersed in solvent or melt gather in microscopic domains and arrange themselves in definite geometry to form clusters. However, it needs to be noted that, these cluster formations have to attain a particular size termed as the critical size to become stable nuclei. Crystal growth is the subsequent increase in size or growth of the nuclei after attainment of critical size. Fast cooling or quenching is hence required during glass formation to inhibit nucleation stage within the liquid melt. In nucleation events, a small number of unit cells combine with each other during short timescales, termed as the nucleation time T_{nuc}. A larger T_{nuc} implies fluctuations during nucleation stage do not allow the formation of crystals within glass melt. This reasoning is applicable for formation of all types of glasses and is irrespective of preparation methods.

Knowledge of thermodynamic and kinetic control parameters of glass formation can help us in design of functional glasses. It may be noted that both aspects mentioned above change with change in chemical composition and volume of the glass melt. The variation in specific volume as a function of temperature is shown in Fig. 9.7. On gradually lowering the temperature of a liquid, a linear decrease in specific volume is observed till the freezing point (melting point or T_{m}). At T_{m}, liquid can undergo either crystallization where liquid is converted to solid or super-cooling in which the material remains in liquid state even below freezing temperature. Crystallization of liquid melt is characterized by a sharp drop in specific volume of the substance at T_{m}, whereas glass formation or super-cooling is manifested by a gradual decrease in slope of the specific volume. The temperature range over which variation in slope of the cooling curve occurs corresponds to glass transition phenomenon and the onset of such change is the glass transition temperature (T_{g}). For convenience, glass transition temperature is expressed as fictive temperature (T_{f}) which is the temperature at the point of intersection of the straight-line regions of cooling curves corresponding to the liquid and the glass. Similar behavior is also observed if the specific volume V is replaced by other thermodynamic parameters such as enthalpy (H) and entropy (S).

It should however be noted that the glass transition temperature is a function of the rate of cooling of liquid melt. Faster or slower cooling of a melt leads to smaller or larger area on the cooling curve in the super-cooled region. Accordingly, there is an increase or decrease in the glass transition temperature. In other words, T_{g} of a particular glassy material is not an intrinsic property. This is because when a melt is quenched to form a glass, the different degrees of freedom undergo relaxation. When the glass is heated, the different degrees of freedom start relaxing. Since the

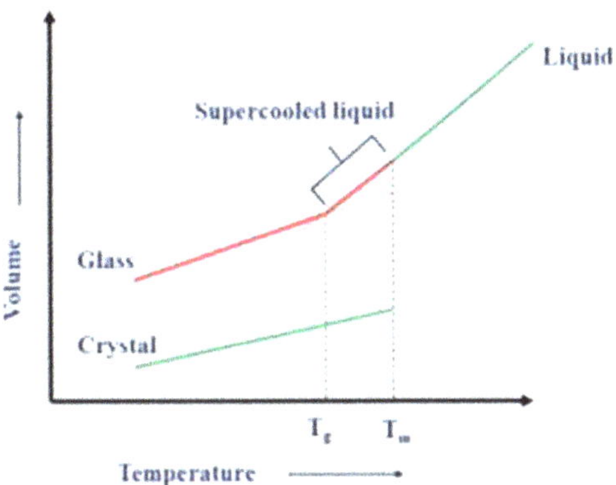

Fig. 9.7 Variation of specific volume as a function of temperature for a liquid melt

relaxation rates are different for different degrees of freedom, the T_g values depend on the degree of freedom (property) which is monitored to determine the T_g. Glass transition temperature (T_g) of any material is related to the cooling rate (q) by Eq. 9.1:

$$q = q_0 \exp\left[-1/c\left(1/T_g - 1/T_m\right)\right] \tag{9.1}$$

where T_m the melting point, q_0 and c are constants [19]. Experimentally, the determined values of T_g are also not unique as it depends upon the heating or cooling rates at which measurements are carried out. Hence, an independent parameter K_g (which describes the tendency to form glass) has been derived by Hruby [20] for characterizing glasses (Eq. 9.2).

$$K_g = \left(T_c - T_g\right)/\left(T_m - T_c\right) \tag{9.2}$$

where T_g is glass transition temperature, T_m is melting temperature and T_c is crystallization temperature. High ($T_c - T_g$) and low ($T_m - T_c$) value indicate high K_g values which imply inhibition of nucleation and subsequent crystallization within glass melt, thereby increasing the glass forming tendency of a melt.

Glass transition can also be defined in terms of viscosity of the melt. According to this definition, at glass transition temperature, the viscosity of a melt liquid attains a value of around 10^{13} poise. Near vicinity of glass transition, viscosity, or the relaxation time suddenly becomes so large, and the equilibrium between thermal state of any material and its surrounding is destroyed. This commonly occurs at around two-third of T_m for most of the oxide glasses. Variation of viscosity of glassy materials as a function of temperature can be explained according to either Arrhenius relaxation law or Vogel-Fulcher law and are represented by Eqs. 9.3 and 9.4. Arrhenius relaxation law states that viscosity (as well as relaxation time) undergoes exponential growth with decrease in temperature according to the relation (Eq. 9.3):

$$\eta = \eta_0 \exp(A/T) \tag{9.3}$$

where η_0 is constant and A is the activation energy for viscous flow of material.

The second equation of viscosity variation with temperature is Vogel-Fulcher, or Vogel-Fulcher-Tammann-Hesse law (Eq. 9.4) according to which:

$$\eta = \eta_0 \exp[B/(T - T_0)] \tag{9.4}$$

where η_0, B, and T_0 are constants. Divergence in case of the second equation is more than the Arrhenius law. This phenomenon is explained by defining a temperature dependent activation energy $A = BT/(T - T_0)$. For the last few decades, glasses have been characterized based on the above-mentioned nature of temperature dependence of glass transition process, namely strong glasses and fragile glasses. The glasses which exhibit high viscosity above melting temperature are termed as strong glasses and they are found to follow the Arrhenius relaxation law. Consequently, glassy materials following Vogel-Fulcher relaxation law are termed as fragile glasses as they exhibit low viscosity above melting temperature. It may be noted that this classification is based upon the flow characteristics of the melt and not on the mechanical properties of glass.

Thermodynamically, glass transition phenomenon is a second-order phase transition and hence second derivative of the free energy with respect to temperature, i.e., $(\partial^2 G/\partial T^2)$p or Cp is discontinuous. Experimentally, it is observed that Cp is discontinuous during glass transition. However, this simple model cannot explain the change in T_g values depending on the thermal history of glass samples.

The phenomenon of glass transition is explained satisfactorily by free volume theory. The basic assumption of this theory is that a glassy material is composed of hard spheres and the total volume of glass is divided into two parts, namely the part which is occupied by the hard spheres (V_{occ}) and the other part in which the hard spheres or molecules are free to move (V_f). According to this theory, a glassy material is defined as the one having V_f independent of temperature, whereas, for liquids, both V_{occ} and V_f decrease and undergo redistribution with reduction in temperature. Based on this theory, glass transition takes place when V_f decreases below a critical value (V_{fg}). The fractional free volume (R) is related to the expression (Eq. 9.5):

$$R = V_{fg}/V_g = T_g \cdot \Delta\alpha_T \tag{9.5}$$

where V_{fg} = free volume of glass
V_g = volume of glass
and $\Delta\alpha_T = \alpha_{T1} - \alpha_{T_g}$ where α_{Tl} and α_{Tg} represent cubical expansion coefficients of liquid and glass, respectively.

In general, it has been observed that for most of the glasses about 10% of the total volume remains free at T_g. Free volume theory has been adapted for an extensive applicability by incorporation of percolation theory [21]. This theory has extended the free volume theory by accounting for exchange of V_f between nearest neighbors of liquid cells without any concurrent change in volumes of the solid cells. Although many theories exist, the phenomenon of glass transition could not be explained fully on the basis of any single theory, till today [22].

9.6 Kinetics of In Situ Crystallization

In situ crystallization of ceramic phase is a menace for glass formation. On the other hand, nucleation and crystallization of desired phases are mandatory for preparation of functional glass-ceramic materials. In view of this, the kinetics of in situ crystallization within glass matrix becomes an important aspect for synthetic chemists. A proper understanding of in situ crystallization becomes mandatory for design of functional glasses and glass-ceramics. The in situ crystallization kinetics of any glassy material can be understood based on time temperature transformation (TTT) diagrams. The TTT diagrams or isothermal transformation diagrams (as they are called) are plots of temperature versus time (either linear or logarithmic scale). These diagrams are experimentally generated from evaluation of percentage of phase transformation as a function of time. Experimental data at different temperatures are then plotted to yield TTT diagram. This plot is quite informative and helps to understand the crystallization kinetics of glass melts.

Annealing of glass materials is generally carried out to improve their functionality. It is well known that extent of diffusion in glasses becomes maximum near the glass transition temperature. Hence, annealing glasses at temperatures very close to the glass transition temperature reduces thermal stress generated due to fast cooling or quenching. An important strategy for glass-ceramic preparation involves both time and temperature optimization for crystallization of desired phase with required extent of crystal growth as crystal size many times decides the functionality of synthesized material.

The above discussion helps us to realize the importance of annealing and time temperature transformation diagram for developing functional glass or glass-ceramic material. In simpler terms it is a plot of percentage composition of both glassy and crystalline domains of any substance with variation of time. These diagrams are valid only for a particular composition of glass melt. Variation in composition of glass melt will definitely alter the TTT diagram. Let us understand in simple terms, the interpretation of such diagram by using Fig. 9.8. It is already discussed that above melting temperature (T_m), glass melt behaves as a pure liquid and below glass transition temperature (T_g) it behaves as a pure glassy material. Crystallization is not possible in a liquid or glass without undergoing any reaction. Thus, it is clear that the chances of crystallization of discrete phases within glass melt will be limited within the range from T_g to T_m. Experimental data obtained from time variation in glass and crystalline domains are plotted within the range of temperatures at which studies were carried out. (shown in purple color). According

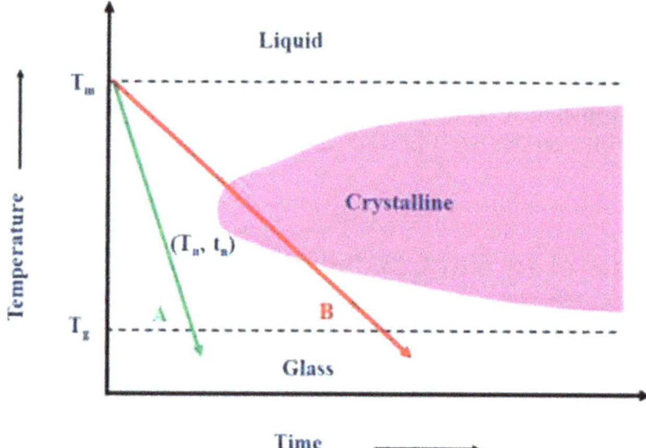

Fig. 9.8 Schematic plot of time temperature transformation (TTT) diagram. Arrows represent variation of temperature of the melt under two different cooling rates

to this diagram, for glass formation, the cooling rate of glass melt should not pass through the crystallization domain. Let us assume that the melt is cooled though two different pathways A and B. The pathway A does not cross through the crystalline domain and hence 100% glassy phase can be obtained by cooling at this rate. If pathway B is followed, the formation certain extent of crystalline phases within the formed glass is expected. The slowest cooling rate which helps in the formation of glassy phase is termed as the critical cooling rate for that glass composition. The critical cooling rate that is necessary for preparation of glass can be obtained by drawing a tangent at the nose (T_n, t_n) of the crystalline domain where T_n is the temperature coordinate and t_n the time coordinate of the nose respectively. Mathematically, it is given by (Eq. 9.6):

$$dT/dt_c = (T_m - T_n)/t_n \tag{9.6}$$

The TTT diagram can be utilized for the generation of functional glass-ceramics. Knowledge of precipitation of varying amount of ceramic phase(s) in glass melt with time variation can play a key role in the development of glass-ceramics with tailor-made properties.

9.7 Structural Aspects of Glasses and Glass-Ceramics

Fundamental understanding of structural arrangements of atoms or molecules in any material is of prime importance. This understanding guides researchers to fine-tune physico-chemical properties of materials as well as to induce suitable functionality.

This structural understanding becomes relatively easy for crystalline materials as understanding of arrangement of unit cell is only adequate to gain an insight of structural aspects throughout the material. However, as mentioned earlier, glasses lack periodic arrangement of atoms or molecules over long range and hence complete structural analysis of glasses or glass-ceramics is really challenging. Generally, the nature of local or short-range order present in these materials is investigated and the information is used to gain a thorough insight about the extended structure. For example, crystalline quartz has regular arrangement of SiO_4 tetrahedra, whereas vitreous silica exhibits randomly connected SiO_4 tetrahedra with order over the length scale of few angstroms. Extent of randomness, size of the crystalline domain, effects of one domain upon the neighboring one, etc., are some of the concerns which need to be answered by structural modeling of glassy system. These challenging aspects need to be addressed while undertaking structural analysis of glasses and glass-ceramics.

Structural studies on glasses can be mainly classified into two types depending upon the length scales in which experimental studies are carried out. They are namely (a) macroscopic structure and (b) microscopic structure. Structural studies undertaken by diffraction of X-rays, neutrons, electrons, etc., are sensitive to the variation of few angstroms (microscopic domain), whereas techniques such as optical and scanning electron microscopy are useful for the detection of structural in-homogeneities of the order of thousands of angstroms (macroscopic domain). The microscopic structural domain has been again subdivided into medium-range order involving a length scale of roughly 10–30 Å and a short-range order involving a length scale of few angstroms. Detailed structural aspects of representative systems obtained from different characterization techniques will be discussed in the following section.

9.8 Characterization Techniques

Functional glasses and glass-ceramics are generally characterized for their structural aspects, thermo-physical properties, mechanical, optical, and chemical properties, etc. In this section, we will focus on some of the important techniques used for the characterization of glasses and glass-ceramics.

9.8.1 Structural Analysis

(a) **Powder X-ray diffraction**: Powder X-ray diffraction (XRD) is one of the most commonly used technique to confirm formation of glassy phases. Since crystalline materials have long-range periodicity of atoms or group of atoms, they are characterized by sharp peaks in the diffraction patterns (Fig. 9.9). Ceramic components in glass-ceramic gives sharp peaks and these peaks are then matched with standard databases to garner information about the nature of embedded crystalline phase. However, glassy samples are characterized by single/multiple broad humps in their diffraction pattern indicating lack of long-range order in the glass (Fig. 9.9).

A powder X-ray diffraction is essentially a plot of intensity as a function of 2θ values (Fig. 9.9). Since glassy materials have random atomic arrangement, there is a considerable distribution in bond lengths and bond angles between the central atom and neighboring atoms giving rise to a broadband of low intensity. However, if a small amount of phase is crystallized within the glass melt, sharp peaks of high intensity (as compared to glassy phase) are observed in the XRD pattern. This helps to identify the nature of phases and also plausible reason for crystallization of phases. In certain cases, the position of first diffraction peak gives idea about medium-range order in glass and this information combined with theoretical simulations can give structural information on extended length scales.

(b) **XAFS spectroscopy**: X-ray absorption fine structure (XAFS) spectroscopy is an important tool for understanding structural aspects of functional materials. XAFS investigates the modulation of the probability of X-ray absorption at energies in and around the binding energies of core electron of any given atom present in any material. Modulation of X-ray absorption depends upon the chemical environment of a particular atom over long or short range depending upon the energies. X-ray absorption spectrum consists of three ranges: edge region, X-ray absorption near edge structure (XANES), and extended X-ray absorption fine structure (EXAFS), as can be seen from Fig. 9.10. Value of edge gives information regarding the oxidation state of investigated ion. Near edge region (XANES) covering over 50–100 eV energy range around the edge

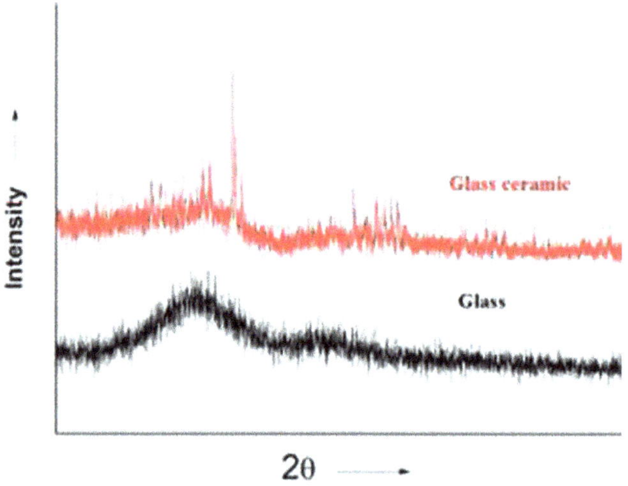

Fig. 9.9 X-ray diffraction patterns of representative glass and a glass-ceramic material

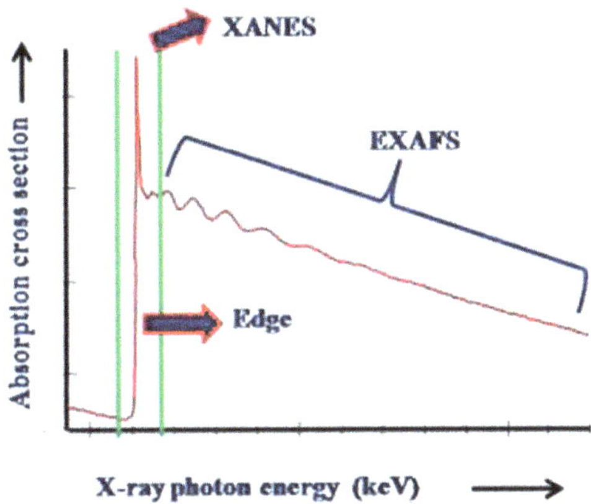

Fig. 9.10 Diagram showing three regions of X-ray absorption spectrum

gives information about the band structure, hybridization of molecular orbitals, and extent of long-range order.

EXAFS comprises of Kronig structure, i.e., the oscillatory function spreading over 100 eV past the absorption edge which gives us information about the local structure, coordination number, nearest neighbor distance, and any disorder prevalent in the sample. Compilation of the obtained data and proper processing (beyond the scope of this chapter) is required to simulate the experimental data. Structural models are then calculated comparing with standard (for crystalline phase) or manual modeling (for glassy phase) and fitted with the experimentally obtained and processed data to impart a complete knowledge of structure existing in the sample.

(c) X-ray photoelectron spectroscopy

X-ray photoelectron spectroscopy is one of the techniques based on X-rays which can be used for understanding the short-range order around a particular atom/ion in glass and glass-ceramics. In this technique, the material is irradiated with X-rays having energy of few keV and the kinetic energy of the emitted electrons is monitored. The process is schematically as shown in Fig. 9.11.

Once energy (hν) of X-ray is higher than the binding energy (BE) of electrons in a particular level, electrons get ejected out. This ejected electron is called photoelectron and kinetic energy (KE) of the photoelectron can be expressed by the relation (Eq. 9.7)

$$\mathrm{KE} = \mathrm{h}\nu - \mathrm{BE} + \phi \tag{9.7}$$

Here, ϕ is the work function, and the value of which can be obtained from calibration of the machine. From the measured values of kinetic energy and ϕ, binding energy can be calculated which is related to the oxidation state of the corresponding element and its coordination environment. For example, bridging

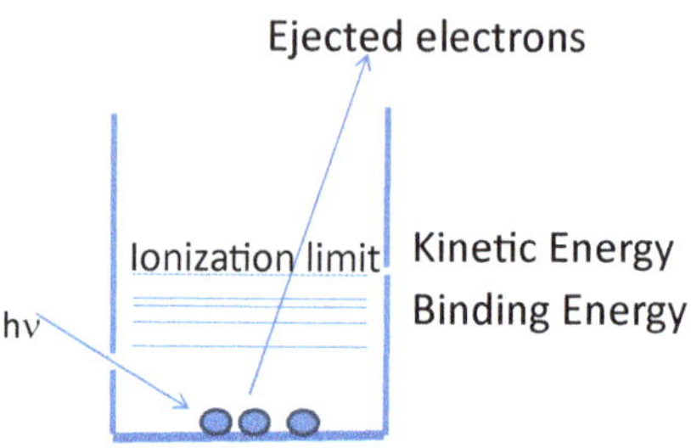

Fig. 9.11 Principle of XPS technique. Shaded circles represent electrons

and non-bridging oxygen atoms present in borosilicate and phosphate glasses can be easily distinguished from O1s XPS patterns. Presence of negative charge with the non-bridging oxygen (NBO) atom leads to lower binding energies for O1s electrons for NBO compared to BO. Wang and Zhang used XPS technique confirm the network modifying and network forming action of Pb^{2+} in binary lead silicate glasses as a function of composition [23].

(d) **Vibration spectroscopic techniques**: Infrared and Raman spectroscopic techniques belong to the category of vibrational techniques and are extensively used for the characterization of glasses and glass-ceramics. Vibration of groups or bonds which involve change in the dipole moment results in absorption of radiation and this forms the basis of IR spectroscopy. For molecules/bonds or a group to be Raman active, there must be change in the polarisability ellipsoid up on absorption of light. Vibrational spectra of glasses are analyzed by assuming that there exists considerable distribution in bond angles and bond lengths of different structural units constituting glass network/structure. Further, the structural units present are capable of being vibrationally excited independent of the surrounding amorphous matrix or other groups present in the glass structure. Typical FTIR spectrum corresponding to borosilicate glass is shown in Fig. 9.12. The pattern essentially consists of three peaks with maxima around 1400, 996, and 464 cm^{-1}. There is a shoulder peak around 700 cm^{-1} present along with the broad peak at 996 cm^{-1}. Based on earlier studies [24], it is clear that the peak at 1400 cm^{-1} is due to the vibration of BO_3 structural units. Corresponding vibration for BO_4- structural units overlap with the asymmetric stretching vibrations of Si–O–Si linkages, thereby appearing as a broad asymmetric peak around 996 cm^{-1}. Symmetric vibration of Si–O–Si linkages appears around 700 cm^{-1} and the peak at 464 cm^{-1} arises due to the bending modes of Si–O–Si linkages. All the peaks are broad indicating a significant extent of disorder (distribution of bond lengths and bond angles) existing in the glass sample.

As the composition of glass changes, all the above modes of vibration undergo change. Weakening and strengthening of the bonds can be clearly seen from the shift in absorption peak maximum and changes in the line shapes. Thus, FTIR spectra can be used to identify and monitor the structural changes taking place with variation in composition of glass samples.

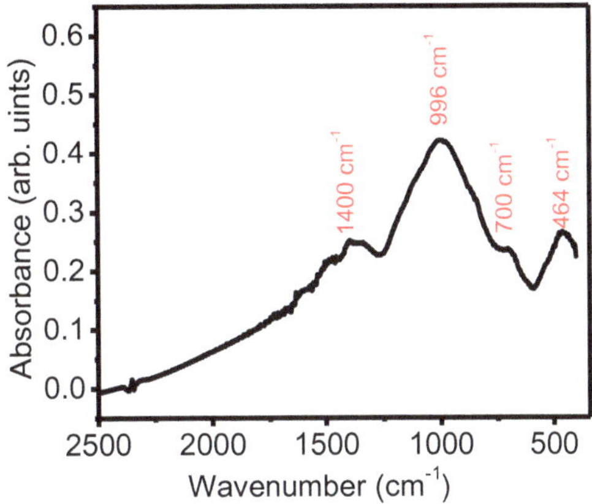

Fig. 9.12 FTIR pattern of sodium borosilicate glass with composition $(SiO_2)_{0.50}(B_2O_3)_{0.30}(Na_2O)_{0.20}$

Raman spectroscopy: The technique also gives significant information regarding short and medium-range order existing in glasses and glass-ceramics. In this technique, atoms are excited by photon emitted from laser into a virtual state (lifetime $\sim 10^{-15}$ s) which relaxes instantaneously to ground level. The emitted photons may have higher frequency (anti Stoke) or lower frequency (Stoke) as compared to the incident photon. The relative shift is termed as Raman shift which provides information about vibrational, rotational, and certain other low frequency transitions for group of atoms/structural units. Because of the presence of short-range order in glasses, vibrations in a glass are limited to a localized domain unlike in crystalline materials. Many structural groups present in silicate, phosphate, and borate glasses are Raman active. For example, a representative Raman spectrum of PbO–P_2O_5 glass containing equimolar amounts PbO and P_2O_5 is shown in Fig. 9.13. The pattern is characterized by sharp peaks around 1150 and 687 cm^{-1}. The peak at 1150 cm^{-1} is more intense with shoulder peaks placed around 1210 and 1060 cm^{-1}. The intense peaks around 1150 cm^{-1} are arising due to the symmetric stretching vibrations of Q^2 structural units of P (P structural units with 2 bridging oxygen atoms).[25,26] The corresponding asymmetric stretching mode is less Raman active (more IR active) and give rise to a shoulder peak around 1210 cm^{-1}. The other shoulder peak around 1060 cm^{-1} is arising due to pyro-phosphate type of structural units ($P_2O_7^{2-}$) present in the glass. P–O–P bridges present in the glass are characterized by a peak around 687 cm^{-1} [25, 26].

In addition to this, Raman spectrum can also give information regarding the medium-range order existing in glass. For example, the presence of boroxol ring in many borate-based glasses is unequivocally confirmed by the observation of Raman peak around 808 cm^{-1}. Recently, Yadav et al. [27] have reviewed Raman studies on a variety of oxide-based glasses.

It may be noted that a complete correlation of theory and experimental data becomes necessary to fully understand and visualize the structural features of a glassy material. Another approach followed by researchers worldwide is calculation of Raman spectra of simple glass samples based on modeling such as molecular

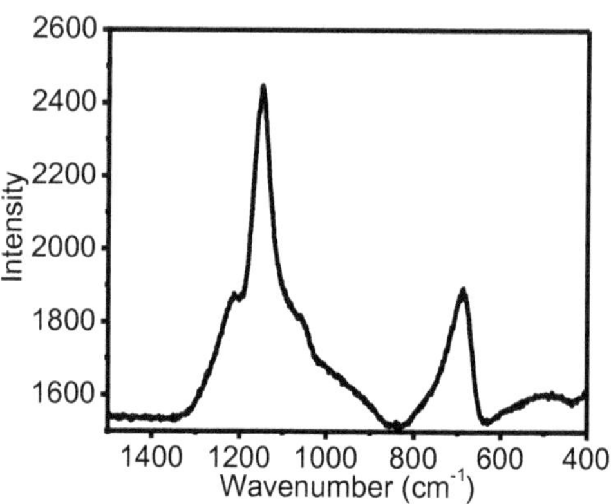

Fig. 9.13 Representative Raman spectrum of PbO–P_2O_5(1:1 mol ratio) glass sample

dynamics or reverse Monte Carlo simulations followed by comparison of the same with the experimentally obtained data. This approach helps to gain information about both short and even intermediate-range order existing in the glass and shall enable researches for establishing structure property correlations.

(e) **Solid-state NMR technique**: Solid-state nuclear magnetic resonance (NMR) is an important tool for detailed characterization of structural units present in glasses and glass-ceramics. Contrary to solution NMR, where sharp lines are observed, NMR line shapes of solids are characterized by broad lines due to anisotropic and orientation-dependent interactions present in solids. Unlike in solutions, these interactions do not get averaged out in the solid state. Advent of newer technologies has now enabled researchers to average these interactions to a very small value leading to relatively sharp NMR lines from solids [28, 29]. Alternatively, broad lines observed from solids contain detailed information regarding the anisotropic interactions existing in solids or glass. The internal interactions existing in glass or in the solid state can be classified into three categories, namely (a) magnetic interaction of the nucleus with the surrounding electron cloud (chemical shift interaction) (b) magnetic dipole –dipole interaction among nuclei, and (c) interaction between electric quadrupole moment (for nuclei having spin, I greater than or equal to 1) and the surrounding electric field gradient. These interactions are briefly described below.

(i) **Chemical shift interaction**: Chemical shift arises because of the effective magnetic field felt around a nucleus and is brought about by the polarization of electron cloud around the nucleus caused by the applied magnetic field. This interaction is sensitive to the configuration of valence electrons, which is governed by the nature of chemical bonding. The interaction is represented by a second rank tensor and can be diagonalized for specific principal axis system with components σ_{11}, σ_{22} and σ_{33}. For nucleus with axial symmetry, the precession frequency (ω_p) of nuclei can be expressed by the relation (Eq. 9.8)

$$\omega_p(\theta) = \gamma B_0[1 - \sigma_{iso} - \Delta\sigma(3\cos^2\theta - 1)/3] \tag{9.8}$$

Where $\omega_p(\theta)$ represents the orientation of principal axis with respect to the applied magnetic field direction, $\sigma_{iso} = 1/3(\sigma_{11} + \sigma_{22} + \sigma_{33})$, $\Delta\sigma = \sigma_{33} - \sigma_{11}$. The symbols γ and B_0 are the gyro-magnetic ratio and strength of applied magnetic field respectively. It may be noted that when $\theta = 54.7°$, $(3\cos^2\theta - 1)$ becomes zero and dependence of chemical shift anisotropy term on Larmor frequency gets averaged out to a small value.

(ii) **Magnetic dipole-dipole interaction**: In this type of interaction, the magnetic moment of one nucleus interacts with the magnetic moment of the other (neighboring nucleus). The nuclei can be either same (homonuclear dipolar interaction) or different (heteronuclear dipolar interaction). The Hamilton corresponding to dipole-dipole interaction is also having an angular dependence of $(3\cos^2\theta - 1)$, where θ is angle between the inter-nuclear distance vector and applied magnetic field. Hence, this interaction also gets averaged out when inter-nuclear bond vector is having an angle 54.7° with the applied magnetic field.

(iii) **Nuclear-quadrupole interaction**: Nuclei which are having spin value I > ½ are known as quadrupole nuclei and charge distribution within the nucleus is asymmetric in nature. This results in nuclear electric quadrupole moment, which is represented as *e*Q where "*e*" is the charge and Q is the nuclear-quadrupole moment. Interaction of nuclear-quadrupole moment with the surrounding electric field gradient (EFG) leads to broadening of NMR line shapes. Hamiltonian corresponding to quadrupolar interaction consists of both first and second-order terms. The former is having an angular dependence of $3\cos^2\theta - 1$, whereas the latter is having a complex angular dependence. Therefore, the quadrupolar interaction get only partially averaged out when inter-nuclear vector is at an angle 54.7° with the direction of applied magnetic field.

From the above discussion, it is clear all the three types of interactions are having an angular term of $3\cos^2\theta - 1$. Hence, by keeping the sample at an angle of 54.7° with respect to the applied magnetic field, the above-mentioned interaction can be minimized leading to narrowing of NMR line shapes. This aspect is discussed further in detail in the following section.

(f) **Magic angle spinning nuclear magnetic resonance (MAS NMR) technique**. The simplest and most popular experimental method for getting high-resolution NMR patterns from solids is MAS NMR technique discovered by Andrew and Lowe [30, 31]. The technique involves rotating the samples at high spinning speeds at an angle of 54.7° with respect to the applied magnetic field. At sufficiently fast spinning speeds, all the interaction vectors get aligned along a direction which is 54.7° with respect to the applied magnetic field. Under this condition, as mentioned above, the angular term $3\cos^2\theta - 1$ becomes very

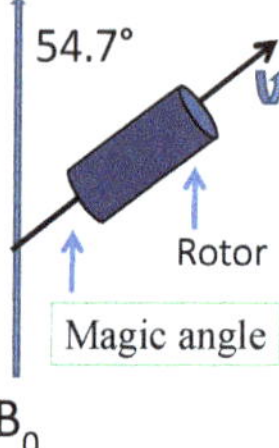

Fig. 9.14 Schematic representation of MAS NMR technique

small resulting in sharp NMR peaks. This is schematically as shown in Fig. 9.14.

There are many other experimental strategies to reduce or simplify the line width/shape of solid-state NMR patterns, details of which are reported elsewhere [29].

NMR parameters

In a typical NMR experiment, the important information that can be obtained directly from the spectrum are the chemical shift and coupling constants. Chemical shift is expressed by the following relation (Eq. 9.9):

$$\delta = \frac{(\omega - \omega_0)}{\omega_0} \times 10^6 \tag{9.9}$$

Where ω and ω_0 represent resonance frequency of the nuclei in the sample and in reference respectively. Values of δ are independent of the applied magnetic field and can be used to monitor the structural changes taking place around a particular nucleus. Parameters like dipolar and quadrupolar coupling constants can be obtained by fitting the NMR line shapes obtained under static conditions so that the interactions are not averaged out.

MAS NMR studies on representative glasses and glass-ceramics

As mentioned earlier, unlike crystals, glasses do not have long-range ordering and there is distribution in bond length and bond angles around different ions/atoms constituting the glass. As a result of this, ^{29}Si MAS NMR pattern of silica glass is characterized by a broad peak, whereas crystalline sample of silica (quartz) is characterized by a sharp peak around −109 ppm. Corresponding ^{29}Si MAS NMR patterns are shown in Fig. 9.15a, b.

It may be noted that although both silica glass and quartz crystal contain Q^4 structural units of silicon (silicon structural units with 4 bridging oxygen atoms), there is significant distortion around Si structural units in silica glass. This results in broad peak in the corresponding ^{29}Si MAS NMR pattern (Fig. 9.15a). Quartz powder is highly crystalline as revealed by the XRD pattern and there exists perfect ordering around Si structural units in quarts, resulting in sharp ^{29}Si MAS NMR peak (Fig. 9.15b).

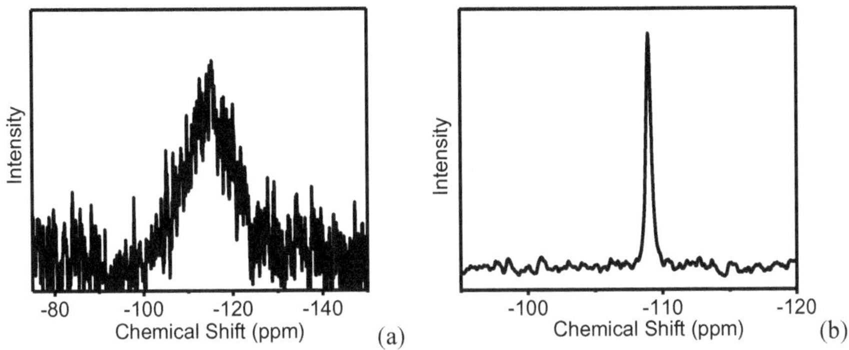

Fig. 9.15 ^{29}Si MAS NMR patterns of pure silica glass (**a**) and quartz crystals (**b**). Spinning speed was 10 kHz

Silicate-based glasses

Single component silica glass forms only at high temperatures due to the high melting point of silica. However, with the addition of Na_2O, PbO, etc., the melting temperatures drastically reduces. A representative ^{29}Si MAS NMR pattern of binary sodium silicate glass containing around 24% sodium oxide is shown in Fig. 9.16a. The pattern is mainly characterized by a broad asymmetric peak which can be de-convoluted into two peaks around −108 and −99 ppm. Based on earlier ^{29}Si MAS NMR studies, peaks around −108 and −99 ppm are attributed to Q^4 and Q^3 structural units of silicon [32]. Here, Q^4 represents silicon structural units having 4 bridging oxygen atoms and Q^3 represents silicon structural units having 3 bridging oxygen atoms. These structural units are schematically as shown in Fig. 9.16b, c.

Borosilicate glasses

Although binary silicate glasses are ideal candidates for understanding the structural aspects, such glasses have minimum technological applications. Borosilicate

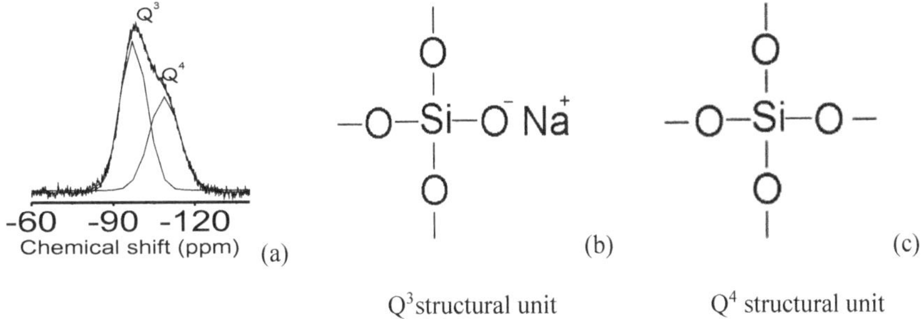

Fig. 9.16 ^{29}Si MAS NMR pattern (**a**) for sodium silicate glass containing around 25% Na_2O. The structural units Q^3 and Q^4 are schematically as shown in Fig. 9.16 (**b** and **c**) respectively

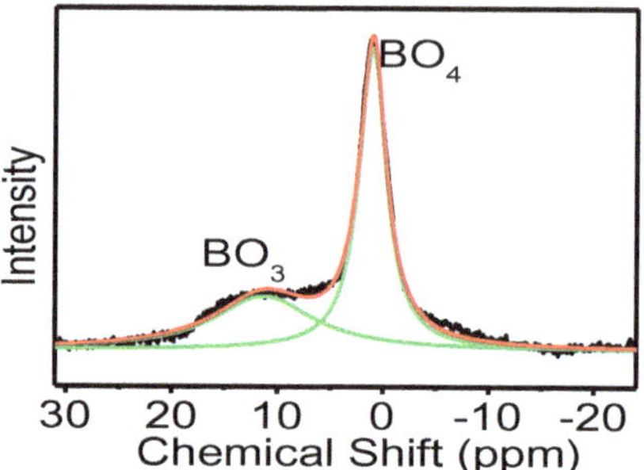

Fig. 9.17 Representative ^{11}B MAS NMR pattern of sodium borosilicate glass with composition $(SiO_2)_{0.42}(B_2O_3)_{0.21}(Na_2O)_{0.22}(BaO)_{0.15}$. Spinning speed is 10 kHz

glasses have a wide range of applications because of its improved physico-chemical as well as thermo-physical properties. ^{11}B MAS NMR is an ideal technique to monitor structural units present in such glass samples. Although boron exists in a variety of structural forms/units, BO_3 and BO_4 are the main structural units in borosilicate glasses. Due to higher symmetry of BO_4 structural units compared to BO_3 units, the former gives rise to sharp peak and the latter is characterized by a broad peak. A representative ^{11}B MAS NMR pattern of sodium borosilicate glass containing both BO_3 and BO_4 structural units is shown below (Fig. 9.17) [33].

In the case of glass-ceramics, in addition to the broad peak due to glassy phase, sharp peaks characteristic of crystalline phase is also observed in the ^{29}Si MAS NMR patterns. Figure 9.18 shows the ^{29}Si MAS NMR patterns of magnesium aluminosilicate (MAS) glass-ceramics prepared by addition of around 8 mol% Al_2O_3. In this glass-ceramics, a fluoro-aluminosilicate phase, namely potassium fluorophlogopite phase is formed, which is having a tile-like structure and is responsible for the machine-able characteristics of the material. The potassium fluorophlogopite phase consists of five different silicon structural units and is characterized by 5 sharp peaks in ^{29}Si MAS NMR pattern. These peaks are superimposed over a broad peak due to the residual glassy phase. Relative concentration of the crystalline phase increases with increase in Al_2O_3 content in the glass-ceramics [34].

In the recent past number of NMR techniques have been reported which include variable angle spinning NMR, double rotation NMR, multiple quantum MAS NMR, etc.,[29] which gives valuable information regarding the short and medium-range order in glasses and glass-ceramics.

9.8.2 Thermo-Physical Analysis

(a) **Differential thermal analysis (DTA)**: DTA technique revolves around the simultaneous heating and cooling of a standard (reference) sample and sample

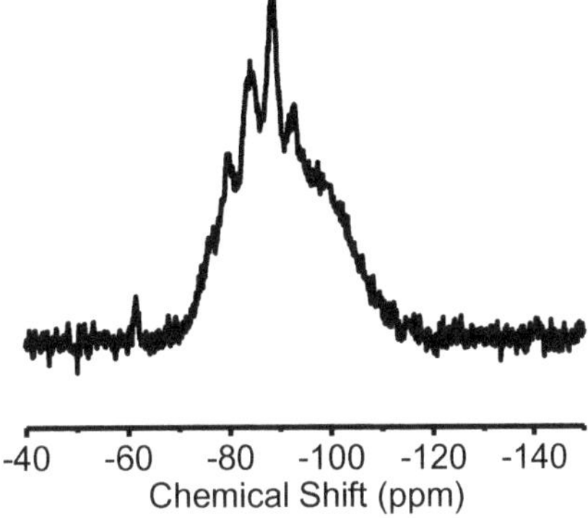

Fig. 9.18 ^{29}Si MAS NMR patterns of magnesium aluminosilicate glass-ceramics prepared by adding 8 mol% Al_2O_3. The spinning speed was 5 kHz

of interest under identical time span. Endothermic or exothermic reactions in the sample material due to absorption or evolution of heat lead to a decrease or increase in temperature, respectively, compared to the standard, which is chosen in such a manner that any sort of thermal event remains absent under experimental conditions. The temperature difference or ΔT is then plotted as a function of time or temperature in a DTA plot. As the change in enthalpy is responsible for a change in temperature, ΔH can also be plotted against temperature in a DTA plot. The major phenomena that can be observed from DTA plots of glasses and glass-ceramics are glass transition temperature characterized by weak and broad endothermic peak, crystallization of a phase characterized by sharp exothermic peak, and melting of residual glass (sometimes crystal) characterized by an endothermic peak. Generally, graphite, alumina, silica, nickel, or platinum crucibles are used for DTA measurements depending upon the temperature range and nature of the material. Disc-shaped thermocouples are used conveniently in DTA to provide optimum thermal contact with bottom of the sample and standard holder material. The only concern for a new experimenter is the choice of suitable heating rate for performing DTA experiments for glassy materials as very fast heating rate leads to overlap of more than one signal, whereas slow heating rate may result in submerged signal. Generally, a heating rate of 10 K min^{-1} is the optimum rate for observing glass transition temperature in borosilicate glasses.

(b) **Differential scanning calorimetry (DSC)**: DSC technique provides information similar to that of DTA technique. However, DSC is more quantitative mode of measurement as compared to DTA. The main reason for this is that temperature control of the sample and reference pans are more accurate in DSC as compared to DTA. The importance of DSC lies in the fact that it can measure enthalpy change quite accurately during any phase transition occurring with the sample. DSC is mainly of two types: temperature compensated DSC and modulated temperature power compensated DSC. In temperature compensated DSC, during any endothermic event sample is provided with increased heat flow to neutralize the temperature lag, whereas heat is given to the standard for any exothermic event. This principle allows for the calculation of heat flow in or heat flow out as a function of temperature leading to the quantification of the net heat associated with physical or chemical process. In spite of this logic being attractive, users are presently shifting to the newer modulated temperature power compensated DSC. Power compensated DSC uses two independent micro furnaces for heating the sample and standard separately. Initially, the temperature of both the sample and standard are kept at the same value through independent heating. During an endothermic process, power is increased in the furnace corresponding to the sample pan to neutralize the temperature difference, and the increase in power is measured. The procedure is reversed for an exothermic process. Thus, here, the primary signal for calculation of thermal event is the adjusted power which gives us the enthalpy changes involved in the process. The major advantages of this technique are the rapid process of heating or cooling and better resolution for any thermal events.

(c) **Thermomechanical analysis (TMA)**: TMA technique is used for the estimation of coefficient of thermal expansion, dilatometric softening, glass transition temperature, etc. In this method, sample is placed on a holder with a push rod assembly and heated at a constant rate over a particular temperature range in a dilatometer. Expansion of the glass or glass-ceramic sample exerts a force on the push rod as a result of which the push rod starts moving upward. The movement of push rod associated with a scale accounts for the total expansion of the solid sample. However, care is taken to subtract the expansion of the dilatometer assembly itself with temperature. The coefficient of thermal expansion can be calculated based on the equation given below (Eq. 9.10).

$$\alpha_L = (\partial L/\partial T)/L \tag{9.10}$$

where α_L is the coefficient of thermal expansion,
L is the length of the test sample and T is the temperature.

Upon attaining glass transition temperature, test material becomes soft and hence the slope of the curve increases nonlinearly. The temperature at which the onset of nonlinearity is taking place is generally considered as the glass transition temperature. More scientifically, one can say that slope of the curve increases drastically by almost three to four times post glass transition because of softening. The intersection points of the tangents drawn from both these domains are considered as the glass transition temperature. Upon heating above the glass transition temperature, decrease in length of the sample is observed. This is attributed to the softening of glass and subsequent dip of push rod assembly into softened glass which is having lower viscosity. The temperature which marks the onset of the negative expansion of the glass sample is termed as the dilatometric softening temperature.

(d) **Micro-hardness characterization**: Hardness of any material is defined as its resistance to permanent deformation by any induced force. Hardness is measured for bulk samples by Vickers indentation technique and for thin films by Knoop indenter method. In the Vickers method, a diamond indenter is placed over the surface of a glassy material. Suitable loads are applied on the surface of glass through the indenter till an impression or deformation mark is created on the surface. Typical deformation area or area having dents (impressions) is observed under optical microscope, and the hardness is calculated based on the following formula (Eq. 9.11):

$$H = F/A_p \tag{9.11}$$

where H is the hardness of the material,

F is the applied load, and A_p is the projected deformed surface area.

Upon performing the experiments with a Vicker pyramidal indenter, the formula is modified to Eq. 9.12,

$$H \approx 1.8544F/d^2 \tag{9.12}$$

where d is the average length of the diagonal and F is the force applied to the diamond indenter. It must however be noted that based on hardness values, glasses can be classified as normal and anomalous glasses. Normal glasses exhibit flow behavior upon indentation, whereas anomalous glasses exhibit densification. Flow behavior occurs when the pressure applied by diamond indenter becomes sufficient to lower the viscosity of glass up to a certain depth from the glass surface and the material flows [25]. In contrast, the presence of heteroatoms as network modifiers within the glass network inhibits mobility, induces twisting and bending of bonds thereby resulting in densification of glassy structure. Another phenomenon observed is recovery of dent marks in glasses upon sufficient heating.

In the following section, brief description is given regarding applications of glasses and ceramics.

9.9 Application of Glasses and Glass-Ceramics

Glasses and glass-ceramics are extensively used in various technological applications ranging from domestic household materials to materials for nuclear waste immobilization. Although people started making glasses as early as 1500 BC, the extensive use of glass for modern technology was realized in the last century. As we are living in an age of high-speed communication, glass fibers are the obvious material of choice for the above applications. It is necessary that the phonon energy of the glass must be quite low to get optimum luminescence efficiency from lanthanide ions. Erbium-doped fiber amplifiers are extensively used for 1.5 μm window of telecommunication. Such fibers are made by doping lanthanide ions like Er^{3+} in glass and drawing the glass into fibers at higher temperatures. The major problem observed in the fabrication of efficient lanthanide ions doped fiber amplifiers is the clustering of lanthanide ions and associated reduction in the extent of luminescence. One way to circumvent this problem is to dope lanthanide ions in suitable host with nanosize dimensions prior to its incorporation in glasses.

Another important application of glass is for their use in glass to metal seals. Our earlier studies have demonstrated that lead silicate glasses with SiO_2 to PbO ratio 6.9 and containing different amounts of alkali and alkaline earth metal oxides are potential candidates for making compression type glass to metal seals. These glasses have got wide application as feed through materials for various pipelines of plants and holder of bulbs. Photograph of typical glass to metal seal used as feed through material is shown in Fig. 9.19a. In addition to this, glasses are extensively used in nuclear technology for immobilization of high-level nuclear waste. For example, detailed studies on physico-chemical properties of barium borosilicate and sodium borosilicate glasses confirmed that the former is a better candidate for immobilization of nuclear waste to be generated from the proposed advanced heavy water reactors (AHWR). Fabrication of glass-based nuclear waste matrices and their property validation is important for catering the needs of nuclear waste immobilization technology [35, 36].

Glass-ceramics, on the other hand, possesses both the properties of glass and ceramics and they have wide applications catering to the modern needs. Research on glass-ceramics led to the development of products with minimal thermal expansion between room temperature to around 700 °C range. Earlier reports [37, 38] suggest that glass-ceramics-based β-spodumene solid solution and β-quartz solid solution can be obtained with low or even negative thermal expansion (shrinkage type of glass phases). Apart from these materials, glass-ceramics-based on β-eucryptite ($LiAlSiO_4$), Cr-doped mullite ($Al_6Si_2O_{10}$), lithium metasilicate (Li_2SiO_3), etc., developed by Corning Incorporated, USA, are worthwhile to mention as they are used in optical applications in glass fiber-based telecommunication technology [39]. These materials made path breaking developments and

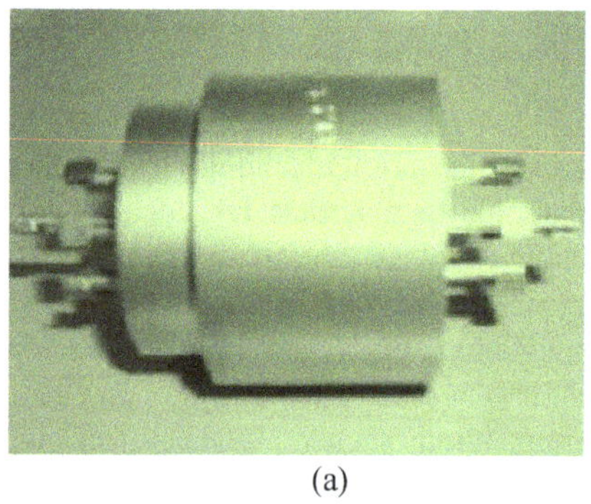

(a)

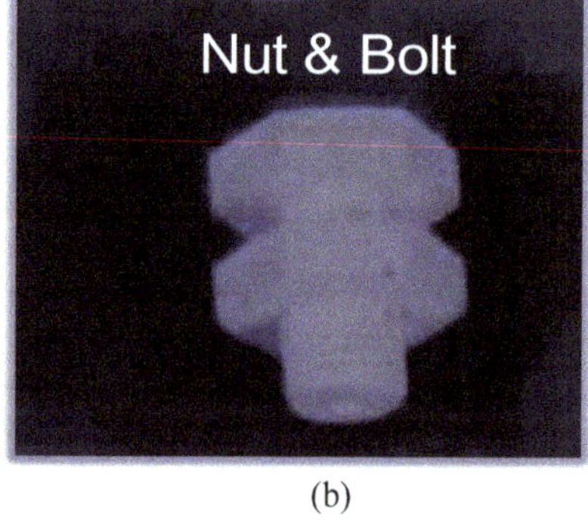

(b)

Fig. 9.19 Photographs of (**a**) compression type glass to metal seals (**b**) bolt and nut fabricated from MAS glass-ceramics

gave several epoch making technologies which revolutionized the communication technology. Reports [40, 41] are also there regarding development of spinel-based transparent glass-ceramic materials with embedded TiO_2 precipitate for production of photovoltaic substrates. Phosphate-based glass-ceramics have been found to be promising in the field of bone replacement and artificial and restorative dentistry [42, 43]. As glass-ceramics have properties of both glasses and ceramics, they are very widely used for making machine-able glass-ceramics. Magnesium aluminosilicate glass when subjected to programmed crystallization with addition of Al_2O_3 leads to glass-ceramics with machine-able properties. A variety of tools can be made by adjusting the Al_2O_3 content and temperature programming. A photograph of nut and bolt fabricated from the machine-able glass-ceramics is shown in Fig. 9.19b.

9.10 Summary

Different aspects of glasses namely synthesis, characterization, properties, etc., are discussed in this chapter. Thermodynamic and kinetic aspects of glass formation are explained in detail to understand the factors that are responsible for glass formation. Conversion of glass to glass-ceramics to achieve both the advantages of glasses and ceramics in the same material and widen the scope of such materials for different applications have been discussed in detail. Representative applications of glasses and glass-ceramics, mentioned at the end of the chapter, clearly establish the importance of carrying out extensive research and development work in this area.

Acknowledgements Authors of this chapter thank former and present colleagues in BARC namely Drs. G. P. Kothiyal, V. K. Shrikhande, S. K. Kulshreshtha, C. P. Kaushik, A. K. Tyagi, Madhumita Goswami, Raman Mishra, Dayamoi Banerjee, P. Sengupta, Mrs. Shobha, Dr. Vidya Thorat, Ms. A. Dhara for their active collaboration and scientific interactions.

References

1. https://web.archive.org/web/20110415194738; http://www.glassonline.com/infoserv/history.html
2. Woodard KL (March 2000) Profiles in ceramics: S. Donald Stookey. Am Ceram Soc Bull 79 (3):34–38
3. Höland W (ed) (2013) G. H. Beall, Handbook of advanced ceramics. Elsevier, Amsterdam, pp 371–381
4. Hench LL, Freiman SW (1971) Advances in nucleation and crystallization in glasses. Beall GH (eds) Structure, properties, and nucleation of glass-ceramics. Columbus, Ohio USA. The Am Ceram Soc special publ no 5:251–261
5. Tashiro T, Wada M (1963) Glass-ceramics crystallized with Zirconia. In: Advances in glass tech. Plenum Press, New York, p 18
6. Cambell J, Suratwala TI (2000) J Non-Cryst Solids 263&264:318

7. Ehrmann PR, Carlson K, Cambell JH, Click A, Brow RK (2004) Neodymium fluorescence quenching by hydroxyl groups in phosphate laser glasses. J Non-Cryst Solids 349:105
8. Mattox DM (1998) Handbook of physical vapour deposition processing (PVD): film formation, adhesion, surface preparation and contamination control. Noyes Publications, Oxford, UK, pp 29–54
9. Soukup RJ, Ianno NJ, Huguenin-Love JL (2007) Solar Energy Mater Solar Cells 91:1383
10. Pribil G, Hubi˘cka Z, Soukup RJ, Ianno NJ (2001) J Vac Sc Technol A19:1571
11. Soukup RJ, Ianno NJ, Pribil G, Hubi˘cka Z (2004) Surf Coat Technol 177&178:676
12. Li L, Lu Q, Fu RKY, Chu PK (2009) Nucl Instrum Methods Phys Res Sect B 267:1696
13. Binnewies M, Locmelis S, Meyer B, Polity A, Hofmann DM, Von H (2009) Prog Solid State Chem 37:57
14. Pierson HO (1999) Handbook of chemical vapor deposition (CVD): principles technology and applications. Noyes Publications, pp 217–226, 295–339
15. Kothiyal GP, Kumar R, Goswami M, Shrikhande VK, Bhattacharya D, Roy M (2007) J Non-Cryst Solids 353:1337
16. Aegerte MA, Mennig M (2004) Sol-Gel technologies for glass producers and users. Kluver Academic Press, New York
17. Young L (1961) Anodic oxide films. Academic Press, New York
18. Primak W (1958) Phys Rev 110:1240
19. Gibbs JH, Di Marzio EA (1958) J Chem Phys 28:373
20. Hruby A (1972) J Phys B 22:1187
21. Turnbull D, Cohen MH (1961) J Chem Phys 34:120
22. Ngai KL (2007) J Non-Cryst Solids 353:709
23. Wang PW, Zhang L (1996) J Non-Cryst Solids 194:129
24. Ross SD (1972) Inorganic infrared and Raman spectra. McGraw-Hill, New York, pp 148, 260
25. Liu HS, Shih PY, Chin TS (1996) Phys Chem Glasses 37:227
26. Rouse GB, Miller PJ, Risen WM (1978) J Non-Cryst Solids 28:193
27. Yadav AK, Singh P (2015) RSC Adv 5:67583
28. Boolchand P (2000) Insulating and semiconducting glasses. In: Series of directions in condensed matter physics, vol 17. World Scientific, Singapore
29. Eckert H (1992) Prog NMR Spectroscopy 24:159
30. Andrew ER, Bradbury A, Eades RG (1959) Nature 183:1802
31. Lowe IJ (1959) Phys Rev Lett 2:285
32. Shrikhande VK, Sudarsan V, Kothiyal GP, Kulshreshtha SK (2001) J Non-Cryst Solids 283:18
33. Sudarsan V, Shrikhande VK, Kothiyal GP, Kulshreshtha SK (2002) J Phys: Condens Matter 14:6553
34. Goswami M, Kothiyal GP, Sudarsan V, Kulshreshtha SK (2005) Glass Technol 46:341
35. Thorat VS, Mishra RK, Kumar A, Sudarsan V, Tyagi AK, Kaushik CP (2019) Bull Mater Sci 42:211
36. Dhara Prakash A, Singh M, Mishra RK, Valsala TP, Tyagi AK, Sarkar A, Kaushik CP (2019) J Non-Cryst Solids 510:172
37. Beall GH (1971) The Am Ceram Soc 61:251
38. Tashiro T, Wada M (1963) Adv in Glass Tech 9:18. Plenum Press, New York
39. Beall GH, Chyung K, Pierson JE (1998) Gratings Proc Int Cong On Glass XVIII (CD-ROM), San Francisco
40. Beall GH, Pinckney LR (1999) J Am Ceram Soc 82:5
41. Beall GH, Pinckney LR (1997) J Non-Cryst Solids 219:219
42. Chevalier J, Gremillard L (2009) J Eur Ceram Soc 29:1245
43. Kasuga T, Nogami M (2004) Trans Mat Res Soc Japan 29:2933

Chapter 10
Synthesis of Materials by Ion Exchange Process: A Mild Yet Very Versatile Tool

V. Grover

Abstract The technological advances of the society have been intricately related to development of novel and improvised materials and methodologies. Conventional synthesis routes involving higher temperatures and longer reaction duration tend to yield the thermodynamically stable products that have the limitation on introducing newer functionalities. The synthesis of the materials with desired properties requires novel routes that can take place at milder conditions. Synthesis by ion-exchange is one such low temperature preparative route that can be utilised to design rational synthesis to obtain materials with desired structures and morphologies. It has become a technique of choice to synthesize novel three dimensional layered structures that possess exchangeable cations. It has been used to synthesize nano-materials, not just de novo, but also as a *post-synthetic* procedure to obtain hitherto inaccessible phases and complex hetero-structures. These have various applications as next generation catalysts, electrical, optical, opto-electronic and magnetic materials. Understanding of mechanism of ion exchange synthesis process would also aid in better fundamental understanding and would ultimately help in planning, control and execution of the synthesis processes in systematic and a logical manner. The chapter discusses the history, fundamentals and applications of "preparation of materials by ion exchange synthesis" with relevant examples.

Keywords Synthesis · Ion-exchange · Kinetically-stable · Layered materials · Nano-heterostructures

---

V. Grover (✉)
Chemistry Division, Bhabha Atomic Research Centre, Mumbai 400085, India
e-mail: vinita@barc.gov.in

V. Grover
Homi Bhabha National Institute, Mumbai 400094, India

A. K. Tyagi and R. S. Ningthoujam (eds.), *Handbook on Synthesis Strategies for Advanced Materials*, Indian Institute of Metals Series,
https://doi.org/10.1007/978-981-16-1807-9_10

10.1 Introduction

In the modern world that is continually growing, there is always an unending quest for newer functional materials for various advanced technological applications. The synthesis or the preparation of materials lies at the core of improvising existing materials/technologies and inventing newer ones. Synthesis of inorganic solids in the desired phase and to the specifications has always been a very challenging and rewarding task for materials scientists. Designing inorganic materials involve a thorough knowledge of many factors such as structures, stabilities, reaction mechanism, and properties. There is no doubt that today, a wealth of knowledge is available in this context, however planning a synthesis and choosing appropriate synthetic methods to achieve a material with desired structure and properties is no easy task. A wide range of synthesis methods are today available in the toolkit of synthetic material chemists which are well documented in this book and elsewhere. These include ceramic solid-state synthesis, gel combustion, ultrasonication, microwave synthesis, and even the higher-end methods including Laser-based synthesis, to name a few. Each method has its own features which enable the experimenter to choose one among them depending on the desired end product. Synthesis of materials by ion exchange is an ambient conditions synthesis process. The ion exchange product effectively depends on various thermodynamics and kinetic parameters, understanding of which aids in better fundamental understanding of the processes that are involved. The nature of reaction during an ion-exchange process can be defined as a non-equilibrium process that is predominantly kinetically driven, and this factor can be controlled and tailored to yield desired products which might be difficult to obtain otherwise.

10.1.1 History of Ion Exchange Process

Ion exchange, as a process, has been well known to chemists for its applications in ion-exchange chromatography and also for its usefulness in separating various components/elements from a mixture. The concept behind this process has also been used by material scientists for obtaining novel materials for its versatility, procedural ease, and ability to yield hitherto unknown phases. Ion exchange is defined as the reaction in which the exchangeable or the free mobile ions of one solid can be exchanged for different ions with different materials which could be soild, liquid, or gas. The ions that are being exchanged may have similar or different charges. Structurally, to facilitate the exchange, the exchanger must have relatively open network structure that should be able to carry the ions and allow them to pass through it. Both natural and synthetic ion exchangers are, which could either organic or inorganic, and are used for various applications.

The major industrial application of ion exchange in the treatment of water. The calcium and magnesium ions present in water are responsible for its hardness that form insoluble precipitates with soap. The hard water is softened by exchanging these ions with sodium ions by passing through cation exchanger containing sodium ions. Earlier natural aluminosilicates were used for this purpose but are now replaced by synthetic resins. Ion exchange process can be used to separate lanthanide ions, which are usually difficult to separate by other separation processes. Other areas where ion exchange is widely used are separation of amino acids, separation of metal ions, removal of trace elements from potable sea water, etc.

Such an important and technically useful process has equally interesting historical references attached to it. The usage of ion exchange has been first reported in the Old Testament of the Holy Bible which mentions that Moses could obtain potable water from the brackish water. This could have been made possible by removal of salt-bearing minerals that contain sodium, magnesium, and calcium by ion exchange [1]. Then about a thousand years later, it was explained by Aristotle in *Aristotle's Problematica* that when sea water is passed through certain sand, it loses its salt content [1]. However, the understanding behind this process was lacking as Aristotle assigned this to differences in densities of salt water and fresh water. Even the scientist credited with the concepts of ions and ionic compounds, Jons Berzelius, observed that the solutions of common salt, when passed through sand was free from saline impregnation after the first pass, could not identify the mechanism. The agricultural chemists H.S. Thompson and J. Thomas Way [2, 3] are generally credited with the discovery of the process of ion exchange. Based on their experiments they came up with a reasonably accurate description of the process. Some of their observations were that ion exchange process sis definitely different from the adsorption and also exchange of some ions was more facile than others [2–4]. The materials responsible for this phenomenon were recognized later on chiefly (by Lemberg and later by Wiegner) such as clay glauconites, humic acid, and zeolites. Their discovery was later on utilised by many geochemists, but the actual application was by Gans for industrial water softening using natural and synthetic aluminium silicates [5]. The first analytical application of this process was for the separation of ammonium ions from urine by Folin and Bell in 1917 [6]. The ultimate revolution came after Adam and Holmes discovered that ion-exchange characteristics of crushed phonographs and that prompted the synthesis of first organic ion exchanger [7]. They were synthesized by reaction of formaldehyde with phenol or its derivatives [7] and exhibited stable ion exchange as compared to any other known exchangers. These synthetic ion exchangers played a very significant role in "Manhattan Project" [8]. The discovery of Promethium ($Z = 61$) has in fact been attributed to the process ion exchange [9]. The resins developed by Adams and Holmes were further researched and improvised by companies in various countries such as Germany, U.S.A, and England after the World War-II. "Ion exchange chromatography" thus emerged out to be an important manifestation of the ion exchange process. In labs, ion exchangers are used as an aid in analytical and preparation chemistry. The aim of scientific research with ion exchange membranes extend far into physiological chemistry, biochemistry, and biophysics. However, the most widely used application is still the purification and also demineralization of water which has been a perennial challenge since Moses and Aristotle, and a task that has been made more compelling by the growth of population and industries.

10.2 Physico-Chemical Description of Ion-Exchange Process

The *ion exchange reaction* occurs when the exchanger is placed in a solution containing a counter-ion, which is different from the ion present in the exchanger [10]. By definition, ion exchange reactions are reversible and are completed when the equilibrium state is achieved. There may be partial or complete exchange of counter-ions during the ion exchange reaction. If the material (the ion exchanger) does not possess a very strong preference to one of the exchanged ions, then both the ions shall be present in both phases in the equilibrium state. This also implies that a relatively stronger affinity to a particular ion will result in complete removal from one of the phases [10]. The same is expected, if solution has much higher concentration of one ion than the other ion. The equilibrium distribution of ions between the solid material and the solution is also the measure of selectivity exhibited by the ion exchanger. Fig. 10.1 shows the schematics of an ion exchange process.

The process of ion exchange in aqueous medium has been extensively understood in context of classical exchangers such as zeolites, hydrous oxides, mica, etc. The ease and existence of ion exchange in these materials have been understood on the basis of presence of channels, cages that are interconnected, and layers that can accommodate ions. Some of these are "slow" ion exchangers but are more elective whereas others are fast exchangers but are consequently less selective. The hydrous oxides and poorly cross-linked sulfonic acid resins belong to latter category, while

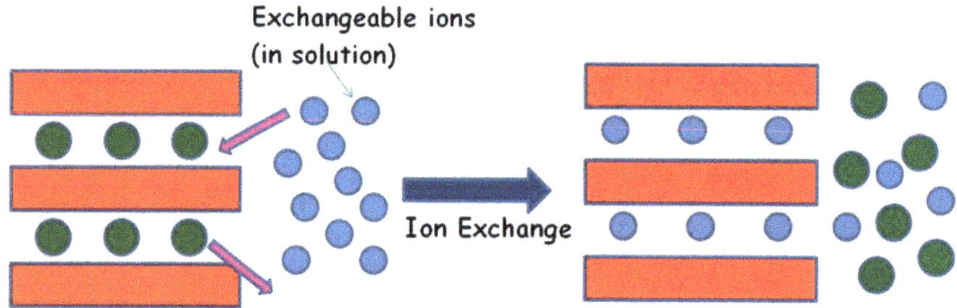

Fig. 10.1 Schematics of ion-exchange process

cross-linked zeolites may allow for selective ion-exchange and belong to former category. The ion transport in "faster" ion exchange materials may be akin to movement in aqueous solution while in "slower" exchange materials is more like ions hopping to vacant sites through anion-bound "restrictive" sites.

The substitution of ions in a solid with those present in the solution in its contact is a widely used strategy employed for modification of the composition and the properties of the materials. Ion substitution reactions have been actually responsible for various re-equilibration processes like those occurring in rocks [11], and also the displacement reactions in metal oxides [12].

The fact that a set of ions can be exchanged in a solid has been very nicely adopted by synthetic chemists for the synthesis of the materials which cannot be easily obtained by other synthetic routes. In this synthesis, the ion exchanger acts as one of the reactants. An example of ion exchange synthesis of more technological relevance is the processing of thin-film semiconductors [13]. A broad range of structures have been obtained via ion exchange of thin-film semiconductors materials for applications in optical waveguides, alloys for infrared photo detectors, etc [13]. The cation exchange procedure is, in general, simple though the anion exchange reactions have also been demonstrated [14–16]. However, in case of anion exchange reactions, the anions have relatively larger size and as the result the lower diffusivity. This leads to sluggish reaction kinetics that yields poorer morphology retention and as the result, often, hollow nanostructures are obtained. Cation exchange reactions, on the other hand, are simple, exhibit rapid reaction rates, and are capable of excellent morphology retention. This makes it an extremely attractive synthetic route to materials scientists.

10.3 Thermodynamics and Kinetics Concept of Ion Exchange

The products obtained from ion exchange synthesis are dictated by two primary factors: thermodynamics and kinetic [17]. While the thermodynamic parameters suggest the feasibility and spontaneity of the ion exchange, the kinetic parameters throw light on the mechanistic pathways of the process. In context of this, it is worthwhile here to discuss the implications as well as significance of both thermodynamic and kinetic factors.

Consider an arbitrary ion exchange reaction between two binary solids:

$$AX + B^{+} \rightarrow BX + A^{+} \tag{10.1}$$

where A and B are the exchangeable species.

This can be split into four reactions:

$$AX \rightarrow A + X \tag{10.2}$$

$$A \rightarrow A^+ + e^- \tag{10.3}$$

$$B^+ + e^- \rightarrow B \tag{10.4}$$

$$X + B \rightarrow BX \tag{10.5}$$

The energy terms involved in Eqs. (10.2) and (10.5) involve crystal energies that comprise of lattice energies and surface energies [18]. Equations (10.3) and (10.4) describe the solvation and desolvation energies of the corresponding cations. This shows that the energy balance of an exchange process is dictated by the ease of dissociation of the exchangeable cation from the parent host lattice and association of the incoming ion and its subsequent desolvation. The stronger solvation of outgoing cation compared to incoming cation, favours the progress of desired reaction. The fact that strain and dislocation energies of the reactant and the product phases also influence the thermodynamics and kinetics of the ion exchange reaction makes the topotactic exchange reactions more favourable than the non-topotactic reactions. For the uninitiated, topotactic reactions are those where the product lattice is related to the reactant lattice structure. Hence ultimately, the energetics and the thermodynamics of the cation exchange process is the sum-balance of lattice energy, solvation energy, interfacial energy, dislocation energy, and dissociation energy [18]. This gives an idea about not just the feasibility of the exchange process but also about the nature of end products e.g. whether the reaction shall yield solid solution or completely exchanged products. This shall depend upon the interfacial energies and also the solubility limits of the two phases into each other.

The knowledge about relative energy terms can be helpful in designing the synthesis. e.g. in a mutual exchange of a monovalent and divalent ions viz. M^+ and M^{2+}, the preferential solvation of divalent ion would control the outcome. In case of isovalent exchange, $M1^+$ and $M2^+$, the lattice energies and mutual solubilities would be more important factor as compared to the solvation energy. For the cases where the energy terms do not differ much, the reaction is driven by law of mass action and can be appropriately designed.

It must be noted, that the exchange reactions carried out under ambient conditions are seldom thermodynamically driven. In such cases, the kinetic factors assume a lot of significance. They depend significantly on the activation energy of each step (the sub-reactions mentioned earlier). Their activation energies are correlated since the product of one is the reactant for the other, and hence they must occur in a concerted manner. The exchange reactions have been described as possessing the "reaction zone". It is the zone where *dissolution* and *re-precipitation* occur together and simultaneously over a small length scale. This reaction zone is actually kind of a link that allows communication between the reactant and the product phase and thus helps in passing on and retention of the crystallographic information from parent phase to product phase [11]. The reaction zone is highly influenced by vacancies, dislocations, and interstitial diffusion, and these manifest in the crystal phase as well as morphology, porosity, etc. of the product. In case of bulk solids, the ion exchange may occur over hours sometimes days and may require elevated temperatures. This has some repercussion in terms of the product obtained which shall be discussed in the section dealing with ion exchange for nano-materials.

10.4 Utilizing Ion Exchange Reactions as Synthesis Process

Synthesis using ion exchange technique can be utilised in two modalities.

10.4.1 By Providing a Heterogeneous Medium Wherein the Desired Product Can Be Easily Separated in a One Pot-Synthesis from by-Products Without Much Reaction Work-Up

To explain this modality, let us consider the synthesis of a product X1Y2 from the reagents X1Y1 and X2Y2. The reaction can be represented as:

$$\mathrm{X1Y1 + X2Y2 \rightarrow X1Y2 + X2Y1}$$

Now, if the reaction is made to happen in a homogeneous media, it is not much useful as a synthetic route. All the four ionic species could be present in ionic/non-ionised form and the solution would contain the ions $X1^+$, $X2^+$, $Y1^-$, $Y2^-$ and the possible products X1Y1, X2Y2, X1Y2 and X2Y1. One of the products needs to be removed by some process such as precipitation or chromatography that makes the reaction non-homogeneous. This is an additional step that would depend upon the respective properties of the reactants and might not be feasible in many cases. However, if we use a cation or an anion exchanger (A) as follows:

$$\mathrm{AX1 + X2Y2 \rightarrow AX2 + X1Y2 (Cation\ exchanger)}$$

Or

$$\mathrm{AY2 + X1Y1 \rightarrow AY1 + X1Y2 (Anion\ exchanger)}$$

This ensures easy separation of the product X1Y2 and if done in columnar mode (reactor), may lead to complete conversion as well. In addition the regeneration of the exchanger is also possible, if desired, as follows:

$$AX2 + X1Y1 \rightarrow AX1 + X2Y1(\text{Cation exchanger})AY1 + X2Y2 \rightarrow AY2 + X2Y1(\text{Anion exchanger})$$

Thus, the ion exchange material maintains the heterogeneity of the system at all stages and also may be recovered, if desired, for a particular process. Another advantage is the purity of the product obtained.

One of the initial patents on this method was obtained on industrial synthesis of alkali nitrates in 1938 [10].

$$2AK + Ca(NO_3)_2 = A_2Ca + 2KNO_3$$

The regeneration can be performed using solution of chlorides as follows

$$A_2Ca + 2KCl = 2AK + CaCl_2$$

10.4.2 The Synthesis of Novel/New Phases of Technologically Important Compounds

It is increasingly being realized that progress in society depends on progress in materials. This can be achieved by synthesizing novel materials with desirable properties. During the past few decades, most of it could be achieved by serendipity and a lot by design too [19]. With the advances in materials science, understanding of structure and structure- property correlations, scientists are making huge leaps in being able to synthesize materials with desired properties up to some extent. Designing a material requires a thorough understanding of the property that is being looked for, and the structure that might be able to yield it, the next logical step is choosing the right synthesis route. The conventional synthesis routes yield thermodynamically stable products which are, no doubt, important and can be scaled up. However, the range of compositions and structures accessible by solid-state conventional synthesis is directed and decided by competition between thermodynamics and kinetic factors involved [20]. The requirement of solid-state synthesis that needs diffusion of ions calls for elevated temperatures which more often than not, yield thermodynamically stable phases. Interestingly, the combination of properties that would be required in the next generation materials would need the novel materials possessing structure that are not always known to occur naturally, in other words, the call of the hour is synthesizing metastable materials. The zeolites and other microporous solids, newer forms of carbon, variety of intercalation compounds, high Tc oxide superconductors, various organic and inorganic nano-composites, and the layered materials with anionic pillars are all metastable materials, to name a few [19]. Obviously, the common high-temperature solid-state synthesis method is not the correct choice for the synthesis of many of these desired solids.

Among soft chemical routes, synthesis by ion-exchange route can be adopted to yield the phases and the crystal structures hitherto inaccessible by other synthesis routes. This particular route involves fabrication of a thermodynamically stable phase in the first step, which is then transformed to the desired phase (metastable or stable) by a low-temperature ion exchange. The ion exchange may occur at low temperature or elevated temperatures depending on kinetics on exchange reaction involved. It provides a facile and versatile preparative method for exploring new and previously unknown compositions of materials with a certain crystal framework. Also, contrary to the common belief that ion exchange would essentially require fast mobility of ionic species involved, appreciable exchange rates have been demonstrated even with the diffusion coefficients in the range of $\sim 10^{-11}$ cm^2 s^{-1}, thus establishing ion exchange as a viable synthetic technique [19]. Further, ion exchange has been known to occur in both stoichiometric as well as non-stoichiometric solids. This is contrary to another common belief that non-stoichiometry and mobile ion vacancy would be a pre-requisite for facile ion exchange.

At this point, it is pertinent to discuss some typical examples of structural classes that are amenable to ion exchange reactions without disrupting the bonding between the layers. **Layered oxides** are a class of solid oxides most facile to ion exchange synthesis. Layered metal oxides consists of stacks of electrically neutral layers or charged layers of metal oxides interspersed with cations/ anions or the metal oxide sheets alternating with covalently bonded interlayers. The examples of compounds with neutral stacks of layers are $MoO_3.2H_2O$ and V_2O_5 [21, 22]. The examples with ionic intergrowths include structures based on A_xMO_2 and A_2MO_3-type which possesses edge- shared MO_6 octahedral layers with alkali cations residing between the layers [23]. Alkali metal niobates (KNb_3O_8 [24] and $K_4Nb_6O_{17}$ [25]) and titanates $[M_2Tib_nO_{2n+1}(M: Na, K)]$ [26–30] and $Na_4Ti_nO_{2n+2}$ [31, 32] also fall under this category. The titanate structures have slabs of layers that shear at every nth plane. Ternary layered oxides of transition metals, such as Mn and V have also been synthesized [33–35].

There is another class of layered oxides that possesses either cation or anion ordering. Brownmillerite ($A_2B_2O_5$) is the class of oxygen-deficient layered oxides that have alternate oxygen octahedra and tetrahedra. These may contain ordering of cations. The anion-ordered layered oxides consist of oxychlorides, oxyfluorides, oxyhydrides, etc. Structural motifs of different layered structures are depicted in Fig. 10.2.

Layered perovskites are special class of layered oxides that contain intergrowths of perovskite (ABX_3) with other structures. They consist of two-dimensional perovskite slabs which are interspersed with cations/cationic structural units. It consists of structural families: Dion-Jacobson phases (DJ) $[\text{general formula}: A'(A_{n-1}B_nO_{3n+1})]$, Ruddlesden Popper phases (RP) $[\text{general formula}: A'_2(A_{n-1}B_nO_{3n+1})]$ and Aurivillius phases $[\text{general formula}: M_2O_2(A_{n-1}B_nO_{3n+1})]$. These structures contain ABX_3 as the basic building block wherein the cation B is octahedrally co-ordinated and A

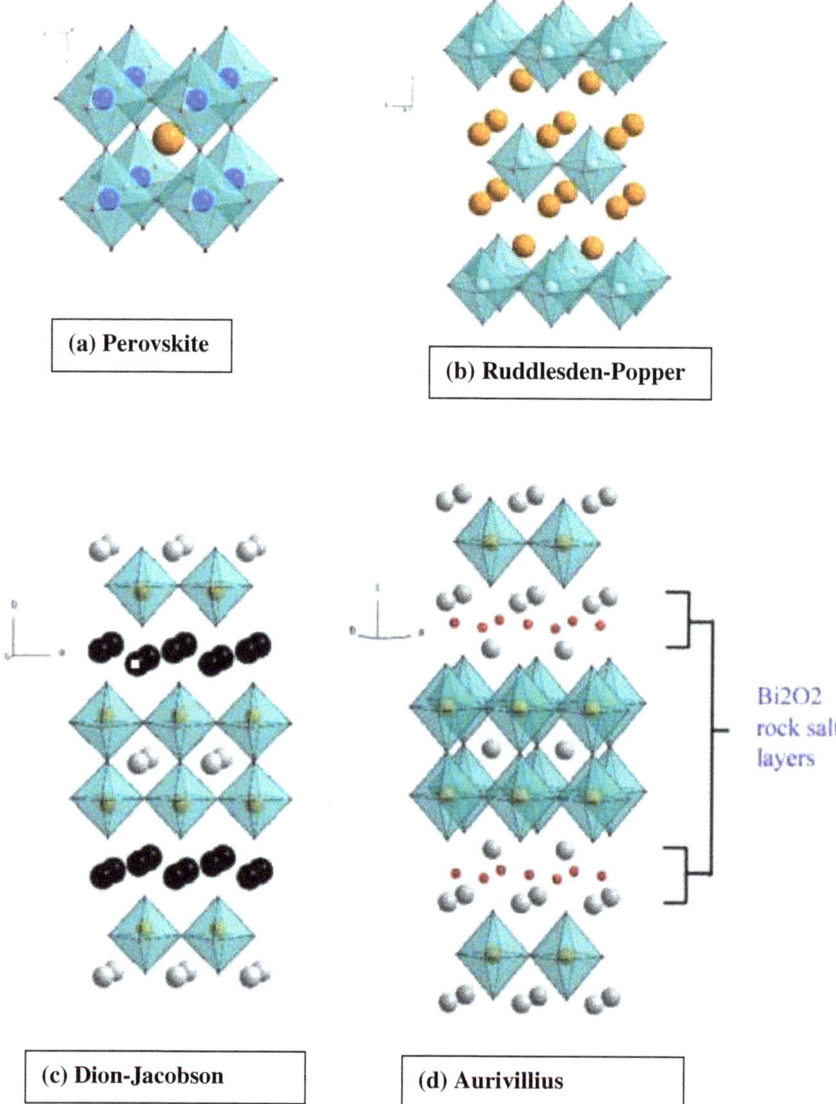

Fig. 10.2 Structural motifs for representative perovskite and perovskite based layered oxides. *Source* https://www.princeton.edu/~cavalab/tutorials/public/structures/perovskites.html **a** Perovskite, **b** Ruddlesden-Popper, **c** Dion-Jacobson, **d** Aurivillius

occupies the cuboctahedral site in ABX_3 slabs. DJ phases differ from RP in the sense that they consist of half the alkali metal cations in the interlayers. In addition, the RP phases have the rock salt arrangement of A–O in the interlayer spacing. The aurivillius phases have $M_2O_2^{2+}$ occupying the interlayer galleries between the perovskite slabs. These are structural classes that have exchangeable cations.

As discussed earlier, that compared to cation exchangers, anion exchangers are relatively less. One typical example of structural family that can exchange anions is layered double hydroxides (LDH). LDH is the famous class that is the structural analogue of Brucite, $Mg(OH)_2$, and has general formula $\left[M_{1-x}^{2+}M_x^{3+}(OH)_2\right]A_{x/n}^{n-} \cdot mH_2O$, where n is the charge on anionic species and x is the $M^{3+}/\left(M^{2+}+M^{3+}\right)$. The structure has positively charged metal hydroxide layers held together by anions and were first discovered in 1940s during the study of the co-precipitation of alkali metal cations with the help of bases [36]. The interlayers may consist of various anions such as carbonates, hydroxides and nitrates. Intercalation of various dyes and organic molecules in LDH has also been reported in the literature [37]. Layered hydroxides of rare earth ions are also known [38, 39].

10.5 Methodology of Ion exchange Reaction

The methodology for ion exchange reaction is simple. The ion exchange reaction is generally carried out by dispersing the reactant (parent) compound in either the molten salt (of the desired exchangeable ion) or the aqueous solution of salt of desired ion. The solution is kept stirred for better reaction rates. The temperature chosen, in most of the cases, is generally room temperature to 70–80 °C. The duration of reaction depends on material. For bulk solids, it may take from few hours to few days whereas for cation exchange in nano-materials, it may take few minutes to few hours. The exchange reactions may sometimes be carried out under hydrothermal conditions as well.

10.6 Layered Compounds and Ion Exchange

The fact that the layered oxides have charged layers (or sheets) held together by charged species intercalated between them, unlike Van der Waal solids, they have the potential to show the ion exchange where interlayer charged species can act as the exchangeable species and metal oxide block may act as host [20]. In addition, the presence of electrostatic forces may also make operations such as pulling apart of the layers to obtain nanosheets or stacking up to obtain complex structure feasibly. The fact that these ion exchange reactions may occur under milder conditions as compared to solid-state reactions may lead to metastable phases or the kinetically stable phases which would otherwise not be accessible. Designing novel metastable phases based on layered oxides employs general chemistry principles involving charge electro-neutrality and the acid–base chemistry fundamentals. The higher charge-to-radius ratio of smaller cations (or divalent cations) can effectively replace and drive larger cations out by lowering the electrostatic energy. This yields iso-structures with smaller cations in the interlayer galleries.

Ion exchange reaction is generally carried out in aqueous solution or molten salts. The reaction of $Tl_2Ti_4O_9$ with KCl–KNO_3 flux to obtain isostructural $K_2Ti_4O_9$ was the first report of ion exchange in layered oxides [27]. Similarly $Na_2Ti_3O_7$ can be acid-exchanged to $H_2Ti_3O_7$ which can again be reverted to the parent compound by reacting with NaOH [40]. Hydrolytic proton exchange of layered titanate $K_2Ti_4O_9$ is an important reaction that brings about the significance of soft-chemical synthesis to obtain metastable phases [41]. It yielded two different products: A partially hydrolyzed $K(H_2O)$–$Ti_4O_8(OH)$ and a fully hydrolysed $(H_2O)_2Ti_4O_7(OH)_2$. The dehydration at 500 °C of both yielded different products wherein the former led to a novel octatitatnate and latter yielded TiO_2(B) which is a metastable modification of TiO_2 having monoclinic structure. This signifies the gentle conversion of one structure into another with retention of chemical bonding. The alkali metal-ions in the inter-layer galleries of layered perovskites can be ion-exchanged with divalent metal cations as well as complex metal halides [42, 43]. Gopalakrishnan et al. synthesized the series $ALaNb_2O7$ (where A = K, Rb, Cs) which is $n = 2$ member of the layered perovskite family described by $A\left[A'_{n-1}B_nO_{3n+1}\right]$ conventional solid state synthesis. The counterparts with Li, Na and NH_4 were prepared by ion-exchange of $RbLaNb_2O_7$ with the corresponding molten nitrates. The topotactic proton exchange of the A atom $ALaNb_2O_7$ (A = K, Rb, Cs) yielded hydrated $HLaNb_2O_7$. The hydrate was found to readily lose water to give an anhydrous structure similar to $RbLaNb_2O_7$ [44]. Similarly, Li ion exchange of $LiNbWO_6$ and $LiTaWO_6$ in acid solution was performed in the same group yielding $HNbWO_6$ and $HTaWO_6$ that adopt cubic ReO_3 structure [45]. Here the ion exchange led to structural transformation from trirutile to ReO_3-type structure. The initial exchange products were identified as hydrated $HNbWO_6 \cdot H_2O$ and $HTaWO_6 \cdot H_2O$ with ReO_3 structures (Fig. 10.3). Both the hydrated products yielded $HMWO_6$ at 400 °C and $MWO_{5.5}$ at 600 °C both of which are ReO_3–type structures. The structural transformation involve tetragonal array of rutile to hcp array which is then followed by cation rearrangement to yield NbO_3 network similar to that in $LiNbO_3$. This is then followed by a transformation involving 60° rotation of octahedra that converts the hexagonal close packed anion array into 3/4 cubic closed packed array found in ReO_3. [46].

Most of these examples involved in ion exchange reactions, interchange cations and anions that are weakly bonded in a framework structure. Another interesting approach to obtain desired metastable phase is by transforming a layered structure into a three-dimensionally bonded solid that proceeds by topochemical condensation reaction. In this case, following the ion exchange, the selected terminal ligands can be removed along a particular crystallographic plane (e.g., removing O^{2-} anion in combination with H_2 to form H_2O). This bond formation in fact acts as a sort of "zipper" to form bonds between parallel planes leading to condensation of the structure. An example is conversion of $KTiNbO_5$ to $HTiNbO_5$ by exchange of K^+

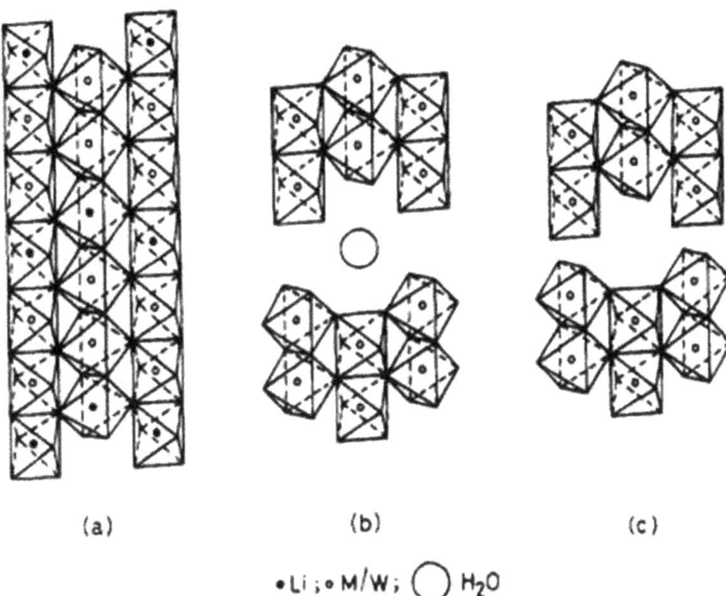

Fig. 10.3 Schematic representation of **a** rutile-like $LiMWO_6$, **b** $HMWO_6.nH_2O$ and **c** $HMWO_6$ (where M = Nb, Ta). Reprinted from Ref [19] with permission from (Chem. Mater. 7 (1995) 1265). Copyright (1995) American Chemical Society

with proton and that can undergo dehydration to yield $TiNbO_{4.5}$ (or $Ti_2Nb_2O_9$) [47].

Wiley and co-workers have been pioneers in topochemical ion exchange in layered perovskites [48]. Several monovalent and divalent ion-exchanges have been carried out by them in two layer DJ phase, $ALaNb_2O_7$ compounds where A could be H, Li, Na, K, NH_4 and Ag. The ion exchange was performed by heating the parent material in molten alkali nitrates. This has recently been done on $ANdNb_2O_7$ compounds and also oxyfluorides such as $RbLaNb_2O_6F$ [49].

The tendency to readily exchange inter-layer alkali metal cations with protons in both RP and DJ series yield strong Brönsted acids when the perovskite slab is formed by Nb. This can also lead to variety of intercalation compounds with different organic bases [44, 50]. These can further be exfoliated into single layers [51] and may also be stuffed with species, such as polyhydroxy aluminates, in the interlayer spaces yielding interesting molecular composites [52].

The fact that RP and DJ series are composed of same perovskite slab, $[A_{n-1}B_nO_{3n+1}]$, but at the same times possess different cation densities in the interlayers suggest that it should, in principle, be possible to vary the interlayer density. In this direction, Uma et al. [53] had synthesized a new series of layered perovskites, $A_{2-x}La_2Ti_{3-x}Nb_xO_{10}$ (A = K, Rb). Here the interlayer alkali-metal ion density varies from 2 (when $\boldsymbol{x}$ **= 0**) to **1** (when $\boldsymbol{x}$ **= l)** per formula unit. Thus, the end members, $A_2La_2Ti_30_{10}$ **(*x*= 0)** and $ALa_2Ti_3NbO_{10}$ (x = l), are the $\boldsymbol{n = 3}$ members of the

Ruddlesden-Popper series and the Dion-Jacobson series respectively and the solid solutions between them, $A_{2-x}La_2Ti_{3-x}Nb_xO_{10}$ can be considered as the bridge between the two series of layered perovskites. They observed RP type phase for $x = 1$ member while DJ type phase was observed for $0.0 \leq x \leq 0.75$ and this structural variation was also manifested in their Brönsted acid behavior as determined by intercalation of organic bases. Many other compounds such as $RbLaNb_2O_7$ [46], $CsCa_2Nb_3O_{10}$ [54], $RbLa_2Ti_2NbO_{10}$ [55], $RbSrNb_2O_6F$ [56], $NaLaTiO_4$ [57], $K_2SrTa_2O_7$ [58] and $Rb_2La_2Ti_3O_{10}$ [59, 60] have also been extensively studied. Layered perovskites have been shown to exhibit enhanced photocatalytic property [61]. In this regard, the role of partial alkali-metal exchange and interlayer hydration on catalytic activity has also been explored [62].

The ion exchange of various known compounds thus has the capability to yield new compounds with different inter-layer cations not feasible by conventional synthetic routes. The new structures obtained by exchanging the inter-layer cation can be further tuned by various other strategies such as intercalation of different molecules including organics, dyes, etc. to obtain different functionalities, tuning the interlayer spacing by intercalating different sized ions/neutral molecules, exfoliating the new structure to obtain building blocks for further engineering.

10.7 Synthesis by Ion-Exchange for Nano-Materials

The physico-chemical properties of the bulk solids are found to change substantially as the crystallite size is decreased to nano-dimensions. Advances in synthetic techniques and procedures that allow for the preparation of monodispersed nano-particles of diverse shapes and sizes have helped in developing a qualitative understanding of the fundamental scaling laws. This understanding has led to development of more efficient materials for various next-generation technological applications as photonic, optoelectronic, and catalytic materials.

The direct preparative routes for nanocrystal synthesis require a fine control of processes such as nucleation, growth, and surface binding kinetics at temperatures that are sufficient to obtain crystalline phase-pure nano-materials [63, 64]. However, the application of these nanomaterials as building blocks for future nano-devices need not only the phase pure materials but also a very systematic tuning of the composition and morphology which generally becomes difficult in the de-novo synthesis. In this context, the post-synthetic transformations have been found quite useful for obtaining complex nano-structures with desired composition and phase. There are a variety of post-synthetic options that are being adopted for modifications of nano-particles such as intercalation-deintercalation and exfoliation etc.

Ion-exchange has been found to be quite useful "post-synthetic" or the secondary procedure for fine control of the desirable characteristics in nano-materials. In addition to providing nano-structures with desired composition and phase, it also provides access to metastable compositions and structures that are otherwise inaccessible via direct synthesis from molecular precursors.

In case of bulk materials, long processing time is one very obvious limitations to ion exchange technique, and any elevation in temperature (to reduce the processing time) is constrained by material stability because the elevated temperatures would also enhance the counter-ionic diffusivity (ion other than which is being exchanged). In addition, the elevated temperatures, lead to the formation of products favored by thermodynamics and thus limits the probability of obtaining any metastable or non-equilibrium product that could be possible. These limitations can be overcome in nano-materials because of the availability of the large surface area and lesser interfacial strain that results in enhanced exchange rates [18]. For example, up to 100 times faster reaction rates have been reported for Ag^+ exchange of CdSe nano-material than in the bulk form, which is similar to molecular exchange reactions as compared to solid-state reactions [65]. In the ion exchange reactions, as in the bulk solids, the ions of the parent nano-material diffuse out into the solution and gets solvated and concurrently, the ions that are to be substituted get introduced into the lattice by inward diffusion. The diffusion rates are higher for nano-materials than the bulk because of larger surface-to-volume ratios. The reaction barrier required for nucleation may also be reduced on the nanoscale. This could be attributed to high curvature surfaces and the low co-ordination facets (present on nanoparticle surface) that serve as high-energy sites which are typically ideal for nucleation. The energetics is also altered as the surface energy has increasingly dominant contribution as the size of the nanoparticle decreases. This also helps in retaining the morphology of the initial nanocrystal template. The possibility of morphology retention along with stoichiometric control over composition enables the synthesis of nanocrystals with properties (morphology, composition, and phase) unattainable by other routes.

Due to the fast kinetics involved in nanoscale cation exchange, it can also be employed as a simple one-step room-temperature synthesis process for obtaining several products. For example, CdX (X = S, Se, Te) nanocrystals can be converted into a range variety of products, e.g. Ag, Cu, Pb, Zn, and Hg chalcogenides [66–69]. Herein, the size and shape-controlled template nanocrystals could be synthesized by the process such as hot injection in the first step [70, 71]. Following which the facile process of cation exchange allows to obtain other nanocrystals, which otherwise are not as easy to synthesize directly with such exquisite size and shape control. Thus, the ability of nanocrystals to undergo facile and rapid ion exchange at ambient temperature throws wide open an gamut of applications at the nano-scale relative to that observed in bulk systems.

It must be mentioned that the kinetics of exchange greatly depends on nano-crystallite sizes, shapes, and compositions. It was discussed earlier that ion exchange process can effectively be explained on the basis of "reaction zone". The reaction zone is typically a several atomic layer thick interface that represents a spatial local variation in the chemical composition as well as charge distribution relative to either of the bulk phases during the course of reaction. The nature, stability, and propagation of the reaction zone directly affects the morphology retention between the parent and product phase [66]. A typical example is the probability of morphology retention in the ion exchange reaction to obtain Ag_2Se

from CdSe nanorods. This was observed to be strongly dependent on the rod diameter. The small diameter (~3.5 nm) CdSe nanorods were observed to be transformed into spherical Ag_2Se nano-particles whereas larger diameter (~5 nm) CdSe nanorods could retain their structural morphology upon conversion to Ag_2Se. It is possible to understand this on the basis of reaction zone of the order of 3 nm in diameter. For thinner nanorods, the dimensions of reaction zone are comparable to that of nanorod size and hence the entire nanorod lies in reaction zone such that during the exchange, both cations and anions can rearrange and may result in the thermodynamically preferred, spherical morphology. On the other hand, in thicker nanorods, the larger size of nanorod as compared to reaction zone causes only a portion to experience significant structural distortions during the exchange process keeping majority of the lattice structure intact thus aiding the retention of rod-like morphology in the product phase as well. *Thus, there has to be an optimization.* While nano-size leads to enhanced kinetics, if the morphology retention is desired, the nano-particle should be large enough so that it is robust enough to prevent any reorganization in anionic framework during the cation exchange process. This was proved by the reversible exchanges between CdSe and Ag_2Se and vice versa.

The post-synthesis done by cation exchange also helps in introducing complexity in the nanostructures. Such complex morphological features lead to interesting electric, optical and optoelectronic behavior of the resultant structures. The complexities are achievable by partial "exchange" or partial "transformations" [72–74]. These partial exchanges may yield homogeneous products which are equivalent to solid solutions. The cation exchange process, though, not usually related to obtains solid solutions, also provide much better control over stoichiometry and hence compositions as compared to widely used hot injection synthesis for nano-materials [18, 75]. Further, whereas hot injection yields only stable thermodynamic phases, the cation exchange reactions may be designed at milder temperatures and ensuring strong thermodynamic driving conditions. Controlled doping via low amount of cation exchange under these conditions can yield controlled stoichiometry and also help in attaining the desired metastability [18]. In some cases, however, the product phase nucleates at a particular facet and grows in a specific direction which is generally topotactic towards the interior of the nanocrystal. This kind of nucleation and growth leads to hetero-structures with sharp interfaces and hetero-junctions [76–79]. Such nanostructures with hetero-junctions allow for band engineering vital for designing various opto-electronic devices required for high-efficiency light-emitting diodes and photovoltaic cells. A case in study is cation exchange of PbX (X: S, Se, Te) with Cd^{2+} ions. The PbSe-CdSe core-shell nanostructures obtained by this route show improved oxidative stability and more importantly very high quantum yields of the excitonic emission from the core were observed. The core shell nano-particles obtained via this route also possess lesser interfacial defects because of milder reaction conditions and topotaxial nature of product formation [80]. Sometimes the unique features of the nanoscale exchange process also yield different complex hetero-structures. For example, in Cu^+ exchange with CdS, the Cu_2S phase nucleates on end facets and hence the product phase grows inwards. The two end faces of CdS (wurtzite) are non-equivalent and hence there is a preferential growth of Cu_2S from one end. This yields binary CdS–Cu_2S nanorods representing the heterojunction [81] (Fig 10.4). On the other hand, in exchange of Ag^+ with CdS, there is a huge interfacial strain between Ag_2S and CdS phases and consequently, equally spaced regions of Ag_2S spread along the CdS rod are obtained [72]. Interestingly, both Cu^+ and Ag^+ can be exchanged with divalent ions in the presence of soft bases, this can be utilized to obtain heterojunction of CdS nanostructure with Pb^{2+} and Zn^{2+} by multi-step exchange processes. Similarly, a range of complex morphologies have been observed for partial exchange of CdSe with Pb^{2+} [80, 82].

In addition to templated morphologies, cation exchange processes may also yield metastable crystallographic phases as directed by parent (or the reactant) phase. The metastable phases, as known, are generally obtained by non-equilibrium processes wherein the kinetics can take over the thermodynamic criteria. Ion exchange, by nature, is a non-equilibrium process. The outcome of the ion exchange is primarily kinetically controlled thus giving access to metastable phases and compositions. The most stable phase for CdSe is hexagonal wurtzite, however, cation exchange of PbSe nanorods yield CdSe in zinc blende structure [82]. Reactant providing a structural template for product phase was demonstrated by Li et al. wherein zinc-blende CdSe yielded structurally related tetragonal/ FCC phase of Cu_2Se upon exchange with Cu^{2+} ions which further yielded zinc blende-type ZnSe. On the other hand, hexagonal wurtzite CdSe yielded hexagonal phase Cu_2Se and ZnSe

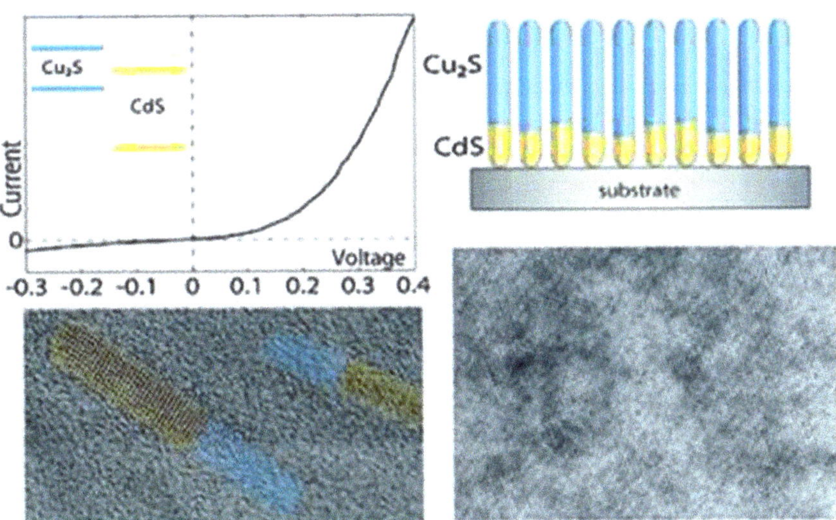

Fig. 10.4 Ion exchange demonstrated as a device fabrication technique. Figure shows the array of nanoscale heterojunctions obtained by facile Cu^+ exchange of a self-assembled CdS nanorods. Reprinted with permission from Ref 81 (J. B. Rivest, S. L. Swisher, L. K. Fong, H. Zheng and A. P. Alivisatos, ACS Nano, 2011, 5, 3811–3816). Copyright (2011) American Chemical Society

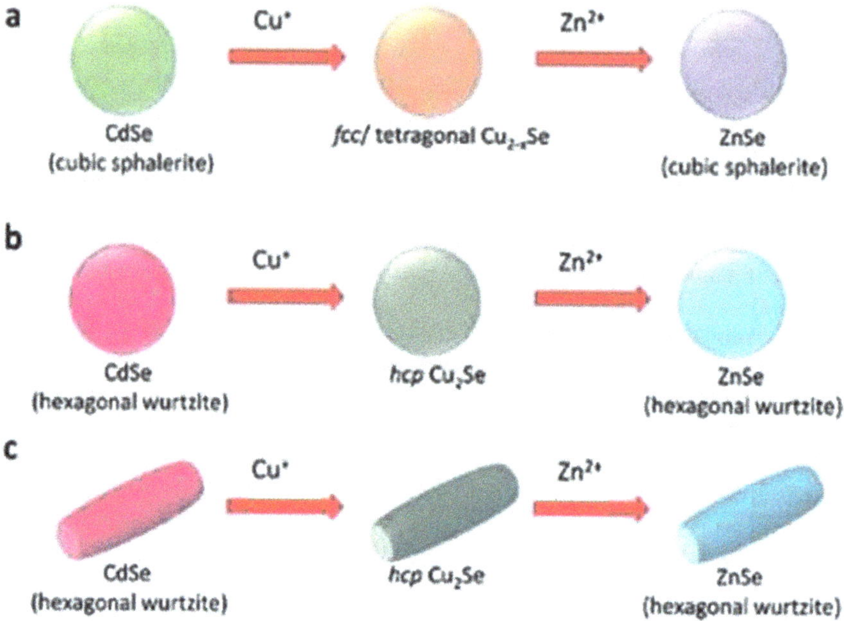

Fig. 10.5 Schematic describing the cation exchange of CdSe acting as both shape/morphology template and crystallographic template. The product nanocrystals adopt the crystallographic phase closely related to the starting lattice. Reprinted with permission Ref 83 (H. Li, M. Zanella, A. Genovese, M. Povia, A. Falqui, C. Giannini and L. Manna, Nano Lett., 2011, 11, 4964–4970). Copyright (2011) American Chemical Society

nanoparticles. Even the retention of morphology was depicted in the latter case (Fig 10.5) [83].

Cation exchange can also be employed as a low temperature, mild, low monetary cost, relatively environmental friendly device fabrication route that also gives access to novel better performing materials. First reported thin-film solar cell composed of Cu_2S–CdS was made from cation exchange some 50 years ago. Cation exchange may be used to access layered materials that ensure fast Li-ion mobility essential for battery applications. For example, $Cu_{2-x}(S_ySe_{1-y})$ nanocrystals, (with $S_{0.5}Se_{0.5}$), was obtained by cation exchange and evaluated for electrode material for Li-ion batteries [84]. These nanocrystals could show lithiation/ delithiation reaction in a partially reversible manner [84].

Cation exchange reaction is an excellent method to obtain hollow nanostructures. Nanostructures possessing hollow interiors exhibit attractive properties and functionalities but their synthesis is rather more complicated as compared to conventional nano-materials [85–88]. Their hollow interiors bestow them with properties such as low thermal expansion, low density, high surface to volume ratios, low refractive indices and these make them attractive candidates for an array of applications ranging from energy storage to drug delivery [89, 90]. Hollow nano-materials are generally prepared by template synthesis. The template could be hard template such as anodic alumina membrane or soft template like polymers, surfactants etc. In case of hard template synthesis, removal of template as well as stability of template is difficult task. The template might not be stable in acidic/alkaline medium or might have high solubility/reactivity. Similarly removal of template at high temperature might lead to non-retention of desired morphology and structure. Another technical barrier with the template method is that it is difficult to produce particles with a smaller size because of the intrinsic size limit of the template itself such as polystyrene or silica spheres. The ion exchange approach has been demonstrated as a versatile and rather easier method to obtain hollow nanostructures [91–93]. In order to obtain ZnO hollow microspheres, $Zn_5(CO_3)_2(OH)_6$ microspheres possessing uniform size and morphology were synthesized as the precursors for the subsequent ion exchange reaction. These were then added to the KOH solution and subsequently aged for sometime (~8 h) at room temperature. Figure 10.6 displays the conversion of filled microspheres to corresponding hollow ZnO microspheres. When $Zn_5(CO_3)_2(OH)_6$ microspheres were calcined at 400 °C, these however, yielded ZnO with similar solid structure [93]. Sometimes these chemical conversions are also carried out by gas phase reactions. For example, synthesis of hollow microspheres of CuO was obtained from CuS, by first synthesizing CuS microspheres and then heating them at air at 700 °C [94]. Heating the precursor structures in air initially led to the formation of a layer of copper oxide as a shell and oxygen continues to diffuse inside that causes increments oxidation. CuS has higher diffusion outwards as compared to inward diffusion of CuO. That causes a cavity in the centre. In addition, the release of SO_2 renders the shell porous. Hence, in the end the CuO microspheres with porous shells are obtained. The as-prepared hollow structures makes available more reactive sites by giving access to increased surface area and the better mechanical stability makes them suitable for applications such as energy and catalysis. Various other hollow structures such as polyhedra of various dimensions, nanotubes, sulphide hollow microspheres have also been reported by cation exchange process.

Nanoparticle synthesis by ion exchange is thus one of the advanced and favored synthetic methods due to possibility that it offers in terms of synthesizing novel nano-particles with complex structures such as segmented, core-shell, hollow microstructures, etc. However, this still in the developing stage. This needs more studies on bulk materials, their kinetics and thermodynamics to develop a complete understanding of the mechanism and the process to generalize for preparation of novel functional materials. As exemplified by many examples discussed here as well as in literature, ion exchange is promising technique and has the potential to become new paradigm for solid-state materials discovery.

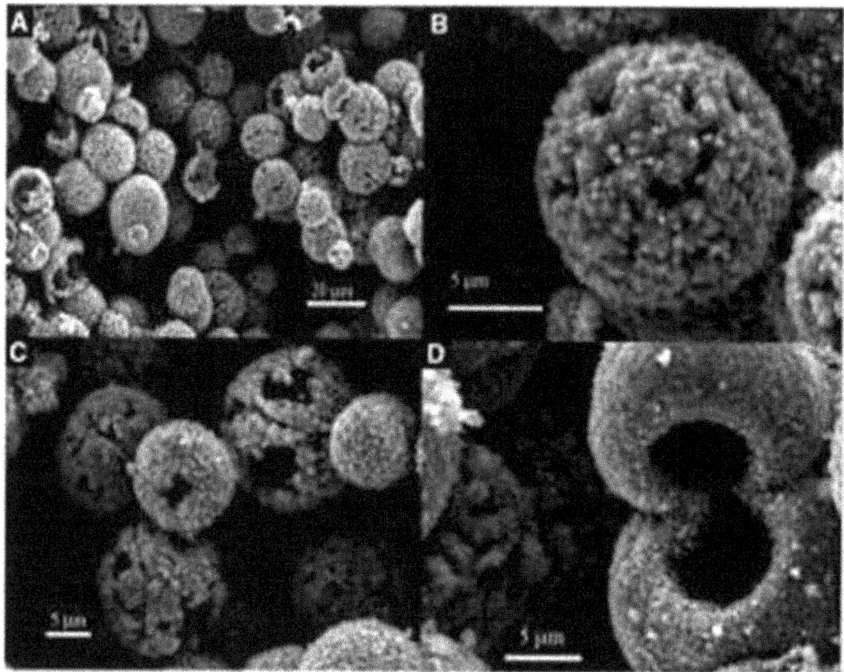

Fig. 10.6 SEM micrographs of the as-synthesized hollow ZnO microsphere obtained by the room-temperature treatment of $Zn_5(CO_3)_2(OH)_6$ microspheres (A) product morphology; (B) and (C) depict the detailed surface view of hollow ZnO microspheres; (D) micrograph of twin hollow ZnO microspheres. Reprinted with permission from Ref 93 (C. Yan and D. Xue, J. Phys. Chem. B, 2006, 110, 11076). Copyright (2006) American Chemical Society

10.8 Conclusion

Ion exchange, as a technique, was documented more than hundred years ago and has now become an integral part of new technical and industrial processes. It was adopted as synthetic process to obtain myriad new compounds, both layered and others, for various applications. With passage of times, it has been employed for obtaining novel nano-materials with phases, previously not known, and complex microstructures that are significant for device applications. This possibility of obtaining various metastable phases with enhanced functionalities and pre-designed microstructures through cation exchange has provided material scientists a very vital tool in their synthesis kit. These metastable materials obtained from ion exchange routes have wide applications in fields such as catalysis, energy, optical materials, electronics, etc. The design of materials for various high-end applications requires a wider understanding of the interfacial processes occurring at nanoscale during the ion exchange process. With the advent of advanced characterization and imaging techniques, such as high-resolution microscopy, especially *in situ*, there is a scope of revolutionizing mechanistic and fundamental understanding of solid-state transformations on the nanoscale. Due to the boundless options that that ion exchange synthesis has to offer, it has opened up a huge landscape for synthetic chemists for rational design of useful functional materials.

References

1. Lucy CA (2003) Evolution of ion-exchange: from Moses to the Manhattan Project to modern times. J Chromatogr A 1000:711
2. Thompson HS (1850) On the absorbent power of soils. J R Agric Soc Engl 11:68
3. Way JT (1850) On the power of soils to absorb manure. J R Agric Soc Engl 11:313
4. Way JT (1852) On the power of soils to absorb manure. J R Agric Soc Engl 13:123
5. Kunin R (1958) Ion exchange resins. John Wiley and Sons, New York
6. Folin O, Bell R (1917) Applications of a new reagent for the separation of ammonia: I. The colorimetric determination of ammonia in urine. J Biol Chem 29:329
7. Adams BA, Holmes EL (1935) Adsorptive properties of synthetic resins. J Soc Chem Ind (London) 54:1–6T
8. Settle FA (2002) Peer Reviewed: analytical chemistry and the manhattan project. Anal Chem 74:36A
9. Marinsky JA, Glendenin LE, Coryell CD (1947) The chemical identification of radioisotopes of neodymium and of element 61. J Am Chem Soc 69:2781
10. Kumar S, Jain S (2013) History, introduction, and kinetics of ion exchange materials. J Chem 957647
11. Putnis A (2002) Mineral replacement reactions: from macroscopic observations to microscopic mechanisms. Mineral Mag 66:689
12. Schmalzried H (1981) Solid state reactions. Verlag Chemie, Weinheim, Germany
13. Fedorov VA, Ganshin VA, Korkishko YN (1993) Ion exchange in II–VI crystals: Thermodynamics, kinetics, and technology. Phys Status Solidi A 139:65
14. Dloczik L, Engelhardt R, Ernst K, Fiechter S, Sieber I, Konenkamp R (2001) Hexagonal nanotubes of ZnS by chemical conversion of monocrystalline ZnO columns. Appl Phys Lett 78:3687
15. Dloczik L, Konenkamp R (2003) Nanostructure transfer in semiconductors by ion exchange. Nano Lett 3:651
16. Park J, Zheng H, Jun Y-W, Alivisatos AP (2009) Hetero-epitaxial anion exchange yields single-crystalline hollow nanoparticles. J Am Chem Soc 131:13943
17. Beberwyck BJ, Surendranath Y, Alivisatos AP (2013) Cation exchange: a versatile tool for nanomaterials synthesis. J Phys Chem C 117(39):19759
18. Rivest JB, Jain PK (2013) Cation exchange on the nanoscale: an emerging technique for new material synthesis, device fabrication, and chemical sensing. Chem Soc Rev 42:89
19. Gopalakrishnan J (1995) Chimie douce approaches to the synthesis of metastable oxide materials. Chem Mater 7:1265
20. Uppuluri R, Sen Gupta A, Ross AS, Mallouk TE (2018) Soft chemistry of ion-exchangeable layered metal oxides. Chem Soc Rev 47:2401
21. Krebs B (1970) The crystal structure of $MoO_3,2H_2O$: a metal aquoxide with both co-ordinated and hydrate water. J Chem Soc D 50:51
22. Bachmann HG, Ahmed FR, Barnes WH, Kristallogr Z (1961) The crystal structure of vanadium pentoxide. Cryst Mater 115:110
23. Delmas C, Fouassier C, Hagenmuller P (1980) Structural classification and properties of the layered oxides. Phys B + C 99:81

24. Gasperin M (1982) Structure du triniobate(V) de potassium KNb3O8, un niobate lamellaire. Acta Crystallogr, Sect B: Struct Crystallogr Cryst Chem 38:2024
25. Gasperin M, LeBihan MT (1982) Mecanisme d'hydratation des niobates alcalins lamellaires de formule $A_4Nb_4O_{17}$ (A = K, Rb, Cs). J Solid State Chem 43:346
26. Andersson S, Wadsley AD (1961) The crystal structure of $Na_2Ti_3O_7$, Acta Crystallogr 14:1245
27. Dion M, Piffard Y, Tournoux M (1978) The tetratitanates $M_2Ti_4O_9$ (M = Li, Na, K, Rb, Cs, Tl, Ag). J Inorg Nucl Chem 40:917
28. Wadsley AD, Mumme WG (1968) The crystal structure of $Na_2Ti_7O_{15}$, and an ordered intergrowth of $Na_2Ti_6O_{13}$ and `$Na_2Ti_8O_{17}$', Acta Crystallogr, Sect B: Struct Crystallogr Cryst Chem 24:392
29. Watanabe M, Bando Y, Tsutsumi M (1979) A new member of sodium titanates, $Na_2Ti_9O_{19}$. J Solid State Chem 28:397
30. Marchand R, Brohan L, Tournoux M (1980) TiO_2(B) a new form of titanium dioxide and the potassium octatitanate $K_2Ti_8O_{17}$. Mater Res Bull 15:1129
31. Olazcuaga R, Reau J-M, Devalette M, Le Flem G, Hagenmuller P (1975) Les phases Na4XO4 (X = Si, Ti, Cr, Mn, Co, Ge, Sn, Pb) et K_4XO_4 (X = Ti, Cr, Mn, Ge, Zr, Sn, Hf, Pb). J Solid State Chem 13:275
32. Werthmann R, Hoppe R, Anorg Z (1984) Über Oxotitanate der Alkalimetalle. Zur Kenntnis von $Na_4Ti_5O_{12}$. Allg Chem 519:117
33. Wadsley AD (1957) Crystal chemistry of non-stoichiometric pentavalent vandadium oxides: crystal structure of $Li_{1+x}V_3O_8$. Acta Crystallogr 10:261
34. Okada K, Marumo F, Iwai S (1978) The crystal structure of $Cs_6W_{11}O_{36}$ Acta Crystallogr, Sect B: Struct Crystallogr Cryst Chem 34:50
35. Delmas C, Fouassier C, Anorg Z (1976) Les Phases K_xMnO_2 ($x \leq 1$). Allg Chem 420:184
36. Feitknecht W, Gerber M (1942) Zur Kenntnis der Doppelhydroxyde und basischen Doppelsalze II. Über Mischfällungen aus Calcium-Aluminiumsalzlösungen, Helv Chim Acta 25:131
37. Lee JH, Young DY, Kim E, Ahn TK (2014) Fluorescein dye intercalated layered double hydroxides for chemically stabilized photoluminescent indicators on inorganic surfaces. Dalton Trans 43:8543
38. Klevtsova RF, Klevtsov PV (1967) X-ray diffraction study of a new modification of yttrium hydroxychloride $Y(OH)_2Cl$. J Struct Chem 7:524
39. Geng F, Matsushita Y, Ma R, Xin H, Tanaka M, Izumi F, Iyi N, Sasaki T (2008) General synthesis and structural evolution of a layered family of $Ln_8(OH)_{20}Cl^4.nH_2O$ (Ln = Nd, Sm, Eu, Gd, Tb, Dy, Ho, Er, Tm, and Y). J Am Chem Soc 130:16344
40. Izawa H, Kikkawa S, Koizumi M (1982) Ion exchange and dehydration of layered [sodium and potassium] titanates, $Na_2Ti_3O_7$ and $K_2Ti_4O_9$. J Phys Chem 86:5023
41. Tournoux M, Marchand R, Brohan L (1986) Layered $K_2Ti_4O_9$ and the open metastable TiO_2(B) structure. Prog Solid State Chem 17:33
42. Hyeon K-A, Byeon S-H (1999) Synthesis and structure of new layered oxides, $M^{II}La_2Ti_3O_{10}$ (M = Co, Cu, and Zn). Chem Mater 11:352
43. Kodenkandath TA, Lalena JN, Zhou WL, Carpenter EE, Sangregorio C, Falster AU, Simmons WB, O'Connor CJ, Wiley JB (1999) Assembly of Metal−Anion arrays within a perovskite host. Low-temperature synthesis of new layered copper−oxyhalides, (CuX) $LaNb_2O_7$, X = Cl, Br. J Am Chem Soc 121:10743
44. Gopalakrishnan J, Bhat V (1987) $AILaNb_2O_7$: A new series of layered perovskites exhibiting ion exchange and intercalation behaviour. Mat Res Bull 22:413
45. Bhat V, Gopalakrishnan J (1988) $HNbWO_6$ and $HTaWO_6$: Novel layered oxides related to the rutile structure. Synthesis and investigation of ion-exchange and intercalation behaviour. Solid State Ionics 26:25
46. Gopalakrishnan J (1986) Low-temperature synthesis of novel metal oxides by topochemical reactions. Proc Indian Natl Sci Acad A 52:48

47. Rebbah H, Desgardin G, Raveau B (1979) Les oxydes $ATiMO_5$: Echangeurs cationiques. Mater Res Bull 14:1125
48. Guertin SL, Josepha EA, Montasserasadi D, Wiley JB (2015) Thermal stability and high temperature polymorphism of topochemically-prepared Dion–Jacobson triple-layered perovskites. J Alloys Compd 647:370
49. Kobayashi Y, Tian M, Eguchi M, Mallouk TE (2009) Ion-Exchangeable, electronically conducting layered perovskite oxyfluorides. J Am Chem Soc 131:9849
50. Jacobson AJ, Johnson JW, Lewandowski JT (1987) Intercalation of the layered solid acid HCa2Nb3O10 by organic amines. Mater Res Bull 22:45
51. Treacy MMJ, Rice SB, Jacobson AJ, Lewandowski JT (1990) Electron microscopy study of delamination in dispersions of the perovskite-related layered phases $K[Ca_2Na_{n-3}Nb_nO_{3n-1}]$: evidence for single-layer formation. Chem Mater 2:279
52. Hardin S, Hay D, Millikan M, Sanders JV, Turney JW (1991) A molecular composite between partially hydrolyzed aluminum cations and a layered calcium niobate perovskite. Chem Mater 3:977
53. Uma S, Raju AR, Gopalakrishnan J (1993) Bridging the Ruddlesden–Popper and the Dion–Jacobson series of layered perovskites: synthesis of layered oxides, A2–xLa2Ti3–xNbxO10 (A = K, Rb), exhibiting ion exchange. J Mater Chem 3:709
54. Dion M, Ganne M, Tournoux M, Ravez J (1984) Structure cristalline de la pérovskite feuilletée ferroélastique $CsCa_2Nb_3O_{10}$. Rev Chim Miner 21:92
55. Gopalakrishnan J, Uma S, Bhat V (1993) Synthesis of layeerd perovskite oxides, $ACa_{2-x}La_xNb_{3-x}Ti_xO_{10}$ (x: K,Rb,Cs) and characterisation of new solid acids, $HCa_{2-x}La_xNb_{3-x}Ti_xO_{10}$ ($0 \leq x \leq 2$) exhibiting variable bronsted acidity. Chem Mater 5:132
56. Choy JH, Kim JY, Kim SJ, Sohn JS, Han OH (2001) New Dion−Jacobson-Type Layered Perovskite Oxyfluorides, $ASrNb_2O_6F$ (A = Li, Na, and Rb). Chem Mater 13:906
57. Blasse G, van del Heuvel GPM (1974) Vibrational spectra and structural considerations of compounds $NaLnTiO_4$. J Solid State Chem 10:206
58. Ollivier PJ, Mallouk TE (1998) A "Chimie Douce" Sythesis of Perovskite-Type SrTa2O6 and $SrTa_{2-x}Nb_xO_6$. Chem Mater 10:2585
59. Gondrand M, Joubert JC (1987) Nouveaux oxydes à structure en couches dérivant de celle de la pérovskite: le titanate double, $Na_2Gd_2Ti_3O_{10}$: cristallochimie et réactions d'échange. Rev Chim Miner 24:33
60. Gopalakrishnan J, Bhat V (1987) $A_2Ln_2Ti_3O_{10}$ (A = potassium or rubidium; Ln = lanthanum or rare earth): a new series of layered perovskites exhibiting ion exchange. Inorg Chem 26:4299
61. Takata T, Shinohara K, Tanaka A, Hara M, Kondo JN and Domen K (1997) A highly active photocatalyst for overall water splitting with a hydrated layered perovskite structure. J Photochem Photobiol. A: Chemistry 106, 45-49
62. Mitsuyama T, Tsutsumi A, Sato S, Ikeue K, Machida M (2008) Relationship between interlayer hydration and photocatalytic water splitting of $A'_{1-x}Na_xCa_2Ta_3O_{10}{\cdot}nH_2O$ (A′=K and Li). J Solid State Chem 181:1419
63. Yin Y, Alivisatos AP (2005) Colloidal nanocrystal synthesis and the organic-inorganic interface. Nature 437:664
64. Park J, Joo J, Kwon SG, Jang Y, Hyeon T (2007) Synthesis of monodisperse spherical nanocrystals. Angew Chem Int Ed 46:4630
65. Chan EM, Marcus MA, Fakra S, ElNaggar M, Mathies RA, Alivisatos AP (2007) Millisecond kinetics of nanocrystal cation exchange using microfluidic X-ray absorption spectroscopy. J Phys Chem A 111:12210
66. Son DH, Hughes SM, Yin Y, Alivisatos AP (2004) Cation exchange reactions in Ionic nanocrystals. Science 306:1009
67. Dloczik L, Koenenkamp R (2004) Nanostructured metal sulfide surfaces by ion exchange processes. J Solid State Electrochem 8:142

68. Mews A, Eychmuller A, Giersig M, Schooss D, Weller H (1994) Preparation, characterization, and photophysics of the quantum dot quantum well system cadmium sulfide/mercury sulfide/cadmium sulfide. J Phys Chem 98:934
69. Camargo PHC, Lee YH, Jeong U, Zou Z, Xia Y (2007) Cation exchange: A simple and versatile route to inorganic colloidal spheres with the same size but different compositions and properties. Langmuir 23:2985
70. Murray C, Norris D, Bawendi M (1993) Synthesis and characterization of nearly monodisperse CdE (E = sulfur, selenium, tellurium) semiconductor nanocrystallites. J Am Chem Soc 115:8706
71. Peng X, Manna L, Yang W, Wickham J, Scher E, Kadavanich A, Alivisatos AP (2000) Shape control of CdSe nanocrystals. Nature 404:59
72. Sadtler B, Demchenko DO, Zheng H, Hughes SM, Merkle MG, Dahmen U, Wang L-W, Alivisatos AP (2009) Selective facet reactivity during cation exchange in cadmium sulfide nanorods. J Am Chem Soc 131:5285
73. Luther JM, Zheng H, Sadtler B, Alivisatos AP (2009) Synthesis of PbS nanorods and other ionic nanocrystals of complex morphology by sequential cation exchange reactions. J Am Chem Soc 131:16851
74. Miszta K, Dorfs D, Genovese A, Kim MR, Manna L (2011) Cation exchange reactions in colloidal branched nanocrystals. ACS Nano 5:7176
75. Wills AW, Kang MS, Wentz KM, Hayes SE, Sahu A, Gladfelter WL, Norris DJ (2012) Synthesis and characterization of Al- and In-doped CdSe nanocrystals. J Mater Chem 22:6335
76. Peng X, Schlamp MC, Kadavanich AV, Alivisatos AP (1997) Epitaxial growth of highly luminescent CdSe/CdS Core/Shell nanocrystals with photostability and electronic accessibility. J Am Chem Soc 119:7019
77. Kim S, Fisher B, Eisler H, Bawendi M (2003) Type-II Quantum Dots: CdTe/CdSe(Core/Shell) and CdSe/ZnTe(Core/Shell) Heterostructures. J Am Chem Soc 125:11466
78. Talapin DV, Nelson JH, Shevchenko EV, Aloni S, Sadtler B, Alivisatos AP (2007) Seeded growth of highly luminescent CdSe/CdS nanoheterostructures with rod and tetrapod morphologies. Nano Lett 7:2951
79. Sitt A, Della Sala F, Menagen G, Banin U (2009) Multiexciton engineering in seeded core/shell nanorods: Transfer from Type-I to Quasi-type-II Regimes. Nano Lett 9:3470
80. Pietryga JM, Werder DJ, Williams DJ, Casson JL, Schaller RD, Klimov VI, Hollingsworth JA (2008) Utilizing the lability of lead selenide to produce heterostructured nanocrystals with bright, stable infrared emission. J Am Chem Soc 130:4879
81. Rivest JB, Swisher SL, Fong LK, Zheng H, Alivisatos AP (2011) Assembled Monolayer Nanorod Heterojunctions. ACS Nano 5(5):3811
82. Casavola M, van Huis MA, Bals S, Lambert K, Hens Z, Vanmaekelbergh D (2012) Anisotropic cation exchange in PbSe/CdSe Core/Shell nanocrystals of different geometry. Chem Mater 24:294
83. Li H, Zanella M, Genovese A, Povia M, Falqui A, Giannini C, Manna L (2011) Sequential cation exchange in nanocrystals: Preservation of crystal phase and formation of metastable phases. Nano Lett 11:4964
84. Dilena E, Dorfs D, George C, Miszta K, Povia M, Genovese A, Casu A, Prato M, Manna L (2012) Colloidal $Cu_{2-x}(S_ySe_{1-y})$ alloy nanocrystals with controllable crystal phase: synthesis, plasmonic properties, cation exchange and electrochemical lithiation. J Mater Chem 22:13023
85. Yan C, Xue D (2008) Formation of Nb_2O_5 nanotube arrays through phase transformation. Adv Mater 20:1055
86. Yan C, Nikolova L, Dadvand A, Harnagea C, Sarkissian A, Perepichka DF, Xue D, Rosei F (2010) Multiple $NaNbO_3/Nb_2O_5$ heterostructure nanotubes: A new class of ferroelectric/semiconductor Nanomaterials. Adv Mater 22:1741
87. Cao H, Qian X, Wang C, Ma X, Yin J, Zhu Z (2005) High symmetric 18-Facet polyhedron nanocrystals of Cu7S4 with a hollow nanocage. J Am Chem Soc 127:16024

88. Khanal A, Inoue Y, Yada M, Nakashima K (2007) Synthesis of silica hollow nanoparticles templated by polymeric micelle with core–shell–corona structure. J Am Chem Soc 129:1534
89. Lee J-H, Huh Y-M, Jun Y, Seo J, Jang J, Song H-T, Kim S, Cho E-J, Yoon H-G, Suh J-S, Cheon J (2007) Artificially engineered magnetic nanoparticles for ultra-sensitive molecular imaging. Nat Med 13:95
90. Wang B, Chen JS, Wu HB, Wang ZY, Lou XW (2011) Quasiemulsion-templated formation of α-Fe_2O_3 hollow spheres with enhanced lithium storage properties. J Am Chem Soc 133:17146
91. Prashant J, Lilac A, Shaul A, Alivisatos AP (2010) Nanoheterostructure cation exchange: anionic framework conservation. J Am Chem Soc 132:9997
92. Yan C, Xue D (2006) Conversion of ZnO Nanorod Arrays into ZnO/ZnS Nanocable and ZnS Nanotube Arrays via an in Situ Chemistry Strategy. J Phys Chem B 110:25850
93. Yan C, Xue D, (2006) Morphosynthesis of hierarchical hydrozincite with tunable surface architectures and hollow zinc oxide. J Phys Chem B 110:11076
94. Liu J, Xue D (2008) Thermal oxidation strategy towards porous metal oxide hollow architectures. Adv Mater 20:2622

Chapter 11
Polyol Method for Synthesis of Nanomaterials

Priyanka Ruz and V. Sudarsan

Abstract Polyol method is a widely used synthesis technique for the preparation of large number of inorganic compounds ranging from metal nanoparticles to alloys, oxides, sulphides, tellurides, fluorides, etc. Improved solubility of commonly available starting materials is the prime reason for wide applicability of the method for nanomaterials synthesis. Careful selection of precursors, relative amounts of polyols, ligands, etc., provide wide tune-ability in sizes and physico-chemical properties of synthesised materials. Morphologies such as spheres, nanorods, platelets and flowers can be synthesised by this method. In this chapter, initially different types of polyols and their physico-chemical properties are discussed briefly. This is followed by the details of synthesis of metals, alloys, oxides, sulphides, selenides, tellurides and fluorides in nano-size dimensions. Finally, the chapter ends with future scope and challenges in this method of synthesis.

Keywords Polyol method · Nanostructure · Size and shape · Capping agent

11.1 Introduction

The unique optical, catalytic and electronic properties of nanomaterials compared to their bulk counterparts have created considerable interest in the area of both fundamental research and development of technologies. Nano-science and nano-technology have evolved continuously over the past three decades and currently play pivotal roles in different industrial sectors such as electronics and information technology, pharmaceuticals, petrochemicals, energy and environment.

P. Ruz · V. Sudarsan (✉)
Chemistry Division, Bhabha Atomic Research Centre, Mumbai 400085, India
e-mail: vsudar@barc.gov.in

P. Ruz
e-mail: pdas@barc.gov.in

V. Sudarsan
Homi Bhabha National Institute, Mumbai 400094, India

A. K. Tyagi and R. S. Ningthoujam (eds.), *Handbook on Synthesis Strategies for Advanced Materials*, Indian Institute of Metals Series,
https://doi.org/10.1007/978-981-16-1807-9_11

Table 11.1 List of selected polyols along with their physico-chemical properties

Polyol	Chemical formulae	[a]B.P. (°C)	[b]η (cP)	[c]E_T
Ethylene Glycol (EG)	$HO–CH_2–CH_2–OH$	197	16.1	0.79
Di-ethylene Glycol (DEG)	$HO–CH_2–CH_2–O–CH_2–CH_2–OH$	245	30.2	0.713
Triethylene Glycol (TEG)	$HO–(CH_2–CH_2–O)_2–CH_2–CH_2–OH$	285	49.0 (20 °C)	0.682
Tetra-ethylene Glycol (TTEG)	$HO–(CH_2–CH_2–O)_3–CH_2–CH_2–OH$	327	44.9	0.664
1,3-propane diol (PDO)	$HO–CH_2–CH_2–CH_2–OH$	213	52 (20 °C)	–
1,4-butane diol (BD)	$HO–CH_2–CH_2–CH_2–CH_2–OH$	228	65	–
1,5-pentane diol (PD)	$HO–CH_2–CH_2–CH_2–CH_2–CH_2–OH$	242	128 (20 °C)	–
Glycerol (GLY)	$HO–CH_2–CH(OH)–CH_2–OH$	290	934	0.812

[a]Boiling Point
[b]Viscosity at 25 °C
[c]E_T: Normalized values of empirical parameter of solvent polarity

For the above-mentioned applications, it is absolutely necessary to have suitable synthesis procedures for preparing nanomaterials in larger scales. Significant interest has grown among researchers all over the world in the area of development of facile synthetic methods for nanomaterials in different sizes and shapes [1]. Methodologies involved in the preparation of nanoparticles can be broadly classified into two categories, namely bottom-up and top-down approaches. Bottom-up solution-based methodologies are very promising because of their inherent versatility and user-friendliness. These synthesis methods allow excellent control over the structure, morphology and chemical nature of the nanoparticles which in turn tune their physico-chemical properties. Among different solution-based methods for nanomaterials synthesis, the most widely used is the "polyol method". In this method, nanoparticles/nanomaterials are formed in a medium of high boiling poly-alcohol (polyol), and the polyol acts both as a solvent for synthesis and as a stabilizing ligand on the nanoparticles to prevent particle aggregation. The simplest member of polyol family is ethylene glycol. Among different types of polyols (listed in Table 11.1), the most widely used ones for the synthesis of nanoparticles are ethylene glycol (EG), glycerol (GLY) and n-butane diol (BD).

The term 'polyol synthesis' was first introduced in 1989 by Fiévet, Lagier and Figlarz to assign the solution phase synthesis of small size metal particles starting from their respective oxides, hydroxides or salts in polyols [2, 3]. Using this route, they reduced ions of copper, cobalt, nickel and platinum to the zero-valent state and the synthesized metal particles were in the micron size range. The lower degree of agglomeration of such particles and tune-ability in their size and shape encouraged researchers for using the process in the synthesis of other metal particles as well as alloys and intermetallics [4–7].

Choice of polyol during metal nanoparticle synthesis is determined by the reduction potential of metal and boiling point of poly-alcohol. For example, nanoparticles of noble metals such as Pd, Pt can be prepared in EG (b.p. ~197 °C), whereas transition metal nanoparticles (Co, Ni and Fe) require higher temperature for reduction and the polyol TEG (B.P. ~285 °C) appears to be a better choice compared to EG. The polyol serves both as a solvent and as a reducing agent. Apart from that, polyol process also provides several other advantages. First of all, polyols are advantageous because of their water equivalent nature. From the tabulated values of normalized empirical solvent polarity, it is clear that the polarity of these polyols are lower than that of water ($E_T = 1$). However, the lower polarity is taken care of by the presence of OH groups giving rise to good coordinating properties to these polyols and making them water equivalent solvent. Hence, inexpensive metal salts such as halides, nitrates or sulphates can be used as starting materials in polyol method. Coordination of polyol molecules with metal ions resulting in intermediate metal complex formation and subsequent reduction to zero-valent metals has been investigated by experimental as well as theoretical approaches [8]. For example, intermediate cobalt complexes have been isolated from the reaction mixture during polyol synthesis of Co nanoparticle [9], and the structures of the intermediates were determined by X-ray diffraction technique (XRD) (Fig. 11.1).

The experimental observations and theoretical predictions match well in case of Cu or Ni complexes formed during synthesis of copper or nickel nanoparticles using EG or GLY [10, 11]. In order to have an idea about the mechanism of

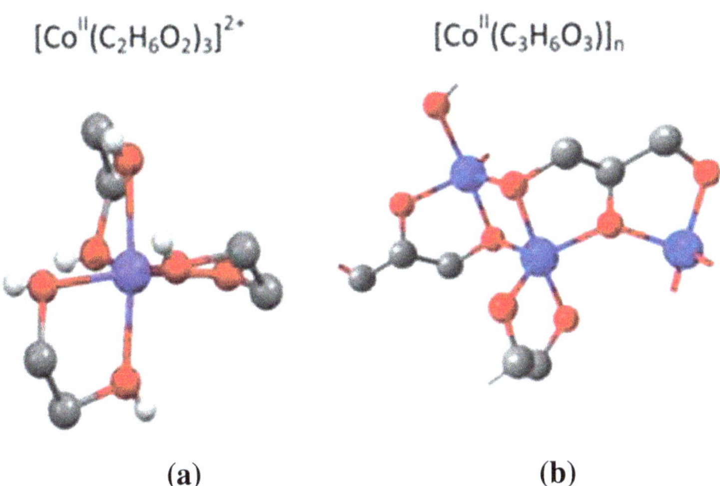

Fig. 11.1 Structure of Co complexes formed with (**a**) EG and (**b**) GLY. Spheres with different colours represent different atoms; blue, red, dark grey and pale grey for Co, O, C and H, respectively. H atoms attached to C and O atoms are omitted in (**b**). Reprinted with permission from Ref. [9]. Copyright (2018) Royal Society of Chemistry

reduction by polyol, some researchers have focused on theoretical modelling of interaction of polyol with as-synthesized metal nanoparticles [12, 13].

High boiling points of the polyols are added advantage as it facilitates relatively high temperature synthesis (200–320 °C) without applying high pressures. High temperature used during polyol synthesis enhances the chance of obtaining well-crystallized materials. It is worth mentioning here that the maximum temperature that can be obtained during polyol synthesis is not only dependent on the boiling point of the polyol but also restricted by its thermal stability. For most of the polyols, thermal decomposition occurs ~30–50 °C below the boiling point.

Chelating properties of the polyols control the nucleation, growth and extent of agglomeration of as-synthesized particles. Highly viscous nature of polyols (Table 11.1) compared to water ($\eta = 0.89$ cP) also influences nucleation and growth rate of nanoparticles [14]. Polyol method is successful in producing well-dispersed crystalline nanoparticles with controlled size and shape since it uses the magic solvent poly-alcohol with exceptional properties such as high boiling point, reducing and chelating nature as explained above. The experimental setup needed for polyol synthesis can be easily fabricated due to its simple design. Besides, polyols are less toxic compared to other conventional solvents and their biocompatible and biodegradable properties are well established. Hence, polyols are regarded as green solvent making them relevant for nanoparticle synthesis in industrial scale. Just after the pioneering invention of polyol process by Fiévet, Lagier and Figlarz, large-scale production of Co- and Ni-based particles were carried out using polyol method by the Eurotungstene Company for manufacturing cemented carbides [15]. The polyol method has high demand in electronic industry. Silver nanoparticles prepared through polyol route show high electrical conductivity and oxidation resistance which make them most suitable raw material for inkjets in electronics [16, 17].

In the following section of this chapter, details of synthesis of different types of nanoparticles and nanomaterials using polyol method have been described. Furthermore, modifications of conventional polyol process to extend synthesis method for challenging nanostructures such as less noble metal nanoparticles with controlled size and shape, water-soluble upconversion nanoparticles with proper surface functionalization for bioimaging, etc., have been discussed. The chapter comprises four main topics:

i. Polyol synthesis of metal nanoparticles
ii. Polyol synthesis of nanostructured metal oxides
iii. Polyol synthesis of nanostructured chalcogenides
iv. Polyol synthesis of metal fluoride nanoparticles.

Among the above-mentioned four topics, the first topic has been discussed under different sub-headings depending on the nature of metals. For the other three topics, specific examples have been selected for demonstrating the preparation of nanoparticle with varying sizes and shapes.

11.2 Polyol Synthesis of Monometallic Nanoparticles

Metal nanoparticles are the very first category of materials that were synthesized by the polyol route. As already described, polyol-mediated synthesis has many advantages over other solution based synthesis methods. In this method, simple low-cost metal salts are usually used as the raw materials. Polyols are reductive in nature at elevated temperature which helps in instantaneous reduction of dissolved metal cations. As-synthesized metal nanoparticles are surface stabilized by chelating polyols preventing particle agglomeration. Thus, polyol method can be regarded as a one-pot synthesis route for synthesis of finely dispersed crystalline nanoparticle with tailored morphologies. Different types of metal nanoparticles have been synthesized by polyol route, and they have been categorized in two classes, namely (i) noble nano-metals and (ii) less noble nano-metals.

11.2.1 Noble Nano-metals

Size- and shape-controlled noble metal nanoparticles are attracting the scientific community due to their unique physiochemical properties and technological importance. They have immense application in the areas of catalysis, fuel cell technology, optoelectronics, sensors, oil refining, photodynamic therapy of cancerous cells, etc. In recent years polyol synthesis method has been extensively studied for the formation of nanoparticles of noble metals (M = Pt, Pd, Rh, Ru, Ir, Au and Ag), and some of the examples are listed in Table 11.2.

Synthesis of sub-micrometer-sized Pd particles with uniform size and shape was reported by Figlarz group using Pd(II) tetraammine complex as starting material in ethylene glycol medium [36]. Pd particles with nano-size dimensions (<10 nm) were synthesized by Bonet et al. who used PVP as a stabilizer to avoid particle aggregation in EG [33]. Xia's group suitably modified polyol process in order to get efficient size and shape control during the synthesis of noble metal nanoparticles [23, 28, 32]. This involved (i) using suitable capping agents, (ii) controlling the reduction step and (iii) oxidative etching. Lee et al. [19] have obtained triangular and hexagonal shaped Pd plates through a modified polyol route where glycerol acted as reducing agent and PVP, a nonionic polymer, as surface capping agent. PVP has a highly polar amide group in its pyrrolidone ring, and its alkyl backbone is totally nonpolar. Thus, PVP has high solubility both in water and in nonpolar solvents. The same stabilizer can be used to get Pt nanoparticles with narrow size distribution by polyol method [33]. During the preparation of Pt nanoparticles by polyol/modified polyol method hexachloroplatinic acid is preferentially used as precursor material and the reaction temperature is kept above 100 °C (also mentioned in Table 11.2) [21, 23, 37–40]. A similar method has also been used for the controlled synthesis of gold nanoparticles of different sizes and shapes. In 1995 Silvert et al. [41] have adopted polyol process to prepare monodisperse,

Table 11.2 Examples of noble metal nanoparticles prepared by polyol route

M	Precursor	Polyol	Additives	Reaction temp (°C)	Shape	Size (nm)	References
Pd	$Na_2[PdCl_4]$	EG	PVP/HCl/$FeCl_3$	85	Triangular/hexagonal nanoplates	10–30	Ref. [18]
	$Na_2[PdCl_4]$	GLY	PVP/H_2O	100	Triangular/hexagonal nanoplates + dodecahedra	57–68	Ref. [19]
	H_2PdCl_4	TTEG	PVP/CTAB	MW[b]	Cubes/bars	15–20	Ref. [20]
Pt	$H_2PtCl_6{\cdot}xH_2O$	EG	PVP	190	Spheres	3.2–6.4	Ref. [21]
	$H_2PtCl_6{\cdot}xH_2O$	EG/ H_2O	PVP/KBr	116	Tripods	5–10	Ref. [22]
			PVP/KBr + HCl		Tetrapods/hexapods/octopods	10–20	
	$H_2PtCl_6{\cdot}xH_2O$	EG	PVP/SDS/Fe(III)/ Fe(II)	110	Nanowire	L:500 D:5	Ref. [23]
Au	$HAuCl_4{\cdot}xH_2O$	EG	PVP	155	Triangular/hexagonal/plates	Several hundreds of nm	Ref. [24]
	$HAuCl_4{\cdot}xH_2O$	EG	NaOH	RT[c]	Multispiked	75	Ref. [25]
	$HAuCl_4{\cdot}xH_2O$	DEG	PVP	B.P.	Decahedra/truncated tetrahedra/ icosahedra	50–100	Ref. [26]
Ag	$AgNO_3$	EG	PVP	120	Quasi spheres	15–21	Ref. [27]
	$AgNO_3$	EG	PVP/Pt seeds	160	Nanowire	L: up to 50 µm D: 30–40 nm	Ref. [28]
		EG	PVP	MW	Spheres	80–100	Ref. [29]
			PVP + Na_2S (1 mM)		Cubes	70–90	
			PVP + Na_2S (2 mM)		Nanowires	L: 5–20 µm D: 60–100 nm	

(continued)

Table 11.2 (continued)

M	Precursor	Polyol	Additives	Reaction temp (°C)	Shape	Size (nm)	References
Rh	$RhCl_3$	EG	PVP/TTAB	185	Cubes	<10	Ref. [30]
	$RhCl_3$	PEG	PVP	190	Octahedra	6.7	Ref. [31]
	Na_3RhCl_6	TEG	Citric acid, ascorbic acid	145	Tetrahedra	18	Ref. [32]
Ru	$RuCl_3{\cdot}xH_2O$	EG	PVP	150	Spheres	2	Ref. [33]
	$RuCl_3{\cdot}xH_2O$	EG	PVP	198		5.4	Ref. [34]
		DEG		245		2.9	
		TEG		285		1.8	
	$RuCl_3{\cdot}xH_2O$	1,2 PD	$CH_3COONa{\cdot}3H_2O$	140	Spheres	6	Ref. [35]

PVP polyvinyl pyrrolidone, *MW* microwave, *RT* room temperature

micron-sized, quasi-spherical Au particles. The reaction was carried out in EG medium with tetrachloroauric acid as starting material and PVP as stabilizing ligand. The PVP/Au precursor ratio was kept high to ensure the formation of nearly spherical Au nanoparticles by covering Au facets from all sides. For anisotropic growth of Au nanoparticles, concentration of PVP has to be reduced. At a lower ratio of PVP to Au precursor, PVP interacts selectively with different crystallographic planes and thereby modifying the growth rate along different planes leading to deviation from spherical shape [24, 42, 43]. A modified polyol method, using binary surfactants CTAB and PVP, was reported for the synthesis of hexagonal gold nanoplates by Wan et al. [44]. They have also fabricated gold nanoparticles of novel shapes such as star and shield by changing the temperature in the early stage of crystal growth. Gold nanoparticles with smaller size (and of different shapes) were obtained by Song et al. [45] by replacing EG with 1,5-pentanediol, adding $AgNO_3$ in very low concentration to the solution of tetrachloroauric acid and PVP. Representative SEM images recorded from the samples are given in Fig. 11.2.

A variety of capping agents, namely poly(diallyldimethylammonium) chloride (PDDA), polyvinylcaprolactam (PVCL), bovine serum albumin (BSA), etc., have been tried in place of PVP to get nano-sized gold particles [25, 46, 47]. The polyol method for silver nanoparticle synthesis was modified by Xia's group for fine control of morphology, and this was achieved by varying the types of surface stabilizers, the relative concentration of Ag precursor/stabilizer and reaction temperature [48–51]. The results have been summarized in a review article entitled 'Shape-Controlled Synthesis of Metal Nanostructures: The Case of Silver' [52]. Different techniques such as IR, Raman, and XPS have been used to study the selective adsorption of PVP during Ag nanoparticle formation. During growth of Ag nanoparticle, PVP is selectively adsorbed on the {100} facets thus forcing Ag atoms to sit on {111} facets, and this favours the formation of Ag nanocubes [53]. Seed-mediated growth procedure has been applied to synthesize nanowires and nanorods of Ag in EG solution in the presence of PVP [28]. Somorjai et al. [30] obtained Rh nanocubes of size <10 nm by using TTAB (trimethyl(tetradecyl)ammonium bromide) as surface stabilizing agent in EG solution as mentioned in Table 11.2. In this method, bromide ions (Br^-), slowly released from TTAB, stabilized the {100} facets inducing formation of Rh nanoparticles with cubical shape. Yan et al. reported that selection of polyol, reduction temperature and amount of surface protecting agent played critical roles in controlling final size of the Ru nanocrystals [See Table 11.2, ref. 34]. Ru nanoparticles with reduced size could be obtained by changing the polyol from EG to DEG and TEG while using ruthenium chloride as precursor and PVP as surface stabilizer. Apart from the addition of surface stabilizer, polyol process may also be modified by oxidative etching (by O_2 or other oxidant such as Fe/Cu salts) which helps to get well-controlled nanostructure geometry specially by removal of twinned defects during early stage of NP growth [54, 55]. Etching usually occurs at the sharp edges, corners, defect sites which have higher surface energies and helps to remove structural anisotropy of the nanoparticles.

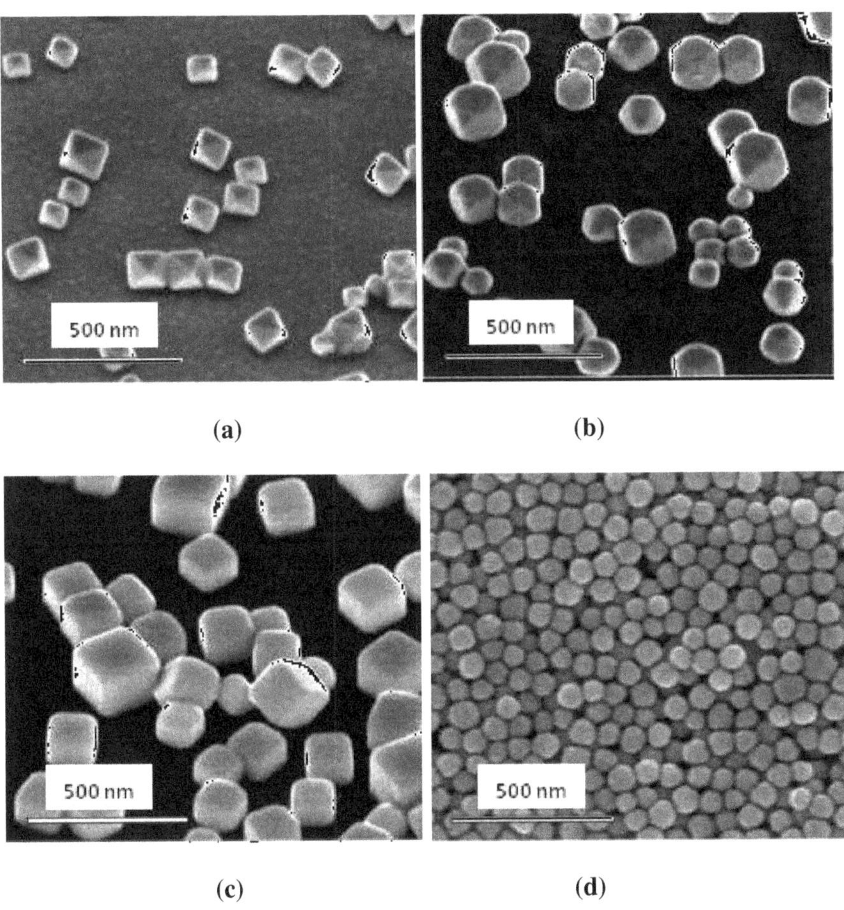

Fig. 11.2 Representative SEM images illustrating Au nanoparticles with different shapes formed by modified polyol synthetic route: (**a**) octahedral, (**b**) cuboctahedra, (**c**) cubes and (**d**) higher polygon. Adapted with permission from Ref. [45]. Copyright (2006) American Chemical Society

Experiments were carried out to prepare supported noble metal nanoparticles by polyol method for different applications. For example, Banerjee et al. have synthesized Pd-doped multi-walled carbon nanotubes (CNTs) by polyol method ($PdCl_2$ as precursor and EG as reducing and capping agent) and found a drastic increase in its hydrogen storage capacity when compared with the sample prepared by wet impregnation method [56]. Polyol route resulted in Pd nanoparticles with smaller sizes and better dispersion on CNTs which showed improved spillover of hydrogen and thereby enhancing its hydrogen uptake.

11.2.2 Less Noble Nano-metals

Synthesis of less noble metal nanoparticles by polyol method faces a lot of challenges (mainly because of oxidation) which can be overcome by adding stronger reducing agents such as sodium borohydride, hydrazine and sodium hypophosphite than polyol [57, 58]. Engels et al. [59] have successfully prepared zero-valent Cu NPs with size ~10 nm by deploying sodium hypophosphite ($NaH_2PO_2{\cdot}H_2O$) as co-reductant in the presence of PVP as anti-agglomerant in EG medium. The nanoparticles thus prepared show resistance to oxidation for several months. Kawasaki et al. [60] have reported formation of highly stable Cu particles (<3 nm) via microwave-assisted polyol method without addition of any surface stabilizer in EG solution. Under microwave irradiation, EG is converted into PEG which gets adsorbed on as-prepared Cu nanoparticle surface and restrict further growth and oxidation of Cu nanoparticles. In addition to use of co-reductant, citrate functionalization was also attempted by Feldmann et al. [61] for the synthesis of Cu nanoparticle. They used sodium citrate in DEG solution of cupric chloride to convert it into copper citrate ($Cu_3(citrate)_2$) which is finally reduced to zero-valent Cu nanoparticle upon $NaBH_4$ addition. The citrate-functionalized Cu nanocrystals show high oxidation resistance as well as colloidal stability. Similar strategy was adopted for the synthesis of finely disperse indium (In) nanoparticle with high air stability [62]. In another work, Feldmann et al. have reported formation of uniform size In^0 nanoparticle (10–15 nm) via a two-step synthesis route. At first low-cost indium precursor ($InCl_3$, $4H_2O$) is reduced to metallic indium by $NaBH_4$ reduction in DEG solution. However, the as-prepared In^0 nanoparticles have tendency to get agglomerated and oxidized in DEG solution. To avoid this, oleylamine-driven phase transfer of In^0 nanoparticles to dodecane was employed during the synthesis [63] (Fig. 11.3).

It is well known that melting point of a metal decreases sharply when its particle size is reduced to nano-size range. Bulk bismuth has low-melting point (271.4 °C), and it is expected that NPs of Bi would have much lower melting point. Xia et al. [64] prepared mono-dispersed liquid droplets of Bi metal with controlled diameter over the range of 100–600 nm. Bismuth acetate was thermally reduced to metallic Bi in boiling EG in the presence of PVP as surface capping agent. The as-prepared Bi atoms nucleate and grow into monodisperse spherical particles as shown in Fig. 11.4. The size of liquid and metallic Bi spherical colloids was controlled by varying the precursor concentration.

Apart from thermal reduction method as mentioned above, photochemical reduction was also used for the synthesis of Bi nanoparticles from $BiCl_3$ precursor [65]. Upon exposure to sunlight, transparent and colourless solution of $BiCl_3$ in DEG changes into black suspension due to photochemical reduction of Bi^{3+} to Bi^0 accompanied by oxidation of DEG. An air contact of black suspension resulted in a clear solution indicating reversible nature of the conversion.

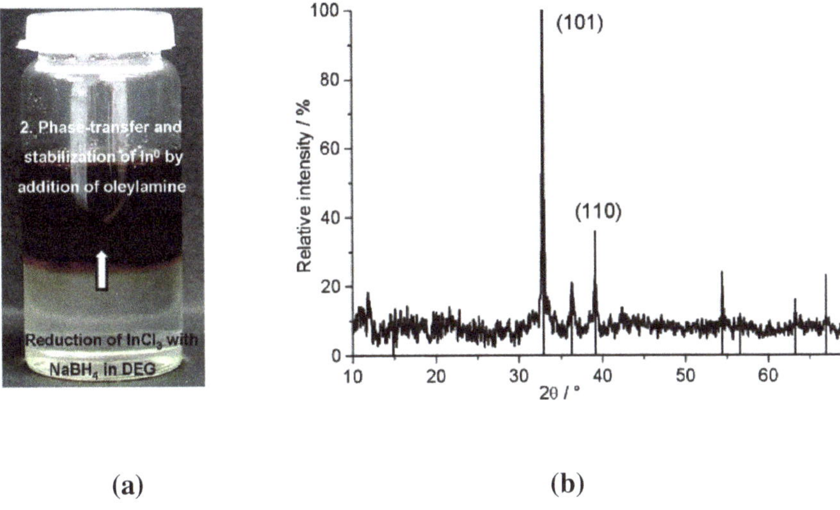

Fig. 11.3 (**a**) Polyol synthesis of indium (In^0) nanoparticle assisted by $NaBH_4$ and subsequent phase transfer to dodecane medium. (**b**) XRD pattern of nanostructured In^0 particles. Adapted with permission from Ref. [63]. Copyright (2009) American Chemical Society

Fig. 11.4 SEM image of spherical colloidal particles of Bi prepared by modified polyol method in EG medium. The inset represents a closer look into the colloidal Bi spheres. Adapted with permission from Ref. [64]. Copyright (2004) American Chemical Society

11.3 Synthesis of Multi-metallic Nanoparticles

11.3.1 Nanoalloys

The polyol synthesis process has been extended beyond metal nanoparticles, and different nanoalloys have been prepared using this solution phase method. The method is widely employed for the synthesis of polymetallic ferromagnetic materials with improved magnetic properties. In 1996 Fievet et al. [7] succeeded in preparing bimetallic Fe–Ni, Fe–Co and trimetallic Fe–Co–Ni compounds from their mixed hydroxides through a modified polyol method where NaOH-polyol combination was used. A high NaOH concentration was maintained to ensure precipitation of metallic Fe from $Fe(OH)_2$ and suppress the formation of oxides. In the course of the reaction, zero-valent nickel and cobalt were formed from Ni(II) and Co(II) hydroxides by polyol and ultimately alloyed with Fe. XRD, EDS and electron microscopic techniques were used to characterize these particles which revealed that binary Fe–Ni particles were spherical in shape with sub-μm size range along with narrow size distribution. Composition of the alloy was also found to be quite homogeneous. Unlike the single-phase Fe–Ni alloy sample, Fe–Co and Fe–Co–Ni powders were found to be multiphasic. Jeyadevan et al. [66] modified the polyol process in such a way that cubic Fe–Co nanoparticles were formed with Fe concentration varying from 20 to 90 mol% with size in the range 35–300 nm. Conventional polyol route is unfavourable for the synthesis of single-phase Fe–Co particle due to the presence of additional α-Co and ferrite phases which are formed during the course of reaction along with Fe–Co. Fe–Co nanoparticles with different compositions were successfully synthesized by reducing Fe and Co salt mixtures at a specified ratio in EG medium at 403 K for 1 h in the presence of NaOH (molar ratio of hydroxyl ion and metal was kept around 40:1). At this temperature and high conc. of OH^-, Fe^0 was easily formed from the iron precursor compared to Co^0 from cobalt precursor. As soon as Fe^0 is formed, it acts as a catalyst and facilitates reduction of cobalt ions leading to Fe–Co nanoalloy formation. Abbas et al. [67] obtained Fe–Co nanoparticle with size ~10 nm by reducing the solution containing both Fe(II) chloride and Co(II) acetate in PEG medium in the presence of NaOH, at 300 °C. Recently, binary nanoalloys of Fe–Ni with varying concentration of Fe were prepared by reduction of Fe(II) chloride and Ni (II) chloride mixture with $NaBH_4$ in EG medium [68]. Fe_xNi_{1-x} ($x = 0.25$, 0.50) nanoparticles are found to exist in face-centred cubic (fcc) structure. Atomic force microscopic (AFM) studies confirmed existence of particles with different size regimes: the smaller particles with 10–20 nm size and relatively bigger particles with size in 30–60 nm range, as shown in Fig. 11.5.

Extensive research has been carried out on the preparation of nanoalloys of noble metals due to their immense applications in the field of heterogeneous catalysis. Park et al. [69] reported glycerol-mediated polyol synthesis of Pd–Pt nanoparticles of octahedral shape and dominantly exposed with {111} facets. The bimetallic Pd–Pt nanoalloy showed excellent catalytic stability and superior electro-catalytic performance for methanol oxidation reaction. Kusada et al. [70] have reported hydrogen storage properties of Ag–Rh alloy though none of the constituent element absorbs hydrogen. This unexpected property of $Ag_{50}Rh_{50}$ alloy is explained in terms of electronic structure similarity between this alloy and Pd. The neighbouring elements in the periodic table, Ag and Rh, have insignificant miscibility even in molten state at very high temperature, and hence, the preparation of Ag–Rh alloy is really challenging. However, through a modified polyol method, Kusada et al. have synthesized Ag–Rh alloy stabilized by PVP at a much lower temperature (<200 °C). The preparation procedure involves mixing equimolar

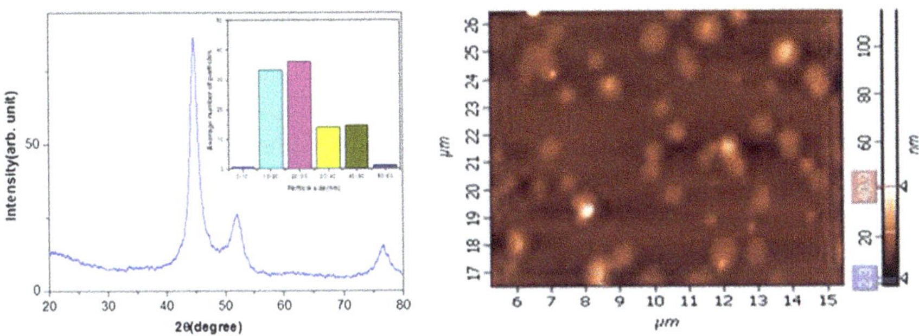

Fig. 11.5 XRD (**a**) and AFM image (**b**) of $Fe_{0.5}Ni_{0.5}$ nanoalloy with corresponding histograms. Adapted with permission from Ref. [68]. Copyright (2016) Elsevier

aqueous solution of $AgNO_3$ and $Rh(CH_3COO)_3$ in EG in presence of PVP at 170 °C. Subsequently, metal ions were reduced to corresponding metals in zero-valent state followed by atomic-level mixing leading to alloy formation. XRD and EDS confirmed the formation of the alloy phase. Hydrogen uptake of $Ag_{50}Rh_{50}$ alloy is almost half that of palladium. Pd-based bimetallic nanoalloy Pd–Rh was prepared by a similar method at much lower temperature ($\sim$100 °C) using $Pd(NO_3)_2$ and $RhCl_3$ as the respective metal precursors in EG medium in the presence of capping agent PVP. The alloy nanoparticles thus prepared was highly crystalline in nature with mean diameter $\sim$3 nm and showed hydrogen uptake at ambient pressure without phase separation.

11.3.2 Core–Shell Nanostructure

Nanoparticles of core–shell structure are of growing interest because of their unique catalytic, optical, electronic or magnetic properties. In case of a metal–metal core–shell nanostructure, the core, initially generated by polyol process, can act as a seed for the growth of the metal shell over the core surface. Tsuji et al. [71] have demonstrated a two-step polyol reduction method for preparing Au@Ag core–shell nanostructures with different shapes. First, reduction of $HAuCl_4.4H_2O$ in EG containing PVP was carried out to prepare Au core having different shapes such as triangular, octahedral, decahedral with {111} facets. Addition of a solution containing silver precursor to the polyol solution of Au nanoparticles led to the formation of a core–shell structure. The epitaxial growth of Ag on Au surface gave rise to various shapes of Au@Ag nanocrystals. Similar type of shape evolution was observed in Au@Ag core–shell nanostructures by Park et al. [72] who reported gradual change in crystal shape from sphere to cuboctahedra to cubes with progress in reaction. Same strategy was followed for the synthesis of Au@Cu core–shell nanoparticles wherein Au and Cu have lattice mismatch unlike Ag–Au system [73].

11.4 Synthesis of Nanostructured Metal Oxides

Apart from metal and alloy nanoparticles, polyol synthesis has been established as a suitable method for the synthesis of different metal oxides. Like metal nanoparticles, oxide nano particles are of immense technological importance in areas such as passive electronic components, catalysis and surface passivation in the form of coating. All these applications require finely dispersed oxide nanoparticles free from agglomeration and uniform size distribution. Oxide nanocrystals with these properties can be obtained by polyol method without any thermal treatment for post-sintering, in most of the cases. However, shape control of the oxide nanoparticles requires prior knowledge regarding the acidic or basic character of their surfaces. Sometimes, additional reagents or ligands need to be mixed in the polyol solution for precise size and shape control of the nanoparticles. Selection of polyol is very important for suppressing the formation of zero-valent metal while preparing metal oxide nanoparticles. For some systems, both zero-valent metal and metal oxide nanoparticles can be obtained as per requirement, just by playing with the reaction temperature. In most of the cases, metal nanostructures are obtained at higher temperatures (>180 °C) whereas nanoparticles of metal oxides are formed in the temperature range of 100–180 °C [58, 74]. Aluminium-doped ZnO (AZO) nanostructures were synthesized by microwave-assisted polyol method using $Zn(CH_3COO)_2 \cdot 2H_2O$ precursor in DEG together with $AlCl_3 \cdot 6H_2O$ [75]. Colloidal stability of AZO nanoparticles was found to be quite good. With the help of NMR-technique, it is confirmed that fourfold coordinated Al^{3+} species exists near the particle surface along with the expected sixfold coordination. The improved electrical conductivity of AZO was supported by the presence of fourfold aluminium environment. Unlike DEG, the use of glycerol as the reducing medium does not support formation of ZnO nanoparticles [76]. In the presence of glycerol (GLY), zinc glycerolate platelets of micron size were obtained at 160 °C. This happened due to the formation of highly stable coordination complex between de-protonated glycerol and Zn^{2+}. The reaction temperature was not sufficient to decompose zinc glycerolate complex into ZnO. However, the complex undergoes conversion into porous ZnO platelets upon heat treatment at 500 °C by loss of glycerol moiety. Ningthoujam et al. [77] prepared crystalline ZnO nanoparticles doped with Li^+ and Eu^{3+} by modified polyol method where EG was used along with urea and the reaction temperature was maintained at 150 °C. At this temperature, urea decomposes to NH_4OH to act as a co-reductant whereas agglomeration of as-synthesized nanoparticles is restricted by surface capping agent, EG. These luminescent nanoparticles when incorporated in functional polymer matrix show potential application in optical/electrical device fabrication. Same strategy was followed for the synthesis of TiO_2 particles with varying concentration of the dopant europium ions [78]. Photoluminescence studies on as-prepared Eu-doped TiO_2 nanoparticles with anatase phase showed better luminescence intensity compared to the samples heat treated beyond 500 °C.

Iron oxide-based magnetic nanoparticles, dispersible in water, were successfully prepared by Thanh et al. [79] based on polyol method at high pressure and temperature. The size as well as morphology of the nanoparticles was tuned by varying the relative concentrations of iron precursor $Fe(acac)_3$, polyol and reaction time. To increase the stability of the nanoparticles, post-synthesis tailoring of the particle surface was done by ligand exchange process with two different ligands, namely DHCA (3,4-dihydroxyhydrocinnamic acid) and tartaric acid. DHCA-treated iron oxide nanoparticles showed very high colloidal stability and better relativity values compared to commercially available magnetic resonance imaging (MRI) contrast agents.

Polyol method is extensively used for the preparation of nanocrystalline metal tungstates which are considered as inorganic functional materials of significant importance in the field of catalysis, photocatalysis, scintillators, fluorescent materials, humidity sensors, etc. A simple polyol method was applied for the synthesis of well-crystallized fluorescent $ZnWO_4$ nanoparticles [80]. These nanoparticles, without any doping, shows intense fluorescence upon excitation with near-UV LED due to the presence of complex anion $[WO_4]^{2-}$. Feldmann et al. [81] obtained faceted microcrystals of β-$SnWO_4$ by MW-assisted polyol synthesis in DEG with $SnCl_2{\cdot}2H_2O$ and $Na_2WO_4{\cdot}2H_2O$ as starting materials. β-$SnWO_4$, thus prepared, with a direct band gap of 2.7 eV shows good catalytic activity for the photodegradation of rhodamine B dye.

Extensive studies were carried out on calcium tungstate ($CaWO_4$)-based phosphor material which emits blue light due to the presence of WO_4^{2-} groups. Dispersible NPs of $CaWO_4$, both lanthanide doped and undoped were prepared using polyol synthesis method [82]. For the synthesis of undoped $CaWO_4$, calcium salt precursor, $Ca(NO_3)_2{\cdot}H_2O$ was dissolved in EG with continuous stirring and $Na_2WO_4{\cdot}2H_2O$ was injected to this solution. After sometime, a white precipitate appeared and the reaction was allowed to continue for two hours. The precipitate was centrifuged and thoroughly washed with methanol and acetone for removal of unreacted species. Lanthanide-doped $CaWO_4$ was synthesized based on the same procedure except that lanthanide precursor was added to calcium salt solution prior to sodium tungstate addition. The NPs, thus prepared, were characterized by XRD and transmission electron microscopy (TEM).

The XRD pattern of $CaWO_4$ is shown in Fig. 11.6a which confirms the formation of $CaWO_4$ nanoparticles with scheelite structure. The crystallite size of $CaWO_4$ nanoparticles was evaluated based on Scherrer formula and found to be around 7 nm. Uniform size distribution was observed for the $CaWO_4$ nanoparticles prepared by polyol method with particle size in the range of 5–10 nm as confirmed by TEM image (Fig. 11.6b).

Upon excitation with 253 nm light, blue emission was observed from undoped $CaWO_4$ nanoparticles whereas Dy^{3+} doped $CaWO_4$ NPs showed strong green emission. (Fig. 11.7a, b). The emission band from undoped sample ranges from 325 to 600 nm with a peak maximum around 425 nm. The origin of this emission is attributed to the charge transfer excitation followed by emission from WO_4^{2-} moiety, involving lower vibrational level of ^1B ($^1T_2$) and $^1A_1$ states of W^{6+} species.

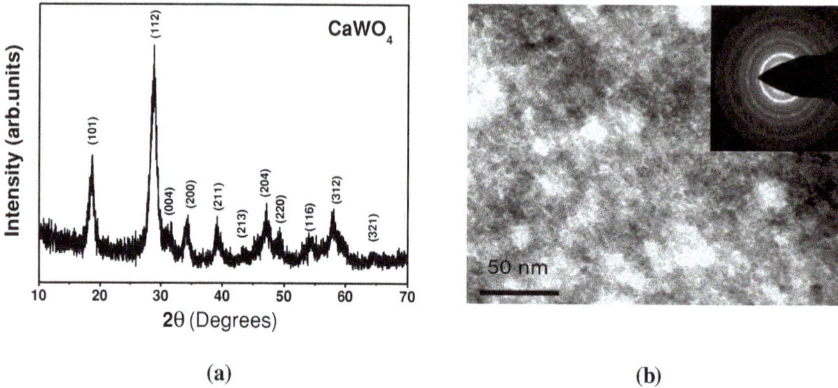

Fig. 11.6 XRD pattern (**a**) and TEM image (**b**) of $CaWO_4$ nanoparticles

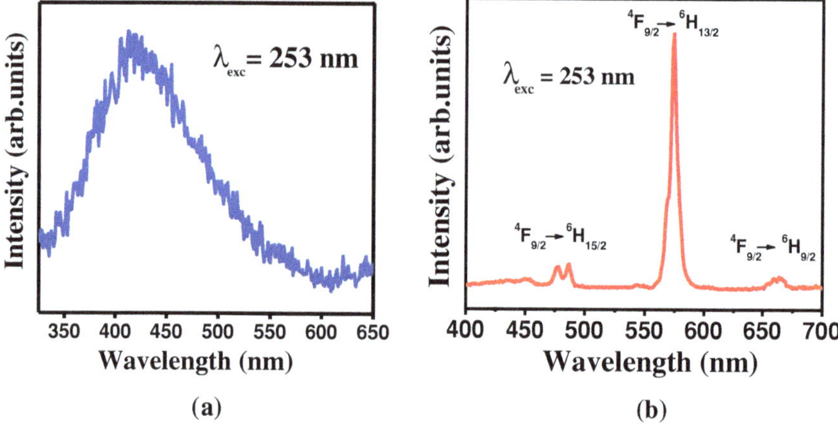

Fig. 11.7 Emission spectra of (**a**) undoped and (**b**) Dy^{3+}-doped $CaWO_4$ nanoparticles after excitation at 253 nm

Emission spectrum of Dy^{3+} doped $CaWO_4$ consists of mainly three peaks at 478, 574, 665 nm which correspond to $^4F_{9/2} \rightarrow {}^6H_{15/2}$, $^4F_{9/2} \rightarrow {}^6H_{13/2}$ and $^4F_{9/2} \rightarrow {}^6H_{9/2}$ transitions, respectively, of Dy^{3+} ions present in the $CaWO_4$ lattice. The decay curve corresponding to ^1B ($^1T_2$) excited state of WO_4^2 tetrahedra in undoped $CaWO_4$ is shown in Fig. 11.8a, b which represents the decay profile of $^4F_{9/2}$ state of Dy^{3+} ions from $CaWO_4$:Dy NPs. For both the samples, decay profiles are bi-exponential in nature with two lifetime components. The lifetime values, evaluated from the corresponding decay curves, are listed in Table 11.3.

Undoped and lanthanide ion-doped $ZnGa_2O_4$ are potential candidates for different types of display devices [83–85]. Undoped $ZnGa_2O_4$ emits blue light upon UV or low-voltage electron excitation due to transition through a self-activation

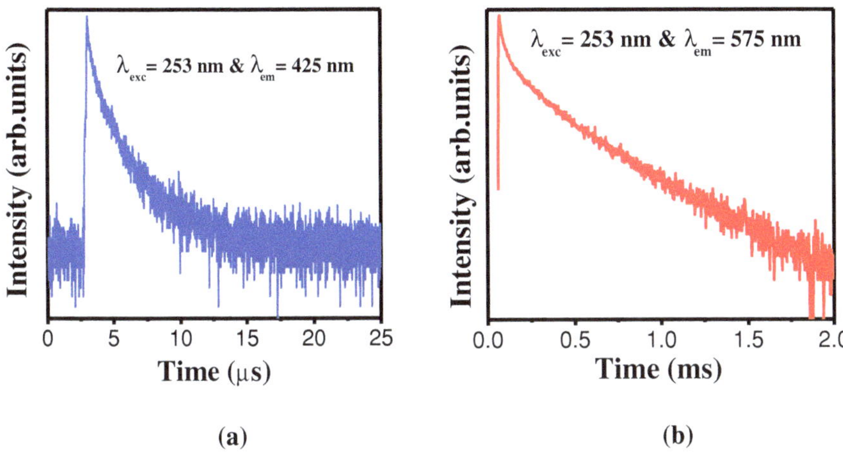

Fig. 11.8 Decay curves corresponding to **(a)** undoped and **(b)** Dy^{3+}-doped $CaWO_4$ nanoparticles

Table 11.3 Lifetime values of ^1B ($^1T_2$) excited state in WO_4^{2-} poly-anion of undoped $CaWO_4$ nanoparticles and $^4F_{9/2}$ excited states of Dy^{3+} in $CaWO_4$:Dy nanoparticle

Sample	^1B ($^1T_2$) lifetime		$^4F_{9/2}$ lifetime	
	τ_1	τ_2	τ_1	τ_2
Undoped $CaWO_4$	0.5	2.6	–	–
Dy^{3+} doped $CaWO_4$	–	–	0.1 ms	0.36 ms

centre involving GaO_6 structural units [86]. Its luminescent properties can be tuned by substituting Zn^{2+} with Cd^{2+}/In^{3+}. Preparation of doped/undoped $ZnGa_2O_4$ nanoparticles, highly dispersible in organic solvents, is required for incorporating them in polymer matrix which in turn is indispensable part for the fabrication of flexible display devices. In^{3+} and lanthanide-doped $ZnGa_2O_4$ nanoparticles were prepared at a lower temperature of 120 °C by a modified polyol method using EG solvent [82]. Initially, gallium (Ga) metal was dissolved using concentrated HCl containing few drops of concentrated HNO_3 and the excess acid was evaporated repeatedly from the solution after adding water. This solution was added to zinc acetate precursor in a double-necked RB flask followed by addition of EG-water mixture. The solution was allowed to reach 100 °C when urea was added. The temperature was subsequently increased to 120 °C to get a white-coloured precipitate. Relative amount of water and EG in the mixed solvent greatly influenced average crystallite size of $ZnGa_2O_4$ nanostructures and can be seen from the XRD patterns shown in Fig. 11.9.

Average crystallite size of the NPs increases with increase in water content in the mixed solvent. The crystallite sizes of the nanoparticles prepared with varying concentration of EG, and water are listed in Table 11.4. Gradual increase in crystallite size is observed with increase in water volume which is mainly due to

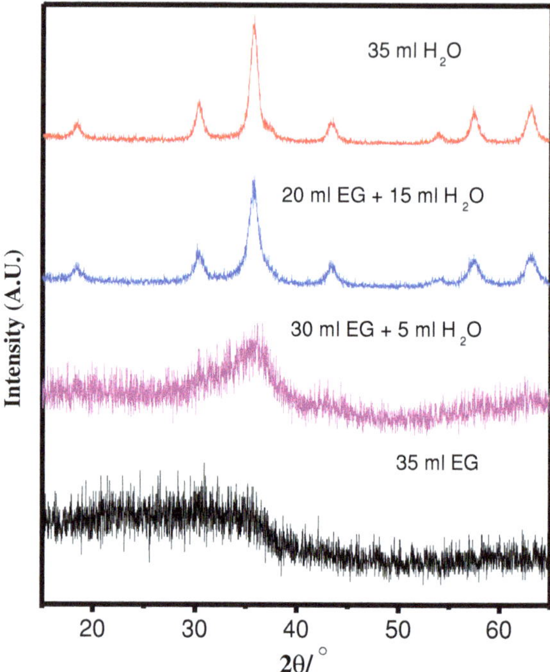

Fig. 11.9 XRD patterns of $ZnGa_2O_4$ nanoparticles synthesized with varying EG-H_2O ratios

Table 11.4 Crystallite size of $ZnGa_2O_4$ nanostructures prepared in EG-H_2O mixed solvents with different ratios keeping relative concentrations of Zn and Ga intact (Zn: 1.432 mmol, Ga: 2.864 mmol)

S. No.	Vol. of H_2O (ml)	Vol. of EG (ml)	Crystallite size (nm)
1	35	0	9.5
2	15	20	6.9
3	5	30	1.2
4	0	35	–

two factors: (1) effective hydrolysis of urea and (2) less control over nanoparticle growth arising from lack of stabilizing agent EG.

Precursor concentration also affects the particle size of $ZnGa_2O_4$ synthesized by this method. At a particular ratio of water and EG in the mixed solvent (10 ml water and 25 ml EG), concentrations of Zn and Ga precursors were varied systematically and XRD patterns of as-prepared NPs are shown in Fig. 11.10. At very low concentration of Ga precursor (0.142 mmol), amorphous phase formation was observed. With increase in precursor concentration, an increase in average crystallite size was observed up to a certain concentration of 1.432 mmol of Ga precursor followed by a gradual decrease in crystallite size and finally loss of crystallinity with further increase in reactant concentrations (22.954 mmol Ga

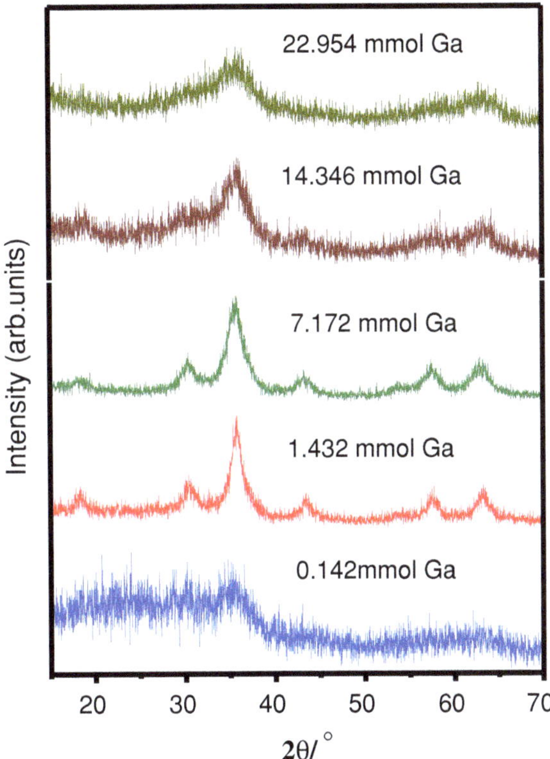

Fig. 11.10 XRD patterns of $ZnGa_2O_4$ particles prepared with different precursors (of Zn and Ga) concentrations

precursor). The calculated crystallite sizes using Scherrer formula are listed in Table 11.5. The observed trend in particle sizes with increasing the concentrations of reactants can be explained based on nucleation and growth of $ZnGa_2O_4$ nanoparticles. At lower concentrations of Zn and Ga precursors, both nucleation and growth processes are slow due to lack of proximity among the reactant ions leading to lower particle size. With increase in precursor concentrations, growth becomes faster compared to nucleation resulting in bigger crystallite size. Above a certain concentration, nucleation is much faster than the growth process and particles with smaller crystallite sizes are observed.

Table 11.5 Crystallite size of $ZnGa_2O_4$ nanoparticles with different precursor concentrations keeping water–EG ratio fixed (10 ml water and 25 ml EG)

S. No.	Concentration of Zn precursor (mmol)	Concentration of Ga precursor (mmol)	Crystalite size (nm)
1	0.071	0.142	–
2	0.716	1.432	5.1
3	3.586	7.172	3.6
4	7.173	14.346	2.8
5	11.477	22.954	–

In^{3+}-doped $ZnGa_2O_4$ nanoparticles with size ~5 nm were prepared based on the same polyol method. The nanoparticles showed strong blue emission with a quantum yield of 10% upon excitation with 270 nm light. These nanoparticles can be easily incorporated in polymethylmethacrylate (PMMA) matrix by in situ polymerization of MMA below 100 °C, thus facilitating their use in the fabrication of polymer-based devices.

Sahu et al. [87] have prepared Tb^{3+}-doped $GdPO_4$ nanorods sensitized with Ce^{3+} by polyol technique. Gd_2O_3, $Tb(NO_3)_3{\cdot}6H_2O$ and $Ce(NO_3)_3{\cdot}6H_2O$ were dissolved in hydrochloric acid, and excess acid was removed by repeated evaporation by adding water. Finally, a transparent solution was obtained. $(NH_4)H_2PO_4$ and EG were added to that solution. Polymer-based films incorporated with nanocrystals of $GdPO_4$:Tb,Ce showed intense green emission under UV irradiation. Luminescence switching (ON–OFF) performance of $GdPO_4$:Tb,Ce nanorods was checked by subsequent addition of potassium permanganate (oxidizing agent) and ascorbic acid (reducing agent) through a redox (Ce^{3+}/Ce^{4+}) reaction. These nanoparticles showed paramagnetic behaviour and are potential candidates for flexible display device and MR imaging.

Recent studies have shown potential of lanthanide orthovanadates ($LnVO_4$) as materials for laser hosts, polarizers, catalysis, biological detection, etc. In this respect, synthesis of size- and shape-controlled $LnVO_4$ and study of their luminescence properties are of prime importance for material scientists. Okram et al. [88] have synthesized nanocrystals of $LaVO_4:Eu^{3+}$ co-doped with metal ions ($Mn^+ = Li^+/Sr^{2+}/Bi^{3+}$) by a simple polyol method in EG medium at 140 °C. The schematic diagram for the preparation of nanostructured $LaVO_4:Eu^{3+}$ starting from lanthanum oxide (La_2O_3) and europium oxide (Eu_2O_3) are depicted in Fig. 11.11. $Li^+/Sr^{2+}/Bi^{3+}$ co-doped $LaVO_4$:Eu samples were prepared following the similar strategy. All the synthesized samples were found to contain both tetragonal and monoclinic phases of $LaVO_4$. Relative ratio of tetragonal to monoclinic phases increased upon increasing the dopant concentrations (Li^+ and Sr^{2+}) as shown in Fig. 11.12. However, the reverse trend was observed for Bi^{3+} co-doped sample. Upon Li^+/Sr^{2+}doping, the lattice of $LaVO_4:Eu^{3+}$ expanded whereas lattice contraction occurred when co-doped with Bi^{3+}. Irrespective of the crystal structure, the authors found an increase in luminescence intensity by co-doping of $Li^+/Sr^{2+}/Bi^{3+}$ in $LaVO_4:Eu^{3+}$ matrix.

Polyol synthesis method was adopted for developing core–shell nanostructures based on metal oxide systems. Core–shell nanostructures of Ag NP @ ZnO were prepared by seed-mediated polyol method in two steps [89]. Initially, Ag nanoparticles were prepared by polyol reduction of $AgNO_3$ salt precursor in EG solvent containing PVP at 160 °C. The as-prepared Ag nanoparticles were used as seed for growing ZnO nanorods over the Ag surface. These core–shell nanostructures with hedgehog-like morphology showed better photocatalytic performance compared to ZnO mainly because of the roles played by Ag cores in generating ZnO with higher specific surface area and hindering the recombination of photogenerated electrons and holes.

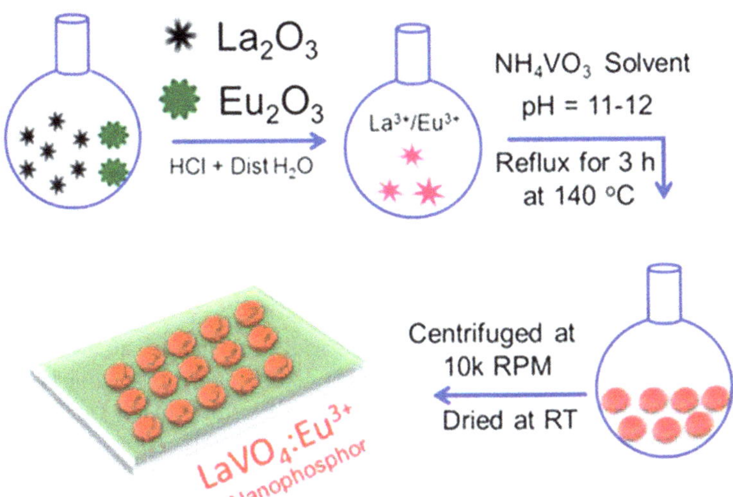

Fig. 11.11 Schematic diagram for the preparation of $LaVO_4:Eu^{3+}$ nanophosphors in ethylene glycol medium. Adapted with permission from Ref. [88]. Copyright (2014) American Chemical Society

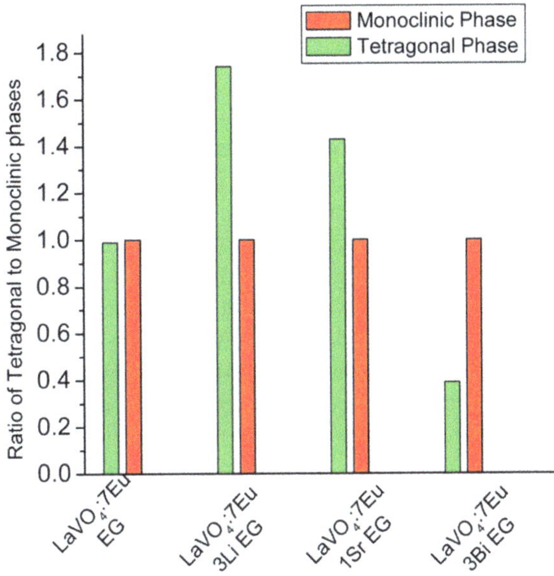

Fig. 11.12 Variation of monoclinic to tetragonal phases in $LaVO_4:Eu^{3+}$ system and in the $LaVO_4:Eu^{3+}$ systems co-doped with Li^+/Sr^{2+}/Bi^{3+}. Adapted with permission from Ref. [88]. Copyright (2014) American Chemical Society

11.5 Synthesis of Nanostructured Metal Chalcogenides

The polyol synthesis method is reported to be extensively used for preparing metal sulphides with nanoscale dimensions. Shen et al. [90] designed a general polyol technique for preparing various types of metal sulphide nanoparticles (Fe_3S_4, Cu_nS, NiS, In_2S_3, Ag_2S, PbS). The metal sulphides were prepared by treating the corresponding metal salt precursors with thiourea in polyol at appropriate reaction conditions. Morphology and crystallinity of the NPs were studied by TEM and powder XRD. Hexagonal NiS nanocrystals with particle size ~35 nm, pseudocubic phase $Cu_{1.8}S$ nanoparticles with size ~65 nm were obtained by this method, whereas the size of cubic phase PbS nanostructures was in the range of 15–120 nm. Apart from thiourea, other sulphide sources such as sulphur powder and Na_2S have also been tried successfully by different research groups for the preparation of metal sulphides via polyol method [91, 92].

Datta et al. [93] reported synthesis of nearly monodisperse In_2S_3 nanoplatelets in high yield by a simple polyol process. $InCl_3$ was dissolved in anhydrous EG along with 1-thioglycerol, and the reaction temperature was set at 70 °C, which was maintained for half an hour with continuous stirring. After that, thiourea was added to the solution and reaction temperature was raised to 140 °C. The reaction was continued at this temperature for 1 h. Then, the reaction mixture was cooled down to room temperature and precipitate was formed by addition of hexane/ethanol. The nanoplatelets were formed in uniform size and shape which has been shown in Fig. 11.13a, and the thickness of the nanoplates was ~7 nm (Fig. 11.13b). Quantum size confinement was proposed for these thin nanoplatelets as large shift in the band gap was observed compared to bulk In_2S_3 in the optical absorption spectrum. The In_2S_3 nanoplatelets prepared by this method were much thinner (~5 times) compared to the bulk In_2S_3 with a Bohr diameter of ~33.8 nm. The In_2S_3 nanoplatelets were almost free from surface defects, and they showed steady band-edge UV luminescence with a peak around 383 nm. Emission maximum remained almost unchanged for these In_2S_3 nanoplatelets upon excitation with different wavelengths indicating their narrow size distribution.

Antimony sulphide (Sb_2S_3) with flower-like structure was prepared via a polyol reflux method by Zhu et al. [94]. Antimony trichloride and L-cysteine were added to EG in the presence of ethylenediamine and refluxed at 175 °C for 50 min. The reaction mixture was allowed to cool down to room temperature and was centrifuged to collect the product. To understand the growth mechanism of micro-flower, a series of experiments were performed by varying the reaction time and keeping all other experimental conditions similar. Some experiments were also tried without adding ethylenediamine in EG. XRD patterns confirmed formation of Sb_2S_3 with orthorhombic stibnite structure. However, Sb_2S_3 crystals with several micrometer sizes were obtained when the reaction was performed without ethylenediamine. When ethylenediamine is present in reaction mixture, a chelate compound of Sb is formed which controls the release of antimony ions, and thus, a fine control over the crystal growth of Sb_2S_3 was obtained resulting in nanosheets

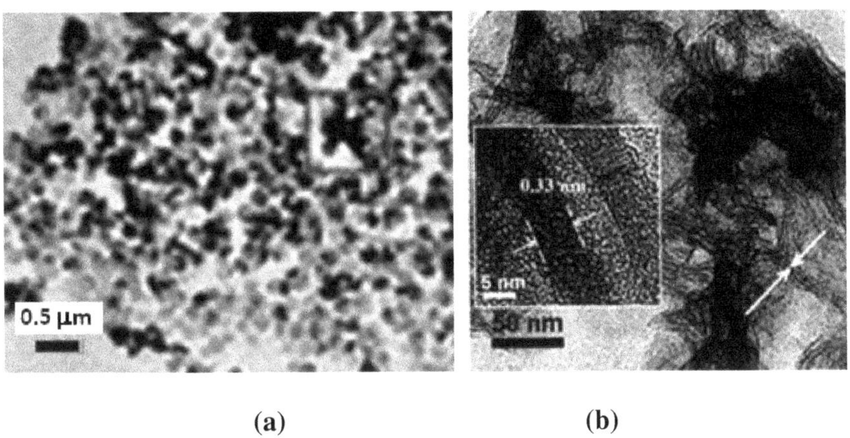

Fig. 11.13 (**a**) TEM image of In_2S_3 nanoplatelets at lower magnification and (**b**) HRTEM image recorded from the region highlighted as red square in (**a**). In the inset of (**b**), two parallel In_2S_3 nanoplatelets are shown with thickness ~7 nm. Lattice fringe spacing was determined as 0.33 nm indicating the presence of (109) plane of tetragonal In_2S_3. Adapted with permission from Ref. [93]. Copyright (2013) Royal Society of Chemistry

of Sb_2S_3. When the reaction was continued for 10 min, nanosheets were accumulated into bulk structures. Upon increasing the reaction time from 10 to 20 min, nanosheets started self-assembling and fluffy spheres of Sb_2S_3 were formed. On further increasing the reaction time beyond 40 min, micro-flower structure formation started by Ostwald ripening process. Scanning electron micrograph of the Sb_2S_3 flowers revealed that they consisted of dozens of nanosheets with ~50 nm thickness. Stacking of such nanosheets leads to the formation of spherical microflowers with 7–10 μm diameter with enough interspaces (Fig. 11.14).

The Brunauer–Emmett–Teller (BET) surface area of the microflowers was measured and found to be 22.787 $m^2\ g^{-1}$. The flower-like Sb_2S_3 showed improved electrochemical behaviour when used as an anode material for Na-ion batteries due to its unique structure which helped in efficient diffusion for Na^+ and e^-.

Transition metal sulphides have emerged as excellent materials for high-performance supercapacitor applications. Ternary Ni–Co sulphides, with varying Ni and Co contents, were prepared employing a one-pot polyol synthesis method in EG medium [95]. Nickel (II) acetate and cobalt (II) acetate were used as precursors of respective metals. The precursors dissolved in EG were transferred to a stainless steel autoclave (Teflon-lined) and kept at 200 °C for 6 h to ensure formation of the ternary sulphide. XRD peaks of $Ni_{1.5}Co_{1.5}S_4$ matches well with those of Co_3S_4 phase indicating that the Co_3S_4 phase is dominant in the mixed sulphide with almost no or negligible fraction of NiS phase (Fig. 11.15a). Because of similar radii of Ni and Co, cobalt ions are partially replaced by nickel ions ultimately maintaining the Co_3S_4 phase in final $Ni_{1.5}Co_{1.5}S_4$ compound. Elemental mapping of the Ni, Co and S in $Ni_{1.5}Co_{1.5}S_4$ sample is shown in Fig. 11.16 which

Fig. 11.14 FE-SEM images (**a**)–(**d**) showing flower-like structure of Sb_2S_3 and EDS profile of representative Sb_2S_3 (**e**). Adapted with permission from Ref. [94]. Copyright (2015) Royal Society of Chemistry

confirmed uniform distribution of the elements in the sample. TEM image demonstrated that the particles are loosely assembled in the sample resulting in porous structure (Fig. 11.15b). It is clear from the TEM image at higher magnification that the sample is composed of interconnected nanoparticles. The supercapacitance performance (pseudo) of bimetallic Ni–Co sulphide was proven to be better compared to that of monometallic Ni and Co sulphides in terms of high specific capacitance, superior rate capability and cyclic life. The relative ratio of Ni and Co ions in the sample is crucial for improved pseudo-supercapacitive performance. $Ni_{1.5}Co_{1.5}S_4$ sample exhibited specific capacitance as high as 1093 F g^{-1} at 1 A g^{-1}. Representative CV curves of $Ni_{1.5}Co_{1.5}S_4$ sample after and before 2000

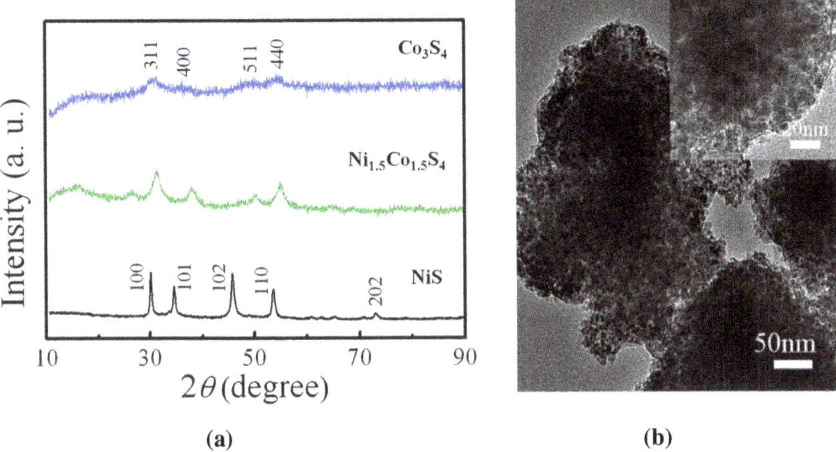

Fig. 11.15 (**a**) XRD patterns of the binary NiS, Co_3S_4 and ternary $Ni_{1.5}Co_{1.5}S_4$; (**b**) TEM image of $Ni_{1.5}Co_{1.5}S_4$ sample. Inset in (**b**) shows TEM image of the sample at higher magnification. Adapted with permission from Ref. [95]. Copyright (2015) Royal Society of Chemistry

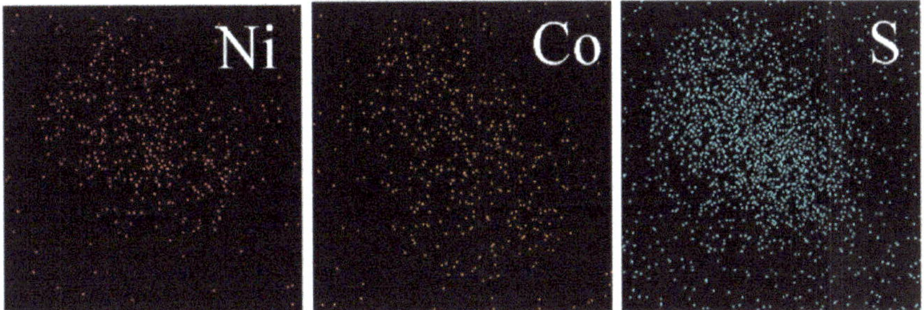

Fig. 11.16 Elemental mapping of $Ni_{1.5}Co_{1.5}S_4$ sample. Adapted with permission from Ref. [95]. Copyright (2015) Royal Society of Chemistry

charge–discharge cycles are shown in Fig. 11.17. There is negligible change in redox peak positions, and both the curves are similar in shape indicating high cyclic stability of $Ni_{1.5}Co_{1.5}S_4$ sample.

Ternary metal sulphides have been demonstrated as potential candidate for photocatalysis. Shang et al. [96] synthesized flower-like $ZnIn_2S_4$ microspheres by polyol-mediated hot-injection technique. $ZnCl_2$ and $InCl_3$ were dissolved in DEG in the presence of capping agent polyacrylic acid (PAA) and heated to 200 °C/220 °C in inert atmosphere. A solution containing thiourea in DEG was injected to the hot solution containing Zn and In precursors, and the temperature was maintained at that value for 2–4 min. The reaction mixture was cooled down to ambient temperature, and precipitate was formed upon addition of ethanol. Electron micrographs of as-synthesized $ZnIn_2S_4$ samples confirmed formation of flower-like

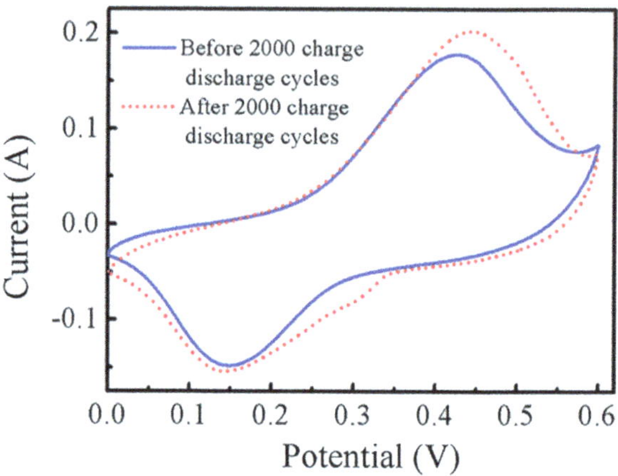

Fig. 11.17 CV curves of $Ni_{1.5}Co_{1.5}S_4$ recorded before and after 2000 charge–discharge cycles. Adapted with permission from Ref. [95]. Copyright (2015) Royal Society of Chemistry

superstructures at both the reaction temperatures as can be seen in Fig. 11.18. When the reaction temperature was fixed at 220 °C, the average diameter of the spherical $ZnIn_2S_4$ flowers was found to be ~500 nm. However, with decrease in synthesis temperature (200 °C), average diameter of the superstructures increased as clear from the SEM image. At lower temperature, the number of nuclei was less leading to relatively large-sized microspheres. The constituent nanosheets of the superstructures are clearly visible from TEM images. The estimated thickness of the nanosheets was ~2–3 nm. Due to ultrathin nature of the nanosheets and its reasonably high specific surface area, the material showed excellent photocatalytic activity for visible light-driven H_2 production.

Apart from sulphides, metal selenides are also interesting class of nanomaterial because of their unique optical and electrical properties. Semiconductors based on metal selenide nanoparticles are widely studied for different applications such as thin film transistors and solar cells. Unlike large number of sulfur precursor available for stable metal sulphide nanoparticle synthesis, precursors for selenium are limited [97]. Liu et al. [98–101] have worked extensively towards the development of synthesis strategies for mono/bi-metallic selenide nanoparticles at mild experimental conditions with lower preparation cost. A mixed solvent of thioglycolic acid (TGA) and monoethanolamine (MEA) was used for dissolution of Se at room temperature in air. The as-prepared Se solution was used as Se precursor for the preparation of different metal selenide nanostructures such as CuSe nanoplates, Sb_2Se_3 nanorods and $CuInSe_2$ nanocrystals by polyol method using triethylene glycol (TEG) in the presence of PVP at ~180 °C. Selenium solution prepared in (TGA + MEA) mixed solvent is highly stable even in presence of air. The dissolution of Se powder happens by reduction of Se via the thiol group of TGA followed by complexing with the amine group of MEA. Single crystalline nature of

Fig. 11.18 SEM images of $ZnIn_2S_4$ samples prepared at (**a**) 220 °C and (**b**) 200 °C. TEM images of $ZnIn_2S_4$ sample prepared at 220 °C with (**c**) low and (**d**) high magnifications. Adapted with permission from Ref. [96]. Copyright (2013) Royal Society of Chemistry

all the selenide samples was confirmed from HRTEM images as shown in Fig. 11.19.

All the samples, i.e. monometallic selenides CuSe and Sb_2Se_3 and bimetallic selenides $CuInSe_2$ prepared by this modified polyol route, were well dispersible, and their morphologies were well controlled. The average diameter of Sb_2Se_3 nanorods was ~80 nm with length ranging from 400 nm to 1.4 μm whereas the average particle size of CuSe nanoplates was varying from 100 to 500 nm. $CuInSe_2$ nanoparticles formed by this process were much smaller in size (~8 nm) as seen from TEM images. Sizes are even smaller than the Bohr radius of $CuInSe_2$ nanocrystals (10.6 nm). Hence, quantum size confinement was expected in this type of $CuInSe_2$ nanocrystal leading to an increase in its band gap. The $CuInSe_2$ nanoparticles were utilized for preparing ethanol-based colloidal ink and deposited as thin films by a dip-coating process followed by selenization at 550 °C. Selenized $CuInSe_2$ film was used to construct a solar cell with typical cell structure of Mo/$CuInSe_2$/CdS/i-ZnO/Al-ZnO, and its power conversion efficiency was determined as 2.65% under AM 1.5 illumination (Fig. 11.20).

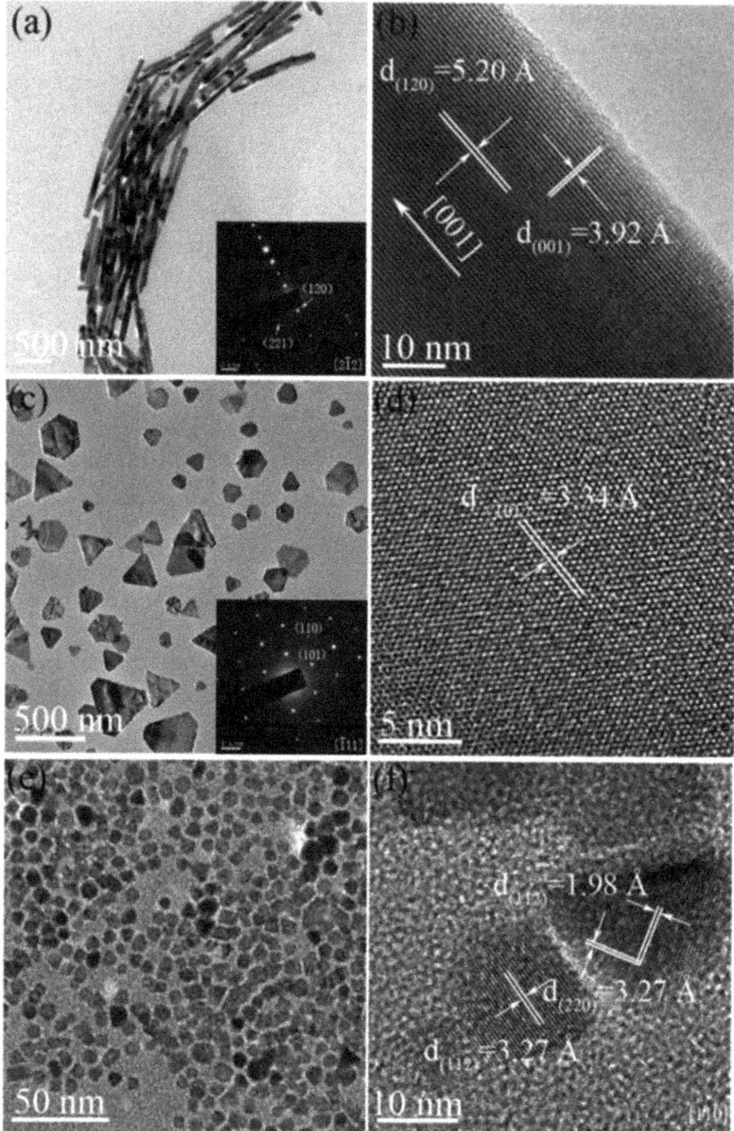

Fig. 11.19 TEM and HRTEM images of Sb_2Se_3 (**a**), (**b**), CuSe (**c**), (**d**), and $CuInSe_2$ (**e**), (**f**) nanoparticles. Insets of (**a**) and (**c**) show the SAED patterns of Sb_2Se_3 nanorod and CuSe nanoplate. Adapted with permission from Ref. [101]. Copyright (2016) Royal Society of Chemistry

The $CuInSe_2$ nanoparticles can also be obtained via microwave-assisted polyol reduction method [102]. However, the particle size obtained by this method was much bigger (~ 10 times) compared to the nanoparticles formed by previous method.

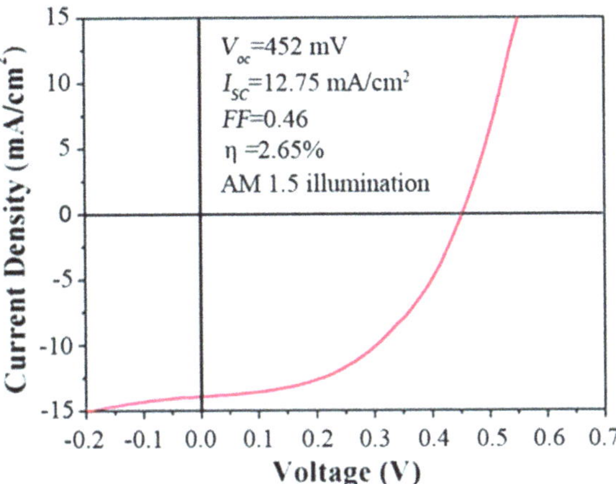

Fig. 11.20 I-V curve of $CuInSe_2$-based solar cell. In the figure, V_{OC}, I_{SC}, FF, η stand for open voltage, short current density, filling factor and efficiency, respectively. Adapted with permission from Ref. [101]. Copyright (2016) Royal Society of Chemistry

Like sulphide and selenides, tellurides are also important class of chalcogenide for alternate energy sources, other than fossil fuel. Energy conversion employing thermoelectric materials are considered as one of the promising alternatives for conversion of waste heat into electric energy. Bismuth telluride is a potential low temperature thermoelectric material [103]. Its ternary analogue, bismuth antimony telluride, $Bi_{0.5}Sb_{1.5}Te_3$ is reported to exhibit better thermoelectric properties, when synthesized in nanoform, compared to Bi_2Te_3 [104]. It is worth mentioning that the thermoelectric performance of a material depends on its figure of merit ZT which is expressed as $ZT = S^2\sigma T/\lambda$, where S, σ and T are Seebeck coefficient, electrical conductivity and temperature, respectively, and λ represents total thermal conductivity. One of the important approaches to improve thermoelectric performance is by lowering the dimension to nanoregime which can generate lots of grain boundaries resulting in reduced thermal conductivities. In addition to that nanostructuring is a good tool for fine-tuning the important parameters such as S, σ and λ by controlling size and morphology of the nanomaterials. In most of the cases, bulk production of thermoelectric nanomaterials is done either by traditional solid-state method or by ball milling, sputtering, etc. Liquid-phase chemical method seems to be an alternative method for preparation of thermoelectric nanomaterials with more complex structure, but suffers from different issues related to scaling up. Transport property measurement of the nanostructured thermoelectric material produced by chemical route becomes difficult as it needs bigger sample size. Anderson et al. [104] reported a modified and scalable polyol process for synthesis of nanostructured thermoelectric $Bi_{0.5}Sb_{1.5}Te_3$ in bulk scale. Nanostructured $Bi_{0.5}Sb_{1.5}Te_3$ was prepared by reducing stoichiometric amounts of bismuth (III) nitrate pentahydrate, antimony (III) chloride, sodium tellurate (VI) dihydrate in TEG using $NaBH_4$ as co-reductant. After addition of $NaBH_4$ the reaction temperature was raised to 250 °C and maintained at least for 1 h. In order to improve the crystallinity, as-synthesized $Bi_{0.5}Sb_{1.5}Te_3$ was annealed at 500 °C for 6 h. XRD patterns of as-synthesized and annealed $Bi_{0.5}Sb_{1.5}Te_3$ samples are shown in Fig. 11.21.

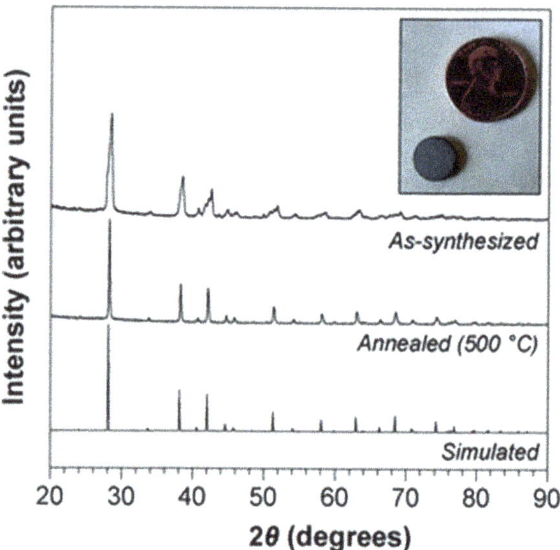

Fig. 11.21 XRD patterns of $Bi_{0.5}Sb_{1.5}Te_3$ samples. Inset shows photograph of sintered pellet of nanostructured $Bi_{0.5}Sb_{1.5}Te_3$ (1 cm diameter × 1 mm thick). Adapted with permission from Ref. [104]. Copyright (2010) Royal Society of Chemistry

The as-synthesized $Bi_{0.5}Sb_{1.5}Te_3$ nano-powders are composed of aggregates of loosely bound platelets, whereas the annealed sample is dense and sintered as can be seen from the SEM image shown in Fig. 11.22. The overall platelet morphology is maintained in the annealed sample.

The Seebeck coefficient (S) of sintered pellets of $Bi_{0.5}Sb_{1.5}Te_3$ nano-powders was measured in the temperature range of 300–500 K, and its temperature dependence is presented in Fig. 11.23. The value of S increases with increase in temperature and reaches S = + 200 μV K^{-1} at 500 K. The Seebeck coefficient of

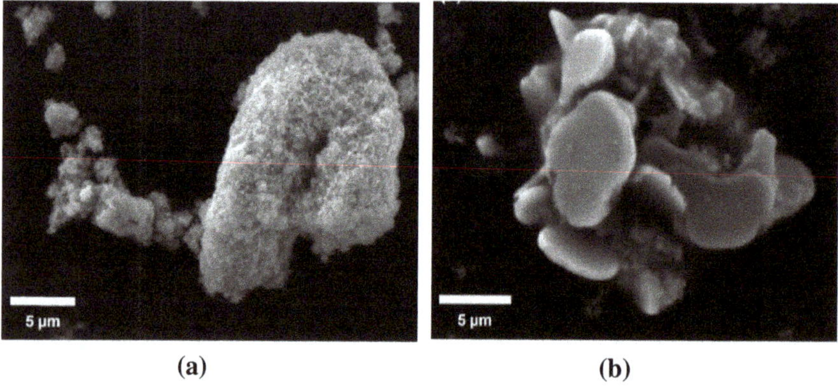

Fig. 11.22 SEM image of $Bi_{0.5}Sb_{1.5}Te_3$ nano-powders (**a**) as-synthesized and (**b**) annealed at 500 °C. Adapted with permission from Ref. [104]. Copyright (2010) Royal Society of Chemistry

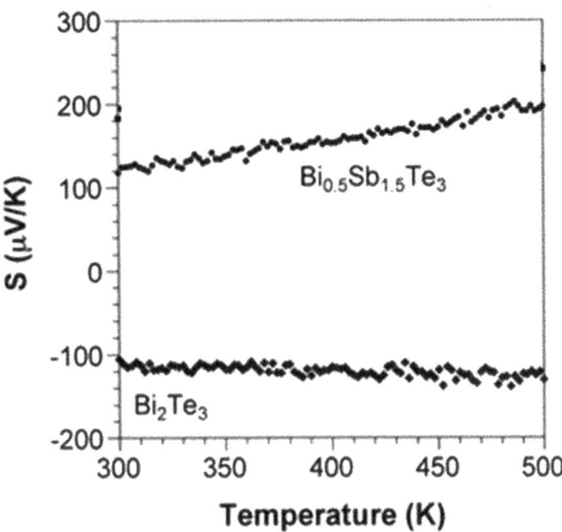

Fig. 11.23 Seebeck coefficient for $Bi_{0.5}Sb_{1.5}Te_3$ and Bi_2Te_3 in temperature range 300–500 K. Adapted with permission from Ref. [104]. Copyright (2010) Royal Society of Chemistry

binary Bi_2Te_3, prepared by same polyol method remains almost constant over the entire temperature range of 300–500 K and much lower than the S value of $Bi_{0.5}Sb_{1.5}Te_3$.

11.6 Synthesis of Metal Fluoride Nanoparticles

Nanostructured rare earth (RE) fluorides are very interesting class of materials for novel applications such as lighting, telecommunication, bioimaging, solar cells and scintillators. Two types of rare earth fluorides, namely binary REF_3 and ternary $AREF_4$ (A = alkali metal), have been established as excellent luminescent host materials for a variety of optically active lanthanides both for down-conversion and for up-conversion processes. Due to relatively higher value of refractive index coupled with low phonon energy for fluoride systems, the probability of non-radiative decay is very low leading to exceptionally high luminescence quantum yields compared to conventional oxide-based hosts. In this section of the chapter, we mainly focus on the polyol synthesis of rare earth fluoride nanostructures with their probable applications. For example, CeF_3, a RE-based luminescent nanomaterial has potential application as laser host material. Wang et al. [105] prepared this material along with $CeF_3:Tb^{3+}$ and $CeF:Tb^{3+}/LaF_3$ (core/shell) nanostructures using simple and facile polyol method in DEG medium. During these synthesis processes, NH_4F was used as the source of fluoride. Nanocrystals of CeF_3 and $CeF_3:Tb^{3+}$ were obtained in a single step whereas the core/shell nanoparticles were formed via two steps. $CeF_3:Tb^{3+}$ core was prepared separately by dissolving $Ce(NO_3)_3.6H_2O$ and $Tb(NO_3)_3$ in DEG at 100 °C followed by

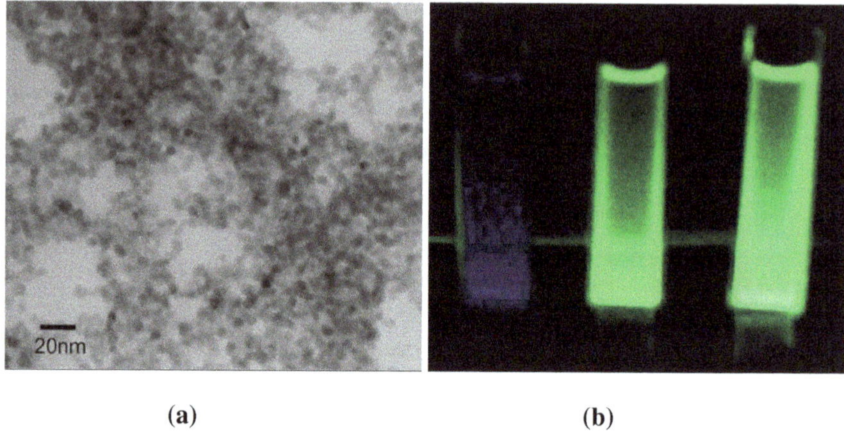

Fig. 11.24 TEM image of CeF_3 nanospheres (**a**) and photographs of colloidal dispersions of CeF_3, $CeF_3:Tb^{3+}$ and $CeF_3:Tb^{3+}/LaF_3$ core–shell NPs upon UV irradiation at 254 nm (**b**). Adapted with permission from Ref. [105]. Copyright (2006) American Chemical Society

addition of NH_4F at an elevated temperature of 200 °C in argon atmosphere. To ensure the formation of $CeF_3:Tb^{3+}$, reaction temperature was maintained at 200 °C for 1 h with vigorous stirring of the reaction mixture. To prepare the LaF_3 shell, a solution of $La(NO_3)_3$ dissolved in DEG at 100 °C was mixed with already prepared $CeF_3:Tb^{3+}$ core solution. The temperature of the reaction mixture was raised to 200 °C under Ar flow and NH_4F was added. The solution was heated at 200 °C for 1 h to precipitate core/shell nanoparticles. XRD patterns revealed hexagonal structure of CeF_3, $CeF_3:Tb^{3+}$ and $CeF_3:Tb^{3+}/LaF_3$ NPs with average crystallite sizes of 8, 8.5 and 11.5 nm, respectively. The nanoparticles of CeF_3 and $CeF_3:Tb^{3+}$ were spherical with average diameter of ~7 nm as illustrated from TEM images (Fig. 11.24a). The mean size of $CeF_3:Tb^{3+}/LaF_3$ core–shell NPs was increased to 11 nm with broadening in particle size distribution. The nanocrystals were dispersed in ethanol, and transparent colloidal solutions were obtained for all the samples. The solutions were irradiated under UV lamp (254 nm), and purple-blue emission was obtained from the solution containing CeF_3 nanoparticles, characteristic of Ce^{3+} 5d-4f transition (320 nm). The other two solutions gave bright green emission which was due to $^5D_4$-$^7F_5$ transition of Tb^{3+}at 542 nm.

Wei et al. [106] have reported synthesis of upconversion luminescence nanoparticles (UCNPs), namely $NaYF_4$:Yb,Er/Tm with size 10–30 nm using a simple polyol method. Oxides of yttrium, ytterbium, erbium and thulium were used as sources of RE^{3+} which were converted into corresponding chlorides by dissolving in hydrochloric acid. During the synthesis, two types of fluorine source were employed: NaF and NH_4F. When NH_4F was used, addition of NaCl was mandatory to provide sodium ions. In case of NaF as fluoride source, sodium ions came from NaF, restricting further addition of NaCl. Three types of polyols (glycol, DEG and glycerol) with different boiling points were used as reaction medium as well as surface capping agent to control particle growth and inhibit particle agglomeration. The product obtained by polyol method, irrespective of the reaction medium, was α- $NaYF_4$:Yb,Er/Tm. It is well known that β form of $NaYF_4$, also known as hexagonal $NaYF_4$, is a better host compared to cubic α- $NaYF_4$. In order to convert α-$NaYF_4$ phase to β form, the products of polyol method were treated solvothermally. Nanoparticles formed in DEG were smaller in size compared to those prepared in glycol or glycerol, which was explained by the researchers in terms of steric repulsion in polyol solution. DEG being bulkier polyol compared to short-chain polyols (glycol/glycerol), it is expected to provide considerable steric hindrance. As a result, the amount of reactants is reduced in the vicinity of initial seeds leading to smaller particle size of the nanocrystal. Li and co-workers [107] have developed a versatile one-pot polyol synthesis method for the preparation of functional-surface tuneable UCNPs by using difunctional ligands containing a carboxyl group, such as 11-aminoundecanoic acid (ADA), folic acid (FA) and poly (ethylene glycol)bis (carboxymethyl)ether (PEGBA), as schematically depicted in Fig. 11.25. This novel approach is based on exchange of surface ligand of DEG-coated upconversion NPs with carboxylic ligands due to stronger complexing

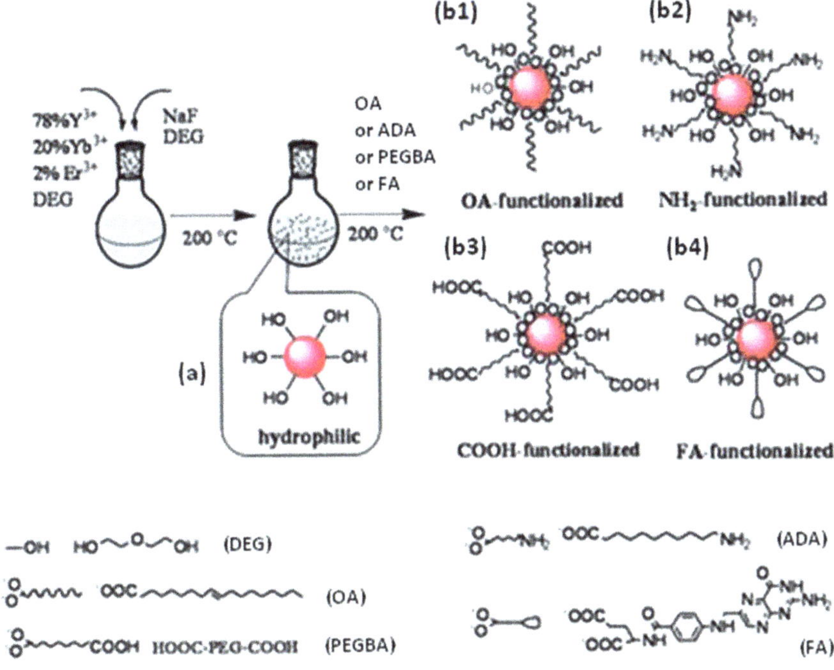

Fig. 11.25 Schematic diagram for the one-pot polyol synthesis of surface-functionalized UCNPs. Untreated UCNPs-DEG (hydrophilic) (**a**); amphiphilic UCNPs-OA: OA-functionalized UCNPs (**b1**); UCNPs-ADA: NH_2-functionalized UCNPs (**b2**); UCNPs-PEGBA: COOH-functionalized UCNPs (**b3**); UCNPs-FA: FA-functionalized UCNPs (**b4**). Adapted with permission from Ref. [107]. Copyright (2010) Royal Society of Chemistry

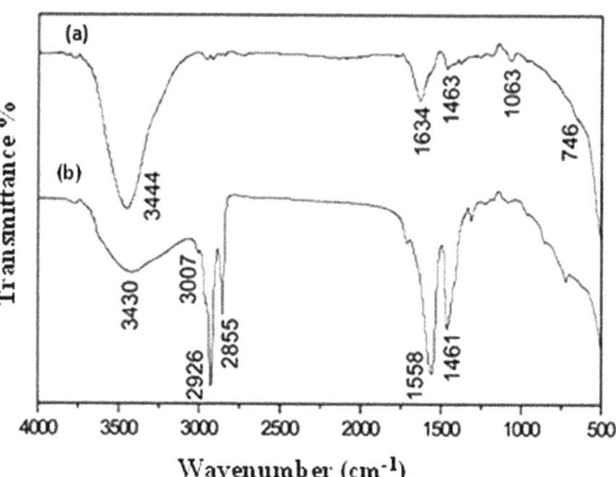

Fig. 11.26 FTIR spectra of DEG coated UCNPs (**a**) and OA-functionalized UCNPs (**b**). Adapted with permission from Ref. [107]. Copyright (2010) Royal Society of Chemistry

ability of the -COOH group with lanthanide ions compared to the –OH group of DEG.

After preparation of $NaYF_4$-based nanoparticles in DEG medium at 200 °C, the reaction temperature was reduced to 100 °C wherein a DEG solution containing OA/ADA/PEGBA/FA was injected. The solution was heated at 200 °C for 1 h to ensure formation of surface-functionalized UCNPs. The successful preparation of surface-functionalized UCNPs via this one-pot polyol synthesis method was verified by FTIR and ^{1}H NMR spectroscopy. As an example, the FTIR spectra of oleic acid-functionalised (OA-functionalized) UCNPs and DEG-coated UCNPs are shown in Fig. 11.26 which confirmed successful ligand exchange between DEG and OA. The characteristic peak at 1063 cm^{-1} is due to C–O bond of DEG which is absent in the FTIR spectrum of UCNPs-OA. Moreover, a new peak is observed at 3007 cm^{-1} in the FTIR spectrum of OA-functionalized UCNPs which is attributed to the = C–H stretching vibration. Among the above-mentioned surface-functionalized UCNPs, FA-functionalized UCNPs have been demonstrated for targeting folate-receptors (FR) over expressing cancer cells with no background fluorescence and can be considered as one of the promising luminescent labels for bioimaging.

11.7 Conclusions and Future Scope

In this chapter, synthesis of a variety of nanostructured materials based on polyol method has been described in detail. In addition to the role of polyol as a solvent and a stabilizing ligand, polyol also acts as a reducing agent during the synthesis of metal and alloy nanoparticles. For metal ions which require strong reducing agents for conversion to zero-valent state, additional reagents can be employed during the synthesis. In most of the cases, nanomaterials with narrower particle size distribution have been obtained by this method. The method can be used for the preparation of nanomaterials with both hydrophobic and hydrophilic surfaces. The examples mentioned in the chapter clearly establish the fact that the methodology can be extended for the synthesis of wide range of inorganic compounds with large extent of tunability in physico-chemical properties. However, a lot of improvements are required to extend the scope of polyol synthesis to industrial scale. Based on extensive literature survey, it is understood that the following aspects of polyol synthesis need to be addressed:

(i) Research efforts should be made towards the establishment of formation mechanism of nanoparticles by polyol method to avoid the choice of polyols on empirical basis during synthesis. This seems to be a tedious job due to wide range of polyols and metal precursors. Already, some theoretical investigations have been carried out for fundamental understanding of the polyol synthesis for some specific polyol-metal cases. However, in-depth theoretical studies along with extensive experimental investigations specially using in situ spectroscopic tools are required to make a reliable database so that size and shape of the final product can be predicted based on the data.
(ii) Like the choice of polyol, in most of the polyol synthesis, capping agent is also selected through an empirical approach due to the lack of proper understanding of the factors that decide the morphology of the final product. Hence, it is obvious that a lot of research scope is opened for the comprehensive understanding of the roles played by capping agent in controlling the size and shape of nanoparticles.
(iii) Thermal decomposition of the polyols near the boiling point is a serious problem which restricts the reaction temperature to a certain upper limit using a particular polyol. Moreover, polyol synthesis faces big challenges of solvent recyclability as far as scaling up is considered.

If the above-mentioned issues are solved, a revolutionary change is expected in the field of polyol synthesis, and this synthetic method can be successfully applied for developing various inorganic functional nanomaterials with different technological importance.

References

1. Yang P, Zheng J, Xu Y, Zhang Q, Jiang L (2016) Colloidal synthesis and applications of plasmonic metal nanoparticles. Adv Mater 28:10508–10517
2. Fiévet F, Lagier JP, Figlarz M (1989) Preparing monodisperse metal powders in micrometer and submicrometer sizes by the polyol process. MRS Bull 14:29–34
3. Fiévet F, Lagier JP, Blin B, Beaudoin B, Figlarz M (1989) Homogeneous and heterogeneous nucleations in the polyol process for the preparation of micron and submicron size metal particles. Solid State Ion 32–33:198–205

4. Viau G, Ravel F, Acher O, Fievet-Vincent F, Fievet F (1994) Preparation and microwave characterization of spherical and monodisperse $Co_{20}Ni_{80}$ particles. J Appl Phys 76:6570–6572
5. Viau G, Ravel F, Acher O, Fievet-Vincent F, Fievet F (1995) Preparation and microwave characterization of spherical and monodisperse Co-Ni particles. J Magn Magn Mater 140–144:377–378
6. Viau G, Fievet-Vincent F, Fievet F (1996) Nucleation and growth of bimetallic CoNi and FeNi monodisperse particles prepared in polyols. Solid State Ion 84:259–270
7. Viau G, Fievet-Vincent F, Fievet F (1996) Monodisperse iron-based particles: precipitation in liquid polyols. J Mater Chem 6:1047–1053
8. Love CP, Torardi CC, Page CJ (1992) Two new barium-copper-ethylene glycol complexes: synthesis and structure of $BaCu(C_2H_6O_2)_n(C_2H_4O_2)_2$ (n = 3, 6). Inorg Chem 31:1784–1788
9. Fiévet F, Ammar-Merah S, Brayner R, Chau F, Giraud M, Mammeri F, Peron J, Piquemal J-Y, Sicard L, Viau G (2018) The polyol process: a unique method for easy access to metal nanoparticles with tailored sizes, shapes and compositions. Chem Soc Rev 47:5187–5233
10. Carroll KJ, Reveles JU, Shultz MD, Khanna SN, Carpenter EE (2011) Preparation of elemental Cu and Ni nanoparticles by the polyol method: an experimental and theoretical approach. J Phys Chem C 115:2656–2664
11. Poul L, Ammar S, Jouini N, Fiévet F, Villain F (2003) Synthesis of inorganic compounds (metal, oxide and hydroxide) in polyol medium: a versatile route related to the sol-gel process. J Sol-Gel Sci Technol 26:261–265
12. Milek T, Zahn D (2014) Molecular simulation of Ag nanoparticle nucleation from solution: redox-reactions direct the evolution of shape and structure. Nano Lett 14:4913–4917
13. Liu B, Greeley J (2013) A density functional theory analysis of trends in glycerol decomposition on close-packed transition metal surfaces. Phys Chem Chem Phys 15:6475–6485
14. Lide DR (2004) Handbook of chemistry and physics on CD-ROM, version 2005. CRC Press, Boca Raton
15. Bonneau M, Chardon N, Andersson S, Muhammed M (1996) US5529804, Sandvik AB and Eurotungstene Poudres S.A.
16. Rajan K, Roppolo I, Chiappone A, Bocchini S, Perrone D, Chiolerio A (2016) Silver nanoparticle ink technology: state of the art. Nanotechnol Sci Appl 9:1–13
17. Alshehri AH, Jakubowska M, Młożniak A, Horaczek M, Rudka D, Free C, Carey JD (2012) Enhanced electrical conductivity of silver nanoparticles for high frequency electronic applications. ACS Appl Mater Interfaces 4:7007–7010
18. Xiong Y, McLellan JM, Chen J, Yin Y, Li Z-Y, Xia Y (2005) Kinetically controlled synthesis of triangular and hexagonal nanoplates of palladium and their SPR/SERS properties. J Am Chem Soc 127:17118–17127
19. Lee YW, Han SB, Ko AR, Kim HS, Park KW (2011) Glycerol-mediated synthesis of Pd nanostructures with dominant {111} facets for enhanced electrocatalytic activity. Catal Commun 15:137–140
20. Yu Y, Zhao Y, Huang T, Liu H (2010) Microwave-assisted synthesis of palladium nanocubes and nanobars. Mater Res Bull 45:159–164
21. Hei H, He H, Wang R, Liu X, Zhang G (2012) Controlled synthesis and characterization of noble metal nanoparticles. Soft Nanosci Lett 2:34–40
22. Ma L, Wang C, Gong M, Liao L, Long R, Wang J, Wu D, Zhong W, Kim MJ, Chen Y, Xie Y, Xiong Y (2012) Control over the branched structures of platinum nanocrystals for electrocatalytic applications. ACS Nano 6:9797–9806
23. Chen J, Xiong Y, Yin Y, Xia Y (2006) Pt nanoparticles surfactant-directed assembled into colloidal spheres and used as substrates in forming Pt nanorods and nanowires. Small 2:1340–1343
24. Lee JH, Kamada K, Enomoto N, Hojo J (2008) Polyhedral gold nanoplate: high fraction synthesis of two-dimensional nanoparticles through rapid heating process. Cryst Growth Des 8:2638–2645

25. Li C, Shuford KL, Chen M, Lee EJ, Cho SO (2008) A facile polyol route to uniform gold octahedra with tailorable size and their optical properties. ACS Nano 2:1760–1769
26. Seo D, Il Yoo C, Chung IS, Park SM, Ryu S, Song H (2008) Shape adjustment between multiply twinned and single-crystalline polyhedral gold nanocrystals: decahedra, icosahedra, and truncated tetrahedra. J Phys Chem C 112:2469–2475
27. Silvert P-Y, Herrera-Urbina R, Duvauchelle N, Vijayakrishnan V, Elhsissen KT (1996) Preparation of colloidal silver dispersions by the polyol process. Part 1-Synthesis and characterization. J Mater Chem 6:573–577
28. Sun Y, Yin Y, Mayers BT, Herricks T, Xia Y (2002) Uniform silver nanowires synthesis by reducing AgNO3 with ethylene glycol in the presence of seeds and poly(vinyl pyrrolidone). Chem Mater 14:4736–4745
29. Liu Z, Tian S, Yang J, Liu H, Liu X, Jia H, Xu B (2016) Sulfur ion-induced shape evolution of Ag nanocrystals by microwave-assisted polyol process. Mater Lett 164:647–650
30. Zhang YW, Grass ME, Kuhn JN, Tao F, Habas SE, Huang WY, Yang PD, Somorjai GA (2008) Highly selective synthesis of catalytically active monodisperse rhodium nanocubes. J Am Chem Soc 130:5868–5869
31. Biacchi AJ, Schaak RE (2011) The solvent matters: kinetic versus thermodynamic shape control in the polyol synthesis of rhodium nanoparticles. ACS Nano 5:8089–8099
32. Xie S, Zhang H, Lu N, Jin M, Wang J, Kim MJ, Xie Z, Xia Y (2013) Synthesis of rhodium concave tetrahedrons by collectively manipulating the reduction kinetics, facet –selective capping, and surface diffusion. Nano Lett 13:6262–6268
33. Bonet F, Delmas V, Grugeon S, Herrera Urbina R, Silvert P-Y, Tekaia-Elhsissen K (1999) Synthesis of monodisperse Au, Pt, Pd, Ru and Ir nanoparticles in ethylene glycol. Nanostruct Mater 11:1277–1284
34. Yan X, Liu H, Liew KY (2001) Size control of polymer-stabilized ruthenium nanoparticles by polyol reduction. J Mater Chem 11:3387–3391
35. Viau G, Brayner R, Poul L, Chakroune N, Lacaze E, Fiévet-Vincent F, Fiévet F (2003) Ruthenium nanoparticles: size, shape, and self-assemblies. Chem Mater 15:486–494
36. Ducamp-Sanguesa C, Herrera Urbina R, Figlarz M (1993) Fine palladium powders of uniform particle size and shape produced in ethylene glycol. Solid State Ion 63–65:25–30
37. Song H, Kim F, Connor S, Somorjai GA, Yang P (2005) Pt nanocrystals: shape control and langmuir−blodgett monolayer formation. J Phys Chem B 109:188–193
38. Wang Y, Ren J, Deng K, Gui L, Tang Y (2000) Preparation of tractable platinum, rhodium, and ruthenium nanoclusters with small particle size in organic media. Chem Mater 12:1622–1627
39. Chen J, Herricks T, Geissler M, Xia Y (2004) Single-crystal nanowires of platinum can be synthesized by controlling the reaction rate of a polyol process. J Am Chem Soc 126:10854–10855
40. Chen J, Herricks T, Xia Y (2005) Polyol synthesis of platinum nanostructures: control of morphology through the manipulation of reduction kinetics. Angew Chem Int Ed 44:2589–2592
41. Silvert P-Y, Tekaia-Elhsissen K (1995) Synthesis of monodisperse submicronic gold particles by the polyol process. Solid State Ion 82:53–60
42. Li C, Cai W, Cao B, Sun F, Li Y, Kan C, Zhang L (2006) Mass synthesis of large, single-crystal Au nanosheets based on a polyol process. Adv Funct Mater 16:83–90
43. Kan C, Wang C, Li H, Qi J, Zhu J, Li Z, Shi D (2010) Gold microplates with well-defined shapes. Small 6:1768–1775
44. Zhu J, Kan C, Li H, Cao Y, Ding X, Wan J (2011) Synthesis and growth mechanism of gold nanoplates with novel shapes. J Cryst Growth 321:124–130
45. Seo D, Ji CP, Song H (2006) Polyhedral gold nanocrystals with Oh symmetry: from octahedra to cubes. J Am Chem Soc 128:14863–14870
46. Lee SJ, Park G, Seo D, Ka D, Kim SY, Chung IS, Song H (2011) Coordination power adjustment of surface-regulating polymers for shaping gold polyhedral nanocrystals. Chem - Eur J 17:8466–8471

47. Lee SJ, Scotti N, Ravasio N, Chung IS, Song H (2013) Bovine serum albumin as an effective surface regulating biopolymer for morphology control of gold polyhedrons. Cryst Growth Des 13:4131–4137
48. Sun Y, Xia Y (2002) Large-scale synthesis of uniform silver nanowires through a soft, self-seeding, polyol proces. Adv Mater 14:833–837
49. Sun Y, Mayers B, Xia Y (2003) Metal nanostructures with hollow interiors. Adv Mater 15:641–646
50. Wiley BJ, Xiong Y, Li ZY, Yin Y, Xia Y (2006) Right bipyramids of silver: a new shape derived from single twinned seeds. Nano Lett 6:765–768
51. Xiong Y, Siekkinen AR, Wang J, Yin Y, Kim MJ, Xia Y (2007) Synthesis of silver nanoplates at high yields by slowing down the polyol reduction of silver nitrate with polyacrylamide. J Mater Chem 17:2600–2602
52. Wiley B, Sun Y, Mayers B, Xia Y (2005) Shape-controlled synthesis of metal nanostructures: the case of silver. Chem -Eur J 11:454–463
53. Sun Y, Xia Y (2004) Mechanistic study on the replacement reaction between silver nanostructures and chloroauric acid in aqueous medium. J Am Chem Soc 126:3892–3901
54. Xiong Y, Cai H, Wiley BJ, Wang J, Kim MJ, Xia Y (2007) Synthesis and mechanistic study of palladium nanobars and nanorods. J Am Chem Soc 129:3665–3675
55. Zheng Y, Zeng J, Ruditskiy A, Liu M, Xia Y (2014) Oxidative etching and its role in manipulating the nucleation and growth of noble-metal nanocrystals. Chem Mater 26:22–33
56. Banerjee S, Dasgupta K, Kumar A, Ruz P, Vishwanadh B, Joshi JB, Sudarsan V (2015) Comparative evaluation of hydrogen storage behavior of Pd doped carbon nanotubes prepared by wet impregnation and polyol methods. Int J Hydrogen Energy 40:3268–3276
57. Zhang HX, Siegert U, Liu R, Bin Cai W (2009) Facile fabrication of ultrafine copper nanoparticles in organic solvent. Nanoscale Res Lett 4:705–708
58. Park BK, Jeong S, Kim D, Moon J, Lim S, Kim JS (2007) Synthesis and size control of monodisperse copper nanoparticles by polyol method. J Colloid Interface Sci 311:417–424
59. Engels V, Benaskar F, Jefferson DA, Johnson BFG, Wheatley AEH (2010) Nanoparticulate copper - Routes towards oxidative stability. Dalton Trans 39:6496–6502
60. Kawasaki H, Kosaka Y, Myoujin Y, Narushima T, Yonezawa T, Arakawa R (2011) Microwave-assisted polyol synthesis of copper nanocrystals without using additional protective agents. Chem Commun 47:7740–7742
61. Kind C, Weber A, Feldmann C (2012) Easy access to Cu^0 nanoparticles and porous copper electrodes with high oxidation stability and high conductivity. J Mater Chem 22:987–993
62. Kind C, Feldmann C (2011) One-pot synthesis of In^0 nanoparticles with tuned particle size and high oxidation stability. Chem Mater 23(22):4982–4987
63. Hammarberg E, Feldmann C (2009) In^0 nanoparticle synthesis assisted by phase-transfer reaction. Chem Mater 21:771–774
64. Wang Y, Xia Y (2004) Bottom-up and top-down approaches to the synthesis of monodispersed spherical colloids of low melting-point metals. Nano Lett 4(10):2047–2050
65. Luz A, Feldmann C (2009) Reversible photochromic effect and electrochemical voltage driven by light-induced Bi^0-formation. J Mater Chem 19:8107–8111
66. Kodama D, Shinoda K, Sato K, Konno Y, Joseyphus RJ, Motomiya K, Takahashi H, Matsumoto T, Sato Y, Tohji K, Jeyadevan B (2006) Chemical synthesis of sub-micrometer-to nanometer-sized magnetic FeCo dice. Adv Mater 18:3154–3159
67. Abbas M, Nazrul Islam M, Parvatheeswara Rao B, Ogawa T, Takahashi M, Kim C (2013) One-pot synthesis of high magnetization air-stable FeCo nanoparticles by modified polyol method. Mater Lett 91:326–329
68. Kumar A, Meena SS, Banerjee S, Sudarsan V, Yusuf SM (2016) Fe-Ni solid solutions in nano-size dimensions:Effect of hydrogen annealing. Mater Res Bull 74:447–451
69. Lee YW, Ko AR, Han SB, Kim HS, Park KW (2011) Synthesis of octahedral Pt-Pd alloy nanoparticles for improved catalytic activity and stability in methanol electrooxidation. Phys Chem Chem Phys 13:5569–5572

70. Kusada K, Yamauchi M, Kobayashi H, Kitagawa H, Kubota Y (2010) Hydrogen-storage properties of solid-solution alloys of immiscible neighboring elements with Pd. J Am Chem Soc 132:15896–15898
71. Tsuji M, Miyamae N, Lim S, Kimura K, Zhang X, Hikino S, Nishio M (2006) Crystal structures and growth mechanisms of Au@Ag core-shell nanoparticles prepared by the microwave-polyol method. Cryst Growth Des 6:1801–1807
72. Park G, Seo D, Jung J, Ryu S, Song H (2011) Shape evolution and gram-scale synthesis of gold@silver core–shell nanopolyhedrons. J Phys Chem C 115:9417–9423
73. Tsuji M, Yamaguchi D, Matsunaga M, Alam MJ (2010) Epitaxial growth of Au@Cu core −shell nanocrystals prepared using the pvp-assisted polyol reduction method. Cryst Growth Des 10:5129–5135
74. Dawood F, Leonard BM, Schaak RE (2007) Oxidative transformation of intermetallic nanoparticles: an alternative pathway to metal/oxide nanocomposites, textured ceramics, and nanocrystalline multimetal oxides. Chem Mater 19:4545–4550
75. Avadhut YS, Weber J, Hammarberg E, Feldmann C, Schmedt auf der Günne J (2012) Structural investigation of aluminium doped ZnO nanoparticles by solid-state NMR spectroscopy. Phys Chem Chem Phys 14:11610–11625
76. Dong H, Feldmann C (2012) Porous ZnO platelets via controlled thermal decomposition of zinc glycerolate. J Alloys Compd 513:125–129
77. Ningthoujam RS, Gajbhiye NS, Ahmed A, Umre SS, Sharma SJ (2008) Re-dispersible Li+ and Eu3+ Co-doped nanocrystalline ZnO: Luminescence and EPR studies. J Nanosci Nanotechnol 8:3059–3062
78. Ningthoujama RS, Sudarsan V, Vatsa RK, Kadam RM, Jagannath, Gupta A (2009) Photoluminescence studies on Eu doped TiO_2 nanoparticles. J Alloys Compd 486:864–870
79. Hachani R, Lowdell M, Birchall M, Hervault A, Mertz D, Begin-Colin S, Thanh NTK (2016) Polyol synthesis, functionalisation, and biocompatibility studies of superparamagnetic iron oxide nanoparticles as potential MRI contrast agents. Nanoscale 8:3278–3287
80. Feldmann C, Roming M, Trampert K (2006) Polyol-mediated synthesis of nanoscale CaF_2 and CaF_2:Ce,Tb. Small 2:1248–1250
81. Warmuth L, Feldmann C (2019) β-$SnWO_4$ with morphology-controlled synthesis and facet-depending photocatalysis. ACS Omega 4(8):13400–13407
82. Boddu S (2011) Synthesis and characterization of lanthanide ions doped nanomaterials. A PhD thesis, Homi Bhabha National Institute, Mumbai
83. Kim JS, Kwon AK, Kim JS, Park HL, Kim GC, do Han S (2007) Optical and structural properties of ZnGa2O4: Eu3+ nanophosphor by hydrothermal method. J Lumin 122&123:851–854
84. Lee JH, Park HJ, Yoo K, Kim BW, Lee JC, Park S (2007) Characteristics of nano-sized $ZnGa_2O_4$ phosphor prepared by solution combustion method and solid state reaction method. J Eur Ceram Soc 27:965–968
85. Bae JS, Shim KS, Moon BK, Choi BC, Jeong JH, Yi S, Kim JH (2005) Photoluminescence characteristics of $ZnGa_2O_{4-x}Mx:Mn^{2+}$ (M=S, Se) thin film phosphors grown by pulsed laser ablation. Thin Solid Films 479:238–244
86. Jeong IK, Park HL, Mho SI (1998) Photoluminescence of $ZnGa_2O_4$ mixed with $InGaZnO_4$. Solid State Commun 108:823–826
87. Sahu NK, Singh NS, Ningthoujam RS, Bahadur D (2014) Ce^{3+}-Sensitized $GdPO_4:Tb^{3+}$ nanorods: an investigation on energy transfer, luminescence switching, and quantum yield. ACS Photonics 1:337–346
88. Okram R, Yaiphaba N, Ningthoujam RS, Singh NR (2014) Is Higher ratio of monoclinic to tetragonal in $LaVO_4$ a better luminescence host? redispersion and polymer film formation. Inorg Chem 53:7204–7213
89. Yin X, Que W, Fei D, Shen F, Guo Q (2012) Ag nanoparticle/ZnO nanorods nanocomposites derived by a seed-mediated method and their photocatalytic properties. J Alloys Compd 524:13–21

90. Shen G, Chen D, Tang K, Liu X, Huang L, Qian Y (2003) General synthesis of metal sulfides nanocrystallines via a simple polyol route. J Solid State Chem 173:232–235
91. Wang H, Li X, Yan K, Liu G, Song W, Shen T, Zou D (2017) Low-temperature synthesis of near-monodisperse globular MoS_2 nanoparticles with sulphur powders. Nano: Brief Rep Rev 12(1–13):1750091
92. Castillón-Barraza FF, Farías MH, Coronado-López JH, Encinas-Romero MA, Pérez-Tello M, Herrera-Urbina R, Posada-Amarillas A (2011) Synthesis and characterization of copper sulfide nanoparticles obtained by the polyol method. Adv Sci Lett 4:1–6
93. Datta A, Mukherjee D, Witanachchi S, Mukherjee P (2013) Low temperature synthesis, optical and photoconductance properties of nearly monodisperse thin In_2S_3 nanoplatelets. RSC Adv. 3:141–147
94. Zhu Y, Nie P, Shen L, Dong S, Sheng Q, Li H, Luo H, Zhang X (2015) High rate capability and superior cycle stability of a flower-like Sb_2S_3 anode for high-capacity sodium ion batteries. Nanoscale 7:3309–3315
95. Chen H, Jiang J, Zhao Y, Zhang L, Guo D, Xia D (2015) One-pot synthesis of porous nickel cobalt sulphides: tuning the composition for superior pseudocapacitance. J Mater Chem A 3:428–437
96. Shang L, Zhou C, Bian T, Yu H, Wu LZ, Tung CH, Zhang T (2013) Facile synthesis of hierarchical $ZnIn_2S_4$ submicrospheres composed of ultrathin mesoporous nanosheets as a highly efficient visible-light-driven photocatalyst for H_2 production. J Mater Chem A 1:4552–4558
97. Webber DH, Buckley JJ, Antunez PD, Brutchey RL (2014) Facile dissolution of selenium and tellurium in a thiol–amine solvent mixture under ambient conditions. Chem Sci 5:2498–2502
98. Liu H, Jin Z, Wang W, Wang Y, Du H (2012) Well-dispersed, size-tunable chalcopyrite $CuInSe_2$ nanocrystals and its ink-coated thin films by polyhydric solution chemical process. Mater Lett 81:173–176
99. Liu H, Jin Z, Wang X, Zheng X, Wang Y, Du H, Cui L (2012) Morphological growth and phase formation of $CuInSe_2$, nanocrystals by an ambient pressure polylol-based solution synthesis. CrystEngComm 14:8186–8192
100. Liu H, Jin Z, Wang W, Li J (2011) Monodispersed sphalerite $CuInSe_2$ nanoplates and highly (112) oriented chalcopyrite thin films by nanoplates ink coating. CrystEngComm 13:7198–7201
101. Liu H, Zhang J, Zheng X, Liu G, Hu H, Pan W, Liu C, Hao Q, Chen H (2016) Polyol process synthesis of metal selenide nanomaterials and their photovoltaic application. Cryst Eng Comm 18:6860–6866
102. Grisaru H, Palchik O, Gedanken A, Palchik V, Slifkin MA, Weiss AM (2003) Microwave-assisted polyol synthesis of $CuInTe_2$ and $CuInSe_2$ nanoparticles. Inorg Chem 42:7148–7155
103. Akshay VR, Suneesh MV, Vasundhara M (2017) Tailoring thermoelectric properties through structure and morphology in chemically synthesized n-type bismuth telluride nanostructures. Inorg Chem 56(11):6264–6274
104. Anderson ME, Bharadwaya SSN, Schaak RE (2010) Modified polyol synthesis of bulk-scale nanostructured bismuth antimony telluride. J Mater Chem 20:8362–8367
105. Wang ZL, Quan ZW, Jia PY, Lin CK, Luo Y, Chen Y, Fang J, Zhou W, O'Connor CJ, Lin J (2006) A facile synthesis and photoluminescent properties of redispersible CeF_3, CeF_3:Tb^{3+}, and CeF_3: Tb^{3+}/LaF_3(core/shell) nanoparticles. Chem Mater 18:2030–2036
106. Wei Y, Lu F, Zhang X, Chen D (2008) Polyol-mediated synthesis and luminescence of lanthanide-doped $NaYF_4$ nanocrystal upconversion phosphors. J Alloys Compd 455:376–384
107. Zhou J, Yao L, Li C, Li F (2010) A versatile fabrication of upconversion nanophosphors with functional-surface tunable ligands. J Mater Chem 20:8078–8085

Chapter 12
Synthesis of Nanostructured Materials by Thermolysis

Bheeshma Pratap Singh, Ramaswamy Sandeep Perala, Manas Srivastava, and Raghumani Singh Ningthoujam

Abstract The thermolysis synthesis for the different nanomaterials such as metal, metal oxides, hollow nanostructures, bimetallic, metal organic frameworks, and carbon dots is provided. Controlled shape and size engineering of particles has been performed using appropriate polyols. The polyols such as ethylene glycol (EG), polyethylene glycol (PEG), and glycerol are frequently used for the nanomaterial processing. The microwave assisted synthesis gives advantages such as fast heating, quick reaction rate, high yields of the product, and less reaction time as compared to the conventional heating techniques. Hydro/solvothermal routes are employed to obtain range of nanomaterials with controlled morphology and crystallinity compared to the other wet-chemical techniques. Sonochemical as well as ultrasonic spray pyrolysis approaches are also utilized for the synthesis of nanomaterials. Ultrasonication produces acoustic cavitation. The cavitation process leads to the formation of bubbles. During the collapse of bubbles, the tremendous amount of energy/high temperature and high pressures are liberated in very short time, and this can be used for synthesis of nanomaterials.

Keywords Nanomaterials · Polyol · Thermolysis · Microwave · Sonochemical · Hydrothermal · Solvothermal

Abbreviations

Nanoparticles	NPs
Microwave	MW
Ethylene glycol	EG
Polyethylne glycol	PEG

B. P. Singh (✉) · R. S. Perala · M. Srivastava · R. S. Ningthoujam (✉)
Chemistry Division, Bhabha Atomic Research Centre, Mumbai 400085, India
e-mail: bheeshmapratap@gmail.com

R. S. Ningthoujam
e-mail: rsn@barc.gov.in

R. S. Ningthoujam
Homi Bhabha National Institute, Mumbai 400094, India

A. K. Tyagi and R. S. Ningthoujam (eds.), *Handbook on Synthesis Strategies for Advanced Materials*, Indian Institute of Metals Series,
https://doi.org/10.1007/978-981-16-1807-9_12

Diethylene glycol	DEG
Tri-ethylene glycol	TEG
Butanediol	BD
Glycerol	GLY
Benzenehexa chloride	Bhc
Tri-ethanol-amine	TEA
Cetyl trimethyl ammonium bromide	CTAB
Sodium dodecyl sulfate	SDS
Polyethylenimine	PEI
Metal Organic Framework	MOF
Hollow Carbon Sphere	HCS

12.1 Introduction

Nanomaterials are regarded as backbone of nanotechnology and have shown promising applications in materials science, bio-physics, chemistry, biotechnology, etc. Synthesis plays a vital role in obtaining desired nanostructured materials. Moreover, synthesis of the nanomaterials with the environment friendly route, use of less toxic and less explosive reactants, and less reaction processing time is highly required. Precise control of shape and size of the nanoparticles (NPs) is necessary in order to tailor electrical, optical, magnetic, and catalytic properties. Nanostructured materials synthesized via thermolysis can provide the various shape and size of the particles which exhibit the different properties as compared to their bulk counterparts, and thus, they are useful in many applications such as optoelectronic devices, photocatalytic activity, data storage, bio-imaging, drug delivery, magnetic resonance imaging (MRI), fuel cell, photovoltaics, photodetector, super capacitors and hyperthermia, etc. [1–6].

Synthesis protocols play the pivotal role in controlling the morphology of the NPs, which finally determines their properties. Herein, thermolysis approach is presented for the synthesis of nanomaterials and their different applications. Thermolysis is a process of synthesis of compounds above room temperature but below the boiling points of the solvents. Generally, solvents used into thermolysis routes should have high boiling point. It may also include the process such as thermal decomposition, substitution, and addition of the ions or elements but we consider the case for thermolysis only when there is an involvement of heating in the presence of a solvent. The scope of thermolysis is very vast, and this can be also used in the preparation of organic and inorganic compounds. However, in this chapter, synthesis of inorganic materials such as metals, alloys, oxides, metal organic frameworks, etc., will be presented. Figure 12.1 represents the schematic of various routes of thermolysis, namely polyol, hydro/solvothermal, microwave (MW), sonochemical, and their applications.

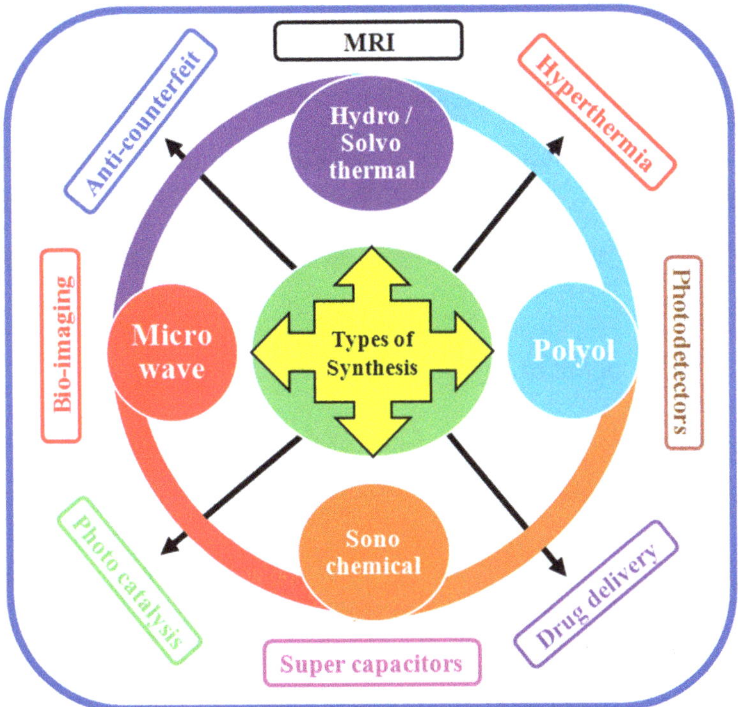

Fig. 12.1 Schematic representations depicting the various routes of thermolysis for synthesis of nanoparticles and their applications

12.1.1 Types of Solvents

On the basis of polarity, solvents can be categorized into two types—polar and non-polar solvents. Generally, polar solvents can dissolve polar molecule, whereas non-polar solvents can dissolve non-polar molecule. There are solvent molecules such as dimethyl sulfoxide (an organosulfur compound with the formula $(CH_3)_2SO$), N,N-dimethylformamide, which can dissolve polar or non-polar molecules. In addition, there are solvent molecules such as surfactant (oleic acid), which can dissolve polar molecule (water) and non-polar molecule (hexane, oil) simultaneously. Here, water-oil is immiscible, but miscible in the presence of oleic acid.

12.1.2 Polar or Hydrophilic Solvents

The solvent molecule having polarity due to presence of a functional group such as OH, which can mix with water molecule easily is considered as polar or hydrophilic solvent. The general structure of the solvents can be expressed as R-OH. Generally, this solvent has high boiling point; and in general, its boiling point increases with increase of OH contents in a molecule. The polar solvents with their structural formula and corresponding boiling points are represented in Table 12.1 [7].

12.1.3 Non-polar or Hydrophobic Solvents

Solvent molecule of non-polar character does not have static charges such as positive and negative charge separation, and in general, it is immiscible with water, and these are considered as non-polar or hydrophobic solvents. In this type of solvents, homogenous distribution of charges results in small dielectric constant. Moreover, non-polar solvents can dissolve non-polar substances such as oils, fats, and grease. Examples of non-polar solvents are summarized in Table 12.2 [8].

Using the hydrophilic solvents during the thermolysis process, agglomerated particles/compounds are prepared. Moreover, suitable use of capping agents such as sodium dodecyl sulfate (SDS), poly vinyl alcohol (PVA) which can exhibit less agglomeration, and nearly, monodispersed type compounds are prepared [9]. Moreover, in hydrophobic type solvents, octadecene, diethylether, and silicon oil have been used. By using these solvents and hydrophobic surfactant such as oleic

Table 12.1 Polar solvents and their corresponding boiling points

Solvents (Polar)	Chemical formula	Boiling point (°C)
Ethylene glycol (EG)	$HO–C_2H_4–OH$	197
Diethyleneglycol (DEG)	$HO–C_2H_4–O–C_2H_4–OH$	244
Tri-ethyleneglycol (TrEG)	$HO–(C_2H_4–O)_2–C_2H_4–OH$	291
Tetraethyleneglycol (TEG)	$HO–(C_2H_4–O)_3–C_2H_4OH$	314
Polyethyleneglycol (PEG)	$HO–(C_2H_4–O)_n–C_2H_4OH$	350
Glycerol (GLY)	$C_3H_8O_3$	290
Butanediol (BD)	$C_4H_{10}O_2$	235
Pentanediol (PD)	$C_5H_{12}O_2$	242
Methanol	$CH_3–OH$	64.6
Ethanol	$CH_3CH_2–OH$	78.5
n-Propanol	$C_3H_7–OH$	97
Acetic acid	$CH_3CO–OH$	117.9
n-Butanol	$C_4H_9–OH$	118
Water	$H–OH$	100

Table 12.2 Non-polar solvents and their corresponding boiling points

Solvents (Non-polar)	Chemical formula	Boiling point (°C)
Hexane	$CH_3(CH_2)_4CH_3$	69
Benzene	C_6H_6	80.1
Diethyl ether	$CH_3CH_2OCH_2CH_3$	118
Carbon tetrachloride	CCl_4	76.8
Diphenyl ether	$C_6H_5–O–C_6H_5$	257
Octadecene	$CH_3C_{15}H_{30}CH = CH_2$	315
Toluene	$C_6H_5CH_3$	111

acid, nearly monodispersed nanoparticles are prepared. Using polar solvents during the compound preparation, the thermolysis process can be categorized as (a) polyol, (b) hydrothermal and/or solvothermal, (c) microwave (MW) assisted synthesis, and (d) sonochemical synthesis.

12.2 Polyol Synthesis Route

Metal nanoparticles of Ag, Au, Pd, and Ni can be prepared by polyol route. Here, EG, PEG, and GLY solvents are used as reaction medium as well as reducing reagent. It restricts an extra addition of reducing agent. The presence of OH group in solvent helps in reducing metal ions. The polyol synthesis exhibits a range of features as compared to traditional sol–gel route such as it comprises high boiling points solvents, reducing properties for the metal synthesis, high solubility of metal salts comparable to water, and vast operating temperature choice. It also favors the synchronizing behavior toward surface functionalization, colloids of highly stable NPs, and vast range of flexibility of polyols ranging from low molecular weight EG to high molecular weight PEGs. Furthermore, it is easier to detach the polyols from the surface of NPs after synthesis by repeated washing with water, and thus, particles can be prepared at large scale. This route has been regarded as green, bio-compatible, and viable solvents, which is also extremely applicable for large-scale commercial synthesis of NPs [7].

12.2.1 Metal NPs

The solution processed polyol mediated synthesis has been emerged as the most effective successful route to prepare the metallic NPs with less agglomeration. Zero valent noble metals Au, Ag, Pd, Cu, and high electropositive Co and Ni metals are prepared. Also, this approach is further extended for the metals and its alloys comprised of particles with very less agglomeration, precise control over morphology, and narrow size arrangement [10]. The precise control over shape and size shows tunability of opto-electronic, magnetic, catalytic, data storage capacities, etc. Xia and co-workers have synthesized the single crystal of Ag NPs via polyol mediated synthesis route. SilverNPs are prepared via reduction of silver nitrate salt precursor in the presence of ethylene glycol at higher temperature at the manifestation of polyvinyl pyrrolidone (PVP). Here, ethylene glycol acts as reaction medium as well as reducing agent while PVP as a capping agent. It was observed that addition of NaCl into the solution mixture increases the etching and oxidation of the twinned particles promoting the formation of monodispersed single crystal of Ag NPs [11]. In recent years, Pt NPs are extensively studied for their catalytic properties [12]. Yang and his co-workers reported the monodispersed Pt NPs are prepared via polyol route. It is reported that by mixing of $AgNO_3$ to the polyol, the morphology of Pt NPs changes. Here, nitrate anion significantly reduces the degree of reduction of Pt(II) and Pt(IV) via ethylene glycol. Generally, PVP is used as regulating and stabilizing agent for the selective growth of nanocrystal in well-defined shape. It is reported that monodispersed Pt nanocrystals with different shapes comprising of cubes, cuboctahedra, and octahedral can be prepared by polyol method. In the synthesis process, $AgNO_3$ solution containing Ag ion shows a vital part for the controlled morphology of Pt NPs [13]. Xia and his co-workers reported the uniform cuboctahedral PdNPs can be prepared by using simple polyol synthesis route. In the typical synthesis of Pd NPs, $Na_2[PdCl_4]$ metal precursor is mixed thoroughly into EG heated at 100 °C in the presence of PVP [14]. The reaction mixture is largely established by cuboctahedra composed with twinned particles. The chloride as well as O_2 anions are accountable for the oxidative etching of twinned NPs. The twinned NPs are stable under Ar atmosphere. Moreover, these twinned NPs are vanished by revealing the sample to air. Thus, twinned NPs are more sensitive due to the higher concentration of surface defects as compared to cuboctahedra. This demonstrates the occurrence of an oxidant in reaction medium affording a way to control the formation of well-defined morphology of PdNPs. Monodispersed gold nanoboxes with highly truncated cubic shape have been synthesised employing silver nanocubes as a sacrificial template in which silver nanocubes react with an aqueous $HAuCl_4$ solution by following reaction process:

$$3\mathrm{Ag}\ (s) + \mathrm{HAuCl_4(aq)} \rightarrow \cdots \mathrm{Au(s)} + 3\mathrm{AgCl(aq)} + \mathrm{HCl(aq)}$$

These polyols mediated silver and gold NPs treasure their applications in diverse zones of claims in photonics, catalysis, and surface enhanced Raman spectroscopy (SERS)-based sensing [15]. For the comparatively less-noble metals, the reducing capacity of the polyol reaches to its higher limit at elevated temperature (>230 °C). Even though if reduction process continues to happen, quick thermal decomposition of the polyol obstructs the nucleation of NPs and impedes their separation subsequent to synthesis. Moreover, by lowering the temperature, less-noble metals such as Co^0 and Ni^0 can be deoxidized by the polyol. At elevated temperature (>150 °C) reducing ability of polyol favors the reduction of these metallic ions to their zero-valence metal from Co^{2+}/Ni^{2+} to Co^0/Ni^0. Meanwhile at the lower temperature (<150 °C), the bonding of ions and molecules to metal ion stability of the polyols as well as the existence of OH-groups entails the re-oxidation to Co^{2+}/Ni^{2+} [16].

12.2.2 Metal Alloys

The bimetallic NPs show the better data storage capacity as compared to monometallic counterparts. In the synthesis of FePt metal alloys, platinum and iron acetyl-acetonates reduce in the presence of ethylene glycol. It has been suggested that in presence of 1,2-hexadecanediol as the reducing reagent, great quality FePt NPs are obtained via the reduction of $C_{10}H_{16}FeO_4$ and $C_{10}H_{16}O_4Pt$. The monodispersed FePt NPs having *fcc* phase show disordered structure. Upon annealing the sample at 650 °C under inert argon atmosphere, structural phase change occurs via disordered *fcc* to ordered L10 phase. The elemental configuration is precisely tuned via tailoring the comparative amounts of metal precursors of Fe (II) and Pt(II) [17]. Also for the preparation of monodispersed FePd NPs, altered polyol mediated synthesis has been implemented. Paladium acetyl acetonate Pd $(acac)_2$ is effortlessly dissolved in diphenyl ether. The 1,2-hexadecanediol is used as a reducing agent, while oleic acid and oleylamine is used as stabilizer. The solution mixture is heated below the boiling point of diphenyl ether and refluxed [18]. The as-prepared FePdNPs are annealed for its structural ordering from disordered *fcc* phase to L10 ordered phase in vacuum [18, 19]. Fievet et al. have also reported the synthesis of various other metal alloys such as $Ni_{1-x}Co_x$, FeNi, Co_xCu_{1-x}, and FeCoNi [20]. Metal carbides such as Co_3C, Co_2C, Ni_3C have been prepared and they exhibit high coercivity in magnetic study. These carbides-based alloys have shown promising applications in data storage [7].

12.2.3 Metal Oxides

This synthesis protocol is effectively extended to synthesize the metal oxides NPs. Highly water-dispersible Fe_3O_4NPs are prepared via thermal decomposition of iron acetylacetonate in the presence of PEG comprising PVP and polyethylenimine (PEI). Fe_3O_4 NPs layered with PEG/PVP and/or PEG/PEI displayed an excellent dispersion constancy in water. The surface of prepared Fe_3O_4NPs can be modified by the use of suitable polymer additives [21]. Monodispersed magnetite Fe_3O_4NPs are prepared using tri-ethylene glycol (TEG). TEG reduces $Fe(acac)_3$, and finally, magnetite is obtained [22]. Polyol mediated submicron-size monodispersed ZnO NPs are prepared via hydrolysis of zinc acetate dehydrate and diethylene glycol [23]. Feldman and Jungk have reported the highly crystalline monodispersed Cu_2O, TiO_2 and Nb_2O_5NPs via polyol technique. Also Y_2O_3, Cr_2O_3, $ZnCo_2O_4$, $ZnO:In^{3+}$, CeO_2, Mn_2O_4 NPs have been successfully synthesized for their different aspect of applications [24]. Very vast ranges of oxide nano-materials have been prepared via polyol for their various applications. Oxides such as MnO_2, Mn_3O_4, Cu_2O, $NiCo_2O_4$ are used as catalysts, while photocatalytic properties have been reported for ZnO, Cu_2O and $BiVO_4$ [7]. Also magnetic oxides NPs such as Fe_2O_3, Fe_3O_4, Gd_2O_3, and spinel ferrites such as $MgFe_2O_4$, $CoFe_2O_4$, $ZnFe_2O_4$ are reported for their biomedical applications. Dye sensitized photovoltaic cells for broad band-gap diluted semiconductors, namely ZnO, TiO_2 have reported. Also high-power batteries applications of the oxides such as V_2O_5, MnO, Mn_2O_3, SnO_2, $CoMn_2O_4$, $LiFePO_4$, and $LiMnPO_4$ have been reported [25, 26]. The involvements of hydrolysis and reduction reactions are major chemical reactions during the polyol synthesis. The micron-size metal particles with tailored morphology are attained in the deficiency of water, while water entails the reduction and increases the hydrolysis which thus results information of metal oxides through polymerization. Poul et al. have reported the layered hydroxide acetate structures of zinc, cobalt, and nickel metals. Presence of high degree of water contents favors the hydrolysis and condensation reaction for layered geometry formation. The use of acetate precursors as compared to its chlorides and sulfates counterparts promotes to the precipitation of metal, oxides, and hydroxides. The precipitation process majorly depends upon hydrolysis ratio (water to metal molar ratio is termed as hydrolysis ratio) [27]. Furthermore, Prevot et al. have extended the polyol synthesis route for the synthesis of layered double hydroxides $[Ni_2Al(OH)_6]Ac{\bullet}nH_2O$, and $[Co_2Al(OH)_6]Ac{\bullet}nH_2O$ using acetate precursors in polyol [28]. Molybdates and tungstates AMO_4 (A = Ca, Ba, Sr, Mg, M = Mo, W) derivative compounds are successfully synthesized via polyol mediated route at ambient temperature. $CaWO_4$ and $CaMoO_4$ show the intrinsic photoluminescence without any impurity ions doping in its bluish green region (400–540 nm). After incorporating the Eu^{3+} and Sm^{3+} ions in these host matrices, sharp characteristics f-f transitions of these activators (Eu^{3+} and Sm^{3+}) exhibit red color emission. $MoO_4{}^{2-}$ and $WO_4{}^{2-}$ act as efficient emission centers which absorbs the UV light and transfer to the corresponding activator ions, and strong emission in red region is observed due to energy transfer from $MoO_4{}^{2-}$ and $WO_4{}^{2-}$ to Eu^{3+} and Sm^{3+} [29–32]. Highly crystalline and luminescent phosphates and fluorides-based compound are obtained [33, 34].

Moreover, the polyol mediated synthesis technique is extensively employed for synthesizing nanostructured chalcogenides. In chalcogenides system, more often sulfide-based compounds are discussed. Usually, CH_4N_2S, hydrogen sulfide, or soluble metal sulfides such as Na_2S are utilized as precursors for the source of sulfur. The polyol mediated synthesized metal chalcogenides have been activated in contemporary years for its explicit assets and applications in vast domain. The $CuInSe_2$ and Cu_2ZnSnS_4 are used in photovoltaic, while In_2S_3 and SnS are used in photo-detectors. Also, $ZnIn_2S_4$, Bi_2Te_3, and Sb_2S_3 are used as photocatalysts, thermo-electrics, and in lithium-ion batteries, respectively [7].

12.2.4 Core@Shell Nanomaterials

In polyol mediated synthesis of core-shell nanoparticles such as ferrimagnetic $CoFe_2O_4$ as a core and antiferromagnetic CoO as shell with high degree of crystallinity, giant exchange bias effect is observed. This exchange bias coupling is impressively employed in memory and switching devices [35]. Two-step polyol mediated synthesis under N_2 gas bubbling is reported for core@shell Cu@Ag NPs. Since oxidation of copper takes place in the existence of oxygen in the solution, hence N_2 gas is passed to suppress the oxidation of copper via core@shell formation. Reduction of Ag^+ is faster as compared to copper oxidation and its reduction to Ag^0 leads suppression of nucleation and growth of Ag NPs [36]. Nguyen et al. have reported the star shaped $Fe_{3-x}O_4$–Au core@shell structured nanomaterials using polyol synthesis. Fe_3O_4 core comprised with gold shell has been prepared via polyol method in which hydroquinone is used as reducing agent. The prepared core@shell materials are utilized for the chemical sensors [37]. The bi-functional behavior of highly luminescent core@shell $CaMoO_4$:Eu@$CaMoO_4$ and its hybrid formation with Fe_3O_4 for its hyperthermia and bio-imaging applications has been reported [38]. Highly luminescent β-$NaY_{0.8}Eu_{0.2}F_4$@γ-Fe_2O_3 core-satellites NPs are prepared for its dual mode magnetic resonance as well as bio-imaging applications [39].

12.2.5 Carbon Dots

Recently, carbon dots have attracted a noticeable attention owing to their superior high quantum yield, less toxicity, and high chemical stability. These properties may be very useful in bio-imaging and optoelectronics application. Carbon dots may be utilized in bio-imaging. Also, narrow emission band of red color with blue and green component covers its utility in full color display. Feldman and his co-workers reported the preparation of ultra-small carbon dots ~3–5 nm via straightforward and effective polyol route. The presence of $MgCl_2 \cdot 6H_2O$ salts promotes the nucleation of the carbon dots via thermal decomposition of glycerol, diethylene glycol, and PEG 400. Further modification of prepared carbon dots with $TbCl_3$/$EuCl_3$, an efficient Förster resonance energy transfer takes place via the carbon dots to the lanthanide ions resulting the sharp characteristics emission of terbium ions (green) and europium ions (red) with a very high quantum yields ~85%. The synthesis strategy and full color display for as-prepared carbon dots and modified with Mg^{2+}, Tb^{3+}, and Eu^{3+} have been presented in Fig. 12.2. The highly efficient multicolored C-dots prepared via polyol route can be very noteworthy for molecular imaging and optoelectronics [40].

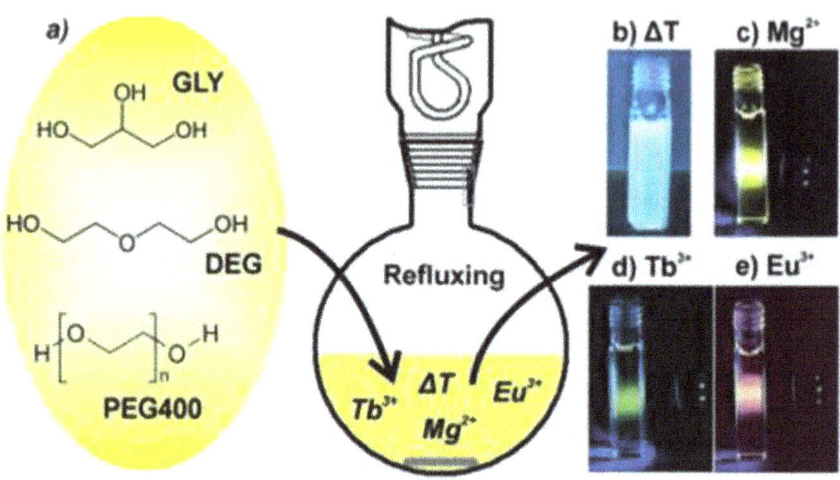

Fig. 12.2 Polyol-assisted synthesis of C-dots: **a** Protocol used for the synthesis; **b** T/PEG (λexc = 366 nm) exhibits blue color emission; **c** Mg^{2+}/PEG; gives yellow color emission; **d** Tb^{3+}/PEG; provides green color emission and **e** Eu^{3+}/PEG; reveals red color emission under blue-LED excitation λ_{max} = 465 nm). Reproduced with permission from RSC publisher [40]

12.3 Microwave Synthesis (MW) Route

Microwave (MW) synthesis approach is an emerging green chemistry route under which materials can be prepared to an atomic level. MW route offers to prepare the multifunctional nanomaterials which establish their profound applications in area of energy production, nano-biomedicine, nano-electronics, etc. In this approach, the choice of green solvents as well as energy efficiency is major parameters to control the NPs growth. Also, MW heating is reflected as a more competent mode to regulate the heating during reaction process since it requires less energy consumption as compared to traditional routes. Furthermore, use of ionic liquids with solvent free as well as nontoxic precursors approach promotes the lucrative green synthesis of NPs. The synthesis process for the nanomaterials has been dependent on innumerable factors, namely temperature, pressure, solvents, synthesis time, etc. Shape and sizes of particles are precisely controlled by MW irradiation [41]. The microwave radiation refers to a part of an electromagnetic spectrum which has radiation frequencies vary between 300 MHz and 300 GHz. The MW approach accelerates the reactions because polar solvent molecules absorb MW energy and thus produce tremendous extent of thermal energy via disturbing the alignment of the molecules with respect to the external field.

In the MW-assisted synthesis, heat up of the reaction mixture is considerably quicker at ambient pressure as compared to convection-based orthodox heating process. The chemical reactions, which are not feasible under the conventional heating, can be executed under like conditions with the MW heating. The experimental findings could not be well explained by the consequence of MW heating only, promoting the clarification of the continuation of the non-thermal MW properties [42].

12.3.1 Principle Behind MW Heating

The MW heating includes two-fold key mechanisms, namely dipolar polarization as well as ionic conduction. Generally, MW can produce heat to material with polar molecules and through ionic conduction. During the MW heating, polar solvents like water molecules attempt to orient with the fast changing alternating electric field which leads to heat generation by the process of rotation, friction, and collision. This type of heat generation leads to dipolar polarization. Moreover, ionic movement into the solution will be dependent on the alignment of the electric field. The constant movement of ions into solution with changing directions leads to rise in local temperature via friction and collision (Fig. 12.3b) [43]. In case of conducting and semiconducting, nano-material heat generation takes place by formation of electric current with involvement of electrons and ions. Energy dissipation in these systems is mainly because of resistance of the material, and this process is governed via ionic conduction namely.

12.3.2 Conventional Versus MW Heating Process

Electric furnace and/or oil bath are frequently used in the traditional heating. Firstly, reactor chambers are heated followed by subsequent heating of reactants via heat convection and/or conduction mechanism. Reactor plays a mediator role, and it transfers the thermal energy via outside heat resource to the solvent and finally on the way to reactants. Sample core proceeds much higher time duration to reach the set temperature value which causes inefficient and inhomogeneous reactions (Fig. 12.3a). It may thus not much effective in mass production. Moreover, in MW heating only target materials heating takes place and homogenous heating can be done. In this type of heating, entire heating of furnace and oil bath does not takes place, and thus, it saves time and energy [41].

12.3.3 MW Effect on Rate of Reaction

Specific reactant energy (E_R) is required to initiate any molecular reaction between A and B. For the proceedings of the reaction, the reactants must attain the activation energy to acquire transition state (TS). Thus, activation energy is indicated by E_{TS}-E_R, and this much energy is engaged by the reactants from the reaction surroundings. Finally, formation of products with lower energy E_P is obtained. Use of MW irradiation does not modify the activation energy. It offers an excessive impetus to complete the reaction more rapidly as compared to conventional heating. Since the MW transfers the energy to the molecule in10^{-9}s while the molecular relaxation takes time in a period of 10^{-5}s, rapid MW energy transfer as compared to

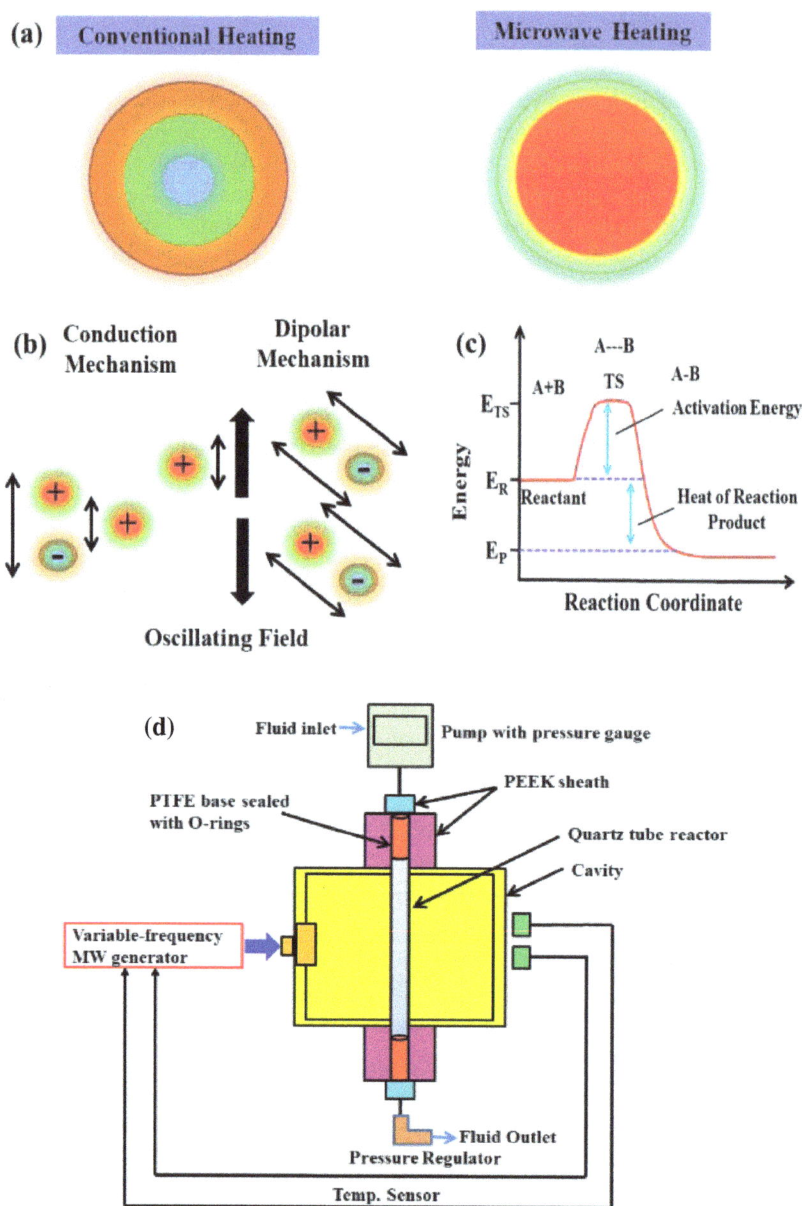

Fig. 12.3 **a** Conventional and MW heating, **b** dipolar and ionic conduction mechanism, **c** effect of MW on reaction medium, and **d** MW reactor for synthesis of nanomaterials. This is redrawn from references [41–44]

molecular relaxation leads to higher instantaneous temperature within the reaction medium which finally speeds up the reaction rate (Fig. 12.3c) [42].

The un-interrupted flow-type reactor with appropriate pressure allows uniform heating of the reaction. This reactor is required during this synthesis for its fast and mass production synthesis of nanomaterials. The constant flow from MW reactor is outfitted to a high-pressure connection. A uniform electromagnetic field alongside a tube-shaped reactor is produced at the midpoint of a cylindrical MW cavity (Fig. 12.3d). The applied MW reactor system comprised of a MW generator has operating frequency and power 2.5 GHz $\pm$ 200 MHz and 100 W. The reactor system automatically records the change in resonance frequency shift. A radiation thermometer is employed for the temperature measurement of the reaction solution via slit of cavity. The pressure to the reaction mixture is maintained via a pump and guided toward the quartz reactor tube. The pressure regulator is applied to vary the pressure required. The quartz reactor tube is linked with the sheath that provides the necessary pressure to sustain $\sim$10 MPa and finally closed through a Kalrez O-ring [44].

The role of solvent during synthesis shows a vital part in the synthesis of nanomaterials for the green approach synthesis. Generally, MW heating ability infers the capability of a solvent to change the MW energy into thermal energy at a specified MW temperature and frequency. The heating ability has been generally governed by dielectric loss tangent($\tan \delta = \varepsilon''/\varepsilon'$) where ε' and ε'' are dielectric constants (see in Table 12.3) [45].

The loss tangent mainly relies on the temperature and MW frequency. The penetration depth profile is strongly dependent on temperature and frequency. The

Table 12.3 Solvents and their corresponding loss tangent at 20 °C

Solvents	tan δ
Ethylene glycol	1.349
Ethanol	0.939
2-propanol	0.800
Methanol	0.660
1,2-dichlorobenzene	0.279
N-methyl-2-pyrrolidone	0.280
1-butyl-3-methylimidazolium hexafluorophosphate	0.190
Acetic acid	0.172
N,N-dimethyl-formamide	0.159
1,2-dichloroethane	0.130
Water	0.119
Chlorobenzene	0.990
Acetone	0.053
Tetrahydrofuran	0.046
Dichloromethane	0.042
Toluene	0.04
Hexane	0.02

penetration depth states to the point at which nanomaterials preserves 37% of the primary irradiation source power. The penetration depth decreases as loss tangent increases. The solvents with low loss tangent show high penetration depth, while high loss tangent possess small penetration depth. The MW penetration depth profile through a solvent is given as

$$D_p = \left(\frac{\lambda_o}{2\pi}\right)\left(\frac{\sqrt{\varepsilon'}}{\varepsilon''}\right)$$

λ_o is the wavelength of the MW radiation, and its value is observed to be 0.122 m at 2.45 GHz. H_2O is frequently utilized solvent for the preparation of nanomaterials via MW. The penetration depth values at 22 °C for water solvent at 2.45 GHz are 1.8, cm while its value is 0.34 cm at 5.79 GHz MW treatment [41]. Tangent loss for the different solvents used in the MW-assisted route at 2.45 GHz is mentioned in Table 12.3 [46]. During the MW supported inorganic nanostructure materials, H_2O (tan $\delta = 0.119$) and alcohols are widely utilized for their good MW heating. Polyol such as EG has loss tangent tan $\delta = 1.349$ and high boiling point (~ 197.9 °C) with strong reducing capability, allow fairly elevated temperatures for the synthesis of inorganic nanostructured materials. The dielectric behavior of solvents generally alters considerably with respect to temperature. Solvent such as ethanol exhibits decent MW absorbing property with loss tangent 0.94 at room temperature. Moreover, loss tangent decreases from 0.27 to 0.08 as temperature rises from ~ 100 to 200 °C. Heating involved during MW synthesis using ethanol solvent is mainly due to the dipolar polarization phenomena. The MW absorbing property decreases as temperature increases. The rise in temperature decreases the viscosity and frictions among the molecules [47].

In case of ionic liquids such as 1-butyl-3-methylimidazolium hexafluorophosphate, heating takes mainly due to the ionic conduction phenomena. The MW absorbing property increases with rise in temperature in case of the ionic liquids [8].

12.3.4 Synthesis of Metal NPs

The metallic nanostructures with different morphologies, namely sphere, nanosheets, nanorods, nanowires, nanotubes, and dendrites have been quickly synthesized via MW heating. The morphology as well as dimension of metallic NPs might be precisely organized by varying the reaction mixture condition such as precursors concentration, proper selection of the solvent, surfactant, and reaction temperature [48]. Green MW approach is employed for the synthesis of noble metals, namely Au, Ag, Pd, and Pt, nano-particles using red wine and/or grape pomace extract which play the role of green solvent source, reducing as well as stabilizing reagent [49]. Reducing agent free MW-assisted Au NPs is prepared using $HAuCl_4$ in aqueous phase. Synthesized gold NPs show aggregation and uncontrolled growth

[50]. MW-assisted hydrothermal approach is applied for the folic acid targeted Au NPs using $HAuCl_4$ and NaOH. The folic acid targeted Au nanoparticles is used for detection of Hela Cells since tumor cells over express as the folate receptors to the cancer cells [51]. Rapid MW supported approach is reported for the synthesis of monodispersed Ag NPs. During the synthesis, basic amino acid like L-lysine or L-arginine is used as a reducing reagent while starch is used as a shielding agent. In a typical synthesis process, 0.4 mmol of soluble starch and 0.16 mmol of L-Lysine or L-arginine are added to 4.0 mL of deionized water followed by addition of 20 mmol aqueous solution of $AgNO_3$ and stirred. The temperature of reaction mixture is kept at 150 °C under MW irradiation. Monodispersed Ag NPs are obtained in a very short period of time $\sim$10 s [52].

Colloidal Pt NPs are prepared by an aqueous solution comprising H_2PtCl_6 and 3-thiophenemalonic acid under MW irradiation (power $\sim$300 W) for 8 min. duration. NPs size has been organized through adjusting the molar ratio of the reaction precursors [53]. Pt nanoclusters with a porous interconnected nanostructure are attained via MW heating to an aqueous solution containing K_2PtCl_4 and 2-[4-(2-hydroxyethyl)-1-piperazinyl]-ethane sulfonic acid for 12 s only. The Pd NPs have been prepared under MW heating in quick time 20 s only, using $PdCl_2$, glucose, and PEG as a capping agent in aqueous solution [54]. MW-assisted single crystal Cu nanowires at 120 °C for 2 h is prepared by using $CuCl_2$, hexadecylamine, ascorbic acid into water. Ascorbic acid as well as hexadecylamine play significant role in controlling morphology and aspect ratio of Cu nanowires [55]. MW approach is further extended for the preparation of luminescent silicon QDs. During synthesis of Si nano-wires, glutaric acid is used at 185 °C for 15 min. Highly luminescent, dispersible and photo as well as pH stable Si quantum dots show their potentiality for the relevance in cellular imaging [56].

12.3.5 Metal Oxides

Substantial work has been dedicated to MW-assisted synthesis of numerous metal oxides owing to its remarkable properties, extraordinary constancy as well as extensive applications in various arenas. MW-assisted syntheses of metal oxide nanomaterials involve a water-soluble metal salt as the metal source, an alkaline component and a surfactant to govern the morphology and dimension of the nanomaterials. MW-assisted synthesized metal oxides such as ZnO, SnO, SnO_2, TiO_2, Fe_3O_4, Co_3O_4, CuO, MnO_2, ZrO_2, WO_3, MoO_3, CeO_2, Nd_2O_3, and Y_2O_3 are reported [57–70]. Also metal tungstates such as AWO_4 (A = Ca, Sr, Ba, Fe, Co, Ni, Mn, Zn, Ag/In), gallate $ZnGa_2O_4$, spinel metal ferrites AFe_2O_4 (A = Mg, Zn, Ni, Mn, Co), metal aluminates MAl_2O_4(M = Zn, Co), perovskites $BiFeO_3$, $ATiO_3$(M = Ba, Sr, Pb), metal molybdates $AMoO_4$ (M = Ba, Ca), and metal vanadates MVO_4 (M = Bi, Ce, Y, La) are also explored [41].

Fe_3O_4 is a significant magnetic functional material and has been found its profound applications in diverse areas of research such as contrast agent in MRI, hyperthermia, gene separation, drug delivery, etc. [71, 72] Under MW irradiation, Fe_3O_4 NPs have been generally prepared by means of Fe(III) salt and/or salts of Fe (III) and Fe(II) dissolved in the water and a reducing agent present in an aqueous phase [73]. Also, surfactant free Fe_3O_4 NPs are prepared using an oxidized iron foil such as Fe_2O_3 into deionized water. During synthesis, no additive like alkali, acid, or surfactant is used. Homemade MW oven at 700 W fixed operating powers for 30 min is applied during synthesis. The as-synthesized Fe_3O_4NPs exhibited magnetization of Ms ~ 51.29 emu g^{-1} [74]. The MW-assisted approach for preparation of Fe_3O_4NPs is reported by using an aqueous solution containing $FeCl_3$, $FeSO_4$, and ammonia [73]. Also, iron oxide/oxyhydroxide NPs have been synthesized under MW irradiation. The well-ordered growth as well as assembly of NPs are commonly observed due to gentle reaction of the reactants (iron salt and sodium hydroxide) [75].

12.3.6 Metal Chalocogenides

Functional metal sulfides nanomaterials have shown their significant uses in optoelectronics and nanomedicines as a carrier drug, bio-imaging, etc. [76, 77]. MW approach assists the fast production of metal sulfide nanostructures and has triggered a great deal of interest owing to its small processing time. Generally, water-soluble metal precursors are used as sulfur source during MW-assisted sulfide synthesis in aqueous phase. Surfactant is used to control the precise size and morphology of particles. Nanostructured metal sulfides are prepared with different shapes and sizes of particles via MW-assisted approach in aqueous solution such as PbS, CuS, CdS, ZnS, Ag_2S, Bi_2S_3, HgS, $AgInS_2$, $AgIn_5S_8$, $CuInS_2$, and $CdIn_2S_4$ [78–88]. The most common sulfur sources used during the metal sulfide synthesis in aqueous solution comprised of Na_2S, CS_2, $Na_2S_2O_3$, NH_2CSNH_2, $NH_2NHCSNH_2$, CH_3CSNH_2, and 3-mercaptopropanoic acid [41]. The synthesis of metal telluride nanomaterials is still tricky related to its metal selenides counterparts. CdTe is an important class of semiconductor material which has wide application in solar cell and in infrared optical window. The p-n junction solar cell is formed via sandwiching CdTe with CdS. The conventional aqueous solution way to prepare the CdTe nanocrystals generally takes prolong time duration. During MW synthesis, NaHTe is normally used as the source of tellurium. High-quality nanocrystal of CdTe is reported via MW-assisted synthesis. An aqueous solutions consisting of $CdCl_2$, NaHTe, 3-mercaptopropionic acid, and NaOH are used. During synthesis, high yield of the product is simply attained by tuning the reaction time and temperature. This approach permits the fast synthesis of CdTe nanocrystals with wide spectral range covering from green to NIR at lower temperature ranges [89]. MW-assisted metal selenides nanostructured materials are less reported as compared to nanostructured metal oxides and sulfides. It is perhaps due to high cost metal precursors, a lesser amount of availability of selenide sources, and difficulties in the synthesis. The quickness of MW-assisted route is more notable for metal selenides with considerably less reaction time period up to minutes. MW synthesis of CdSe, PbSe, and Cu_2Se NPs is reported by taking metal acetates and/or sulfide precursors under refluxing conditions with Na_2SeSO_3 into aqueous solution. Crystal structure of CdSe NPs is reliant on the reaction time duration under MW irradiation. CdSe nanocrystal exhibits cubic phase for MW heating of 10 min; while for prolong duration time of 30 min of MW irradiation, hexagonal CdSe nano-crystal is obtained [90].

12.3.7 Core@Shell Structure

As for as controlled size and shape of particles, synthesis of nanomaterials and tailoring their properties at nano-scale are concerned, and core-shell strategy has been employed by simple coating of organic and/or inorganic nanomaterials. Core@shell formation of the nanomaterials decreases the surface energy of the arrangement. Outer shell coated over inner core materials exhibit significant role to advance the reactivity and oxidation ability of the interior core manufacturing. Core@shell anomaly comprises inner inorganic core covered by outer shell forms the heterogeneous system [91]. Core@shell nanostructured materials are synthesized via MW-assisted route in aqueous solution. Au@Pd and Au@Pt structures are attained through the reduction of Au(III) complexes. The reduction of chloro-complexes of Au and nano-powders of Pd and Pt takes place under hydrothermal, and MW irradiation leads to the realization of bimetallic core@shell structure [92]. Also Se@C core@shell nanostructure is obtained using starch and H_2SeO_3into deionized water under MW-assisted hydrothermal process. The prepared product comprised of Se nano-rod as the core and amorphous carbon as the shell materials [93]. Pd@Pt core@shell nanostructures in aqueous solution are synthesized using K_2PtCl_4, $PdCl_2$, and CTAB under MW irradiation. Morphology of the core@shell nanostructures is precisely organized by changing the Pt and Pd molar ratio of precursors [94]. Rapid, modest, and one step MW supported approach to formulate the gold coated with silica shell NPs(Au@SiO_2) has been reported. This approach circumvents the time wasting orthodox routes (Fig. 12.4).

The MW-assisted scheme deals with the uniform SiO_2 coating over the colloidal AuNPs, exploiting silane as the coupling reagent. Monodispersed AuNPs (particle size $\sim$16 nm) are obtained by employing citrate reduction approach. The tetra-ethoxysilane TEOS is used during the silica coating as coupling reagent under the MW irradiations. Dynamic light scattering results the excellent dispersion of the prepared NPs in aqueous medium. Additionally, surface functionalization of silica-coated AuNPs (Au@SiO_2) is perfumed via conjugation of different functional assemblies as amine (–NH_2), carboxylic (–COO) and alkyl groups [95].

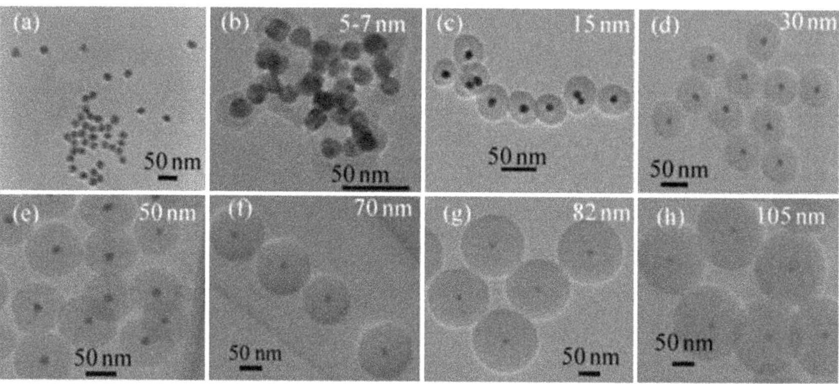

Fig. 12.4 TEM Micrographs of **a** Au NPs **b–h** Au@SiO_2NPs with various silica shell thicknesses with varying concentration of TEOS 1, 2, 3, 5, 10, 15, and 20 nm. Reproduced with permission from Elsevier publisher [95]

12.3.8 Hollow-Type Structure

In recent times, the hollow-archetype structures comprised of inorganic and/or organic composites have fascinated a lot of attention owing to their low density, enormous surface area, and induced large porosity in the sample surface. These assets of hollow structured materials could find their powerful applications in diverse research zones such as catalysis, super capacitors, sensors, drug delivery and microcapsule reactors, etc. [96–99]. The conventional heating-based engineering tactics are broadly introduced to synthesize the hollow-type designs of inorganic nano-materials. Moreover, these routes involve precious metal precursors and long duration reaction time. Thus, MW-assisted approach has been admitted for the synthesis such hollow-types design due to its remarkable highlights such as uniform volumetric heating, fast reaction rate, and energy consumption [100]. The hollow porous carbon sphere is synthesized via MW-assisted method and possesses the following such as amorphous phase of porous structure, uniform size, high pore volume, and highly dispersed, which are used as anode materials in lithium-ion batteries [101].

12.4 Hydro- and/or Solvothermal Approach

Hydro- and/or solvothermal process is a wet-chemical synthesis approach commonly implemented for preparing nanomaterials with precise particle size, shape, and composition. During the synthesis, high vapor pressure is generated by the consequence of heating of the reaction mixture in a sealed vessel above the ambient temperature and pressure. The discrete feature of hydro and/or solvothermal

Table 12.4 Characteristics of usually employed solvents during hydrothermal and/or solvothermal approaches [102]

Solvent	Formula	Critical temperature (°C)	Critical pressure (MPa)
Water	H_2O	374	22.09
Ethylenediamine	$H_2N–C_2H_4–NH_2$	319.88	62.09
Methanol	CH_3OH	239.19	8.09
Ethanol	C_2H_5OH	241.09	6.09
Toluene	C_7H_8	320.59	4.19
Ethanolamine	$HO–C_2H_4–NH_2$	398.24	8

treatment is the deployment of capping ligands to control the growth of the particles during synthesis and prevent the agglomeration of the nanomaterials in solutions. The chelating ligands associated with surface of particles can provide versatile functional groups for bio-conjugation. The limitation of the hydro- and/or solvothermal method suffers scalability of the product preparations as well as limiting operating temperature at 300 °C into a Teflon line autoclave (Table 12.4).

Nanomaterials production through hydrothermal/solvothermal approach in aqueous solution involves crystal nucleation followed by growth. The nanomaterials can be prepared into desired shape, size, and morphology of particles by tuning the reaction parameters such as temperature, pH, precursors concentrations, and surfactants as additives. The mechanism governing the controlled shape, size, and morphology of particles via varying the reaction conditions is large due to nucleation and growth rates which rely on supersaturation. The term supersaturation is distinct as the proportion of the real concentration to the saturated precursor concentration into the solution [103, 104].

When the solute precursor solubility exceeds its maximum value into the solution, resulting solution becomes supersaturated thus promoting the nucleation. The synthesis process is completely irreversible one in which precipitation of the solute leads to the nucleation of macroscopic size crystals [105]. After nucleation, the sequential growth of crystals takes place. The growth of crystals involves the integration of growth units. The growth units correspond to the similar crystal entities exhibiting the identical and/or dissimilar structure from the precursor solution resulting in an increase in sizes. The schematic representation of the involved mechanisms of crystal growth via hydrothermal/solvothermal methods has been shown in Fig. 12.5. Since plentiful availability, less toxicity, and an elevated dielectric constant are required, and water has been frequently utilized solvent in the hydrothermal/solvothermal routes. In this technique, the values of critical temperature and pressure of water are ~374 °C and ~22 MPa. The variation in temperature and pressure infers the changes in the property of water. This change is even more profound above its critical point value. At room temperature, the dielectric constant value for water is 78. The high value of dielectric constant promotes the dissolution of polar salts into it. In the critical zone, the dielectric constant of water decreases (~10) via increase in temperature and decrease in

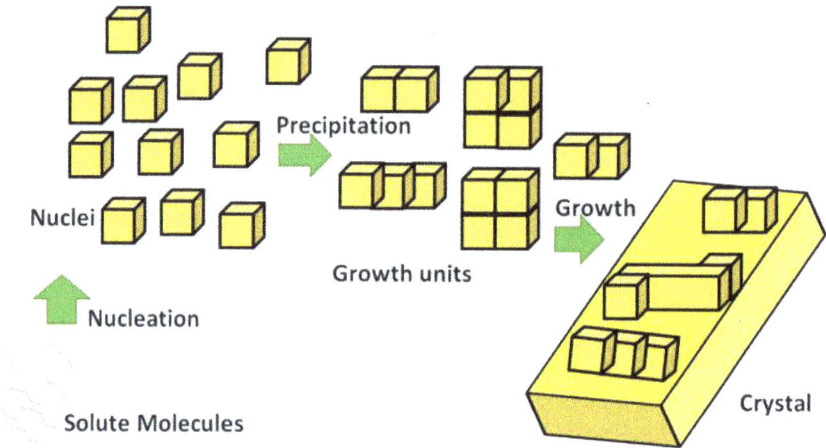

Fig. 12.5 Schematic representation of crystal growth mechanism for the hydro- and/or solvothermal processes. This is redrawn from references [103–106]

pressure. The dielectric constant value drops considerably in the range ~2 to 10 above its critical point [106]. The striking change in the dielectric constant results in an extremely reduced solubility of solute resulting into supersaturation in the solution and thus nucleation and growth of crystal takes place. The low dielectric constant of water allows the dissolution of organic compounds in the supercritical water. Similar tendency has been also observed for solvothermal systems. Non-aqueous phase organic solvents have been extensively used in the solvothermal synthesis. The organic solvents such as methanol, 1, 4-butanediol, toluene, and amines are usually used in the solvothermal route. The solvothermal synthesis can be processed at comparatively lower temperature and pressure as compared to the hydrothermal approach. Moreover, sensitive precursors to water can be easily tackled into solvothermal route. However, morphology and crystal phase of the prepared compound can be easily tailored through this route [107].

12.4.1 Synthesis of Nanomaterials via Hydrothermal and/or Solvothermal Approaches

Nanomaterial with manageable morphology, size, crystallinity of particles, and easy way of surface functionalization has attracted widespread research courtesy due to their unique optoelectronic, magnetic, and thermo-mechanical properties. Nanomaterials possess high surface to volume ratio in particles, and quantum confinement effect in semiconductor mainly governs the properties of the nanomaterials. In quantum confinement effect, the particle sizes are squeezed below to its Bohr radius. In order to synthesize the nanomaterials, the hydrothermal and/or solvothermal synthetic routes are reflected as the most favorable tactics. In these techniques, controlled shape and size of particles, and exceedingly crystalline with high yield of nanomaterials at low cost is produced. Furthermore, these routes can be interconnected with MW to get the high quality nanocrystals with enhanced reproducibility.

12.4.2 Metal Oxides NPs

Metal oxide NPs are significantly useful materials due to their exceptional properties. These metal oxide NPs are typically utilized in a variety of research fields such as in catalysis, ceramic, optoelectronics, and so forth. In solvothermal technique, smaller size metal oxide NPs are prepared at comparatively lower temperature as compared to its hydrothermal approach [108]. Various metals oxides NPs such as Al_2O_3, CuO, γ-Fe_2O_3, $CoFe_2O_4$, CeO_2, NiO, ZrO_2, TiO_2, $BaTiO_3$, and $SrTiO_3$ are synthesized via hydro- and/or solvothermal approaches. Furthermore, the hydrothermal and solvothermal approaches are employed for the synthesis of $ZnGa_2O_4$, $BaZrO_3$, and $LiNbO_3$ [107, 109]. Persistent luminescence properties of Cr doped $ZnGa_2O_4$ have been reported for the hydrothermally synthesized sub 10 nm particles [110]. TiO_2 is widely studied metal oxide NPs for its photocatalytic properties. The oxide NPs prepared via these routes exhibit high crystallinity, precisely controlled particle shape, size which determines their advanced applications. Semiconductor metal chalcogenides nanomaterials are widely used in optoelectronic devices and in photovoltaic. The chalcogenides nanometrials comprised of ZnS, ZnSe, ZnTe, CdS, CdSe, CdTe, CuInSe, $Cu_{2-x}Se$, $AgInS_2$, and $AgGaS_2$ are reported via hydrothermal approach [107]. Ultra-small ($\sim$4 nm) hexagonal ZnS nanosphere has been synthesized using tetra-pyridine–di-thio-cyanato-zinc precursor into ethylene glycol via solvothermal route. During the synthesis, the increase in reaction temperature (160, 180, 200 °C) fetch bigger size ZnS nanospheres (200, 350, and 450 nm) [111]. Solvothermal synthesis of monodispersed Mn-doped ZnS nanospheres is reported. During the synthesis, $ZnCl_2$, $MnCl_2$, and sulfur powders are used as the precursors in oleic acid [112]. The hydro-solvothermal route has been further extended to synthesize ZnSe nanocrystal in form of quantum dots, nanorods, nanoplates, and in bulk form. ZnSe as well as ZnTe NPs are prepared through a solvothermal technique. During the synthesis, less toxic metal precursors are utilized as compared to the conventional chemical vapor deposition technique. The metal precursors $Zn(CH_3COO)_2$, $ZnSO_4$, and/or Zn powders, Se, Te powders and/or Na_2SeO_3, and Na_2TeO_3 are used as Zn, Se, and Te sources during synthesis. The frequent solvents used during the synthesis generally contain ammonia, triethylamine, ethylenediamine, and hydrazine as well. The pH of the reaction mixture is maintained at $\sim$10 [107, 113]. Highly ordered CdS sphere is synthesized via solvothermal route using $Cd(NO_3)_2$ as metal precursors, thiourea as S source, while polyvinylpyrrolidone (PVP) functions as a capping agent. Surfactant PVP provides the needed nucleation sites during nanocrystal growth. Higher concentration of PVP entails more accessible nucleation sites promotes the NPs growth and consequently leads to small size NPs. This method has been also extended for the preparation of other sulfides like HgS, Ag_2S, and Bi_2S_3. During the hydrothermal and/or solvothermal synthesis of CdS NPs, thioglycolic acid and dithioglycol have been used as S sources [107, 114]. The compounds based on nitrides, arsenides, and phosphide having direct band-gap semiconductors are extensively applied in light emitting diodes, lasers, photo-detectors, optical amplifiers, etc. In light emitting diode (LED) application, commercially available GaN chip is used for illumination purpose [115]. Xie and his co-workers first reported the solvothermal synthesis of GaN compound. Polycrystalline GaN has been synthesized via solvothermal route using $GaCl_3$ and Li_3N precursors in benzene at 280 °C for 6–12 h. The absence of quantum confinement in GaN is attributed to its large size ($\sim$32 nm) as compared to its Bohr exciton radius ($\sim$11 nm). The innovative benzene solvent tactics has been executed at relatively lower temperature than that of traditional routes [116]. Moreover, GaN NPs have been synthesized using anhydrous ammonia as a solvent. Anhydrous ammonia has been condensed in a quartz tube containing the reactants Ga metal, GaI_2, and/or GaI_3 as Ga sources while NH_4I, NH_4Cl, or NH_4Br are added to produce the GaN. During the ammonia thermal synthesis of GaN, temperature plays a decisive role. Cubic phase amorphous GaN can be prepared when the growth temperature is $\sim$300 °C. When growth temperature is kept at 440 °C, hexagonal phase GaN is obtained [117]. Furthermore, other metal nitrides such as NbN, ZrN, HfN, and Ta_3N_5 are synthesized using $NbCl_5$, $ZrCl_4$, $HfCl_4$, and $TaCl_5$ as the metal precursor along with $LiNH_2$ as the nitriding reagent. By using benzene solvothermal reaction, highly crystalline Ta_3N_5, ZrN, HfN, and NbNNPs of changeable sizes can be synthesized [118].

12.4.3 Hydrothermal Treatment for Hollow Structures

Hollow structured nanomaterials such as hollow carbon sphere and mesoporous carbon spheres are of prime interest and widely applied in catalysis, bio-imaging, and drug delivery, lithium-ion batteries, fuel cells, sensors, etc. [119]. Low temperature hydrothermal approach is further used for the preparation of carbonaceous NPs. Highly biocompatible as well as economically viable precursors such as sugar, glucose, cyclo-dextrins, fructose, sucrose, cellulose, and starch have been used during hydrothermal synthesis of carbon NPs [120]. Generally, hydrothermally synthesized carbon NPs are spherical in nature. Reaction temperature plays a decisive factor in formation of carbon spheres. Carbon spheres of different diameter $\sim$0.2, 0.5, 0.8, 1.1, and 1.5 μm are produced using 0.5 M glucose as the carbon source and setting the reaction time and temperature of 2, 4, 6, 8, and 10 h and 160 °C, respectively. Growths of these NPs are consequence of dehydration, condensation, polymerization as well as aromatization processes. The reaction temperature range generally occurs from 150 to 350 °C, while synthesis time is generally set in between 4 and 24 h. Based on the precursors used during the synthesis, the diameters of the synthesized NPs may diverge from 100 nm to few microns. The hollow carbon NPs have been synthesized using mesoporous or nonporous silica templates. Ultimately, selective etching of carbon NPs deposited onto mesoporous and/or nonporous silica templates promotes the formation of hollow carbon spheres (shown in Fig. 12.6) [121].

Wet aptitude of the surface of templates plays an extremely vital role. When the surface of the templates is not wetted by the precursor, then nonporous carbon NPs is obtained. Likewise, this route is comprehensively used for the synthesis of composite carbonaceous NPs. The composite carbonaceous NPs can be prepared via simple mixing of the metal precursors such as Ag, Pd, Se, and/or metal oxides as Fe_3O_4 and SnO_2 with a carbon source [122]. Silicon and germanium are significant semiconductor metals mainly due to their vast applications in optoelectronic devices. In order to synthesize crystalline silicon and germanium NPs via

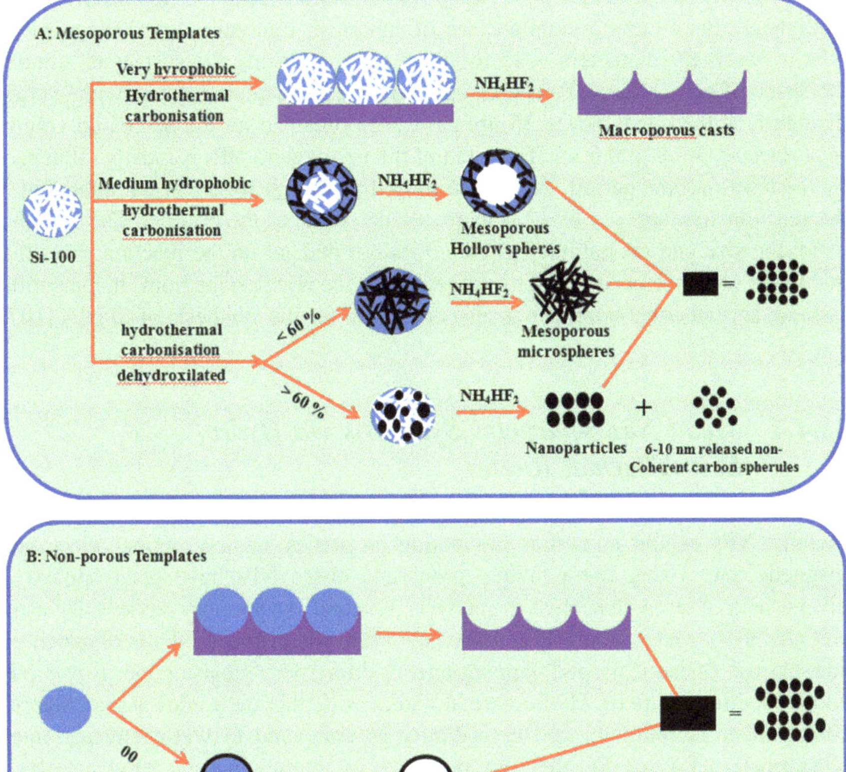

Fig. 12.6 (A) and (B) represent the schematic representation of the hydrothermal carbonization process using silica templates with different polarities, resulting in the formation of various carbon morphologies. This is redrawn from Refs. [120, 121]

hydro- and/solvothermal approach, a very high reaction temperature is kept. Solvothermal decomposition of tetra-ethyl germanium results the formation of the reduced crystallinity of germanium NPs. The decomposition can be executed at higher temperature in the presence or absence of surfactants into organic solvents and/or super-fluidic solvents such as CO_2. The surfactants involved during synthesis lead the nucleation and crystal growth via reverse micelles formation. Also surfactants permit not only the precise control over the morphology of the germanium NPs, but also stabilize the synthesized NPs [123]. Germanium nanocubes of 100 nm edge length via hexane solvothermal route have been prepared using heptaethylene glycol monododecyl as a surfactant. The same molar ratio of $GeCl_4$ and phenyl-$GeCl_3$ is employed as germanium precursors. Herein, Na metal sprinkled in toluene is used as the reducing reagent. The reaction mixture has been heated at 280 °C for 72 h in a hydrothermal reactor. Highly crystalline as-prepared germanium nanocubes with diamond cubic structure is observed. The synthesized germanium nanocubes consist of multi-slighter nanocubes that are linked to the surface-adsorbed surfactant NPs [124]. When the surfactant is replaced by pentaethylene glycol ether, a combination of spherical, triangular, and hexagonal Ge NPs is observed (diameters ~15 to 70 nm). By adjusting the surfactant amount, the shape of germanium NPs can be tailored. The germanium spheres with average diameters of the range ~6 to 35 nm have been obtained for the decreased volume of surfactant ~1.8 to 0.6 ml. The yield of the germanium NPs is mostly influenced by the reaction time period. Prolong heating time during solvothermal treatment of the reaction mixture ~4 to 12 h increases the yield of the product. On the other hand, the size and crystallinity of NPs does not depend on the reaction time. It is due to the presence of the capping reagents into the reaction medium. In the similar fashion, solvothermal approach is also employed for the synthesis of Si NPs [107].

12.4.4 Metal Nano-particles Synthesis via Hydro/ SolvoThermal Routes

Metallic NPs exhibit numerous fascinating properties, namely optical, electronic, magnetic, etc. Using these unique properties, these NPs have been applied in diverse area of research such as in catalysis, sensors, and memory devices. By using wet chemistry routes, controlled shape-size synthesis of metal NPs is obtained via suitable use of reducing and capping agents. Moreover, precise control over NPs and crystallinity is poor. Hydrothermal and/or solvothermal routes assure superior control over morphology and crystallinity as compared to wet-chemical routes. Ultra-small ~1.7 nm Pt NPs are prepared by employing an ethylene glycol solvothermal route in basic medium. Here, ethylene glycol acts as solvent as well as a reducing agent. $H_2PtCl_6{\cdot}6H_2O$ is used as Pt metal precursor during synthesis. The reaction mixture temperature is kept at 160 °C for 3 h [125]. The EG solvothermal route is further extended for the preparation of Ag NPs. In a typical synthesis of Ag

NPs, $AgNO_3$ is added into suitable amounts of toluene, EG, and dodecylthiol and kept in the reactor at 160–170 °C. The prepared Ag NPs exhibit an ordered spherical shape (average diameter ~10 nm). Thiol used during the synthesis acts as a complexing reagent, while ethylene glycol functions as reducing agent. By adjusting ethylene glycol to thiol ratio ~3 to 1.5, the morphology of the Ag NPs shows remarkable change in the morphology. Ag NPs show spherical to rectangular shape morphology with a somewhat decreased diameter ~6 to 10 nm [126]. DMF solvothermal approach has been employed to grow Ag and Au NPs. In addition to EG and DMF, other reducing agents such as $NaBH_4$, N_2H_4, NH_2OH, and ethanol have also been used to prepare metal NPs [127]. DMF solvothermal method has been utilized for the synthesis of Pt-Ni alloy system. Platinum (II) 2,4-pentanedionate and nickel (II) 2,4-pentanedionate are used as precursors for Pt and Ni metals, respectively. During the synthesis, 30 mM $Pt(acac)_2$ and 10 mM $Ni(acac)_2$ concentrations have been dissolved in DMF. The reaction mixture has been placed in an electrical furnace at 200 °C for one day. Temperature below 200 °C, decomposition of $Ni(acac)_2$ has not been completed, as-prepared platinum-nickel alloy NPs show high crystallinity. The average diameter of alloy NPs is ~10–13 nm. By tuning the metal precursors ratio, different sizes NPs are obtained for the bimetallic Pt–Ni alloy. The similar approach can be utilized for the synthesis of Pt–Co and Pt–Fe bimetallic alloys NPs [107].

12.4.5 Metal Organic Framework (MOF) NPs

MOFs are promising functional nanomaterials which attracted a great deal of attention owing to their potential applications in hydrogen storage, catalysis, bio-imaging, drug delivery, sensors, etc., gas separation and proton exchange membranes, etc. MOFs consist of a chain of hybrid nonporous crystalline nanomaterials comprised of metals and organic linkers with pore sizes of ~0.4 to 6 nm [128–131]. MOFs are usually prepared through hydrothermal and/or solvothermal routes. Gentle temperature and pressure are used during MOF NP synthesis to keep away from decomposition of organic ligands. MOF NPs for biomedical use are highly demanding. The surfactant free MOFs NPs growth is performed via an easy solvothermal route. Solvothermal synthesis of Fe(III) MOF NCs with an organic framework Fe3-(μ3-O)$Cl(H_2O)_2(BDC)_3$ has been prepared via taking equal-molar $FeCl_3$ and BDC in DMF under MW heating at 150 °C [132]. The MW heating promotes smaller NPs. The NPs thus obtained are highly crystalline octahedral with an average diameter ~200 nm. Furthermore, MOF NPs are targeted and functionalized with dye molecules for cellular imaging. Also, these MOFs can be used as anticancer pro-drugs for tumor therapies. Though these functionalized NP_S are not stable in a physiological atmosphere, post-synthesis coating of the MOF NPs with silica layer resolves the stability issue. Surfactant-assisted hydrothermal synthesis has been employed under the presence and absence of microwave heating to synthesize the paramagnetic Gd(III) MOF NPs. As-synthesized MOF NPs show high crystallinity. When the reaction is carried out at room temperature, only amorphous GdMOF NPs can be obtained because the nucleation rate is too rapid and dominates the NP growth. The MOF NPs obtained under microwave heating have a formula of $[Gd_2(bhc)(H_2O)_6]$ MOF NPs. The synthesized Gd(III) MOFs can be applied to magnetic resonance imaging [133]. Similar hydrothermal approach has been utilized to synthesize Tb (III), Eu (III), and Mn(II) MOF NPs. MW heating, temperature of reaction mixture, and ratio of water to surfactant ratio impart the noteworthy effects over morphology of the synthesized MOFNPs. Lower synthesis temperature and a higher ratio of water to surfactant lead to the development of nanorods, while greater temperature and MW heating promotes the formation of shorter nanorods or octahedral NPs [107].

12.5 Sonochemical Synthesis

Nanostructured materials open a new realm of diverse applications such as optoelectronics, catalysis, sensing, water splitting, and in medical diagnostics. Nanomaterials exhibit distinct properties as compared to its bulk. The characteristics of the nanostructured materials are heavily dependent on synthesis process by which they have been prepared. Therefore, it induces an interest among the materials scientists to develop the easily adaptable synthesis methodologies to prepare the varieties of nanostructured materials. Among the various synthesis route available such as photo and wet chemistry, hydrothermal and flame pyrolysis for the materials synthesis, ultrasound-based nanomaterials gained a lot of attention for its multi-dimensional applications [134]. Ultrasonic wave serves as an efficient energy source for synthesis of organo-metallic and inorganic materials which found their profound applications in organo-metallic chemistry and industrial manufacturing [135]. This synthesis approach provides high temperature as well as high-pressure synthesis conditions. It covers a vast range of reaction condition which is not feasible with other pre-existing techniques. Sonochemistry involves the utilization of ultrasound energy with frequency (ν) range $\sim$15 kHz to 10 MHz. This approach provides unique pathways to prepare the compounds in a simple and easily accessible at high temperature and pressure. Ultrasonic waves predominantly cause cavitation in an aqueous medium followed by formation, evolution, and breakdown of micro bubbles. An ultrasonic energy initiates the formation and growth of bubbles to a solution containing chemical precursors followed by diffusion of the precursor vapor. After reaching the bubble size to its critical limit, the bubble collapses and generates shockwaves.

At specific conditions, overgrowth of a micro bubble can take place and later breakdown which finally liberating the energy deposited in the bubble at a very short time duration. The chemical reaction process takes place under fast heating and cooling rate $>10^{10}$ K s^{-1}. The cavitational collapse leads to a very high temperature (5000 K) and pressure (1000 bar). This collapse process is almost adiabatic in its preceding stages which are mainly accountable for the extreme environments representative of sonochemistry [136]. Remarkably, such unusual environments are not resulting straightway from ultrasound since acoustic wavelengths have far longer value as compared to the molecular sizes. Therefore, there is an absence of any straight molecular-level interface between ultrasound and the elementals pieces. The phenomenon of acoustic cavitation occurs via development, evolution, and implosive breakdown of bubbles in liquid phase. The cavitation process under highly intense ultrasound is accountable for the chemical effects of ultrasound [137]. Under ultrasound irradiation, sinusoidal expansion and compression of acoustic waves form the cavities and allow the bubble oscillation. The oscillating bubbles collect the ultrasonic energy and finally grow to a critical size. Under these thrilling environments, emission of light occurs. The emission of light under sonochemical irradiation is termed as sono-luminescence [138]. The reaction involved in ultrasonic system comprised of a power source, a piezoelectric transducer with electrodes, a Ti horn bearing stainless steel collar, reaction mixture container, and a bath chamber. Highly intense ultrasonic titanium horn with piezoelectric transducer is preferred for the lab-scale sonochemical reactions (Fig. 12.7). Ultrasonic horn provides 10–100 W acoustic powers during sonication to the liquid solution. The cavitation procedure happens over a wide spread range of frequencies (10 Hz to 10 MHz). The common alternate to ultrasonic horn in the synthesis set up into laboratory, and ultrasonic cleaning baths are used. The power density of ultrasonic cleaning baths is governed by a small input power as compared to power produced by an ultrasonic horn. Cleaning baths are frequently trivial for many sonochemical reactions. Moreover, it can be helpful for the study of the substantial effects of ultrasound on very reactive metals such as lithium or magnesium producing emulsions which promotes the suspensions of solids and exfoliating layer structured nanomaterials.

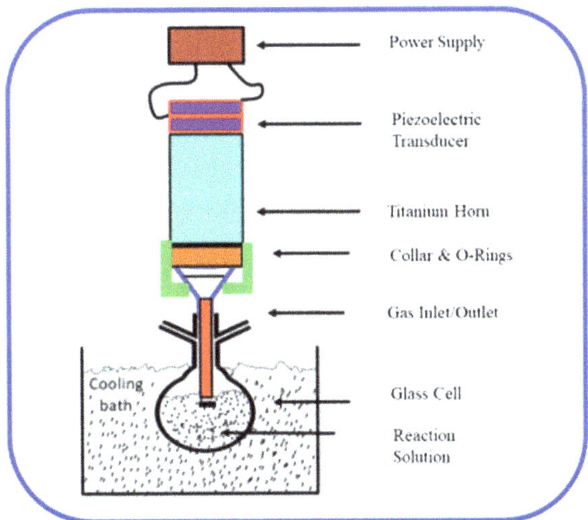

Fig. 12.7 Typical high intensity ultrasonic set up for nanostructured synthesis. This is redrawn from references [136–138]

Metal, metal oxide, sulfide, and alloy NPs are prepared by sonochemical synthesis approach using aqueous and non-aqueous solutions. The structural, opto-electronic, magnetic, and electrochemical assets of metallic NPs have the emphasis of contemporary research accomplishments. Under sonochemical route synthesis, particle undergoes to exhibit different morphologies such as spherical, rod, wires, hollow, layered type structure 2D materials, etc. [139].

12.5.1 Metal NPs Synthesis via SonoChemical Route

In literature, sonochemical synthesis of nanostructured metallic noble NPs as Au, Ag, Pt, and Pd is explored. The sonochemical reduction of noble metal salts has advantages above other existing old-fashioned reduction methods such as sodium boro-hydride, hydrogen, and alcohol, etc. In this synthesis, involvement of reducing agent is not required. Hence, the reaction rates during the synthesis are rationally fast which finally leads to the formation of metal NPs [140–142]. The water molecules under sonolysis generate $H^{\bullet}$ radicals which function as reductants. Moreover, 2-propanol as organic additive is incorporated to generate a secondary radical species. The secondary radical species can considerably endorse the reduction rate. The detailed process of metal preparation via sonochemical route has been demonstrated as [140]:

$$H_2O \rightarrow H\cdot + OH\cdot \tag{12.1}$$

$$H\cdot + H\cdot \rightarrow H_2 \tag{12.2}$$

$$H\cdot + OH\cdot \rightarrow H_2O \tag{12.3}$$

$$RH + OH\cdot\,(\text{or } H\cdot) \rightarrow R\cdot + H_2O \text{ or } (H_2) \tag{12.4}$$

$$OH\cdot + OH\cdot \rightarrow H_2O_2 \tag{12.5}$$

$$\text{Au (III), Ag(I), Pt (III), Pd (III)} + H\cdot \text{ or } R\cdot \rightarrow \text{Au(0), Ag(0), Pt(0), Pd(0)} \tag{12.6}$$

$$nM(0) \rightarrow M_n(M = \text{Noble metal}) \tag{12.7}$$

The sonochemical reduction of noble metals involved tedious reduction process. The consequence of several factors, namely time, concentration, and ultrasound frequency, with diverse organic additives to control shape, and size of particles has been reported. It is reported that particle size obtained during sonolysis is inversely reliant on alcohol concentration as well as alkyl chain stretch. It has been thoroughly linked to the circumstance that alcohols adsorbed on the surface of metallic NPs stabilizes and furthermore confine the nucleation rate [140]. Highly monodispersed gold nanodecahedra under ultrasonic irradiation with elevated yield and markedly enlarged reproducibility is reported. Gold nanodecahedra is obtained via sonochemical reduction of $HAuCl_4$ over already synthesized gold seeds into DMF solution. Unexpectedly, in the absence of ultrasound, inferior yield of gold nanodecahedra by elevated polydispersity is observed via thermal reduction [143, 144].

An analogous synthesis method has utilized for the preparation of silver nanoplates. The ultrasound-assisted Ostwald ripening procedure promotes the development of silver nanoplates via silver NPs produced at an initial stage of reaction [145]. The synthesis of Pd and Pt NPs via reduction of K_2PdCl_4 and H_2PtCl_6 precursors under sonochemical approach has been reported. The consequence of inert atmosphere on the particle size of Pd and Pt NPs has been reported. Under the Ar atmosphere, particle size of Pd is observed to be 3.5 nm, while its size is 2 nm in presence of N_2 atmosphere. Comparatively, narrower sizes of Pt NPs are obtained under Xe atmosphere. This was mainly due to the acoustic cavitation leading to formation of hot spot temperature [146]. Also, highly stable Pd NPs are prepared via ultrasonic irradiation (frequency 50 kHz for three hours) to $Pd(NO_3)_2$ solution into ethylene glycol and PVP. Ethylene glycol reduces Pd(II) to metal Pd while co-ordination of Pd atom to the carbonyl functional group of PVP promotes the steadiness of Pd NPs [147].

Fe_3O_4 magnetic NPs are synthesized by sonochemical route. Iron pentacarbonyl and polyethylene glycol have taken together in hexadecane. The decomposition of $Fe(CO)_5$ occurs in the presence of polyethylene glycol, and monodispersed ultra-small NPs $\sim$3 nm particles are obtained. The reaction was carried out in dark condition. The presence of a black slurry in reaction vessel leads to $Fe(CO)_5$ decomposition [148].

12.5.2 Metal Chalcogenides

Metal sulfide NPs show emerging character owing to their extensive use in miscellaneous fields such as lasers, optoelectronic, thermoelectric devices, and in infrared spectroscopy cells. The nano-crystalline metal chalcogenides exhibit superior performance as compared to their bulk counterparts. Bismuth sulfide nanorods are synthesized through a facile sonochemical method using bismuth nitrate and sodium thiosulfate in an aqueous solution. Different complexing agents such as triethanol amine, sodium tartrate, and ethylenediamine tetra-acatetic acid are used during synthesis. These complexing agents promote the different diameter and length of Bi_2Se_3 nanorods [149].

In another report, hexagonal phase cadmium sulfide and cadmium selenide NPs are observed using cadmium acetate as metal precursor. The NPs have been obtained under reduced atmosphere $H_2/Ar(5/95)$ via ultrasonication. The hydrogen used during the reaction process acts as reducing agent, while high temperature is attained via collapse of bubbles followed via reduction of precursor [150].

Similarly, synthesis of PbSe and ZnSe NPs using sonochemical technique has been reported by Gedanken and his co-workers [151].

Metal selenides demonstrate extensive applications in diverse field such as in optical filters, optical recording materials, photovoltaic, sensors as well as laser materials [152]. The CdSe semiconductor nanocrystal finds its profound application in photoconductor. Hollow sphere of CdSe via sonochemical approach has been reported. Uniform and regular spheres of CdSe nanocrystal with average diameter of ~ 120 nm are obtained [153]. Ultrasonically synthesized hexagonal CdSe nanocrystals are reported using cadmium acetate and metal selenium in a reduced atmosphere of H_2/Ar. The precise experiments demonstrate that the hydrogen used here functions as a reducing agent. An extremely extraordinary temperature is attained through the bubble breakdown which accelerates the reduction of Se [154]. HgSe is an important semiconductor material which is utilized in photoconductors, solar cell, IR detector, tunable lasers, etc. At room temperature, HgSe is prepared under ultrasonic irradiation using mercury acetate and sodium selenosulfate in aqueous phase. The HgSe NPs with various sizes are obtained using the complexing agents such as ethylenediamine (EDA), ammonia, and tri-ethanolamine (TEA). The high concentration of TEA promotes the growth of tiny particles [155]. A range of metal chalcogenides such as MoS_2, Cu_3Se_2, Cu_7Te_4, Cu_4Te_3, $AgBiS_2$, etc., have been synthesized via sonochemical approach. During the typical ultrasonic synthesis, these metal chalcogenides necessitate the metal precursors into aqueous solution, and a chalcogen source such as thiourea is used for sulfur, while selenourea is used for Se sources. Under ultrasonic irradiation, in situ generation of H_2S or H_2Se reacts with metal salt precursors in aqueous phase to generate metal chalcogenide NPs [156–159].

12.5.3 Metal Carbides

Sonochemical approach has been further extended for the preparation of metal carbides such as Fe_3C, PdC, Mo_2C. At room temperature, PdC NPs are synthesized via the reduction of $Na_2PdCl_4$with interstitial carbon in aqueous solution. The concentration of carbon in Pd particles has been optimized by altering the nature and the concentration of organic additives. The PdC synthesis is composed of firstly formation of Pd cluster growth during Pd nanoparticle formation. In second step, organic additives have been adsorbed onto Pd cluster, and lastly, diffusion of carbon atoms takes place into Pd metal lattice. Higher carbon chain such as hexanol, ethanol, methanol, and precise concentration of isopropyl alcohol (IPA) promotes the enormous number of C atoms in Pd metals [160]. Iron carbide NPs have been prepared via sonochemical approach using Ferrocene $Fe(C_5H_5)_2$ as a metal precursor. The absence of oxygen in the precursor prevents the formation of Fe_3O_4NPs. Amorphous phase iron carbide is obtained via sonolysis of ferrocene in diphenylmethane. The modifications of reaction parameters such as ultrasound frequency, tip diameter, and immersion depth throughout the experiment promote the formation of monodispersed nanoparticles ~6 to 12 nm [161]. For the preparation of molybdenum carbide NPs, first slurry of molybdenum hexacarbonyl is prepared through ultrasonication (operating frequency ~20 kHz). Here, hexadecane is used as a solvent because it has small vapor pressure. The acquired nanomaterial shows porous character with aggregate of 2 nm sized Mo_2C NPs [162].

12.5.4 Bimetallic NPs/Metal Alloys/Metal Composites

Sonochemical route has been implemented for the synthesis of colloidal bimetallic NPs. Bimetallic NPs have shown the fruitful application as catalyst as well as in optoelectronic device applications. Suslick and his co-workers for the first time revealed the utilization of ultrasound to produce bimetallic NPs [163, 164]. Bimetallic NPs composed of gold and palladium is reported using sonochemical synthesis. Ultrasonic irradiation promotes the reduction by sodium dodecyl sulfate (SDS) of Au (III) and Pd (II) ions from $NaAuCl_4{\cdot}2H_2O$ and Na_2PdCl_4 in an aqueous medium. The surfactant SDS used here improves the reducing rate as well as the stability of the nanomaterials. The obtained nanocrystals exhibit the spherical shape particles with diameter ~8 nm [165]. An outstanding soft magnetic property has been reported for an amorphous phase ferromagnetic alloys comprised of Fe and Co. The boosted magnetic behavior of alloys has been used in magnetic data storage as well as in power transformers [166]. Furthermore, sonochemical decomposition synthesis is employed for the preparation of amorphous phase alloy of $Co_{20}Ni_{80}$ and $Co_{50}Ni_{50}$. The explosive organic metal precursors such as $Co(NO)(CO)_3$ and $Ni(CO)_4$ in decalin are utilized at 273 K under argon pressure (100–150 kPa). Synthesized alloys NPs exhibit superparamagnetic behavior [167]. Under ultrasonic irradiation, the Fe/Co alloy is synthesized using $Fe(CO)_5$ and $Co(NO)(CO)_3$ precursors in diphenylmethane solution in an inert argon atmosphere. The NPs having a metal alloy core along with a coated shell are obtained. The alloy NPs reveal exceptional storage constancy with magnetic performance [168]. Pt–Ru bimetallic system has been extensively applied in catalysis as well as in fuel cells. The colloidal synthesis of Pt–Ru bimetallic NPs in an aqueous phase has been obtained via sonochemical reduction of Pt(II) and Ru(III). The synthesis is executed at the ultrasound frequency ~210 kHz, while temperature is set to 20 °C. By the use of SDS as a stabilizing as well as capping agent, particle size of 5–10 nm NPs is obtained. Furthermore, using PVP, ultra-small bimetallic NPs in the range of ~5 nm are obtained [169].

12.5.5 Metal Oxide NPs

Metal oxides display a vigorous character in the arena of interdisciplinary branch of sciences, namely physics, chemistry, and materials science. In the technological point of view, these metal oxides exhibit profound applications such as in sensor, fuel cells, piezoelectric devices, microprocessors and microelectronic circuits, etc. Moreover, these metal oxides are employed in catalysis and as surface coating to prevent the corrosion as well. Owing to their vast range applicability, the synthesis of these metal oxide NPs has drawn much attention to the research community. Mesoporous tin oxide SnO_2NPs have been reported via sonochemical method using tin ethoxide as a precursor and CTAB as surfactant which controls its structural morphology [170]. The synthesis of lanthanide metal oxides such as Y_2O_3, Er_2O_3, CeO_2, Sm_2O_3 and La_2O_3, via sonochemical route has been also reported. These metal oxides are synthesized via nitrate salts of the corresponding rare earth metal ions as precursors, SDS as surfactants and urea as a precipitating reagent, respectively. Molar ratio of rare earth metal ions, SDS, and urea (1:2:30) is maintained during the synthesis [171]. Surfactant free zinc oxide nanoparticles are prepared via sonochemical route using zinc acetate and 1,4-butanediol. During synthesis process, 1,4-butandiol functions both as solvent and capping agent. For desired product formation, reaction is executed under ultrasonic irradiation [172]. The shape-selective ZnO NPs synthesis via sonochemical route is reported. The prepared ZnONPs exhibit different shapes such as nanorods, nanocups, nanodisks, nanoflowers, and nanospheres. It is observed that precursor concentration, power density of ultrasonic source, sonication time, kind of hydroxide anion mediators, and the capping agent are crucial aspects in the shape-selective ZnO nanomaterial preparation [173]. Furthermore, other metal oxides TiO_2, CeO_2, MoO_3, V_2O_5, In_2O_3, $ZnFe_2O_4$, $PbWO_4$, $BiPO_4$, and $ZnAl_2O_4$ have been successfully synthesized. Sonochemically prepared titania nanoprticles show superior photocatalytic activity to commercially available Degussa P25. The improved photocatalytic activity is ascribed to the highly crystalline nature of titania generated by rapid hydrolysis via ultrasound irradiation [174–181].

12.5.6 Sonochemical Preparation of Hollow and Layered Structures

The deposition of inorganic NPs on solid substrates such as silica or carbon nanotubes via sonochemical approach is employed to obtain hollow nano-structures. Hollow spheres of MoS_2 and MoO_3 have been reported via sonochemical technique [182]. MoS_2/SiO_2 composite under ultrasonic irradiation is obtained using isodurene slurry of molybdenum hexacarbonyl, sulfur, and silica nanospheres under an inert Ar atmosphere. Analogous process is carried out in the presence of air, and the absence of sulfur promotes the formation of a MoO_3/SiO_2 nano-composite.

Successive HF treatment leaches out the silica spheres which results in the formation of MoS_2 and MoO_3 hollow spheres. (Fig. 12.8 a, b and c). Remarkably after heat treatment, hollow MoO_3 nanospheres exhibit hollow single crystal (Fig. 12.8d).

Sonochemical synthesis of hollow hematite has been reported. During the synthesis, carbons NPs are utilized as an instinctively detachable template. The mechanism behind the formation of the hollow hematite formation is employed in situ combustion of the carbon NPs. However, sonochemical decomposition of Fe $(CO)_5$ resulting an amorphous iron NPs which form shells in the vicinity of carbon NPs. The fast oxidation of the high surface area iron shells under air exposure ignites the interior carbon particles. The combustion of the carbon NPs produces sufficient heat to crystallize the iron oxide shells consequently leads to hollow α-Fe_2O_3 cores [183]. Sonochemical synthesis of porous Co_3O_4 nanotubes has been reported. Carbon nanotubes (CNTs) have been utilized as a sacrificial template

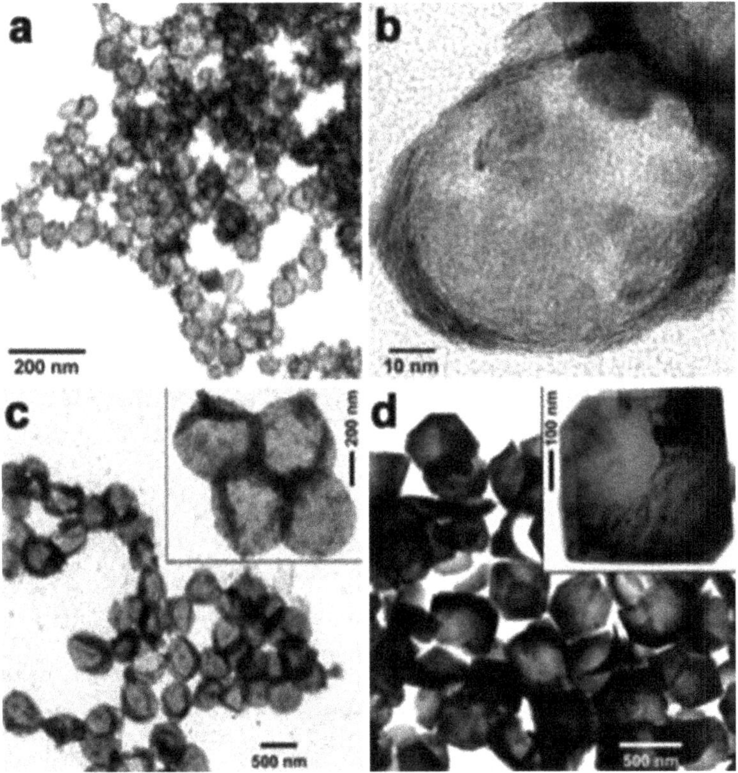

Fig. 12.8 TEM images of sonochemically prepared hollow MoS_2 nano-spheres **a** and **b** and hollow MoO_3 **c** after leaching of the silica template but before thermal annealing and **d** after thermal annealing and formation of hollow single crystals (**d**). Reproduced with permission from ACS publisher [182]

during the synthesis leads to the formation of CoO_x/CNTs composite. The nano-composite is further calcined into air to be on fire which finally removes the carbon nanotubes leads to porous nanotubes of CoO_x in Co_3O_4. The synthesized porous Co_3O_4 nanotubes act as an excellent electrode material employed in lithium batteries [184]. Hollow FePt spheres are reported via sonochemical deposition technique. Sonochemically synthesized FePt bimetallic particles are deposited on polyelectrolyte layers modified silica spheres. Consequent HF curing yields hollow FePt spheres. Fascinatingly, annealing of hollow FePt spheres exhibits magnetic properties with modified character from soft to hard magnet [185]. Polymer spheres such as polystyrene and poly-methyl-methacrylate are also used as a template material to create hollow structured materials [186]. In this situation, polymer cores are separated by either thermal pyrolysis or extraction with organic solvents from composites.

Ultrasound is proved to be formidable means for the chemical preparation of mono and multilayered graphene. The pristine graphite has been oxidized by Hummer's approach to synthesize the grapheme oxide. In grapheme oxide, inter-layer distance is increased as compared to graphite which results in weaker Vander Waals force. Subsequent to gentle sonication, single-layered graphene oxides have been synthesized that can be further reduced to graphene. The direct liquid-phase exfoliation of graphite via sonication gives the easy processing of graphene. In order to attain high yields of exfoliated graphene from graphite, the surface energy of the solvent must be equal to the surface energy of graphite (40–50 mJ m^{-2}). Sonication of graphite in appropriate solvents such as N-methyl-pyrrolidone (NMP)) promotes the formation of single layer and few layer graphene [187].

Ultrasound is repeatedly used to separate single-walled CNTs, which typically form bundles because of the presence of the Van der Waals force. This methodology has been utilized for the synthesis of other layered material such as $MoSe_2$, $MoTe_2$, MoS_2, WS_2, $TaSe_2$, $NbSe_2$, $NiTe_2$, BN and Bi_2Te_3. These materials have been exfoliated in the liquid phase to prepare monolayer nanosheets [188].

12.5.7 Sonochemical Preparation of Protein and Polymer Pano and Microstructures

Ultrasonic approach has been further transformed to synthesize biomaterial and polymers as well. The sonochemical synthesis of protein microspheres is obtained via sonication of a protein solution which contains serum albumins (Fig. 12.9).

Highly biocompatible and stable microspheres have meticulous attention in a variety of biomedical applications applied as contrast agents for MRI, sonography, optical coherence tomographym, and drug delivery carriers [189]. The collective effort from emulsification termed as a physical effect and cavitation corresponding to chemical effect consequently promotes the microsphere formation. The protein microspheres produced via ultrasonic emulsification attain significantly enhanced

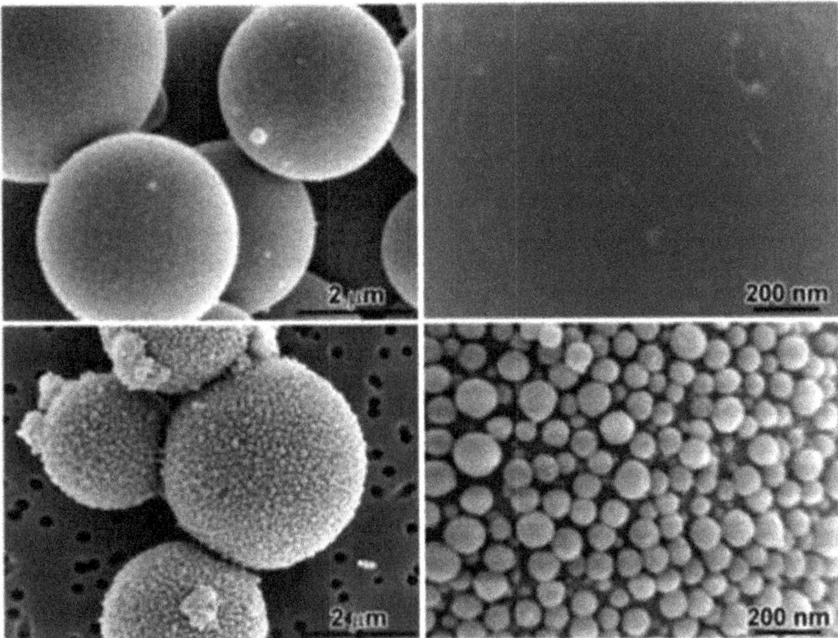

Fig. 12.9 SEM images of sonochemically prepared protein microspheres before and after nanoparticle functionalization by layer-by-layer adhesion: upper left, native microspheres as-prepared by sonication of bovine serum albumin and upper right, close-up of the surface; lower left, silica-coated microspheres using a RGD polylysinepeptide to reverse surface charge and lower right, close-up of its surface. Reproduced with permission from ACS publisher [189]

stability via covalent disulfide cross-linking of cysteine. Hydroperoxyl radicals are mainly generated, and these radicals are primarily accountable for this cross-linking [190, 191]. Hydroperoxyl radicals HO_2. generated during the sonolysis of water promotes induce cross-linking of the disulfide bonds linking with cysteine amino acid. The protein microspheres have been tailored via conjugation of selective cancer-cell ligands such as folate, RGD peptides, and mercaptoethane sulfonate to its surface [192].

12.5.8 Core @Shell Nanomaterials

An easy and effective sonochemical route has been employed for ZnO@CdS core@shell synthesis. ZnO nanorods is synthesized via oxidation of ZnO/Zn particles in air at high temperature [193]. In a typical synthesis process under ultrasonic irradiation, ZnO nanorods, cadmium chloride, and thiourea in an aqueous solution result the ZnO nanorods/Cds NPs formation [194]. Synthesis of Fe_3O_4@SiO_2 core@shell NPs via sonochemical route is reported [195]. It is observed that Fe(II) and Fe(III) metal precursors get precipitated under ultrasonic irradiation which exhibit narrower Fe_3O_4 NPs size distribution as compared to precipitation obtained via mechanical stirring. It is reported that an alkaline TEOS hydrolysis into water alcohol mixture with Fe_3O_4 NPs increases the homogeneity and reduces the agglomeration of $Fe_3O_4@SiO_2$ NPs significantly. The thickness of silica shell is easily tailored under ultrasound irradiation. Core/shell hetero structured $SnO_2@CdS$ has been prepared using SnO_2 nanobelt as a support. In the typical preparation, first CdS NPs are deposited over SnO_2 surface. The CdS NPs thus formed during the synthesis are of spherical type, and their sizes are in the order of 10–20 nm. Single crystal SnO_2 nanobelt having rutile structure is prepared via thermal evaporation metallic tin powders at 800 °C. In a typical synthesis process, SnO_2 nanobelts, cadmium chloride, and thiourea are mixed into 100 ml deionized water irradiated with ultrasound (100 W, 40 kHz) for 1–3 h [196]. Under ultrasound irradiation, the reduction of $AuCl_4^-$ ions to Au^0 (gold metal) and $AgNO_3^+$ ions to Ag^0 (silver metal) takes place in the presence of alcohols at room temperature [197]. Under ultrasound irradiation, cavitation takes place which generates primary and secondary radicals which reduces the gold chloride and silver nitrates into their respective metal ions. The presence of hydrogen atom along with alcohol radicals reduces the gold and silver ions to produce Au–Ag bimetallic core@shell NPs. It is also reported that in the presence of polymer, reaction of primary radicals with the polymer promotes the formation of polymeric radicals. It assists in reduction of the analogous metal ions to produce metal NPs [198, 199].

12.5.9 Ultrasonic Pyrolysis (USP)

Ultrasound produces chemical reactions in sonochemical synthesis. Moreover, in USP (ultrasonic pyrolysis), an ultrasound has not engaged directly during synthesis. Ultrasound acts as phase separator into one micro-droplet reactor to other. In sonochemical route, highly intense low-frequency ultrasound ~20 kHz has been generally used; whereas in USP, high frequency ~2 MHz with low intensity ultrasound is employed. Ultrasound is employed for nebulising the precursor solution which assist to generate micron-sized droplets during USP synthesis. Droplets thus generated by ultrasonic nebulisation have been heated under a gas flow before dispensed to chemical reaction. Large-scale production of ultrafine particles finds its utility in industry. Also, this technique is quite useful in film deposition. During this technique, creation of aerosols under ultrasonic nebulizer followed by thermal decomposition occurs [199, 200]. As compared to traditional techniques, USP has shown many merits such as easy and continuous operation, high purity reaction products, etc. This technique is equally applicable for small- and large-scale mass production with high reliability as well as reproducibility. Spherical NPs are prepared via the USP route. Liquid droplet formation takes place under ultrasonic nebulisation in USP route. In the USP process, firstly generated liquid droplets are subjected to the heated zone having Ar, N_2 and O_2 as carrier gases followed by evaporation of solvents at droplet surface. The droplets rapidly contract and attain supersaturation after heating. The precipitation of solute takes place at droplet surface. Moreover, thermal decomposition can generate intermediate compounds in the form of porous or hollow structured particles [173]. In a typical USP experimental set up, a vessel having transducer at the base fixed with a mist connected to the tubular furnace. The collector chambers have been positioned at the furnace outlet (Fig. 12.10). The USP procedure requires droplet creation, diffusion of the solutes, evaporation of solvents, and precipitation followed by decomposition and densification (Fig. 12.10b). For the film deposition, silicon and glass substrates have been positioned inside the furnace.

USP technique is established as a multifaceted approach for the synthesis of fine powders of metals, metal alloys, and ceramic materials. For the synthesis of the metals NPs such as Ag, Pd, Au, Cu, Ni, Co, and their alloys as Ag–Pd, USP technique is extensively utilized. In order to prepare metal or alloys particles, generally it engages the USP decomposition of metal salt precursor solution in an

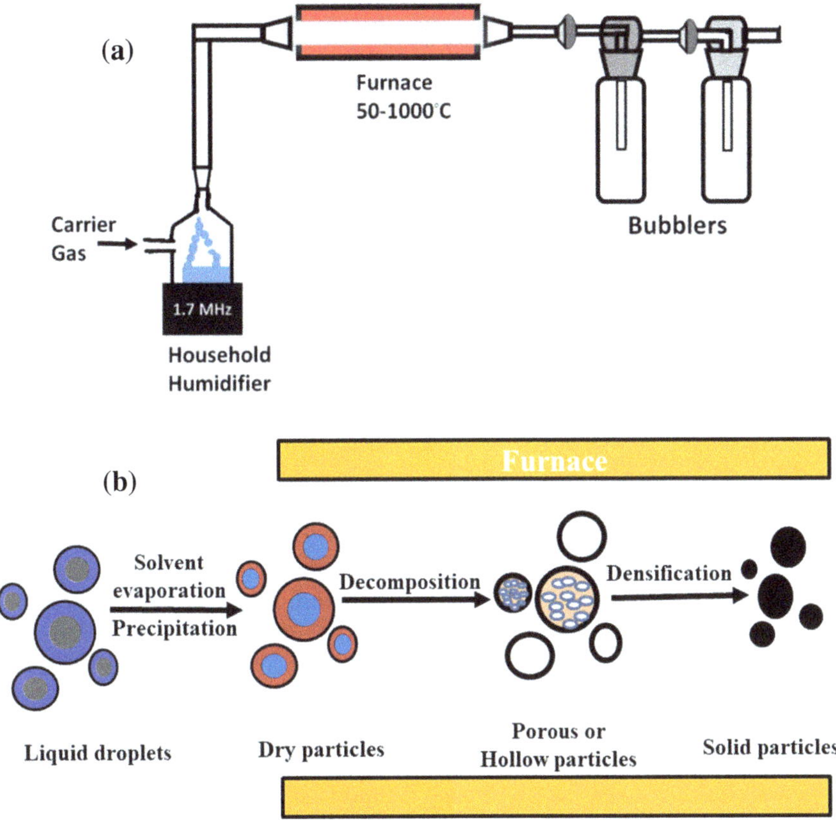

Fig. 12.10 Schematic illustration of **a** typical USP apparatus and **b** a schematic diagram of simplified USP process. This is redrawn from references [173, 199, 200]

inert atmosphere (N_2 or Ar flow) [201–206]. Highly crystalline nanowires of Zn, Cd, Co, and Pb nanowires are produced by USP route using methanolic solutions of corresponding metal acetates [207]. Micron-sized metal oxides and chalcogenides synthesis are reported via USP technique. During metal oxides synthesis, precursors of metal nitrates, chlorates, and acetates have been engaged; while for the preparation of metal chalcogenide, chalcogen sources are mixed into the precursor solution [140]. Suslick and Skrabalak employed silica as a sacrificial module to synthesize porous MoS_2 structure by USP approach. Silica NPs are tightly packed into an evaporating droplet and can afford an in situ nanostructure scaffold [208].

The exploitation of a silica template is more utilized for the preparation of metal oxides NPs. The USP preparation of diverse shapes of titania nanostructures, comprising porous, hollow, and ball-in-ball architectures has been reported. Silica-titania composite porous titania as well as silica-titania composite covered with cobalt oxide NPs are reported by employing the USP synthesis route. Under USP process, an aqueous solution containing silica NPs and titanium complexes forms silica-titania nano-composite. A porous titania microsphere is obtained after etching of the silica-titania composite using HF. At initial stage of etching fetches a ball-in-ball structure consists of silica core covered with porous titania shell. Moreover, complete etching result in the vanishing of silica core and, finally, porous spherical shells of titania is obtained [209].

The formation of porous carbon under the USP technique has been reported. In this context, this technique is applied for the synthesis of several carbon nanostructures which occurs by alkali halocarboxylate decomposition (Fig. 12.11) [210].

As compared to tiresome multistep conventional approaches for porous carbon preparation, one step USP approach removes the issue of use of high cost template materials. Based on the variety of alkali halo-carboxylates such as lithium chloroacetate, sodium chloroacetate, potassium chloroacetate, lithium dichloroacetate, sodium dichloroacetate, and potassium dichloroacetate, a miscellaneous range of nanostructures has been synthesized via USP process [210].

12.6 Conclusions and Future Prospects

In summary, thermolysis (i.e., polyol, hydro/solvothermal, MW and sonochemical) approaches for the nanomaterial synthesis have been briefly described. Since precise control over shape, size, and dimension greatly alter the physiochemical properties. Therefore, the prepared nanomaterials encounter diverse area of interests into materials research community for the range of applications such as photoconductors, bio-imaging, laser materials, photocatalysis, biosensors, supercapacitors, etc. Polyol synthesis has unquestionably turned out to be a handy synthesis approach which assures the preparation of great quality nanomaterials. Ethylene glycol, diethylene glycol, glycerol, and butanediol are extensively applied as polyol solvents for the preparation of nanomaterials. Currently, a variety of nanomaterials such as elemental metals, metal oxides, and metal chalcogenides have been

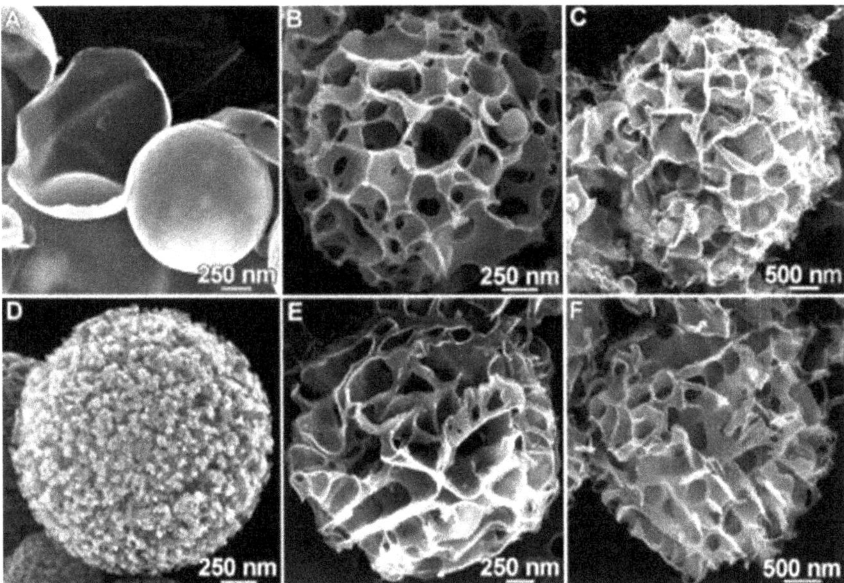

Fig. 12.11 SEM images of USP porous carbons. Reaction conditions: 1.5 M solutions, 700 °C, Ar at 1.0 slpm. Product from (A) lithium chloroacetate, (B) sodium chloroacetate, (C) potassium chloroacetate, (D) lithium dichloroacetate, (E) sodium dichloroacetate, and (F) potassium dichloroacetate. Reproduced with permission from ACS publisher [210]

synthesized via polyol mediated approach. Highly water dispersible and chelating properties of the polyols are extremely beneficial. During the synthesis, polyol act as solvent, reducing agent as well as capping agent which lead to surface functionalization as well as high colloidal stability of the NPs. Consequently, NPs with regular size distribution with less extent of agglomeration have been obtained. During the metal NPs synthesis in the polyol medium, the reductive character of the polyols is an added advantage which permits a straight reduction via heating of metal precursors. Also, higher boiling point of polyols fetches highly crystalline compounds during the synthesis. Due to these merits, polyol synthesis is broadly employed to attain materials with a range of applications as photocatalysis, MRI, sensing, photovoltaic, batteries, thin film, diluted transparent semiconductor, pigments, bio-imaging, and in drug delivery, etc. The polyol synthesis entails the preparation of heterostructured nanomaterials. Also, PEG exhibits control of the nucleation and growth of particles and particles coated with PEG show high bio-compatibility which permits the best possible cellular uptake of the pegylated NPs. However, there are also obvious restrictions to polyol synthesis. The fast thermal decomposition close to the boiling point of the polyols limits the availability of temperature choice of reactions.

The striking features of the MW-assisted synthesis exhibits rapid and uniform heating, fast reaction rate, higher yields, and shorter reaction time as compared to the usual heating techniques. Rapid synthesis of inorganic materials in aqueous phase under MW heating is observed. This unique approach leads to its cost effectiveness, energy saving, and a high product yield. Furthermore, the MW reactors with accurate power, temperature, and pressure controlled probes have been used for nanomaterials synthesis. However, detailed mechanism involved into MW synthesis and its consequence on reaction rate, nucleation, and growth in aqueous phase has been still matter of debate, and future research will be focussed in this direction. Precisely designed MW reactors will give an idea about the non-thermal heating issues involved under MW synthesis.

Hydrothermal and/or solvothermal routes are extensively utilized for the preparation of inorganic nanomaterials comprising metals, metal oxides, meal chalcogenides, and transition metals. These techniques are further extended for the synthesis of metal ions and linking organic ligands for metal organic framework (MOF)NPs.

The miscellaneous range of applications of ultrasound irradiation has been investigated in the synthesis of nanostructured materials. Sonochemical as well as ultrasonic spray pyrolysis techniques are used during the nanomaterials synthesis. Ultrasonic irradiation leads to acoustic cavitation. Bubbles are generated during cavitation, which can efficiently gather the ultrasound energy. After collapse of bubbles, massive amount of energy is released leading to localized hot spots with extremely high temperatures and pressures which are mainly responsible for chemical effects of ultrasound. By employing this route, elemental metals, metal alloys, metal oxides, metal sulfides, and metal carbides are synthesized. This synthesis approach is subsequently utilized for the synthesis of carbons, polymers and biomaterials, etc. During the synthesis of nano-composites, template-based strategy is used to obtain hollow structured materials via subsequent removal of template. Scale-up of materials and energy efficiency is the prime hindrance of sonochemical synthesis. Moreover, laboratory apparatus for sonochemical reactors is commercially available even though large scale apparatus remains quite rare. Future study will be to design precisely controlled sonochemical reactors for the large-scale production of the nanomaterials under ultrasonication.

In USP route, the ultrasound acts to nebulize the precursor solutions resulting production of micron-size droplets to facilitate the chemical reactions inside the reactor. The micro-reactors permit the simplistic control over elemental structure at the micron-size gauge. USP synthesis route is explored for the preparation of hetero-structure and/or composite materials. The synthesis of nanomaterials via USP generated nano-composites is vigorous as well as proficient. Moreover, processing of metal oxides via USP route is still inadequate mainly owing to the deficiency of appropriate precursors for metal oxides and uncontrolled fast sol-gel reactions. In view of the present restrictions, systematic development of new metal oxide precursors and precise control over the reaction parameters will be prime focus of the future study for the further advancement in USP technique.

Acknowledgements B.P.S. acknowledges the award through the Inspire Faculty (IFA17-MS-109) provided by DST, Government of India. Also, M.S. acknowledges the award through CSIR-SRA(9146-A) provided by CSIR, New Delhi, India.

References

1. Xia Y, Peidong Y, Sun Y, Wu Y, Mayers M, Gates B, Yin Y, Kim F, Yan H (2003) One dimensional nanostructures: synthesis, characterization, and applications. Adv Mater 15:353
2. Tsuji M (2017) Microwave-assisted synthesis of metallic nanomaterials in liquid phase. Chemistry 2:805
3. Jaiswal S, Smriti D (2017) Microwave-assisted eco-friendly synthesis and antimicrobial evaluation of Aryl-Triazole-1, 3, 4-Thiadiazols. J Med Res Innovat 1:17
4. Kumar VV, Kumar A (2020) Facile synthesis of flowerlike $LiFe_5O_8$ microspheres for electrochemical supercapacitors. Inorg Chem Commun 112:
5. Bharti A, Cheruvally G (2018) Surfactant assisted synthesis of Pt-Pd/MWCNT and evaluation as cathode catalyst for proton exchange membrane fuel cell. Int J Hydrogen Energy 31:14729
6. Soni AK, Joshi R, Singh BP, Kumar NN, Ningthoujam RS (2019) Near-infrared- and magnetic-field-responsive $NaYF_4:Er^{3+}/Yb^{3+}@SiO_2@AuNP@Fe_3O_4$ nanocomposites for hyperthermia applications induced by fluorescence resonance energy transfer and surface plasmon absorption. Appl ACS Nano Mater 2:7350
7. Dong H, Chen YC, Feldmann C (2015) Polyol synthesis of nanoparticles: status and options regarding metals, oxides, chalcogenides, and non-metal elements. Green Chem 17:4107
8. Baghbanzadeh M, Carbone L, Cozzoli PD, Kappe CO (2011) Microwave-assisted synthesis of colloidal inorganic nanocrystals. Angew Chem Int Ed 50:11312
9. Thorat ND, Shinde KP, Pawar SH, Barick KC, Betty CA, Ningthoujam RS (2012) Polyvinyl alcohol: an efficient fuel for synthesis of superparamagnetic LSMO nanoparticles for biomedical application. Dalton Trans 41:3060
10. Fiévet F, Vincent FF, Lagier JP, Dumont B, Figlarz M (1993) Controlled nucleation and growth of micrometre-size copper particles prepared by the polyol process. J Mater Chem 3:627
11. Wiley B, Herricks T, Sun Y, Xia Y (2004) Polyol synthesis of silver nanoparticles: use of chloride and oxygen to promote the formation of single-crystal, truncated cubes and tetrahedrons. Nano Lett 4:1733
12. El-Sayed MA (2001) Some interesting properties of metals confined in time and nanometer space of different shapes. Acc Chem Res 34:257
13. Song H, Kim F, Connor S, Somorjai GA, Yang P (2005) Pt nanocrystals: shape control and Langmuir-Blodgett monolayer formation. J Phys Chem, B 109:188
14. Xia X, Choi S, Herron JA, Lu N, Scaranto J, Peng HC, Wang J, Mavrikakis M, Kim MJ, Xia Y (2013) Facile synthesis of palladium right bipyramids and their use as seeds for overgrowth and as catalysts for formic acid oxidation. J Am Chem Soc 135:15706
15. Sun Y, Xia Y (2002) Shape-controlled synthesis of gold and silver nanoparticles. Science 298:2176
16. Toneguzzo P, Viau G, Acher O, Guillet F, Bruneton E, Fievet F (2000) CoNi and FeCoNi fine particles prepared by the polyol process: Physico-chemical characterization and dynamic magnetic properties J Mater Sci 35:3767
17. Jezequel D, Guenot J, Jouini N, Fievet F (1995) Suhinicrometer zinc oxide particles: Elaboration in pniyoi medium and morphological characteristics. J Mater Res 10:77
18. Liu C, Wu X, Klemmer T, Shukla N, Yang X, Weller D, Roy AG, Tanase M, Laughlin D (2004) Polyol process synthesis of monodispersed FePt nanoparticles. J Phys Chem B 108:6121

19. Watanabe K, Kura H, Sato T (2006) Transformation to L10 structure in FePd nanoparticles synthesized by modified polyol process. Sci Technol Adv Mater 7:145
20. Viau G, Fievet-Vincent F, Fievet F (1996) Nucleation and growth of bimetallic CoNi and FeNi monodisperse particles prepared in polyols. Solid State Ionics 84:259
21. Zhang B, Tu Z, Zhao F, Wang J (2013) Superparamagnetic iron oxide nanoparticles prepared by using an improved polyol method. Appl Surface Sci 266:375
22. Cai W, Wan J (2007) Facile synthesis of superparamagnetic magnetite nanoparticles in liquid polyols. J Colloid Interface Sci 305:366
23. Jézéquel D, Guenot J, Jouini N, Fiévet F (1995) Suhinicrometer zinc oxide particles: Elaboration in pniyoi medium and morphological characteristics. J Mater Res 10:77–83
24. Feldmann C, Jungk HO (2001) Polyol-mediated preparation of nanoscale oxide particles. Angew Chem Int Ed 40:359–362
25. Feldmann C (2003) Polyol-mediated synthesis of nanoscale functional materials. Adv Funct Mater 13:101
26. Ammar S, Helfen A, Jouini N, Fiévet F, Rosenman I, Villain F, Molinié P, Danot M (2001) Magnetic properties of ultrafine cobalt ferrite particles synthesized by hydrolysis in a polyol medium. J Mater Chem 11:186
27. Poul L, Jouini N, Fiévet F (2000) Layered hydroxide metal acetates (metal= zinc, cobalt, and nickel): elaboration via hydrolysis in polyol medium and comparative study. Chem Mater 12:3123
28. Prévot V, Forano C, Besse J (2005) Hydrolysis in polyol: new route for hybrid-layered double hydroxides preparation. Chem Mater 17:6695
29. Schmitt P, Brem N, Schunk S, Feldmann C (2011) Polyol-Mediated synthesis and properties of nanoscale molybdates/tungstates: color, luminescence, catalysis. Adv Funct Mater 21:3037
30. Parchur AK, Ningthoujam RS (2011) Preparation and structure refinement of Eu^{3+} doped CaMoO4 nanoparticles. Dalton Trans 40:7590
31. Maheshwary, Singh BP, Singh J, Singh RA (2014) Luminescence properties of Eu^{3+}-activated $SrWO_4$ nanophosphors-concentration and annealing effect. RSC Adv 4:32605
32. Maheshwary BP, Singh RA (2015) Color tuning in thermally stable Sm^{3+}-activated $CaWO_4$ nanophosphors. New J Chem 39:4494
33. Feldmann C, Roming M, Trampert K (2006) Polyol-Mediated Synthesis of Nanoscale CaF_2 and CaF_2:Ce,Tb. Small 2:1248
34. Parchur AK, Prasad AI, Rai SB, Tewari R, Sahu RK, Okram GS, Singh RA, Ningthoujam RS (2012) Observation of intermediate bands in Eu^{3+} doped YPO_4 host: Li^+ ion effect and blue to pink light emitter. AIP Adv 2:032119
35. Flores N, Franceschin G, Gaudisson T, Beaunier P, Yaacoub N, Grenèche JM, Valenzuela R, Ammar S (2018) Giant exchange-Bias in polyol-made $CoFe_2O_4$@CoO Core–shell like nanoparticles. Part Part Syst Charact 35:1800290
36. Tsuji M, Hikino S, Sano Y, Horigome M (2009) Preparation of Cu@Ag core–shell nanoparticles using a two-step polyol process under bubbling of N_2 gas. Chem Lett 38:518
37. Nguyen TT, Lau-Truong S, Mammeri F, Ammar S (2020) Star-shaped $Fe_{3-x}O_4$-Au core-shell nanoparticles: from synthesis to SERS application. Nanomaterials 10:294
38. Parchur AK, Ansari AA, Singh BP, Syed FN, Rai SB, Ningthoujam RS (2014) Enhanced luminescence of $CaMoO_4$:Eu by core@shell formation and its hyperthermia study after hybrid formation with Fe_3O_4: cytotoxicity assessment on human liver cancer cells and mesenchymal stem cells. Integer Biol 6:53
39. Mnasri W, Tahar LB, Beaunier P, Haidar DA, Boissière M, Sandre O, Ammar S (2020) Polyol-made luminescent and superparamagnetic β-$NaY_{0.8}Eu_{0.2}F_4$@γ-Fe_2O_3 core-satellites nanoparticles for dual magnetic resonance and optical imaging. Nanomaterials 10:393
40. Dong H, Kuzmanoski A, Gobi DM, Popescu R, Gerthsen D, Feldmann C (2014) Polyol-mediated C-dot formation showing efficient Tb^{3+}/Eu^{3+} emission. Chem Commun 50:7503–7506

41. Kumar A, Kuang Y, Liang Z, Sun X (2020) Microwave chemistry, recent advancements and eco-friendly microwave-assisted synthesis of nanoarchitectures and their applications: a review. Mater Today. NANO 11:
42. Zhu YJ, Chen F (2014) Microwave-assisted preparation of inorganic nanostructures in liquid phase. Chem Rev 114:6462
43. Collins MJ Jr (2010) Future trends in microwave synthesis. Med Chem 2:151
44. Nishioka M, Miyakawa M, Daino Y, Kataoka H, Koda H, Sato K, Suzuki TM (2013) Single-mode microwave reactor used for continuous flow reactions under elevated pressure. Ind Eng Chem Res 52:4683
45. Kappe CO (2004) Controlled microwave heating in modern organic synthesis. Angew Chem Int Ed 43:6250
46. Bilecka I, Niederberger M (2010) Microwave chemistry for inorganic nanomaterials synthesis. Nanoscale 2:1358
47. Robinson J, Kingman S, Irvine D, Licence P, Smith A, Dimitrakis G, Obermayer D, Kappe CO (2010) Understanding microwave heating effects in single mode type cavities-theory and experiment. Phys Chem Chem Phys 12:4750
48. Tsuji M, Hashimoto M, Nishizawa Y, Kubokawa M, Tsuji T (2005) Microwave-assisted synthesis of metallic nanostructures in solution. Chem Eur J 11:440
49. Baruwati B, Varma RS (2009) High value products from waste: grape pomace extract – a three-in-one package for the synthesis of metal nanoparticles. Chem Sus Chem 2:104
50. Hernandez CV, Mariscal MM, Esparza R, Yacaman MJ (2010) A synthesis route of gold nanoparticles without using a reducing agent. Appl Phys Lett 96:
51. Zhang ZW, Jia J, Ma YY, Weng J, Sun YA, Sun LP (2011) Microwave-assisted one-step rapid synthesis of folic acid modified gold nanoparticles for cancer cell targeting and detection. Med Chem Comm 2:1079
52. Hu B, Wang SB, Wang K, Zhang M, Yu SH (2008) Microwave-assisted rapid facile "green" synthesis of uniform silver nanoparticles: self-assembly into multilayered films and their optical properties. J Phys Chem C 112:11169
53. Luo YL (2007) A simple microwave-based route for size-controlled preparation of colloidal Pt nanoparticles. Mater Lett 2007:61
54. Mehta SK, Gupta SJ (2011) Time-efficient microwave synthesis of Pd nanoparticles and their electrocatalytic property in oxidation of formic acid and alcohols in alkaline media. Appl Electrochem 41:1407
55. Liu Y-Q, Zhang M, Wang F-X, Pan G-B (2012) Facile microwave-assisted synthesis of uniform single-crystal copper nanowires with excellent electrical conductivity. RSC Adv 2:11235
56. He Y, Zhong YL, Peng F, Wei XP, Su YY, Lu YM, Su S, Gu W, Liao LS, Lee S-T (2011) One-pot microwave synthesis of water-dispersible, ultraphoto- and pH-stable, and highly fluorescent silicon quantum dots. J Am Chem Soc 133:14192
57. Jing ZH, Zhan JH (2008) Fabrication and gas-sensing properties of porous ZnO nanoplates. Adv Mater 20:4547
58. Wu DS, Han CY, Wang SY, Wu NL, Rusakova IA (2002) Microwave-assisted solution synthesis of SnO nanocrystallites. Mater Lett 53:155
59. Wang HE, Xi LJ, Ma RG, Lu ZG, Chung CY, Bello I, Zapien JA (2012) Microwave-assisted hydrothermal synthesis of porous SnO_2 nanotubes and their lithium ion storage properties. J Solid State Chem 190:104
60. Zhang PL, Yin S, Sato T (2009) Synthesis of high-activity TiO_2 photocatalyst via environmentally friendly and novel microwave assisted hydrothermal process. Appl Catal B 89:118
61. Muraliganth T, Murugan AV, Manthiram A (2009) Facile synthesis of carbon-decorated single-crystalline Fe_3O_4 nanowires and their application as high performance anode in lithium ion batteries. Chem Commun 131:7360
62. Meher SK, Rao GR (2011) Effect of microwave on the nanowire morphology, optical, magnetic, and pseudocapacitance behavior of Co_3O_4. J Phys Chem C 115:25543

63. Zhu HT, Zhang CY, Tang YM, Wang JX (2007) Novel synthesis and thermal conductivity of CuO nanofluid. J Phys Chem C 111:1646
64. Ruong TT, Liu YZ, Ren Y, Trahey L, Sun YG (2012) Morphological and crystalline evolution of nanostructured MnO_2 and its application in lithium–air batteries. ACS Nano 6:8067
65. Bondioli F, Ferrari AM, Leonelli C, Siligardi C, Pellacani GC (2001) Microwave-hydrothermal synthesis of nanocrystalline zirconia powders. J Am Ceram Soc 84:2728
66. Hariharan V, Radhakrishnan S, Parthibavarman M, Dhilipkumar R, Sekar C (2011) Synthesis of polyethylene glycol (PEG) assisted tungsten oxide (WO_3) nanoparticles for l-dopa bio-sensing applications. Talanta 85:2166
67. Wei GD, Qin WP, Zhang DS, Wang GF, Kim RJ, Zheng KZ, Wang LL (2009) Synthesis and field emission of MoO_3 nanoflowers by a microwave hydrothermal route. J Alloys Compd 481:417
68. Liao XH, Zhu JM, Zhu JJ, Xu JZ, Chen HY (2001) Preparation of monodispersed nanocrystalline CeO_2 powders by microwave irradiation. Chem Commun 937
69. Zawadzki M (2008) Preparation and characterization of ceria nanoparticles by microwave-assisted solvothermal process. J Alloys Compd 451:297
70. de Moura AP, de Oliveira LH, Paris EC, Li MS, Andrés J, Varela JA, Longo E, Viana Rosa IL (2011) Photolumiscent properties of nanorods and nanoplates Y_2O_3: Eu_{3+}. J Fluoresc 21:1431
71. Joshi R, Singh BP, Ningthoujam RS (2020) Confirmation of highly stable 10 nm sized Fe_3O_4 nanoparticle formation at room temperature and understanding of heat generation under AC magnetic fields for potential application in hyperthermia. AIP Adv 10
72. Joshi R, Perala RS, Srivastava M, Singh BP, Ningthoujam RS (2019) Heat generation from magnetic fluids under alternating current magnetic field or induction coil for hyperthermia-based cancer therapy: basic principle. J Radiat Cancer Res 10:156
73. Hong RY, Pan TT, Li HZ (2006) Microwave synthesis of magnetic Fe_3O_4 nanoparticles used as a precursor of nanocomposites and ferrofluids. J Magn Magn Mater 303:60
74. Miao F, Hua W, Hu L, Huang KM (2011) Magnetic Fe_3O_4 nanoparticles prepared by a facile and green microwave-assisted approach Mater Lett 65:1031
75. Parsons JG, Luna C, Botez CE, Elizalde J, Torresdey JLG (2009) Microwave-assisted synthesis of iron(III) oxyhydroxides/oxides characterized using transmission electron microscopy, X-ray diffraction, and X-ray absorption spectroscopy. J Phys Chem Solids 70:555
76. Figueroa LA, Alvarez OM, Cruz JS, Contreras RG, Torres LSA, de la F. Hernández J, Arrocena MCA (2017) Nanomaterials made of non-toxic metallic sulfides: a systematic review of their potential biomedical applications. Mater Sci Eng C 76:1305
77. Tiwari A, Dhoble SJ (2016) Stabilization of ZnS nanoparticles by polymeric matrices: syntheses, optical properties and recent applications. RSC Adv 6:64400
78. Zhao Y, Liao XH, Hong JM, Zhu JJ (2004) Synthesis of lead sulfide nanocrystals via microwave and sonochemical methods. Mater Chem Phys 87:149
79. Mu CF, Yao QZ, Qu XF, Zhou GT, Li ML, Fu SQ (2010) Controlled synthesis of various hierarchical nanostructures of copper sulfide by a facile microwave irradiation method. Colloids Surf A 371:14
80. Zhu JJ, Zhou MG, Xu JZ, Liao XH (2001) Rapid synthesis of nanocrystalline SnO_2 powders by microwave heating method. Mater Lett 47:25
81. Zhao Y, Hong JM, Zhu JJ (2004) Microwave-assisted self-assembled ZnS nanoballs. J Cryst Growth 270:438
82. Xing RM, Liu SH, Tian SF (2011) Microwave-assisted hydrothermal synthesis of biocompatible silver sulfide nanoworms. J Nanopart Res 13:4847
83. Liao XH, Wang H, Zhu JJ, Chen HY (2001) Preparation of Bi_2S_3 nanorods by microwave irradiation. Mater Res Bull 36:2339

84. Shao MW, Kong LF, Li Q, Yu WC, Qian YT (2003) Microwave-assisted synthesis of tube-like HgS nanoparticles in aqueous solution under ambient condition. Inorg Chem Commun 6:737
85. Zhang WJ, Li DZ, Chen ZX, Sun M, Li WJ, Lin Q, Fu XZ (2011) Microwave hydrothermal synthesis of AgInS2 with visible light photocatalytic activity. Mater Res Bull 46:975
86. Zhang WJ, Li DZ, Sun M, Shao Y, Chen ZX, Xiao GC, Fu XZ (2010) Microwave hydrothermal synthesis and photocatalytic activity of AgIn5S8 for the degradation of dye. J Solid State Chem 183:2466
87. Bensebaa F, Durand C, Aouadou A, Scoles L, Du X, Wang D, Page YL (2010) A new green synthesis method of $CuInS_2$ and $CuInSe_2$ nanoparticles and their integration into thin films. J Nanopart Res 12:1897
88. Apte SK, Garaje SN, Bolade RD, Ambekar JD, Kulkarni MV, Naik SD, Gosavi SW, Baeg JO, Kale BB (2010) Hierarchical nanostructures of $CdIn_2S_4$ via hydrothermal and microwave methods: efficient solar-light-driven photocatalysts. J Mater Chem 20:6095
89. Li L, Qian HF, Ren JC (2005) Rapid synthesis of highly luminescent CdTe QDs in the aqueous phase by microwave irradiation with controllable temperature. Chem Commun 528
90. Zhu JJ, Palchik O, Chen SG, Gedanken A (2000) Microwave Assisted Preparation of CdSe, PbSe, and $Cu_{2-x}Se$ Nanoparticles J Phys Chem B 104:7344
91. Gawande MB, Goswami A, Asefa T, Guo H, Biradar AV, Peng DL, Zboril R, Varma RS (2015) Core–shell nanoparticles: synthesis and applications in catalysis and electrocatalysis. Chem Soc Rev 44:7540
92. Belousov OV, Belousova NV, Sirotina AV, Solovyov LA, Zhyzhaev AM, Zharkov SM, Mikhlin YL (2011) Formation of bimetallic Au-Pd and Au-Pt nanoparticles under hydrothermal conditions and microwave irradiation. Langmuir 27:11697
93. Yu JC, Hu XL, Li Q, Zheng Z, Xu YM (2006) Synthesis and characterization of core-shell selenium/carbon colloids and hollow carbon capsules. Chem Eur J 12:548
94. Zhang H, Yin YJ, Hu YJ, Li CY, Wu P, Wei SH, Cai CX (2010) Pd@Pt Core−shell nanostructures with controllable composition synthesized by a microwave method and their enhanced electrocatalytic activity toward oxygen reduction and methanol oxidation. J Phys Chem 114:11861
95. Bahadur NM, Watanabe S, Furusawa T, Sato M, Kurayama F, Siddiquey IA, Kobayashi Y, Suzuki N (2011) Rapid one-step synthesis, characterization and functionalization of Silica coated gold nanoparticles. Colloid Surface Physicochem Eng Aspects 392:137
96. Yu J, Yu X (2008) The greenhouse gas emissions and fossil energy requirement of bioplastics from cradle to gate of a biomass refinery. Environ Sci Technol 42:
97. Ming B, Li J, Kang F, Pang G, Zhang Y, Liang C, Xu J, Wang X (2012) Microwave–hydrothermal synthesis of birnessite-type MnO_2 nanospheres as supercapacitor electrode materials. J Power Sources 198:428
98. Frank C (2000) Hollow capsule processing through colloidal templating and self-assembly. Chem Eur J 6:413
99. An K, Hyeon T (2009) Synthesis and biomedical applications of hollow nanostructures. Nano Today 4:359
100. Li X, Sun P, Yang T, Zhao J, Wang Z, Wang W, Liu Y, Lu G, Du Y (2013) Template-free microwave-assisted synthesis of ZnO hollow microspheres and their application in gas sensing. Cryst Eng Comm 15:2949
101. Zou S, Xu X, Zhu Y, Cao C (2017) Microwave-assisted preparation of hollow porous carbon spheres and as anode of lithium-ion batteries. Microporous Mesoporous Mater 251:114
102. Li J, Chen Z, Wang RJ, Proserpio DM (1999) Low temperature route towards new materials: solvothermal synthesis of metal chalcogenides in ethylenediamine. Coord Chem Rev 192:707
103. Sue K, Suzuki M, Arai K, Ohashi T, Ura H, Matsui K, Hakuta Y, Hayashi H, Watanabe M, Hiaki T (2006) Size-controlled synthesis of metal oxide nanoparticles with a flow-through supercritical water method. Green Chem 8:634

104. Andelman T, Tan MC, Riman RE (2010) Thermochemical engineering of hydrothermal crystallisation processes. Mater Res Innov 14:9
105. Kashchiev D (1982) On the relation between nucleation work, nucleus size, and nucleation rate. J Chem Phys 76:5098
106. Uematsu M, Franck EU (1981) Static dielectric constant of water and steam. J Phys Chem Refer Data 9:1291
107. Li J, Wu Q, Wu J. Synthesis of nanoparticles via solvothermal and hydrothermal methods. Handbook of nanoparticles. https://doi.org/10.1007/978-3-319-13188-7_17-1
108. Li Y, Duan X, Liao H, Qian Y (1998) Self-regulation synthesis of nanocrystalline $ZnGa_2O_4$ by hydrothermal reaction. Chem Mater 10:17
109. Niederberger M, Pinna N, Polleux J, Antonietti M (2004) A general soft-chemistry route to perovskites and related materials: synthesis of $BaTiO_3$, $BaZrO_3$, and $LiNbO_3$. Nanoparticles Angew Chem Int Ed 43:2270
110. Srivastava BB, Kuang A, Mao Y (2015) Persistent luminescent sub-10 nm Cr doped $ZnGa_2O_4$ nanoparticles by a biphasic synthesis route. Chem Commun 51:7372
111. Jiang L, Yang M, Zhu S, Pang G, Feng S (2008) Phase evolution and morphology control of ZnS in a solvothermal system with a single precursor. J Phys Chem C 112:15281
112. Zhang Y, Li Y (2004) Synthesis and characterization of monodisperse doped ZnS nanospheres with enhanced thermal stability. J Phys Chem B 108:17805
113. Cozzoli PD, Manna L, Curri ML, Kudera S, Giannini C, Striccoli M, Agostiano A (2005) Shape and phase control of colloidal ZnSe nanocrystals. Chem Mater 17:1296
114. Li XH, Li JX, Li GD, Liu DP, Chen JS (2007) Controlled synthesis, growth mechanism, and properties of monodisperse CdS colloidal spheres. Chem A Eur J 13:8754
115. Ningthoujam RS, Gajbhiye NS (2015) Synthesis, electron transport properties of transition metal nitrides and applications. Prog Mater Sci 70:50
116. Xie Y, Qian Y, Wang W, Zhang S, Zhang Y (1996) A benzene-thermal synthetic route to nanocrystalline GaN. Science 272:1926
117. Purdy AP (1999) Ammonothermal synthesis of cubic gallium nitride. Chem Mater 11:1648
118. Mazumder B, Chirico P, Hector AL (2008) Direct solvothermal synthesis of early transition metal nitrides. Inorg Chem 47:9684
119. Hu B, Wang K, Wu L, Yu SH, Antonietti M, Titirici MM (2010) Engineering carbon materials from the hydrothermal carbonization process of biomass. Adv Mater 22:813
120. Titirici M-M, Antonietti M, Baccile N (2008) Hydrothermal carbon from biomass: a comparison of the local structure from poly-to monosaccharides and pentoses/hexoses. Green Chem 10:1204
121. Titirici MM, Thomas A, Antonietti M (2007) Replication and coating of silica templates by hydrothermal Carbonization. Adv Funct Mater 17:1010
122. Titirici M-M, Antonietti M (2010) Chemistry and materials options of sustainable carbon materials made by hydrothermal carbonization. Chem Soc Rev 39:103
123. Wang WZ, Huang JY, Ren ZF (2004) Synthesis of germanium nanocubes by a low-temperature inverse micelle solvothermal technique. Langmuir 21:751
124. Wang WZ, Poudel B, Huang JY, Wang DZ, Kunwar S, Ren ZF (2005) Synthesis of gram-scale germanium nanocrystals by a low-temperature inverse micelle solvothermal route. Nanotechnology 16:1126
125. Rowsell JLC, Yaghi OM (2005) Strategies for hydrogen storage in metal–organic frameworks. Angew Chem Int Ed 44:4670
126. Li J-R, Kuppler RJ, Zhou H-C (2009) Selective gas adsorption and separation in metal–organic frameworks. Chem Soc Rev 38:1477
127. Carpenter MK, Moylan TE, Kukreja RS, Atwan MH, Tessema MM (2012) Solvothermal synthesis of platinum alloy nanoparticles for oxygen reduction electrocatalysis. J Am Chem Soc 134:8535
128. Sadakiyo M, Yamada T, Kitagawa H (2009) Rational designs for highly proton-conductive metal−organic frameworks. J Am Chem Soc 131:9906

129. Lee J, Farha OK, Roberts J, Scheidt KA, Nguyen ST, Hupp JT (2009) Metal-organic framework materials as catalysts. Chem Soc Rev 38:1450
130. Rieter WJ, Taylor KM, An H, Lin W (2006) Nanoscale metal-organic frameworks as potential multimodal contrast enhancing agents. J Am Chem Soc 128:9024
131. McKinlay AC, Morris RE, Horcajada P, Férey G, Gref R, Couvreur P, Serre C (2010) BioMOFs: metal-organic frameworks for biological and medical applications. Angew Chem Int Ed 49:6260
132. Pashow KMLT, Rocca JD, Xie Z, Tran S, Lin W (2009) Postsynthetic modifications of iron-carboxylate nanoscale metal-organic frameworks for imaging and drug delivery. J Am Chem Soc 131:14261
133. Taylor KML, Jin A, Lin W (2008) Surfactant assisted synthesis of nanoscale gadolinium metal–organic frameworks for potential multimodal imaging. Angew Chem Int Ed 47:7722
134. Taylor KML, Rieter WJ, Lin W (2008) Manganese-based nanoscale metal-organic frameworks for magnetic resonance imaging. J Am Chem Soc 130:14358
135. Rieter WJ, Taylor KML, Lin W (2007) Surface modification and functionalization of nanoscale metal-organic frameworks for controlled release and luminescence sensing. J Am Chem Soc 129:9852
136. Suslick KS, Flannigan DJ (2008) Inside a collapsing bubble: sonoluminescence and the conditions during cavitation. Annu Rev Phys Chem 59:659
137. Suslick KS (1990) Sonochemistry. Science 247:1439
138. Suslick KS, Doktycz SJ (1990) Advances in sono-chemistry, vol 1. Mason TJ (ed), JAI Press, New York, pp 197
139. Frenzel H, Schultes H (1934) Luminescence in water carrying supersonic waves. Z Phys Chem 27b:421; Suslick KS (1995) MRS Bull 20:29
140. Bang JH, Suslick KS (2010) Applications of ultrasound to the synthesis of nanostructured materials. Adv Mater 22:1039
141. Dhas NA, Raj CP, Gedanken A (1998) Synthesis, characterization, and properties of metallic copper nanoparticles. Chem Mater 10:1446; Su CH, Wu PL, Yeh CS (2003) Sonochemical synthesis of well-dispersed gold nanoparticles at the ice temperature. J Phys Chem B 107:14240
142. Caruso RA, Kumar MA, Grieser F (2002) Sonochemical formation of gold sols. Langmuir 18:7831
143. Iglesias AS, Santos IP, Juste JP, González BR, García de Abajo FJ, Marzán LML (2006) Synthesis and optical properties of gold nanodecahedra with size control. Adv Mater 18:2529
144. Santos IP, Iglesias AS, García de Abajo FJ, Marzán LML (2007) Environmental optical sensitivity of gold nanodecahedra. Adv Funct Mater 17:1443
145. Jiang LP, Xu S, Zhu JM, Zhang JR, Zhu JJ, Chen HY (2004) Ultrasonic-assisted synthesis of monodisperse single-crystalline silver nanoplates and gold nanorings. Inorg Chem 43:5877
146. Fujimoto T, Terauchi S, Umehara H, Kojima I, Henderson W (2001) Sonochemical preparation of single-dispersion metal nanoparticles from metal salts. Chem Mater 13:1057
147. Nemamcha A, Rehspringer JL, Khatmi D (2006) Synthesis of palladium nanoparticles by sonochemical reduction of palladium (II) nitrate in aqueous solution. J Phys Chem B 110:383
148. Khalil H, Mahajan D, Rafailovich M, Gelfer M, Pandya K (2004) Synthesis of zerovalent nanophase metal particles stabilized with poly (ethylene glycol). Langmuir 20(16):6896
149. Wang H, Zhu JJ, Zhu JM, Chen HY (2002) Sonochemical method for the preparation of bismuth sulfide nanorods. J Phys Chem B 106:3848
150. Li HL, Zhu YC, Chen SG, Palchik O, Xiong J, Koltypin Y, Gofer Y, Gedanken A (2003) A novel ultrasound-assisted approach to the synthesis of CdSe and CdS nanoparticles. J Solid State Chem 172:102
151. Zhu JJ, Aruna ST, Koltypin Y, Gedanken A (2000) A novel method for the preparation of lead selenide: pulse sonoelectrochemical synthesis of lead selenide nanoparticles. Chem Mater 12:143

152. Wang WZ, Geng Y, Yan P, Liu FY, Xie Y, Qian YT (1999) A novel mild route to nanocrystalline selenides at room temperature. J Am Chem Soc 121:4062
153. Zhu JJ, Xu S, Wang H, Zhu JM, Chen HY (2003) Sonochemical synthesis of CdSe hollow spherical assemblies via an In-Situ template route. Adv Mater 15:156
154. Li H, Zhu Y, Chen S, Palchik O, Xiong J, Koltypin Y, Gofer Y, Gedanken A (2003) A novel ultrasound-assisted approach to the synthesis of CdSe and CdS nanoparticles. J Solid State Chem 172:102
155. Wang H, Xu S, Zhao X-N, Zhu J-J, Xin X-Q (2002) Sonochemical synthesis of size-controlled mercury selenide nanoparticles. Mater Sci Eng, B 96:60
156. Mdleleni MM, Hyeon T, Suslick KS (1998) Sonochemical synthesis of nanostructured molybdenum sulfide. J Am Chem Soc 120:6189
157. Xie Y, Zheng X, Jiang X, Lu J, Zhu L (2002) Sonochemical Synthesis and Mechanistic Study of Copper Selenides $Cu_{2-x}Se$, β-CuSe, and Cu_3Se_2 Inorg Chem 41:387
158. Li B, Xie Y, Huang J, Liu Y, Qian Y (2000) Sonochemical synthesis of nanocrystalline copper tellurides Cu_7Te_4 and Cu_4Te_3 at room temperature. Chem Mater 12:2614
159. Pejova B, Grozdanov I, Nesheva D, Petrova A (2008) Size-dependent properties of sonochemically synthesized three-dimensional arrays of close-packed semiconducting $AgBiS_2$ Quantum Dots. Chem Mater 20:2551
160. Okitsu K, Nagata Y, Mizukoshi Y, Maeda Y, Bandow H, Yamamoto TA (1997) Synthesis of palladium nanoparticles with interstitial carbon by sonochemical reduction of tetrachloropalladate (II) in aqueous solution. J Phys Chem B 101:5470
161. Nikitenko S, Koltypin Y, Palchik O, Felner I, Xu XN, Gedanken A (2001) Synthesis of highly magnetic, air stable iron-iron carbide nanocrystalline particles by using power ultrasound. Angew Chem Intd Ed 40:4447
162. Hyeon T, Fang M, Suslick KS (1996) Nanostructured molybdenum carbide: sonochemical synthesis and catalytic properties. J Am Chem Soc 118:5492–5493
163. Suslick KS, Hyeon T, Fang M, Ries JT, Cichowlas AA (1996) Sonochemical synthesis of nanophase metals, alloys and carbides. Mater Sci Forum 225–227:903–912
164. Bellissent R, Galli G, Hyeon T, Migliardo P, Parette P, Suslick KS, Noncryst J (1996) Magnetic and structural properties of amorphous transition metals and alloys. Solids 205:656
165. Mizukoshi Y, Okitsu K, Maeda Y, Yamamoto TA, Oshima R, Nagata Y (1997) Sonochemical preparation of bimetallic nanoparticles of gold/palladium in aqueous solution. J Phys Chem B 101:7033
166. Egami T (1984) Magnetic amorphous alloys: physics and technological applications. Rep Prog Phys 47:1601
167. Shafi K, Gedanken A, Prozorov R (1998) Sonochemical preparation and characterization of nanosized amorphous Co–Ni alloy powders. J Mater Chem 8:769
168. Li Q, Li H, Pol VG, Bruckental I, Koltypin Y, Moreno JC, Nowik I, Gedanken A (2003) Sonochemical synthesis, structural and magnetic properties of air-stable Fe/Co alloy nanoparticles. New J Chem 27:1194
169. Vinodgopal K, He Y, Ashokkumar M, Grieser F (2006) Sonochemically prepared platinum-ruthenium bimetallic nanoparticles. J Phys Chem B 110:3849
170. Srivastava DN, Chappel S, Palchik O, Zaban A, Gedanken A (2002) Sonochemical synthesis of mesoporous tin oxide. Langmuir 18:4160
171. Wang Y, Yin L, Gedanken A (2002) Sonochemical synthesis of mesoporous transition metal and rare earth oxides. Ultrason Sonochem 9:285
172. Bhatte K, Fujita S, Arai M, Pandit A, Bhanage B (2011) Ultrasound assisted additive free synthesis of nanocrystalline zinc oxide. Ultrason Sonochem 18:54
173. Jung S-H, Oh E, Lee K-H, Yang Y, Park CG, Park W, Jeong S-H (2008) Sonochemical preparation of shape-selective ZnO nanostructures. Cryst Growth Des 8:265–269
174. Yu JC, Yu J, Ho W, Zhang L (2001) Preparation of highly photocatalytic active nano-sized TiO_2 particles via ultrasonic irradiation. Chem Commun:1942
175. Zhang D, Fu H, Shi L, Pan C, Li Q, Chu Y, Yu W (2007) Synthesis of CeO_2 Nanorods via Ultrasonication Assisted by Polyethylene Glycol. Inorg Chem 46:2446

176. Krishnan CVC, Burger J, Chu C (2006) Polymer-assisted growth of molybdenum oxide whiskers via a sonochemical process. J Phys Chem B 110:20182
177. Mao C-J, Pan H-C, Wu X-C, Zhu J-J, Chen H-Y (2006) Sonochemical route for self-assembled V_2O_5 bundles with spindle-like morphology and their novel application in serum albumin sensing. J Phys Chem B 110:14709
178. Dutta DP, Sudarsan V, Srinivasu P, Vinu A, Tyagi AK (2008) Indium oxide and europium/dysprosium doped indium oxide nanoparticles: sonochemical synthesis, characterization, and photoluminescence studies. J Phys Chem C 112:6781
179. Geng J, Zhu JJ, Lu DJ, Chen HY (2006) Hollow $PbWO_4$ Nanospindles via a Facile Sonochemical Route. Inorg Chem 45:8403
180. Geng J, Hou WH, Lv YN, Zhu JJ, Chen HY (2005) One-Dimensional $BiPO_4$ nanorods and two-dimensional BiOCl lamellae:fast low temperature sonochemical synthesis, characterization, and growth mechanism. Inorg Chem 44:8503
181. Dutta DP, Ghildiyal R, Tyagi AK (2009) Luminescent properties of doped zinc aluminate and zinc gallate white light emitting nanophosphors prepared via sonochemical method. J Phys Chem C 113:16954
182. Dhas NA, Suslick KS (2005) Sonochemical preparation of hollow nanospheres and hollow nanocrystals. J Am Chem Soc 127:2368
183. Bang JH, Suslick KS (2007) Sonochemical synthesis of nanosized hollow hematite. J Am Chem Soc 129:2242
184. Du N, Zhang H, Chen BD, Yu JB, Ma XY, Liu ZH, Liu ZH, Zhang YQ, Yang DR, Huang XH, Tu JP (2007) Porous Co_3O_4 nanotubes derived from $Co_4(CO)_{12}$ clusters on carbon nanotube templates: a highly efficient material for Li-battery applications. Adv Mater 19:4505
185. Wang J, Loh KP, Zhong YL, Lin M, Ding J, Foo YL (2007) Bifunctional FePt core-shell and hollow spheres: sonochemical preparation and self-assembly. Chem Mater 19:2566
186. Yin J, Qian X, Yin J, Shi M, Zhou G (2003) Preparation of ZnS/PS microspheres and ZnS hollow shells. Mater Lett 57:3859
187. Hernandez Y, Nicolosi V, Lotya M, Blighe FM, Sun Z, De S, McGovern IT, Holland B, Byrne M, Gun'Ko YK, Boland JJ, Niraj P, Duesberg G, Krishnamurthy S, Goodhue R, Hutchison J, Scardaci V, Ferrari VAC, Coleman JN (2008) High-yield production of graphene by liquid-phase exfoliation of graphite. Nat Nanotechnol 3:563
188. Coleman JN, Lotya M, O'Neill A, Bergin SD, King PJ, Khan U, Young K, Gaucher A, De S, Smith RJ, Shvets IV, Arora SK, Stanton G, Kim HY, Lee K, Kim GT, Duesberg GS, Hallam T, Boland JJ, Wang JJ, Donegan JF, Grunlan JC, Moriarty G, Shmeliov A, Nicholls RJ, Perkins JM, Grieveson EM, Theuwissen K, McComb DW, Nellist PD, Nicolosi V (2011) Two-dimensional nanosheets produced by liquid exfoliation of layered materials. Science 331:568
189. Toublan FJJ, Boppart S, Suslick KS (2006) Tumor targeting by surface-modified protein microspheres. J Am Chem Soc 128:3472
190. Suslick KS, Grinstaff MW (1990) Protein micro-encapsulation of non-aqueous liquids. J Am Chem Soc 112:7807
191. Grinstaff MW, Suslick KS (1991) Air-filled protein aceous microbubbles: synthesis of an echo-contrast agent. Proc Natl Acad Sci USA 88:7708
192. Xu H, Zeiger BW, Suslick KS (2013) Sonochemical synthesis of nanomaterials. Chem Soc Rev 42:2555
193. Kitano M, Kitano T, Hamabe S, Maeda T, Okabe J (1990) Growth of large tetrapod-like ZnO crystals: experimental considerations on kinetics of growth. Crys Growth 102:965
194. Gao T, Li Q, Wang T (2005) Sonochemical synthesis, optical properties, and electrical properties of core/shell-type ZnO nanorod/CdS nanoparticle composites. Chem Mater 17 (4):887
195. Morel AL, Nikitenko SI, Gionnet K, Wattiaux A, Him JLK, Labrugere C, Chevalier B, Deleris G, Petibois C, Brisson A, Simonoff M (2008) Sonochemical approach to the

synthesis of $Fe_3O_4@SiO_2$ core−shell nanoparticles with tunable properties. ACS Nano 2 (5):847
196. Gao T, Wang T (2004) Sonochemical synthesis of SnO_2 nanobelt/CdS nanoparticle core/shell heterostructures. Chem Commun 22:2558
197. Anandan S, Grieser F, Kumar MA (2008) Sonochemical synthesis of Au-Ag core-shell bimetallic nanoparticles. J Phys Chem C 112(39):15102
198. Salkar RA, Jeevanandam P, Aruna ST, Koltypin Y, GedankenJ A (1999) The sonochemical preparation of amorphous silver nanoparticles. Mater Chem 9:1333
199. Ranasinghe JC, Dikkumbura AS, Hamal P, Chen M, Khoury RA, Smith HT, Lopata K, Haber LH (2019) Monitoring the growth dynamics of colloidal gold-silver core-shell nanoparticles using in-situ second harmonic generation and extinction spectroscopy. J Chem Phys 151:
200. Kodas TT, Hampden-Smith M (1999) Aerosol processing of materials. Wiley-VCH, New York
201. Suh WH, Suslick KS (2005) Magnetic and porous nanospheres from ultrasonic spray pyrolysis. J Am Chem Soc 127:12007
202. Pluym TC, Powell QH, Gurav AS, Ward TL, Kodas TT, Wang LM, Glicksman HD (1993) Solid silver particle production by spray pyrolysis. J Aerosol Sci 24:383
203. Majumdar D, Kodas TT, Glicksman HD (1996) Gold particle generation by spray pyrolysis. Adv Mater 8:1020
204. Kim JH, Germer TA, Mulholland GW, Ehrman SH (2002) Size-monodisperse metal nanoparticles via hydrogen-free spray pyrolysis. Adv Mater 14:518
205. Nagashima K, Wada M, Kato A (1990) Preparation of fine Ni particles by the spray-pyrolysis technique and their film forming properties in the thick film method. J Mater Res 5:2828
206. Pluym TC, Kodas TT, Wang L-M, Glicksman HD (1995) Silver-palladium alloy particle production by spray pyrolysis. J Mater Res 10:1661
207. Vivekchand SRC, Gundiah G, Govindaraj A, Rao CNR (2004) A new method for the preparation of metal nanowires by the nebulized spray pyrolysis of precursors. Adv Mater 2004(16):153
208. Skrabalak SE, Suslick KS (2005) Porous MoS_2 synthesized by ultrasonic spray pyrolysis. J Am Chem Soc 127:9990
209. Suh WH, Jang AR, Suh Y-H, Suslick KS (2006) Porous, hollow, and ball-in-ball metal oxide microspheres: preparation, endocytosis, and cytotoxicity. Adv Mater 2006:18
210. Skrabalak SE, Suslick KS (2006) Porous carbon powders prepared by ultrasonic spray pyrolysis. J Am Chem Soc 128:12642

Chapter 13
Hot Injection Method for Nanoparticle Synthesis: Basic Concepts, Examples and Applications

Abhishek Kumar Soni, Rashmi Joshi, and Raghumani Singh Ningthoujam

Abstract Highly monodispersed nanoparticles produced by the hot injection method are discussed. The hot injection synthesized nanoparticles are remarkable materials with size-dependent properties leading to advanced developments in nanoscience and nanotechnology. The basic classical theory, nucleation and growth of the nanocrystals are discussed in the framework of hot injection method. Kinetics of the hot injection method has been explored with the help of Ostwald ripening process and crystal growth mechanism. A comparison of hot injection synthesis method with other methods has been explored. A detailed description of monodispersed nanocrystals has been given by taking various important examples such as Co, Ag, Au, CdSe, PbSe, PbS, SnS_2, FeS_2, $CuInS_2$, Cu_2ZnSnS_4, Cu_2NiSnS_4 and ferrites, nanomaterials. Applications of the hot injection synthesized nanoparticles in different areas and their uses in device fabrication have also been specified. This chapter suggests the utility of the hot injection method in the formation of size and shape-dependent colloidal nanoparticles with desirable optical, electrical and magnetic properties.

Keywords Hot injection method · Nanoparticles · Nucleation and growth · Monodisperse · Nanocrystals · Quantum dots

List of Abbreviations

CQDs	Carbon quantum dots
CTAB	Cetyltrimethylammonium bromide
CZTS	Copper zinc tin sulfide
DDT	Dodecanethiol
DTR	Diffusion transfer reversal
E°	Reduction potential

A. K. Soni · R. Joshi · R. S. Ningthoujam (✉)
Chemistry Division, Bhabha Atomic Research Centre, Mumbai 400085, India
e-mail: rsn@barc.gov.in

R. Joshi · R. S. Ningthoujam
Homi Bhabha National Institute, Mumbai 400094, India

A. K. Tyagi and R. S. Ningthoujam (eds.), *Handbook on Synthesis Strategies for Advanced Materials*, Indian Institute of Metals Series,
https://doi.org/10.1007/978-981-16-1807-9_13

E_g	Band gap
FTO	Fluorine doped tin oxide
H_c	Coercivity
HOMO	Highest occupied molecular orbital
LUMO	Lowest unoccupied molecular orbital
M_r	Retentivity
MSCs	Magic size clusters
OA	Oleic acid
OAM	Oleylamine
ODE	1-Octadecene
PVP	Polyvinylpyrrolidone
QDs	Quantum dots
SAM	Self-assembly monolayer
SDS	Sodium lauryl sulfate
SERS	Surface enhanced Raman scattering
SLBL	Successive layer by layer
TOP	Trioctylphosphine
TOPO	Trioctylphosphine oxide
TREG	Triethylene glycol

13.1 Introduction

Nanoscience and nanotechnology are related to study and application of extremely small particles (1–100 nm in one of dimensions). This links chemistry, biology, physics, materials science and engineering. The size and shape of the nanoparticles can tune the optical, electrical, thermal, catalytic and magnetic properties. Thus, the application of these size-dependent nanoparticles in various fields viz. medical, energy science, optoelectronics, display and photonic, optics, biomedical sciences, chemical industries, space industry and sensing opens new hope and challenge to fabricate these nanoparticles in large scale with desired properties [1–8]. The main aim is to develop the nanoparticles via suitable synthesis techniques, which can efficiently control monodispersity, particle sizes and shapes for the desired applications. Colloidal synthesis of the nanocrystals swiftly preserves extensive era in the development of size-dependent optical active nanomaterial. Nowadays, the current challenge for nanocrystal growth and nucleation in nanoparticle synthesis is to understand the influence of reaction parameters, precursor used as raw materials and their concentration, temperature and types of the ligand. The nucleation is an initial phase of the nanoparticle synthesis which involved basically self-gathering or/association process. Nucleation is a stochastic process because it may differ for two identical systems occurring at different time [9]. The size of nanoparticles and their uniformities are strongly related to the chemical and physical properties in an ensemble of nanocrystals. For the colloidal synthesis, burst nucleation and diffusion-controlled growth are the two significant factors to control the size distribution of the nanoparticles. Hot injection and heat up (or thermolysis) methods involve the decomposition of organometallic compounds/complex at higher temperature in the presence of long-chain fatty acids (acting as surface passivation over nanoparticles) and high-boiling point solvents. These two methods are the two main strategies for the synthesis of uniform nanocrystals of various materials including metals, alloys, oxides, selenides, tellurites, phosphides, sulfides, core-shell as well as composites. Heat up method suffers some disadvantage such as laborious cooling process, volatile solvent removal, controlling thickness(layers) of shell coating and anisotropy in shape as well [10–12]. The benefit of the hot injection synthesis method is that it requires one-pot synthesis which allows successive layer-by-layer (SLBL) synthesis of multi-shell nanoparticles. Hot injection synthesis of inorganic nanoparticles deals through molding with the role of the precursor and next to renovation in the product development. First, Murray et al. have reported a general method for the synthesis of cadmium chalcogenide nanocrystals in 1993, which later named as the "hot injection" method [10]. The method was practically enable to make CdSe nanocrystals with moderate size uniformity in which mean size was controllable from 2 to 12 nm. Many reports have been carried out in the earlier years, which were showing size-dependent tuning of the band gap as well as the progress of the band-edge structures. This can be done via quantum dots or quantum confinement effect of semiconductor nanocrystals produced by hot injection method [13–15]. The size of particles in nanometer range changes magnetic, optical and catalytic properties as compared to bulk [16, 17]. The phenomenon of the quantum dots (quantum crystallites) is related to the band structure formation with changing the crystallite size. Classical model for uniform microparticles delivers a useful basis for the systematic understanding of nanocrystal sizes and reaction conditions in which burst nucleation, diffusion controlled growth and Ostwald ripening are the important processes for the nanocrystal realizations. LaMer and Dinegar have studied the condition for nucleation and growth of sulfur sols for dilute and concentrated solutions of precursors [18]. In hot injection method, a cold stock solution containing the precursors is injected quickly into the hot solution containing surfactant and high-boiling point solvent. The kinetics of the hot injection method is such that there is no or little induction time between injection and precipitation. This is an essential step in the hot injection synthesis process which characteristically determines the formation of the monodispersed nanocrystals. Understanding about monitoring and handling of the size-dependent nanoparticles and kinetics of their growth in the hot injection method is very interesting. Peng et al. have reported the colloidal semiconductor nanocrystals growth via "Focusing" of size distributions in which concentration of the monomer can change size distribution [19]. Here, monomer is a minimal building unit of a crystal. These monomers in solution can be dissolved or form crystal particles. The supersaturation occurs when concentration of monomers ($[M]$) is more than that of equilibrium monomers concentration of the bulk solid ($[M]_0$). Monomers can either form crystal particles or dissolve back into the solution. For

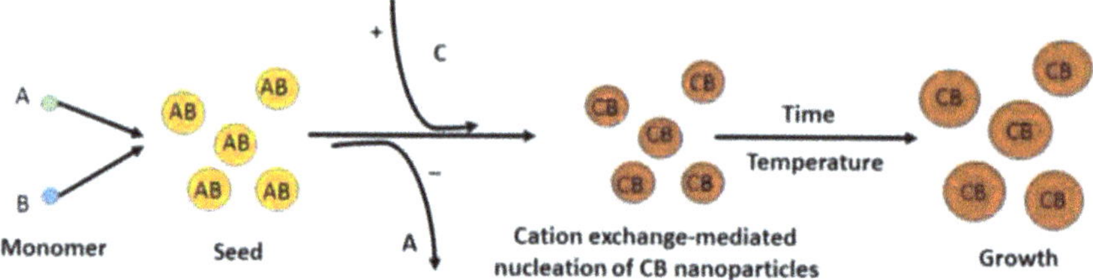

Fig. 13.1 Formation of cation-exchange mediated CB nanocrystals and growth

higher monomer concentration, the rate of growth of smaller nanocrystals is faster than that of the larger nanocrystals, and thus the size distribution is focussed down to one that is nearly monodisperse. For lower monomer concentration (below the critical threshold), smaller nanocrystals are exhausted and larger nanocrystals grow. Therefore, the distribution broadens or defocusing was achieved in this case which may promote the Ostwald ripening process. Control in the nucleation rate by generating intermediate nuclei through the modified hot injection and cation-exchange process for PbSe quasi seeds has been reported by Kovalenko et al. [20]. Figure 13.1 shows the cation-exchange mediated nucleation of CB nanocrystals. Here, seed AB nanocrystals are already formed. After injection of C precursor, C exchanges with A to form CB nanocrystals. With increase of time of reaction or temperature, chance of growth of particles is enhanced. The nucleation and growth can be separated by quenching process [21]. Here, cold precursor(s) are injected into the coordinating solvent at high temperature. Within a second to few seconds, nucleation starts. Further growth of particles can be stopped by pouring hot solution into cold ethanol (20 °C), which causes quenching.

In this chapter, basic concepts encountered for understanding various nanoparticles prepared by the hot injection method are mentioned. This includes following concepts such as kinetics of nucleation and growth of nanoparticles, Oswald ripening process, quantum dots, quantum confinement, role of surfactants, difference between the hot injection method and other methods to prepare monodispersed nanoparticles. The advantages and disadvantages of the hot injection method are provided. Many examples of nanomaterials prepared by hot injection and heat up methods are mentioned. Applications of nanomaterials prepared by hot injection method in various areas such as solar cells, high-density storage devices, laser devices and biomaterials are provided.

13.2 Basic Concepts

In the classical nucleation theory, homogeneous nuclei can be formed with the help of thermodynamics via understanding the total free energy of the nanoparticles [22]. Total free energy (ΔG) of the nanoparticle is defined as the sum of surface free energy ($\Delta G_s = 4\pi r^2\gamma$) under consideration of spherical particles and bulk free energy ($\Delta G_b = \frac{4}{3}\pi r^3\Delta G_v$) and can be written as,

$$\Delta G = \Delta G_s + \Delta G_b$$

$$\Delta G = 4\pi r^2\gamma + \frac{4}{3}\pi r^3\Delta G_v$$

where $\Delta G_v = \frac{-k_b T\ln(S)}{v}$ is a crystal free energy, k_B is Boltzmann's constant, S is the supersaturation of the solution, v is its molar volume, γ is surface energy, and r is the radius of spherical particle. The value of ΔG_v is temperature dependent. Critical free energy can be calculated by differentiating ΔG with respect to r and setting it to zero value, which can be represented as,

$$\frac{\mathrm{d}\Delta G}{\mathrm{d}r} = 0$$

The critical radius (r_{cri}) for minimum size at which a particle can exist in a solution without actuality redissolved which can be obtained by utilizing the above equation [22].

$$r_{\mathrm{cri}} = \frac{2\gamma v}{k_b T\ln(S)} = \frac{-2\gamma}{\Delta Gv}$$

Figure 13.2 shows the total free energy plot with contribution of ΔG_s and ΔG_b as a function of particle size for an arbitrary system. When $r < r_{\mathrm{cri}}$, system tries to get lower free energy by sacrificing monomers or smaller size particles [22, 23]. This can happen by deposition of monomers or smaller sized particles over particles already present in solution, leading to larger-sized particles [22, 23]. The nucleation occurs when concentration of monomers becomes supersaturated (S), which is defined as ratio between the monomer concentration and the monomer concentration in equilibrium with solubility of the bulk solid phase. The rate of nucleation and type of precursor used during the reaction can determine the shape of particles [20].

Rempel et al. reported the formation of the small nanocrystals having narrow size distributions via either modulating temperature or by targeted introduction of additives with the help of the precursor activation chemistry [24]. The impurity atoms present in the system may produce a heterogeneous nucleation at a particular nucleation sites via displacements, vessel surfaces and imperfections. The heterogeneous nucleation depends on the contact angle of the nucleus with respect to the substrate, and hence, these nanoparticles are not in spherical shapes. The value of r_{cri} is identical for both the nucleation processes, but the critical volume of nucleus for heterogeneous nucleation is smaller than that of the homogeneous nucleation. Additionally, the total free energy essential for heterogeneous nucleation in the nanoparticle growth process is equivalent to the product of homogeneous

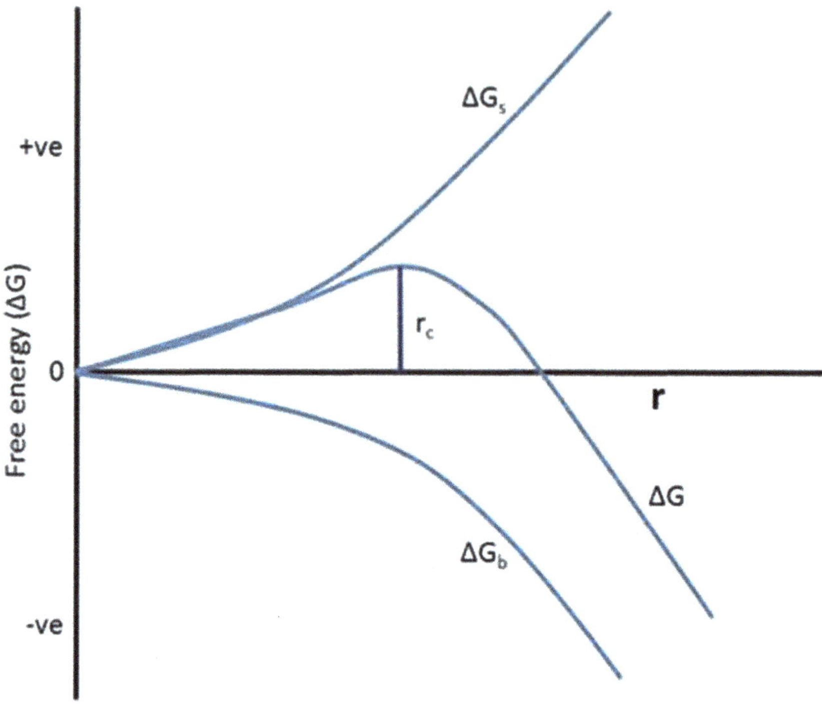

Fig. 13.2 Total free energy plot with contribution of ΔG_s and ΔG_b as a function of particle size for an arbitrary system

nucleation and a function of contact angle (θ), which can be precisely represented as follows [9, 24],

$$\Delta G_t(\text{Heterogeneous}) = f(\theta)\Delta G_t(\text{Homogeneous})$$

13.2.1 *Kinetics of the Hot Injection Method*

The hot injection method is a rapid technique for the preparation of monodispersed nanocrystals of metals, alloys, core-shell, composites, etc. The synthesized nanoparticles with tunable size and shape can be fabricated for a particular system or application. The kinetics of reaction can be controlled by changing the reaction parameters such as injection temperature, waiting time, use of seed or catalytic sites, reducing or oxidizing agent or reaction atmosphere, and hence, growth of the nanocrystals can be controlled. In the further sections, some kinetics of the hot injection method with Ostwald ripening process and growth mechanism of the nanoparticles will be described.

13.2.2 Ostwald Ripening Process

Ostwald ripening is an important phenomenon in solid or liquid which explain the evolution of an inhomogeneous structure growth over time. This phenomenon can affect the nanoparticles growth and nucleation in the hot injection synthesis. The Wilhelm Ostwald in 1896 has described this phenomenon to understand the mechanism of the nanoparticle growth process. In chemistry, Ostwald ripening explains the process such as smaller particle having higher surface energy than larger particle which can provide higher solubility due to higher total Gibbs energy. This leads to formation of energetically favorable larger particles. The system tends to lower its overall energy thus molecules on the surface of small particles tend to detach from the particle and simultaneously defuse in the solution. Thus, an increment in the free particle concentration can be achieved in this process. The free molecules have a tendency to condense on the surface of the larger particles due to the supersaturation of the free molecules in solution. Thus, smaller particles decrease and larger particles parallel grow, and hence, the overall particle size will increase effectively. The solubility of the molecules or nanocrystals is very sensitive to the size of nanocrystal, and it motivates Ostwald ripening process in the solution. If the size distribution is primarily quite extensive, the size uniformity can be enhanced via Ostwald ripening phenomenon as the lesser particles dissolve away to a certain step. Ostwald ripening is a slow process compared to size focusing. Size focussing produces narrow size distributions, which require the maintenance of a high supersaturation of monomers during particle growth [25]. For example, $NaAF_4$ (A = rare earth atom) nanocrystals can be prepared using metal trifluoro acetates as thermally labile precursors for F^- and Na^+/A^{3+} ions. Here, rate of productions of monomers $NaYF_4$ is high. This is a requirement of size focusing. In case of Oswald's ripening, high concentration of monomers is not required. Particle growth is due to depletion of small size particles.

13.2.3 Growth Mechanism

Growth of the crystal particles (particularly spherical particles) can take place through two steps processes. In the first step, transport of the monomers from the bulk solution onto the crystal surface is happened and could be described by the Fick's law of diffusion [22],

$$j = -D\frac{\mathrm{d}[M]}{\mathrm{d}x}$$

where j is the monomer flux, D is the diffusion constant, and M and x are the monomer concentration and the radial distance from the center of the spherical particle, respectively. The second step are the reaction of the monomers at the

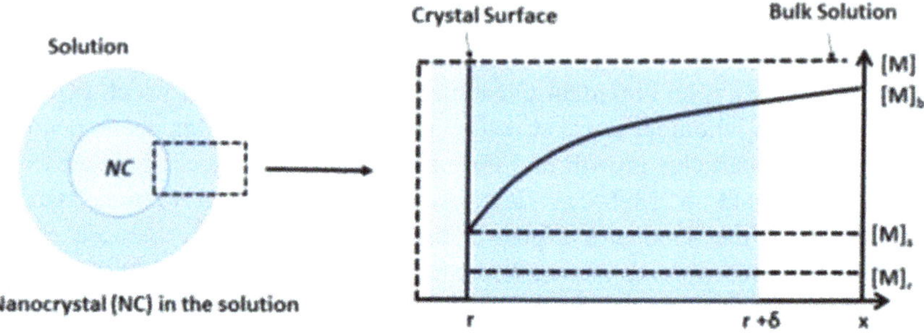

Fig. 13.3 Schematic illustration of diffusion layer structure near the surface of a nanocrystal (left) and plot for the monomer concentration as a function of distance x (right). Redrawn from Ref. [22]. The shaded area indicates the diffusion layer

surface. The concentration gradient around the surface with diffusion layer and their schematic of the particle was sketched in Fig. 13.3. Diffusion layer structure near the surface of a nanocrystal (left) and plot for the monomer concentration as a function of distance x (right) is shown in Fig. 13.3. The graph shows the monomer concentration $[M]_s$ at the surface of the crystal ($x = r$). At the point ($x = r + \delta$), the concentration reaches the value equal to bulk concentration of the solution, i.e., $[M]_b$. The total flux of the monomer is equal to the monomer consumption rate by the surface reaction of the particle. With the help of growth of the particle, the consumption rate of monomer can be obtained and is equal to total flux of the monomer due to the mass balance. The monomer consumption rate can be written as following equation [26–28],

$$J = 4\pi r^2 k\left([M]_s - [M]_r\right)$$

where k and $[M]_r$ are the reaction constant and solubility of the spherical particle of radius, respectively. The growth rate can be expressed as following equation [26, 29],

$$\frac{\mathrm{d}r}{\mathrm{d}t} = \frac{DV_m([M]b - [M]r)}{r - D/k}$$

where V_m is the molar volume of the monomer in crystal, and another term has their common meanings. A schematic profile of size distribution and nucleation process with growth periods was depicted in Fig. 13.4. In the lower part of Fig. 13.4, the relative standard deviation of the size distribution σ_r (size) was shown. The general growth rate equation for a spherical particle can be expressed as follows [30],

$$\frac{\mathrm{d}r^*}{\mathrm{d}\tau} = \frac{S - \exp(1/r^*)}{r^* + K\exp(\alpha/r^*)}$$

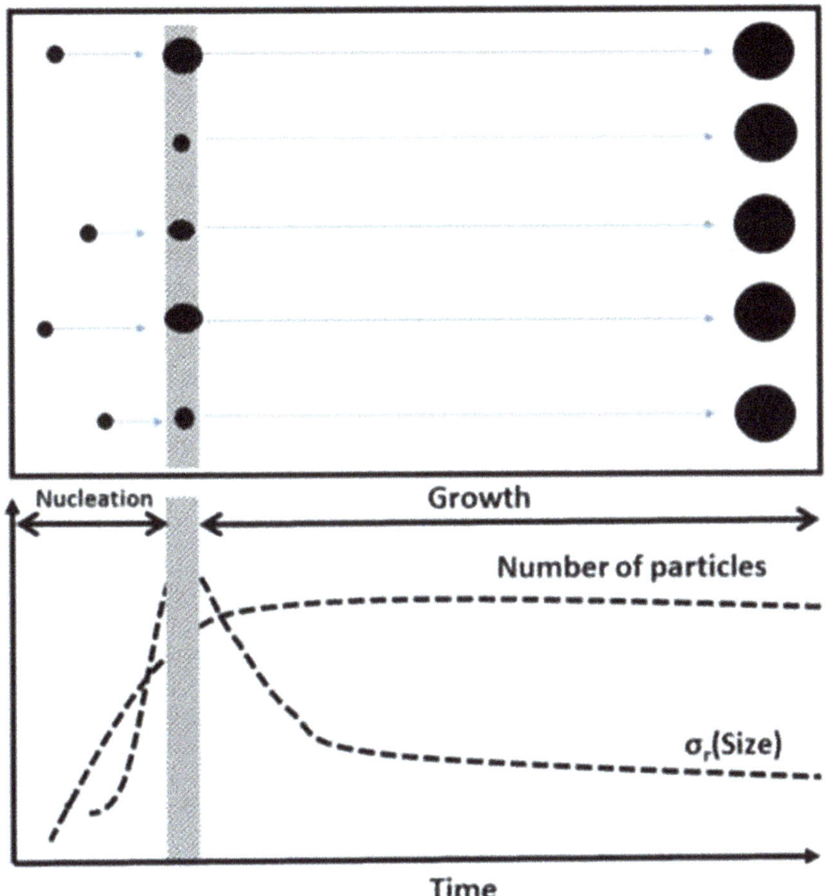

Fig. 13.4 Schematic illustration of the size distribution control process. The thick vertical line (shaded) indicates the point in time at which the nucleation process is terminated, dividing the nucleation and growth periods. In the lower part of the figure, the time evolution of the number of particles and the relative standard deviation of the size distribution, σ_r(size), are shown. Redrawn from Ref. [26]

where r^* and τ are the dimensionless particle radius and time, respectively. K is the dimensionless parameter describing the processes involved in the reaction. S and α are the oversaturation of the monomer in solution and transfer coefficient, respectively.

The chemical potential of a particle and its radius can be related to each other with the help of the Gibbs–Thomson relation. Chemical potentials of the monomers in the solution and the particle with the radius "r' can be taken as μ_b and $\mu(r)$, respectively. The difference between these two chemical potentials of monomers in particle and solution can be written as follows,

$$\Delta\mu = \frac{2\gamma V_m}{r}$$

The equation clearly shows that the chemical potential is dependent on the radius of the particle. Strictly, smaller particles have a higher chemical potential. Figure 13.5 shows the schematics of chemical potential versus reaction coordinate for particles of three different sizes (r_1, r_2 and r_3). The radius of the particles is taken as increasing order ($r_3 > r_2 > r_1$). For the particles of small radius, the chemical potential of its monomers is more than that of the monomers in the solution. Therefore, for smaller particles, dissolution in reaction is foremost. Furthermore, for the particles of larger radius, the precipitation reaction will be dominant on the surface since chemical potential of its monomers is less that of the monomers in the solution.

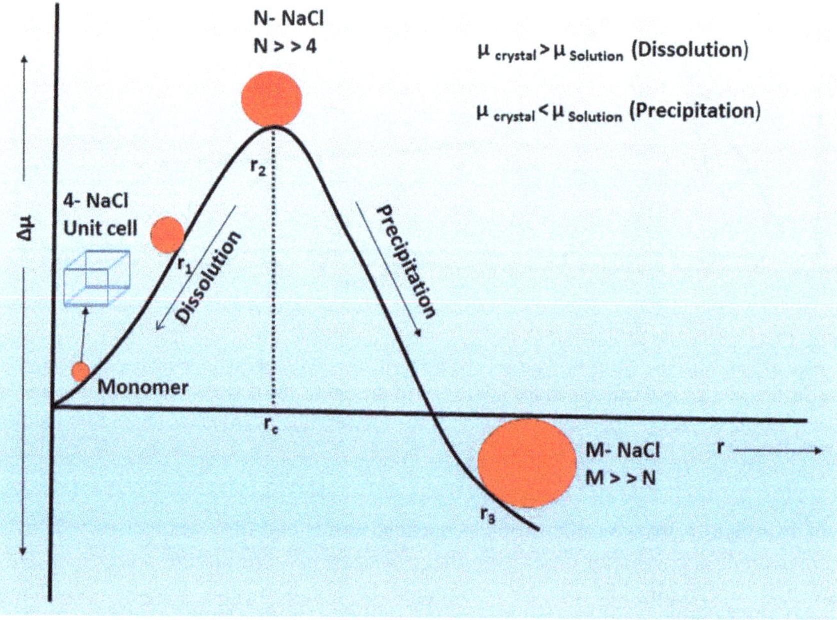

Fig. 13.5 Schematics of chemical potential versus reaction coordinate for NaCl particles of three different sizes (r_1, r_2 and r_3, where $r_3 > r_2 > r_1$)

13.2.4 Quantum Dots and Quantum Confinement

Quantum dots are the semiconductor particles of spherical nanometer sizes; whose sizes are less than the exciton Bohr's radius [31]. Since size is so small, one quantum dot acts as H-like atom where a positive nucleus is surrounded by a negative electron with radius (*R*). They obey a principle of quantum confinement. They are considered as 0 (zero) dimensional particles. Thus, movement of electrons in all three *x*-, *y*- and *z*- dimensions are confined. Their band gap (E_g) between the lowest energy level (valence band) and the highest energy level (conduction band) is more than the bulk value ($E_{g(R\rightarrow\infty)}$). Also, they show the many discrete absorption bands. The relation between E_g and $E_{g(R\rightarrow\infty)}$ is given below:

$$E_g = E_{g(R\rightarrow\infty)} + \frac{h^2}{8R^2}\left(\frac{1}{m_e^*} + \frac{1}{m_h^*}\right)$$

where m_e^* and m_h^* are the effective masses of electron and hole, respectively. For example, bulk PbSe has a large exciton Bohr's radius of 46 nm with band gap of 0.26 eV [32]. When particle size is less than 46 nm, it shows quantum confinement effect, and those particles are considered as quantum dots (QDs). These can show the sharp absorption and emission peaks. Depending on particle size, emission peak can vary from 900 to 4100 nm range. The luminescence quantum yields can be achieved up to 90%. Interestingly, QDs have one emission peak, but have many discrete absorption peaks (Fig. 13.6) [32].

When particle changes the shape from spherical to wire, plane, confinement effect decreases even though one of its dimensions is of nanometer scale. They are considered as 1- and 2-dimensional particles. Instead, the polarizability factor along axis and plane arise in those 1- and 2-dimensional particles. Such particles show the

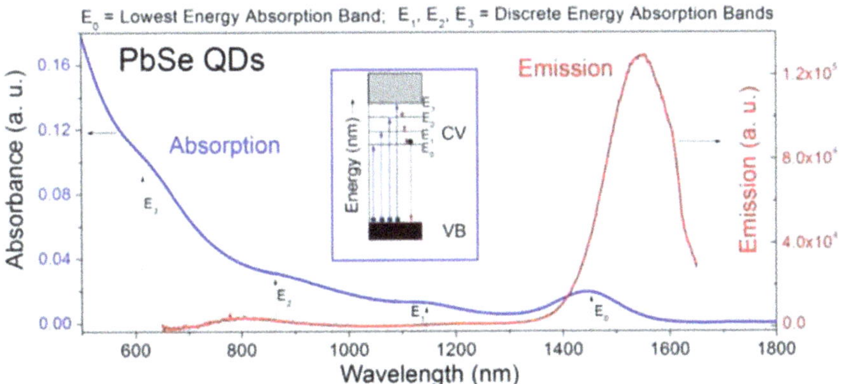

Fig. 13.6 Absorption and emission spectra of the PbSe QDs along with their energy level diagram (inset). Wavelength of excitation is 633 nm (He–Ne Laser)

unusual optical and electronic properties. Generally, spherical particles show the isotropic effect, whereas wire- or plane-typed particles can have anisotropic effect. The development of the quantum dots, wires and rods with various sizes and shapes can be obtained via surface engineering. By changing dielectric medium, the quantum confinement effect can be generated. Particle prepared in ligand free or liquid medium free (say solid state synthesis) may not be able to show the quantum confinement effect since size is more than bulk exciton Bohr's radius, but in different dielectric medium, it can show it [31]. For example, bulk SnO_2 is *n*-type semiconductor with the Bohr's radius of 2.7 nm and band gap of 3.6 eV. SnO_2 nanoparticles with size 2.3–3.1 nm show the blue shift in band gap by 0.27 eV. They are considered as the QDs [33]. When size of SnO_2 is 5 nm, it does not show the formation of exciton. The particle sizes of SnO_2 in 5–50 nm are not considered as QDs even though they belong to nanomaterials. In such particles, surface-related properties can be observed. However, when SnO_2 particles are dispersed in a medium A = TiO_2 or SiO_2 or Y_2O_3 matrix [34–37], it can generate the exciton even though particle size is about 5 nm. This was shown by Ningthoujam and his group [34–37]. It means that exciton (e-h pair) can be confined in the SnO_2: A nanostructures/nanohybrids/core-shell.

The absorption and generated emissions within QDs are dependent on the size of particles. QDs have high absorption coefficient and zero dimensions which can produce the sharper density of states than the higher dimensional particles [38]. Each quantum dot has a capability to convert incoming light (excitation light) into the light which can have small or larger Stokes shift. The dispersion capability of the quantum dots is very high in different solvents (hexane, silicone-based polymeric fluids, iodide and other polar medium) [39–41]. The applicability of the colloidal quantum dots in many fields such as light-emitting diodes, display and solid-state lighting technologies, bioimaging and drug delivery has been reported in the various literatures [42–44]. In this study, hot injection method was used to achieve the monodispersible quantum dots with tailoring optical properties. The various quantum dots, viz. CdSe, PbSe, PbS, $AgInS_2$/ZnS, Zn-In-S:Ag, perovskite $CsPbBr_3$, CdTe, CZTS (Cu_2ZnSnS_4), etc., were synthesized by using the hot injection and thermolysis techniques [45–48].

13.2.5 Use of the Surfactants for Nanoparticle Synthesis

The surfactant can reduce the surface tension between two liquids or between a liquid and a solid during the synthesis process. The surfactant used in the synthesis of nanoparticles prevents agglomeration of the nanoparticles by acting as a capping regent. The surfactant plays an important role in the colloidal chemical synthesis of the nanocrystals. In the hot injection synthesis method, the surfactant provides the uniform colloidal stability in the liquid for the nanoparticles so that monodispersed nanosized particles can stay for a long time. Generally, polar group of the surfactants is used in the reaction medium to bind the surface of the nanocrystals. For

CdSe nanoparticles synthesis, the prominent surfactants, namely tri-*n*-octylphosphine oxide (TOPO), tri-*n*-octylphosphine (TOP), fatty acids, hexadecylamine and phosphonic acids, have been used [19, 48, 49]. In hot injection method, the surfactants comprising functionalities (e.g., alcohols, acids, amines and thiols) can interact with the surface of the particles to stabilize growth of the particles and also to protect the particles from the sedimentation, agglomeration or losing their surface properties. For Ag nanoparticles in different synthesis techniques, surfactants such as starch, glucose, polysaccharides, β-D-glucose and citrate ions have been used [50–54]. Poly (vinyl alcohol), poly (vinylpyrrolidone), poly (ethylene glycol), poly (methacrylic acid) and polymethylmethacrylate as a polymeric compound were reported as a protective agent in the fabrication of the nanoparticles.

The biopolymer chitosan has been used to attach metal or magnetic nanoparticles due to its eco-friendliness [55, 56]. However, its stability is poor in many solvents. It needs to cross-link between chitosan polymers. Since chitosan molecule has functional group of amine ($-NH_2$), it can have a positive charge on $-NH_3$ group in acidic medium. However, cross-linker (sodium tri-polyphosphate) can be used to bridge between $-NH_3$ group (+ve) of chitosan and –O (−ve) of sodium tri-polyphosphate.

13.2.6 Difference Between Hot Injection and Other Methods to Prepare Monodispersed Nanoparticles

To prepare monodispersed nanoparticles, the following methods can be adopted.

A Electrostatic repulsion method
B Reverse micelle formation method
C Thermolysis method
D Hot injection method

A **Electrostatic repulsion method**

In electrostatic repulsion method, small size particles (a few nanometers 2–10 nm) have the same charge (−ve or +ve) and are able to stabilize in a liquid (closed system) for a few hours to months (preferably in the absence of UV-light or radiation or other) [57]. Like charge particles are able to disperse in a medium (liquid). Usually, the stable colloidal particles have a large value of surface charge potential (zeta potential, $\zeta \geq -30$ or +30 mV). Particles with very small sizes as well as high zeta potential can repulse each other. In such condition, they are considered as monodispersed particles. Even ligand-free particles can be stable. When size of particles increases, the particles are tending to settle to the bottom of container/beaker. This is because of gravitational forces, which dominate the repulsion forces among like charged particles. Then, agglomeration or aggregation follows.

In agglomeration, individual particles can be separated by external forces such as ultrasonication or addition of long-chain amphiphilic molecules. Some kind of van der Waals forces exists between particles. On other hand, it is difficult to separate individual particles in the case of aggregation. Here, some kind of dipolar interactions (magnetic or electric poles) is present in particles, e.g., Au/Ag/Cu (2–10 nm) and Fe_3O_4 (10 nm) [57, 58]. Rate of nanoparticles settlement is dependent on type of metal ions present in a liquid medium and pH of medium in a fixed temperature [58]. Figure 13.7 shows the schematic diagrams of (a) formation of dispersed particles in a liquid, (b) settlement of larger particles in a liquid, (c) agglomeration of particles and (d) aggregation of particles.

B **Reverse micelle formation method**

In reverse micelle formation method, three phases such as water, oil (hexane, benzene, cyclohexane) and surfactant (amphiphilic molecules such as oleic acid (OA), oleylamine (OAM), cetyltrimethylammonium bromide (CTAB), sodium lauryl sulfate (SDS)) are required, and this mixture acts as a medium for the formation of nanoparticles [59, 60]. Sometime, co-surfactants such as a long-chain hydrocarbon alcohol are added to reduce surface tension of medium. Surfactant acts as interface between water and oil layers. In reverse micelle formation, amount of water is very less than oil or surfactant, i.e., droplets of water molecules in oil medium. Water droplet or aqueous region is surrounded by hydrophilic head group of surfactant molecules (Fig. 13.8a), and hydrophobic tail of surfactant molecules is away and free to interact with oil medium. This is called water to oil (W/O) microemulsion. In this, nanoparticles in the form of metals, alloys, semiconductors, insulators or composites can be formed (Fig. 13.8a, b). Thus, water droplet region is also considered as nanoreactor. As ratio of water to surfactant molecules increases, size of nanoparticles can be increased.

Examples:

(i) Formation of metal nanoparticles (Ag, Au, Pd, Pt, Ni, Cu)
(ii) Formation of alloys nanoparticles (FePd, FePt)
(iii) Formation of metal oxides/sulfides/selenides/telulides (Fe_3O_4, SnO_2, TiO_2, PbS, PbSe, PbTe)
(iv) Formation of composites (Ag/Fe_3O_4, Au/Ferrites).

Usually, smaller size particles (2–5 nm) can be stable and monodispersed. Particles are in dynamic state or Brownian motion. With increase of size of particles, the stability of particles in the oil medium decreases. It results in agglomeration of particles. This is due to the following possible mechanism: the formation of fusion among micelles, attraction among particles and acceleration due to gravitation. Since it contains aqueous medium, stability of particles in chemical or thermal terms decreases with times (days to months).

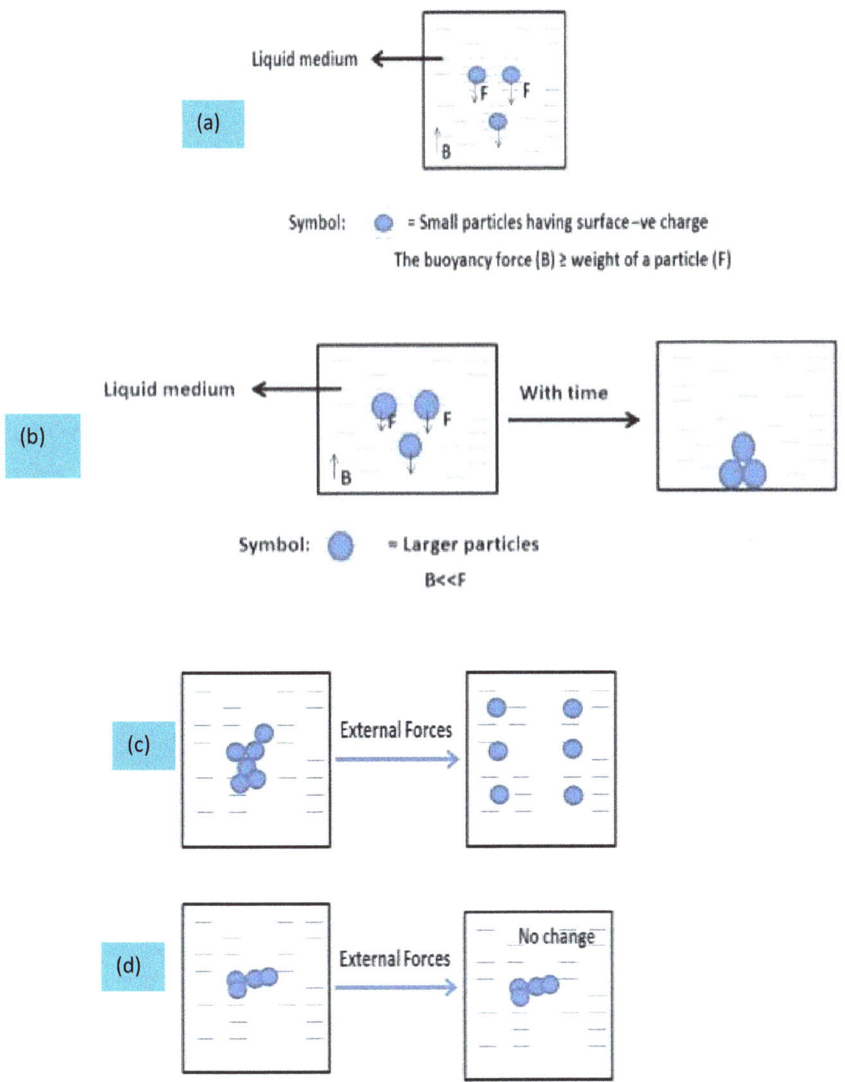

Fig. 13.7 Schematic diagrams of **a** formation of dispersed particles in a liquid, **b** settlement of larger particles in a liquid, **c** agglomeration of particles and **d** aggregation of particles

C Thermolysis method

Thermolysis means the decomposition of precursors in a liquid medium at the elevated temperature [60, 61]. To break a chemical bond of metal–ligand, it requires energy, which can be provided by external heating. After decomposition, it remains as metals, alloys, oxides, sulfides, telullides or sulfides, etc. However, water

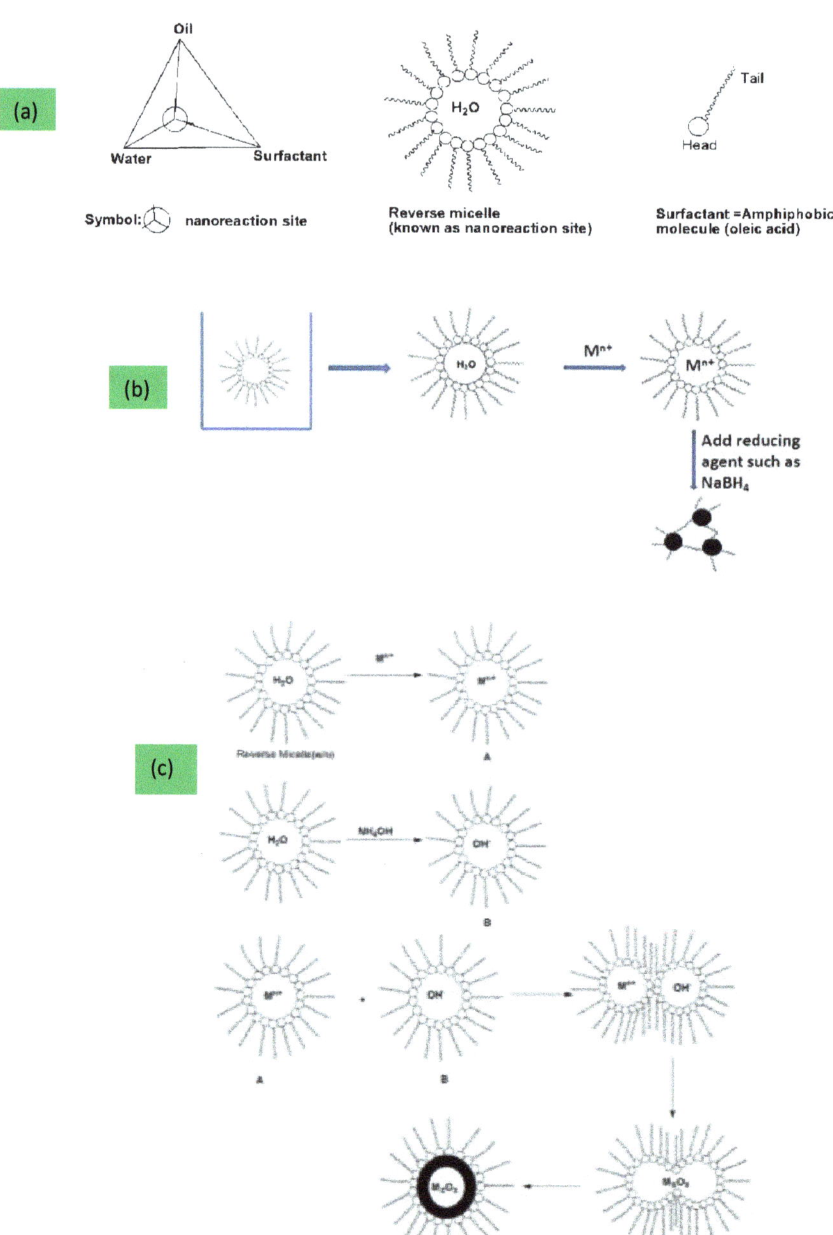

Fig. 13.8 **a** Reverse micelle formation, **b** formation of metal nanoparticles and **c** formation of metal oxides

medium cannot be used in high-temperature decomposition reaction. In such condition, high-boiling point organic solvents are necessary. To reduce agglomeration among particles after the preparation of particles, the following arrangement is required:

(i) A liquid medium having a high boiling point as well as long-chain hydrocarbons
(ii) Complexing agents
(iii) Additives (if requires)
(iv) Inert atmosphere
(v) Precursors for the formation of particles

Examples of a liquid medium having a long-chain hydrocarbon as well as a high boiling points are 1-octadecence (CH_3–$(CH_2)_{15}CH{=}CH_2$; b.p. 315 °C), diphenylether ($(C_5H_5)_2O$; b.p. 258 °C), dibenzylether ($(C_6H_5$–$CH_2)_2O$; b.p. 298 °C), etc. This liquid medium may or may not be taking part during the formation of nanoparticles. Since this has high boiling point, size and shape of particles can be engineered by changing temperature and duration of heating time.

Examples of complexing agents are oleic acid (b.p. 360 °C), olylamine (b. p. 364 °C), trioctylphosphine (TOP, b.p. 290 °C), trioctylphosphine oxide (TOPO, b.p. 408 °C), etc. They can make complexation with *s*, *p*, *d* block elements (Fig. 13.9).

Examples of additives are a long-chain hydrocarbon with one or more functional groups (1,2-hexadecanediol, 1,2-hexanediol, tris(diethylamino)phosphine, etc.) which can act as reducing or oxidizing agent or catalyst to change size/shape of particles during the reaction. It may take part in heat transfer process.

In this process of reaction, water has to be removed. In order to remove water, system needs to be heated at elevated temperature 120–150 °C for a few hours. In case, if water is present, agglomeration is associated. Round-bottom flask (RB flask with 100–200 ml capacity) is taken. The precursors (metal salt, complexing agent, additive (if required), liquid medium) are taken into RB flask. Here, complexing agent (about 5 ml or g) and metal salt (1 g) are approximately molar ratio of (10–100):1. Amount of additive depends on the purpose of reaction. Liquid medium (solvent + complexing agent) will have 20–50 ml. However, amount of precursors can be increased. It is to be noted that higher amount of precursors may not get uniform heating, and it results in the formation of polydispersed particles. RB flask should have option for two or more connectors, which can provide inlet of gas flow or removing of gases, water during the reaction. The inlet gas will be argon. First, the reaction system is pumped to remove unwanted gases (water, methanol, ethanol, HCl, HNO_3, etc.) at 120–150 °C. After removing gases (2–4 h duration at 120–150 °C), the Ar gas is passed. The heating temperature increases slowly to reach the desired reaction temperature (200–300 °C) for the formation of unform size particles. In this way, monodispersed particles can be formed, and also higher stability in term of thermal or chemical as compared to that of electrostatic repulsion or reverse micelle formation method is observed.

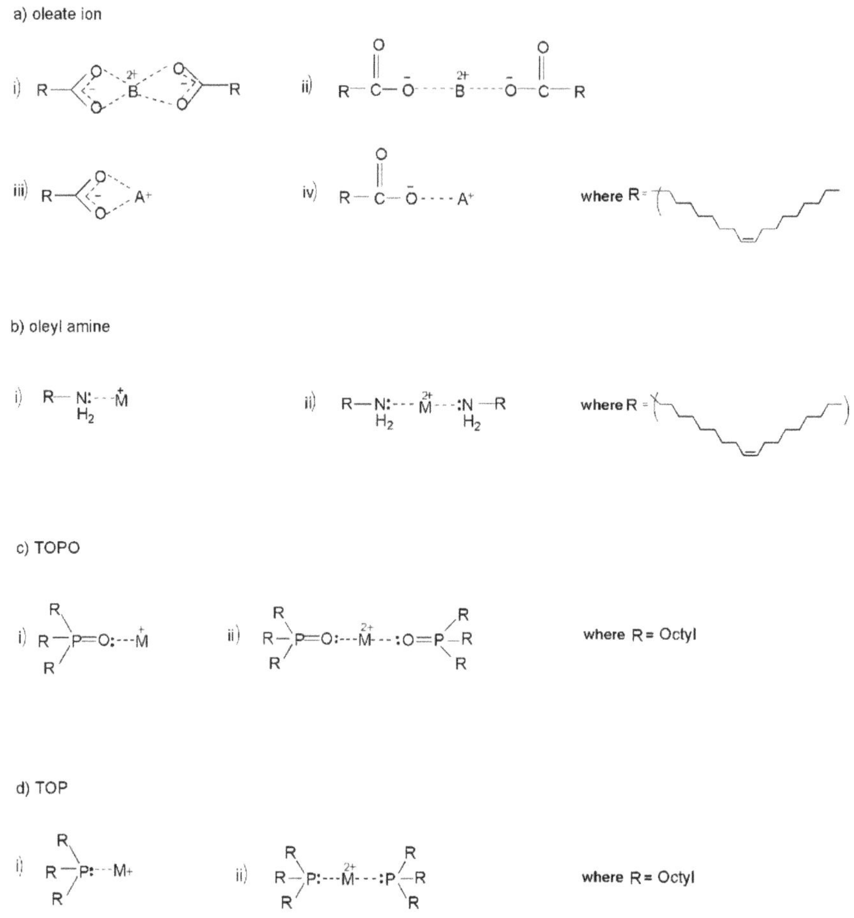

Fig. 13.9 Possible ways of interaction between dipositive and unipositive cations with **a** Oleate ion, **b** Oleylamine, **c** TOPO and **d** TOP

Examples

(i) Formation of metal nanoparticles (Ag, Au, Pd, Pt, Ni, Cu)
(ii) Formation of alloys nanoparticles (FePd, FePt)
(iii) Formation of metal oxides/sulfides/selenides/telulides (Fe_3O_4, SnO_2, TiO_2, PbS, PbSe, PbTe)
(iv) Formation of composites (Ag/Fe_3O_4, Au/Ferrites).

D **Hot injection method**

In hot injection method, nucleation takes place instantaneously at a particular temperature higher than room temperature or sometimes higher than 100 °C [63]. The requirements in the hot injection are similar to thermolysis [62]. The synthesis procedures for the formation of different types of nanoparticles using hot injection technique are given below:

(a) A-type single species (metal nanoparticles Ag, Au, Pd, Pt, Cu): A complexing agent and additives (if required) are heated at 120–150 °C in a liquid medium. Initially, evacuation of gases from RB flask takes place to make system free from residual gases/moisture, followed by passing Ar gas. To perform this, Schlenk lines (Fig. 13.10) are used, and this has option for use of vacuum line as well as gas line in alternate way. In this way, moisture and unwanted gases will be removed through vacuum line at initial stage. Meanwhile, a cold solution of metal-complexing ligand will be kept ready, and the solution will be prepared in inert environment which may need glove box. At a particular temperature, a few ml of metal-complex solution will be injected instantaneously. Injection time will be 1 s or less. In such way, instantaneous nucleation occurs. These nuclei can be frozen instantaneously by pouring hot particles/mixture into ethanol (which was kept in 20–22 °C). In this way, nucleation can be separated from the growth of particles. The surface of particles is passivated by surfactant molecules (say oleic acid), which controls further surface oxidation or reduction. After sometimes (waiting time or delay time after injection of metal-complex solution in RB flask), growth of particles will start if heating is continuing. Depending on waiting time, reaction injection temperature as well as additives, the different sizes and shapes can be prepared.

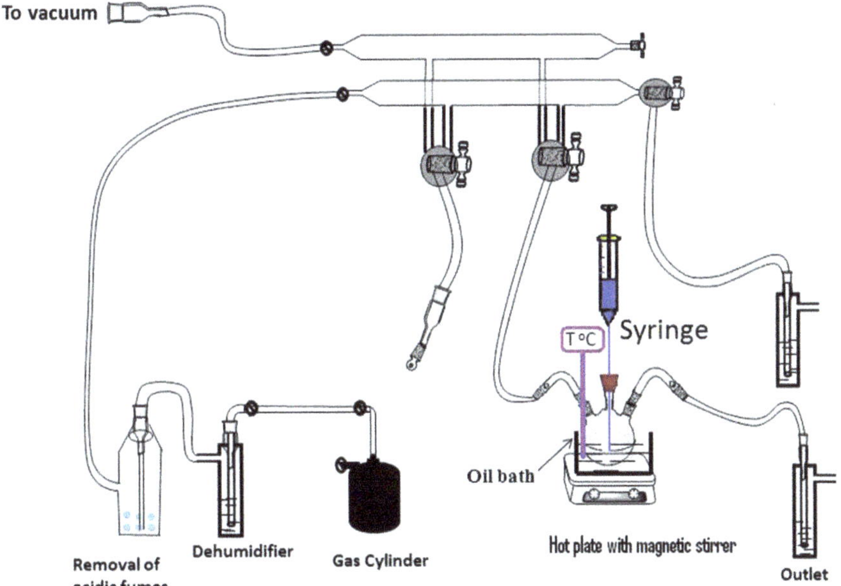

Fig. 13.10 Typical apparatus of the Schlenk lines

(b) AB-type single species (alloys such as AgAu, optical materials such as PbSe, CdTe, PbS): In this, metal ions (say Au, Pb or Cd) with lower reduction potential will make complexation with surfactant ligand. If required, the additives will be added. Initially, vacuum line will be operated at elevated temperatures (120 °C) for a few hours and followed by passing inert gas. Then, the temperature will be rising to higher or lower. Then, another species (say Ag, Se, Te, S) complexed with ligand will be injected into metal ion complexation system instantaneously. In this way, AuAg, PdSe, PbTe, PbS nanoparticles will be formed. Depending on thermodynamic stability of the system, different sizes and shapes will be formed.

(c) Pseudo type ($A_{1-x}B_xC$): A few examples are $PbSe_{1-x}Te_x$, $PbSe_{1-x}S_x$, etc. This can be prepared by a similar method as mentioned above.

(d) Core-shell type (PbSe@CdTe): In this, core particles will be formed, and then, precursors of shell will be added into system. With increase of time (at a particular temperature) or temperature, shell will be deposited over core particles. Thus, core-shell type can be formed.

(e) Seed-mediated A type: Here, seed will act as catalyst for the formation of A type species. In this way, the different sizes and shapes of the particles of A type can be prepared. Usually, seeds are noble metals, which can enhance shape anisotropy as well as reduce reaction temperature. In some cases, alumina micronsize particles are used as catalyst in preparation of magnetic nanoparticles. They can be separated each other by magnet.

(f) See mediated AB type: Here, seed will act as catalyst for the formation of AB-type species. Examples: Noble nanoparticles of Ag, Au, Pd are used as seed for the formation PbSe with different sizes and shapes [64].

Notably, metal nanoparticles such as Co, Ni, Pt, Pd prepared by hot injection method need a longer duration (a few minutes) for nucleation as compared to optical materials such as PbSe, ZnTe, PbS (a few seconds). In some cases, metal precursors are injected into a solvent or another metal solution at higher temperature (T_1) and followed by heat treatment at high temperature (T_2) for the formation of particles. This is similar to combination of hot injection and thermolysis methods. Thus, FePt, FePd, CoPt, CoPd were usually prepared by combination method. Figure 13.11 shows the various steps involved in the (a) thermolysis, (b) hot injection and (c) combination of the thermolysis and hot injection approaches.

13.3 Advantages and Disadvantages of Hot Injection Method

The preparation of nanoparticles prepared by the hot injection method has the following advantages:

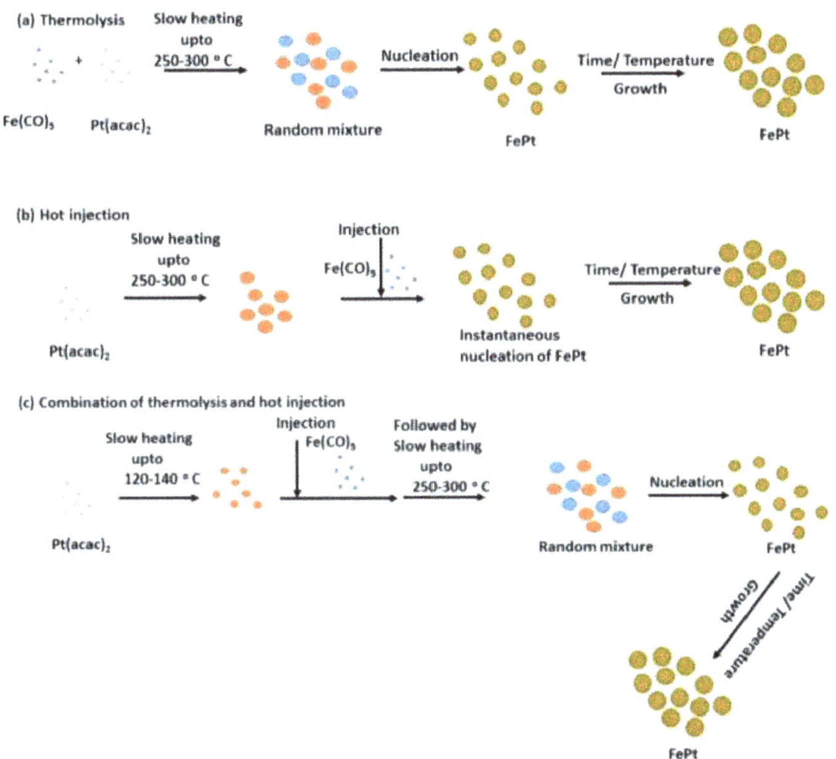

Fig. 13.11 Various steps involved in **a** the thermolysis, **b** hot injection and **c** combination of the thermolysis and hot injection

A Separation of nucleation and growth stages of particles
B Narrow size distribution
C Good control of particle size and shape
D High-temperature reaction leads to the formation of highly crystalline compounds or materials of the desired phase
E The preparation of core-shell particles (layer by layer)
F The formation of superlattices from nanoparticles

However, the hot injection method also has the following disadvantages:

A Expensive and often use of toxic or harmful chemicals
B Difficult to scale-up for the commercial production
C Set up the vacuum lines and inert gas atmosphere

13.4 Nanoparticles Synthesized by the Hot Injection Method

The nanoparticles have been synthesized using various methods and can be used in various applications [65–69]. One way to synthesize nanoparticles in controlled manner (with respect to size and shape) is the hot injection method. Nanoparticles prepared by the hot injection method on the basis of physical properties can be grouped in the following ways: metallic, magnetic, optical and multifunctional nanoparticles.

13.4.1 Metallic Nanoparticles

13.4.1.1 Silver (Ag) Nanoparticles

The hot injection method is well-known approach for the synthesis of the Ag nanoparticles. Among all the metals, Ag has the highest thermal and electrical conductivity with good optical reflectivity. Ag can be used as antibacterial agent on both gram-negative and gram-positive bacteria. Recently, strong effort has been made to bind the Ag nanoparticles to the DNA of bacterial cells [70–72]. The synthesis of the Ag nanoparticles under various conditions was first studied by Rothenberg and his group [73]. The Ag nanoparticles have been synthesized by different synthesis techniques such as photochemical, laser ablation, electron irradiation, gamma irradiation, chemical reduction, biological synthetic and microwave methods. For practical application of the nanoparticles, the study of transformation of the metals in the physicochemical environment is needed [74]. Kim et al. have reported that for rapid nucleation, the injection rate and the reaction temperature are important factors in terms of tuning particle size and attaining monodispersity [75]. A detailed mechanism of Ag nucleation and growth produced from γ-irradiation of silver perchlorate was studied by Henglein and Giersig [76]. Growth of the Ag nanoparticles for monodispersity occurs during surface reduction through electron transfer. The electron transfer can reduce Ag^{+} ions onto the particle surface available in the solutions [22].

$$Ag^{+} + e^{-} \rightarrow Ag^{0}$$

The Ag^{+} ion reduction has a positive standard reduction potential (E°) of around +0.799 V in water. The Ag^{+} ion reduction into metallic Ag^{0} in the solution can be carried out by using reliable reducing agent. The sodium borohydride (E ° = −0.481 V) and sodium citrate (E° = −0.180 V) are able to reduce Ag^{+} ion to metallic Ag^{0} [77]. Seed-mediated Ag nanoparticles (size about ~10 nm) have been synthesized on a nanosized solid pre-synthesized $CaCO_3$ (size about ~52 nm) through hot injection approach [78]. Figure 13.12 shows the scheme for the

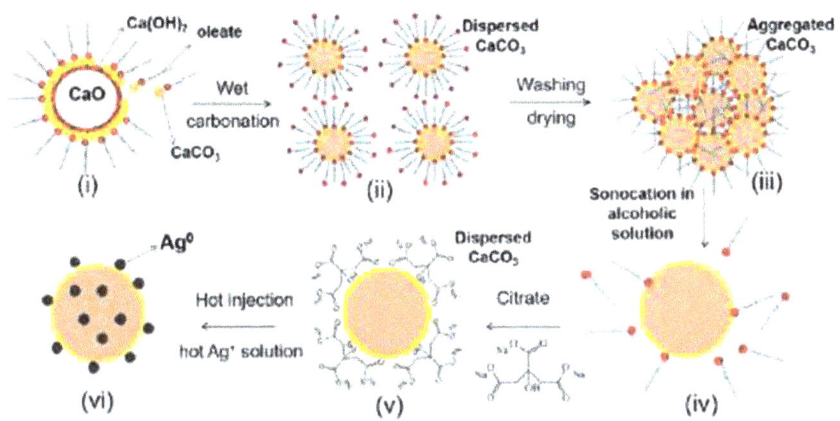

Fig. 13.12 Scheme for the different phases of the formation of Ag@$CaCO_3$ composite nanocrystals (In situ modification of the $CaCO_3$ nanocrystal). Reproduced with permission from ACS Publications [78]

different phases of formation of Ag@$CaCO_3$ composite nanocrystals (In situ modification of the $CaCO_3$ nanocrystal). Figure 13.12 (i–iii) shows synthesis and in situ modification of $CaCO_3$ nanocrystals; (iv) re-dispersion of the oleate-modified $CaCO_3$ nanocrystals (in water/ethanol = 4:1 v/v) containing trisodium citrate; (v) stabilization of the $CaCO_3$ nanocrystals by adsorption of the citrate molecules on the nanocrystals surface; (vi) hot injection of the $CaCO_3$ dispersion into hot $AgNO_3$ solution and the formation of the tiny Ag nanocrystals on $CaCO_3$ nanocrystals. The purpose of the oleate ions is such that it can stabilize the $CaCO_3$ nanocrystals and provide dispersion with clustering, nucleation and growth as well as controlling the surface properties of the $CaCO_3$ nanocrystals. Different steps in the seed-mediated formation of tiny Ag nanocrystals on the $CaCO_3$ surface have been represented through scheme (not to scale) presentation view (Fig. 13.13). Stabilization by oleate ions and reduction on the surface of $CaCO_3$ by using citrate is the two important phases in this scheme.

Nanobars and nanocubes shape formation of Ag were prepared via poly (vinyl pyrrolidone) (PVP) and Br^- ions which can selectively bind to {100} facets of Ag and slowdown their growth rate to favor the formation of shaped particles [79]. Bharti et al. have reported in situ approach to synthesize noble plasmonic Ag nanoparticles from aqueous PVP solution of metal salt using radiolysis of water via synchrotron monochromatic X-ray irradiation without any chemical reducing agent [80]. Ag nanoparticles prepared by the reverse micellar system exhibited a wide collection of shapes and depend strongly on the reaction temperature [81]. The effect of the localized surface plasmon resonance is such that the peak wavelength

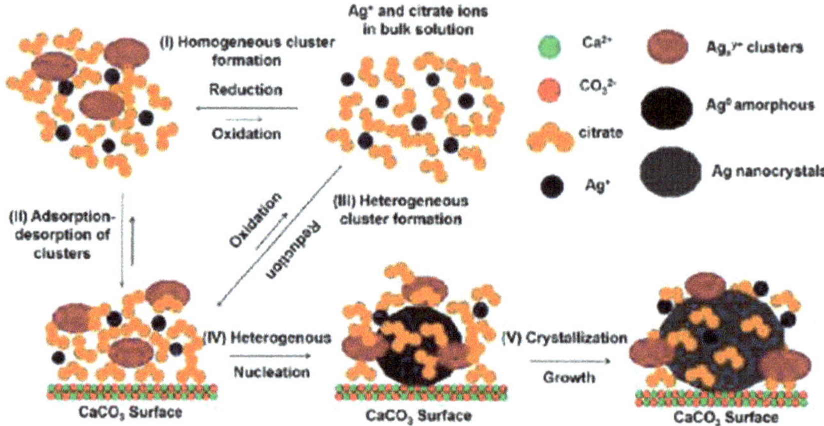

Fig. 13.13 Scheme showing the different steps in the seed-mediated formation of tiny Ag nanocrystals on the $CaCO_3$ surface. **i** homogeneous clusters formation; the complexation and redox reaction of Ag^+ and citrate ions in the solution induces the homogeneous formation of tiny clusters (Ag_x^{y+}) stabilized by citrate ions; **ii** adsorption/desorption of Ag_x^{y+} clusters on the $CaCO_3$nanocrystal surface; **iii** heterogeneous formation of Ag_x^{y+} clusters on the $CaCO_3$nanocrystals surface, possibly facilitated by the citrate monolayer adsorbed on the $CaCO_3$nanocrystals surface; further citrate electron transfer reduction of Ag^+ to metallic silver Ag^0 on the surface; **iv** heterogeneous nucleation; densification and formation of stable amorphous Ag^0 nuclei on the $CaCO_3$ surface; and **v** growth and crystallization of the amorphous Ag^0 nuclei to form spherical nanocrystals. Reproduced with permission from ACS Publications [78]

increases with an increase in the mean diameter of Ag nanoparticles. The growth of the Ag nanoparticles beyond the critical radius (r_{cri}) is accompanied by a shift of the plasmon resonance toward longer wavelengths. It is observed that the mean diameter of Ag nanoparticles was dependent upon the average esterification degree of sucrose fatty acid esters forming reverse micelles [77]. Nanoparticle shapes and specific adsorption sites are important for surface plasmon resonance enhancement, and the surface-enhanced Raman scattering (SERS) performance is a dominant factor in the particle size observation [82].

13.4.1.2 Gold (Au) Nanoparticles

Au nanoparticles in the range 9–27 nm were prepared by injecting $HAuCl_4 \cdot 3H_2O$-surfactant (Oleylamine or PVP, 1-octadecanethiol) in solvent (toluene or ethylene glycol) at 120 or 170 °C for 2 h [83]. The nanoparticles show surface plasmon resonance peak in 530–560 nm and can be dispersed/grown over silicon substrate.

13.4.2 Magnetic Nanoparticles

13.4.2.1 Cobalt (Co) Particles

Timonen et al. have reported the preparation of magnetic cobalt nanoparticles by injecting [$Co_2(CO)_8$] dissolved in ortho-dichlorobenzene(*o*-DCB, b.p. 180 °C) into a solution of TOPO and oleic acid in *o*-DCB at 180 °C temperature under N_2 atmosphere [61]. After the injection, it was found that the temperature was dropped to 143 °C, which is a characteristic observance of hot injection method. Cobalt is a ferromagnetic metal and exhibits size-dependent properties at the nano level. It has been observed that the number of nuclei measured in the hot injection synthesis of cobalt nanoparticles depends on temperature kinetics after the injection than on the injection itself. The injection provides the direct supersaturation which is not sufficient high to cause burst nucleation. Therefore, the nucleation is hindered until adequate monomers are produced from the precursor. Temperature kinetic modification can be used to control the number of nuclei formed in the delayed nucleation. The kinetically controllable nucleation led to a technologically relevant synthesis of the monodispersed cobalt nanoparticles. Cobalt nanoparticles have also been prepared with different shapes and sizes such as nanorods and nanotubes which depend on preparation condition [84]. The synthesis parameters viz. injection temperature and time metal to surfactant ratio and reaction time can impact in mean and median particle diameter of the cobalt nanoparticle [85].

13.4.2.2 FePt and FePd Nanoparticles

The important materials in terms of high–density magnetic data storage and electrical conductivity were prepared by combination of hot injection and thermolysis methods. Examples are FePt and FePd nanomaterials [86–89]. Initially, A = Pt or Pd acetyl acetonate is reduced to Pt or Pd nanoclusteres in the presence of oleic acid and high-boiling point organic solvent. Then, at higher temperature 100–140 °C, [B $(CO)_5$] (B = Fe or Co) solution is injected to Pt or Pd clusters/seed (Fig. 13.11). When temperature rises to 250–300 °C, nanocrystals of AB with size of about 3–10 nm are formed. During the synthesis, some of $B(CO)_5$ is lost due to evaporation. In order to reduce loss, the cold condenser can be provided. At room temperature, as prepared sample FePd with size 3–4 nm shows cubic phase with super paramagnetic nature. With further annealing at 600 °C, it shows tetragonal phase with high ferromagnetic nature. The coercivity (H_c), retentivity (M_r) and saturation magnetization at 300 °C are found to be 1300 Oe, 34 emu/g and 94 emu/g, respectively.

13.4.2.3 Ferrite (AB_2O_4) Nanoparticles

Cobalt-antimony and ferrite nanoparticles ($CoSb_2$ and $CoFe_2O_4$) have also been taken a keen attention among the researchers [90, 91]. The $CoSb_2$ nanoparticles were synthesized via combination of hot injection and thermolysis methods in which non-hydrated Sb-oleate dissolved in 1-octadecene is directly injected into a hot solution of the hydrated Co-oleate in 1-octadecene at a specified temperature [90]. Also, the highly monodispersible water soluble cobalt ferrite ($CoFe_2O_4$) and manganese ferrite ($MnFe_2O_4$) nanoparticles were synthesized by combination of the hot injection and thermolysis processes of pivalate clusters as single-source precursors [91]. The pre-synthesized precursors $[Fe_2CoO(O_2C^tBu)_6(HO_2C^tBu)_3]$ and $[Fe_2MnO(O_2C^tBu)_6(HO_2C^tBu)_3]$ were decomposed in the mixture of polyvinylpyrrolidone (PVP) (used as a capping agent) and triethylene glycol (TREG) (solvent) at 285 °C temperature to prepare $CoFe_2O_4$ and $MnFe_2O_4$ nanoparticles, respectively.

13.4.3 Optical Nanoparticles

13.4.3.1 CdSe Nanoparticles

The typical synthesis of CdSe nanoparticles via hot injection involved injection of room temperature (cold) solution of the precursor molecules into the solution at about 300 °C (hot) of apolar coordinating solvent trioctylphosphine oxide (TOPO) [10]. The solution of precursor contains $CdMe_2$(Me = methyl) and Se in trioctyl phosphine (TOP) [10]. As advanced method, cadmium acetate $[Cd(Ac)_2]$ and CdO were used alternatively as cadmium compounds/precursor, because $(Me)_2Cd$ is extremely toxic and pyrophoric. Upon addition of the precursor via injection, an instantaneous formation of nuclei of CdSe nanoparticles is resulted. CdSe is a compound of II–VI semiconductor composed of Cd^{2+} and Se^{2-} ions. Rapid hot injection method for the synthesis of CdSe quantum dots was prepared by varying the Cd:Se molar ratio [92]. The work suggested that the Cd:Se molar ratio can affect the nanoparticle size and quantum yield. Experimental study shows that Cd:Se molar ratio can enhance the photoluminescence properties. The particle formed at highest stabilizer concentration exhibits low photoluminescence peak intensity. The CdSe nanocrystals prepared with the wurtzite structure have been found highly anisotropic in nature [93]. The growth rate is fast along the *c*-axis which results a rod-like faceted shape along the *c*-axis. The wurtzite crystal structure is favored toward the minimum surface energy. Figure 13.14 shows the TEM images of quantum rods of the different CdSe samples, (a–c) low-resolution TEM images of three quantum-rod samples with different sizes and aspect ratios and (d–g) high-resolution TEM images of four representative quantum rods. Figure 13.15 shows the three-dimensional orientation of CdSe quantum rods observed by TEM images. The capability to control shape of the nanoparticles further provides a good

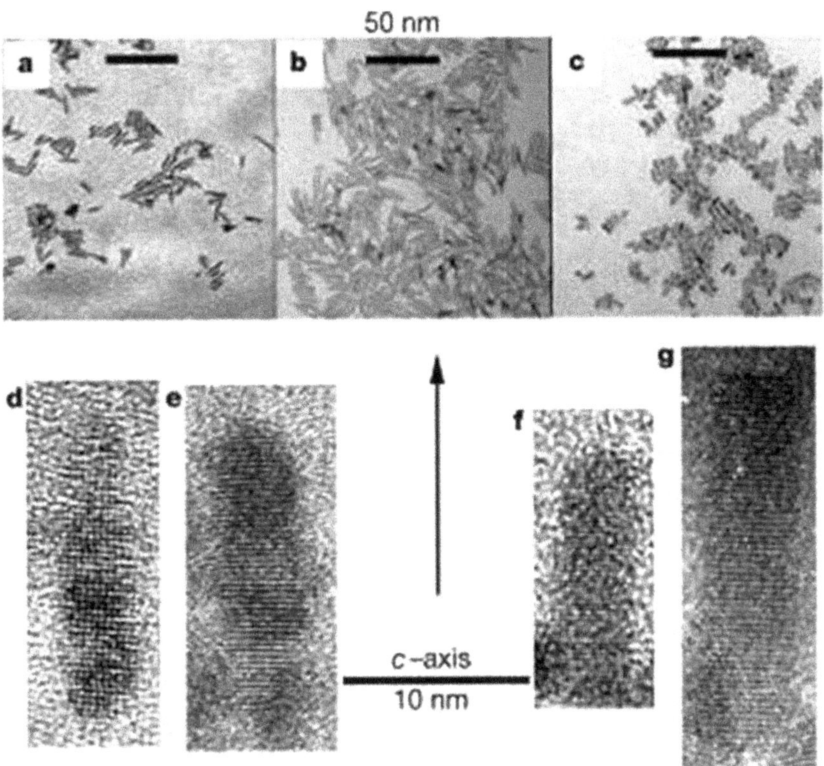

Fig. 13.14 TEM images of quantum rods of the different CdSe samples. **a–c** low-resolution TEM images of three quantum-rod samples with different sizes and aspect ratios, **d–g** high-resolution TEM images of four representative quantum rods, **d** and **e** are taken from the sample shown in **a**; **f** and **g** are taken from the sample shown in **c**. Reproduced with permission from Nature Publishing Group [93]

chance to further study in the quantum confinement process. The CdSe quantum dots prepared by the thiol capped ligands can act as hole-acceptor for improvement in the fluorescence intensity [94]. The highly luminescent (quantum efficiency of 29%) and defect-free CdSe nanoparticles by using a combination of the reverse micelle method and post-growth annealing processes were reported by Kim et al. [95]. They reported that the annealing temperature can control the radius of the CdSe nanoparticles. The band-edge emission shifts to longer wavelength with increase of annealing temperature, and this suggests that size of quantum dots increases with annealing. The shift in the absorbance spectra toward relative to the bare sample has been reported. Also, emission intensity increases with annealing indicating decrease in defect or trap levels in CdSe quantum dot [29].

In each unit cell of the CdSe (E_g = 1.74 eV), the single-electron molecular orbitals in the nanocrystals are formed from the highest occupied molecular orbital

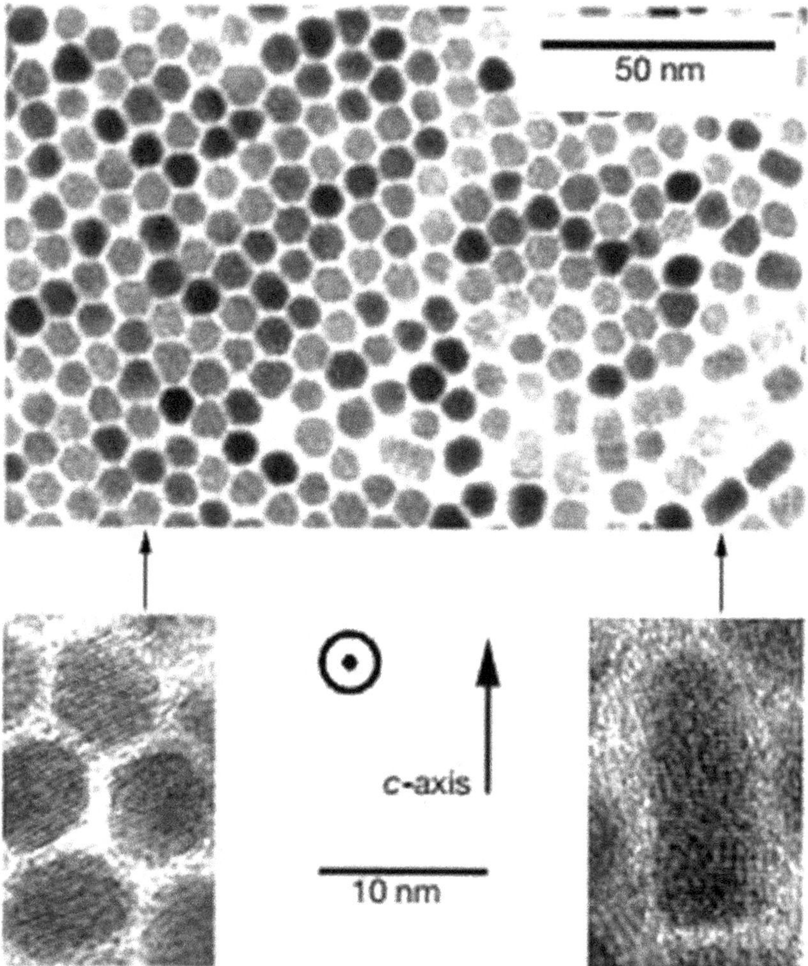

Fig. 13.15 Three-dimensional orientation of CdSe quantum rods observed by TEM. Reproduced with permission from Nature Publishing Group [93]

(HOMO) and lowest unoccupied molecular orbital (LUMO) [96]. The particle in sphere model is an easy model for the electronic structure of CdSe nanocrystals. The carriers are confined by a potential which is infinite at the interface. The energy levels of CdSe quantum dots attached to an Au (111) substrate, via a rigid sulfur-terminated oligo (cyclohexylidene) self-assembly monolayer (SAM) have been studied by Bakkers et al. [97]. They described the individual energy levels of nano-crystalline quantum dots with the help of resonance tunneling spectroscopy. They used scanning tunneling spectroscopy in finding the single-particle electron levels and Coulomb repulsion energies in a colloidal CdSe quantum dot having diameter of about 4.3 ± 0.4 nm. The advantage of the preparation of different size CdSe nanocrystals in the luminescence is such that it can emit the wavelength of about 450–650 nm, which nearly covers the visible color spectrum [98]. The surface ligands can affect the hot carrier thermalization. The crystal structure of the CdSe:0.5en (en = ethylenediamine) has been demonstrated in 2003 by Deng et al. [99]. Figure 13.16 shows a close view of the CdSe structure as a three-dimensional (3D) network with 2D CdSe monolayers connected via bidentate ethylenediamine molecules. It was observed that Cd and Se atoms were associated, and their fair positions in the inorganic layers of CdSe:0.5en precursor were nearly the similar as the atoms of the wurtzite CdSe crystal. The photoluminescence study was performed in the colloidal CdSe nanocrystals prepared by the hot injection method in liquid paraffin [100]. The photoluminescence spectra were carried out at room temperature (295 K) as well as low temperature (15 K), and the emission peak behavior was monitored. The shifting in the peak at higher energy range with decreasing the nanocrystal size shows a characteristic observation of quantum size effect. The Raman spectra of the CdSe nanocrystals were found to be same as the bulk CdSe. The Raman peak shift at lower frequency was observed due to the size dependence and could be related with the phonon quantum confinement effect [100].

13.4.3.2 PbSe and PbS Nanoparticles

Lead selenide (PbSe) nanoparticles (diameter 3–13 nm) have been synthesized by hot injection method. Here PbO, oleic acid and 1-octadecene (ODE) were heated at 150 °C to get colorless solution in vacuum to remove unwanted water and other gases, and it was further heated at 180 °C [101]. Then Ar gas is passed into reaction medium or solution. The selenium-trioctylphosphine solution was injected into the hot solution. After only 10 s of injection at a temperature of 150 °C, monodispersed nanocrystals of about 3.5 nm size were achieved. With further longer-time treatments for 800 s, it produced 100% chemical yields with 9.0 nm size of nanocrystals. They reported that the as prepared PbSe exhibited narrow size distribution without any post synthetic size selection. The luminescence quantum yields with various sizes of particles are in the range of 35–89%. Chalcogenide nanocrystals (such as PbS, PbSe, PbTe) have been fabricated in many reports [102–106]. PbSe is a compound of IV–VI semiconductors, which is a direct band gap semiconductor with band gap about 0.28 eV. PbSe nanocrystals show the extreme quantum confinement in which the electron, hole and exciton have large Bohr radii of around ~23, ~23, ~46 nm, respectively [107]. Initially, lead chalcogenide semiconductor nanocrystals were reported inside polymer and oxide glass hosts [108]. A heteroepitaxial growth of mondispersed PbS nanocrystals onto the surface of TiO_2 nanoparticles via hot injection method was done by Acharya et al. [109]. They observed a size-dependent alignment of the energy levels within the PbS/TiO_2 system. In this approach, a photoinduced electron transfer across PbS/TiO_2 has been detected when the nanocrystal size is below 7 nm. But, when the diameter of

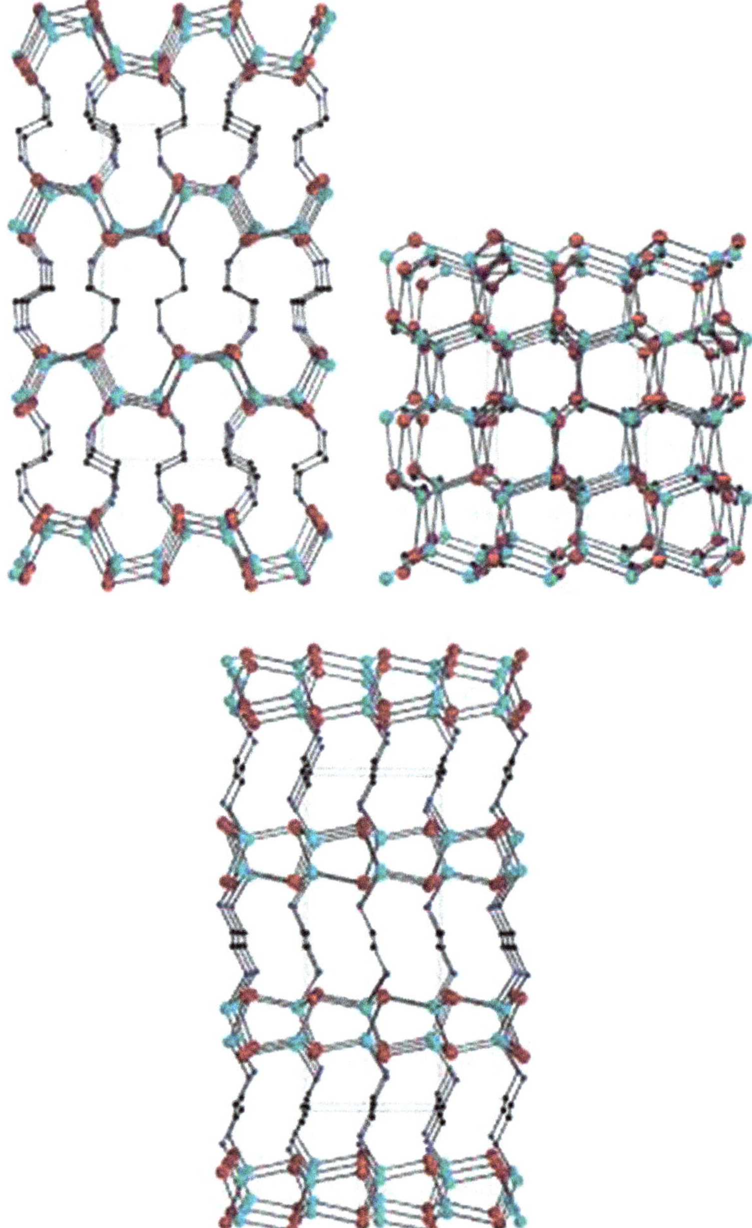

Fig. 13.16 Structural views of CdSe.0.5en down the **b** (top, left), **c** (top, right) and **a** (bottom) axes of the orthorhombic cell. Red, cyan, blue and black balls correspond to Cd, Se, N and C atoms, respectively. Hydrogen atoms are omitted for clarity. Here "en' stands for ethylene diamine. Reproduced with permission from ACS Publications [99]

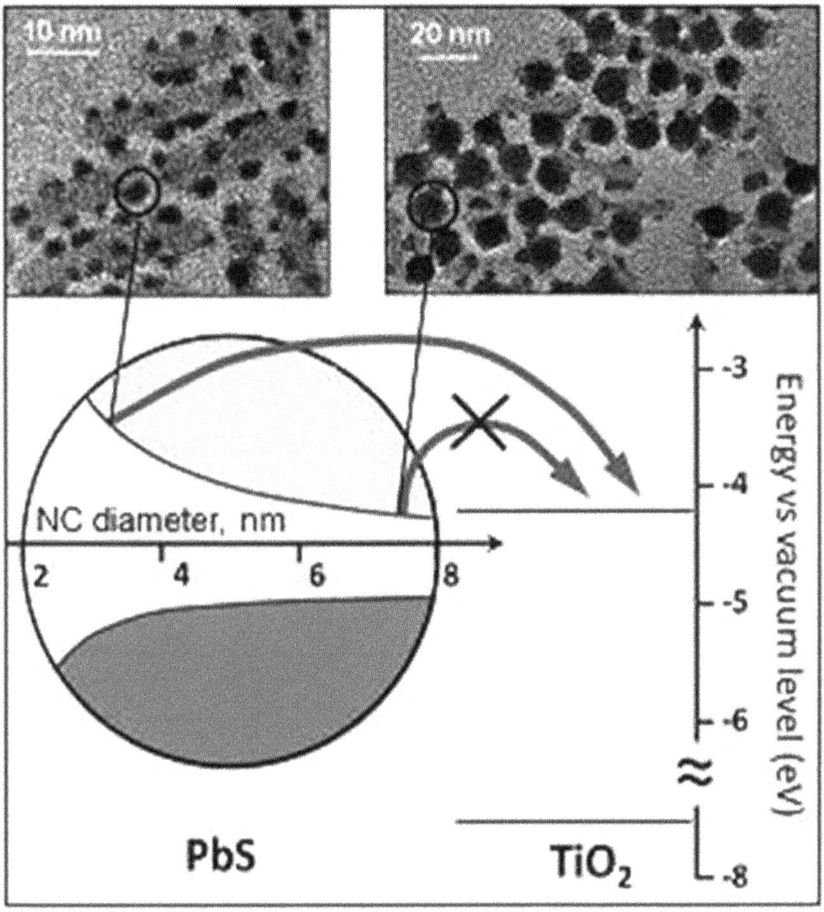

Fig. 13.17 Energy level diagram showing the relative alignment of the conduction and valence band edges in the PbS/TiO_2 hetrostructures. Reproduced with permission from ACS Publications [109]

the nanocrystal is greater than 7 nm, the PbS/TiO_2 exhibits an alignment of the conduction and valence band and also the excited carriers remain within the PbS material. Figure 13.17 shows the energy level diagram with relative alignment of the conduction and valence band edges in the PbS/TiO_2 heterostructures. A photoinduced electron transfer from PbS to TiO_2 domain is allowed only when the size is less than 7 nm, which may induce an interesting optoelectronic properties in the prepared PbS/TiO_2 nanocrystals for solar cell application. Zhang et al. have reported thin film fabricated by using PbS quantum dots film photovoltaic applications [110]. They have studied the effect of halide treatment on PbS quantum dots thin film. Depending on size of QDs, the emission can be observed in the wide

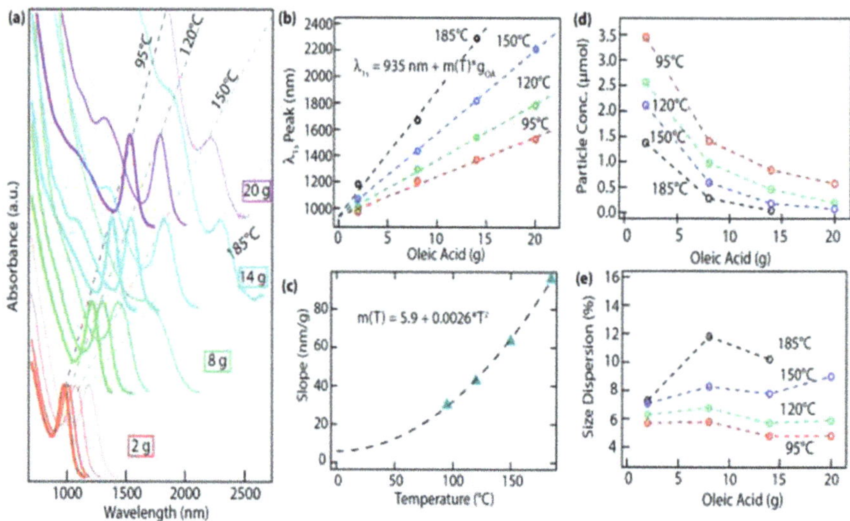

Fig. 13.18 **a** Absorption spectra of PbS quantum dots synthesized at different injection temperatures with various oleic acid (OA) amounts. Red traces are for synthesis with 2 g of OA, green traces are for 8 g, blue traces are for 14 g, and purple traces are for 20 g oleic acid. The injection temperature was varied from 95 to 185 °C (the different injection temperatures have faded colors). **b** The peak of the 1S exciton as a function of grams of oleic acid. The traces are a best-fit linear function, and the slope is plotted versus temperature in panel **c**. **d** Particle concentration as a function of the amount of oleic acid added. **e** Size dispersion (%) synthesized at various conditions. Reproduced with permission from ACS Publications [113]

region of the electromagnetic spectrum. PbS and PbSe nanocrystals show strong electrons-holes quantum confinement which can enhance the photo-conversion efficiency of solar cells due to the multiple exciton generation [111, 112]. A relation between concentration of ligand (oleic acid, g_{OA}) and the lowest energy excitation peak position (wavelength of the 1S exciton, λ_{1S}) of the PbSe quantum dot at various hot injection temperature has been studied by Zhang et al. [113]. They studied PbSe quantum dot for solar cell fabrication by the same size PbSe quantum dots at four different injection temperatures (95, 120, 150 and 185 °C). Figure 13.18a shows the absorption spectra of the hot injection synthesized PbS quantum dot at different injection temperatures with different oleic acid amounts. From Fig. 13.18b, it is clearly seen that the lowest energy exciton peak position and the oleic acid amount display a linear relationship whose slope depends on T_{inj} (Fig. 13.18b). The larger quantum dots can be obtained by the higher T_{inj} which indicates steeper slope values. Thus, a general relation was developed with the help of this experimental conclusion which has represented by the following equation,

$$\lambda_{1S} = 935\,\text{nm} + m(T_{inj}) \cdot g_{OA}$$

where $m(T_{inj}) = 5.9 + 0.0026 \cdot T_{inj}^2$ is the temperature-dependent slope value as given in Fig. 13.18c. Particle concentration decreases with increase of amount of oleic acid at a particular injection temperature (Fig. 13.18d). Also, particle concentration decreases with increase of injection temperature at a particular concentration of oleic acid. Narrower size dispersion was achieved with lower T_{inj} and higher oleic acid.

Ningthoujam et al. have separated magic size clusters (MSCs) of PbSe from their counterpart QDs using oleylamine as reducing agent employing hot injection method [32]. Otherwise, it is difficult to remove MSCs from QDs while using commonly used synthesis method. The absorption and emission peaks of MSCs are assigned to 600 and 780 nm, respectively, and these values do not change when size of particles is less than 1.6 nm. In case of QDs, absorption peak shifts from 940 to 2400 nm with increase of size from 2.8 to 10 nm. Peak varies with change in decimal value of particle size. The luminescence quantum yield is found to be 80% when size of particle is 4.1 nm. Their lifetime values are found to be in the range of 1.0–1.3 μs. It is to be noted that most QDs (e.g., TiO_2, SnO_2) compounds show the lifetimes in a few ns.

13.4.3.3 CuSe Nanoparticles

Copper selenide (CuSe) is a p-type semiconductor and always play a good part in the solar cell fabrication, ionic conductor and gas sensor [86, 114, 115]. CuSe and $Cu_{2-x}Se$ crystals were synthesized via hot injection method using copper and selenium–triethylene glycol (TEG) solution and triethylenetetramine (TETA) as the reducing agent and polyvinylpyrrolidone (PVP) as the capping agent at 230 °C temperature [87]. Figure 13.19 shows the SEM images of copper selenide powders (a) CuSe and (b) $Cu_{2-x}Se$. The SEM micro-images reflect that the formed particles

Fig. 13.19 SEM images of copper selenide powders synthesized using the thermal decomposition (hot injection) process **a** CuSe, **b** $Cu_{2-x}Se$. Reproduced with permission from Elsevier Publishers [87]

are big in size and lie in the micro-meter range. The observed morphology for the synthesized particles is of hexagonal flakes type with the size about 0.90–0.88 μm and thickness of about 0.3 μm. $CuSe_2$ was obtained when the TETA amount is low up to 0.4 ml in the copper selenide. However, on increasing the TETA amount of about 0.7 ml in the reaction medium, the CuSe was achieved. Some co-existence of the crystalline phase has also been reported on increasing the TETA content.

13.4.3.4 SnS_2 Nanoparticles

SnS_2 is an *n*-type semiconductor having optical band gap around 2.2–2.4 eV [116, 117]. The electronic, chemical and physical properties of the SnS_2 are remarkable which makes the material to be useful in different technology-based applications. The facile hot injection method has been used to synthesize the tin disulfide [118]. For this, the SnO was added into the oleic acid and heated at 320 °C temperature to get the tin precursor. The sulfur precursor was prepared by mixing thioacetamide with oleyamine at 280 °C. The Sn precursor was injected quickly into the S precursor at 280 °C, and it was kept for 30 min under Ar atmosphere. The obtained precipitate was washed and collected by centrifugation to get final product. Figure 13.20 shows the synthesis of SnS/SnS_2 at 280 °C for (a) 10 min, (b) 30 min and SnS_2 for (c) 60 min as reaction time. Figure 13.20d shows schematic diagram of growth mechanism of SnS_2 crystals. The layered-like structure can be seen in the TEM images for 10 min which shows the existence of the amorphous phase of the

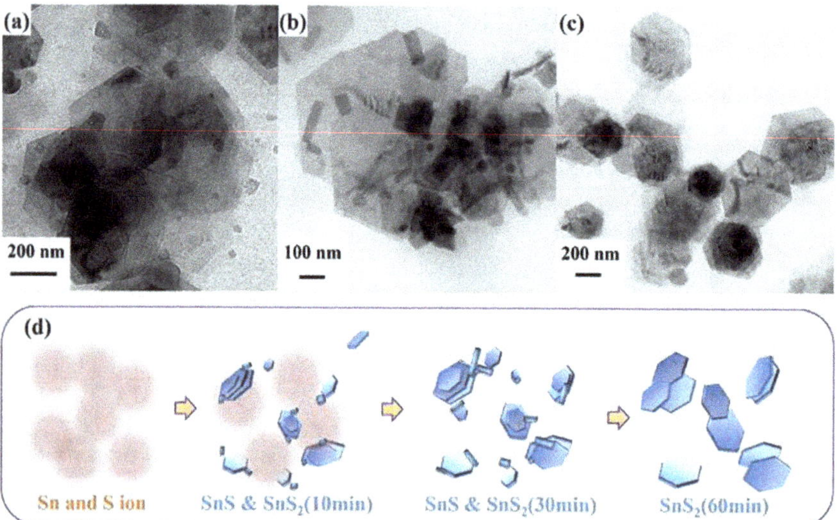

Fig. 13.20 Synthesized of SnS/SnS_2 at 280 °C for **a** 10 min, **b** 30 min, and SnS_2 for **c** 60 min as reaction time. **d** Schematic diagram of growth mechanism of SnS_2 crystals. Reproduced with permission from Elsevier Publications [118]

SnS/SnS_2. On increasing time, the amorphous phase gradually changes into the square (SnS) and hexagonal (SnS_2)-shaped crystals. It was observed that the amorphous phase disappears when the time increases up to 30 min. At the end of 60 min, complete formation of hexagonal shaped particles is reported with a size and thickness of about 460–530 nm and 22 nm, respectively. The observed formation mechanism demonstrates development of layer-by-layer growth of SnS_2 along (001) due to Ostwald ripening process.

13.4.3.5 FeS_2 Nanoparticles

FeS_2 nanocrystals of cuboid and spherical shapes (size about 80 and 30 nm, respectively) were obtained using trioctylamine (TOA) and oleylamine (OLA) as solvents to dissolve sulfur via hot injection method [119]. Here, Fe precursor was prepared by mixing of $FeCl_2 \cdot 4H_2O$, octadecylamine (ODA) and TOA in a three-necked flask at temperature around 150 °C. To form spherical shaped-nanoparticles of FeS_2, the S precursor prepared from elemental sulfur in OLA was injected into the hot solution of Fe precursor. In case of the formation of cuboid-shaped nanoparticles of FeS_2, the S precursor prepared from elemental sulfur in TOA at 70 °C was injected into the hot solution of Fe precursor. This was further heated up to 220 °C under N_2 atmosphere to complete the growth of the iron pyrite nanocrystals. The cooling was done at normal room temperature, and when the temperature reaches up to 100 °C, a small amount of the chloroform was added. Then after, the iron pyrite nanocrystal was purified by mixed solvent precipitation using chloroform and methanol. Nanocrystals were stored in the chloroform. The mechanism proposed for the formation of FeS_2 can be achieved by using the following reaction,

$$\mathrm{Fe(complex)_2} \rightarrow \mathrm{FeS}(\text{in the presence of the } 1/8\,\mathrm{S_8})$$

$$\mathrm{RCH_2NH_2} \rightarrow \mathrm{RCSNH_2} + \mathrm{H_2S}(\text{in the presence of the } 1/4\,\mathrm{S_8})$$

$$3\,\mathrm{FeS} + \mathrm{RCSNH_2} + \mathrm{H_2S} \rightarrow \mathrm{Fe_3S_4} + \mathrm{RCH_2NH_2} + \mathrm{S}$$

$$\mathrm{Fe_3S_4} + \mathrm{RCSNH_2} + \mathrm{H_2S} \rightarrow 3\mathrm{FeS_2} + \mathrm{RCH_2NH_2}$$

As per the reaction given above, FeS phase can be obtained, when the S source solution was injected rapidly into the iron source solution. The second reaction can be possible for formation of the hydrogen sulfide and amide which are the active sulfur sources compared to the elemental sulfur. The third and fourth reaction can be happened which may transform the FeS into the FeS_2 phase through the Fe_3S_4 in the presence of amide and hydrogen sulfide. Figure 13.21 shows the TEM images of the growth process of cuboid and spherical FeS_2 nanocrystals for different durations. The formation of the nanocubes and edged nanospherical FeS_2 nanocrystals can be obtained in this synthesis process. For cube growth mechanism, TOA was adopted which is combined with elemental sulfur. After injection, reaction between S and ODA (primary amine) takes place. This is a slow reaction

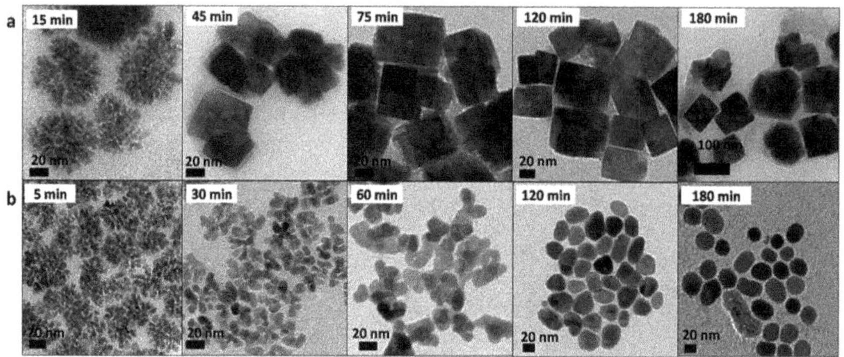

Fig. 13.21 TEM images of the growth process of **a** cubic and **b** spherical FeS_2 nanocrystals for different durations. Reproduced with permission from Elsevier Publications [119]

and actively formed sulfur source occurs. Low nuclei concentration was generated, and hence, bigger seeds can be produced (Fig. 13.21a). Afterward, the process is happened in such a way that sulfur was combined with OLA as S source. This can be formed as active sulfur source. This also may create a high nuclei concentration which resulted in a fast growth with overall smaller seed development (Fig. 13.21b).

13.4.3.6 $CuInS_2$ Nanocrystals

Copper-indium sulfide chalcopyrite ($CuInS_2$) ternary compound was synthesized by hot injection method where thiourea solution was injected into a hot copper-indium solution [120]. In a typical synthesis for $CuInS_2$, the precursors CuCl and $InCl_3$ were dissolved in oleylamine and oleic acid in a three-neck flask under air atmosphere. The solution temperature was increased to 150–240 °C under Ar atmosphere until a clear solution was formed. At the same time, thiourea (CH_4N_2S) was separately dissolved in oleylamine keeping temperature around 50–150 °C. The thiourea solution was injected in the solution, and temperature was kept at 300 °C for 2–4 h. The reaction was cooled down at normal room temperature, and ethanol was added to the solution for precipitating $CuInS_2$ nanoparticles. Pan et al. have reported the synthesis of Cu-In-S ternary nanocrystals with tunable structure and composition using hot injection method [121]. They have described a hot injection method based on $Cu(dedc)_2$ and $In(dedc)_3$ precursors, where dedc represents diethyldithiocarbamate. Oleylamine as the activation agent and oleic acid or dodecanthiol as the capping agent. The nearly monodisperse Cu-In-S particles were prepared, and these particles can be potential material for low-cost, high-efficiency solar cell device fabrication.

13.4.3.7 Cu_2SnSe_3 Nanoparticles

The Cu_2SnSe_3 was prepared by hot injection method in which the selenium precursor was injected into a hot solution at a given reaction temperature [122]. Initially, the stock solution was prepared by taking $CuCl_2$, $SnCl_2$, OLA and dodecanethiol (DDT) in three-neck flask and stirred by alternate arrangement of vacuum and N_2 gas passing at room temperature. Subsequently, the mixed solution was retained at 60 °C under vacuum for 30 min and then heated to 180 °C. The color of the solution changed from blue to milky at 60 °C and to brown at the temperature of over 150 °C. Another stock solution was made by taking OLA, DDT and Se powder into a three-neck flask at room temperature, cycled between vacuum and N_2 three times to eliminate O_2 in the flask and then stirred under a N_2 atmosphere for dissolving Se powder to get stock solution. The second stock solution was quickly injected into the reaction medium of the first stock solution. The color of the solution changed instantly from light brown to dark brown, which indicates the progress of nucleation and subsequent growth of CTSe nanostructures. The reaction was allowed for heating at 180 °C for 30 min and then allowed to cool at room temperature. The obtained brown–black product was centrifuged and collected. The upper unwanted yellow color solution was discarded, and the remaining products were washed with chloroform and ethanol.

13.4.3.8 Cu_2ZnSnS_4 · (CZTS) Nanocrystals

A novel quaternary semiconductor CZTS nanocrystal with average size ~4 to 9 nm was fabricated by using the hot injection method [88]. Oleylamine (OLA) is used for CZTS preparation both as a solvent and the stabilizer. $CuCl_2$, $ZnCl_2$ and $SnCl_4$ and oleylamine were mixed in a three-neck round-bottom flask, stirred and heated to 170 °C for 1 h under inert atmosphere of argon. At 120 °C temperature, the color of the mixture slowly transformed from dark blue to a brown-yellow. Then the brown-yellow solution was heated up to 230 °C, and the pre-prepared thioacetamide-oleylamine solution was rapidly injected. The CZTS nanoparticles were precipitated and dispersed in ethanol to form the ink solution. CZTS is a direct band gap semiconductor with about 1.4–1.6 eV. The band gap of the CZTS is favorable for solar cell application and can be used as photovoltaic material [89, 123]. Colloidal CZTS nanoparticles were synthesized via hot injection method by many researchers [88, 123–126]. CZTS can be found in the kesterite or stannite as well as wurtzite phase via monitoring the preparation strategy [88, 127]. In a typical synthesis of the wurtzite, CZTS nanocrystals can be obtained via hot injection method reported by Lu et al. [124]. Firstly, $CuCl_2$, $ZnCl_2$ and $SnCl_4$ were dissolved in dodecanethiol (DDT) at 120 °C to produce a homogenous precursor solution of metal thiolates. Afterward, the hot solution was injected into DDT and oleylamine or oleic acid in a three-neck flask at 240 °C. The mixture color became brownish and was preserved for 1 h to permit the progress of CZTS nanocrystals formation. Finally, the products were precipitated by ethanol and re-dispersed in cyclohexane.

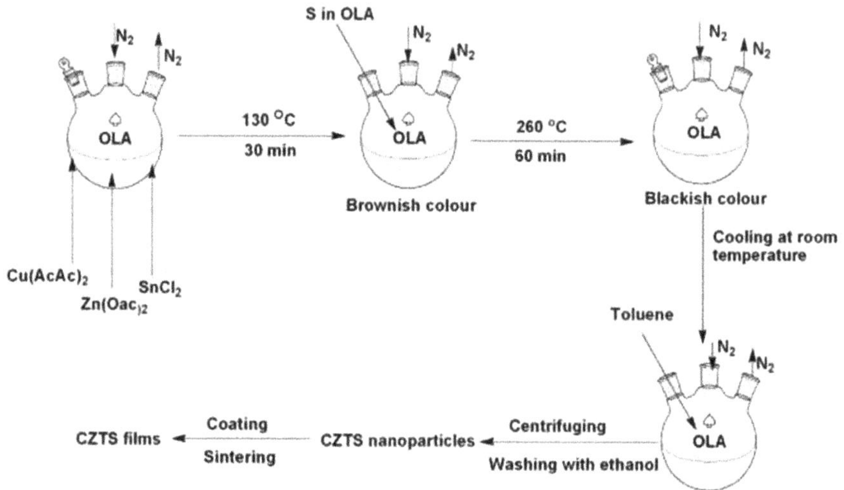

Fig. 13.22 Schematic flow diagram of colloidal CZTS synthesis and film preparation

CZTS compound has some advantages over other compound such as used as material for solar radiation due to suitable optical band gap, non-toxic, large absorption coefficient and large abundance ratio of elements, which make the CZTS derived thin film for solar cell application [128]. The colloidal CZTS nanocrystals were synthesized in oleylamine through hot injection method, and then, the films on glass substrates were fabricated by Zhou et al. [128]. The schematic illustration for the same has been shown in Fig. 13.22. A mixture of $Cu(AcAc)_2$, $Zn(OAc)_2$, $SnCl_2$ and oleylamine (OLA) was added into a three-neck round-bottom reaction flask connected to a N_2 gas cylinder. After being heated at 130 °C temperature for 30 min, the solution becomes brownish. The brownish solution was injected with sulfur–OLA solution and then heated at the reaction temperature 240–280 °C for 1 h. A color change from dark brownish into blackish was found, and the mixture solution was cooled down to room temperature. Then, precipitates of nanoparticles were collected by centrifugation and washing with ethanol. After that, the precipitate was dispersed in toluene. CZTS films were deposited on soda-lime glass slides by drop-casting, from the CZTS dispersed nanoparticles.

13.4.3.9 Cu_2NiSnS_4 Nanoparticles

Kamble et al. have reported for the first time quaternary chalcopyrite Cu_2NiSnS_4 (CNTS) nanoparticles synthesized by hot injection method [129]. For typical modified synthesis of the nanoparticles, the stoichiometric amounts of $CuCl_2$, $SnCl_2$and $Ni(NO_3)_2$ were mixed with oleylamine in a four-neck flask at 220 °C in N_2 atmosphere. In a separate pot, sulfur was added to OLA, and this mixture was kept stirring to get orange-red solution. When the temperature was reached 120 °C, the sulfur solution was added to the flask, and temperature was maintained at 240 °C for 1 h. Afterward, the solution was cooled down to room temperature, and the precipitate was dispersed in acetone. The solution was washed and centrifuged, and this process was repeated several times to eliminate all the impurities. The obtained precipitate was dried in vacuum oven at 600 °C to get CNTS nanoparticles powder. After that, the CNTS nanoparticle powder was dispersed in toluene. The films were grown by coating over fluorine doped tin oxide (FTO) substrate using doctor blade technique. The prepared CNTS films were further annealed at 450 °C in Ar atmosphere for 1 h. They reported the optical band gap of the CNTS nanoparticles at around 1.38 eV, which are in the anticipated range (1.1–1.5 eV) for solar cell application.

13.5 Applications of the Hot Injection Synthesized Nanoparticles

13.5.1 Solar Cells

Ningthoujam et al. have performed photoconductivity study of PbSe QDs with size of 5 nm prepared by hot injection method [32]. The device architectures of solar cells with active layers have been shown in Fig. 13.23. Since PbSe is of n-type semiconductor, p-type semiconductor-MEHPPV (2-Methoxy-5-(2′-ethyl-hexyloxy)-1,4-phenylene vinylene) was interfaced with PbSe to form *p-n* junction. MEHPPV:PbSe QDs mixture was deposited over transparent indium tin oxide (ITO) film. A possible mechanism for photoconductivity in a hybrid of MEH-PPV and PbSe was explained via using energy level diagram (Fig. 13.23c). On back side, aluminum film was deposited. It makes solar cells with area of 0.06 cm^2. Pure MEHPPV does not show the conductivity in the dark. On other hand, MEHPPV:PbSe shows the conductivity in dark indicating absorption of near infrared light from sun, and conductivity is enhanced significantly in the presence of visible light indicating absorption of light in both regions of visible as well as NIR.

Bucherl et al. have reported the nucleation and growth of chalcogenide semiconductor nanocrystals by using hot injection method and shown their use in device fabrication for solar cell application [130]. They focus to develop the device by using substrate-based film followed by making of hot injection synthesized ink solution. Figure 13.24 shows the schematic overview of sintered nanocrystals for hot injection synthesized ink solution-based solar cell fabrication. (a) Nanocrystals are synthesized via hot injection. (b) Nanocrystals are re-dispersed in an organic solvent to formulate an ink solution. (c) Inks are deposited onto a metal-coated substrate. (d) As deposited nanocrystals films are then annealed in an inert atmosphere with chalcogen vapor to form a dense crystalline *p*-type absorber layer. The device can be completed by depositing an n-type buffer layer monitored by a window layer and a grid of conductive top contacts.

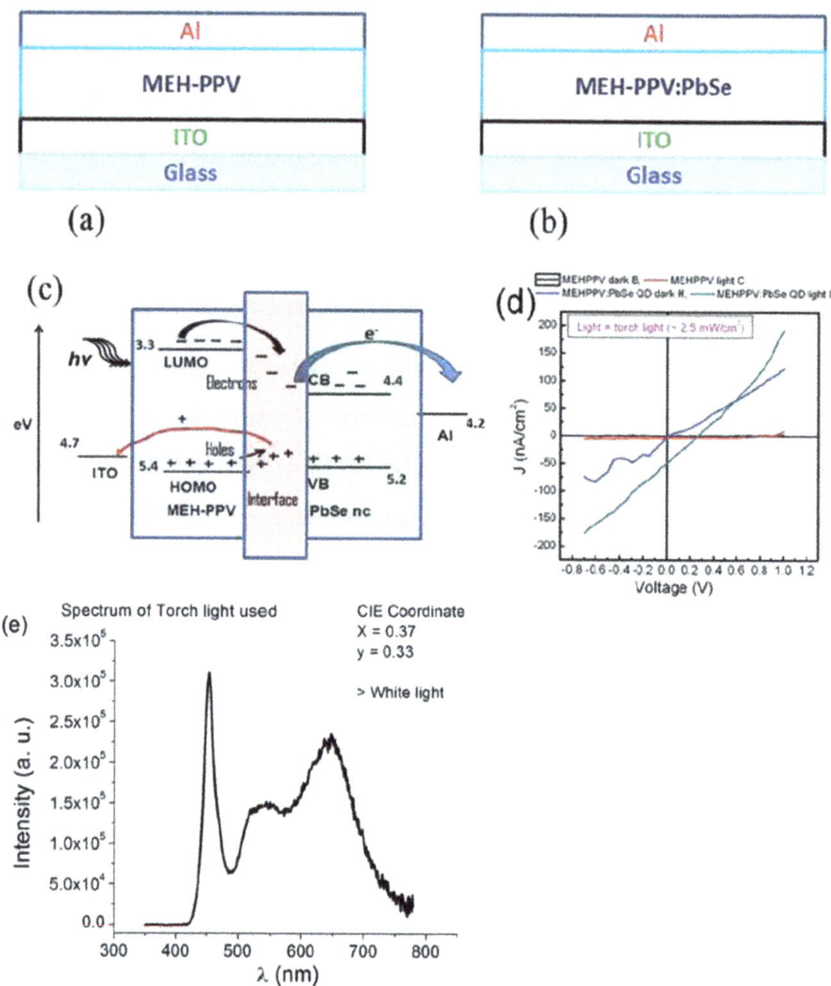

Fig. 13.23 Device architectures of solar cells with active layers: **a** plain MEH-PPV and **b** blend of MEH-PPV and PbSe QDs using olylamine (OAM). **c** A possible mechanism for photoconductivity in a hybrid of MEH-PPV and PbSe. The size of PbSe QDs is 5 nm. **d** *J–V* characteristics of ITO/MEHPPV/Al and ITO/MEHPPV:PbSe QDs/Al devices in the dark and under illumination of white light of 2.5 mW cm^{-2}. **e** Spectrum of torch light used to provide white light on the solar cells. Reproduced with permission from RSC Publications [32]

Heteroepitaxial growth of monodispersed PbS nanocrystals onto the surface of TiO_2 nanoparticles via hot injection method can be used as a photovoltaic response medium with the help two-electrode cell filled with polysulfide electrolyte [109]. Solar cells fabricated from the same size PbS quantum dots synthesized at different conditions demonstrate similar power conversion efficiency (small fluctuation between ~4.7 and ~5%) [107]. CZTS thin film (prepared by hot injection CZTS

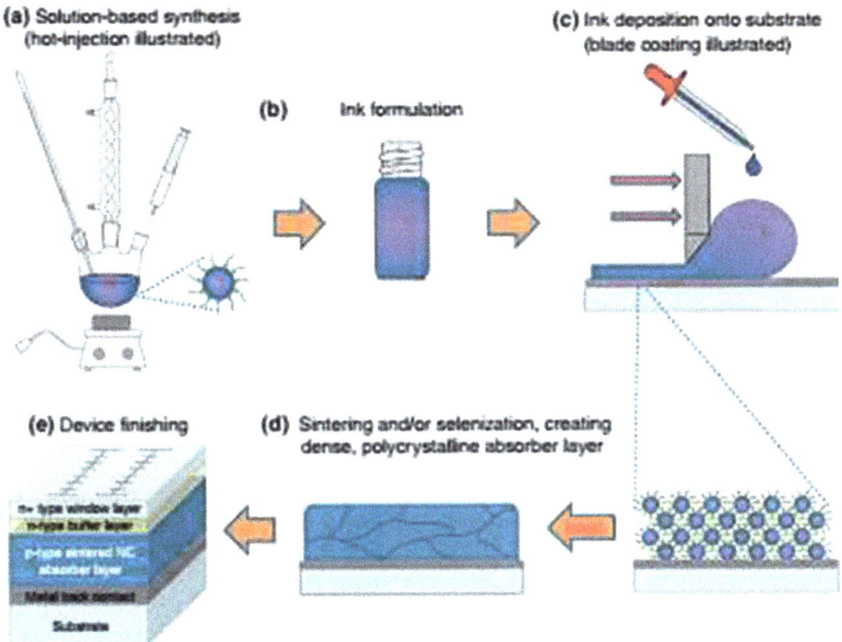

Fig. 13.24 Schematic overview of sintered nanocrystal ink solar cell fabrication. Reproduced with permission from Elsevier Publications [130]

nanoparticles) formation exhibited enhanced conductivity under light irradiation in the current potential (I-V) graph under an illumination intensity of 100 mW cm^{-2} (Xenon Lamp) [124]. It was observed that the electrical conductivity of the CZTS thin film under light irradiation is nearly double as that of the CZTS film in darkness. The conductivity of the thin film enhances due to the illumination which provides an excitation of electrons in the valence band to the conduction band, and hence, holes are created in the CZTS. The excitation of the electron in CZTS enhances the conductivity of the thin film, which further generates photocurrent. The production of photocurrent can be used in the solar energy conversion systems (solar cell) of the wurtzite CZTS materials [124].

The $Co_{0.85}Se$ nanocrystals were synthesized through the hot injection method using oleylamine as solvent [131]. The photo response behavior of the prepared $Co_{0.85}Se$ nanocrystals was studied by making thin film prepared by drop-casted nano-ink on soda-lime glass. I–V performance was investigated for $Co_{0.85}Se$ film using a Keithley 2400 source meter under simulated 100 mW cm^{-2} power obtained by Xenon lamp source and in the dark condition. Figure 13.25 shows the current–potential characteristics of the$Co_{0.85}Se$ nanocrystals film under the dark (black) and

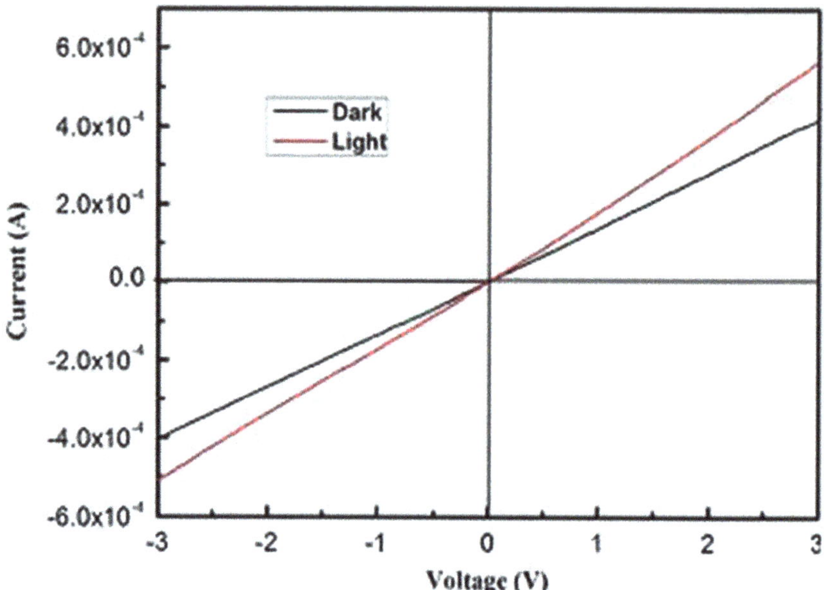

Fig. 13.25 Current–potential characteristics of the $Co_{0.85}Se$ nanocrystals film under dark (black) and illumination conditions (100 mW cm^{-2}) (red). Reproduced with permission from Elsevier Publications [131]

illumination (red) conditions. The performance of the developed $Co_{0.85}Se$ thin film device shows an increment in the current about 1.35 times under illumination compared to the dark state and exhibits wide prospective in photo-electrical conversion application. The photoconductivity response is observed because incident light excites electron and hole pair ions in the film which increases the conductivity [131].

13.5.2 High-Density Data Storage Devices

The ordered face-centered tetragonal (fct, $L1_0$) phase of CoPt, FePt, FePd, CoPd nanoparticles (3–10 nm) and their composites with Cr, Ag, Au, Cu, C in the form of films have been used as extremely high-density magnetic recording (EHDR) media. It can store data of 1 Tb/in^2 [132]. The coercivity (H_c) of 24 kOe, saturation magnetization (M_s) of 750 emu/cm^3, remanence (M_r) of 510 emu/cm^3 and magnetic anisotropy constant (K_u) of 1.2 × 10^7 erg/cm^3 can be achieved.

13.5.3 Laser Devices and Optical Telecommunications

The QDs and quantum wells can change absorption and emission wavelength depending on particle size and shapes and have high absorption or emission cross section. Their emitted light can be wave guided using proper matrix and can be used as laser [133–135]. QDs are used in the data storage/transport in computer in short time. In optical telecommunications, QDs having laser action in 1.33–1.55 μm are used, where loss of energy is very less and interference from overtones of water is also very less [136, 137]. The data exchange with 100 Gbit/s has been reported. Some of QDs examples are PbSe, PbS, InP/AlGaInP, InGaAs/AlGaAs, etc.

13.5.4 Photodetectors

QDs have found applications in the detection of photons from visible to IR region [138, 139]. Sensitivity of QDs at different temperatures varies as compared to the bulk and it gives the better sensitivity per amount of light absorption. Some of the examples are PbSe, HgCdTe, InGaAs, GaAs/AlGaAs, InAs/GaAs and other composites.

13.5.5 Biomaterials

The bright color, Ag@$CaCO_3$, has been developed by using hot injection method treated with the viscose fibers, which exhibit good UV protection properties and admirable antibacterial activity against Escherichia coli (Gram-negative gamma proteobacterium), and consequentially shows potential multifunctional materials for biomedical applications [78]. Enzyme immunoassay method for the detection of protein using biocatalytic Ag nanoparticles as an enhanced substrate based for SERS was reported by Chen et al. [82]. They suggested the use of the SERS-enhanced biocatalytic production of Ag nanoparticles in sensitive immunoassay. Ferrite nanoparticles prepared by the hot injection method have always being considered a good choice in many fields such as drug delivery, hyperthermia and magnetic resonance imaging [91, 140–143]. In this regards, highly monodispersed and water soluble nanoparticles based on the cobalt, iron and ferrite nanoparticles were reported by many researchers [90, 91]. On the other hand, metal nanoparticles such as Au are used in therapy of cancer through photothermal heating, contrast agent, drug and gene delivery systems [144–147]. Au has high atomic mass as compared to soft tissue cells. Because of this, Au nanoparticles can absorb X-ray radiation significantly as compared to soft tissues. This can give the enhanced contrast during radiation dose provided in tumors.

13.5.6 Imaging, Labeling and Sensing

QDs conjugated with suitable molecules/ligands are used in imaging, labeling and sensing of DNA, proteins and biomolecules, etc. [148, 149]. These have been tested in in vitro and in vivo conditions. QDs have many advantages over dyes because of possession of large absorption cross section and sharp emission peak, high luminescence quantum yield and tuneable emission peaks from visible to NIR. Their energy transfer process from QDs to biomolecules/acceptor or vice versa can be used in sensing of biomolecules. Deep tissue imaging can be performed using suitable QDs. Many nanoparticles (particularly Gold) are used extensively as contrast agent in radiation dose therapy, drug and gene delivery systems [148, 149].

13.5.7 Catalysis

Sono-chemical synthesis study of colloidal Ag catalysts for reduction of complexing Ag in diffusion transfer reversal (DTR) system shows that the precursor $AgNO_3$ is better than $Ag(S_2O_3)_2^{3-}$ [150].

13.5.8 Thermoelectric Devices

Song et al. have reported the hot injection synthesis and characterization of monodispersed ternary Cu_2SnSe_3 nanocrystals of small size with an average diameter of 5 nm for thermoelectric applications [151].

13.6 Conclusions

This chapter includes introduction of nanoparticles, basic concepts of hot injection method, kinetics and growth mechanism for the formation of monodispersed particles. High-quality quantum dots are prepared by hot injection method, and many examples are provided. The roles of surfactant, solvent, additives, temperature and duration of heat treatment are important to control the shape and size of nanoparticles for a particular study. In addition, many synthesis methods through electrostatic repulsion, reverse micelle and thermolysis are mentioned to get monodispersed particles. Their merits and demerits are discussed in detail. Many examples of metal, alloys, composites, magnetic materials and quantum dots (semiconducting nanoparticles) are mentioned. Their applications in various areas of solar cells, high-density data storage devices, laser devices, optical telecommunications, photodetectors, biomaterials, imaging, labeling, sensing, catalysis and thermoelectric devices are highlighted.

Acknowledgements Dr. Abhishek Kumar Soni acknowledges Science and Engineering Research Board (SERB), Government of India for providing the National Post-doctoral Fellowship (N-PDF file No. PDF/2016/003419).

References

1. Colvin VL, Schlamp MC, Alivisatos AP (1994) Light-emitting-diodes made from cadmium selenide nanocrystals and a semiconducting polymer. Nature 370:354–357
2. Hoffman AJ, Mills G, Yee H, Hoffmann MR (1992) Q-sized cadmium sulfide: synthesis, characterization, and efficiency of photoinitiation of polymerization of several vinylic monomers. J Phys Chem 96:5546–5552
3. Hamilton JF, Baetzold RC (1979) Catalysis by small metal clusters. Science 205:1213–1220
4. Mansur HS, Grieser F, Marychurch MS, Biggs S, Urquhart RS, Furlong DN (1995) Photoelectrochemical properties of 'Q-state' CdS particles in arachidic acid Langmuir-Blodgett films. J Chem Soc Faraday Trans 91:665–672
5. Hsu CW, Zhen B, Qiu W, Shapira O, DeLacy BG, Joannopoulos JD, Soljacic M (2014) Transparent displays enabled by resonant nanoparticle scattering. In: Lasers and electro-optics (CLEO), conference, pp 1–2. IEEE Xplore. ISBN: 978-1-55752-999-2
6. Anikeeva PO, Halpert JE, Bawendi MG, Bulovic V (2009) Quantum dot light-emitting devices with electroluminescence tunable over the entire visible spectrum. Nano Let 9:2532–2536
7. Wang Y, Herron N (1991) Nanometer-sized semiconductor clusters: materials synthesis, quantum size effects, and photophysical properties. J Phys Chem 95:525–532
8. Schmid G (1992) Large clusters and colloids. Metals in the embryonic state. Chem Rev 92:1709–1727
9. Sear RP (2007) Nucleation: theory and applications to protein solutions and colloidal suspensions. J Phys Condens Matter 19:033101
10. Murray C, Norris DJ, Bawendi MG (1993) Synthesis and characterization of nearly monodisperse CdE (E = sulfur, selenium, tellurium) semiconductor nanocrystallites. J Am Chem Soc 115:8706–8715
11. Abel KA, Boyer JC, Andrei CM, van Veggel FC (2011) Analysis of the shell thickness distribution on $NaYF_4$/$NaGdF_4$ core/shell nanocrystals by EELS and EDS. J Phys Chem Lett 2:185–189
12. Zhang F, Che R, Li X, Yao C, Yang J, Shen D, Hu P, Li W, Zhao D (2012) Direct imaging the upconversion nanocrystal core/shell structure at the subnanometer level: shell thickness dependence in upconverting optical properties. Nano Lett 12:2852–2858
13. Bawendi MG, Steigerwald ML, Brus LE (1990) The quantum mechanics of larger semiconductor clusters ("quantum dots"). Annu Rev Phys Chem 41:477–496
14. Brus L (1986) Electronic wave functions in semiconductor clusters: experiment and theory. J Phys Chem 90:2555–2560
15. Bosnick KA, Jiang J, Brus LE (2002) Fluctuations and local symmetry in single-molecule rhodamine 6G Raman scattering on silver nanocrystal aggregates. J Phys Chem B 106:8096–8099
16. Bell AT (2003) The impact of nanoscience on heterogeneous catalysis. Science 299:1688–1691
17. Batlle X, Labarta AL (2002) Finite-size effects in fine particles: magnetic and transport properties. J Phys D 35:R15–R42

18. LaMer VK, Dinegar RH (1950) Theory, production and mechanism of formation of monodispersed hydrosols. J Am Chem Soc 72:4847–4854
19. Peng X, Wickham J, Alivisatos AP (1998) Kinetics of II-VI and III-V colloidal semiconductor nanocrystal growth:"focusing" of size distributions. J Am Chem Soc 120:5343–5344
20. Kovalenko MV, Talapin DV, Loi MA, Cordella F, Hesser G, Bodnarchuk MI, Heiss W (2008) Quasi-seeded growth of ligand-tailored PbSe nanocrystals through cation-exchange-mediated nucleation. Angew Chem 47:3029–3033
21. Park J, Joo J, Kwon SG, Jang Y, Hyeon T (2007) Synthesis of monodisperse spherical nanocrystals. Angew Chem 46:4630–4660
22. Thanh NT, Maclean N, Mahiddine S (2014) Mechanisms of nucleation and growth of nanoparticles in solution. Chem Rev 114:7610–7630
23. van Embden J, Chesman AS, Jasieniak JJ (2015) The heat-up synthesis of colloidal nanocrystals. Chem Mater 27:2246–2285
24. Rempel JY, Bawendi MG, Jensen KF (2009) Insights into the kinetics of semiconductor nanocrystal nucleation and growth. J Am Chem Soc 131:4479–4489
25. Rinkel T, Nordmann J, Raj AN, Haase M (2014) Ostwald-ripening and particle size focussing of sub-10 nm $NaYF_4$ upconversion nanocrystals. Nanoscale 6:14523–14530
26. Klabunde KJ, Richards RM (2009) Nanoscale materials in chemistry. A. John Wiley & Sons, Inc. Publicayion, Hoboke, New Jersey; Sugimoto T (2001) Monodispersed particles. Elsevier, New York
27. Mullin JW (2001) Crystallization, 4th edn. Butterworth-Heinemann, Oxford
28. Sugimoto T (1987) Preparation of monodispersed colloidal particles. Adv Colloid Interface Sci 28:65–108
29. Kwon SG, Hyeon T (2011) Formation mechanisms of uniform nanocrystals via hot injection and heat up methods. Small 7:2685–2702
30. Talapin DV, Rogach AL, Shevchenko EV, Kornowski A, Haase M, Weller H (2002) Dynamic distribution of growth rates within the ensembles of colloidal II−VI and III−V semiconductor nanocrystals as a factor governing their photoluminescence efficiency. J Am Chem Soc 124:5782–5790
31. Ningthoujam RS (2012) Enhancement of luminescence by rare earth ions doping insemiconductor host (chapter 7). In: Rai SB, Dwivedi Y (eds) Nova Science Publishers Inc, USA, pp 145–182
32. Ningthoujam RS, Gautam A, Padma N (2017) Oleylamine as a reducing agent in syntheses of magic-size clusters and monodisperse quantum dots: optical and photoconductivity studies. Phys Chem Chem Phys 19:2294–2303
33. Zhu H, Yang D, Yu G, Zhang H, Yao K (2006) A simple hydrothermal route for synthesizing SnO_2 quantum dots. Nanotechnology 17:2386
34. Ningthoujam RS, Vatsa RK, Vinu A, Ariga K, Tyagi AK (2009) Room temperature exciton formation in SnO_2 nanocrystals in SiO_2: Eumatrix: quantum dot system, heat-treatment effect. J Nanosci Nanotechnol 9:2634–2638
35. Ningthoujam RS, Sudarsan V, Kulshreshtha SK (2007) SnO_2: Eu nanoparticles dispersed in silica: a low-temperature synthesis and photoluminescence study. J Lumin 127:747–756
36. Ningthoujam RS, Sudarsan V, Godbole SV, Kienle L, Kulshreshtha SK, Tyagi AK (2007) SnO_2:Eu^{3+} nanoparticles dispersed in TiO_2 matrix: improved energy transfer between semiconductor host and Eu^{3+} ions for the low temperature synthesized samples. Appl Phys Lett 90:
37. Ningthoujam RS (2010) Generation of exciton in two semiconductors interface: SnO_2:Eu–Y_2O_3. Chem Phys Lett 497:208–212
38. Leatherdale CA, Woo WK, Mikulec FV, Bawendi MG (2002) On the absorption cross section of Cd senanocrystal quantum dots. J Phys Chem B 106:7619–7622
39. Seo YS, Raj CJ, Kim DJ, Kim BC, Yu KH (2014) Effect of CdSe/ZnS quantum dots dispersion in silicone based polymeric fluids. Mater Lett 130:43–47

40. Sayevich V, Gaponik N, Plötner M, Kruszynska M, Gemming T, Dzhagan VM et al (2015) Stable dispersion of iodide-capped PbSe quantum dots for high-performance low-temperature processed electronics and optoelectronics. Chem Mater 27:4328–4337
41. Spirin MG, Brichkin SB, Razumov VF (2015) Luminescent properties of CdSe quantum dots in dispersion media with different polarity. High Energy Chem 49:426–432
42. Probst CE, Zrazhevskiy P, Bagalkot V, Gao X (2013) Quantum dots as a platform for nanoparticle drug delivery vehicle design. Adv Drug Deliv Rev 65:703–718
43. Dai X, Zhang Z, Jin Y, Niu Y, Cao H, Liang X, Chen L, Wang J, Peng X (2014) Solution-processed, high-performance light-emitting diodes based on quantum dots. Nature 515:96–99
44. Qasim K, Lei W, Li Q (2013) Quantum dots for light emitting diodes. J Nanosci Nanotechnol 13:3173–3185
45. Salaheldin AM, Walter J, Herre P, Levchuk I, Jabbari Y, Kolle JM, Brabec CJ, Peukert W, Segets D (2017) Automated synthesis of quantum dot nanocrystals by hot injection: mixing induced self-focusing. Chem Eng 320:232–243
46. Zhang J, Gao J, Miller EM, Luther JM, Beard MC (2013) Diffusion-controlled synthesis of PbS and PbSe quantum dots with in situ halide passivation for quantum dot solar cells. ACS Nano 8:614–622
47. Xuan TT, Liu JQ, Yu CY, Xie RJ, Li HL (2016) Facile synthesis of cadmium-free Zn-In-S: Ag/ZnS nanocrystals for bio-imaging. Sci Rep 6:24459
48. Qu L, Peng ZA, Peng X (2001) Alternative routes toward high quality CdSe nanocrystals. Nano Lett 1:333–337
49. Talapin DV, Rogach AL, Kornowski A, Haase M, Weller H (2001) Highly luminescent monodisperse CdSe and CdSe/ZnS nanocrystals synthesized in a hexadecylamine−trioctylphosphine oxide−trioctylphospine mixture. Nano Lett 1:207–211
50. Vigneshwaran N, Nachane RP, Balasubramanya RH, Varadarajan PV (2006) A novel one-pot ‘green’ synthesis of stable silver nanoparticles using soluble starch. Carbohydr Res 341:2012–2018
51. Tai CY, Wang YH, Liu HS (2008) A green process for preparing silver nanoparticles using spinning disk reactor. AIChE J 54:445–452
52. Raveendran P, Fu J, Wallen SL (2003) Completely “green” synthesis and stabilization of metal nanoparticles. J Am Chem Soc 125:13940–13941
53. Pillai ZS, Kamat PV (2004) What factors control the size and shape of silver nanoparticles in the citrate ion reduction method? J Phys Chem B 108:945–951
54. Komarneni S, Li D, Newalkar B, Katsuki H, Bhalla AS (2002) Microwave−polyol process for Pt and Ag nanoparticles. Langmuir 18:5959–5962
55. Reddy DH, Lee SM (2013) Application of magnetic chitosan composites for the removal of toxic metal and dyes from aqueous solutions. Adv Coll Interface Sci 201–202:68–93
56. Jia Z, Yujun W, Yangcheng L, Jingyu M, Guangsheng L (2006) In situ preparation of magnetic chitosan/Fe_3O_4 composite nanoparticles in tiny pools of water-in-oil microemulsion. React Funct Polym 66:1552–1558
57. Xu H, Liu X, Su G, Zhang B, Wang D (2012) Electrostatic repulsion-controlled formation of polydopamine–gold Janus particles. Langmuir 28:13060–13065
58. Singhal P, Pulhani V, Musharaf Ali SKM, Ningthoujam RS (2019) Sorption of different metal ions on magnetic nanoparticles and their effect on nanoparticles settlement. Environ Nanotechnol Monit Manag 11(2019):
59. Luwang MN, Ningthoujam RS, Srivastava SK, Vatsa RK (2011) Preparation of white light emitting YVO_4:Ln^{3+} and silica-coated YVO_4:Ln^{3+}(Ln^{3+} = Eu^{3+}, Dy^{3+}, Tm^{3+}) nanoparticles by CTAB/n-butanol/hexane/water microemulsion route: energy transfer and site symmetry studies. J Mater Chem 21:5326–5337
60. Luwang MN, Ningthoujam RS, Singh NS, Tewari R, Srivastava SK, Vatsa RK (2010) Surface chemistry of surfactant AOT-stabilized SnO_2 nanoparticles and effect of temperature. J Colloid Interface Sci 349:27–33

61. Timonen JV, Seppälä ET, Ikkala O, Ras RH (2011) From hot-injection synthesis to heating-up synthesis of cobalt nanoparticles: observation of kinetically controllable nucleation. Angew Chem 50:2080–2084
62. Humphrey JJ, Sadasivan S, Plana D, Celorrio V, Tooze RA, Fermín DJ (2015) Surface activation of Pt nanoparticles synthesised by "hot injection" in the presence of oleylamine. Chem Eur J 21:12694–12701
63. de Mello Donegá C, Liljeroth P, Vanmaekelbergh D (2005) Physicochemical evaluation of the hot-injection method, a synthesis route for monodisperse nanocrystals. Small 1:1152–1162
64. Yong KT, Sahoo Y, Choudhury KR, Swihart MT, Minter JR, Prasad PN (2006) Shape control of PbSe nanocrystals using noble metal seed particles. Nano Lett 6:709–714
65. Guzelian AA, Katari JB, Kadavanich AV, Banin U, Hamad K, Juban E, Heath JR (1996) Synthesis of size-selected, surface-passivated InP nanocrystals. J Phys Chem A 100:7212–7219
66. Cao Y, Banin U (2000) Growth and properties of semiconductor core/shell nanocrystals with InAs cores. Am Chem Soc 122:9692–9702
67. Lassenberger A, Grünewald TA, van Oostrum PDJ, Rennhofer H, Amenitsch H, Zirbs R, Lichtenegger HC, Reimhult E (2017) Monodisperse iron oxide nanoparticles by thermal decomposition: elucidating particle formation by second-resolved in situ small-angle x-ray scattering. Chem Mater 29:4511–4522
68. Hikmet RAM, Talapin DV, Weller H (2003) Study of conduction mechanism and electroluminescence in CdSe/ZnS quantum dot composites. J Appl Phys 93:3509–3514
69. Kim H, Park B, Cho K, Kim JH, Lee JW, Kim DW, Kim S (2005) Transport of charge carriers in HgTe/CdTe core-shell nanoparticle film. Jpn J Appl Phys 44:5703
70. Deng Z, Zhu H, Peng B, Chen H, Sun Y, Gang X, Jin P, Wang J (2012) Synthesis of PS/Ag nanocomposite spheres with catalytic and antibacterial activities. ACS Appl Mater Interfaces 4:5625–5632
71. Tran QH, Nguyen VQ, Le A-T (2013) Silver nanoparticles: synthesis, properties, toxicology, applications and perspectives. Adv Nat Sci Nanosci Nanotechnol 4:033001
72. Lu L, Sun RW-Y, Chen R, Hui C-K, Ho C-M, Luk JM, Lau GKK, Che C-M (2008) Silver nanoparticles inhibit hepatitis B virus replication. Antiviral Ther 13:253–262
73. Wang J, Boelens HF, Thathagar MB, Rothenberg G (2004) In situ spectroscopic analysis of nanocluster formation. Chem Phys Chem 5:93–98
74. Cozzoli PD, Comparelli R, Fanizza E, Curri ML, Agostiano A, Laub D (2004) Photocatalytic synthesis of silver nanoparticles stabilized by TiO_2 nanorods: a semiconductor/metal nanocomposite in homogeneous nonpolar solution. J Am Chem Soc 126:3868–3879
75. Kim D, Jeong S, Moon J (2006) Synthesis of silver nanoparticles using the polyol process and the influence of precursor injection. Nanotechnology 17:4019
76. Henglein A, Giersig M (1999) Formation of colloidal silver nanoparticles: capping action of citrate. J Phys Chem B 103:9533–9539
77. Pacioni NL, Borsarelli CD, Valentina Rey, Alicia VV (2015) Silver nanoparticle applications. In: Alarcon EI, Griffith M, Udekwu KI (eds) Engineering materials. Springer International Publishing, Cham. https://doi.org/10.1007/978-3-319-11262-6_2
78. Barhoum A, Rehan M, Rahier H, Bechelany M, Van Assche G (2016) Seed-mediated hot-injection synthesis of tiny Ag nanocrystals on nanoscale solid supports and reaction mechanism. ACS Appl Mater Interfaces 8:10551–10561
79. Sun Y, Mayers B, Herricks T, Xia Y (2003) Polyol synthesis of uniform silver nanowires: a plausible growth mechanism and the supporting evidence. Nano Lett 3:955–960
80. Bharti A, Bhardwaj R, Agrawal AK, Goyal N, Gautam S (2016) Monochromatic x-ray induced novel synthesis of Plasmonic nanostructure for photovoltaic application. Sci Rep 6
81. Noritomi H, Umezawa Y, Miyagawa S, Kato S (2011) Preparation of highly concentrated silver nanoparticles in reverse micelles of sucrose fatty acid esters through solid-liquid extraction method. ACES 1:299

82. Chen J, Luo Y, Liang Y, Jiang J, Shen G, Yu R (2009) Surface-enhanced Raman scattering for immunoassay based on the biocatalytic production of silver nanoparticles. Anal Sci 25:347–352
83. Benkovicova M, Vegso K, Siffalovic P, Jergel M, Luby S, Majkova E (2013) Preparation of gold nanoparticles for plasmonic applications. Thin Solid Films 543:138–141
84. Ma WW, Yang Y, Chong CT, Eggeman A, Piramanayagam SN, Zhou TJ, Song T, Wang JP (2004) Synthesis and magnetic behavior of self-assembled Co nanorods and nanoballs. J Appl Phys 95:6801–6803
85. Zacharaki E, Kalyva M, Fjellvåg H, Sjåstad AO (2016) Burst nucleation by hot injection for size controlled synthesis of ε-cobalt nanoparticles. Chem Cent J 10:10
86. Xu J, Zhang W, Yang Z, Ding S, Zeng C, Chen L, Wang Q, Yang S (2009) Large-Scale synthesis of long crystalline $Cu_{2-x}Se$ nanowire bundles by water-evaporation-induced self-assembly and their application in gas sensing. Adv Funct Mater 19:1759–1766
87. Yang CT, Hsiang HI, Tu JH (2016) Copper selenide crystallites synthesized using the hot-injection process. Adv Powder Technol 27:959–963
88. Méndez-López A, Morales-Acevedo A, Acosta-Silva YJ, Ortega-López M (2016) Synthesis and characterization of colloidal CZTS nanocrystals by a hot-injection method. J Nanomater, Article ID 7486094, 7 pages
89. Steinhagen C, Panthani MG, Akhavan V, Goodfellow B, Koo B, Korgel BA (2009) Synthesis of Cu_2ZnSnS_4 nanocrystals for use in low-cost photovoltaics. J Am Chem Soc 131:12554–12555
90. Ahn JP, Kim MS, Kim SH, Lee BH, Kim DK, Park JS (2016) Hot-injection thermolysis of cobalt antimony nanoparticles with Co (II)-Oleate and Sb (III)-Oleate. J Korean Ceram 53:367–375
91. Abdulwahab KO, Malik MA, O'Brien P, Timco GA (2014) A direct synthesis of water soluble monodisperse cobalt and manganese ferrite nanoparticles from iron based pivalate clusters by the hot injection thermolysis method. Mater Sci Semicond Process 27:303–308
92. Williams JV, Kotov NA, Savage PE (2009) A rapid hot-injection method for the improved hydrothermal synthesis of CdSe nanoparticles. Ind Eng Chem Res 48:4316–4321
93. Peng X, Manna L, Yang W, Wickham J, Scher E, Kadavanich A, Alivisatos AP (2000) Shape control of CdSe nanocrystals. Nature 404:59–61
94. Wuister SF, de Mello Donega C, Meijerink A (2004) Influence of thiol capping on the exciton luminescence and decay kinetics of CdTe and CdSe quantum dots. J Phys Chem B 108:17393–17397
95. Kim JI, Kim J, Lee J, Jung DR, Kim H, Choi H, Lee S, Byun S, Kang S, Park B (2012) Photoluminescence enhancement in CdS quantum dots by thermal annealing. Nanoscale Res Lett 7:482
96. Nirmal M, Brus L (1999) Luminescence photophysics in semiconductor nanocrystals. Acc Chem Res 32:407–414
97. Bakkers EPAM, Hens Z, Zunger A, Franceschetti A, Kouwenhoven LP, Gurevich L, Vanmaekelbergh D (2001) Shell-tunneling spectroscopy of the single-particle energy levels of insulating quantum dots. Nano Lett 1:551–556
98. Dabbousi BO, Rodriguez-Viejo J, Mikulec FV, Heine JR, Mattoussi H, Ober R et al (1997) (CdSe) ZnS core−shell quantum dots: synthesis and characterization of a size series of highly luminescent nanocrystallites. J Phys Chem B 101:9463–9475
99. Deng ZX, Li L, Li Y (2003) Novel inorganic–organic-layered structures: crystallographic understanding of both phase and morphology formations of one-dimensional CdE (E = S, Se, Te) nanorods in ethylenediamine. Inorg Chem 42:2331–2341
100. Valcheva E, Yordanov G, Yoshimura H, Ivanov T, Kirilov K (2014) Low temperature studies of the photoluminescence from colloidal CdSe nanocrystals prepared by the hot injection method in liquid paraffin. Colloids Surf A 461:158–166
101. Yu WW, Falkner JC, Shih BS, Colvin VL (2004) Preparation and characterization of monodisperse PbSe semiconductor nanocrystals in a noncoordinating solvent. Chem Mater 16:3318–3322

102. Sashchiuk A, Langof L, Chaim R, Lifshitz E (2002) Synthesis and characterization of PbSe and PbSe/PbS core–shell colloidal nanocrystals. J Cryst Growth 240:431–438
103. Lifshitz E, Bashouti M, Kloper V, Kigel A, Eisen MS, Berger S (2003) Synthesis and characterization of PbSe quantum wires, multipods, quantum rods, and cubes. Nano Lett 3:857–862
104. Wehrenberg BL, Wang C, Guyot-Sionnest P (2002) Interband and intraband optical studies of PbSe colloidal quantum dots. J Phys Chem B 106:10634–10640
105. Murray CB, Sun S, Gaschler W, Doyle H, Betley TA, Kagan CR (2001) Colloidal synthesis of nanocrystals and nanocrystal superlattices. IBM J Res Dev 45:47–56
106. Du H, Chen C, Krishnan R, Krauss TD, Harbold JM, Wise FW, Thomas MG, Silcox J (2002) Optical properties of colloidal PbSe nanocrystals. Nano Lett 2:1321–1324
107. Efros AL, Efros AL (1982) Interband absorption of light in a semiconductor sphere. Sov Phys Semicond 16:772–775
108. Bakueva L, Musikhin S, Hines MA, Chang T-WF, Tzolov M, Scholes GD, Sargent EH (2003) Size-tunable infrared (1000–1600 nm) electroluminescence from PbS quantum-dot nanocrystals in a semiconducting polymer. Appl Phys Lett 82:2895–2897
109. Acharya KP, Hewa-Kasakarage NN, Alabi TR, Nemitz I, Khon E, Ullrich B, Anzenbacher P, Zamkov M (2010) Synthesis of PbS/TiO2 colloidal heterostructures for photovoltaic applications. J Phys Chem C 114:12496–12504
110. Zhang Z, Yang J, Wen X, Yuan L, Shrestha S, Stride JA, Conibeer GJ, Patterson RJ, Huang S (2015) Effect of halide treatments on PbSe quantum dot thin films: stability, hot carrier lifetime, and application to photovoltaics. J Phys Chem C 119:24149–24155
111. Yang J, Elim HI, Zhang Q, Lee JY, Ji W (2006) Rational synthesis, self-assembly, and optical properties of PbS−Au heterogeneous nanostructures via preferential deposition. J Am Chem Soc 128:11921–11926
112. Nozik AJ (2002) Quantum dot solar cells. Physica E Low Dimens. Syst Nanostruct 14:115–120
113. Zhang J, Crisp RW, Gao J, Kroupa DM, Beard MC, Luther JM (2015) Synthetic conditions for high-accuracy size control of PbS quantum dots. J Phys Chem Lett 6:1830–1833
114. Lévy-Clément C, Neumann-Spallart M, Haram SK, Santhanam KSV (1997) Chemical bath deposition of cubic copper (I) selenide and its room temperature transformation to the orthorhombic phase. Thin Solid Films 302:12–16
115. Lakshmikumar ST, Rastogi AC (1994) Selenization of Cu and In thin films for the preparation of selenide photo-absorber layers in solar cells using Se vapour source. Sol Energy Mater Sol Cells 32:7–19
116. Chen D, Chen W, Ma L, Ji G, Chang K, Lee JY (2014) Graphene-like layered metal dichalcogenide/graphene composites: synthesis and applications in energy storage and conversion. Mater Today 17:184–193
117. Panda SK, Antonakos A, Liarokapis E, Bhattacharya S, Chaudhuri S (2007) Optical properties of nanocrystalline SnS_2 thin films. Mater Res Bull 42:576–583
118. Huang PC, Wang HI, Brahma S, Wang SC, Huang JL (2017) Synthesis and characteristics of layered SnS_2 nanostructures via hot injection method. J Cryst Growth 468:162–168
119. Trinh TK, Pham VTH, Truong NTN, Kim CD, Park C (2017) Iron pyrite: phase and shape control by facile hot injection method. J Cryst Growth 461:53–59
120. Vahidshad Y, Ghasemzadeh R, Irajizad A, Mirkazemi SM (2013) Synthesis and characterization of copper indium sulfide chalcopyrite structure with hot injection method. J Nanostruct 3:145–154
121. Pan D, An L, Sun Z, Hou W, Yang Y, Yang Z, Lu Y (2008) Synthesis of Cu−In−S ternary nanocrystals with tunable structure and composition. J Am Chem Soc 130:5620–5621
122. Song JM, Liu Y, Niu HL, Mao CJ, Cheng LJ, Zhang SY, Shen YH (2013) Hot-injection synthesis and characterization of monodispersed ternary Cu_2SnSe_3 nanocrystals for thermoelectric applications. J. Alloys Compd 581:646–652

123. Wang W, Winkler MT, Gunawan O, Gokmen T, Todorov TK, Zhu Y, Mitzi DB (2014) Device characteristics of CZTSSe thin-film solar cells with 12.6% efficiency. Adv Energy Mater 4
124. Lu X, Zhuang Z, Peng Q, Li Y (2011) Wurtzite Cu 2 $ZnSnS_4$ nanocrystals: a novel quaternary semiconductor. Chem Commun 47:3141–3143
125. Singh A, Geaney H, Laffir F, Ryan KM (2012) Colloidal synthesis of wurtzite Cu_2ZnSnS_4 nanorods and their perpendicular assembly. J Am Chem Soc 134:2910–2913
126. Engberg S, Li Z, Lek JY, Lam YM, Schou J (2015) Synthesis of large CZTSe nanoparticles through a two-step hot-injection method. RSC Adv 5:96593–96600
127. Kameyama T, Osaki T, Okazaki KI, Shibayama T, Kudo A, Kuwabata S, Torimoto T (2010) Preparation and photoelectrochemical properties of densely immobilized Cu_2ZnSnS_4 nanoparticle films. J Mater Chem 20:5319–5324
128. Zhou M, Gong Y, Xu J, Fang G, Xu Q, Dong J (2013) Colloidal CZTS nanoparticles and films: preparation and characterization. J Alloy Compd 574:272–277
129. Kamble A, Mokurala K, Gupta A, Mallick S, Bhargava P (2014) Synthesis of Cu_2 $NiSnS_4$ nanoparticles by hot injection method for photovoltaic applications. Mater Lett 137:440–443
130. Bucherl CN, Oleson KR, Hillhouse HW (2013) Thin film solar cells from sintered nanocrystals. Curr Opin Chem Eng 2:168–177
131. Liu Y, Liu F, Huang C, Lv X, Lai Y, Li J, Liu Y (2013) Hot-injection synthesis of $Co_{0.85}Se$ nanocrystals for photo-electrical application. Mater Lett 108:110–113
132. Xu YF, Yan ML, Sellmyer DJ (2007) FePt nanocluster films for high-density magnetic recording. J Nanosci Nanotechnol 7:206–224
133. Sokolovskiĭ GS, Vinokurov DA, Deryagin AG, Dudelev VV, Kuchinskiĭ VI, Losev SN, Lyutetskii AV, Pikhtin NA, Slipchenko SO, Sokolova ZN, Tarasov IS (2008) Switching between generation of two quantum states in quantum-well laser diodes. Tech Phys Lett 34:708–710
134. Lüdge K, Schöll E, Viktorov E, Erneux T (2011) Analytical approach to modulation properties of quantum dot lasers. J Appl Phys 109:103–112
135. Saravanamoorthy SN, Peter AJ, Lee CW (2017) Optical peak gain in a PbSe/CdSe core-shell quantum dot in the presence of magnetic field for mid-infrared laser applications. Chem Phys 483:1–6
136. Shutts S, Elliott SN, Smowton PM, Krysa AB (2015) Exploring the wavelength range of InP/AlGaInP QDs and application to dual-state lasing. Semicond Sci Technol 30:
137. Ezra YB, Lembrikov BI, Zarkovsky S (2017) Applications of quantum dot (QD) lasers in optical communications for datacenters. In: Transparent optical networks (ICTON), 2017 19th international conference. IEEE Xplore, pp 1–4
138. Martyniuk P, Rogalski A (2008) Quantum-dot infrared photodetectors: status and outlook. Prog Quant Electron 32:89–120
139. Prabhakaran P, Kim WJ, Lee KS, Prasad PN (2012) Quantum dots (QDs) for photonic applications. Opt Mater Exprs 2:578–593
140. Majewski P, Thierry B (2007) Functionalized magnetite nanoparticles—synthesis, properties, and bio-applications. Crit Rev Solid State Mater Sci 32:203–215
141. Mout R, Moyano DF, Rana S, Rotello VM (2012) Surface functionalization of nanoparticles for nanomedicine. Chem Soc Rev 41:2539–2544
142. Lee HJ, Jang KS, Jang S, Kim JW, Yang HM, Jeong YY, Kim JD (2010) Poly(amino acid)s micelle-mediated assembly of magnetite nanoparticles for ultra-sensitive long-term MR imaging of tumors. Chem Commun 46:3559–3561
143. Lee SW, Bae S, Takemura Y, Shim IB, Kim TM, Kim J, Lee HJ, Zurn S, Kim CS (2007) Self-heating characteristics of cobalt ferrite nanoparticles for hyperthermia application. J Magn Magn Mater 310:2868–2870
144. Pissuwan D, Niidome T, Cortie MB (2011) The forthcoming applications of gold nanoparticles in drug and gene delivery systems. J Control Release 149:65–71
145. Yao C, Zhang L, Wang J, He Y, Xin J, Wang S, Xu H, Zhang Z (2016) Gold nanoparticle mediated phototherapy for cancer. J Nanomater

146. Abadeer NS, Murphy CJ (2016) Recent progress in cancer thermal therapy using gold nanoparticles. J Phys Chem C 120:4691–4716
147. McQuaid HN, Muir MF, Taggart LE, McMahon SJ, Coulter JA, Hyland WB, Jain S, Butterworth KT, Schettino G, Prise KM, Hirst DG, Botchway SW, Currel FJ (2016) Imaging and radiation effects of gold nanoparticles in tumour cells. Sci Rep 6:19442
148. Medintz IL, Uyeda HT, Goldman ER, Mattoussi H (2005) Quantum dot bioconjugates for imaging, labelling and sensing. Nat Mater 4:435
149. Li J, Zhu JJ (2013) Quantum dots for fluorescent biosensing and bio-imaging applications. Analyst 138:2506–2515
150. Zhang JP, Chen P, Sun CH, Hu XJ (2004) Sonochemical synthesis of colloidal silver catalysts for reduction of complexing silver in DTR system. Appl Catal A 266:49–54
151. Sharma S, Gajbhiye NS, Ningthoujam RS (2010) Synthesis and self-assembly of monodisperse Co_xNi_{100-x} (x = 50, 80) colloidal nanoparticles by homogenous nucleation. J Colloid Interface Sci 351:323–329

Chapter 14
Synthesis of Advanced Materials by Electrochemical Methods

Manoj Kumar Sharma

Abstract Electrochemical synthesis or electro-synthesis is a method to synthesize chemical compounds using electrochemical techniques that involves either application of potentials or currents. The setup for electrochemical synthesis is simple consisting of a potentiostat or a galvanostat and an electrochemical cell with electrodes, solvents, supporting electrolytes, precursor chemicals. The advantages of electrochemical synthesis over other synthesis methods are (1) non-requirement of chemical reductants or oxidants as the precursor chemicals in solution undergo electron transfer directly on the electrode surface, (2) large redox potential range of several volts is easily accessible by selecting appropriate combination of electrode materials, solvents and supporting electrolytes. Such a large potential range of several volts involves very high energy either comparable or more than the most of chemical bonds and activation energy involved in chemical reactions, leading to a controlled generation of highly energetic intermediates under mild experimental conditions, (3) higher selectivity and percentage yield because of a very precise control over potential and current, (4) evade the much-needed separation and waste treatment of redundant end products from the desired product, thus, helping in an effectual control over environmental pollution, (5) understanding of the reaction mechanisms/kinetics gained by the analysis of the electrochemical parameters like current and potential measured during the electrochemical synthesis. Electrochemical methods are used to synthesize a large number of organic and inorganic compounds with many applications in almost all disciplines of science and technology. Shape, size, and dimensions of the material can be easily controlled by using either template-assisted or template-free electrochemical methods.

Keywords Electrochemical synthesis · Electrodes · Potential · Current · Electrochemical window · Nanomaterials · Conducting polymers · Semiconductors · Graphene

---

M. K. Sharma (✉)
Fuel Chemistry Division, Bhabha Atomic Research Centre, Mumbai 400085, India
e-mail: mkumars@barc.gov.in

Homi Bhabha National Institute, Mumbai 400094, India

A. K. Tyagi and R. S. Ningthoujam (eds.), *Handbook on Synthesis Strategies for Advanced Materials*, Indian Institute of Metals Series,
https://doi.org/10.1007/978-981-16-1807-9_14

14.1 Introduction

Though the development of the electrostatic generator and scientific understanding of theories governing the electricity began around sixteenth century, but the application of electric current to drive the chemical reaction began in early nineteenth century when English chemist William Nicholson and German chemist Johann Wilhelm Ritter carry out the electrolysis of water to produce hydrogen and oxygen molecules [1]. Figure 14.1 shows the Ritter's electrochemical cell used for electrolysis of water.

Concurrently, Alessandro Volta invented the first battery to produces electric current from the redox chemical reaction [1]. The voltaic pile invented by Volta is shown in Fig. 14.2.

The redox chemical reactions in the two half-cells drive the flow of electric current between them. This phenomenon leads to an idea among the researchers that the desired chemical reactions can be carried out by the flow of electrical current in the half-cells. The advent of new sophisticated advance instruments and the improved understanding of the theories governing the thermodynamics and kinetics of the electrochemical reactions lead to the development of numerous electrochemical synthesis schemes for the manufacturing/production of many materials of interest.

In 1807, Humphry Davy produced metallic potassium by carrying out the electrolysis of molten caustic potash (KOH) and potassium became the first element to be discovered by electrolysis. Later, Davy used the electrolysis to isolate and produce sodium metal. Davy proposed in his lectures 'On Some Chemical Agencies

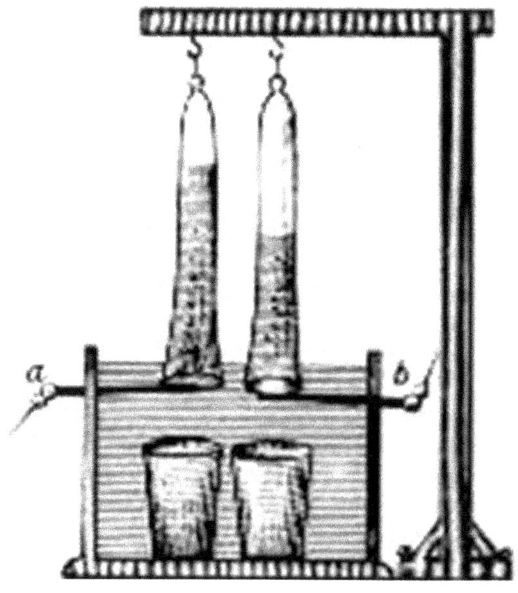

Fig. 14.1 Ritter's electrochemical cell for electrolysis of water [reproduced with permission of Elsevier (Ref. [1]a, Hydrogen from Water Electrolysis in Storing Energy, 2016)]

Fig. 14.2 A Voltaic Pile [reproduced with the permission of American Physical Society (Ref. [1]b, APS News, March 2006, vol. 15, No. 3)]

of Electricity' that passage of the electric currents through chemicals can lead to decomposition of many chemical compounds into their basic elements. For his contribution to electrolysis, Institut de France awarded him the Napoleon Prize in 1807. Davy also isolated some of the alkaline earth metals from their salts using electrolysis.

In the beginning, the attempts to improve the purification/production of elements by electrochemical techniques often lead to the discoveries of new elements. Elements like iridium, osmium, palladium, and rhodium were discovered by

William Hyde Wollaston and Smithson Tennant while designing an efficient electrochemical synthesis route for the production of very pure platinum metal.

Discovery and isolation of new elements were not the only initial application of electrolysis, but the first telegraph, developed by Samuel Thomas von Soemmering in 1809, used the principle of electrolysis. The device used by him had 26 wires (each wire representing one letter of German alphabet) which terminated in a container of acid and the passage of the current through these wires caused the acid in the containers to decompose chemically leading to formation of bubbles. The message was read by monitoring the wire's terminal, in receiving station, at which bubble formation took place. The messages were sent communicating one letter at a time.

Michael Faraday too studied the effect of passage of electric current through the chemical compounds and found that the major effect was either electrochemical decomposition or deposition. Faraday discovered that the amount of the decomposed/deposited substance was directly proportional to the amount of electricity passing through the solution (Faraday's first law). These findings led to the development of Faraday's two laws of electrochemistry:

- The amount of a substance deposited/produced on each electrode of an electrolytic cell is directly proportional to the amount of electricity passing through the cell.
- The quantities of different elements deposited/produced by a given amount of electricity are in the ratio of their chemical equivalent weights.

Chemical manufacturers, material scientists, and associated researchers always wish to come up with new advance technologies to improve the efficiency of the chemical processes and their wish list include the increased selectivity, decreased energy consumption, no or minimum production of toxic chemical as by-products, either no or easy separation steps in case of formation of chemical by-products, minimize the drainage of valuable chemicals in industrial waste stream, etc. Very often, many synthesis routes and technologies are available with their own advantages and disadvantages, but still there is always a scope of little R&D to either improve the existing technology or come up with new innovative synthesis schemes. Electrochemical synthesis can be considered as an innovative method because it has the capability to cover many of the agendas in the wish list of the chemical manufactures/material scientists. The advantages of electrochemical synthesis over other synthesis methods are (1) non-requirement of chemical reductants or oxidants as the precursor chemicals in solution undergo electron transfer directly on the electrode surface, (2) large redox potential range of several volts is easily accessible by selecting appropriate combination of electrode materials, solvents and supporting electrolytes. Such a large potential range of several volts involves very high energy either comparable or more than most of chemical bonds and activation energy involved in chemical reactions, leading to a controlled generation of highly energetic intermediates under mild experimental conditions, (3) higher selectivity and percentage yield because of a very precise control over potential and current, (4) evade the much-needed separation and waste treatment of redundant end products from the desired product, thus, helping in an effectual control over environmental pollution, (5) understanding of the reaction mechanisms/kinetics gained by the analysis of the electrochemical parameters like current and potential measured during the electrochemical synthesis. However, little or no knowledge of electrochemical methods among chemical manufactures, organic chemists, inorganic chemists, and material scientists prevents the induction of electrochemical synthesis and associated R&D in the large-scale production of smart functional material. This chapter will present how the electrochemistry can be used to synthesize a large number of organic and inorganic compounds with innumerable applications in all disciplines of science and technology. Shape, size and dimensions of the material can be easily controlled by using either template-assisted or template-free electrochemical methods. Since like other synthesis methods, electrochemical methods have their own limitations, therefore, this chapter will also talk about the factors to be taken under consideration prior to carrying out electrochemical synthesis.

Commercial applications of electrochemical methods are diverse and include chlor-alkali industry, metal extraction, separation and purification, pulp and paper industry, water and effluent treatment for destruction of toxic chemicals in domestic & industrial waste stream, air pollution control by either electrocatalytic production of oxygen or electrochemical conversion of toxic pollutants to non-toxic chemicals, pollution monitoring, groundwater remediation, etc.

Researchers have been attempting to synthesize new low cost advanced functional and smart materials for innumerable applications such as sensors, catalysts, electrochemical energy storage, actuators, molecular electronic devices, electrochromic displays, photo electrodes for solar cells, imaging and photothermal therapeutic agents, biomaterials, etc. The large-scale synthesis of these materials at low cost is very essential if researchers wish to transfer the benefits from lab to society. These advanced functional and smart materials include conducting polymers (polyaniline, polypyrrole, polythiophene, etc.) and their composites, metal nanoparticles (Au, Pt, Pd, Ag, etc.), semiconductors (TiO_2, ZnS, CdS, PbS, CdSe, etc.), intermetallic (Pt–Pd, Au–Ag, Pt–Au, etc.), nanostructured carbonaceous materials (graphene, graphene oxide, etc.), metal-organic frameworks (MOFs), porous template membranes (porous anodic alumina) etc. Numerous synthesis methods/schemes are available in literatures; chemical, electrochemical, physical vapor deposition, chemical vapor deposition, plasma techniques, ultrasonication, molecular beam epitaxy, photochemical, hydrothermal, radiation-induced, microwave-assisted, microbe-assisted, etc. These different methods and techniques have their own advantages and disadvantages. Reducing the size of a material to nanometer-scale will lead to evolution of unique chemical and physical properties which are not only distinct from the bulk material but also superior to bulk materials [2]. The shape and size of nanomaterials influence their properties [3]. Therefore, the syntheses of shape- and size-controlled nanomaterials still allure researchers. Both template-assisted (track-etched membranes, porous anodic alumina, surfactants, emulsions, etc.) and template-free methods of synthesis are available to control the size, shape of the product. Track-etched membranes and porous anodic alumina membranes of various pore sizes & pore lengths are commercially available templates and frequently used to synthesize nanowires, nanorods, nanotubes of conducting polymers, metals, alloys, semiconductors, etc. by electrochemical route. Theoretical predictions and experimental results show that one-dimension (1D) nanomaterials have superior properties [4]. Porous anodic alumina membranes of various pore sizes & pore lengths are effortlessly prepared by electrochemical methods. The template-assisted synthesis often requires the additional steps of removing the template from the synthesized materials, thus, the non-template methods are preferred over the template methods [5]. Fabrication of nano-dimension materials also helps in miniaturization of devices.

Many electrochemical (potentiostatic, galvanostatic, potentiodynamic) and combined electrochemical-chemical methods are used to synthesize a variety of advanced materials with size ranging from nanometers to bulk in various geometries in solution as well as on solid electrodes for wide range of applications. Various solid conducting substrates are used as the working electrodes. The electrochemical synthesis of nanostructured advanced material on the solid conducting substrates is also desired for the development of miniaturized device. It has been observed that the performance of the deposited material may get influenced by the properties of the substrates, therefore, deposition of advanced material on different solid supports is probed to achieve the desired performance of the deposited materials, as observed in case of many catalysts [6]. The material synthesized by chemical routes is either uniformly dispersed in the solution or get precipitated as solid powder in the solution. Extreme experimental conditions like high temperature, either very high or low (near vacuum) pressure as required in some of the synthesis methods are not very often required during electrochemical synthesis. The materials electrodeposited on the substrate are mostly pure and does not require extraction/separation from the initial mixture solution of monomers, oxidants, solvents, etc. Electrochemical method can be easily coupled with various characterization techniques such as visible, IR, Raman, NMR, EPR, ellipsometry, AFM, XRD, EXAFS, etc. for in situ characterization of the intermediate and final products during the electrochemical synthesis process [7].

Electrochemical synthesis/electrolysis involves the transfer of electron from electrode to the chemical compounds in the solution or vice versa across the electrode/solution interface to bring the chemical changes. Either the current or the voltage applied to the working electrode drive the flow of electron across the electrode/solution interface. The electrochemical setup is simple consisting of a potentiostat or a galvanostat and a glass cell with either two or three-electrode configuration, solvents, supporting electrolytes, precursor chemicals. Schematic diagram of an electrochemical cell in Fig. 14.3 shows counter, working and reference electrodes dipped in an electrolyte solution.

The choice of two-electrodes and three-electrodes configuration depends on whether the applied potential on the working electrode needs to be controlled or not. The third electrode in three-electrode configuration is a reference electrode, and the potential on the working electrode is controlled and measured with respect to the reference electrode. The salient feature of a reference electrode is well-defined

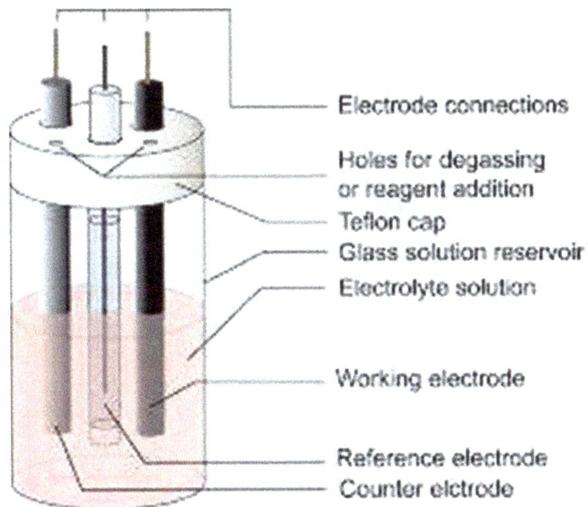

Fig. 14.3 Schematic representation of an electrochemical cell [Reproduced with the permission of American Chemical Society (Ref. [8]b, J. Chem. Edu., **95** (2018) 197–206), https://pubs.acs.org/doi/10.1021/acs.jchemed.7b00361; further permissions related to the material excerpted should be directed to the ACS]

stable electrochemical potential at constant temperature. Though controlling current of the working electrode is much easier and required less sophisticated instruments, but the control of the working electrode's potential provides enhanced selectivity over the redox reaction. A vast choice of materials for working, counter and reference electrodes as well as a vast choice of solvents (aqueous and non-aqueous, protic and aprotic) and supporting electrolytes are available in the arsenal of an electrochemist. The overall aim is to obtain as much possible wider electrochemical window (the potential range between which solvents, supporting electrolytes, electrodes are neither oxidized nor reduced) by proper selection of electrodes' material and supporting electrolytes-solvents system. Influence of electrolyte, electrode material, and solvent is illustrated in Fig. 14.4.

The width of the electrochemical window is very often restricted by either oxidation or reduction of supporting electrolyte and solvent. The working electrode itself may undergo oxidative dissolution at anodic potential or hydrogen evolution occurs in aqueous solution at cathodic potential, thus, further restricting the width of the electrochemical window. The metals, semiconductor oxides (indium tin oxide, titanium oxide), graphite, glassy carbon, highly-oriented pyrolytic graphite (HOPG), carbon-paste electrodes, mercury, etc. have been conventionally used working electrodes, but in recent years chemically-modified electrodes using nanostructured materials of metals, semiconductors, conducting polymers, carbon nanotubes, graphene, fullerene, metal-organic frameworks. These chemically-modified electrodes have higher heterogeneous electron transfer rate constant value, higher sensitivity, higher selectivity, and many a time wider electrochemical windows than the conventional electrodes. The boron-doped diamond electrode has the largest electrochemical window of about 4 V in aqueous solution. Platinum wires and coils, glassy carbon rods, graphite rods are normally used as counter electrodes, and sometimes are confined in glass tube with vycor frit to prevent the counter

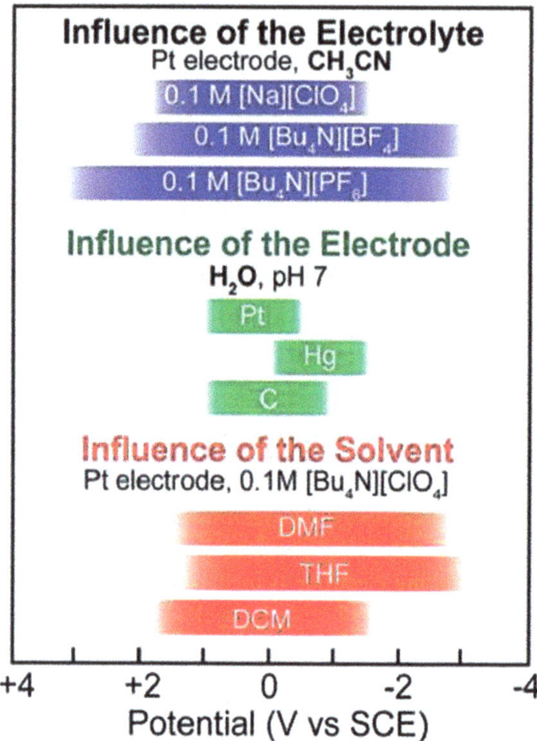

Fig. 14.4 Influence of electrolyte, electrode material, and solvent on electrochemical window [Reproduced with the permission of American Chemical Society (Ref. [8]b, J. Chem. Edu., **95** (2018) 197–206), https://pubs.acs.org/doi/10.1021/acs.jchemed.7b00361; further permissions related to the material excerpted should be directed to the ACS]

redox reaction (specially in case of bulk electrolysis) at its surface. The reference electrode can be categorized as aqueous and non-aqueous reference electrodes based on the solvent used in the electrochemical cell. Since the absolute standard for electrochemical potential measurement is unavailable, the standard hydrogen electrode (SHE) is chosen as the reference electrode and its potential is defined as 0 V at unit H^+ ion activity and 1 bar pressure of hydrogen gas. The most commonly used reference electrodes in aqueous solutions are silver/silver chloride in saturated KCl, saturated calomel electrode (Hg/Hg_2Cl_2 in saturated KCl). In non-aqueous solution, either Ag wire dipped in $AgNO_3$ solution or double-junction Ag/AgCl electrode is used as reference electrode. Many times, a Pt wire dipped in the non-aqueous solution along with a ferrocene-ferricinium redox couple as an internal reference is also used as non-reference electrode. The more detailed information about the working, counter, reference electrodes, and electrochemical cells are available in the literatures [8]. To obtain the best results, electrochemical parameters (applied voltage, current, time, etc.), cell design (electrodes' materials, electrodes' dimension, electrodes' separation, etc.), choice of chemical reagents (solvents, supporting electrolytes, etc.,) need to be optimized.

Though organic solvents have more wider electrochemical windows than the aqueous solvents, but the disadvantages like low vapor pressure, high flammability, toxicity, etc. of organic solvents discourage their uses. However, new solvents like room temperature ionic liquids (RTILs) and deep eutectic solvents (DES) have wider electrochemical windows of about 4–5 V similar to that of organic solvents, but at the same time do not suffer from the disadvantages like low vapor pressure, high flammability, toxicity, etc. [9]. The wider electrochemical window is not only the prerequisite, but the higher value of heterogeneous electron transfer rate constant for a redox process is another factor which too decides the choice of the working electrode and supporting electrolytes-solvents system. The heterogeneous electron transfer reactions at noble metal electrodes (Pt, Au) generally show higher rate constant than that of the carbonaceous electrodes like graphite and glassy carbon electrodes. However, carbonaceous materials like carbon nanotubes (CNTs) and graphene modified electrodes show much-enhanced rate of electron transfer than graphite, glassy carbon [10]. A large electrochemical window of 4–5 V can be easily achieved by judicious selection of the electrode materials and supporting electrolytes-solvents system. A wider electrochemical window encompassing large potential ranges in both cathodic and anodic regions is preferred as it provides a large workable potential for electrolysis with 100% current efficiency.

Since a large category of materials viz. metallic, intermetallic, semiconductors, conducting polymers, metal-conducting polymer composites, organic compounds, metal hexacyanometallates based metal-organic frameworks (MOFs), graphene, graphene oxide, metal oxides, biomaterials, ceramics, optical materials, etc. can be synthesized by electrochemical methods in size ranging from bulk to nanometers and in different shapes (nanowires, nanorods, nanotubes, and nanometer thick films of large dimension). Very often a single material finds multiple applications, for e.g., conducting polymers can find applications as catalyst, sensors, biosensors, electrochromic devices, electrochemical energy storage devices, biomaterials for tissue engineering, etc. [11]. Therefore, it is not an exaggeration if it is said that the electrochemical methods can be used by researchers for synthesizing advanced materials with innumerable applications in almost all disciplines of science and technology.

However, in this chapter, the main focus will be on synthesis of conducting polymers, semiconductors, metal nanoparticles, graphene, porous anodic alumina templates, and metal hexacyanometallates based metal-organic frameworks (MOFs) by electrochemical methods, as these materials with numerous technological applications draw huge attentions of the scientific communities. The electroreduction of CO_2 to synthesize organic compounds is also discussed briefly. Though the electrochemical techniques are also used in synthesis of ceramics, biomaterials, magnetic materials, treatment of organic/inorganic pollutants in water, but their synthesis will be not be discussed keeping consideration of space.

14.2 Electrochemical Synthesis of Conducting Polymers

Conducting polymers (polypyrrole, polyaniline, polythiophene, etc.) are organic polymers capable of conducting electricity and are known for their controllable electrical conductivity and electrochromic properties associated with multiple redox states [12]. The dependence of electrical conductivity of polyaniline (PANI) base on molar fraction of doped camphorsulfonic and picric acid is shown in Fig. 14.5. Conducting polymers find lots of interest among researchers for their applications in catalysis, molecular electronic devices, gas sensors, biosensors, electrochemical energy storage devices, electrochromic displays, actuators, electrochemical separation, biomaterials for tissue engineering, etc. [13].

Conductive polymers are prepared by oxidative polymerization of monomers [14]. A large number of methods for synthesising conducting polymers are reported in literature, but chemical [15] and electrochemical methods [16] are the most preferred one for synthesis. Hybrid electrochemical/chemical approach of synthesizing conducting polymers is also reported [17]. The chemical method involves addition of chemical oxidant to monomers solution under continuous stirring. The precipitated polymers are filtered and washed. Potassium dichromate, ammonium persulfate or peroxydisulfate, ceric nitrate, hydrogen peroxide, ceric sulfate, etc. are

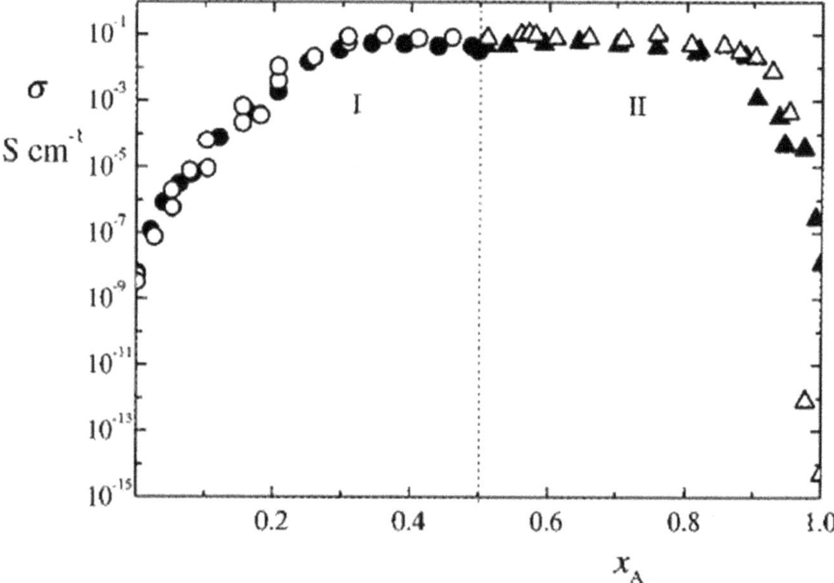

Fig. 14.5 Electrical conductivity, σ, of the PANI base blended in the solid-state with camphorsulfonic acid (full symbols) and picric acid (open symbols) in dependence on molar fraction of acid, x_A, for acid in deficit (case I, $x_A < 0.5$, circles) and acid in excess (case II, $x_A > 0.5$, triangles) [Reproduced with the permission of American Chemical Society (Ref. [19]a, Macromolecules, **31** (1998) 2218–2222)]

commonly used oxidants. The physical and chemical properties of the prepared conducting polymer depends on the concentration of monomer, nature of oxidants, oxidant to monomer ratio, counter-ion concentration, stirring rate, solution temperature, acidity of the solution, etc. [18]. Polymers prepared by chemical methods are often contaminated by chemicals used in the synthesis, thus, purity of the polymer is a major concern. The doped chemical impurities severely affect the properties and performance of the synthesized conducting polymer. Another major apprehension during the chemical synthesis is the coagulation/precipitation of the nanostructured conducting polymer in solution. The stabilizing/protecting agents prevent the coagulation/precipitation and also act as templates for nanoparticles formation in solution, but they can again severely affect the properties of the synthesized polymers [19]. Chemical synthesis of nanostructured conducting polymers of various shapes (nanotubes, nanowires, nanospheres, nanocubes, etc.) and sizes in solution is possible using templates-free, template assisted and sometimes combination of both, but removal of template is always an additional indispensable and inevitable step [20].

Electrochemical methods involve anodic oxidation of monomer on a working electrode surface which are inert conducting substrate like platinum, gold, carbon (graphite, glassy carbon, pyrolytic, vitreous), stainless steel, copper, lead, palladium, semiconductors, optically transparent (ITO, FTO, metal wire mesh) etc. by applying either suitable anodic potential or anodic current [21]. Optically transparent electrodes (OTEs) are prepared by (i) coating a thin film of either indium tin oxide (ITO) or fluoride-doped tin oxide (FTO) on transparent glass (ii) fabricating very fine mesh of metal wire with greater than 80% of transparency. Since the conducting polymers are electrochromic, the electrochemical synthesis of conducting polymers on OTE has the additional advantage of coupling with spectroscopic techniques for in situ optical characterization that helps in understanding the synthesis mechanism and various properties of the conducting polymer. Electrochemical synthesis of conducting polymers involves galvanostatic (constant current), potentiostatic (constant potential), and potentiodynamic (potential scan) methods [22]. Electrochemical formation of polymer film can be carried out either in dynamic (convection-controlled) or static (diffusion-controlled) solution, unlike chemical methods where continuous stirring is always required for mixing the monomers, oxidants, and other additives. Galvanostatic method involves application of constant anodic current on the working electrode to carry out the oxidative polymerization of monomers in solution leading to the formation of conducting polymer film on the working electrode surface. Thickness and morphology of the conducting polymer film can be controlled by optimizing current density, time for which current is applied (pulse time), monomer concentration in the solution, etc. Either a single or multiple current pulse(s) can be used for polymer synthesis, and the numbers of current pulse determine film thickness and morphology. In potentiostatic method, either a single or multiple pulses of anodic potential is applied on working electrode to oxidize the monomer. Thickness and morphology of the conducting polymer film can be controlled by optimizing applied potential, time for which potential is applied (pulse time), numbers of pulse monomer concentration in

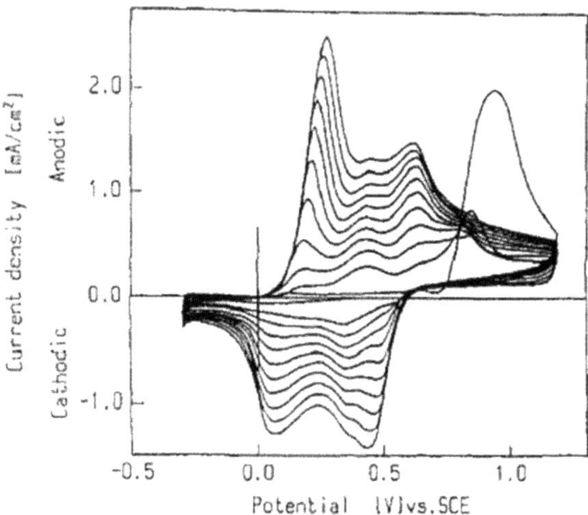

Fig. 14.6 CV for electropolymerization of 0.1 M aniline in 0.1 M H_2SO_4 solution at Pt [Reproduced with the permission of American Chemical Society (Ref. [22]h, Macromolecules, **20** (1987) 1793–1796)]

the solution, etc. Potentiodynamic method involves the scanning of potential on working electrode between cathodic and anodic limits of the chosen electrochemical windows, as shown in Fig. 14.6, to prepare conducting polymer film on the working electrode surface. At anodic potentials, monomers in the solution undergo oxidation leading to polymer formation. During the potential scanning polymer film growth (at anodic potentials) as well as continuous reversible reduction and oxidation of polymer film take place on the working electrode surface. Thickness and morphology of the conducting polymer film can be controlled by optimizing numbers of cyclic voltammetry scan, potential scan rate, monomer concentration in the solution, choice of anodic and cathodic potential limits, etc.

Temperature of the solution also influences the rate of oxidation of monomer. Since the conducting polymer is synthesized on the substrate/working electrode, its separation from the solution is easy and mostly the conducting polymer is of high purity as the monomers and other chemical additives remains in solution. Stable nanometre thick polymer film of any dimension can be easily synthesized on conducting substrate by electrochemical methods.

Electrochemical synthesis of one-dimensional (1D) nanostructures of conducting polymers (nanowires, nanorods, and nanotubes) has been carried out using either template-free method (Fig. 14.7) or nanoporous templates like porous anodic alumina, track-etched membranes, etc. (Fig. 14.8) [23]. The optimized ion-diffusion paths in one-dimensional conducting polymers enhance the conductivity.

The electrodeposition into nanoporous template membrane is simple, economical, and can be easily carried out at ambient temperature-pressure conditions without any requirement for special sophisticated instrumentation and experimental setup, thus, offering many advantages over other techniques (physical vapor deposition, chemical vapor deposition, atomic layer deposition, etc.). The steps involved in the electrochemical synthesis of one-dimensional (1D) nanostructures in template membrane are: (i) one side of the insulating template membrane is

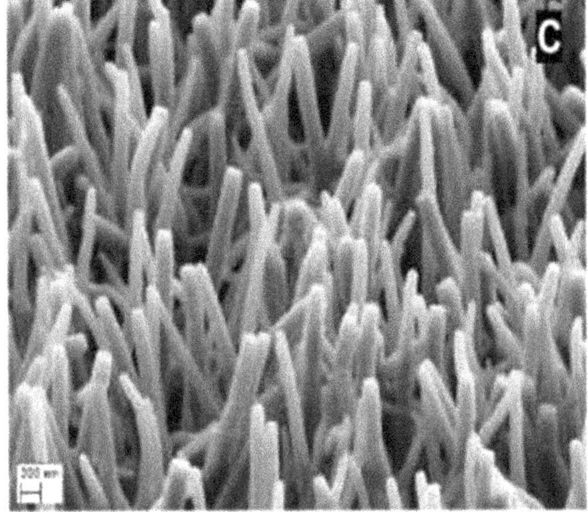

Fig. 14.7 Polypyrrole nanowires synthesized without using template via. potentiostatic deposition [Reproduced with the permission of Elsevier (Ref. [23]h, Electrochem. Commun., **11** (2009) 298–301)]

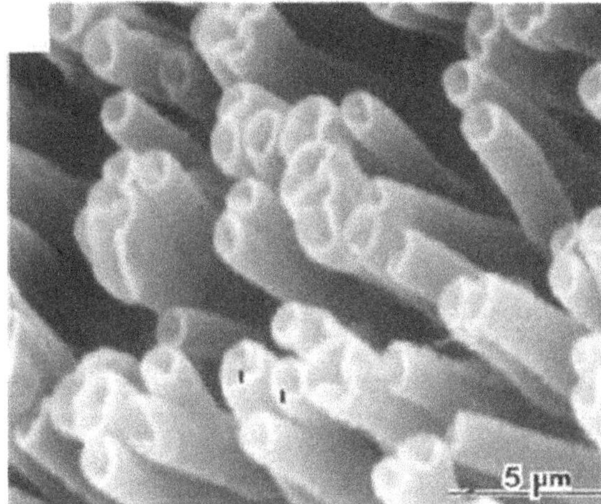

Fig. 14.8 Electrochemically grown polypyrrole fibrils in Nucleopore membrane (pore diameter is 1 μm) [Reproduced with the permission of The Electrochemical Society (Ref. [20]e, J. Electrochem. Soc., **133** (1986) 2206–2207)]

coated with a 500 nm to 1 μm thick conducting metal layer using sputter deposition technique and the coated metal layer serves as conducting substrate for electrodeposition. To ensure the complete uniform metal deposition on the template membrane surface, the thickness of the coated metal layer is sometimes further increased by electrodeposition after the sputtering, (ii) the metal-coated template membrane is fixed on the Au or Cu plate to prepare working electrode/cathode and immersed in the monomer solution, (iii) oxidative polymerization of the monomer to synthesize polymers is carried out by choosing any of the electrochemical techniques viz. galvanostatic, potentiostatic and potentiodynamic. The nanowires grown in the template membrane are attached to the coated metal layer, and protrude like the bristles of brush even after the template dissolution. The free-standing nanowires

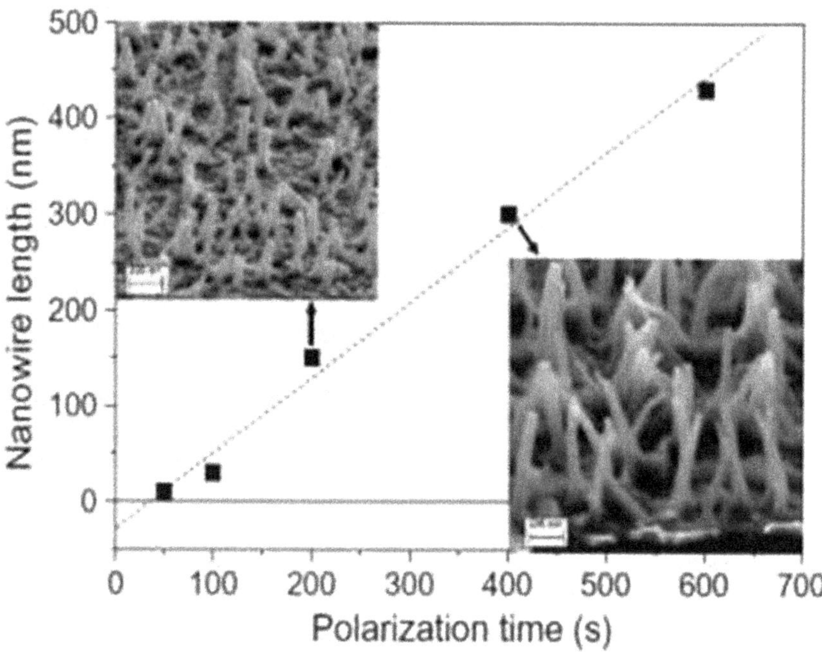

Fig. 14.9 Influence of polarisation time on nanowires length synthesized via. potentiostatic deposition [Reproduced with the permission of Elsevier (Ref. [23]h, Electrochem. Commun., **11** (2009) 298–301)]

are obtained after detaching them from coated metal film. The nanowires of different radii and length are prepared by using nanoporous template membranes with different pore dimension as well as by varying the electrodeposition time. The influence of deposition time on nanowires' length synthesized by potentiostatic method is shown in Fig. 14.9.

The choice of the electrochemical method of synthesis viz. galvanostatic, potentiostatic, and potentiodynamic can influence the properties of conducting polymers. The properties of the polyaniline (PANI) film synthesized on stainless steel substrate by potentiostatic, galvanostatic, and potentiodynamic methods have been compared by electrochemical impedance spectroscopy (EIS) [24]. The EIS spectra of galvanostatically and potentiostatically prepared PANI films were similar in shape. The resistance of aqueous pore (R_{aq}) due to ionic doping–dedoping process is greater than the resistance (R_p) of electronic charge transfer on polymer chain. The magnitudes of R_{aq} and R_p increase with thickness of PANI. In case of potentiodynamically grown PANI, $R_{aq} \leq R_p$ and the impedance parameters depend on the scan rate applied during electrosynthesis. The R_p decreases in magnitude with an increase in thickness of PANI.

Apart from chemical and electrochemical synthesis, other synthesis techniques viz. photochemical, metathesis, concentrated emulsion, plasma, pyrolysis, etc. are reported in the literature [25]. The advantage of plasma process is non-requirement of both oxidant and solvent for polymerization leading to formation of conducting polymer which is extremely pure, non-doped, and does not require separation. The polymer can also be synthesized on non-conducting substrates, unlike electrochemical methods. However, very high temperature of plasma often leads to the degradation of conducting polymer and also doping is required later to make the polymer conducting. Cost-effective photochemical synthesis of conducting polymers requires simpler equipments, but it requires that either the monomers itself work as photosensitizer or a photosensitizer, such as ruthenium complexes or organometallic complexes, is required. Another disadvantage of the photochemical method is that the complete polymerization takes longer time.

Metal nanoparticles are incorporated into the polymer matrix to form metal-polymer nanocomposites having superior properties and performance compared to metal nanoparticles and polymer alone. Incorporation of metal nanoparticles into polymer matrix improves their physical and chemical properties [26]. In case of conducting polymers, incorporation of metal nanoparticles not only improves the physical and chemical properties, but the electronic properties also improve many fold. Many reports show that incorporation of metal nanoparticles in conducting polymer matrix leads to a significant improvement in its electrocatalytic and sensing properties. Therefore, metal-polymer composites find more widespread applications in catalysis, sensing, electronics, etc. than either polymer or metal nanoparticles alone as the blending of polymer with metal nanoparticles often gives rise to a new set of properties. A numerous approach for synthesis of metal-polymer composites either in solution or solid-support are available: (i) first preparation of metal nanoparticles by reducing metal ions followed by the oxidative polymerization of monomers in the same solution, (ii) reduction of metal ions by monomer which undergoes simultaneous oxidative polymerization, (iii) formation of polymer followed by metal ion reduction in the same solution. To get maximum homogeneity, those methods are preferred where polymerization and nanoparticles formation occur simultaneously. Though many synthesis methods (chemical, vapor deposition, electrochemical, γ-radiolysis, UV-radiation, etc.) are reported in the literature [27], electrochemical methods offer some additional advantages over other reported methods. Polymer-metal nanocomposites of various dimensions can be easily formed on the substrate of any shape and size by in situ and ex situ electrochemical routes.

14.3 Electrochemical Synthesis of Metal Nanoparticles

Nanostructures (nanoparticles, nanowires, nanorods, nanotubules, nanocubes, nanoprisms, etc.) of metals, very often noble metals, have found numerous applications encompassing almost every discipline of science and technology. They are used as catalyst, electrocatalyst, chemical sensor, biosensors, imaging and photothermal therapeutic agents in health care, photonics, magnetic materials etc. [28].

Metal nanoparticles are synthesized in either aqueous solution or non-aqueous solution or sometimes at the interface of both by reducing metal ions [29]. Various methods for reduction of metal ions are reported in the literature viz. chemical, electrochemical, microbial, ultrasonication, physical vapor deposition, radiation-assisted (UV light, electron beam, γ-radiation), galvanic replacement, etc. [30]. Thermolysis of organometallic precursor in presence of stabilizer can also be used to synthesize metal nanoparticles [31]. Synthesis of thin films of metal nanoparticles immobilized on substrates like metal, metal oxide, carbonaceous electrodes, semiconductors, polymers, etc. have also been reported, and immobilized metal nanoparticles show very good catalytic activity depending on the nature of the support [32]. Physical methods like molecular beam epitaxy (MBE), physical vapor deposition (PVD), chemical vapor deposition (CVD), etc. and electrochemical deposition are widely used to prepare supported nanoparticles [33]. Electrochemical reduction of metal ions in solution is carried out at the working electrode/cathode in an electrochemical cell with either three-electrode or two-electrode configuration using numerous techniques, viz. (i) galvanostatic method that involves application of constant cathodic current for time duration varying from microseconds to a few tens of seconds, (ii) potentiostatic method that involves application of constant cathodic potential for time duration varying from microseconds to a few tens of seconds, (iii) potentiodynamic method that involves cycling potential between cathodic and anodic potential limits for number of cycles. Electrochemical deposition is a promising technique to synthesize and immobilize the metal nanoparticles onto the conducting substrate surface with easy control of particle size and shape by optimizing the electrochemical parameters like applied current and potential, time duration, number of current and potential pulses, number of cycles, cathodic and anodic potential limits. The template-assisted electrochemical preparation of metal nanoparticles with desired shape and size is also reported in the literature. Electrodeposition of noble metals nanoparticles on conducting substrates, like graphite, glassy carbon, indium tin oxide, etc. reduces the loading of precious noble metals and finds great potential applications in sensing, catalysis, and interface-electrochemistry [34].

The electrochemical synthesis of 1D metal nanostructure (nanowires, nanorods, nanotubes) with desired lengths, diameters, and composition in nanoporous template materials like porous anodic alumina and track-etched membranes is well-established method [35] and is already discussed for electrochemical growth of 1D nanostructures of conducting polymers. Some of the limitation encountered while electrodeposition of 1D nanostructures of metals in porous template membranes are (1) the electrochemical growth of free-standing 1D metal nanostructures in track-etched membrane is impossible due to the growth of the metal nanowires on the sputter-deposited metal layer (2) sputter deposited metal layer closes the opening of the pores on one surface and the availability of open pores on the other surface depends on electrodeposition time. Figure 14.10a shows the three different stages of the growth of 1D nanostructure and reduction current as a function of deposition time during the potentiostatic deposition. Figure 14.10b shows the SEM

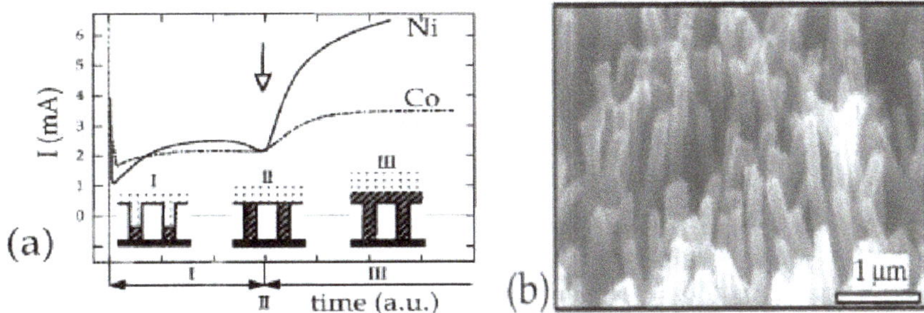

Fig. 14.10 **a** Electrochemical reduction current *I* as a function of time *t* for the potentiostatic deposition. The schematic display three different stages of the growth process. **b** SEM image of the Ni nanowires after dissolution of polycarbonate membrane [Reproduced with the permission of American Chemical Society (Ref. [35]a, J. Phys. Chem. B, **101** (1997) 5497–5505)]

image of potentiostatically deposited Ni nanowires obtained after dissolution of the porous template membrane.

If the electrodeposition time is less, then pores are incompletely filled leading to the availability of a large number of pores with open mouth on the surface of the template membrane. The incomplete filled cylindrical pores with open mouth can be used for encapsulating the materials by functionalizing the 1D metal nanostructures grown within the pores [36].

The use of removable liquid mercury as conducting substrate eliminates the requirement of sputter deposition of a metal coating onto one surface of the template membrane [37]. The experimental set-up (see Fig. 14.11) comprises a two-compartment electrochemical cell and a track-etched membrane that is placed between the two compartments separating mercury from the aqueous solution of $HAuCl_4$. Liquid mercury in contact with one of the surfaces of the track-etched membrane serves as cathode for the electrodeposition of gold in the nanopores. The gold nanorods are not attached to any conducting substrate, therefore, this method synthesizes free-standing gold nanorods encapsulated in a malleable track-etched

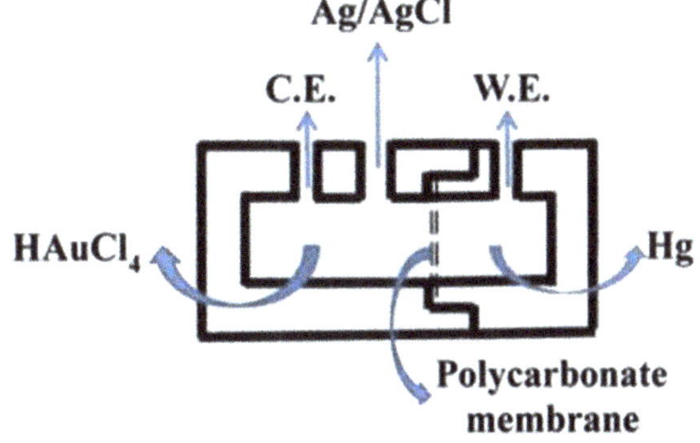

Fig. 14.11 Schematic of electrochemical cell with liquid mercury as working electrode/cathode for synthesis of gold nanorods [Reproduced with the permission of Springer (Ref. [37], J. Nanopart. Res., 14 (2012). https://doi.org/10.1007/s11051-012-1094-z)]

membrane. The gold nanorods are freed from the template membrane after the dissolution in dichloromethane.

This new approach is simple, cost-effective, and saves time. The shape of the electrodeposited nanowires and nanotubes is the replica of the internal architecture of the pores; therefore, to prepare good-quality nanorods, a template porous membrane with well-defined cylindrical pores should be selected. The metal nanowires containing track-etched membrane can be used as smart surfaces/ materials for catalytic and optical applications. The nanowires in the incomplete-filled cylindrical pores can be functionalized to encapsulate desired chemical or biochemical species to develop sensors based on the detection of analyte by surface-enhanced Raman scattering (SERS) technique.

Self-assemblies of surfactants and polymers have been also used as template and protecting agent for preparing nanoparticles of various shapes and sizes [38]. Template-free electrochemical methods for preparing metal nanoparticles are too reported and are preferred over template-assisted methods [39]. Simultaneous reduction of two different metals ions lead to formation of metal alloy (Au–Ag, Au–Pt, etc.) and alloys have properties superior to individual metals, for example, lowered activation barrier for a specific chemical reaction, enhanced selectivity, and better resistance against contamination and poisoning [40]. The controlled deposition of one metal on the surface of other metal is used to synthesize the core-shell metal nanoparticles are synthesized by depositing one metal on the surface of other metal [41]. As discussed earlier, nanocomposites of metals with polymer can be prepared by incorporation of metal nanoparticles into the polymer matrix.

14.4 Electrochemical Synthesis of Semiconductors

Semiconductors find numerous technological applications viz. light emitting diode, photocatalyst, the gain medium in semiconductor lasers etc. [42]. Numerous methods for synthesis of semiconductors of various shapes and size are available, but technological application of these semiconductors requires their immobilization on conducting support. The chemically synthesized semiconductors nanoparticles can be transferred from solution to the substrate to prepare thin film, but their loose adherence to substrate restricts their applications in development of device. Though molecular beam epitaxy and chemical vapor deposition methods are well explored for single-step synthesis of semiconductor nanoparticles and thin films on support, but the control over the average diameter of nanocrystals is difficult due to its dependence on factors intrinsic to the interface obtained upon deposition such as the lattice mismatch between the substrate and the deposited material [43].

Electrodeposition techniques are used for synthesis of semiconductors. The epitaxial electrodeposition of semiconductor nanocrystallites (wurtzite CdSe) on Au (111) surface by galvanostatic method was reported by Hodes and co-workers for the first time [44]. The epitaxially oriented size-selective semiconductor nanocrystals of luminescent CdS, CuI, ZnO, and optically intrinsic thin films of

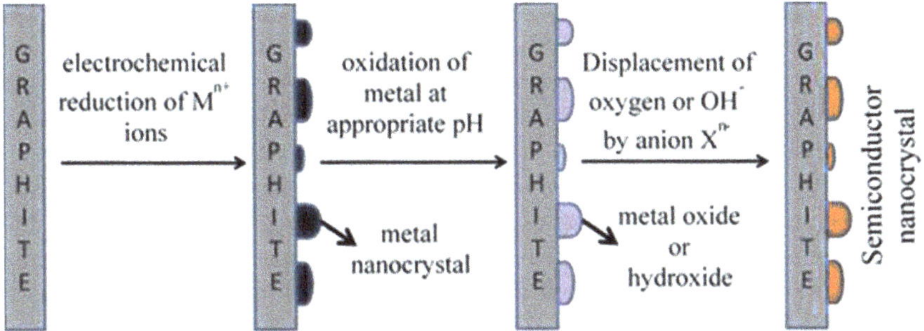

Fig. 14.12 Schematic diagram of the hybrid electrochemical/chemical (E/C) method employed for the synthesis of semiconductor nanocrystals

ZnO are prepared either in solution or on some support (graphite, etc.) using hybrid electrochemical/chemical (E/C) method [45].

The E/C synthesis (schematic shown in Fig. 14.12) involves the following steps: (1) electrochemical reduction of M^{n+} ions from dilute M^{n+} plating solution to deposit metal nanoparticles on the graphite basal plane surface (2) electrochemical or chemical oxidation of deposited metal nanocrystals at appropriate pH to either M $(OH)_n$ or metal oxide nanocrystals; (3) displacement of oxygen or OH^- with an anion by immersing in an aqueous solution of anion to obtain the desired semiconductor nanocrystals of interest.

Nanowires, nanotubes, nanorods of semiconductors with desired diameters and lengths are synthesized in porous anodic alumina and track-etched membranes by electrochemical methods similar to that used for metal nanowires and nanotubes [46]. Self-assemblies of surfactants have been also used as templates and protecting agents for electrochemical synthesis of semiconductor nanoparticles either in solution or on conducting substrate. Template-free electrochemical methods for preparing semiconductor nanoparticles are too reported [47].

14.5 Electrochemical Synthesis of Graphene-Based Materials

As mentioned earlier that the carbonaceous electrodes like graphite and glassy carbon electrodes have smaller value of heterogeneous electron transfer rate constant compared to noble metal electrodes. However, carbonaceous materials like carbon nanotubes (CNTs) and graphene modified electrodes show much enhanced heterogeneous electron transfer rate than glassy carbon and graphite due to their extraordinary electronic transport properties. Graphene is two-dimensional one-atom-thick single layer of sp^2 hybridized carbon atoms arranged in hexagonal lattice and finds potential application in electrocatalysis, sensing, composite materials, electrochemical energy storage device, next-generation electronic devices, biological devices, nanomedicines, etc. due to its fascinating electrical, mechanical, and chemical properties [48]. However, these vast potential applications of graphene cannot be successfully delivered from lab to society unless a simple method for low-cost mass production of high quality, solution-processable, structurally coherent graphene is developed. Several methods for synthesis of graphene are reported in the literature, but suffer from some or the other limitations, thus, keeping the quest alive for developing a simple method to synthesize low-cost, high-quality graphene in bulk quantities. High-quality graphene can be synthesized by mechanical exfoliation and epitaxial growth methods, but, in small quantities [49]. Chemical vapour deposition (CVD) technique is used to synthesize large-area high-quality graphene using catalytic metal substrates, but requirement of sacrificial metal, high temperature, and multistep transfer processes onto the substrates increases the production cost [50]. Hummers and modified-Hummers methods are frequently used for bulk production of graphene oxide (GO) by oxidative exfoliation of graphite in an aqueous solution consisting of concentrated sulphuric acid, sodium nitrate, and potassium permanganate [51]. The GO can be reduced by thermal, photocatalytic, chemical, or electrochemical procedures to produce reduced graphene oxide, thus, partially restoring the electronic properties of graphene [52].

Electrochemical exfoliation of graphite to synthesize graphene does not need harsh chemical oxidants for carrying out intercalation/exfoliation; therefore, it has attracted huge attention of the researchers for bulk synthesis due to its fast, easy, fast and environmentally friendly process [53]. Electrochemical exfoliation involves application of either anodic or cathodic potential on the graphite working electrode in electrolyte solution (schematic shown in Fig. 14.13).

Depending on the polarity of the applied potential, either bulky anions or cations along with other species like water intercalates into the interlayer spacing between graphene sheets leading to expansion of graphite. The subsequent exfoliation of the graphene sheets from the expanded graphite electrode occurs through either expanding gases, produced by electrolysis of intercalated species, between the graphene sheets or mechanical exfoliation (e.g. sonication). Electrochemical exfoliation of graphite has been carried out in aqueous acidic solution (H_2SO_4, H_3PO_4), ionic liquids and neutral pH aqueous solution of inorganic salts such as sodium sulfate, ammonium sulfate, and potassium sulfate [54]. The electrochemical exfoliation of graphite in ionic liquids produce low yield of small lateral size graphene which is often functionalized with ionic liquids resulting in its poor electronic properties. Higher yield of large size high-quality graphene is produced on exfoliation of graphite in aqueous acidic solution, but the overoxidation of graphite produce graphene with considerable fraction of oxygen-containing functional groups resulting in the poor properties of graphene. Electrochemical exfoliation of graphite in aqueous solution of inorganic salts prevents the overoxidation of graphite resulting in formation of graphene with lower oxygen content (C/O ratio of 17.2). The large-scale synthesis of graphene oxide sheets is reported via chemical oxidation of electrochemically exfoliated graphene sheet.

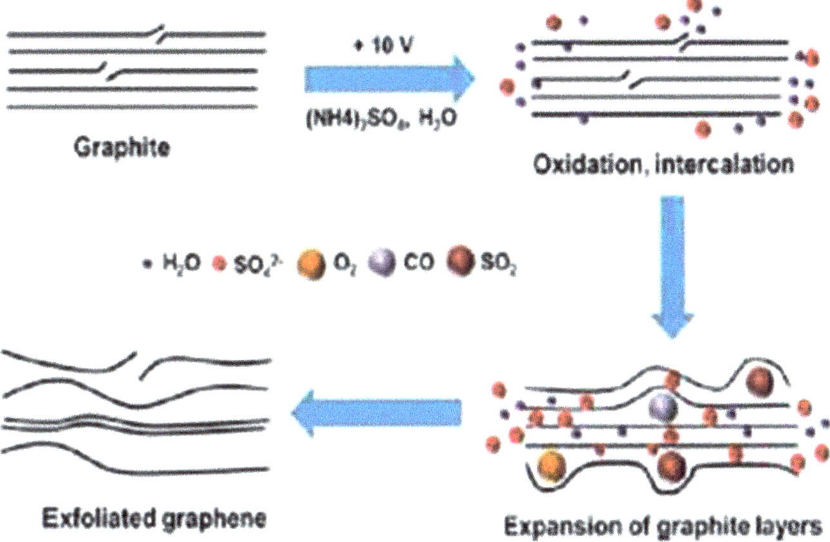

Fig. 14.13 Schematic illustration of the mechanism of electrochemical exfoliation of graphite [Reproduced with the permission of American Chemical Society (Ref. [49]e, J. Am. Chem. Soc., **136** (2014) 6083–6091)]

The electrochemical setup used for exfoliation of graphite is normally a three-electrode cell consisting of a graphite (powder, rod, flakes, sheets, foils, highly orientated pyrolytic graphite) as working electrode, Pt or graphite counter electrode, reference electrode, electrolyte, and either electrochemical workstation or power supply. All the electrodes are immersed into electrolyte with an adequate spacing between them and the constant voltage is applied on the graphite working electrode. Electrochemical exfoliation of graphite can be classified into anodic graphite exfoliation and cathodic graphite exfoliation depending on the positive (anodic), and negative (cathodic) potential, respectively, applied to the working electrodes [55].

The mechanism of graphene synthesis via anodic graphite exfoliation involves oxidation of graphite working electrode by anodic potential which creates positive charge on graphite, followed by the intercalation of bulky anions in the interlayer spacing between the graphene sheets in graphite. The intercalation of bulky anions increases the interlayer spacing leading to subsequent exfoliation of the graphene sheets. In cathodic exfoliation, the negative charge on the graphite favors the intercalation of bulky cations in the interlayer spacing between the graphene sheets in graphite. The intercalation of bulky cations increases the interlayer spacing and helps in subsequent exfoliation of the graphene sheets.

14.6 Electrochemical Synthesis of Highly Ordered Nanoporous Anodic Aluminium Oxide (AAO) Templates

As we have discussed earlier that one-dimension nanomaterials (nanowires, nanorods, nanotubes) of conducting polymers, metals, alloys, semiconductors, etc. very often shows better chemical, physical, electronic properties, thus finding more extensive applications in sensors, catalysis, optical devices, microelectronics, electrochemical energy storage, memory devices and many more. Sometimes new unique properties are observed in one-dimensional nanomaterials which can be exploited for new kind of applications. Gold nanorods/nanowires are used for non-invasive light-activated photothermolysis therapy (optical hyperthermia), which involves nonradioactive decay of the absorbed light into heat [56]. The conductivity of polypyrrole nanowires with very small diameter (10–50 nm) is very high compared to that of polypyrrole fibrils with large diameters [57]. One-dimension nanomaterials can be easily prepared in porous templates membranes. As mentioned earlier, that the shape of the electrodeposited 1D nanostructure is the replica of the internal architecture of the pores, therefore, to prepare good-quality 1D nanostructure, a template porous membrane with well-defined cylindrical pores should be selected. Therefore, preparation of porous template membrane with vertically aligned and highly ordered nanopores is a very essential prerequisite for synthesizing high-quality nanowires with narrow homogeneous distribution of pore radius and length. Electrochemical technique is widely used for synthesizing porous anodic aluminum oxide (AAO) membrane template with vertically aligned and highly ordered array of nanopores [58]. Electrochemical method involves an easy, inexpensive process of oxidizing high purity aluminum metal foil in strong acid solution to produce AAO membrane with perfectly ordered and size-controlled nanopores. The formation of nanopores does not require any lithography or templating. The electrochemical oxidation of aluminum can lead to formation of either a barrier-type non-porous anodic aluminum oxide film or porous anodic aluminum oxide film depending on the nature of an electrolyte, pH of an electrolyte, and anodizing time (schematic diagram shown in Fig. 14.14).

The strong acid solutions used are sulphuric acid, oxalic acid, phosphoric acid, and chromic acid [59]. The localized dissolution of oxide during the AAO film growth results in pore formation. The pore geometry (pore diameter and density) is tunable over wide ranges and can be easily controlled by proper choice of electrolytes, electrolyte concentration, electrolyte pH, anodizing potential/current, anodizing time, and number of anodization steps. Thus, electrochemical oxidation of aluminum at high current efficiency in largely near-neutral electrolytes at ambient temperatures produces compact barrier-type AAO films which are highly uniform in thickness (Fig. 14.14a). Recent studies show the development of the barrier films at current efficiencies of approximately 60% or above. The development of

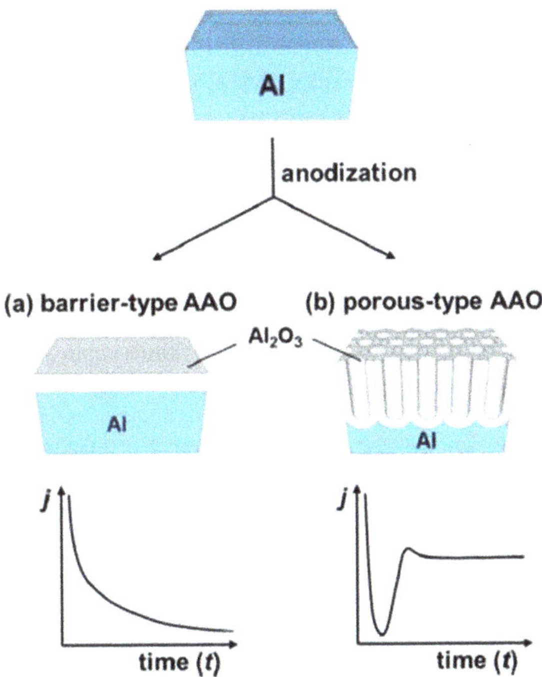

Fig. 14.14 Two different types of AAO film formed by anodization of Al **a** barrier type AAO **b** porous-type AAO along with their current (j)-time (t) transients under potentiostatic conditions [Reproduced with the permission of American Chemical Society (Ref. [58]c, Chem. Rev., **114** (2014) 7487–7556)]

relatively high field across the alumina leads to Al^{3+} egress and O^{2-} or OH^- ingress through the anodic film. In acid electrolytes, or selected alkaline electrolytes, the relatively compact barrier-type AAO films are not produced on aluminum due to anodic polarization. Characterization of the AAO films reveals that the cylindrical pores are normal to the macroscopic aluminum surface and are separated from aluminum surface by a relatively compact barrier layer of scalloped shape. The selective incorporation of electrolyte species in AAO films by anodisation under specific experimental conditions is possible to produce tailor-made AAO films for particular applications. The anodization at high voltage in phosphoric acid produces AAO films with relatively large diameter pores, while AAO films with fine pores are produced by anodizing in sulphuric acid.

The electrochemical setup used for oxidation of aluminium is normally consisted of either electrochemical workstation or power supply connected to a two-electrode electrochemical cell consisting of a high purity aluminium metal as working/anodic electrode and Pt as counter (cathodic) electrode immersed in electrolyte solution. The oxidation is carried out by applying a constant anodic voltage to the aluminium working electrode.

14.7 Electrochemical Synthesis of Transition Metal Hexacyanometallates Based Metal-Organic Frameworks (MOFs)

The transition metal hexacyanometallates based metal-organic frameworks (MOFs) are a type of mixed valence compound with the general formula $M_x^y[M^z(CN)_6]$ where M^y and M^z are transitional metals with different oxidation numbers. The water molecules and ions other than the transition metals may also be present in these mixed valence compounds. Prussian blue (PB) is the oldest reported transition metal hexacyanometallates, and PB along with its analogues have gained lot of interest among researchers for multiple applications such as electrocatalyst, electrochromic devices, ion-exchangers, ion-sensors, electrochemical energy storage devices, magnetic and photomagnetic devices [60]. PB and its analogues are generally synthesized by either chemical or electrochemical method from a solution containing transition metal ions (M^y) and transition metal cyanides $[M^z(CN)_6]^{n-}$ ions [61]. Nanostructured compound of transition metal hexacyanometallates with desired geometries (shape, size, and thickness) can be easily synthesized on conducting substrates by choosing any of the electrochemical methods viz. potentiostatic, galvanostatic or potentiodynamic technique. The desired geometry is easily obtained by optimizing the current density, deposition potential, deposition time, etc. [62]. The oxidation states of transition metal ions in electrochemically deposited compound can be easily altered to obtain desired functional properties (for e.g., electrochromism, ion-exchange) by application of suitable electrical potential.

14.8 Electrochemical Synthesis Organic Compounds from CO_2 by Electroreduction of CO_2

Carbon dioxide is abundantly present in the atmosphere and the concentration of CO_2 in atmosphere has been increasing at alarming rate in a last few decades due to continuously increasing consumption of fossil fuels (coal, petroleum products, woods, etc.). The CO_2 is a green-house gas and alarmingly high levels of CO_2 in atmosphere have raised the concern among the environmentalist because of global warming and associated climate change. Therefore, the low-cost conversion of CO_2 into useful organic compounds is an interesting option drawing the attention of many researchers around the world to deal with the global warming and associated climate change. Literature shows that CO_2 can be electrochemically reduced to many organic compounds, viz. ethanol, methane, ethylene, etc. Ethanol is used in direct ethanol fuel cells (DEFCs) to generate electrical energy, therefore, electroreduction of CO_2 into hydrocarbons and alcohols is a process to synthesize energy storage material [63]. The organic compounds produced by electroreduction of CO_2 and their distribution depends on the working electrode materials, pretreatment of the electrode surface, the electrode potential, temperature, and the electrolyte [64]. Electroreduction of CO_2 at copper electrode produces a wide variety of hydrocarbons and alcohols.

14.9 Concluding Remarks

From the above discussion, we can conclude that electrochemical tools are being increasingly used by researcher for synthesizing different type of bulk and nanostructured materials viz. conducting polymers, metal, alloys, semiconductors, template membranes, mixed valence compounds for numerous applications encompassing almost all disciplines of science and technology. The electrochemical methods are also available for conversion of toxic, environment unfriendly carbon dioxide gas into useful organic compounds (hydrocarbon and alcohols) from which energy can be generated later on. Ethanol is used in direct ethanol fuel cells (DEFCs) to generate electrical energy. The electrochemically prepared nanoporous anodic alumina templates can be used for synthesizing one-dimensional nanomaterials. Organic and inorganic pollutants in waste water can also be electrochemically converted to non-toxic compounds.

References

1. (a) Chisholm G, Cronin L (2016) Storing energy (chapter 16). Letcher TM (ed). Elsevier Inc.; (b) This month in physics history, Volta describes the electric battery. APS News, March 2006 15(3)
2. (a) Eustis S, El-Sayed MA (2006) Chem Soc Rev 35:209–217; (b) Hu J, Odom TW, Lieber CM (1999) Acc Chem Res 32:435–445; (c) El-Sayed MA (2004) Acc Chem Res 37:326–333; (d) Hodes G (2007) Adv Mater 19:639–655; (e) Kharisov BI, Dias HVR, Kharissova OV, Va´zquez A (2014) J Nanopart Res 16:2665
3. (a) El-Sayed MA (2004) Acc Chem Res 37:326–333; (b) Ivanova OS, Zamborini FP (2010) J Am Chem Soc 132:70–72; (c) Berry RS (1998) Nature 393:212–213; (d) Sanchez-Sanchez CM, Solla-Gullon J, Vidal-Iglesias FJ, Aldaz A, Montiel V, Herrero E (2010) J Am Chem Soc 132:5622–5624; (e) Schmidt M, Kusche R, von Isseddorff B, Haberland H (1998) Nature 393:238–240; (f) Hu JT, Odom TW, Lieber CM (1999) Acc Chem Res 32:435
4. (a) Hernandez SC, Chaudhuri D, Chen W, Myung NV, Mulchandani A (2007) Electroanalysis 19:2125–2130; (b) Martin CR (1994) Science 266:1961–1966; (c) Schonenberger C, van der Zande BMI, Fokkink LGJ, Henny M, Schmid C, Kruger M, Bachtold A, Huber R, Birk H, Staufer U (1997) J Phys Chem B 101:5497–5505
5. Guin SK, Sharma HS, Aggarwal SK (2010) Electrochim Acta 55:1245–1257; (b) Tang Y, Liu Y, Yu S, Guo W, Mu S, Wang H, Zhao Y, Hou L, Fan Y, Gao F (2015) Electrochim Acta 161:279–289; (c) Feng K, Park HW, Wang X, Lee DU, Chen Z (2014) Electrochim Acta 139:145–151; (d) Taurino I, Sanzó G, Mazzei F, Favero G, De Micheli G, Carrara S (2015) Scientific reports 5:15277; (e) Lee JK, Samanta D, Nam HG, Zare RN (2018) Nat Commun 9:1562; (f) McCarthy CP, McGuinness NB, Alcock-Earley BE, Breslin CB, Rooney AD (2012) Electrochem Commun 20:79–82

6. (a) Tiana Y, Hu L, Han S, Yuan Y, Wang J, Xu G (2012) Anal Chim Acta 738:41–44; (b) Dong SJ, Qiu QH (1991) J Electroanal Chem 314:223–239; (c) Tang H, Chen J, Nie L, Liu D, Deng W, Kuang Y, Shouzhuo Y (2004) J Colloid Interface Sci 269:26–31; (d) Holdcroft S, Funt BL (1988) J Electroanal Chem 240:89–103
7. (a) LaGraff JR, Gewirth AA (1994) J Phys Chem 98:11246–11250; (b) Li J-T, Zhou Z-Y, Broadwell I, Sun S-G (2012) Acc Chem Res 45:485–494; (c) Chabala ED, Rayment T (1994) Langmuir 10:4324–4329; (d) Tao NJ, Shi Z (1994) J Phys Chem 98:7422–7426; (e) Heineman WR (1983) J Chem Edu 60:305; (f) Webster RD (2004) Anal Chem 76:1603–1610; (g) Abruna HD, Bommarito GM, Yee HS (1995) Acc Chem Res 28:273–279; (h) Zaleski S, Wilson AJ, Mattei M, Chen X, Goubert G, Cardinal MF, Willets KA, Duyne RPV (2016) Acc Chm Res 49:2023–2030
8. (a) Fritz S, Bond AM, Compton RG, Fiedler DA, Inzelt G, Kahlert H, Komorsky-Lovrić Š, Lohse H, Lovrić M, Marken F, Neudeck A, Retter U, Scholz F, Stojek Z (eds) (2010) Electroanalytical methods—guide to experiments and applications. Springer, pp 273–335; (b) Elgrishi N, Rountree K, McCarthy BD, Rountree ES, Eisenhart TT, Dempsey JL (2018) J Chem Edu 95:197–206
9. (a) Hayyan M, Mjalli FS, Hashim MA, AlNashef IM, Mei TX (2013) J Indust Engn Chem 19:106–112; (b) Anthony JL, Bennecke JF, Holbrey JD, Maginn EJ, Mantz RA, Rogers RD, Trulove PC, Visser AE, Welton T (2002) In: Wasserscheid P, Welton T (eds) Ionic liquids in synthesis. Wiley-VCH Verlag GmbH & Co. Chapter 3; (c) Gazlinski M, Lewandowski A, Stepniak I (2006) Electrochim Acta 51:5567. (d) Silvester DS (2011) Analyst 136:4871–4882; (e) Smith EL, Abbott AP, Ryder KS (2014) Chem Rev 114:11060–11082; (f) Zhang Q, De Oliveira Vigier K, Royer S, Jerome F (2012) Chem Soc Rev 41:7108–7146
10. (a) Nugent JM, Santhanam KSV, Rubio A, Ajayan PM (2001) Nano Lett 1:87–91; (b) Yang C, Denno ME, Pyakurel P, Venton BJ (2015) Anal Chim Acta 887:17–37; (c) Choi H-J, Jung S-M, Seo J-M, Chang D-W, Liming D, Baek J-B (2012) Nano Energy 1:534–551; (d) Allen MJ, Tung VC, Kaner RB (2010) Chem Rev 110:132–145
11. (a) Das TK, Prusty S (2012) Polymer-plastics technology and engineering 51:1487–1500; (b) Ates M, Karazehira T, Sezai Sarac A (2012) Current Phys Chem 2:224–240
12. (a) Macdiarmid AG, Chiang J-C, Halpern M, Huang W-S, Mu S-L, Somasiri NLD, Wu W, Yaniger SI (1985) Mol Cryst Liq Cryst 121:173–180; (b) Malinauskas A, Holze R (1998) Synth Met 97:31–36; (c) Patil AO, Heeger AJ, Wudl F (1988) Chem Rev 88:183–200; (d) Chiang J-C, MacDiarmid AG (1986) Synth Met 13:193–205
13. (a) Peng X, Huo K, Fu J, Zhang X, Gao B, Chu PK (2013) Chem Commun 49:10172–10174; (b) Bobacka J (2006) Electroanalysis 18:7–18; (c) Ciric′ G (2013) Marjanovic Synth Met 170:31–56; (d) Leroy D, Martinot L, Debecker M, Strivay D, Weber G, Jerome C, Jerome R (2000) J Appl Polym Sci 77:1230–1239; (e) Jerome C, Martinot L, Jerome R (1999) J Radioanal Nucl Chem 240:969–972; (f) Wang Y, Lu Y, Li H, Liu J, Li S, Qiu Z, Gong M, Wu B, Chu J, Wang X, Zhang R (2018) Polym Bullet 75:3427–3443; (g) Zheng W, Spinks GM, Wallace GG (2016) Materials and energy, 7(WSPC reference on organic electronics: organic semiconductors, volume 2: fundamental aspects of materials and applications), pp 321–343; (h) Nguyena TN, Rohtlaid K, Plesse C, Nguyen GTM, Soyer C, Grondel S, Cattan E, Madden JDW, Vidal F (2018) Electrochim Acta 265:670–680; (i) Yakhmi JV, Saxena V, Aswal DK (2012) Conducting polymer sensors, actuators and field-effect transistors. Functional materials, pp 61–110. Elsevier Ltd., London, UK
14. (a) Sapurina I, Stejskal J (2008) Polym Int 57:1295–1325; (b) Gospodinova N, Terlemezyan L, Mokreva P, Kossev K (1993) Polymers 34:2434–2437; (c) Yu I, Shishov MA, chapter 9, New polymers for special applications. INTECH, Ailton De Souza Gomes
15. (a) DeSurville R, Jozefowicz M, Yu L-T, Perichon J, Buvet R (1968) Electrochim Acta 13:1451–1458; (b) MacDiarmid AG, Chiang J-C, Halperm M, Huang WS, Krawczyk JR, Mammone RJ, Mu SL, Somasiri NL, Wu W (1984) Polymer preprints 25:248–249; (c) Macdiarmid AG, Chiang J-C, Halpern M, Huang W-S, Mu S-L, Somasiri NLD, Wu W, Yaniger SI (1985) Mol Cryst Liq Cryst 121:173–180; (d) Green AG, Woodhead AE (1910) J

Chem Soc (Trans.) 97:2388; (e) Green AG, Woodhead AE (1912) J Chem Soc (Trans.) 101:1117; (f) Langer J (1978) Solid State Commun 26:839

16. (a) Mohilner DM, Adams RN, Argersinger WJ (1962) J Am Chem Soc 84:3618–3622; (b) Kitani A, Kaya M, Sasaki K (1986) J Electrochem Soc 133:1069–1073; (c) Mengoli G, Munari MT, Biacco P, Musiani MM (1981) J Appl Polymer Sci 26:4247; (d) Noufi R, Nozik AJ, White J, Warren LF (1982) J Electrochem Soc 129:2261; (e) Aurian-Blajeni B, Taniguchi I, O'M. Bockris J (1983) J Electroanal Chem Interfacial Electrochem 149:291; (f) Kobayashi T, Yoneyama H, Tamura H (1984) J Electroanal Chem Interfacial Electrochem 161:419; (g) Gottesfeld S, Redondo A, Feldberg SW (1987) J Electrochem Soc 134:271
17. Sharma MK, Ambolikar AS, Aggarwal SK (2013) RSC Adv 3:25674–25676
18. (a) Trchova M, Sedenkova I, Konyushenko EN, Stejskal J, Holler P, Ciric-Marjanovic G (2006) J Phys Chem B 110:9461–9468; (b) MacDiarmid AG (1997) Synth Met 84:27–34; (c) Focke WW, Wnek GE, Wei Y (1987) J Phys Chem 91:5813–5818; (d) Zang J, Li CM, Bao S-J, Cui X, Bao Q, Sun CQ (2008) Macromolecules 41:7053–7057; (e) Skotheim TA, Elsenbaumer RL, Reynold J (1998) Handbook of conducting polymers. Marcel Dekker, New York
19. (a) Stejskal J, Sapurina I, Trchova M, Prokes J, Krivka I, Tobolkova E (1998) Macromolecules 31:2218–2222; (b) Gupta V, Miura N (2005) Electrochem Solid-State Lett 8:A630–A632
20. (a) Zang J, Li CM, Bao S-J, Cui X, Bao Q, Sun CQ (2008) Macromolecules 41:7053–7057; (b) Zhong W, Deng J, Yang Y, Yang W (2005) Macromol Rapid Commun 26:395–400; (c) Wang J, Xu Y, Yan F, Zhu J, Wang J (2011) J Power Sources 196:2373–2379; (d) Mu B, Zhang W, Wang A (2014) J Nanopart Res 16:2432; (e) Penner RM, Martin CR (1986) J Electrochem Soc 133:2206–2207
21. (a) Stilwell DE, Park S-M (1988) J Electrochem Soc 135:2254–2262; (b) Wang K, Huang J, Wei Z (2010) J Phys Chem C 114:8062–8067; (c) Wang K, Huang J, Wei Z (2010) J Phys Chem C 114:8062–8067; (d) Zang J, Li CM, Bao S-J, Cui X, Bao Q, Sun CQ (2008) Macromolecules 41:7053–7057; (e) Hand RL, Nelson RF (1974) J Am Chem Sac 96:850; (f) Carlin CM, Kepley LJ, Bard AJ (1985) J Electrochem Soc 132:353; (g) Oyama N, Ohnuki Y, Chiba K, Ohsaka T (1983) Chem Lett 1759; (h) Gupta V, Miura N (2005) Electrochem Commun 7:995–999; (i) Gupta V, Miura N (2005) Electrochem Solid-State Lett 8:A630–A632; (j) Wang K, Huang J, Wei Z (2010) J Phys Chem C 114:8062–8067; (k) Liu FJ, Huang LM, Wen TC, Li CF, Huang SL, Gopalan A (2008) Synth Met 158:767–774; (l) Mengoli G, Munari MT, Folonari C (1981) J Electroanal Chem 124:237; (m) Geni EM, Tsintavis C, unpublished work; (n) Pfeiffer B, Thyssen A, Wolff M, Schultze JW (1986) Int Workshop-Electrochem Polym Layers, Duisburg, F.R.G., September 15–17; (o) Noufi R, Nozik AJ, White J, Warren LF (1982) J Electrochem Soc 129:226; (p) Aurian-Blajeni B, Taniguchi I, O'M. Bockris J (1983) J Electroanal Chem 149:291; (q) Wang J, Xu Y, Yan F, Zhu J, Wang J (2011) J Power Sour 196:2373–2379
22. (a) Wang K, Huang J, Wei Z (2010) J Phys Chem C 114:8062–8067; (b) Liu FJ, Huang LM, Wen TC, Li CF, Huang SL, Gopalan A (2008) Synth Met 158:767–774; (c) Gupta V, Miura N (2005) Electrochem Commun 7:995–999; (d) Zang J, Li CM, Bao S-J, Cui X, Bao Q, Sun CQ (2008) Macromolecules 41:7053–7057; (e) Kitani A, Kaya M, Sasaki K (1986) J Electrochem Soc 133:1069–1073; (f) Stilwell DE, Park S-M (1988) J Electrochem Soc 135:2254–2262; (g) Focke WW, Wnek GE, Wei Y (1987) J Phys Chem 91:5813–5818; (h) Watanabe A, Mori K, Iwasaki Y, Nakamura Y, Niizuma S (1987) Macromolecules 20:1793–1796
23. (a) Tuan CV, Tuan MA, Hieu NV, Trung T (2012) Curr Appl Phys 12:1011–1016; (b) Peng X, Huo K, Fu J, Zhang X, Gao B, Chu PK (2013) Chem Commun 49:10172–10174; (c) Stejskal J, Sapurina I, Trchova M (2010) Prog Poly Sci 35:1420–1481; (d) Yu X, Li Y, Kalantar-zaheh K (2009) Sens Actuators B 136:1–7; (e) Wang X, Shao M, Shao G, Wu Z, Wang S (2009) J Colloid Interface Sci 332:74–77; (f) Wang Y, Jing X (2008) J Phys Chem B 112:1157–1162; (g) Li GR, Feng ZP, Zhong JH, Wang ZL, Tong YX (2010) Marcromolecules 43:2178–2183; (h) Debiemme-Chouvy C (2009) Electrochem Commun

11:298–301; (i) Gupta V, Miura N (2005) Electrochem Solid-State Lett 8:A630–A632; (j) Niu Z, Yang Z, Hu Z, Lu Y, Han CC (2003) Adv Funct Mater 13:925; (k) Gupta V, Miura N (2005) Electrochem Commun 7:995–999; (l) Wang K, Huang J, Wei Z (2010) J Phys Chem C 114:8062–8067; (m) Zang J, Li CM, Bao S-J, Cui X, Bao Q, Sun CQ (2008) Macromolecules 41:7053–7057

24. Mondal SK, Rajendra Prasad K, Munichandraiah N (2005) Synth Met 148:275–286
25. (a) Millard M (1974) Synthesis of organic polymer films in plasmas. In: Hollahan JR, Bell AT (eds) Techniques and applications of plasma chemistry. Wiley, New York; Shen M (1976) Plasma chemistry of polymers. Marcel Dekker, New York; (b) Hernandez R, Diaz AF, Waltman R, Bargon J (1984) ISE meeting, July 1984. Berkeley, CA, U.S.A.; J Phys Chem 88:3333; (c) Nayak B, Bhakta RC (1983) J Appl Electrochem 13:105; (d) Nedungati PAK, Gupta A, Mukherji SK, Zutshi K (1986) Int workshop-electrochemistry of polymer layers, Duisburg, F.R.G., Sept. 15–17, p 1.4; (e) Uvdal K, Loglund M, Danneton P, Bertilsson L, Strafstrom S, Salaneck WR, MacDiarmid AG, Ray A, Scherr EM, Hjertberg T, Epstein AJ (1989) Synth Met 29:E451
26. (a) Pandey RK, Lakshminarayanan V (2014) Mater Res Bull 50:413–416; (b) Liu FJ, Huang LM, Wen TC, Li CF, Huang SL, Gopalan A (2008) Synth Met 158:767–774; (c) Yang C-C, Wu T-y, Chen H-R, Hsieh T-H, Ho K-S, Kuo C-W (2011) Int J Electrochem Sci 6:1642–1654; (d) Biswas S, Dutta B, Bhattacharya S (2014) J Mater Sci 49:5910–5921
27. (a) Biswas S, Dutta B, Bhattacharya S (2014) J Mater Sci 49:5910–5921; (b) Yang C-C, Wu T-y, Chen H-R, Hsieh T-H, Ho K-S, Kuo C-W (2011) Int J Electrochem Sci 6:1642–1654; (c) Hatchett DW, Josowicz M, Janata J (1999) Chem Mater 11:2989–2994; (d) Ciric-Marjanovic G (2013) Synth Met 170:31–56; (e) Pillalamarri SK, Blum FD, Tokuhiro AT, Bertino MF (2005) Chem Mater 17:5941–5944; (f) Balan L, Burget D (2006) Eur Polym J 42:3180–3189; (g) Granot E, Katz E, Basnar B, Willner I (2005) Chem Mater 17:4600–4609
28. (a) Nørskov JK, Bligaard T, Hvolbæk B, Abild-Pedersen F, Ib Chorkendorff, Christensen CH (2008) Chem Soc Rev 37:2163–2171; (b) Katz E, Willner I, Wang J (2004) Electroanalysis 16:19–44; (c) Štrbac S, Srejic I, Smiljanic M, Rakocevic Z (2013) J Electroanal Chem 704:24–21; (d) Mulvaney P, Yu Y-t (2002) Mater Forum 26:39–43; (e) Lang X, Hirata A, Fujita T, Chen M (2011) Nat Nanotechnol 6:232–236; (f) Skrabalak SE, Chen J, Au L, Lu X, Li X, Xia Y (2007) Adv Mater 19:3177–3184; (g) Tong L, Zhao Y, Huff TB, Hansen MN, Wei A, Cheng J-X (2007) Adv Mater 19:3136–3141; (h) Chah S, Hammond MR, Zare RN (2005) Chem Biol 12:323–328; (i) Eustis S, El-Sayed MA (2006) Chem Soc Rev 35:209–217; (j) Storhoff JJ, Lazarides AA, Mucic RC, Mirkin CA, Letsinger RL, Schatz GC (2000) J Am Chem Soc 122:4640–4650
29. (a) Daniel M-C, Astruc D (2004) Chem Rev 104:293–346; (b) Song Q, Ai X, Wang D, Hong X, Wei L, Yang W, Liu F, Bai Y, Li T, Tang X (2000) J Nanopart Res 2:381–385; (c) Rao CNR, Kulkarni GU, Thomas PJ, Agrawal VV, Saravanan P (2003) J Phys Chem B 107:7391–7395
30. (a) Ma H, Yin B, Wang S, Jiao Y, Pan W, Huang S, Chen S, Meng F (2004) ChemPhysChem 5:68–75; (b) Skrabalak SE, Chen J, Au L, Lu X, Li X, Xia Y (2007) Adv Mater 19:3177–3184; (c) Zhao W, Gonzaga F, Li Y, Brook MA (2007) Adv Mater 19:1766–1771; (d) Yan Y-M, Tal-Vered R, Yehezkeli O, Cheglakov Z, Willner I (2008) Adv Mater 20:2365–2370; (e) Wijnhoven JEGJ, Zevenhuizen SJM, Hendriks MA, Vanmaekelbergh D, Kelly JJ, Vos WL (2000) Adv Mater 12:888–890; (f) Prasad BLV, Stoeva SI, Sorensen CM, Klabunde KJ (2003) Chem Mater 15:935–942; (g) Low CTJ, Ponce-de-Leon C, Walsh FC (2012) Electrochem Commun 22:166–169; (h) Yu Y-Y, Chang S-S, Lee C-L, Wang CRC (1997) J Phys Chem 101:6661–6664; (i) Sakai T, Alexandridis P (2005) J Phys Chem B 109:7766–7777; (j) Sylvestre J-P, Kabashin AV, Sacher E, Meunier M, Luong JHT (2004) J Am Chem Soc 126:7176–7177; (k) Harada M, Einaga H (2006) Langmuir 22:2371–2377; (l) Huang MH, Choudrey A, Yang P (2000) Chem Commun, 1063–1064
31. Amiens C, Chaudret B, Ciuculescu-Pradines D, Colliere V, Fajerwerg K, Fau P, Kahn M, Maisonnat A, Soulantica K, Philippot K (2013) New J Chem 37:3374–3401

32. (a) Wee G, Mak WF, Phonthammachai N, Kiebele A, Reddy MV, Chowdari BVR, Gruner G, Srinivasan M, Mhaisalkar SG (2010) J Electrochem Soc 157:A179–A184; (b) Tang Z, Liu S, Dong S, Wang E (2001) J Electroanal Chem 502:146–151; (c) Chang G, Zhang J, Oyama M, Hirao K (2005) J Phys Chem B 109:1204–1209; (d) Endo K, Sugawara Y, Mishima S, Okada T, Morita S (1991) Jpn J Appl Phys 30:2592–2593; (e) Santra AK, Yang F, Goodman DW (2004) Surf Sci 548:324–332; (f) Kao W-H, Kuwana T (1984) J Am Chem Soc 106:473–476; (g) Ficicioglu F, Kadirgan F, J Electroanal Chem 430:179–182; (h) Seger B, Kongkanand A, Vinodgopal K, Kamat PV (2008) J Electroanal Chem 621:198–204
33. (a) Fernández-Garrido S, Ramsteiner M, Gao G, Galves LA, Sharma B, Corfdir P, Calabrese G, de Souza Schiaber Z, Pfüller C, Trampert A, Marcelo J, Lopes J, Brandt O, Geelhaar L (2017) Nano Lett 17:5213–5221; (b) Jerng SK, Yu DS, Kim YS, Ryou J, Hong S, Kim C, Yoon S, Efetov DK, Kim P, Chun SH (2011) J Phys Chem C 115:4491–4494; (c) Nejati S, Lau KKS (2011) Langmuir 27:15223–15229; (d) Ismach A, Druzgalski C, Penwell S, Schwartzberg A, Zheng M, Javey A, Bokor J, and Zhang Y (2010) Nano Lett 10:1542–1548; (e) Li L, Yang Y, Yang G, Chen X, Hsiao BS, Chu B, Spanier JE, Li CY (2006) Nano Lett 6:1007–1012; (f) Liu J, Zhou Y, Lin Y, Li M, Cai H, Liang Y, Liu M, Huang Z, Lai F, Huang F, Zheng W (2019) ACS Appl Mater Interfaces 11:4123–4130; (g) Seo B, Choi S, Kim J (2011) ACS Appl Mater Interfaces 3:441–446; (h) Pham-Truong T-N, Mebarki O, Ranjan C, Randriamahazaka H, Ghilane J (2019) ACS Appl Mater Interfaces 11:38265–38275; (i) Quinn BM, Dekker C, Lemay SG (2005) J Am Chem Soc 127:6146–6147
34. (a) Moreno-Mañas M, Pleixats R (2003) Acc Chem Res 36:638–643; (b) Kamat PV (2002) J Phys Chem B 106:7729–7744; (c) Sakai N, Fujiwara Y, Arai M, Yu K, Tatsuma T (2009) J Electroanal Chem 628:7–15; (d) Tian Y, Liu H, Zhao G, Tatsuma T (2006) J Phys Chem B 110:23478–23481; (e) Tian N, Zhou Z-Y, Sun S-G, Cui L, Ren B, Tian Z-Q (2006) Chem Commun 4090–4092; (f) Saha MS, Li R, Cai M, Sun X (2007) Electrochem Solid State Lett 10:B130–B133; (g) Lu G, Zangari G (2006) Electrochim Acta 51:2531–2538; (h) Zoval JV, Lee J, Gorer S, Penner RM (1998) J Phys Chem B 102:1166–1175; (i) Kao W-H, Kuwana T (1984) J Am Chem Soc 106:473–476; (j) Mazur M (2004) Electrochem Commun 6:400–403
35. (a) Schonenberger C, van der Zande BMI, Fokkink LGJ, Henry M, Schmid C, Kruger M, Bachtold A, Huber R, Birk H, Staufer U (1997) J Phys Chem B 101:5497–5505; (b) Liu Z, El Abedin SZ, Ghazvini MS, Endres E (2013) Phys Chem Chem Phys 15:11362–11367; (c) Kovtyukhova NI, Mallouk TE (2002) Chem Eur J 8:4355–4363; (d) Nicewarner-Pena SR, Freeman RG, Reiss BD, He L, Pena DJ, Walton ID, Cromer R, Keating CD, Natan MJ (2001) Science 294:137–141; (e) Kline TR, Tian M, Wang J, Sen A, Chan MWH, Mallouk TE (2006) Inorg Chem 45:7555–7565
36. (a) Cusma A, Curulli A, Zane D, Kaciulis S, Padeletti G (2007) Mater Sci Eng C 27:1158–1161; (b) Mbindyo JKN, Mallouk TE, Mattzela JB, Kratochvilova I, Razavi B, Jackson TN, Mayer TS (2002) J Am Chem Soc 124:4020–4026
37. Sharma MK, Ambolikar AS, Aggarwal SK (2012) J Nanopart Res 14:1094
38. (a) Yu Y-Y, Chang S-S, Lee C-L, Wang CRC (1997) J Phys Chem B 101:6661–6664; (b) Mohamed MB, Ismail KZ, Link S, El-Sayed MA (1998) J Phys Chem B 102:9370–9374; (c) Yin B, Ma H, Wang S, Chen S (2003) J Phys Chem B 107:8898–8904; (d) Chang S-S, Shih C-W, Chen C-D, Lai W-C, Wang CRC (1999) Langmuir 15:701–709; (e) Park S, Boo H, Lee SY, Kim H-M, Kim K-B, Kim HC, Chung TD (2008) Electrochim Acta 53:6143–6148; (f) Ma H, Yin B, Wang S, Jiao Y, Pan W, Huang S, Chen S, Meng F (2004) ChemPhysChem 5:68–75
39. (a) Tian Y, Liu H, Zhao G, Tatsuma T (2006) J Phys Chem B 110:23478–23481; (b) Mazur M (2004) Electrochem Commun 6:400–403; (c) Stiger RM, Gorer S, Craft B, Penner RM (1999) Langmuir 15:790–798; (d) Zoval JV, Lee J, Gorer S, Penner RM (1998) J Phys Chem B 102:1166–1175

40. (a) Zhou M, Chen S, Zhao S, Ma H (2006) Physica E 33:28–34; (b) Guerin S, Attard GS (2001) Electrochem Commun 3:544–548; (c) Liu Y-C, Lee H-T, Peng H-H (2004) Chem Phys Lett 400:436–440
41. (a) Robinson DA, White HS (2019) Nano Lett 19:5612–5619; (b) Ting L (2007) Trans Nonferrous Met Soc 17:1343–1346; (c) Kulp C, Chen X, Puschhof A, Schwamborn S, Somsen C, Schuhmann W, Bron M (2010) ChemPhysChem 11:2854–2861
42. (a) Vanheusden K, Seager CH, Warren WL, Tallant DR, Voight JA (1996) J Appl Phys 68:403; (b) Vanheusden K, Warren WL, Seager CH, Tallant DR, Voight JA (1996) J Appl Phys 79:7983; (c) Bachir S, Sandouly C, Kossanyi J, Ronfard-Haret JC (1996) J Phys Chem Solids 57:1869–1879; (d) Pichat P, Herrmann J-M, Jisdier J, Mozzanega M-N (1979) J Phys Chem 83:3122; (e) Kurbatov LN, Kozina GS, Kostinskaya TA, Rudnevskii VS, Lobachev AV, Kuznetsov VA, Kuzmina IP, Shaldin YV, Shternberg AA (1980) Kvantovaya Electronika Moskva 7:378–382; (f) Kamat PV, Tvrdy K, Baker DR, Radich JG (2010) Chem Rev 110:6664–6688; (g) Michalet X, Pinaud FF, Bentolila LA, Tsay JM, Doose S, Li JJ, Sundaresan G, Wu AM, Gambhir SS, Weiss S (2005) Science 307:538–544
43. (a) Hoyer P, Eichenberger R, Weller H (1993) Ber Bunsen Ges Phys Chem 97:630; (b) Hoyer P, Weller H (1994) Chem Phys Lett 221:379–384; (c) Bahnemann DW, Kormann C, Hoffmann MR (1987) J Phys Chem 91:3789–3798; (d) Haase M, Weller H, Henglein A (1988) J Phys Chem 92:482; (e) Sakohara S, Tickanen LD, Anderson MA (1992) J Phys Chem 96:11086–91; (f) Spanhel L, Anderson MA (1991) J Am Chem Soc 113:2826–2833; (g) De Merchant J, Cocivera M (1995) Chem Mater 7:1742; (h) Roth AP, Williams DF (1981) J Appl Phys 52:6685–6692; (i) Roth AP, Williams DF (1981) J Electrochem Soc 128:2684–2686; (j) Maruyama T, Shionoya J (1992) J Mater Sci Lett 11:170–172; (k) Bethke S, Pan H, Wessels BW (1998) Appl Phys Lett 52:138–140; (l) Izaki M, Omi T (1996) Appl Phys Lett 68:2439–2440; (m) Izaki M, Omi T (1996) J Electrochem Soc 143: L53–L55; (n) Peulon S, Lincot D (1996) Adv Mater 8:166–170
44. (a) Golan Y, Margulis L, Rubinstein I, Hodes G (1992) Langmuir 8:749–752; (b) Golan Y, Margulis L, Hodes G, Rubinstein I, Hutchinson JL (1994) Surf Sci 311:L633–L640
45. (a) Nyffenegger RM, Craft B, Shaaban M, Gorer S, Erley G, Penner RM (1998) Chem Mater 10:1120–1129; (b) Gorer S, Hsiao GS, Anderson MG, Stiger RM, Lee J, Penner RM (1998) Electrochim Acta 43:2799–2809; (c) Hsiao GS, Anderson MG, Gorer S, Harris D, Penner RM (1997) J Am Chem Soc 119:1439–1448; (d) Gorer S, Ganske JA, Hemminger JC, Penner RM (1998) J Am Chem Soc 120:9584–9593; (e) Anderson MA, Gorer S, Penner RM (1997) J Phys Chem B 101:5895–5899
46. (a) Lai M, Riley DJ (2006) Chem Mater 18:2233–2237; (b) Zheng MJ, Zhang LD, Li GH, Shen WZ (2002) Chem Phys Lett 363:123–128; (c) Routkevitch D, Bigioni T, Moskovits M, Xu JM (1996) J Phys Chem 100:14037–14047; (d) Sander MS, Gronsky R, Sands T, Stacy AM (2003) Chem Mater 15:335–339
47. (a) Choi K-S, Lichtenegger HC, Stucky GD, McFarland EW (2002) J Am Chem Soc 124:12402–12403; (b) Yoshida T, Tochimoto M, Schlettwein D, Wohrle D, Sugiura T, Minoura H (1999) Chem Mater 11:2657–2667; (c) Xu L, Guo Y, Liao Q, Zhang J, Xu D (2005) J Phys Chem B 109:13519–13522; (d) Cao B, Cai W, Zeng H, Duan G (2006) J Appl Phys 99:073516; (e) Michaelis E, Wohrle D, Rathousky J, Wark M (2006) Thin Solid Films 497:163–169; (f) Ellias J, Tena-Zaera R, Levy-Clement C (2008) J Electroanal Chem 621:171–177; (g) Ellias J, Tena-Zaera R, Levy-Clement C (2007) Thin Solid Films 515:8553–8557
48. (a) Allen MJ, Tung VC, Kaner RB (2010) Chem Rev 110:132–145; (b) Avouris P, Dimitrakopoulos C (2012) Mater Today 15:86–97; (c) Nag A, Mitra A, Mukhopadhyay SC (2018) Sensor Actuat A 270:177–194; (d) Georgakilas V, Tiwari JN, Kemp KC, Perman JA, Bourlinos AB, Kim KS, Zboril R (2016) Chem Rev 116:5464–5519; (e) Roy-Mayhew JD, Aksay IA (2014) Chem Rev 114:6323–6348; (f) Liu M, Zhang R, Chen W (2014) Chem Rev 114:5117–5160; (g) Ambrosi A, Chua CK, Bonanni A, Pumera M (2014) Chem Rev 114:7150–7188; (h) Mao HY, Laurent S, Chen W, Akhavan O, Imani M, Ashkarran AA,

Mahmoudi M (2013) Chem Rev 113:3407–3424; (i) Wu J, Pisula W, Mullen K (2007) Chem Rev 107:718–747; (j) Geim AK, Novoselov KS (2007) Nat Mater 6:183–191; (k) Park S, Ruoff RS (2009) Nat Nanotechnol 4:217–224
49. (a) Avouris P, Dimitrakopoulos C (2012) Mater Today 15:86–97; (b) Lin L, Deng B, Sun J, Peng H, Liu Z (2018) Chem Rev 118:9281–9343; (c) Sun Z, Fang S, Hu YH (2020) Chem Rev. https://doi.org/10.1021/acs.chemrev.0c00083; (d) Yu P, Lowe SE, Simon GP, Zhong YL (2015) Curr Opin Colloid Interface Sci 20:329–338; (e) Parvez K, Wu Z-S, Li R, Liu X, Graf R, Feng X, Mullen K (2014) J Am Chem Soc 136:6083–6091; (f) Geim AK (2011) Rev Mod Phys 83:851–862; (g) Mas-Balleste R, Gomez-Navarro C, Gomez-Herrero J, Zamora F (2011) Nanoscale 3:20–30; (h) Yi M, Shen Z (2015) J Mater Chem A, 3:11700–11715
50. (a) Reina A, Jia XT, Ho J, Nezich D, Son HB, Bulovic V, Dresselhaus MS, Kong J (2009) Nano Lett 9:30–35; (b) Kim KS (2009) Nature 457:706–710; (c) Sutter PW, Flege JI, Sutter EA (2008) Nat Mater 7:406–411; (d) Nag A, Mitra A, Mukhopadhyay SC (2018) Sensor Actuat A 270:177–194
51. (a) Hummers WS, Offeman RE (1958) J Am Chem Soc 80:1339–1339; (b) Chen J, Yao B, Li C, Shi G (2013) Carbon 64:225–229; (c) Zaaba N, Foo K, Hashim U, Tan S, Liu W-W, Voon C (2017) Procedia Eng 184:469–477
52. (a) Shen J, Hu Y, Shi M, Lu X, Qin C, Li C, Ye M (2009) Chem Mater 21:3514–3520; (b) Chen W, Yan L, Bangal P (2010) J Phys Chem C 114:19885–19890; (c) Ekiz OO, Urel M, Guner H, Mizrak AK, Dana A (2011) ACS Nano 5:2475–2482; (d) Williams G, Seger B, Kamat PV (2008) ACS Nano 2:1487–1491; (e) Gomez-Navarro C, Weitz RT, Bittner AM, Scolari M, Mews A, Burghard M, Kern K (2007) Nano Lett 7:3499–3503; (f) Stankovich S, Dikin DA, Piner RD, Kohlhaas KA, Kleinhammes A, Jia Y, Wu Y, Nguyen ST, Ruoff RS (2007) Carbon 45:1558–1565
53. (a) Low CTJ, Walsh FC, Chakrabarti MH, Hashim MA, Hussain MA (2013) Carbon 54:1–21; (b) Abdelkader AM, Cooper AJ, Dryfe RAW, Kinloch IA (2015) Nanoscale 7:6944–6956; (c) Yu P, Lowe SE, Simon GP, Zhong YL (2015) Curr Opin Colloid Interface Sci 20:329–338; (d) Zhong YL, Tian Z, Simon GP, Li D (2015) Mater Today 18:73–78; (e) Parvez K, Wu ZS, Li R, Liu X, Graf R, Feng X, Mullen K (2014) J Am Chem Soc 136:6083–6091
54. (a) Liu N, Luo F, Wu HX, Liu YH, Zhang C, Chen J (2008) Adv Funct Mater 18:1518–1525; (b) Lu J, Yang JX, Wang JZ, Lim AL, Wang S, Loh KP (2009) ACS Nano 3:2367–2375; (c) Parvez K, Li RJ, Puniredd SR, Hernandez Y, Hinkel F, Wang SH, Feng XL, Mullen K (2013) ACS Nano 7:3598–3606; (d) Su CY, Lu AY, Xu YP, Chen FR, Khlobystov AN, Li LJ (2011) ACS Nano 5:2332–2339; (e) Liu JL, Yang HP, Zhen SG, Poh CK, Chaurasia A, Luo JS, Wu XY, Yeow EKL, Sahoo NG, Lin JY, Shen ZX (2013) RSC Adv 3:11745–11750; (f) Parvez K, Wu ZS, Li R, Liu X, Graf R, Feng X, Mullen K (2014) J Am Chem Soc 136:6083–6091
55. (a) Parvez K, Wu ZS, Li R, Liu X, Graf R, Feng X, Mullen K (2014) J Am Chem Soc 136:6083–6091; (b) Rao KS, Senthilnathan J, Liu YF, Yoshimura M (2014) Sci Rep 4:4237–4242; (c) Su CY, Lu AY, Xu Y, Chen FR, Khlobystov AN, Li LJ (2011) ACS Nano 5:2332–2339; (d) Wang J, Manga KK, Bao Q, Loh KP (2011) J Am Chem Soc 133:8888–8891; (e) Mao M, Wang M, Hu J, Lei G, Chen S, Liu H (2013) Chem Commun 49:5301–5303; (f) Abdelkader AM, Kinloch IA, Dryfe RAW (2014) ACS Appl Mater Interfaces 6:1632–1639; (g) Huang H, Xia Y, Tao X, Du J, Fang J, Gan Y, Zhang W (2012) J Mater Chem 22:10452–10456
56. (a) Tong L, Zhao Y, Huff TB, Hansen MN, Wei A, Cheng J-X (2007) Adv Mater 19:3136–3141; (b) Huang X, El-Sayed IH, Qian W, El-Sayed MA (2006) J Am Chem Soc 128:2115–2120; (c) Takahashi H, Niidome T, Nariai A, Niidome Y, Yamada S (2006) Chem Lett 35:500–501
57. (a) Cai Z, Lei J, Liang W, Menon V, Martin CR (1991) Chem Mater 3:960–967; (b) Parthasarathy RV, Martin CR (1994) Chem Mater 6:1627–1632; (c) Cai Z, Martin CR (1989) J Am Chem Soc 111:4138–4139; (d) Martin CR, Partasarathy R, Menon V (1993) Synth Met 55–57:1165–1170

58. (a) Thompson GE (1997) Thin solid films 297:192–201; (b) Lee W, Schwirn K, Steinhart M, Pippel E, Scholz R, Gosele U (2008) Nat Nanotech 3:234–239; (c) Lee W, Park S-J (2014) Chem Rev 114:7487–7556
59. (a) Lee W, Schwirn K, Steinhart M, Pippel E, Scholz R, Gosele U (2008) Nat Nanotech 3:234–239; (b) Li AP, Muller F, Birner A, Nielsch K, Gosele U (1998) J Appl Phys 84:6023–6026; (c) Masuda M, Hasegawa E, Ono S (1997) J Electrochem Soc 144:L127–L130; (d) Masuda H, Fukuda K (1995) Science 268:1466–1468; (e) Masuda H, Satoh M (1996) Jpn J Appl Phys 35:L126–L129; (f) Nielsch K, Choy J, Schwirn K, Wehrspohn RB, Gosele U (2002) Nano Lett 2:677–680; (g) Masuda H, Yada K, Osaka A (1998) Jpn J Appl Phys 37: L1340–L1342; (h) Kushwaha MK (2014) Procedia Mater Sci 5:1266–1273
60. (a) Senthil Kumar S, Joseph J, Phani KL (2007) Chem Mater 19:4722–4730; (b) Baioni AP, Vidotti M, Fiorito PA, Ponzio EA, Cordoba de Torresi SI (2007) Langmuir 23:6796–6800; (c) Milonjic S, Bispo I, Fedoroff M, Los-Neskovic C, Vidal-Madjar C (2002) J Radioanal Nucl Chem 252:497–501; (d) Tani Y, Eun H, Umezawa Y (1998) Electrochim Acta 43:3431–3441; (e) Bhatt P, Yusuf SM, Mukadam MD, Yakhmi JV (2010) J Appl Phys 108:023916 1–6; (f) Pajerowski DM, Andrus MJ, Gardner JE, Knowles ES, Meisel MW, Talham DR (2010) J Am Chem Soc 132:4058–4059; (g) Lu Y, Wang L, Cheng J, Goodenough JB (2012) Chem Commun 48:6544–6546; (h) Okubo M, Honma I (2013) Dalton Trans 42:15881–15884; (i) Pintado S, Goberna-Ferron S, Escudero-Adan EC, Galan-Mascaros JR (2013) J Am Chem Soc 135:13270–13273; (j) de Tacconi NR, Rajeshwar K, Lezna RO (2003) Chem Mater 15:3046; (k) Jayalakshmi M, Scholz F (2000) J Power Sour 91:217–213
61. (a) Neff VD (1978) J Electrochem Soc 125:886–887; (b) Miao Y, Chen J, Wu X, Miao J (2007) Colloids Surf A Physicochem Eng Asp 295:135–138; (c) Liu S-Q, Xu J-J, Chen H-Y (2002) Electrochem Commun 4:421–425; (d) Garcia-Jarero JJ, Navarro-Laboulais J, Vicente F (1996) Electrochim Acta 41:835–841; (e) Yang C, Wang C-H, Wu J-S, Xia X (2006) Electrochim Acta 51:4019–4023; (f) Pintado S, Goberna-Ferron S, Escudero-Adan EC, Galan-Mascaros JR (2013) J Am Chem Soc 135:13270–13273; (g) Abbaspour A, Kamyabi MA (2005) J Electroanal Chem 584:117–123; (h) Senthil Kumar SM, Pillai KC (2006) J Electroanal Chem 589:167–175; (i) Yu H, Song S-W, Lian Y-Y, Liu Z-Y, Qi G-C (2010) J Electroanal Chem 650:82–89; (j) Jiang W, Yuan R, Chai Y-Q, Yin B (2010) Anal Biochem 407:65–71; (k) de Tacconi NR, Rajeshwar K, Lezna RO (2003) Chem Mater 15:3046
62. (a) Yu H, Song S-W, Lian Y-Y, Liu Z-Y, Qi G-C (2010) J Electroanal Chem 650:82–89; (b) Wang X, Zhang Y, Jiang S, Ji X, Liu Y, Banks CE (2011) Int J Electrochem. https://doi.org/10.4061/2011/395724; (c) Sheng Q, Zhang D, Wu Q, Zheng J, Tang H (2015) Anal Methods 7:6896–6903; (d) Kumar SS, Joseph J, Phani KL (2007) Chem Mater 19:4722–4730; (e) Karyakin AA, Puganova EA, Budashov IA, Kurochkin IN, Karyakina EE, Levchenko VA, Matveyenko VN, Varfolomeyev SD (2004) Anal Chem 76:474; (f) Orellana M, Ballestros L, Rio RD, Grez P, Schrebler R, Cordova R (2009) J Solid State Electrochem 13:1303–1308
63. (a) Li J, Kuang Y, Meng Y, Tian X, Hung W-H, Zhang X, Li A, Xu M, Zhou W, Ku C-S, Chiang C-Y, Zhu G, Guo J, Sun X, Dai H (2020) J Am Chem Soc 142:7276–7282; (b) Liu Y, Fan X, Nayak A, Wang Y, Shan B, Quan X, Meyer TJ (2019) PNAS 116:26353–26358; (c) Gao D, Aran-Ais RM, Jeon HS, Cuenya BR (2019) Nat Catal 2:198–210; (d) Luo W, Xei W, Li M, Zhang J, Zuttel A (2019) J Mater Chem A 7:4505–4515
64. (a) Gao D, Aran-Ais RM, Jeon HS, Cuenya BR (2019) Nat Catal 2:198–210; (b) Sanchez OG, Birdja YY, Bulut M, Vaes J, Breugelmans T, Deepak P (2019) Curr Opin Green Sustain Chem 16:47–56; (c) Kortlever R, Shen J, Schouten KJP, Calle-Vallejo F, Koper MTM (2015) J Phys Chem Lett 6:4073–4082; (d) Aran-Ais RM, Gao D, Cuenya BR (2018) Acc Chem Res 51:2906–2917; (e) Larrazabal GO, Martin AJ, Perez-Ramirez J (2017) J Phys Chem Lett 8:3933–3944; (f) Zhu DD, Liu JL, Qiao SZ (2016) Adv Mater 28:3423–3452

Chapter 15
Synthesis of Advanced Inorganic Materials Through Molecular Precursors

G. Kedarnath

Abstract Requirement in various arenas of technologies for sophistication and miniature of devices augmented the quest for novel advanced materials of versatile nature. To meet these demands, there has always been a need to explore and develop new ways of synthesis of advanced materials. Of these synthetic methods, molecular precursor route has an edge over other preparative methods due to the distinct advantages of easy processability and better control over size, shape and quality of the resulting materials by tuning the reactivity of the molecular reactants. Molecular precursor approach can either be multiple or single-source precursor depending on the number of precursor being used for the synthesis of required material. This chapter will give a brief introduction of molecular precursor-based synthesis followed by a discussion on preparation of various advanced material through molecular precursor approach and its comparison with other conventional methods. The subsequent discussion includes characterization, property evaluation and applications of these materials. At the end, the chapter will be concluded with a brief note on future prospective of the molecular precursor route.

Keywords Molecular precursor · Single source molecular precursor · Metal chalcogenide · Hot injection · Criteria · Reaction pathways · Metal nano · Metal oxide · AACVD · Characterization techniques · Advanced materials

15.1 Introduction

Technological demands for the benefits of humankind have always been in a dynamic state and have created insatiable quest for new and advanced inorganic materials with tailor made multifunctional properties. The materials with such

G. Kedarnath (✉)
Chemistry Division, Bhabha Atomic Research Centre, Mumbai 400 085, India
e-mail: kedar@barc.gov.in

G. Kedarnath
Homi Bhabha National Institute, Anushaktinagar, Mumbai 400 094, India

A. K. Tyagi and R. S. Ningthoujam (eds.), *Handbook on Synthesis Strategies for Advanced Materials*, Indian Institute of Metals Series,
https://doi.org/10.1007/978-981-16-1807-9_15

exotic properties are possible due to their size and shape tuneable properties at nanoscale level and find various applications in different arenas [1–8]. Therefore, a need to explore and develop new ways of synthesis of advanced materials is inevitable. Accordingly, a wide range of methods for the synthesis of advanced materials have been devolved over the years which may be broadly categorized as physical (top-down or fabrication approach), chemical (bottom-up or synthesis strategy) or hybrid methods and have been reviewed in the literature [9]. Of these synthetic methods, chemical or bottom-up approach has advantages of low production cost, easy processability and phase purity. The important objective of the approach is to connect organic moieties to inorganic structures for synthesizing materials which are difficult to prepare through thermodynamically controlled chemical syntheses. In general, the approach is a kinetically controlled process for deriving inorganic solids with specific properties from molecular precursors. Among chemical methods, molecular precursor route has an edge over other preparative routes for obvious reasons [10]. In view of this, synthesis of advanced inorganic materials will be discussed in detail in the following sections.

15.2 Advanced Materials Through Molecular Precursor Route

Over the years, several chemical methods have been evolved for the synthesis and deposition of inorganic materials. Of these, molecular precursor approach is versatile in nature and has distinct benefits of easy processability and better-quality control over size, shape and phase purity of the resulting materials by tuning the reactivity of the molecular reactants. Other advantages are the mild decomposition conditions which enable better control over the formation of material while clean decomposition avoids the surface contamination of unreacted reactants when the synthetic procedure involves metal salts or oxides, etc. The term molecular precursor route comes from the molecular form of reactant or precursors being used for the synthesis of nanomaterials. This route uses molecular precursors as the reactants and converts them to the required material by applying suitable reaction conditions. By definition, molecule is a group of atoms bonded together chemically. However, in the synthesis or deposition of materials, the term “molecular” has not been used strictly. Therefore molecular route in general may be referred to a synthetic approach where one of the precursor/reactants required to obtain the target material is metalloorganic or organometallic complex. Molecular precursors can be used as reactants in precipitation/reduction at room temperature, micro-emulsion, sonication, microwave, solvo-/hydro-thermal and thermolysis methods for the synthesis of materials. Although molecular precursors are associated with these thermal or non-thermal-based methods, in general, many reports in the literature consider thermal decomposition of molecular precursors either in high boiling coordinating or non-coordinating or a mixture of the solvents as the molecular precursor method.

Since these reactions are performed at relatively moderate temperatures, the nanoparticles produced are crystalline in nature. Molecular precursor approach can either be single- or multiple-source molecular precursor depending whether all the elements required in the final product are in a single or more than one molecule. Of these molecules, at least one should be organometallic or metalloorganic. However, many reports correlate the advent of molecular precursor approach with the method for the preparation of semiconductor nanocrystallites described by Murray, Norris, and Bawendi [11]. The investigation describes the synthesis of high quality CdSe nanocrystals by injecting solutions of $(CH_3)_2Cd$ and tri-n-octylphosphine selenide (TOPSe) into hot tri-n-octylphosphine oxide (TOPO) in the temperature range 120–300 °C.

15.2.1 Multiple Source Molecular Precursor (MSMP) Method

The method can be defined as a molecular approach where one or more reactants of the synthesis are organometallic or metalloorganic complexes. For instance, in case of a multiple source precursor method for binary materials, metal complex (organometallic or metalloorganic) where metal is bonded to a donor/acceptor atom of an organic moiety can be a metal source and an organic ligand containing chalcogen acts as a chalcogen source. However, in case of ternary materials, the precursors can be either two different single-source molecular precursors (SSMPs) for binary materials or it can be a SSMP for binary material with another metal source. For example, ternary materials like $CuInSe_2$ can either be prepared by thermolysis of $[Cu\{SeC_5H_3(3\text{-}Me)N\}]_4$ and $[In\{SeC_5H_3(3\text{-}Me)N\}_3]$ [12] or $[Cu\{SeC_5H_3(3\text{-}Me)N\}]_4$ and an indium source (e.g. $InCl_3$, $In(OAc)_3$, etc.) or $[In\{SeC_5H_3(3\text{-}Me)N\}_3]$ and a copper source (e.g. CuCl, $Cu(OAc)_2$, etc.). In case of utilizing two different SSMPs for binary materials, the SSMPs can consist of different metals with same chalcogen as in the above case or same metal with different chalcogen like in the case of preparation of CdSSe by thermolysis of cadmium thiolate and a cadmium selenolate complexes.

15.2.1.1 Criteria for the Reactants

In MSMP method, at least one of the reactants acting as metal carrier/source should be organometallic or metalloorganic in nature. The reactants should be sufficiently volatile and reactive in order to obtain materials of good quality. The ideal precursors should be less- or non-toxic, thermodynamically stable but kinetically labile in order to release metal or chalcogen depending whether the precursor is metal or chalcogen source. The precursors should be cheap.

15.2.1.2 Role and Criteria for Surfactants / Passivating Agents

Nanoparticles are not thermodynamically stable due to their higher surface energy owing to larger surface to volume ratio compared to their bulk counterparts. In order to produce stable nanoparticles, the particle growth must be arrested by passivating their surface by adding surface passivating agents during the reaction which avoids the particles to come together.

The surfactants/passivating/stabilizing agents in general are high boiling coordinating solvents with a coordinating head consisting of a donor atom which coordinates to the surface of the nanoparticles and a hydrophobic long alkyl chain. However, in some cases, long chain alkenes like 1-octadecene also passivate the surface of the nanoparticles through their alkene bond and behave similar to that of coordinating agents [13]. The passivating agents prevent agglomeration of the particles by avoiding neighbouring particles coming together during the synthesis of nanoparticles due to steric repulsion of long alkyl chains. This leads to colloidal stability of the nanoparticles in organic solvents for a long time depending on how strongly these passivating agents are coordinated to the surface of the particles. Adsorption and desorption of the passivating agents from the surface of the particles during the growth phase of the particle and the affinity of the donor atom to a specific or selective crystal face facilitate control over size, shape and distribution of the particles.

The passivating agents used in the reaction should have sufficiently high boiling point and better chemical, thermal stability under the reaction conditions. They should be cheap and easily available. They should be less or non-toxic and should not increase the toxicity of the nanoparticles once they bind to their surface.

15.2.2 Single Source Molecular Precursor (SSMP) Method

Single Source Molecular Precursor (SSMP) is defined as a single molecule containing all the desired elements required for the formation of final requisite material either through a synthetic approach or a deposition method. Ideally, the structure of the precursor is designed in such a way that core resembles that of the desired material as much as possible.

15.2.2.1 Criteria for Selection of SSMP

Criteria for selecting a SSMP are (i) the precursor molecule should be synthesized easily in high yields with long shelf life, (ii) the molecule should be readily available at lower cost with good quality and quantity, (iii) the molecule should be air-stable and non-toxic, (iv) the precursor should have good solubility in common organic solvent so that it is easy to purify the precursor, (v) the molecule should decompose cleanly and controllably, (vi) should decompose at moderately lower temperatures to yield volatile and non-toxic by-products, (vii) should yield stable by-products which can be removed or separated easily from the required product.

15.2.2.2 Design and Synthesis of Single Source Molecular Precursor

To meet these criteria, various types of ligands have been designed, synthesized and utilized for the synthesis of SSMP. For instance, sterically hindered ligands [14] have been used for the preparation of SSMP with an intention to reduce the molecular aggregation leading to the formation of monomeric complexes which can be obtained in the pure form and to increase the volatility of the precursor. The usage of bulky ligands however resulted in the incorporation of more carbon impurity into required materials necessitated a search for alternate ligands which include saturation of coordination sphere of metal centre with neutral donor ligands, use of internally functionalized ligands [15] and labile carbon-chalcogen bonds as in chalcogenocarboxylic acids [16].

15.2.2.3 Reaction Pathways for the Decomposition of SSMP

Single source molecular precursor in general consists of M-E (M=metal or E=donor atom) linkage which is required in the final material as the core of the molecule, with different ligands surrounding the metal atom through the coordination of donor atom to metal [17]. Typically, the reaction pathway involves decomposition of the precursor without breaking the M-E linkage but with the removal of ligands, resulting in the final desired material. However, sometimes SSMP may not have M-E bond but still afford M-E linkage in the final product as in the case of some metal adducts with organochalcogen ligands which decomposes to the material with M-E bond in the final product.

The use of SSMP for the preparation of material can be categorized into three different cases with respect to the reaction control over the final products [18]. In the first case, the required elements and their stoichiometric ratio in the SSMP matches with the ratio essential in the final target material while there is no control over the side products of the reaction.

In case of $[R'_mM(ER)_n]$ (R′ and ER are ligands) as a SSMP for the preparation of binary oxide/chalcogenide material where one of the component is a metal and other is a chalcogen (E=O, S, Se or Te), a reaction pathway can be illustrated as shown in Eq. 15.1. Here, the stoichiometric ratio of M and E (1:n) fixed in the SSMP is same as which is required in the material M-E_n and the side products are R and R′.

$$[R'_mM(ER)_n] \rightarrow ME_n + nR + mR' \tag{15.1}$$

In case of ternary system like binary metal oxide, where there are two metal components (M and M′) and two ligands (OR, L), the ligands may be freed as it is or in a modified way as given below in Eq. 15.2. Both the ligands may split further.

$$MM'(OR)_n(L)_m \rightarrow MM'O_n + nR + mL \quad (15.2)$$

In the second scenario, not only the elemental composition of the SSMP have the excess (Eq. 15.4) or required (Eq. 15.3) stoichiometry but also the ligands are designed or selected in such a way that the decomposition of the precursor happens at relatively low temperatures in order to minimize the contamination of the final product with the unwanted side products. In both the cases, the volatile by-product is thermodynamically stable which will be easily removed from the final product. The moderate temperature also helps in obtaining the smaller particles owing to slow down of particle growth. In case of $[M(ER)_n]$ as SSMP for the preparation of binary material, following reactions may be illustrated the path.

$$M(ER)_n \rightarrow ME_n + R_n \quad (15.3)$$

$$M(ER)_n \rightarrow ME + R_nE_{n-1} \quad (15.4)$$

In case of binary metal oxide, where there are two metal components (M and M′) and two ligands (OR, L), the volatile or gaseous ligands are released without any further decomposition.

$$MM'_n(OR)L \rightarrow MM'_nO + R - L \quad (15.5)$$

For example, in Eq. 15.6, volatile H_2 and isobutylene are evolved leaving behind pure product [8].

$$Mg\{[(HC)_3C - O]_2AlH_2\}_2 \rightarrow 4H_2 + 4H_2C = C(CH_3)_2 + MgAl_2O_4 \quad (15.6)$$

In the third situation, the SSMP disintegrates into multi-phase materials with perfect stoichiometry and structure [18]. In such type of cases, multiple phases formed have molecular level interpenetration. Being originated from the same molecule, distribution and the grain sizes of the phases can be easily controlled by parameters such as temperature and pressure (Eq. 15.7).

$$2R'_2M(ER)_2 \rightarrow ME_2 + M + 2R_2E + 2R'_2 \quad (15.7)$$

In case of ternary materials like binary metal oxides, following reaction (Eq. 15.8) represents third case.

$$MM'_n(OR)_m \rightarrow MM'_{n/2}O_{m/2} + {}^{n}\!/_{2}M' + {}^{m}\!/_{2}HOR + {}^{m}\!/_{2}(R - H) \quad (15.8)$$

For example, in the following reaction (Eq. 15.9) tin(II) disproportionate into tin (IV) and tin(0) with the loss of *iso*-butylene and *tert*-butanol [19].

$$BaSn_2[OC(CH_3)_3]_6 \rightarrow 3HOC(CH_3)_3 + 3H_2C = C(CH_3)_2 + Sn + BaSnO_3 \quad (15.9)$$

In addition to the above case, reductive elimination of metal can occur in heavy metal complexes leaving behind the heavy metal (M) and ligand (RE) as in Eq. 15.10.

$$M(ER)_n \rightarrow M + (RE)_n \quad (15.10)$$

15.3 Classification of Molecular Precursor Method Based on Mode of Synthesis

15.3.1 Hot Injection and Heat-Up Method

Both the single or multiple source precursor method, further may be categorized as hot-injection [11, 20] or heat-up [21] depending on the heating procedure. Hot-injection is a versatile synthetic route for the preparation of various highly uniform nanocrystals with tunable size, shape, and surface passivation having general applicability. In hot-injection method, a stock solution containing reactive precursors (metal and non-metal sources in case of MSMP method and a single-source molecular precursor in SSMP method) is rapidly injected into a pre-heated high boiling coordinating solvent inducing a high level of supersaturation in the reaction mixture resulting in immediate formation of nanocrystals. In general, the reaction temperatures are near 200 °C or higher [22]. Reaction parameters such as temperature and time can be tuned for steady growth of the particles formed. Hot injection method uses non-ionic precursors in high-boiling coordinating solvents which allow relatively slow growth of nanoparticles at high temperature resulting defect-free nanocrystals. The former approach also lets the separation of the nucleation and growth phases which helps in achieving monodispersed nanocrystals without any size-selection process.

Besides hot-injection, heat-up method [21] is one of the important approaches which yield highly uniform nanocrystals with narrow size distribution. In this method, the reaction mixture prepared at a lower temperature is heated gradually to the reaction temperature at which nanocrystals are formed. The difference between hot injection and heat-up methods is that both the methods have different reaction pathways such as thermal treatment given to the reaction mixture for obtaining nanoparticles. In heat-up method, the reaction of the precursors takes place slowly during the heating process in contrast to the hot-injection method.

15.3.2 Hot-Injection Mechanism

The classical model for the formation of uniform microparticles affords a useful basis for understanding of nanocrystal formation. The basic concepts of the former model such as burst nucleation, diffusion controlled growth, and Ostwald ripening, can be extended to explain the mechanism of nanocrystal formation.

In general, the formation of nanocrystals is comprised of two steps, viz. nucleation and growth phase. The graphical and pictorial representations of the same have been shown in Figs. 15.1 and 15.2 [23, 24] (where C_m, C_{min} and C_{max} are concentration of monomer, minimum concentration of monomer for nucleation and maximum concentration of monomer for supersaturation, respectively). Nanoparticle synthesis by thermolysis of precursors is a type of precipitation reaction. Precipitation involves the formation of sparingly soluble species under supersaturation condition which dictate nucleation followed by growth processes that affect the size, morphology, and properties of the products. The required supersaturation conditions can be achieved by altering varying parameters related to solubility, temperature and concentration. For instance, supersaturation can be attained by dissolving the solute at higher temperature followed by cooling to low temperatures or by adding the excess reactant during the reaction.

In general there are three types of nucleation processes, viz. homogeneous, heterogeneous and secondary nucleation processes. Homogeneous nucleation is a process in which crystallization is induced with the nuclei generated within the

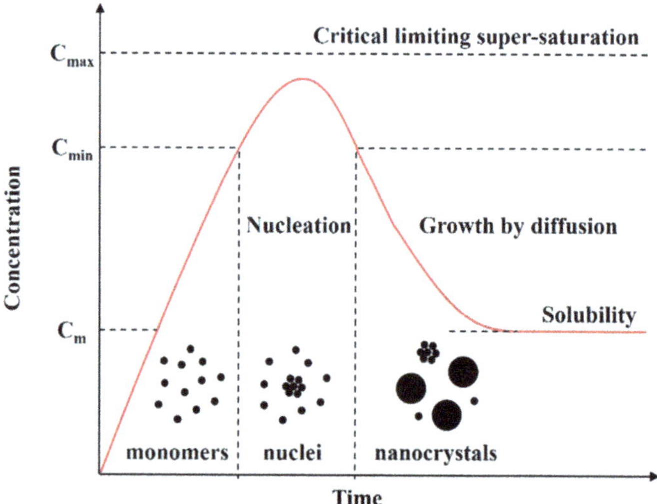

Fig. 15.1 Schematic illustration of the nucleation and growth process of nanocrystals in solution: precursors are initially dissolved in solvents to form monomers, followed by the generation of nuclei and the growth of nanocrystals via the aggregation of nuclei (Reproduced with the permission from American Chemical Society publishers [23])

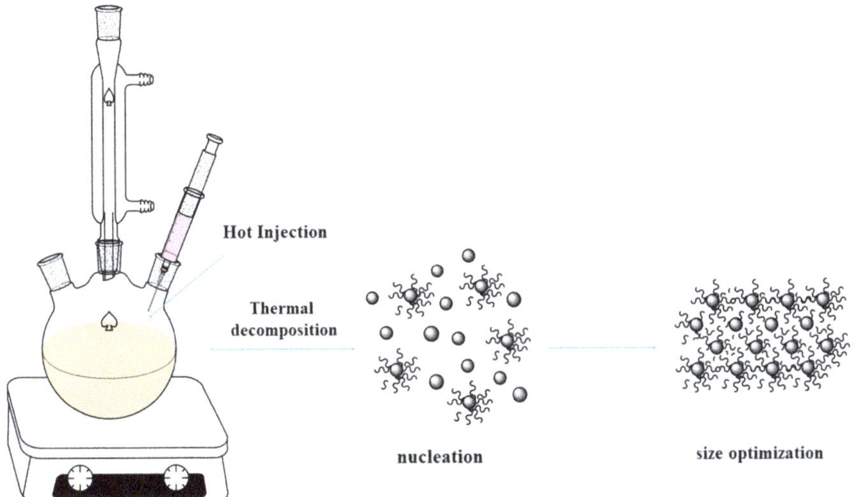

Fig. 15.2 Schematic illustration of hot-injection mechanism involving injection of precursors into surface passivating agents, nucleation followed by reaction cooling leading to the growth phase. (Reproduced with the permission from ACS publishers [24])

solution. In the former type of process, the system consists of a single liquid phase initially and the nucleation takes place spontaneously [25]. The former reaction has very high energy barrier and requires high supersaturation level for the initiation of nucleation process in the solution. However, the supersaturation level is relaxed after the formation of nuclei followed by their growth. As a result, the nucleation process ceases once the concentration is below the critical level. This type of nucleation occurs only for a short time while the supersaturation level remains very high. The thermodynamical driving force for such type of nucleation is the relieving of excess free energy of the monomers in the supersaturated solution.

According to the classical theory of nucleation, the overall free energy of a spherical nucleus ($\Delta G_T(r)$) is the sum of the new energy due to the formation of a new volume, i.e., bulk contribution (($\Delta G_B(r)$) and the free energy due to the new surface created, i.e., surface contribution (($\Delta G_S(r)$) which is expressed as follow (Eq. 15.11) [21].

$$\Delta G_T(r) = \Delta G_B(r) + \Delta G_S(r) \tag{15.11}$$

$$\Delta G_T(r) = \frac{4}{3}\pi r^3 \Delta G_V + 4\pi r^2 \gamma \tag{15.12}$$

where γ is the surface free energy per unit area and ΔG_V is the free energy per unit volume of a crystal which is equal to $-\frac{RT \ln S}{V_m}$. In the above equation, until $S > 1$, the molar volume of the monomer in the crystal (V_m) drives the formation of crystals to ease the excess free energy of the monomers in the supersaturated solution. However, the extra free energy required for the formation of nuclei due to the surface term causes the nuclei smaller than critical radius (r_c) to dissolve back into the solution due to decrease in the free energy while nuclei larger than r_c will grow to form stable particles because of decreases free energy for growth.

The critical radius (r_c) can be obtained by making dG/dr = 0

$$r_c = \frac{2\gamma V_m}{RT \ln S} \tag{15.13}$$

The reaction rate for the formation of nuclei in the Arrhenius form and the formation energy of nuclei, ΔG_N, is equal to $\Delta G(r_c)$.

$$\frac{dN}{dt} = Aexp\left[\frac{-\Delta G_N}{k_B T}\right] = Aexp\left[\frac{-16\pi\gamma^3 V_m^2}{3k_B^3 T^3 N_A^2 (\ln S)^2}\right] \tag{15.14}$$

where N, A, k_B, N_A and T are the number of nuclei, the pre-exponential factor, the Boltzmann constant, Avogadro's number and temperature, respectively.

The growth of spherical crystal particles in the solution takes place in two steps, viz. the transport of the monomers from the solution onto the crystal surface followed by the reaction of the monomers on the surface [22]. Applying Fick's law of diffusion to these two processes, the total flux of the monomer onto the surface of the particle (Eq. 15.15) and the monomer consumption rate by the growth of the particles (Eq. 15.16) can be expressed as:

$$J_I = 4\pi Dr([M]_b - [M]_s) \tag{15.15}$$

$$J_{II} = 4\pi r^2 k([M]_s - [M]_r) \tag{15.16}$$

where D is the diffusion rate and k is the reaction constant, $[M_s]$, $[M_b]$ are monomer concentrations at the surface and bulk, respectively, and $[M_r]$ is the concentration of the spherical particle of radius r. For ensuring mass balance, J_I should be equal to J_{II}. Therefore, equating Eqs. 15.15 and 15.16, $[M_s]$ can be expressed as:

$$[M]_s = \frac{D[M]_b + k[M]_r}{D + kr} \tag{15.17}$$

The monomer consumption is related to rate of the particle volume change as follows:

$$J = \frac{4\pi r^2}{V_m}\frac{dr}{dt} \tag{15.18}$$

By substituting $[M]_s$ from Eq. 15.17 into Eq. 15.15 and using Eq. 15.18, growth rate for particle can be expressed as:

$$\frac{dr}{dt} = \frac{DV_m\left([M]_b - [M]_r\right)}{r - D/k} \tag{15.19}$$

According to Gibbs–Thomson relation, a spherical particle with radius r in nanometre scale has extra chemical potential, $\Delta\mu$ equal to $2\gamma V_m/r$ where γ is the interfacial tension. The relationship between monomer concentration of particle size r ($[M_r]$) and crystal size is established by the Gibbs–Thomson equation which can be expressed as:

$$[M]_r = [M]_\circ \exp\left(\frac{2\gamma V_m}{rRT}\right) \tag{15.20}$$

$$[M]_r \cong [M]_\infty\left(1 + \frac{2\gamma V_m}{rRT}\right) \tag{15.21}$$

where $[M]_\infty$ is a constant.

By substituting Eq. 15.20 into Eq. 15.19, the growth rate equation for spherical particles is given as follow:

$$\frac{dr^*}{d\tau} = \frac{S - \exp\left(\frac{1}{r^*}\right)}{r^* + K} \tag{15.22}$$

where

$$r^* = \frac{RT}{2\gamma V_m} r \tag{15.23}$$

$$\tau = \frac{R^2T^2D[M]_\circ}{4\gamma^2 V_m} t \tag{15.24}$$

$$K = \frac{RT}{2\gamma V_m}\frac{D}{k} \tag{15.25}$$

$2\gamma V_m/RT$ and K are capillary length and Damköhler number (Da), respectively. The former is a measure of size effect on the chemical potential of a particle and Da shows whether the diffusion rate (D) or reaction rate (k) govern the growth reaction [26].

The growth process can be either diffusion-limited or reaction-limited. When $K/r^* < < 1$, the growth rate equation is written as

$$\frac{dr}{dt} = \frac{DV_m}{r}\left([M]_b - [M]_r\right) \tag{15.26}$$

This type of growth is called diffusion controlled growth which is also experimentally verified in precipitation reactions.

Both the hot-injection and heat-up methods follow different reaction pathways, such as thermal decomposition reactions, nonhydrolytic sol–gel reactions, and reductions for producing uniform nanocrystals of metals to metal chalcogenides. This indicates that both the methods have common mechanism independent of the specific reaction conditions.

15.3.3 Molecular Approach

A complementary to the classical nucleation theory is the molecular approach, which considers nucleation and growth of nanoparticles as the reactions between the monomers/precursors and their clusters [27–31] which has been proven both experimentally and numerical simulations. The reaction for the particle formation can be given as below:

$$M + M_n \rightleftarrows M_{n+1} \tag{15.27}$$

where M and M_n are a monomer and a cluster of n monomers, respectively, and k_f and k_d are the formation and dissolution reaction rate constants for cluster M_n, respectively. At equilibrium, the concentration of cluster M_n in terms of the cluster formation energy, Δ G(M_n) can be given as below.

$$[M_n] = [M]exp\left[\frac{-\Delta G(M_n)}{k_B T}\right] \tag{15.28}$$

Let M_c be a cluster of critical size with critical radius r_c, then, the nucleation rate is equal to the formation rate of M_c. For this reaction, M + $M_{c-1} \rightarrow M_c$, the reaction rate is proportional to [M][M_{c-1}].

From this relation, we obtain the nucleation-rate equation as [27]

$$\frac{dN}{dt} = K[M]_o^2 Sexp\left[\frac{-\Delta G(M_{c-1})}{k_B T}\right] \tag{15.29}$$

The above equation derived from the molecular approach is comparable to that derived from the classical nucleation theory. The only exception is that the exponential term is multiplied by the supersaturation level(S) which indicates that irrespective of the argument, starting from the same molecular model, similar results can be obtained [28, 29]. However, S factor does not lead to a substantial deviation from the classical nucleation theory as the change of the exponential term $\exp[-G/k_BT]$ has significantly larger effect on the dependence of the nucleation rate on the supersaturation level compared to multiplication by S.

15.3.4 Heat-Up Mechanism

Mechanism for the formation of nanoparticles using molecular precursors in a heat-up method has been explained using a simulation model [22]. The model is based on experimental evidence that the formation of nanoparticles utilizing molecular precursors is a two-step process which is observed in a synthesis of iron oxide nanocrystal formation from the iron–oleate complex (Fig. 15.3) [32]. The first step involves the thermal decomposition of molecular precursor generating an intermediate species which act as the monomer for the formation of nanocrystals in the second step. The simulation model considers the production of nanoparticles from a molecular precursor in three stages. In the first step, the monomer M is formed from the thermal decomposition of molecular precursor, P (Eq. 15.30). The second step is the nucleation process where 'n' number of monomers generates a nanocrystal N_n (Eq. 15.31). In third stage, the nanocrystal N_n grows by acquiring more number of monomers from the solution (Eq. 15.32)

$$P \rightarrow M \tag{15.30}$$

$$nM \rightarrow N_n \tag{15.31}$$

$$N_n + mM \rightarrow N_{n+m} \tag{15.32}$$

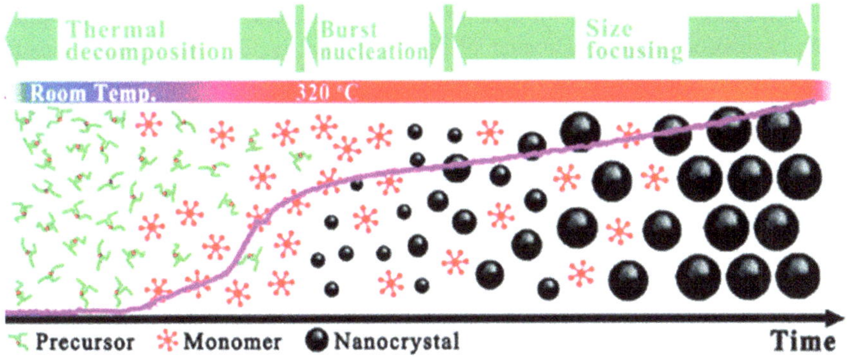

Fig. 15.3 Mechanism of the formation of iron oxide nanocrystals obtained from the solution-phase thermal decomposition of iron − oleate complex via the "heating-up" process (Reproduced with the permission from American Chemical Society publishers [32])

The rate of homogeneous nucleation and the growth rate of the nanocrystals in steps 2 and 3 can be calculated using Eq. 15.14 and Eqs. 15.22–15.25, respectively. This model is an extension of the previous hot-injection model with different approach for supplying monomer. In hot-injection method, the precursors are injected externally at the beginning of the synthesis which either convert swiftly into the monomers or themselves act as the monomers. To describe hot-injection method in the present model, the first step can be viewed as an instantaneous process or considering both the precursor and monomer to be identical. In contrast to hot injection method, the monomers in heat-up method are produced from the precursors during heating.

15.4 Preparation of Advanced Materials Through Molecular Precursor Method

The molecular precursor route both in hot injection and heat-up methods has been adopted for the preparation of unitary (metals), binary (metal oxides, metal chalcogenides), ternary and quaternary materials. A clear distinction between MSMP and SSMP methods appears in the preparation of bimetallic, binary-, ternary- metal chalcogenides and ternary metal oxides through molecular precursor route. This is because; synthesis of unitary metal- and metal oxide-nanostructures through molecular precursor approach involves only one molecular precursor which can only be categorized under SSMPs. Therefore, a discussion of both MSMPs and SSMPs included in both bimetallic and metal chalcogenide nanostructures.

15.4.1 Preparation of Metal Nanoparticles

Solution-based approaches for synthesis of metal nanoparticles and their applications have been subject of matter of many reviews [9, 33]. Way back from the pioneering work of Faraday in 1857, the reduction of metal salts in the presence of stabilizing agents to generate zerovalent metal colloids in aqueous or non-aqueous medium is one of the most traditional approaches for synthesis of metal nanoparticles. The reduction of metal salts to the zerovalent metallic form is an effective and a single step strategy to produce a variety of metal nanostructures. A typical reduction reaction of a metal cation (Eq. 15.33), is given below:

$$M^{n+} + \mathrm{ne}^{-} \rightarrow M^{o} \tag{15.33}$$

The route has been adopted for the synthesis of noble and group 11 metals such as Cu, Ag and Au. Other than this traditional approach, a molecular precursor route involving either reduction or thermal decomposition or a combination of both of them or a reductive elimination (in case of heavy metals) of organometallic/

metalloorganic precursors is an attractive route for the preparation of metal nanoparticles. Metal nanoparticles have been prepared by thermolysis of labile organometallic compounds either by hot-injection or by heat-up method. Although this route is little expensive due to the involved synthesis of organometallic precursors, the advantages associated with the route make it promising for the somewhat large-scale synthesis. The particle size, its distribution and surface states are controlled in a better way than that are synthesized by reduction of metal salts.

Synthesis of metal nanoparticles by the decomposition of metalloorganic precursors as metal source was first introduced by Chaudret in 1990s. The decomposition can be achieved by mild conditions of solution chemistry. The decomposition step involves either reduction or ligand displacement from the metal coordination sphere in the presence of a stabilizer. However, the ease of decomposition depends on the strength of M-C in organometallic complexes and M-X (where X is a main group element such as N, P or O) in metalloorganic complexes [34]. The strength and reactivity of the bond in turn depend on the oxidation state of the metal M. Initial synthetic method for the synthesis of metal nanomaterials using molecular precursors involves thermal decomposition of zero valent metal carbonyl complexes. Carbonyl complexes in general eliminate CO upon heating, sonication or photolysis. This strategy has been adopted by Sulick et. al. [35, 36] and Sun and Alivisatos [37] for the preparation of iron and cobalt nanoparticles using $[Fe(CO)_5]$ and $[Co_2(CO)_8]$, respectively. However, CO being a π acceptor and a σ donor ligand, it is difficult to remove of all CO ligands present in the complex as it requires high temperature or ultrasonication. Adoption of such conditions leads to an uncontrolled growth of nanoparticles. To circumvent these problems, hydrogenation of olefin complexes has been adopted which react readily at room temperature with H_2 resulting in the formation of metal nanoparticles which are suitable capped by stabilizers. The main advantages of this procedure are synthesis of monodispersed nanoparticles due to better control over nucleation step achieved under mild conditions and tuning of kinetics of the nucleation step by changing temperature and the pressure of the reducing gas. For instance, [Ru(COD)(COT)] (COD = 1,5-cyclooctadiene; COT = 1,3,5-cyclooctatriene), is a suitable precursor for the preparation of ruthenium nanoparticles where one of the double bonds of the COD ligand can be easily displaced by H_2 leading to a rapid reaction resulting in nascent Ru(0) nanoparticles [38] which are immediately capped by stabilizers like alcohols. Chaudret and co-workers utilized $[Co(\eta^3\text{-}C_8H_{13})(\eta^4\text{-}C_8H_{12})]$ [39], $[Ni(COD)_2]$ [40], [Ru(COD)(COT)] [38] $[Rh(C_3H_5)_3]$ [41], $[Pd_2(dba)_3]$ [35], $[Pt(dba)_x]$[42] (dba=dibenzylideneacetone) for the synthesis of respective metal nanoparticles. The olefin complexes are rather limited to only a few metal nanoparticles. For example, bis(cyclooctatetraene)iron (0) complex $[Fe(C_8H_8)_2]$ is the only suitable precursor for the synthesis of Fe nanoparticles which were obtained by the decomposition of the former under H_2 at 90 °C [43]. Other than olefin complexes, homoleptic metal alkyl/aryl derivatives containing M-C σ bonds have also been used for the preparation of coinage metal nanoparticles. For instance, mesityl derivatives of copper, $[Cu(C_9H_{11})]$ has been used for the synthesis of copper nanoparticles [44] while $[Pt(CH_3)_2(COD)]$ containing both M-C σ and olefin linkage has been utilized for growing truncated cubes or arrows [45]. Allyl complexes having the formal oxidation state higher than that of olefinic complexes, $[Rh(\eta^3\text{-}C_3H_5)_3]$ [46] and $[Co(\eta^3\text{-}C_8H_{13})(\eta^4\text{-}C_8H_{12})]$ [47] where allyl ligand is easily displaced by H_2 to yield respective metal nanoparticles stabilized by polyvinylpyrrolidone (PVP). For copper and other main group elements like indium, precursors such as $[CpCu(^tBuNC)]$ [48] and $[In(\eta^5\text{-}C_5H_5)]$ [49] have been used for the preparation of copper and indium nanoparticles in the presence of PPh_3 and PVP, respectively. In addition to organometallic complexes, metalloorganic complexes (MOCs) containing amide ligands have also been used for the preparation of metal nanoparticles. These MOCs readily react with H2 even at room temperature and decompose to respective metal nanoparticles along with amines. The amines released during the reaction in turn weakly cap the surface of the particles and control their shape. This strategy has been used to synthesize transition and main group metal nanoparticles by homoleptic amide derivatives of these metals. For example, $\{Fe[N(SiMe_3)_2]_2\}_2$ [50], $\{Co[N(SiMe_3)_2]_2\}$ [51], $\{Cu[N(SiMe_3)_2]_2\}$ [52] have been employed for the synthesis of monodispersed metal nanoparticles. Amidinate derivatives are another class of precursors which react with dihydrogen to undergo decomposition accompanied by hydrogenolysis of amidinate to amine to yield metal nanoparticles [52, 53].

15.4.2 Preparation of Bimetallic Nanostructures

Traditionally, bimetallic alloy nanostructures can be prepared by chemical methods such as chemical reduction or physical process such as microwave synthesis [54, 55]. For instance, bimetallic nanostructures have been synthesized by reduction or disproportionation of inorganic compounds in liquid polyols [56]. Bimetallic nanostructures using molecular precursor method have been achieved by thermal decomposition of two molecular precursors together in one pot. The main advantage of this method is to synthesize metal alloys, whose reduction potentials are different. The chemical composition of bimetallic nanostructures can be adjusted by taking suitable ratio of precursors being used in the synthesis. However, the different reactivities of the molecular precursors lead tonon-homogeneous chemical distribution of the metals in bimetallic nanostructures due to different kinetic decomposition. For instance, the co-decomposition of [Ru(COD)(COT)] and $[Pt(CH_3)_2(COD)]$ in the presence of dppb (1,4-bis(diphenylphosphinobutane)) gives nanoparticles of Ru rich core and a disordered shell of both Ru and Pt, instead of an alloy or a core–shell structure [57]. In contrast, homogeneous PtRu nanoparticles have been prepared by hydrogenation of a mixture of [Ru(COD)(COT)] and Pt_2dba_3(dba=dibenzylideneacetone) in the presence of PVP (polyvinylpyrrolidone) [58]. For the first time, homogeneous FePt nanocrystals were prepared by reducing platinum acetylacetonate with a long chain diol and decomposing $[Fe(CO)_5]$ in the presence of oleic acid and a long chain amine [59].

This successful synthesis of FePt nanocrystals using a combination of reduction and thermal decomposition steps has been adopted for the synthesis of CoPt, FePd, $CoPt_3$ alloy nanoparticles [60, 61].

15.4.3 Preparation of Metal Oxide Nanostructures

Metal oxides have distinctive functionalities that are not known in other solid materials. Nanostructures of the former form an important category of functional materials exhibiting unique properties due to their size and a high density of corner or edge surface sites with potential applications in gas sensors [62], photovoltaics [63], fuel cells [64], lithium-ion batteries [65], hydrogen production by water splitting and its storage [66, 67], water and air purification [68, 69]. In all of these arenas, metal oxide nanostructures are playing a critical role by either improving the efficiency of the storage devices or conversion processes or by ameliorating the design and performance of device. Therefore, the synthesis and application of metal oxide nanostructures with uniform size and morphology have been a subject of matter in nanoscience and nanotechnology.

Over the years, either a number of synthetic approaches have emerged or adapted for the synthesis of metal oxide nanostructures which may be broadly categorized as aqueous or non-aqueous methods. Both the methods adopt either metal salts or molecular precursors containing M–O/M-C/M–N bonds as the starting materials for the production of the metal oxides. In this section, the focus will be on methods using molecular precursors for the synthesis of metal oxide nanostructures. In general, transition metals like Ti, V, Mn, Fe, Co, Ni, Y, Mo, W, Cu, Zn; the f-block elements such as Ce, Nd, Sm and Gd) and main group elements (e.g. In, Sn, Pb and Bi) form solid-state alkoxides. Of these, transition metals with partially-filled d-/f-orbitals containing extra electrons easily coordinate with hydroxyl groups of alcohols to form alkoxides. In contrast, Group I and II metal alkoxides are difficult to prepare due to poor coordination ability while alkoxides of noble metals are not easy to synthesize as these metal ions have tendency to get reduced back to respective metal particles.

Aqueous methods using molecular precursors for metal oxide synthesis either involves the hydrolysis of metal alkoxides followed by condensation [70] or the reaction of M-C/M–N bonds of reactive molecular precursors such as alkyl or amido complexes with water in an exothermic fashion which induce the formation of metal oxide nanostructures at room temperature through the condensation of the transient hydroxides. For instance, luminescent ZnO nanoparticles have been synthesized at room temperature by the oxidation of $[Zn(c\text{-}C_6H_{11})_2]$ in the presence of long chain amine ligands [71]. The size in the range of 3–10 nm and shape in the form of nanodiscs and nanorods have been controlled varying the nature and quantity of additional ligands. The reactive complex, $[Fe(N(SiMe_3)_2)_2]$ has been used for the production of iron oxide nanoparticles soluble in organic solvents [72]. However, the rapid hydrolysis and condensation steps involved in aqueous methods lead to the loss of control over the size and morphology of the metal oxides, often results in amorphous product. Furthermore, the control over the composition and the homogeneity becomes difficult in case of multi component oxides due to the different reactivity of the metal alkoxides.

To overcome the drawbacks of aqueous methods, an effective and alternate method the form of non-aqueous solution has been evolved. The latter method offers crystalline materials at low temperatures with a better control over size and morphology of the product. The method also provides increased reaction variables such as temperature, time, and nature of surfactant. Furthermore, the non-aqueous method adapts the organic chemistry principles involving O-C bond for better designing and synthesis of materials. Non-aqueous method can be either surfactant- or solvent- controlled approach. The synthesis of metal oxides in both the approaches involves the injection of molecular precursors either in hot and high boiling surfactant (e.g. oleylamine (OLA), hexadecylamine (HDA), tri-*n*-octylphosphineoxide (TOPO), etc.) or organic solvents (e.g. ether, alcohols, etc.) providing the oxygen for the metal oxide. These surfactants and organic solvents control the size, shape and other functionalities of the material. Both the approaches have their advantages and limitations.

Non-aqueous method for metal oxide synthesis started way back in 1928 with the investigation on alkyl orthosilicates involving the reaction of silicon tetrachloride with alcohols by Dearing and Reid [73]; however, the method gained momentum only in the 1990s with a focus on metal oxide gels [74, 75] and powders for catalysis [76].

Non-aqueous method for metal oxide nanocrystals using thermal decomposition at high temperatures produces materials with striking properties due to altered defect structure and nature of the surface. Surfactant-controlled route involves the conversion of the molecular precursor into its corresponding metal oxide in the presence of a surfactant using a hot injection method. In the latter method, the precursors are injected into hot surfactant maintained at a required temperature leading to the formation of target material. Surfactants are typically high boiling coordinating solvents with long alkyl chain which prevent agglomeration of nanoparticles through passivation of their surface.

A variety of transition metal oxide and main group metal oxide nanostructures have been prepared either by thermal or solvo-/hydro-thermal decomposition of metal alkoxides. Of them, most studied metal oxide nanostructures with respect to synthesis are ZnO [77, 78] and TiO_2 [79, 80] which have been reviewed in detail by many authors. For example, spherical ZnO nanoparticles have been prepared by thermolysis of zincethylhexanoate in diphenylether and amines [81]. Thermal treatment of zinc glycolates [82] and glycerolate [83] has also been used for the synthesis of various ZnO nanorods, octahedrons and hierarchical structures.

TiO_2 nanomaterials have potential application in various fields [63, 84]. Titanium alkoxides such as titanium glycolate [3] and titanium glycerolate [85] have been used as molecular precursors to prepare TiO_2 nanomaterials. A number of other transition metal oxides have been synthesized by adopting this approach. For instance, Rockenberger et al. [86] have demonstrated the preparation of *c*-

Fe_2O_3, Cu_2O and Mn_3O_4 nanocrystals by injecting the solutions of their respective metal cupferron complexes, M^xCup_x (M=Metal ion; Cup=*N*-nitrosophenylhydroxylamine, $C_6H_5N(NO)O^-$) in octylamine into long chain amines at 250–300 °C. Similarly, nanocrystals of CdO have been prepared by the decomposition of the metal cupferron complexes in tri-n-octylphosphineoxide (TOPO) under solvothermal conditions [87].

A number of relatively cheaper metal oleates have also been used for the synthesis of monodisperse metal oxide nanocrystals such as ZnO, Fe_3O_4, MnO, CoO and NiO nanocrystals using 1-octadecene (ODE), octyl ether, oleyl amine (OLA) and trioctylamine (TOA) have been used as solvents. Pyramidal ZnO nanocrystals have been obtained by the thermolysis of the Zn-oleate complex [88]. Ultra large-scale synthesis of monodispersed iron oxide nanostructures have been obtained by thermolysis of iron-oleate complex in ODE [89]. MnO multipods have been synthesized by thermal decomposition of a manganese oleate complex in oleic acid and OLA at 320 °C [90]. The preparation of NiO and CoO nanoparticles was reported. The dot-like NiO and flower-like CoO nanoparticles have been prepared via thermal decomposition of their respective metal oleates [91].

Main group metal oxides like SnO_2 spherical nanoparticles have also been prepared by thermolysis of tinethylhexanoate in diphenyl ether and amines in the range of 230–250 °C [81]. In addition to above molecular precursors, metal diketonates and reactive metal alkyls have also been used for the synthesis of metal oxide nanostructures using molecular precursor approach.

15.4.4 Preparation of Metal Chalcogenides Nanostructures

Metal chalcogenide structures are of considerable interest due to their applications in optoelectronics [92], IR- detectors [93], solar cells [94], thermoelectrics [95], lithium-ion batteries [96] and bio-imaging, etc. In view of this, synthesis of high quality metal chalcogenide nanostructures with uniform size and morphology is of considerable interest in miniaturization of devices. Although a number of methods have been reported for the synthesis of metal chalcogenide nanostructures, molecular precursor approach is one of the versatile methods for the preparation of high quality nanomaterials.

Synthesis of monodispersed nanocrystals through molecular precursor approach is pioneered by Bawendi et al. [11]. The method involves one pot synthesis of CdSe nanocrystals by hot injection of $(CH_3)_2Cd$ and tri-*n*-octylphosphine selenide (TOPSe) into hot tri-*n*-octylphosphine oxide (TOPO). This synthetic route is one of the most accepted non-aqueous chemical methods for monodispersed metal oxide, chalcogenide, phosphide nanocrystals, etc. The use of TOPSe and $(CH_3)_2Cd$ motivated by the investigations of Steigerwald et al. [97] where organometallic precursors were used for the preparation of CdSe clusters in an inverse micelles method. The method has been developed later by Alivisatos and co-workers [98]. A number of adaptions to Bawendis' method were reported in the literature. These modifications comprise the use of co-surfactants, ligand exchange after synthesis to alter the properties, inorganic passivating shells, alternate metal and chalcogen sources as precursors. For instance, co-surfactant like hexadecylamine (HDA) along with TOPO to increase the quantum yield of CdSe QDs [99] while CdTe QDs recapped with thiols showed improved quantum yield and solubility [100]. Further, these surfactants can also be removed completely and shell of another inorganic material can be grown uncapped nanocrystals to prepare core–shell nanocrystals with enhanced luminescence and quantum yields [101–103].

Unlike group II-VI materials, III-VI materials show great diversity and a variety of stoichiometries in their binary compounds. However, their polytypism, the variety of stoichiometries and limited availability of group III elements, especially indium are a problem as compared to the II-VI materials [104]. Synthesis of group III metal chalcogenides, involves reaction of group III metal salts, metalloorganic, metal complexes as metal sources and $(NH_4)_2E$, $(NH_2)_2CE$, TOPE, OLA-E (E=S, Se or Te), etc. as chalcogen sources in various coordinating- and non-coordinating solvents under various conditions.

Group IV-VI materials are narrow band-gap semiconductors and have attracted considerable interest due to their potential applications in photovoltaics [105], thermoelectrics [95] and infrared detectors [106], etc. Both hot injection and heat-up methods have been used for their synthesis using multiple source precursor method. Most of these methods use OLA as a passivating ligand, while group IV metal- oxide, -acetate, -nitrate, -oleates, bis[bis(trimethylsilyl)amino]M(II)(M=Sn, Ge or Pb) and $(TMS)_2E$ (E=S or Se), elemental sulphur/R_3PE (E=Se or Te) used as metal and chalcogen source for the synthesis of group IV metal chalcogenide nanomaterials. OLA is used as passivating ligand due to its strong affinity for group IV metals.

The intrinsic problems related to the use of toxic and volatile compounds in multiple source molecular precursor method at elevated temperatures underpin the development of SSMP method as alternative routes to synthesize quality nanoparticles. The mechanism for producing these nanoparticles involves the decomposition of the precursor which drives the formation of nanoparticles while growth of the particles stops when supply of the precursor is depleted. A number of thiolate [107], dithiolate (xanthate [108], dithiocarbamate [109], dithiocarboxylate [110], dithiophosphate [111]), thiobiuret [112], didimethylaminoalkylselenolate [113], thiosemicarbazide [114], diselenocarbamate [115], dichalcogenophosphinato [116], dichalcogenoimidodiphosphinato [117], pyridylchalcogenolate [118, 119], pyrimidylchalcogenolate [120], arylchalcogenolate [121], arylchalcogenocarboxylate [16] complexes of group- I, -II, -III, -IV, -V have been prepared using various ligands and have been used as SSMPs for the preparation of metal chalcogenides.

To mention a few, thermolysis of $[Cu\{SeC_5H_3(Me\text{-}3)N\}]_4$ in TOPO or HDA/TOPO affords copper selenide nanoparticles of different shape and size while $[Cu\{TeC_5H_3(R\text{-}3)N\}]_4$ (R=Me or H) decomposes in TOPO at 150 °C gives spherical copper telluride nanoparticles [122]. Highly luminescent CdSe quantum dots (Fig. 15.4) have been prepared by thermolysis of $[Cd(SeCH_2CH_2NMe_2)_2]$ in HDA/TOPO [113]. Both the absorption and emission maxima were red shifted with the

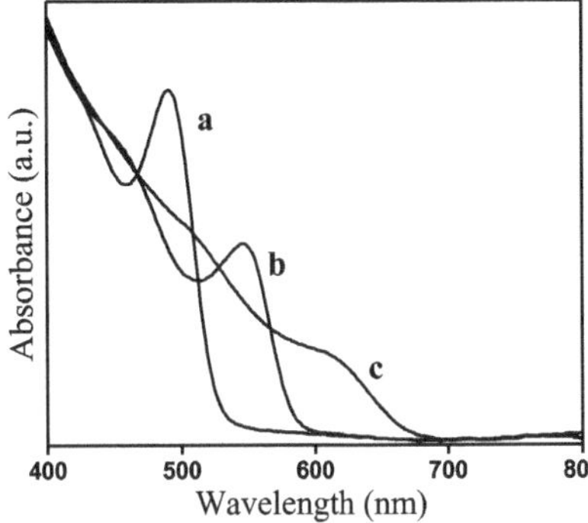

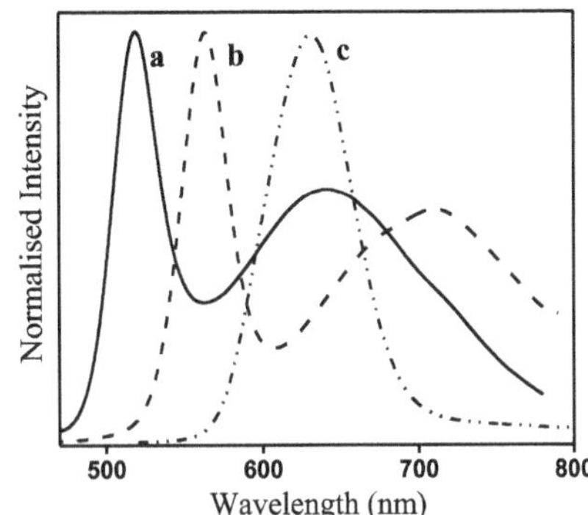

Fig. 15.4 Absorption and emission spectra of CdSe nanoparticles obtained by thermolysis of [Cd$(SeCH_2CH_2NMe_2)_2$] at 187 °C in HDA/TOPO, at **a** 4 **b** 6 **c** 30 min of preparation (Reproduced with the permission from Elsevier publishers [113])

Fig. 15.5 SEM images of SnSe obtained by thermolysis of [$Et_2Sn\{2\text{-}SeC_5H_3(Me\text{-}3)N\}_2$] in oleylamine at 215 °C (Reproduced with the permission from RSC publishers [123])

increasing reaction duration. Emission spectra for the smaller reaction durations (4 and 6 min) showed both the band edge and trapped state emission whereas the emission spectrum of quantum dots isolated at a reaction duration of 30 min displayed only a band edge emission indicating absence of trapped states. High quality SnSe hexagons (Fig. 15.5) with an average thickness of 80 nm have been synthesized by thermolysis of [$Et_2Sn\{2\text{-}SeC_5H_3(Me\text{-}3)N\}_2$] in OLA [123].

Other than synthesis of nanoparticles, SSMPs have also been used for the deposition of nanostructures in the form thin films using aerosol assisted chemical vapour deposition (AACVD) (Fig. 15.6). Such films deposited on suitable substrates with little processing can directly be used as devices. Recently, a number of metal chalcogenide nanostructured thin films have been deposited using SSMPs. For instance, [$Cu\{SeC_5H_3(3\text{-}Me)N\}]_4$ and [$Cu\{TeC_5H_3(Me\text{-}3)N\}]_4$ have been used for the deposition of Cu_5Se_4 and $Cu_{1.85}Te$, thin films respectively [122]. Similarly,

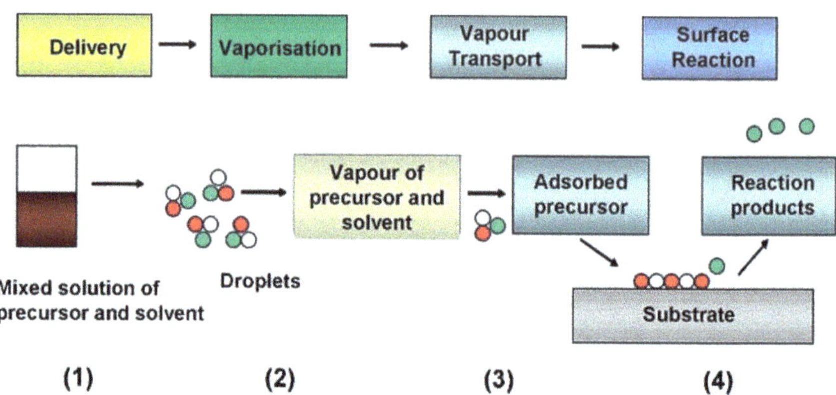

Fig. 15.6 Schematic representation of the AACVD process (Reproduced with the permission from RSC publishers [125])

thin films containing of Sb_2Se_3 or BiSe nano-wires and nano-ribbons have been deposited by AACVD employing [M{Se-C_5H_3(R-3)N}$_3$] (M=Sb or Bi) as SSMPs [124].

15.5 Merits and Demerits of Molecular Precursor Method

Over the years, a number of researchers are advocating the use of molecular precursors for the synthesis of nanocrystals and the deposition of thin films [10, 18]. The main advantage of the molecular method over other methods in thermolysis mode is the separation of the nucleation and growth in a controlled fashion. As a consequence, an excellent control over the crystallite size and shape of the nanocrystals is achieved resulting in monodispersed colloids. However, all the multiple source molecular precursors may not be suitable for the deposition of thin films for various reasons. This problem has been overcome by the advent of single-source molecular precursors (SSMPs) which can be used for both the synthesis of nanocrystals and the deposition of thin films using aerosol assisted chemical vapour deposition (AACVD) technique.

The merits of the single-source precursor approach are manifold. Most of the single-source precursors are also air-stable, less toxic and are therefore easy to handle, purify and characterize. The molecular level mixing and the predetermined stoichiometry of the elements in the precursors allow compositional purity or stoichiometric delivery of the materials at low crystallization temperatures. Designing of the SSMPs by proper selection of ligands are highly useful in achieving lower and cleaner decomposition of the precursors leading to good quality nanomaterials and thin films. The existence of preformed bonds in the single-source molecular precursors affords material with lower defects which are prerequisites for device fabrication. The use of SSPs reduced the environmental impact of material processing. Furthermore, in some cluster-based precursors, the phase of the produced material can be controlled by the core of the molecular precursor.

Other than synthesis of nanocrystals, molecular precursor route can also be used for thin film deposition by AACVD of SSMPs. AACVD through SSMPs has several advantages over conventional chemical vapour deposition (CVD). Traditional CVD requires volatile precursors with moderate vapour pressure for thin film deposition; however, the use of SSMPs by AACVD has helped to overcome this rigid condition. The temperature and flow conditions become simpler with the help of SSMPs using AACVD. SSMPs made it possible to deposit thin films with relatively simple deposition setup due to easy precursor delivery. SSMP allows good control of the composition up to the deposition zone and a simplification of the reactor design and processing conditions. Better homogeneity of the thin films is achieved due to molecular level premixing of the desired elements in SSMPs. Thus, proper designing of single-source precursors can afford a simple and clean route to advanced materials, reduce the need of multiple precursors which are toxic, highly volatile pyrophoric and expensive in nature. Such precursors are not only difficult to handle but also create lot of complications during the gas-phase reactions resulting in the formation of non-stoichiometric films [10, 17].

Despite numerous advantages of MSMPs and SSMPs, there are also several demerits of using these precursors [126]. For instance, volatile molecular precursors used for the synthesis of nanomaterials afford non-stoichiometric and impure materials as these precursors may escape from the reaction flask even before participating in the reaction. MSMPs employed for the deposition of thin films may decompose or loss their identity before reaching the substrate and leads to incorporation of impurities in the thin films. Designing single-source precursors with a moderate vapour pressure for thin film deposition by traditional CVD is a difficult task. Many SSMPs often have low vapour pressures leading to the contamination of the thin films due to incomplete decomposition of the precursors on the substrate. In such cases, increasing the substrate temperatures can result in stoichiometric deficient cleaner films which may have different properties compared to the films with compositional purity. However, techniques such as AACVD and spray pyrolysis for the deposition of thin films using SSMPs have overcome these issues. Compared to conventional CVD, the AACVD involves the production of aerosol droplets of precursors which will be transported with the help of inert carrier gases. AACVD can also be used for involatile or thermally unstable precursors making volatility is no longer a crucial condition for the selection of precursors for deposition of thin films.

A major drawback of the method is the relatively high cost of the molecular precursors as compared to the co precipitation method in the aqueous phase. However, using relatively cheap metal precursors such as metal oleates for metal oxide synthesis may solve the problem. The advantages of SSMP are lost if the thermal decomposition temperature is too high or else strength of the chemical bond between the core elements of single-source molecule are similar or weaker than those between the core and the organic ligands, leading to adopt a non-preferred decomposition path by the precursor. For instance, preparation of nitrides using tetrakis(dialky1-amido) complexes needs an extra NH_3 is added to attain the correct stoichiometry. Another important drawback of SSMP is the carbon incorporation in the films if the decomposition pathway of the precursor leads to any non-volatile by-products due to improper cleavage or due to the use of bulkier ligands. Such type of situation arises due to the improper design of precursors. Furthermore, it is difficult to deposit materials with specific non-stoichiometry.

15.6 Applications of Advanced Materials Prepared Through Molecular Precursor Method

Nanoparticles are unique in nature for two important reasons. The first reason being, increase in the surface to volume ratio with the decrease in size of nanoparticles. This results in the increase in the ratio of number of atoms on the surface compared to those in the interior of the particles. For instance, in quantum dots more than a third of all atoms reside on the surface. Electronic properties of nanoparticles, especially semimetal and semiconductor nanoparticles change with size and shape of the particles. For instance, band gap of the quantum dots increases with the decrease in the size and shape of the nanoparticles.

It is envisaged that these exceptional properties make these nanoparticles as building blocks in advanced electronic, medical and catalytic applications. The full potential of high quality (shape and phase pure monodispersed nanoparticles) nanoparticles prepared by molecular precursors route is yet to be recognized. These high quality nanoparticles have the potential impact on many future technologies spanning over disciplines like chemistry, biology, physics, material science, environmental science, medicine, and electrical engineering.

15.6.1 Optoelectronic and Optical Applications

The advances made in the preparation of size and shape tunable semiconductor nanoparticles using molecular precursor route led to a number of potential optoelectronics applications such as light-emitting diodes [127], low-threshold lasers [128], single-electron devices [129], and quantum computing. Especially, tunability of band gap of semiconductor nanoparticles by varying size and shape opened up applications in tunable diodes and lasers. The degree of control over the size and shape led to another interesting application in the form single electron transistors where a single nanoparticle is placed between two gold electrodes by dithiols [129]. These single-electron nanoparticle devices form the building blocks for emitters enabling quantum encryption.

In addition to optoelectronic applications, defects within the particle due to various reasons act as traps for the carriers which give rise to nonlinear optic effects [130]. Such type of defects may arise due to the large surface area of nanocrystals, the surrounding medium and nature of the capping agent. The nanoparticles also have potential applications in optical amplifiers for telecommunications [131].

15.6.2 Biological and Health Care Applications

The progress made in the preparation of highly luminescent, nearly monodispersed nanocrystals with good reproducibility and easy manipulation of organically passivated quantum dots opened up many applications in biology and biomedicine. They have been used as fluorescent probes for the optical tagging of chemical and biological entities [132] and luminescent probes in combinatorial chemistry and biological screening applications.

15.6.3 Catalysis and Chemical Sensors

The larger surface to volume ratio of the nanoparticles relative to bulk materials, along with size and shape-dependent tunability of band gap makes semiconductor nanoparticles better candidates for sensitizers and catalysts in photochemical reactions [133–135]. For instance, nano-aluminium is highly reactive compared to its bulk counterpart due to increased surface area. Therefore former is used as solid-fuel in rocket propulsion while the latter is used in utensils [136]. The size quantization also affects the redox levels of the conduction and valence bands as the charge carriers migrate to the surface of the particles for oxidation/reduction processes to occur. For example, ZnS nanoparticles have been used for the oxidation of alcohols and the reduction of CO_2 to formic acid [137].

15.6.4 Energy Conversion and Storage

The applications of monodispersed and phase pure colloidal nanoparticles prepared through molecular precursor route in photovoltaic devices are well documented in the literature [24, 138, 139]. Especially, metal oxide and metal chalcogenide nanoparticles have been used in various types of solar cells such as dye-sensitized [140], quantum dot-sensitized solar cells [141], heterojunction quantum dot solar cells [142] and quantum dot solar cells [143], etc.

Taking account of the progress made in several areas with such as faster rate, it may be expected that the impact of nanoscience and nanotechnology on our lives will rise remarkably in the near future.

15.7 Characterization Techniques

The quality of the materials to be used in various applications is very important as they are based on a particular property of the material which in turn is dependent on the synthetic method adopted for the preparation of these materials. Therefore, the proper characterization of a precursor and the materials with respect to their purity is of critical importance to realize the applications of the advanced materials obtained by these precursors. Especially, the purity of a required material generally depends on the purity and decomposition pathway of the molecular precursor under particular conditions. A number of techniques to characterize the materials have been discussed in various chapters of this book. Therefore, in the present section, some of the techniques for evaluating the purity of the molecular precursors and their decomposition pathway will be discussed.

15.7.1 Nuclear Magnetic Resonance (NMR)

Nuclear Magnetic Resonance (NMR) spectroscopy is an important analytical technique to determine the purity and molecular structure of compounds. In some cases, it can also be used for quantitative measurements. NMR can also be used to infer the basic structure of unknown compounds.

The basic principle of NMR involves two important sequential steps. The first step involves the splitting of nuclear spin energy levels of nuclear spin active nuclei in the presence of an external magnetic field owing to alignment of nuclear spins either in the same or opposite direction of the applied field. In the second step, the nuclei to be analysed are swapped with radio frequency, transition of spins in lower energy state to a higher energy state takes place through energy transfer. If the spins in higher energy state return to it lower energy state, energy/radiation is emitted in the radio frequency region. The signal corresponding to this energy transfer measured in number of ways and processed to yield an NMR spectrum for the nucleus to be analysed.

The resonant frequency of the energy transition is dependent on the effective magnetic field at the nucleus in the focus which is affected by electron shielding around the nuclei. The latter is in turn dependent on the chemical environment. The resonant frequency gives the information of chemical environment around the nucleus.

15.7.2 Single Crystal X-ray Diffraction (SCXD)

X-ray diffraction has various applications in the chemical, biochemical, physical and material sciences. Single crystal X-ray diffraction is a technique to determine the molecular structure of organometallic complexes, proteins, and polymers, etc. unambiguously under the proper conditions. The XRD technique is based on Bragg's law. Three important basic steps of single crystal X-ray crystallography are, (i) selection of a crystal with insignificant imperfections and optimum size (approximately larger than 0.1 mm in all directions), (ii) collection of reflections from various planes of the crystal in the presence of monochromatic X-rays and (iii) the collected data with the help of computational technique and chemical information is used to generate and refine a model of the arrangement of atoms within the crystal. The ultimately refined model called the crystal structure of a molecule is deposited with the Cambridge Crystallographic Data Centre (CCDC) which is included in the Cambridge Structural Database (CSD). The mean chemical bond lengths and angles can be determined using X-ray diffraction data. This technique is helpful in predicting tailor made material by assessing the core of the molecular structure. For instance, in a particular molecular structure when the bonds between the elements of the core must be stronger than those between the core and the organic ligands, clean removal of organic groups can be carried out by selective bond breaking. This leads to a tailor made material from the structural unit represented by the core. Such an approach has been adopted in growing a metastable cubic GaS phase using gallium chalcogenide cubane $[(^{t}Bu)GaS]_4$ [144].

15.7.3 *Thermogravimetry*

Thermogravimetric analysis (TGA) is a thermal analytical method which measures the amount of weight change of a material, either as a function of increasing temperature, or as a function of time at a constant temperature, under a flow of nitrogen, helium, air, other gas, or in vacuum. The technique is useful in determining the decomposition temperature, thermal behaviour and decomposition path of the molecular precursor. These details are helpful in the synthesis of nanomaterials and deposition of thin films using molecular precursors. The weight loss details along with XRD pattern of the residue obtained by the decomposition of molecular precursor can be used to determine the composition and phase of the material formed after the decomposition.

15.7.4 *Massspectrometry*

Mass spectrometry is one of the important tools to determine both the structure and decomposition of pathway of a molecule. This technique is especially useful for predicting the plausible molecular structure of a molecule whose single crystal is difficult to grow for single crystal X-ray diffraction analysis.

15.8 Conclusions and Future Prospective

The quest for novel advanced materials gathered momentum due to technological innovations in various fields. For obtaining these materials a number of synthetic methods have been explored. Among these methods, molecular precursor route has been both versatile as well as easily processing method for scaling up the synthesis of high quality materials in terms of size, shape and phase. Both multiple and single-source molecular precursors in hot injection and heat-up mode can be used for nanomaterial synthesis depending on the final product required. This chapter provides some of the basic and advanced information about molecular precursor method and its advantages over other methods. Although there are various advantages of molecular precursor method, the challenges associated with this method for scaling up from lab to commercial scale are being addressed with continuous development in fundamental and engineering level.

Future developments in this method of synthesis for advanced materials should be focussed on precursor chemistry, controlling the reactivities of reagents, heat treatment and management, reproducibility during scale-up. These advances will make sure that the synthetic procedures are not limited to batch methods but progress towards scale-independent processes.

References

1. Huynch W, Peng XG, Alivisatos AP (1999) CdSe nanocrystal rods/poly (3-hexylthiophene) composite photovoltaic devices. Adv Mater 11:923
2. Nirmal M, Brus L (1999) Luminescence photophysics in semiconductor nanocrystals. Acc Chem Res 32:407
3. Chen X, Mao SS (2007) Titanium dioxide nanomaterials: synthesis, properties, modifications, and applications. Chem Rev 107:2891
4. Chen WCW, Nie S (1998) Quantum dot bioconjugates for ultrasensitive nonisotopic detection. Science 281:2016
5. Pankhurst QA, Connolly J, Jones SK, Dobson J (2003) Applications of magnetic nanoparticles in biomedicine. J Phys D Appl Phys 36:R 167
6. Ahmadi TS (1996) Shape-controlled synthesis of colloidal platinum nanoparticles. Science 272:1924
7. Serpone N, Dondi D, Albini A (2007) Inorganic and organic UV filters: Their role and efficacy in sunscreens and suncare products. Inorg Chim Acta 360:794
8. Walter P, Welcomme E, Hallégot P, Zaluzec NJ, Deeb C, Castaing J, Veyssière P, Bréniaux R, Lévêque J-L, Tsoucaris G (2006) Early use of PbS nanotechnology for an ancient hair dyeing formula. Nano Lett 6:2215–2219
9. Cushing BL, Kolesnichenko VL, O'Connor CJ (2004) Recent advances in the liquid-phase syntheses of inorganic nanoparticles. Chem Rev 104:3893–3946
10. Malik MA, Afzaal M, O'Brien P (2010) Precursor chemistry for main group elements in semiconducting materials. Chem Rev 110:4417–4446
11. Murray CB, Norris DJ, Bawendi MG (1993) Synthesis and characterization of nearly monodisperse CdE (E = sulfur, selenium, tellurium) semiconductor nanocrystallites. J Am Chem Soc 115:8706

12. Sharma RK, Kedarnath G, Kushwah N, Pal MK, Wadawale A, Vishwanadh B, Paul B, Jain VK (2013) Indium(III) (3-methyl-2-pyridyl)selenolate: Synthesis, structure and its utility as a single source precursor for the preparation of In_2Se_3 nanocrystals and a dual source precursor with [Cu{SeC5H3(Me-3)N}]4 for the preparation of CuInSe2. J Organomet Chem 747:113–118
13. Lee DC, Pietryga JM, Robel I, Werder DJ, Schaller RD, Klimov VI (2009) Colloidal synthesis of infrared-emitting germanium nanocrystals. J Am Chem Soc 131:3436–3437
14. Bochmann M, Webb KJ, Hursthouse MB, Mazid M (1991) Three-coordinate thiolato complexes of zinc: solution and solid-state structures and EHMO analysis of the bonding pattern of $[Zn(S\text{-tert-}Bu_3C_6H_2\text{-}2,4,6)_2]_2$. J Chem Soc Dalton Trans 2317
15. Dey S, Jain VK, Chaudhury S, Knoedler A, Lissner F, Kaim W (2001) 2-(Dimethylamino) ethaneselenolates of palladium(II): synthesis, structure, spectroscopy and transformation into palladium selenide. J Chem Soc Dalton Trans 723
16. Kedarnath G, Kumbhare LB, Jain VK, Phadnis PP, Nethaji M (2006) Group 12 metal monoselenocarboxylates: synthesis, characterization, structure and their transformation to metal selenide (MSe; M = Zn, Cd, Hg) nanoparticles. Dalton Trans 2714–2718
17. Bloor L, Carmalt CJ, Pugh D (2011) Single-source precursors to gallium and indium oxide thin films. Coord. Chem. Revz. 255:1293
18. Veith M (2002) materials—a one step strategy. J Chem Soc Dalton Trans 2405–2412
19. Veith M, Kneip SJ (1994) New metal-ceramic composites grown by metalorganic chemical vapour deposition. J Mater Sci Lett 13:335
20. Donegá CDM, Liljeroth P, Vanmaekelbergh D (2005) Vanmaekelbergh, Physicochemical evaluation of the hot-injection method, a synthesis route for monodisperse nanocrystals. Small 1:1152–1162
21. van Embden J, Chesman ASR, Jasieniak JJ (2015) The heat-up synthesis of colloidal nanocrystals. Chem Mater 27:2246–2285
22. Kwon SG, Hyeon T (2011) Formation mechanisms of uniform nanocrystals via hot-injection and heat-up methods Formation mechanisms of uniform nanocrystals via hot-injection and heat-up methods. Small 7:2685–2702
23. LaMer VK, Dinegar RH (1950) Theory, production and mechanism of formation of monodispersed hydrosols. J Am Chem Soc 72:4847–4854
24. Carey GH, Abdelhady AL, Ning Z, Thon SM, Bakr OM, Sargent EH (2015) Colloidal quantum dot solar cells. Chem Rev 115:12732–12763
25. Burda C, Chen X, Narayanan R, El-Sayed MA (2005) Chemistry and properties of nanocrystals of different shapes. Chem Rev 105:1025–1102
26. Talapin DV, Rogach AL, Haase M, Weller H (2001) Evolution of an ensemble of nanoparticles in a colloidal solution: theoretical study. J Phys Chem B 105:12278
27. Kashchiev D, Rosmalen GMV (2003) Nucleation in solutions revisited. Cryst Res Technol 38:555
28. Privman V, Goia DV, Park J, Matijevi E (1999) Mechanism of formation of monodispersed colloids by aggregation of nanosize precursors. J Colloid Interf Sci 213:36
29. van Embden J, Sader JE, Davidson M, Mulvaney P (2009) Evolution of colloidal nanocrystals: theory and modeling of their nucleation and growth. J Phys Chem C 113:16342
30. Robb DT, Privman V (2008) Model of nanocrystal formation in solution by burst nucleation and diffusional growth. Langmuir 24:26
31. Laaksonen A, Talanquer V, Oxtoby DW (1995) Nucleation: measurements, theory, and atmospheric applications. Annu Rev Phys Chem 46:489
32. Kwon SG, Piao Y, Park J, Angappane S, Jo Y, Hwang N-M, Park J-G, Hyeon T (2007) Kinetics of monodisperse iron oxide nanocrystal formation by "heating-up" process. J Am Chem Soc 129:12571–12584
33. Qazi UY, Javaid R (2016) A review on metal nanostructures: preparation methods and their potential applications. Adv Nanopart 5:27–43

34. Amiens C, Chaudret B (2007) Organometallic synthesis of nanoparticles. Mod Phys Lett B 21:1133–1141
35. Bradley JS, Hill EW, Behal S, Klein C, Chaudret B, Duteil A (1992) Preparation and characterization of organosols of monodispersed nanoscale palladium. Particle size effects in the binding geometry of adsorbed carbon monoxide. Chem Mater 4:1234
36. Suslick KS, Choe SB, Cichowlas SBA, Grinstaff M (1991) Sonochemical synthesis of amorphous iron. Nature 353:414
37. Sun S, Murray CB (1999) Synthesis of monodisperse cobalt nanocrystals and their assembly into magnetic superlattices. J Appl Phys 85:4325
38. Duteil A, Quéau R, Chaudret B, Mazel R, Roucau C, Bradley JS (1993) Preparation of organic solutions or solid films of small particles of ruthenium, palladium, and platinum from organometallic precursors in the presence of cellulose derivatives. Chem Mater 5:341
39. Osuna J, de Caro D, Amiens C, Chaudret B, Snoeck E, Respaud M, Broto J-M, Fert A (1996) Synthesis, characterization, and magnetic properties of cobalt nanoparticles from an organometallic precursor. J Phys Chem 100:14571
40. Ould Ely T, Amiens C, Chaudret B, Snoeck E, Verelst M, Respaud M, Broto J-M (1999) Broto, Synthesis of nickel nanoparticles. Influence of aggregation induced by modification of poly (vinylpyrrolidone) chain length on their magnetic properties. Chem Mater 11:526
41. Ciuculescu D, Amiens C, Respaud M, Falqui A, Lecante P, Benfield RE, Jiang L, Fauth K, Chaudret B (2007) One-Pot Synthesis of Core− Shell FeRh Nanoparticles. Chem Mater 19:4624
42. Rodriguez A, Amiens C, Chaudret B, Casanove M-J, Lecante P, Bradley JS (1996) Synthesis and isolation of cuboctahedral and icosahedral platinum nanoparticles. ligand-dependent structures. Chem Mater 8:1978
43. de Caro D, Ould Ely T, Mari A, Chaudret B, Snoeck E, Respaud M, Broto J-M, Fert A (1996) Synthesis, characterization, and magnetic studies of nonagglomerated zerovalent iron particles. Chem Mater 8:1987
44. Barrière C, Alcaraz G, Margeat O, Fau P, Quoirin J-B, Anceau C, Chaudret B (2008) Copper nanoparticles and organometallic chemical liquid deposition (OMCLD) for substrate metallization. J Mater Chem 18:3084
45. Axet MR, Philippot K, Chaudret B, Cabie M, Giorgio S, Henry C (2011) TEM and HRTEM evidence for the role of ligands in the formation of shape-controlled platinum nanoparticles. Small 7:235
46. Ibrahim M, Garcia MAS, Vono LLR, Guerrero M, Lecante P, Rossi LM, Philippot K (2016) Polymer versus phosphine stabilized Rh nanoparticles as components of supported catalysts: implication in the hydrogenation of cyclohexene model molecule. Dalton Trans 45:17782–17791
47. Respaud M, Broto JM, Rakoto H, Fert AR, Thomas L, Barbara B, Verelst M, Snoeck E, Lecante P, Mosset A, Osuna J, Ould Ely T, Amiens C, Chaudret B (1998) Surface effects on the magnetic properties of ultrafine cobalt particles. Phys Rev B Condens Matter 57:2925
48. de Caro D, Wally H, Amiens C, Chaudret B (1994) Synthesis and spectroscopic properties of a novel class of copper particles stabilized by triphenylphosphine. J Chem Soc Chem Commun 1891
49. Soulantica K, Maisonnat A, Fromen M-C, Casanove M-J, Lecante P, Chaudret B (2001) Synthesis and self-assembly of monodisperse indium nanoparticles prepared from the organometallic precursor [In(η^5-C_5H_5)]. Angew Chem Int Ed 40:448
50. Kelsen V, Wendt B, Werkmeister S, Junge K, Beller M, Chaudret B (2013) The use of ultrasmall iron(0) nanoparticles as catalysts for the selective hydrogenation of unsaturated C–C bonds. Chem Commun 3416
51. Andersen RA, Faegri KJ, Green JC, Haaland A, Lappert MF, Leung WP, Rypdal K (1988) Synthesis of bis [bis (trimethylsilyl) amido] iron (II). Structure and bonding in M[N $(SiMe_3)_2]_2$ (M= manganese, iron, cobalt): two-coordinate transition-metal amides. Inorg Chem 27:1782

52. Barrière C, Piettre K, Latour V, Margeat O, Turrin C-O, Chaudret B, Fau P (2012) Ligand effects on the air stability of copper nanoparticles obtained from organometallic synthesis. J Mater Chem 22:2279
53. Schette K, Barthel J, Endres M, Siebels M, Smarsly BM, Yue J, Janiak C (2017) Synthesis of Metal Nanoparticles and Metal Fluoride Nanoparticles from Metal Amidinate Precursors in 1-Butyl-3-Methylimidazolium Ionic Liquids and Propylene Carbonate. Chem Open 6:137–148
54. Tekaia-Elhsissen K, Bonet F, Silvert PY, Herrera-Urbina R (1999) Finely divided platinum–gold alloy powders prepared in ethylene glycol. J Alloys Compd 292:96–99
55. Chau JLH (2007) Synthesis of Ni and bimetallic FeNi nanopowders by microwave plasma method. Mater Lett 61:2753–2756
56. Viau G, Fievet-Vincent F, Fievet F (1996) Nucleation and growth of bimetallic CoNi and FeNi monodisperse particles prepared in polyols. Solid State Ionics 84:259–270
57. Lara P, Ayval T, Casanove M-J, Lecante P, Mayoral A, Fazzini P-F, Philippot K, Chaudret B (2013) On the influence of diphosphine ligands on the chemical order in small RuPt nanoparticles: combined structural and surface reactivity studies. Dalton Trans 42:372–382
58. Qi X, Axet MR, Philippot K, Lecante de P, Serp P (2014) Seed-mediated synthesis of bimetallic ruthenium–platinum nanoparticles efficient in cinnamaldehyde selective hydrogenation. Dalton Trans 43:9283–9295
59. Sun S, Murray CB, Weller D, Folks L, Maser A (2000) Monodisperse FePt Nanoparticles and Ferromagnetic FePt Nanocrystal Superlattices. Science 287:1989
60. Chen M, Nikles DE (2002) Synthesis of spherical FePd and CoPt nanoparticles. J Appl Phys 91:8477
61. Shevchenko EV, Talapin DV, Rogach AL, Kornowski A, Haase M, Weller H (2002) Colloidal Synthesis and Self-Assembly of CoPt3 Nanocrystals. J Am Chem Soc 124:11480
62. Jing Z, Zhan J (2008) Fabrication and gas-sensing properties of porous ZnO nanoplates. Adv Mater 20:4547
63. Bai Y, Mora-Seró I, De Angelis F, Bisquert J, Wang P (2014) Titanium dioxide nanomaterials for photovoltaic applications. Chem Rev 114:10095–10130
64. Bi L, Boulfrad S, Traversa E (2014) Steam electrolysis by solid oxide electrolysis cells (SOECs) with proton-conducting oxides. Chem Soc Rev 43:8255–8270
65. Li W, Zeng L, Wu Y, Yu Y (2016) Nanostructured electrode materials for lithium-ion and sodium-ion batteries via electrospinning. Sci China Mater 59:287–321
66. Acar C, Dincer I, Naterer GF (2016) Review of photocatalytic water-splitting methods for sustainable hydrogen production. Int J Energy Res 40:1449–1473
67. Ran J, Zhang J, Yu J, Jaroniecc M, Qiao SZ (2014) Earth-abundant cocatalysts for semiconductor-based photocatalytic water splitting. Chem Soc Rev 43:7787–7812
68. Legrini O, Oliveros E, Braun AM (1993) Photochemical processes for water treatment. Chem Rev 93:671–698
69. Brown GE, Henrich VE, Casey WH, Clark DL, Eggleston C, Felmy A, Goodman DW, Grätzel M, Maciel G, McCarthy MI, Nealson KH, Sverjensky DA, Toney MF, Zachara JM (1999) Metal oxide surfaces and their interactions with aqueous solutions and microbial organisms. Chem Rev 99:77–174
70. Turova NY, Turevskaya EP (2002) The chemistry of metal alkoxides. Kluwer Academic Publishers, Boston
71. Monge M, Kahn ML, Maisonnat A, Chaudret B (2003) Room-Temperature Organometallic Synthesis of Soluble and Crystalline ZnO Nanoparticles of Controlled Size and Shape. Angew Chem Int Ed 42:5321
72. Dumestre F, Chaudret B, Amiens C, Renaud P, Fejes P (2004) Fejes, Superlattices of iron nanocubes synthesized from $Fe[N(SiMe_3)_2]_2$. Science 303:821–823
73. Dearing AW, Reid EE (1928) Alkyl Orthosilicates. J Am Chem Soc 50:3058
74. Vioux A (1997) Nonhydrolytic sol− gel routes to oxides. Chem Mater 9:2292

75. Hay JN, Raval HM (2001) Synthesis of organic− inorganic hybrids via the non-hydrolytic sol− gel process. Chem Mater 13:3396
76. Inoue M (2004) Glycothermal synthesis of metal oxides. J Phys Condens Matter 16:S1291
77. Kołodziejczak-Radzimsk A, Jesionowski T (2014) Zinc oxide—from synthesis to application. Materials 7:2833–2881
78. Klingshirn C, Fallert J, Zhou H, Sartor J, Thiele C, Maier-Flaig F, Schneider D, Kalt H (2010) 65 years of ZnO research–old and very recent results. Phys Stat Sol (B) 247:1424–1447
79. Wang Y, He Y, Lai Q, Fan M (2014) Review of the progress in preparing nano TiO2: An important environmental engineering material. J Environ Sci 26:2139–2177
80. Cargnello M, Gordon TR, Murray CB (2014) Solution-phase synthesis of titanium dioxide nanoparticles and nanocrystals. Chem Rev 114:9319–9345
81. Epifani M, Arbiol J, Dı´az R, Pera´lvarez MJ, Siciliano P, Morante JR (2005) Synthesis of SnO_2 and ZnO Colloidal Nanocrystals from the Decomposition of Tin(II) 2-Ethylhexanoate and Zinc(II) 2-Ethylhexanoate. Chem Mater 17:6468
82. Zhang JW, Zhu PL, Li JH, Chen JM, Wu ZS, Zhang ZJ (2009) Fabrication of octahedral-shaped polyol-based zinc alkoxide particles and their conversion to octahedral polycrystalline ZnO or single-crystal ZnO nanoparticles. Cryst Growth Des 9:2329
83. Zhao J, Zou XX, Zhou LJ, Feng LL, Jin PP, Liu YP, Li GD (2013) Precursor-mediated synthesis and sensing properties of wurtzite ZnO microspheres composed of radially aligned porous nanorods. Dalton Trans 42:14357
84. Campbell AS, Dong C, Maloney A, Hardinger J, Hu X, Meng F, Guiseppe-Elie A, Wu N, Dinu CZ (2014) Systematic Study of the Catalytic Behavior at Enzyme–Metal-Oxide Nanointerfaces. Nano Life 4:145005
85. Jiang X, Wang Y, Herricks T, Xia Y (2004) Ethylene glycol-mediated synthesis of metal oxide nanowires. J Mater Chem 14:695
86. Rockenberger J, Scher EC, Alivisatos AP (1999) A new nonhydrolytic single-precursor approach to surfactant-capped nanocrystals of transition metal oxides. J Am Chem Soc 121:11595
87. Ghosh M, Rao CNR (2004) Solvothermal synthesis of CdO and CuO nanocrystals. Chem Phys Lett 393:493
88. Choi S-H, Kim E-G, Park J, An K, Lee N, Kim AC, Hyeon T (2005) Large-Scale Synthesis of Hexagonal Pyramid-Shaped ZnO Nanocrystals from Thermolysis of Zn− Oleate Complex. J Phys Chem B 109:14792
89. Park J, An K, Hwang Y, Park J-G, Noh H-J, Kim J-Y, Park J-H, Hwang N-M, Hyeon T (2004) Ultra-large-scale syntheses of monodisperse nanocrystals. Nat Mater 3:891
90. Zitoun D, Pinna N, Frolet N, Belin C (2005) Single crystal manganese oxide multipods by oriented attachment. J Am Chem Soc 127:15034
91. Chen Z, Xu A, Zhang Y, Gu N (2010) Preparation of NiO and CoO nanoparticles using M2+-oleate (M = Ni, Co) as precursor. Curr Appl Phys 10:967–970
92. Steckel JS, Snee P, Coe-Sullivan S, Zimmer JP, Halpert JE, Anikeeva P, Kim L-A, Bulovic V, Bawendi MG (2006) Color-saturated green-emitting QD-LEDs. Angew Chem Int Ed 45:5796–5799
93. Pramanik P, Basu PK, Biswas S (1987) Preparation and characterization of chemically deposited tin (II) sulphide thin films. Thin Solid Films 150:269
94. Liu J, Tanaka T, Sivula K, Alivisatos AP, Frechet JMJ (2004) Employing end-functional polythiophene to control the morphology of nanocrystal− polymer composites in hybrid solar cells. J Am Chem Soc 126:6550
95. Zhao LD, Lo SH, Zhang Y, Sun H, Tan G, Uher C, Wolverton C, Dravid VP, Kanatzidis MG (2014) Ultralow thermal conductivity and high thermoelectric figure of merit in SnSe crystals. Nature 508:373–377
96. Wen S, Pan H, Zheng Y (2015) Electronic properties of tin dichalcogenide monolayers and effects of hydrogenation and tension. J Mater Chem C 3:3714–3721

97. Steigerwald ML, Alivisatos AP, Gibson JM, Harris TD, Kortan R, Muller AJ, Thayer AM, Duncan TM, Douglass DC, Brus LE (1988) Surface derivatization and isolation of semiconductor cluster molecules. J Am Chem Soc 110:3046
98. Olshavsky MA, Goldstein AN, Alivisatos AP (1990) Organometallic synthesis of gallium-arsenide crystallites, exhibiting quantum confinement. J Am Chem Soc 112:9438
99. Talapin DV, Rogach AL, Kornowski A, Haase M, Weller H (2001) Highly luminescent monodisperse CdSe and CdSe/ZnS nanocrystals synthesized in a hexadecylamine−trioctylphosphine oxide− trioctylphospine mixture. Nano Lett 1:207
100. Wuister SF, Swart I, van Driel F, Hickey SG, Donegá CDM (2003) Highly luminescent water-soluble CdTe quantum dots. Nano Lett 3:503
101. Talapin DV, Koeppe R, Goetzinger S, Kornowski A, Lupton JM, Rogach AL, Benson O, Feldmann J, Weller H (2003) Highly emissive colloidal CdSe/CdS heterostructures of mixed dimensionality. Nano Lett 3:1677
102. Danek M, Jensen KF, Murray KF, Murray CB, Bawendi MG (1996) Synthesis of Luminescent Thin-Film CdSe/ZnSe Quantum Dot Composites Using CdSe Quantum Dots Passivated with an Overlayer of ZnSe. Chem Mater 8:173
103. Haubold S, Hasse M, Kornowski A, Weller H (2001) Strongly luminescent InP/ZnS core–shell nanoparticles. Chem Commun 2:331
104. Barron AR (1995) MOCVD of group III chalcogenides. Adv Mater Optics Electron 5:245
105. Singh JP, Bedi RK (1991) Electrical properties of flash-evaporated tin selenide films. Thin Solid Films 199:9
106. Wise FW (2000) Lead salt quantum dots: the limit of strong quantum confinement. Acc Chem Res 33:773
107. Nomura R, Inazawa SJ, Kanaya K, Matsuda H (1989) Thermal decomposition of butylindium thiolates and preparation of indium sulfide powders. Appl Organomet Chem 3:195
108. Shah AY, Kedarnath G, Tyagi A, Betty CA, Jain VK, Vishwanadh B (2017) Versatile Precursors for Photo Responsive Germanium Sulfide Nanostructures. Chem Select 2:4598–4604
109. Trindade T, O'Brien P (1996) A single source approach to the synthesis of CdSe nanocrystallites. Adv Mater 8:161
110. Kedarnath G, Jain VK, Ghoshal S, Dey GK, Ellis CA, Tiekink ERT (2007) Zinc, cadmium and mercury dithiocarboxylates: Synthesis, characterization, structure and their transformation to metal sulfide nanoparticles. Eur J Inorg Chem 1566–1575
111. Biswal JB, Garje SS, Nuwad J, Pillai CGS (2013) Bismuth (III) dialkyldithiophosphates: Facile single source precursors for the preparation of bismuth sulfide nanorods and bismuth phosphate thin films. J Solid State Chem 204:348–355
112. Ramasamy K, Malik MA, O'Brien P, Raftery J (2010) Metal complexes of thiobiurets and dithiobiurets: Novel single source precursors for metal sulfide thin film nanostructures. Dalton Trans 39:1460
113. Kedarnath G, Dey S, Jain VK, Dey GK, Varghese B (2006) 2-(N,N-Dimethylamino) ethylselenolates of cadmium(II): Syntheses, structure of [Cd3(OAc)2(SeCH2CH2NMe$_2$)$_4$] and their use as single source precursors for the preparation of CdSe nanoparticles. Polyhedron 25:2383–2391
114. Nair PS, Radhakrishnan T, Revaprasadu N, Kolawole GA, O'Brien P (2002) A single-source route to CdS nanorods. J Chem Soc Chem Commun 564
115. Trindade T, O'Brien P (1997) Synthesis of CdS and CdSe nanocrystallites using a novel single-molecule precursors approach. Chem Mater 9:523
116. Duan T, Lou W, Wang X, Xue Q (2007) Size-controlled synthesis of orderly organized cube-shaped lead sulfide nanocrystals via a solvothermal single-source precursor method. Colloids Surf A 310:86
117. Copsey MC, Chivers T (2005) Formation of Ga_2Te_2 and M_3Te_3 (M = Ga, In) rings from reactions of sodium ditelluroimidodiphosphinate with Group 13 halides. Chem Commun 4938

118. Kedarnath G, Jain VK (2013) Pyridyl and pyrimidyl chalcogen (Se and Te) compounds: A family of multi utility molecules. Coord Chem Rev 257:1409–1435
119. Sharma RK, Kedarnath G, Kushwah N, Pal MK, Wadawale A, Vishwanadh B, Paul B, Jain VK (2013) Indium(III) (3-methyl-2-pyridyl)selenolate: Synthesis, structure and its utility as a single source precursor for the preparation of In2Se3 nanocrystals and a dual source precursor with $[Cu\{SeC_5H_3(Me\text{-}3)N\}]_4$ for the preparation of $CuInSe_2$. J Organomet Chem 747:113–118
120. Tyagi A, Kedarnath G, Wadawale A, Shah AY, Jain VK, Vishwanadh B (2016) Diorganotin (iv) 4,6-dimethyl-2-pyrimidyl selenolates: synthesis, structures and their utility as molecular precursors for the preparation of SnSe2 nano-sheets and thin films. RSC Adv 6:8367–8376
121. Campos MP, Owen JS (2016) Synthesis and surface chemistry of cadmium carboxylate passivated CdTe nanocrystals from cadmium bis(Phenyltellurolate). Chem Mater 28:227–233
122. Sharma RK, Kedarnath G, Jain VK, Wadawale A, Pillai CGS, Nalliath M, Vishwanadh B (2011) Copper(I) 2-pyridyl selenolates and tellurolates: Synthesis, structures and their utility as molecular precursors for the preparation of copper chalcogenide nanocrystals and thin films. Dalton Trans 40:9194–9201
123. Sharma RK, Kedarnath G, Wadawale A, Betty CA, Vishwanadh B, Jain VK (2012) Diorganotin(IV) 2-pyridyl selenolates: synthesis, structures and their utility as molecular precursors for the preparation of tin selenide nanocrystals and thin films. Dalton Trans 41:12129–12138
124. Sharma RK, Kedarnath G, Jain VK, Wadawale A, Nalliath M, Pillai CGS, Vishwanadh B (2010) 2-Pyridyl selenolates of antimony and bismuth: Synthesis, characterization, structures and their use as single source molecular precursor for the preparation of metal selenide nanostructures and thin films. Dalton Trans 39:8779–8787
125. Ritch JS, Chivers T, Afzaal M, O'Brien P (2007) The single molecular precursor approach to metal telluride thin films: imino-bis(diisopropylphosphine tellurides) as examples. Chem Soc Rev 36:1622–1631
126. Marchand P, Carmalt CJ (2013) Molecular precursor approach to metal oxide and pnictide thin films. Coord Chem Rev 257:3202–3221
127. Steckel JS, Zimmer JP, Coe-Sullivan S, Stott NE, Bulovic V, Bawendi MG (2004) Blue luminescence from (CdS) ZnS core–shell nanocrystals. Angew Chem 116:2206
128. Crooker SA, Barrick T, Hollingsworth JA, Klimov VI (2003) Multiple temperature regimes of radiative decay in CdSe nanocrystal quantum dots: Intrinsic limits to the dark-exciton lifetime. Appl Phys Lett 82:2793
129. Klein DL, Roth R, Lim AKL, Alivisatos AP, McEuen PL (1997) A single-electron transistor made from a cadmium selenide nanocrystal. Nature 389:699
130. Wang Y, Herron N (1991) Nanometer-sized semiconductor clusters: materials synthesis, quantum size effects, and photophysical properties. J Phys Chem 95:525
131. Olsson YK, Chen G, Rapaport R, Fuchs DT, Sundar VC, Steckel JS, Bawendi MG, Aharoni A, Banin U (2004) Fabrication and optical properties of polymeric waveguides containing nanocrystalline quantum dots. Appl Phys Lett 85:4469
132. Bruchez M, Moronne M, Gin P, Weiss S, Alivisatos AP (1998) Semiconductor nanocrystals as fluorescent biological labels. Sci 281:2013–2016
133. Hagfeldt A, Grätzel M (1995) Light-induced redox reactions in nanocrystalline systems. Chem Rev 95:49
134. Henglein A, Gutierrez M, Fisher CH (1984) Photochemistry of Colloidal Metal Sulfides 6. Kinetics of Interfacial Reactions at ZnS-Particles. Ber Bunsen-Ges Phys Chem 88:170
135. Henglein A, Fojtik A, Weller H (1987) Reactions on colloidal semiconductor particles. Ber Bunsen-Ges Phys Chem 91:441
136. Tan RQ, He Y, Zhu YF (2003) Hydrothermal preparation of mesoporous TiO2 powder from Ti(SO4)2 with poly (ethylene glycol) as template. J Mater Chem 38:3973
137. Nazeeruddin MK, Baranoff E, Grätzel M (2011) Dye-sensitized solar cells: a brief overview. Solar Energy 85:1172–1178

138. Nikolenko LM, Razumov VF (2013) Colloidal quantum dots in solar cells. Russ Chem Rev 82:429–448
139. Carey GH, Abdelhady AL, Ning Z, Thon SM, Bakr OM, Sargent EH (2015) Colloidal quantum dot solar cells. Chem Rev 115:12732–12763
140. Redmond G, Fitzmaurice D, Grätzel M (1994) Visible Light Sensitization by cis-Bis (thiocyanato)bis(2,2'-bipyridyl-4,4'-dicarboxylato)ruthenium(II) of a Transparent Nanocrystalline ZnO Film Prepared by Sol-Gel Techniques. Chem Mater 6:686–691
141. Yu X-Y, Lei B-X, Kuang D-B, Su C-Y (2011) Highly efficient CdTe/CdS quantum dot sensitized solar cells fabricated by a one-step linker assisted chemical bath deposition. Chem Sci 2:1396–1400
142. Gur I, Fromer NA, Geier ML, Alivisatos AP (2005) Air-Stable All-Inorganic Nanocrystal Solar Cells Processed from Solution. Sci 310:462–465
143. Ning Z, Ren Y, Hoogland S, Voznyy O, Levina L, Stadler P, Lan X, Zhitomirsky D, Sargent EH (2012) All-Inorganic Colloidal Quantum Dot Photovoltaics Employing Solution-Phase Halide Passivation. Adv Mater 24:6295–6299
144. MacInnes AN, Power MB, Barron AR (1992) Chemical vapor deposition of cubic gallium sulfide thin films: a new metastable phase. Chem Mater 4:11–14

Chapter 16
Synthesis of Metal Organic Frameworks (MOF) and Covalent Organic Frameworks (COF)

Adish Tyagi and Siddhartha Kolay

Abstract Since the very beginning, nature has demonstrated its ability to create complex systems with advance functions from atomic-level assembly. Gaining inspiration from nature, an enormous progress has been achieved in constructing crystalline porous material frameworks like metal organic framework (MOF) and covalent organic framework (COF) with predetermined topologies, large surface areas, tunable pore sizes and functionalities. They provide many key features required in industrial applications, like high surface area, uniform nanoporosity, interconnected pore/channel system, accessible pore volume, high adsorption capacity and shape/size selectivity. These features make them an ideal material for gas storage and separation (such as H_2, CH_4, CO_2). In fact, having superior crystallinity, porosity and stability relative to other porous materials, MOFs and COFs affirm their candidature for a wide range of applications. This chapter gives a brief background of porous materials and their classification, followed by a discussion on MOFs, COFs and their properties. The subsequent section includes design and synthesis strategies followed by a detailed discussion related to their applications in various frontline areas.

Keywords MOFs · COFs · Reticular synthesis · Dynamic covalent chemistry · Porous materials

A. Tyagi (✉) · S. Kolay
Chemistry Division, Bhabha Atomic Research Centre, Mumbai 400085, India
e-mail: tyagia@barc.gov.in

S. Kolay
e-mail: siddhart@barc.gov.in

A. Tyagi · S. Kolay
Homi Bhabha National Institute, Anushaktinagar, Mumbai 400094, India

A. K. Tyagi and R. S. Ningthoujam (eds.), *Handbook on Synthesis Strategies for Advanced Materials*, Indian Institute of Metals Series,
https://doi.org/10.1007/978-981-16-1807-9_16

Acronyms used in the chapter

		Organic linker
MOF	Metal organic framework	
COF	Covalent organic framework	
ZSM-5	$Na_nAl_nSi_{96-n}O_{192}.16H_2O$ (0 < n < 7)	
SBU	Structural building unit	
DCC	Dynamic covalent chemistry	
POP	Porous polymeric framework	
MOF-5	$[Zn_4O(C_8O_4H_4)_3]$	1,4-benzenedicarboxylate
MOF-74	Framework of M^{2+} with 2,5-dihydroxybenzene-1,4-dicarboxylate	2,5-dihydroxybenzene-1,4-dicarboxylate
MOF-177	$[Zn_4O(C_{27}H_{15}O_6)_3]$	1, 3, 5-tris(4-carboxyphenyl)benzene or benzene tri benzoic acid
HKUST-1	$Cu_3O_{12}C_{18}H_6$	1,3,5-benzene tricarboxylic acid
ZIF 8	$ZnN_4C_8H_{10}$	2-methylimidazole
COF-5	$C_9H_4BO_2$	
IRMOF	Metal Organic Framework. with 1,4-benzodicarboxylic acid	
IRMOF-3	$[Zn_4O(C_8H_5NO_4)_3]$	2-aminobenzene-1, 4 dicarboxylic acid
IRMOF-11	$[Zn_4O(C_{18}H_{12}O_4)_3]$	4, 5, 9, 10-tetrahydropyrene-2, 7-dicarboxylic acid
MIL	Materials Institute Lavoisier	
IRMOF-20	$[Zn_4O(C_8H_2O_4S_2)_3]$	Thieno[3, 2-b]thiophene-2, 5-dicarboxylic acid
MIL-53(Fe)	3D-[M(μ$_4$-bdc)(μ-OH)]	
Mn-BTC	Mn-MOF with benzene-1,3,5-tricarboxylic acid	
DAAQ-TFP	COF from 2,6-diaminoanthraquinone (DAAQ) and 1,3,5-triformylphloroglucinol (TFP)	2,6-diaminoanthraquinone and 1,3,5-triformylphloroglucinol
MIL-100 (Cr)	$[M_3O(H_2O)_2F_{0.8}(OH)_{0.2})\{C_6H_3(CO_2)_3\}$ 2 H_2O ()]	
MIL-101 (Cr)	3D-$[Cr_3(O)(bdc)_3(F)$-($H_2O)_2]$	
MCM-41	Mobil Composition of Matter No. 41	
COF-320	Synthesized from tetra-(4-anilyl)methane and 4,4′-biphenyldialdehyde	tetra-(4-anilyl)methane and 4,4′-biphenyldialdehyde

16.1 Introduction

Meeting ever-growing energy needs for the over growing population demands an enormous amount of sustainable energy source and their storage medium [1]. Moreover, the power generation through burning of fossil fuels causes significant damage to the environment due to the release of a large amount of CO_2 and SO_2

gases in the burning process. This has led to extensive work not only in searching an alternative non-emissive renewable energy sources like solar, wind and wave or other energy production sources like nuclear power, hydrogen power and electrochemical but also focussing on the storage of the energy for their effective use [2–6]. Porous materials have emerged as the material of choice in the area of energy storage since they already find relevance in the field of gas storage for clean energy production like (H_2, CH_4, etc.), water purification, photocatalysis and heterogeneous catalysis [7, 8]. Porous materials are defined as open framework solids. They are also called *molecularly engineered* materials since their structures can be tuned at molecular level making them adaptable for a wide variety of applications. They are of great technological significance because of their high surface area, tunable pore size and ability to interact with reactants (atoms, ions and molecules) at their surface and bulk. Researchers have designed novel porous materials by gaining insight from nature. Honeycombs with hexagonal pores, hollow bamboo, alveoli in lungs, bones, etc., are few examples of porous structures found in nature. Some common examples of porous materials designed by researchers include activated carbon, mesoporous silica, zeolites, metal organic frameworks, covalent organic frameworks [9–11].

The first porous material ever used by the mankind was porous carbon (activated charcoal). It had great significance since ancient times due to its miraculous ability to cleanse and detoxify. Its actual discovery is likely to be long before its first recorded use ~3750 BC for melting and combining metals in ancient Egypt. Around 400 BC, first written record of antibacterial and water purification use of activated charcoal was found. Apart from this, people have also used the decolorizing property of activated charcoal [12]. With the progress of time, new porous materials were discovered and utilized for the benefit of human mankind like porous hydraulic cements, used as an important building material in Roman antiquity [13].

Next big thing in the field of porous materials was the discovery of zeolites minerals (stilbite) by the Swedish mineralogist Axel Fredrik Cronstedt in 1756, which he described as the "boiling stones" that is why the word zeolite originates from the Greek words "zeo (to boil)" and "lithos (stone)" [14]. Zeolites are "tectosilicate" minerals, constructed from interconnected TO_4 tetrahedra (where T stands for Si, Al or B) through corner sharing, to produce three-dimensional frameworks [15]. Nanopores of zeolites have been used for many decades in applications ranging from catalysis to gas separation to ion exchange [16]. Realizing the importance of natural zeolites, researchers put significant efforts to synthesize zeolites in laboratory, with first artificial zeolite synthesized in 1950 by Milton and Breck by reactive gel crystallization method [17]. In 1972, the miraculous ZSM-5, a high-silica zeolite was developed by Argauer and Landolt which found enormous applications in the field of catalysis, [18]. Following this, various other porous silicates and phosphates were discovered and got utilized in various industrial applications. However, the difficulty in pre-designing the structure and pre- and post-functionalization of pore channels, to perform highly specific and cooperative functions limited their sophisticated use and further led to search for new ultra-porous materials [19]. Also, the need to achieve uniform pore size, shape and volume has been realized over the years as it can lead to superior properties. For instance, a distribution of pore sizes would severely limit the ability of the porous solid to separate molecules of differing sizes leading to poor selectivity [20].

Design and synthesis of porous materials with ordered tunable pore size having specific functionality are one of the main challenges for improving their performance in the area of gas storage, energy storage, electronic properties, etc. [21]. With untiring efforts of the scientific community worldwide, the understanding about the structure of ordered porous materials and how to control and adjust them has increased considerably. In the ongoing quest, researchers prepare porous solids which involve the coordination of metal ion centre with the organic linkers. In fact, these inorganic–organic porous materials were known since very long time. Early examples include metal cyanide complexes (Hofmann-type clathrates, Prussian blue-type structures) Werner complexes and open diamond-like framework constructed from copper nitrate complexes. During 1990s, metal–organic porous materials gained renewed interest but their inability to maintain permanent porosity and avoid collapse of frameworks upon guest removal or guest exchange were major drawbacks which limits the application of this class of materials.

These shortcomings were overcome by the pioneering work of Omar M. Yaghi and co-workers [22] employing reticular chemistry which led to the discovery of a new type of inorganic–organic hybrid porous materials with permanent porosity in 1998 and were classified as metal organic frameworks (MOF). The hybrid material consists of Zn^{2+} ions, and 1,4-benzene dicarboxylates [22] were termed as MOF-5. MOFs can be defined as a class of hybrid framework solid materials comprised of organized organic linkers and metal cations. On a fundamental level, MOFs epitomize the beauty of chemical structures obtained by combining organic and inorganic chemistry, two disciplines often regarded as dissimilar. Due to the ultra-high uniform porosity, adjustable pore size, MOFs exhibit improved performance in the areas of catalysis, adsorption and separation [9, 23]. However, their inadequate chemical and thermal stability due to reversible nature of the coordinate bond generated during synthesis were few shortcomings needs to be overcome. The thermo-chemical stability of porous materials can be improved by replacing the weak coordinate bonds with strong covalent bond. However, building crystalline organic framework structure by linking organic building blocks through strong covalent bonds is a challenging task as they often result in the formation of amorphous cross-linked polymer [24].

O. M. Yaghi and co-workers [25] were able to solve this long standing problem by synthesizing the first crystalline covalent organic framework (COF) in 2005 using basic principles of dynamic covalent chemistry (DCC). COFs have received considerable interest in recent times, due to their ability to merge the advantages of both porous materials and polymers. COFs exhibit well-defined porosity, easy processability so much so that some of the COFs can be dissolved in solvent and then processed using solution based techniques without destroying porosity. Lastly, the availability of various synthetic routes for COFs enables to fine tune multiple functionalities into the porous frameworks or at the porous surface [26–29]. Before proceeding further, it will be essential to classify the porous materials on the basis of pore size and depending upon the building blocks.

16.1.1 Classification of Porous Materials

16.1.1.1 Depending upon the Pore Size

Porous solids are classified by IUPAC into three categories on the basis of their pore sizes: (a) Microporous materials—pore diameter in the range of 2 nm and below, e.g. MOFs, (b) Mesoporous materials—pore diameter in the range of 2–50 nm; e.g. Mesoporous silica and alumina and (c) Macroporous materials having pore diameter more than 50 nm, e.g. metal foams (Fig. 16.1) [20].

16.1.1.2 Depending upon the Building Block Framework

Porous materials can also be classified as (i) purely inorganic, (ii) inorganic–organic hybrid and (iii) purely organic on the basis of the constituted framework material type (Fig. 16.2).

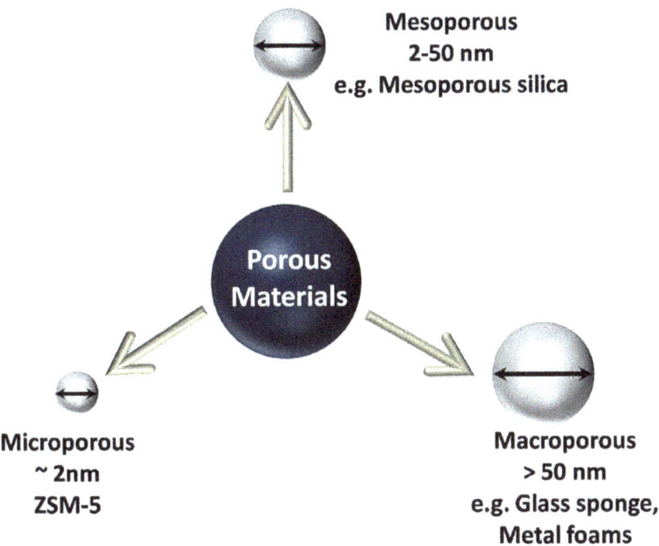

Fig. 16.1 Classification of porous materials on the basis of pore size

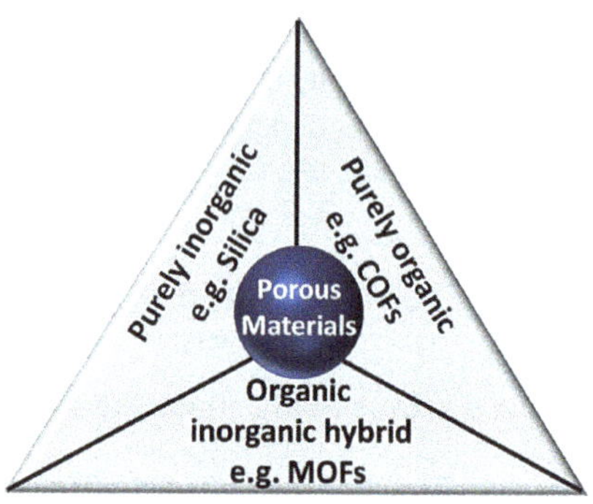

Fig. 16.2 Classification of porous materials on the basis of building block framework

(a) **Inorganic porous frameworks**

This class of materials includes zeolites, aluminophosphates, metallosilicates (titanosilicates, germanosilicates) and metal phosphates. Zeolites are crystalline aluminosilicates which consist of interconnected TO_4 tetrahedra (T = Si, Al) via corner sharing with a general formula $M_{x/n}[(AlO_2)_x(SiO_2)_y]\cdot wH_2O$ (M = Na, Li, K, Ca and Mg), where m is the valency of the metal ion which balances the negative charges on aluminosilicates. The presence of these cations in the frameworks leads to exchange characteristics in zeolites. Generally, zeolites are bronsted acids and their framework contains pore having diameter in the range from 3 to 15 Å [30a]. Variety of materials having different chemical functionality can be realized by substituting Si and Al which are substituted with different elements like P or Fe. However, from applications point of view, high-silica materials are most important due to their exceptionally high thermal and hydrothermal stability under process conditions [30b, c].

Another important material of this class is aluminophosphates (AlPOs) obtained by the replacement of the silicon atoms in zeolites with aluminium and phosphorus. Aluminophosphates are usually formed as neutral, due to the presence of octahedral aluminium sites in the framework. Due to low production costs, high chemical and thermal stability, materials of this class were widely employed as industrial adsorbents and catalyst [20, 31].

(b) **Inorganic–organic hybrid porous frameworks**

Lack of functionalization and control over the structural integrity of inorganic materials (activated carbons and zeolites) limits their ability to carry out specialized functions. The development of inorganic–organic hybrid porous frameworks leads to the discovery of novel unprecedented structures which allows the tailoring the functionality of both pores and surface by judicious choice of inorganic and organic parts. The common example of this class includes metal organic frameworks (MOFs). Developed by the three-dimensional crystalline assembly of metal containing units (secondary building blocks (SBUs)) and organic ligands, MOFs possess flexible structure of well-defined pore sizes, surfaces areas [32]. The metallic units are present at the node separated by organic spacers. The metal ions employed in the construction of MOFs include metals from alkaline earth metals, transition metals, p-block elements, rare earth elements (RE = Ln, Y, Sc), actinides and even mix metals [33–35]. Since MOFs structure and in turn its property depends on the choice of metal ions and organic spacer, a variety of organic ligands of different shape and size have been utilized in the synthesis of MOFs. The commonly used organic ligands are carboxylates, Schiff bases, imidazolate, phosphates, pyrazine and bipyridine.

Construction of MOFs follows the basic rules of reticular synthesis where there is direct correlation between the structure of reactants (building block), the structure of frameworks and porosity. In principle, reticular synthesis can be considered as a process where judiciously designed primary building blocks are stitched together to form secondary building blocks (SBUs), held together by strong bonding interactions leading to a predetermined ordered structure [36] as shown in Fig. 16.3.

The amazing thing about primary building blocks used in the reticular synthesis to generate MOFs is that they maintain their structural integrity throughout the construction process unlike traditional solid-state method where reactants do not maintain their structure during synthesis leading to poor or no correlation between reactants and final frameworks.

The topology of the structures constructed by assembling the SBUs through reticular chemistry can be described by a net, assigned by a three-letter symbol such as pts, rht and soc. The details of nets can be obtained from Reticular Chemistry Structure Database (RCSR) [37] or a computer program TOPOS [38] while new nets can be identified by a mathematical program called SYSTRE [39].

It is important to note that reticular synthesis is different from retro-synthesis as well as from supra-molecular assembly. Unlike retro-synthesis, structural integrity of building block used in reticular synthesis remains intact throughout the process and

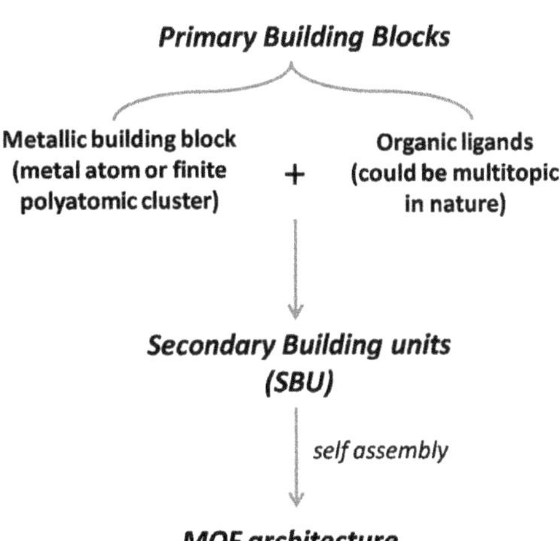

Fig. 16.3 Schematic representation of construction of MOF through primary building blocks

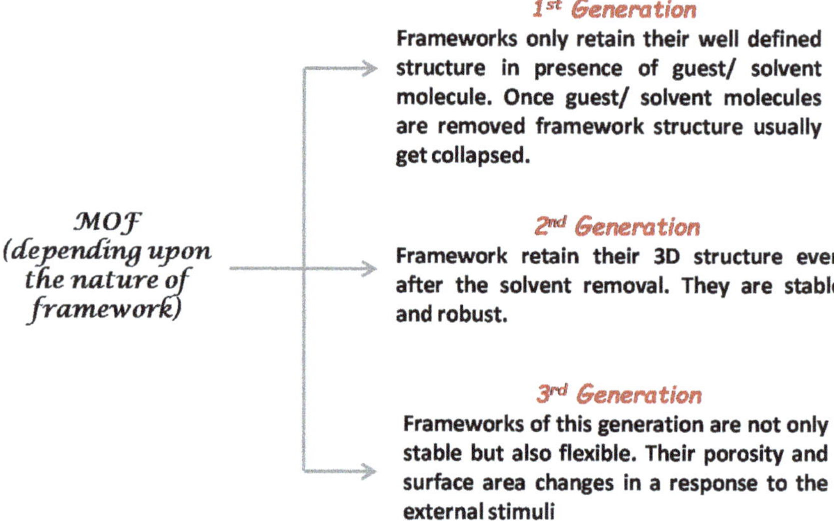

Fig. 16.4 Classification of MOFs depending upon the nature of framework

in contrast to weak interactions responsible for the supra-molecular assemblies, building blocks in reticular synthesis are linked through strong bonds throughout the crystalline network [36, 40, 41]. MOFs can be classified as (i) first generation, (ii) second generation and (iii) third generation depending upon the stability and flexibility of frameworks [42]. The description of different generations of MOFs is discussed in Fig. 16.4. The details of underlying principles involved in the design and synthesis of MOFs are discussed in a separate section.

Due to open framework nature, porosity and tunable functionalities, MOFs find extensive use in gas storage, separation, catalysis, energy storage and various other important applications which will be discussed in a separate section.

Despite the exciting properties, MOFs also suffer from few but serious disadvantages which are as follows:

(i) MOFs lack sufficient thermal and chemical stability, since they are held by relatively weak coordinate bond [43].
(ii) Most of the MOFs are sensitive towards the presence of water [44].
(iii) The porosity of MOFs is difficult to extend to mesoporous region, thus making it inefficient for certain application like storage of bigger sized molecules (e.g. bio-molecules) [44].

(c) Organic porous frameworks
Porous materials which are completely built out of organic building blocks have always attracted the scientific community. Organic porous materials are exclusively prepared from organic building blocks linked through strong covalent bond. Materials of this class possess some unique advantages like:

(i) Organic frameworks show high thermal and chemical stability, since they are prepared by connecting organic building blocks through strong covalent bonds.
(ii) Unlike other frameworks, organic framework materials are made up of only lightweight elements (C, B, O, H, etc.) which appreciably reduce their density and make them suitable for gas storage purpose.

Depending upon the crystallinity of framework, organic framework materials are classified as:

(i) *Porous polymeric framework* (*POPs*): These are class of heavily cross linked microporous polymers synthesized by irreversible C–C coupling reaction like Sonogashira–Hagihara coupling, Suzuki coupling, Friedel–Crafts reaction, acetyl cyclotrimerization, oxidative coupling reaction and phenazine ring fusion reaction [45–49]. POPs are amorphous in nature due to irreversible nature of bond formation mechanism (Fig. 16.5). These materials possess high thermal and chemical stability with high surface area. Examples of this class include CMP-5 and CMP-0 with surface area in the range from 512 to 1018 m^2g^{-1} as reported by Cooper and co-workers [50].
(ii) *Covalent organic frameworks* (*COFs*): Covalent organic frameworks are relatively new class of crystalline organic porous materials generated through reticular chemistry. In simple words, COFs can be explained as the geometric

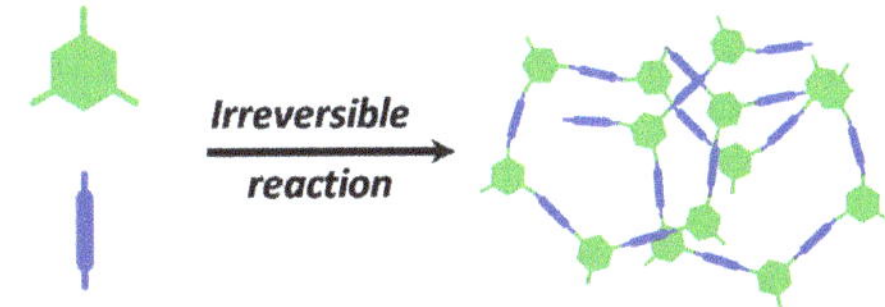

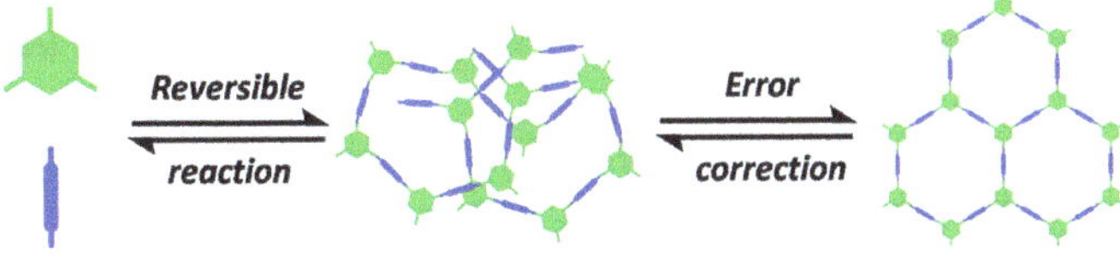

Fig. 16.5 Schematic representations of POPs and COFs by irreversible and reversible reactions, respectively

constructs of molecules which are positioned in a specific spatial orientation to enclose space into which reactivity of the atom and molecules are expressed in ways not possible in discrete molecules [51].

G. N. Lewis, a pioneer in the field of chemical bonding, had beautifully explained how atoms combine to form molecules via strong covalent bonds. With advances in the field of synthetic chemistry, researchers have developed "Retro-synthesis" which discusses the principles to synthesize pre-designed molecules from readily available small molecules. With untiring efforts, researchers learnt how to link the small organic molecules by covalent bonds to form macro molecular structures (polymers). Researchers were able to generate well-defined 2D and 3D supra-molecular structures through self-assembly of building blocks which are held together by weak interactions like van der Waals forces, intra- and inter-molecular hydrogen bonding and C···H interactions. However, any modifications in supra-molecular assemblies without destroying the structure are difficult to achieve because [51]:

(i) Any modification of the constituent building blocks will alter the interaction leading to formation of different assemblies.
(ii) Any chemistry operation on these assemblies can destroy their structural integrity.

Moreover, supra-molecular assemblies have poor thermo-chemical robustness as constituent building blocks are held by weak chemical interactions. Thus, it was considered essential to develop strategies that assemble molecular building blocks through strong covalent interactions instead of weak interactions. In 2005, Yaghi et al. have successfully demonstrated the formation of organic crystalline frameworks by linking the organic molecules in a precise manner through covalent bonds while maintaining their molecular integrity. They named these frameworks as covalent organic frameworks (COFs) [25]. COFs are mainly constructed by organic reactions which are reversible in nature. The reversible bond formation in COFs synthesis provides error checking and proof reading characteristic to the system. This creates an auto repair mechanism through multiple reversible bond formation cycles leading to the formation of stable COFs as final product [52]. The general approach for the synthesis of COFs can be described in following steps [51]:

(i) Target network topology is identified which is subsequently deconstructed into its fundamental units.
(ii) Evaluation of these units according to their connectivity and geometry (e.g. tetrahedral vs square-planar for the connectivity of four).
(iii) Identification of organic molecules (linkers) called building blocks equivalent to these geometric units.
(iv) Construction of COF by the formation of strong covalent bonds between the building blocks using the principle of reticular chemistry.

The widely used reactions for COFs synthesis were:

(i) Boronic acid trimerization [25]
(ii) Boronate ester formation [52]
(iii) Nitrile group trimerisation [26, 53]
(iv) Schiff base reaction [53].

It is worth mentioning that any attempt to synthesize COFs using irreversible reactions always leads to the formation of amorphous porous polymeric frameworks (POPs) because formation of amorphous material using irreversible reaction is thermodynamically favourable. Thus, reversibility during bond formation is an essential criterion for the formation of extended crystalline COFs. Microscopic reversibility can be achieved by controlling the concentration of the by-product (generally water) and pressure.

COFs exhibit high physico-chemical stability and low density due to presence of strong covalent bonds and lightweight constituent atoms. COFs can be pre-designed and modified without destruction of structure. Moreover, their pores can be functionalized and the size can be tuned through proper selection of the building blocks. As a result, COFs open up a new dimension in the material chemistry and are used in the various fields like gas storage, separation, catalysis, photo-conducting materials and sensors.

COFs can be realized in two dimensions (2D) or three dimensions (3D) depending upon the symmetric combinations of organic molecules used for the preparation. 3D COFs are rather rare due to scarcity of higher symmetric organic building units, and most of the COFs reported so far possess 2D structure [25, 53, 54]. In 2D COFs, the covalently bound framework proceeds in horizontal direction and thus exhibits homogeneous porosity in their extended 2D sheets. Stacking of these 2D sheets occurs via π-π interaction in eclipsed conformation which results in the formation of ordered columnar channels. Such channels in 2D COFs could facilitate charge carrier transport within the column, which implies that 2D COFs have potential for wide range of applications in fields such as sensing, separation, storage and catalysis. In contrast, 3D COFs possess extended three-dimensional framework which could led to high specific surface areas comparable to MOFs and are likely ideal candidates for gas storage and separation applications.

Despite such promising properties, real-life applications of COFs are yet to be realized due to the issues related to their chemical stability and scalability. Moreover, in the presence of moisture COFs generally get hydrolyzed to the starting materials. This is because COFs are generally formed by condensation reaction where water is excluded as by-product and its presence leads to reversible back reaction causing hydrolysis of COFs.

The following sections will discuss the design and synthesis strategy, synthetic protocols and applications of MOFs and COFs.

16.2 Design and Synthesis Strategy

The performance of the porous materials depends to a great extent on the proper design and synthesis of the materials with well-defined pore size and controlled functionality. In a broader view, MOFs may be considered as a combination of two central components: connectors and linkers [55, 56]. These two components are considered as starting reagents with which the principle framework of the MOFs is constructed. Apart from them, there are other auxiliary components, such as blocking ligands, counter anions, and non-bonding guests or template molecules. Figures 16.7 and 16.8 give an overview of the different types of connectors and linkers and their combination in the formation of different geometries of the MOFs.

The stability of the frameworks plays a very important role in the successful synthesis of frameworks. The coordination numbers as well as the coordination geometries of the metal ions are two very crucial parameters in deciding the stability of the framework. The nature of the metal ions and its oxidation state has a strong influence on the shape of the pore. Thus, depending on the nature of metal ions and its oxidation state various geometries like linear, T-shaped, Y-shaped, tetrahedral, square-planar, square-pyramidal, trigonal–bipyramidal, octahedral, trigonal-prismatic, and pentagonal-bipyramidal (Fig. 16.6) are possible.

Transition metal ions are commonly used as versatile connectors in the construction of MOFs. Linkers afford a wide variety of linking sites with tuned binding strength and directionality. The role of connectors and linkers in the formation of stable framework is well-explained by Kitagawa et al. [55] and Zhou et al. [56].

Depending upon the charge on the linker molecules, linkers are broadly classified into four categories: (a) inorganic ligands (e.g. halides, CN^-, SCN^-, cyanometallate ($[M(CN)_x]^{n-}$), (b) neutral organic ligands like pyridine (py), 4,4'-bipyridine (bpy), (c) anionic organic ligands like various carboxylates with suitable spacer and (d) cationic organic ligands like poly pyridinium ions. Figure 16.7 gives a schematic representation of the various types of linkers. The suitable combination of connector(s) and linker(s) gave different structural motif.

The nature of bonding and overall charge plays a very important role in deciding the stability of frameworks [55]. Overall framework has to be neutral. Since in MOFs, most of the connectors are cationic metal ions, the positive charge of the metal ions got balanced by negatively charged linkers such as carboxylates. Though the above-mentioned four types of linkers are used in metal–organic framework synthesis, it has been observed that organic ligands with carboxylate functional groups with suitable spacer (type C) are most commonly used in MOFs synthesis. It is also seen that apart from the anionic linkers some inorganic anions, like BF_4^-, NO_3^-, PF_6^-, SiF_6^- and N_3^-, are also used in MOFs synthesis as a counter-anion of the metal salts, and these inorganic anions exist either as free guests or as counter-ions mostly to neutralize the cationic connector. Besides charge neutralization they also help in increasing the stability of the frameworks and modify the channel shape through hydrogen bonding with their O and F atoms.

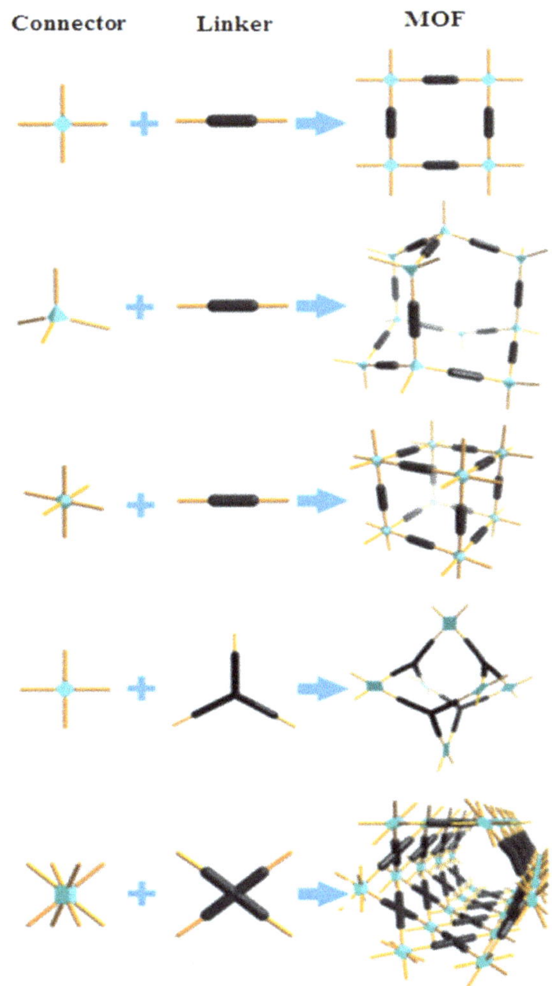

Fig. 16.6 Different geometries of connectors and linkers and their combination forming different MOFs [56] [Copyright: Royal Society of Chemistry]

In general, the bonding interaction in MOFs can be divided into three classes (Fig. 16.8), namely (a) coordination bond (CB) where a pure coordination bonding interaction takes place between the connectors and the linkers with donation of electron pair from linker to the connector, (b) combination of coordination bond and hydrogen bond (CB + HB): both coordinate and H-bonding are present in the framework. The presence of H-bonding in addition to coordination bonds imparts additional stability in the framework compare to coordinate bonding only, (c) co-ordination bonding along with other interaction like d_{π}-p_{π}, d_{π}-d_{π} or δ-bonding.

Sometimes, the mutual orientation of the two aromatic rings may undergo π-π interaction or the π-cloud of one aromatic ring interacts with CH-moieties of other unit. It has been observed that 1D and 2D motifs often aggregate through these

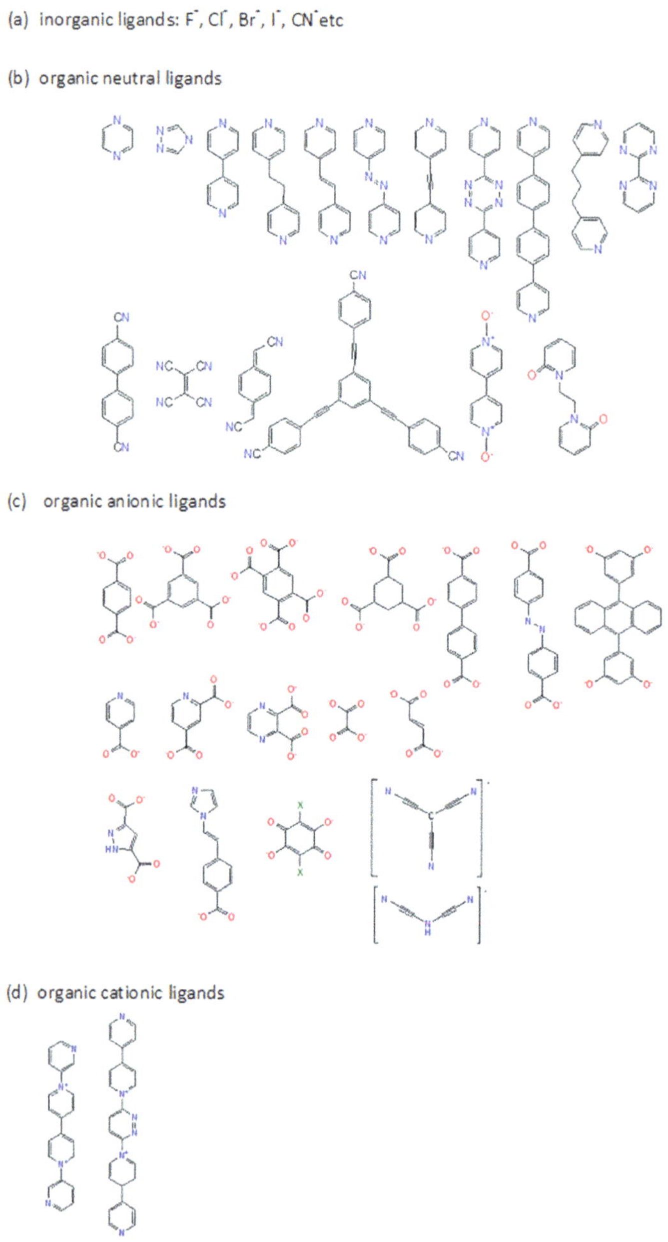

(a) inorganic ligands: F^-, Cl^-, Br^-, I^-, CN^-etc

(b) organic neutral ligands

(c) organic anionic ligands

(d) organic cationic ligands

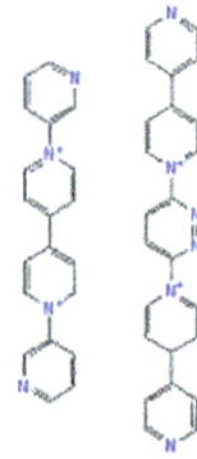

Fig. 16.7 Different types of linkers used in MOFs synthesis

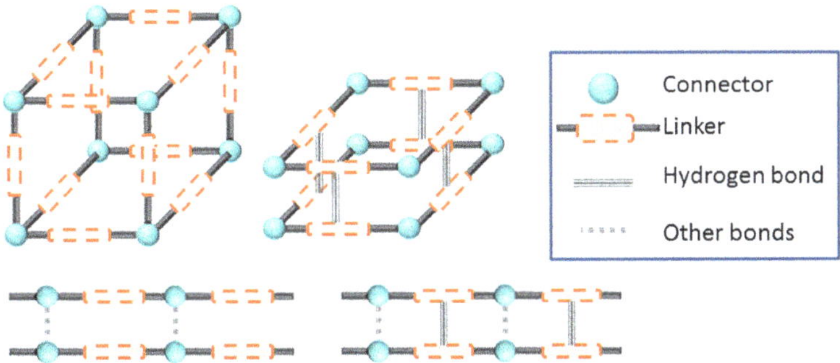

Fig. 16.8 Types of interactions in porous framework structure

additional weak bonds to give 3D frameworks. This type of weak interaction plays a very important role in the interaction of the guest molecules with the motifs.

Unlike MOFs, in covalent organic frameworks (COFs) the symmetry of the building blocks governs the topological structures of the pores [26]. For example, COFs with hexagonal pores can be generated by the combination of a C3-symmetrical building blocks or a C3- and C2-symmetrical building block, while the combination of C4- and C2- or C6- and C2-symmetrical building blocks results tetragonal and trigonal pores, respectively. Figure 16.9 represents the formation of various topological structures of the pores in combination of various building blocks.

In COFs, there are no metal ions, and hence, the possibility of frameworks extension via the coordination bonding of the organic ligands with the metal ions which is one of the main modes of interaction in MOFs is not possible [52]. Thus, in COFs the extension of the frameworks takes place via various condensation reactions. Depending upon the type of functional groups present in the building blocks, various condensation reactions such as boronic acid trimerization, boronate ester formation, Schiff base reaction and nitrile trimerization are reported for the construction of COFs (Figs. 16.10 and 16.11).

Most of the afore-mentioned chemical reactions, being kinetically irreversible in nature lead to the formation of polymeric compound instead of crystalline framework material [25, 26]. COFs are highly porous crystalline and stable materials which can be prepared by reversible cross-linking of rigid organic building blocks. In this aspect, concept of dynamic covalent chemistry (DCC) pertaining to reversible bond formation (reticular chemistry) find relevance for the synthesis of crystalline COFs. Contrary to conventional covalent bond formation, DCC regulates the thermodynamic equilibrium during bond formation via self-correction and thus leading to the formation of the most thermodynamically stable crystalline structures. Figure 16.12, gives the various steps involved in the formation of crystalline COFs.

However, reversibility is not the sole factor that can assure the long range ordering. The topology of the building blocks has important role to play in deciding the crystallinity of COFs. The symmetry of building blocks meets the requirement of constructing the regular pores while DCC accounts for the reversible formation

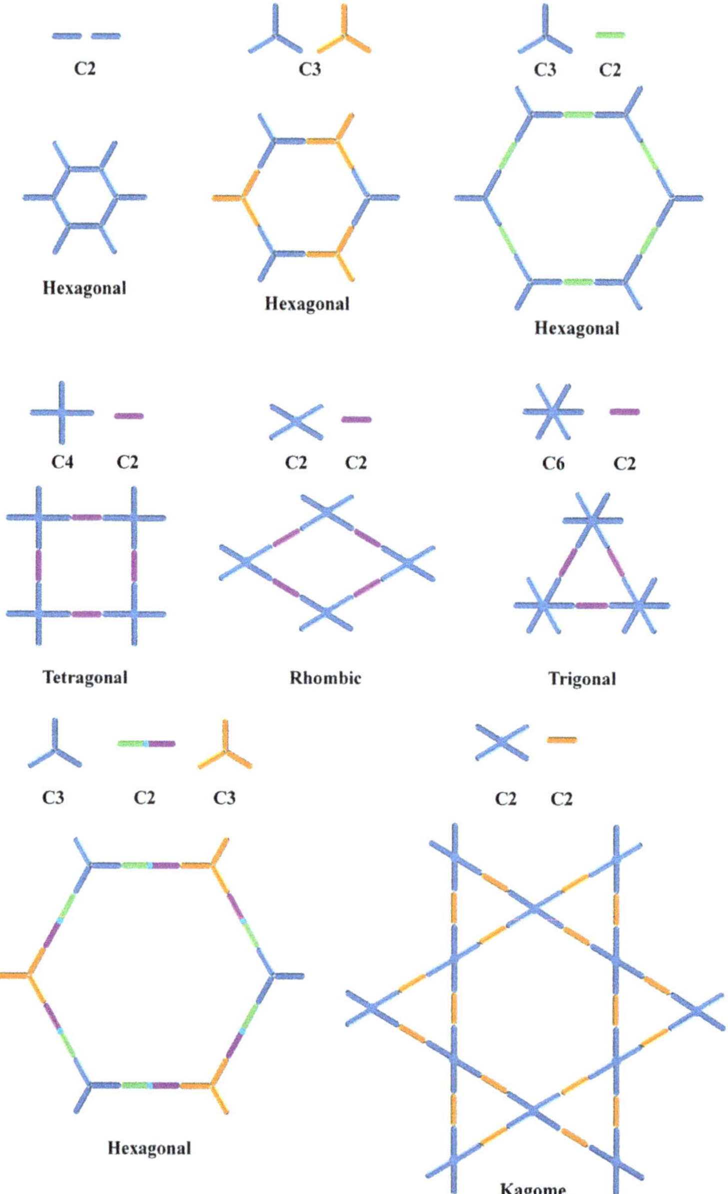

Fig. 16.9 Fusion of building blocks with different geometries to design COFs [26]

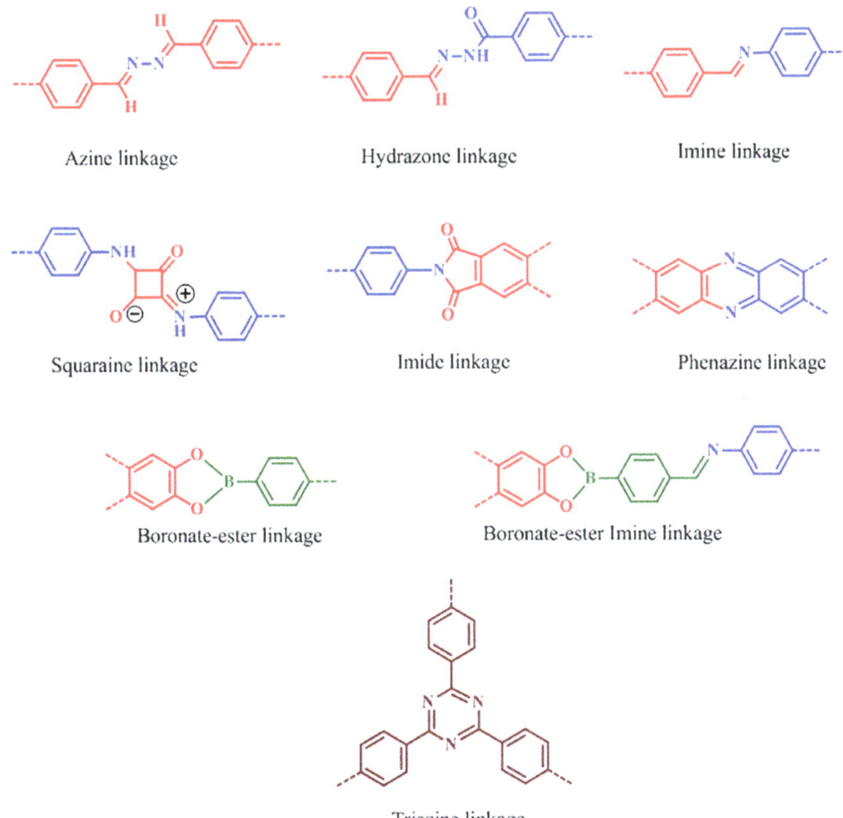

Fig. 16.10 Various linkages amendable for the preparation of COFs [52]

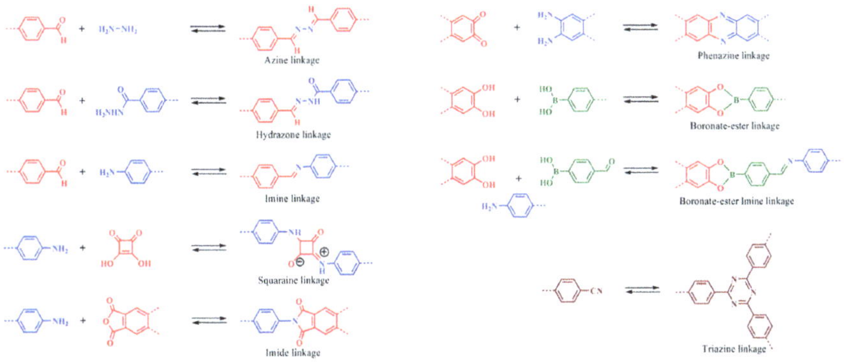

Fig. 16.11 Schematic representation of the reactions for the preparation of COFs [52]

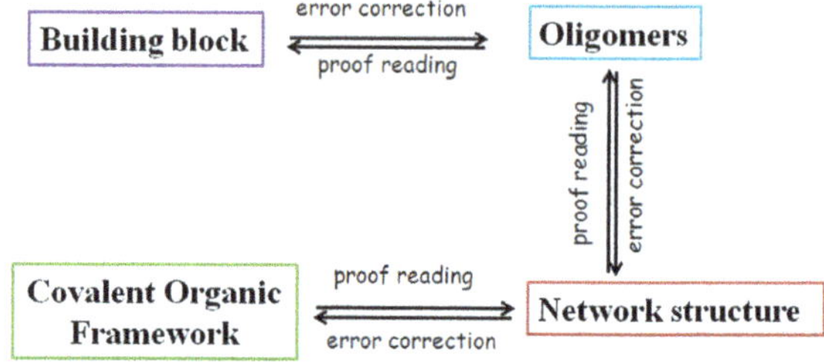

Fig. 16.12 Schematic representation of steps in dynamic covalent chemistry [58]

of covalent bonds through self-correction. Thus, choice of building block should be in accordance with the reactivity of the functional group that triggers dynamic covalent bond formation. In addition, different types of organic linkers on basis of symmetry should be selected so that the topology of the pore of the framework materials can be controlled. For instance, Schiff base condensation reaction is the more profuse chemistry used for the synthesis of imine-based COFs. This is attributed to the control provided over bond formation, breaking and reformation of bond which ultimately facilitate crystallization process in COFs formation [57].

Apart from DCC strategy and topology of building blocks, several other factors such as reaction conditions, temperature, amount of the reactants, water and catalyst amount influence the reaction. These important parameters needed to be concerned during synthesis of thermodynamically stable highly porous and crystalline COFs. Moreover, the solvent combination is also vital parameter to control the porosity and crystallinity of the COFs. To sum up, there are handful reversible reactions available that fulfils the criteria for the formation of thermodynamically stable crystalline architectures.

16.2.1 Synthetic Methods

As already mentioned, the important as well as the crucial step in MOFs/COFs synthesis is to find out the suitable conditions that lead to the formation of building blocks without decomposition of the organic linker. Shape and size of the channel of the frameworks depend to a large extent on the synthetic methods employed. Figure 16.13 gives an overview of the different synthetic methods conventionally used for MOFs synthesis.

The parameters affecting the framework synthesis can be divided mainly into two categories: compositional parameters (molar ratios of starting materials, pH, solvent, etc.) and process parameters (reaction time, temperature, pressure, etc.). Temperature is one of the main parameters in the preparation of porous frameworks. Various synthetic routes like solvothermal, non-solvothermal, microwave heating, electrochemistry, mechanochemical and sonochemistry (Fig. 16.14) are frequently used. A brief knowledge about the various synthetic routs is mentioned below.

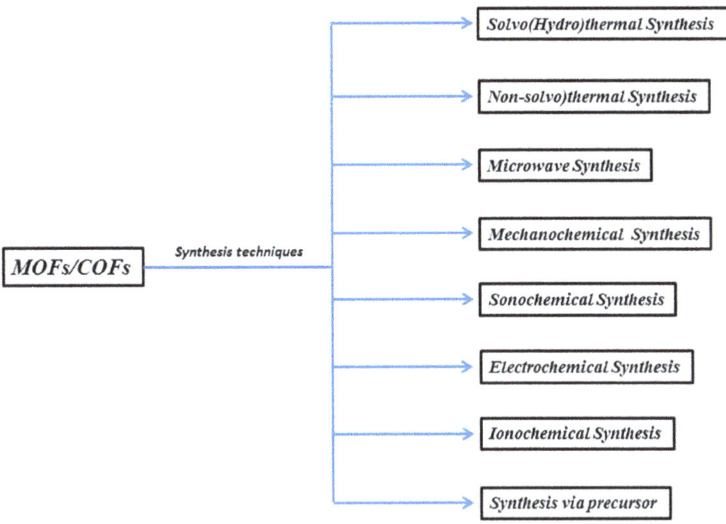

Fig. 16.13 Overview of MOF synthesis under different conditions leading to different product

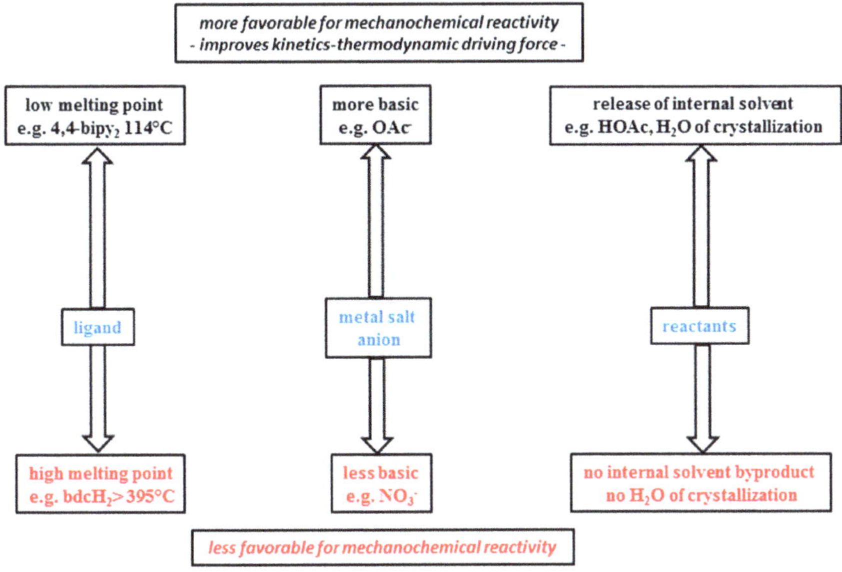

Fig. 16.14 Combination of favourable and unfavourable conditions for mechanochemical reactivity

16.2.1.1 Solvothermal Synthesis

This is one of the highly popular and commonly used techniques for the synthesis of porous crystalline frameworks. Here the reaction takes place in close vessels under autogeneous pressure above the boiling point of the solvent. Generally, this is a comparatively low temperature and high pressure reaction and the reaction is carried out in auto-clave. The method involves mixing of the reactants in a single solvent or combination of solvents in different ratios. The solvent or mixture of solvents of different polarity has been identified as one of the important parameters directly related to induce crystallinity in the frameworks [59–61]. The organic precursor employed is usually soluble in organic solvents which are one of the reasons to make this method quite popular in synthesis of highly porous crystalline frameworks. This synthetic route is widely used in the synthesis of porous MOFs and COFs. Since COFs are mainly prepared by condensation reactions, water is generated as side product and occupies the space above the reaction mixture. The water further cools down and again mixed with the organic solvent which the system can utilize during the process of self-healing and thus governs the reversibility of the reaction. Though the technique is most widely used one, the major drawbacks are long reaction time (several hours), temperature and solvent selection.

16.2.1.2 Non-solvothermal Synthesis

The non-solvothermal reaction can further be classified in two categories: (1) taking place at room temperature and (2) taking place at elevated temperatures. The synthesis at low temperature takes place either by precipitation followed by recrystallization or by slow evaporation of the solvent. The methods are well known to grow simple molecular or ionic crystals because of the possibility of tuning the reaction conditions and hence the nucleation rate and crystal growth. For crystal growth, concentration of reactants plays an important role. Reactants concentration has to be optimized such that it should not exceed critical nucleation concentration, and it can be done by varying temperature or more commonly by slow evaporation of solvent. Other methods routinely employed to obtain crystalline MOFs are layering of solutions or slow diffusion of reactants. Some prominent MOFs obtained at room temperature by just mixing the starting materials are MOF-5, MOF-74, MOF-177, HKUST-1 or ZIF-8. There is also an increasing interest in creating protocols for COFs synthesis under mild conditions such as room temperature and hence improve COFs processability on surfaces which is necessary for their use in practical applications. Bein et al. [62] have synthesized thin films of COF-5 and BTD-COF via vapour-assisted conversion method. This method involves the drop casting of the boronic acid precursors on glass slide, followed by incubation in a desiccators containing mesitylene–dioxane solvent mixture (1:1) in a separate glass vials. With time, the solvent vapours diffuse out slowly from the glass vials. As a result, the solvent molecule gets in contact with the reactant molecules drop casted over the glass substrate and COF thin films get formed on the glass slide. Due to slow diffusion kinetics of the solvent vapours, the COF crystallite formation reaction was complete in 72 h. Top view scanning electron microscopy (SEM) reveals continuous coverage of inter-grown particles on the substrate. Surface area of the synthesized COF thin film is commensurate to that of its solvothermal counterpart. On a similar ground, Ballesté et al. [63] have established micro-fluid-based synthetic method where the reaction between constituent building blocks leading to imine-based COFs takes place under controlled diffusion conditions at room temperature. This method involves the mixing of droplets of reactants and acetic acid injected through the separate nozzles inside the channel. COFs will get formed within few seconds and can be collected simultaneously from the outlet. Material obtained using this method is observed to have crystalline fibre like morphology. The dynamic nature of the protocol allows the direct printing of the COFs onto the surfaces. Variation of the reaction temperature has a strong influence on the product formation, crystallinity, reaction rates as well as morphology of crystals. A prolonged reaction time may lead to degradation of the framework.

16.2.1.3 Microwave-Assisted Synthesis

Microwaves (MW) are also used in the synthesis of porous frameworks [64]. This route is considered as more efficient than other conventional routes for small-scale synthesis because of less reaction time, clean products and high yield. When a sample (solid/polar solvent) is subjected to microwave, an electric current is generated due to the mobile electron or ions. In solid sample, this electric current leads to the heating of the sample due to the resistance of the solid sample. In case of solution, the presence of polar solvent and hence the polar molecules orient themselves according to the electromagnetic field. Since under microwave an oscillating field is produced, the molecules change their orientation accordingly. Thus by applying suitable frequency, collision between the molecules takes place. As a result, the kinetic energy of the molecule and hence the temperature of the sample increase. Due to the direct interaction of the radiation with the sample, MW-heating presents a very high energy-efficient method of heating. Cooper et al. [65] have first introduced the microwave heating method for rapid synthesis of crystalline porous frameworks, and the framework was purified easily with few washing steps only. At present, different synthetic protocols are under development for utilizing the microwave heating for synthesis of framework materials on bulk scale. The bottle neck of this technique is the choice of appropriate solvents and selective energy input.

16.2.1.4 Mechanochemical Synthesis

Mechanical grinding has been identified as another efficient methodology where mechanical force is used for the preparation of frameworks at room temperature [66]. Mechanical force helps in breaking the intra-molecular bonds and forming new chemical bonds. The method can be used in a solvent free condition and hence is environment friendly and is taking place at room temperature. The first example of such synthesis was reported in 2006 [67] where three-dimensional microporous $[Cu(INA)_2]$ (INA = isonicotinic acid) was obtained by grinding mixture of copper acetate and isonicotinic acid for 10 min. In contrast to conventional synthesis, here metal salts can be replaced by metal oxides as starting material and in that case water is formed as the only side product. One such example is the synthesis of pillared-layered MOFs using ZnO. The success of the technique depends on various parameters like counterpart of the metal ions, ligands basicity and ligands melting points. Figure 16.14 gives an overview of the various favourable conditions for this approach.

From Fig. 16.14, it is seen that ligands having low melting point and metal salt with basic counterpart like acetate, combination of reactants where internal solvent like acetic acid, water of crystallization will be released during the course of the reaction favour the success of the technique. Ligand with high melting point and metal ions with less basic counter-ions like nitrate and the combination of ligand and metal system where no solvent will be released during the milling, the chance of the reaction decreases. The technique is also used for the synthesis of COFs. In a typical synthesis, the reactants are mixed and ground in mortar–pestle in the presence of very small amount of solvent, typically water. The paste so obtained is further subjected to high temperature (90–120 °C) typically for 12–24 h followed by several washing with suitable solvent in order to obtain highly porous and crystalline framework materials. The use of mechanical force for the synthesis of highly crystalline and porous imine-based COFs was reported by Banerjee et al. [66]. Further, they have demonstrated the synthesis of COFs in bulk scale as well as the processability of COFs in different shape and size without compromising the porosity and crystallinity of the materials. One of the major limitations with this methodology is the exfoliation of the frameworks into sheets during the grinding process which ultimately affect the porosity and crystallinity of the frameworks.

16.2.1.5 Sonochemical Synthesis

Sonochemical technique is also used for the synthesis of MOFs. This is a fast, energy-efficient, environment friendly, room temperature method. A fundamental principle of sonochemistry and its use for the synthesis of nanomaterials is well-documented in the literatures [68, 69]. Here a high energy ultrasound with a frequency between 20 kHz and 10 MHz is applied to the reaction mixture.

16.2.1.6 Electrochemical Synthesis

This is also well-established and extensively used technique for the synthesis of MOFs with the exclusion of anions, such as nitrate, perchlorate or chloride. In this process rather than using metal salts, metal ions are continuously introduced through anodic dissolution to the reaction medium containing dissolved linker molecules and a conducting salt. This technique was successfully utilized in the synthesis of Zn_2^+ , Cu_2^+ and Al_3^+-based MOFs by Gascon et al. [70].

16.2.1.7 Ionothermal Synthesis

In this method, the ionic liquids simultaneously serve the purpose of both solvent and template or structure directing agent in the formation of solids [71]. This strategy is well known to prepare COFs containing triazine core with high porosities and surface areas by trimerization reaction of simple, cheap and abundant aromatic nitriles [72]. The triazine-based materials synthesized by this methodology are often observed to have nearly same properties as that of zeolites and metal organic framework (MOFs). Typically, it is high-temperature reaction where solid reactant mixed with $ZnCl_2$ was taken in a quartz tube followed by heating at 400 °C to afford the COFs. The molten $ZnCl_2$ acts as a catalyst for trimerization reaction which is otherwise reversible at this temperature. Aromatic nitriles show good solubility in this ionic melt due to strong Lewis acid base interactions. The trimerization reaction can be monitored simply by recording FTIR spectra of the reaction mixture at different reaction times and temperature. The disappearance of strong absorption of carbonitrile band at 2218 cm^{-1} and appearance of intense band at 1352 cm^{-1} corresponding to the formation of triazine ring are indicative of completion of trimerization reaction.

16.2.1.8 Synthesis of Mono Layers on Surface

Owing to their robust structures, accessible functionalization sites, tunable optical and electronic properties, the thin layer of self-organized molecular networks (2D) is ideally suited for many high end applications [54, 61(a),73]. However, the growth of thin film on desirable substrate is highly challenging task as the frameworks are inherently insoluble and thus inhibit the processability. Being unprocessable powders, the porous frameworks cannot be interfaced to electrodes or fabricated into device form unless there is a way of synthesis of oriented 2D layered framework on desirable surface. The thin layer of the porous framework could provide better opportunity of understanding the structural details. Porte et al. [74] have reported the first example of surface covalent organic frameworks (SCOFs) by growing SCOF-1 (dehydration of 1,4-benzenediboronic acid (BDBA) with boronic acid) and SCOF-2 (condensation of BDBA with 2,3,6,7,10,11-hexahydroxytriphenylene (HHTP)) on the clean Ag(111) surface. Later, Dichtel et al. [74c] have demonstrated a simple solvothermal condensation method where oriented COFs thin films were directly grown over single layer graphene (SLG) supported over various substrates. The thickness, crystallinity and morphology of the film were controlled with the reaction time. In due course, Wan et al. [75] have emerged with an innovative idea where highly ordered surface COFs layer was fabricated over 2D materials. To achieve this, they introduced a small amount of $CuSO_4.5H_2O$ as a water regulator which further acts as an equilibrium manipulating agent in a dehydration reaction of BDBA into a closed system. The product obtained was highly ordered molecular network which otherwise known to form disordered network at high temperature [74a, 76]. The reversible release of water from $CuSO_4.5H_2O$ during the heating cooling cycles controls the reversibility of the reaction, thus improving the crystallinity of surface covalent organic framework (SCOFs) drastically.

16.2.1.9 Interfacial Synthesis

Interfacial growth of COFs thin-film methodology is a unique approach. Adopting this methodology, a self-standing 2D COFs membrane can be easily produced. The idea behind this interfacial synthesis is reaction at the interface formed by two different immiscible solvent containing respective reactants. Likewise other methods, the thin films obtained are not defect-free but this limitation can be minimized through varying the different parameters such as concentration of the reactants and temperature. Recently, Dey et al. [77] have demonstrated the synthesis of highly crystalline and porous thin films of COFs via interfacial synthetic method.

Synthesis via Precursor Approach

Apart from the above-mentioned various techniques, precursor approach is also extensively used for MOFs preparation. In this approach, pre-built polynuclear coordination complexes having structure and functions similar to or identical with the inorganic bricks are used and the ligands were substituted with new ligands to form the desired frameworks. The steps involved in the process may well be understood with the below-mentioned example (Fig. 16.15), where $[Zr_6]$ methacrylate oxocluster was used as the precursor and the new framework was

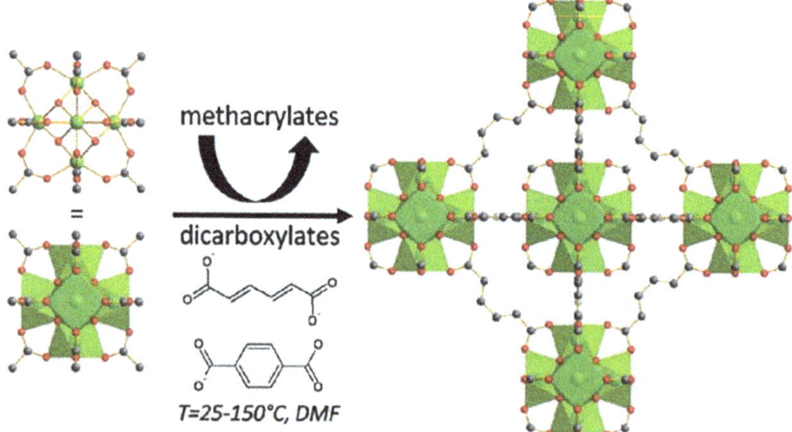

Fig. 16.15 Synthesis of Zirconium dicarboxylate from $[Zr_6]$ methacrylate oxocluster [78] [Copyright: Royal Society of Chemistry]

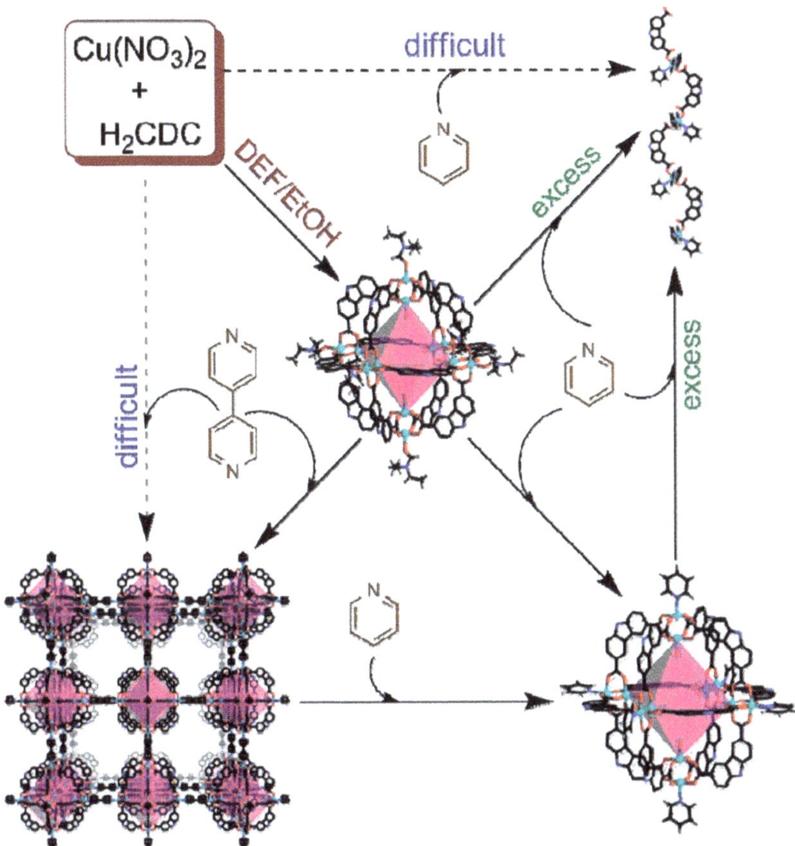

Fig. 16.16 Synthesis of MOFs by substituting the axial solvent molecules [79] [Copyright: American Chemical Society]

synthesized by replacing monocarboxylate by dicarboxylate [78]. Another way of such synthesis is the replacement of coordinating solvent with linker. $[Cu_2(CDC)_2(DMA)(EtOH)]_6$(CDC = 9H-carbazole-3,6-dicarboxylate; DMA = N, N-dimethyl acetamide) has two axial ethanol molecule per copper centre. This axial ethanol molecule may be replaced by some other linkers. Figure 16.16 represents one such example where new MOFs are formed by replacing the axial ethanol [79]. The direct syntheses of these MOFs are rather difficult.

16.2.2 Methods for Post-Synthetic Functionalization of MOFs

In general, the presence of functional group dictates the application of the framework. The direct synthesis of MOFs with functionalized organic linkers is limited because of the direct coordination of the functional group with the metal centre and hence prevents the formation of the frameworks. Moreover, MOFs contain organic linkers, the frameworks can be functionalized even after their formation, and the method is known as post-synthetic modification (PSM). There are a variety of ways by which the framework can be functionalized. Based on the nature of the bond formed or broken during PSM, the process is broadly classified into three categories, namely (A) covalent modification; (B) dative modification and (C) post-synthetic de-protection as discussed below.

In covalent modification, a new covalent bond is formed (Fig. 16.17). This method of PSM is the most powerful and versatile ways to introduce a large number of functional group. One such example is schematically shown in Fig. 16.19 [80] where acetylation takes place due to the reaction of acetic anhydride with IRMOF-3 via the breakage of N–H bond and formation of new N–C bond.

In dative PSM, a new metal–ligand coordination boned is formed (Fig. 16.18). Therefore, in dative PSM, either a ligand is added to the framework that coordinate

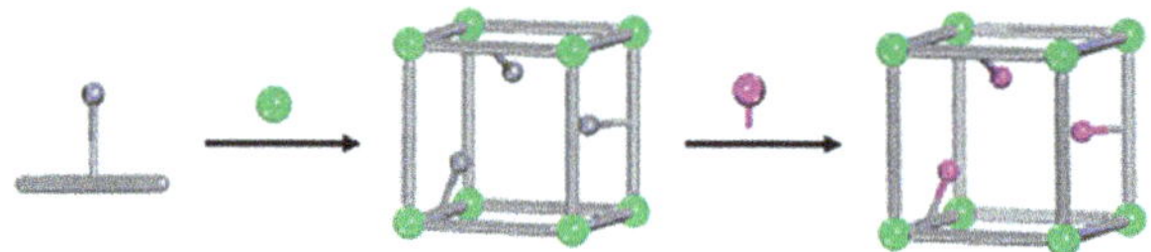

Fig. 16.17 Covalent post-synthetic modification [81] [Copyright: American Chemical Society]

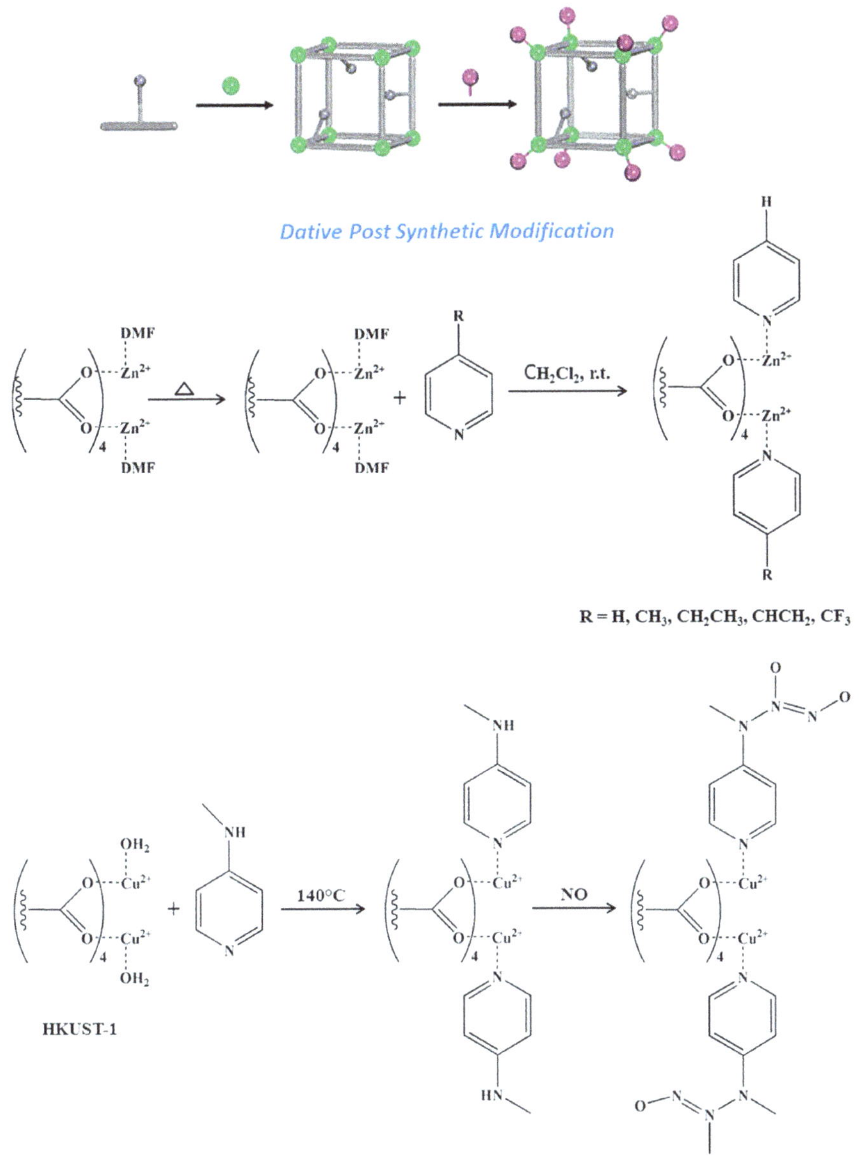

Fig. 16.18 Dative post-synthetic modification [81] [Copyright: American Chemical Society]

to the SBU of the MOF or a metal source is added which then bound to the organic linker of the MOF via dative bonds.

Examples of dative PSM are given in Fig. 16.20. In both the cases, on heating the frameworks under vacuum, the axially coordinated solvents (DMF and H_2O) could be removed and on subsequent treatment of the coordinatively unsaturated

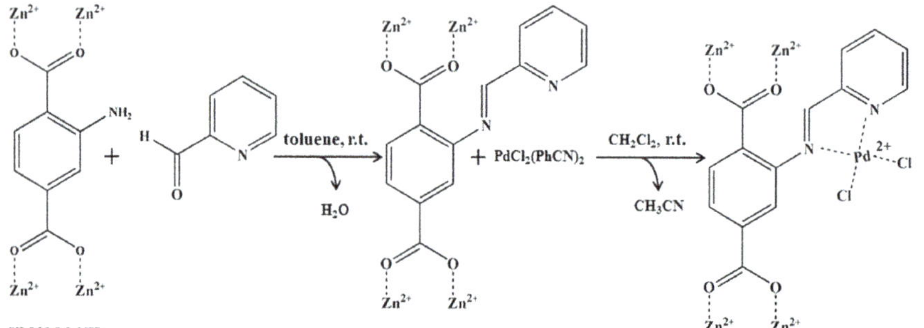

Fig. 16.19 Example of combine covalent and dative post-synthetic modification [81]

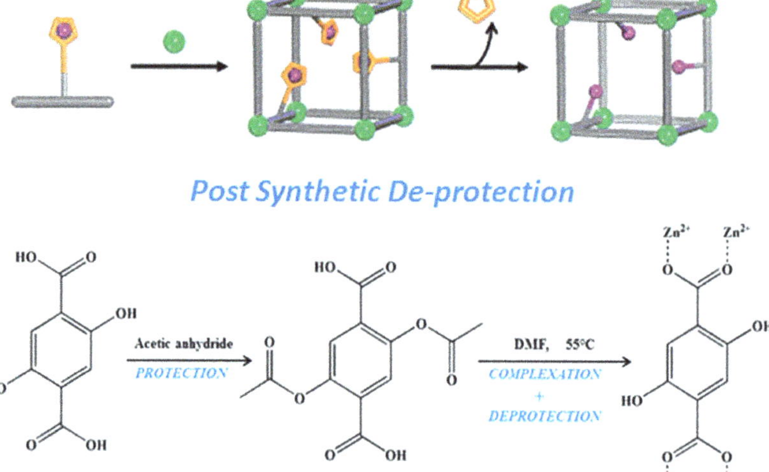

Fig. 16.20 Example of post-synthetic deprotection [81] [Copyright: American Chemical Society]

frameworks with pyridine derivatives new frameworks could be formed [82, 83]. It has been observed that the direct synthesis by solvothermal technique was not as fruitful as by dative PSM due to a significant loss of pyridine during synthesis in case of Zn(II) and not reactivity of pyridine in case of KHUST-1.

Figure 16.19 represents one example where both covalent and dative post-synthetic modification take place. The first step represents the covalent modification, and the second step represents the dative modification.

Sometimes during framework synthesis functional group needs to be protected by converting it into some suitable derivative so that it does not coordinate the metal centre during synthesis. After framework formation, the deprotection of the functional group is carried out and the modification is known as post-synthetic deprotection modification.

In principle, both the covalent or dative bonds can be broken during post-synthetic deprotection. Figure 16.20 reflects one of the serendipitous finding by Yamada and Kitagawa, where in-situ deprotection of an organic linker was observed, resulting in a functionalized MOF [84]. The protection of hydroxyl group was carried out via acetylation of 2,5-dihydroxyterephthalic acid with acetic anhydride and deprotection takes place during MOF formation.

16.2.3 Activation

The final step before utilizing the porosity of MOFs is the removal of guest molecules from the pore keeping the structural integrity and porosity. The guest molecules may be solvents or any other chemicals used during synthesis. The process is known as "activation". High boiling solvents like N, N-dimethylformamide (DMF), N, N-diethylformamide (DEF) or dimethyl sulfoxide (DMSO) are conventionally used in MOFs synthesis. These high boiling solvents generate high capillary force and high surface tension during activation. This high surface tension leads to the partial or full collapse of the frameworks in many cases during activation. In many cases, it has been observed that the measured surface area is less than the one calculated from single crystal X-ray data and is due to incomplete activation. Therefore, the activation is a very crucial step in MOFs chemistry. The following four techniques are the most common and widely used strategies for MOFs activation: (a) conventional heating under vacuum; (b) solvent exchange; (c) supercritical CO_2 ($scCO_2$) processing; and (d) freeze-drying [85]. Figure 16.21 illustrates the various physical phenomena occured during these steps.

Conventional activation: Activation by heating under vacuum which is similar to the technique used for activation of zeolites and carbons is the simplest technique. The essential criteria for applying this technique are that the frameworks have to be thermally stable. Though this strategy for activation has been successfully applied in some cases like the activation of Cr-MIL-101 [86] and UIO-66 [87], this strategy finds minimal utility for accessing the full porosity of many MOFs. It has been observed that in many cases the framework losses the crystallinity and porosity upon activation by this technique. This observation can be well-understood based on the diagram of Fig. 16.21. As the framework passes through liquid-to-gas phase, a significant amount of surface tension and capillary forces generates and this force breaks the weak coordination bonds resulting the loss of crystallinity and porosity.

Activation by solvent exchange: The most commonly used activation technique is the activation by solvent exchange. Normally during synthesis of MOFs, higher boiling solvents like DMF and DEF are used and remain trapped in the pores. These higher boiling solvents are replaced with lower boiling solvents like acetone and methanol which are removed by mild heating under vacuum. One such example of showing activation by solvent exchange is the activation of MOF-5. The framework

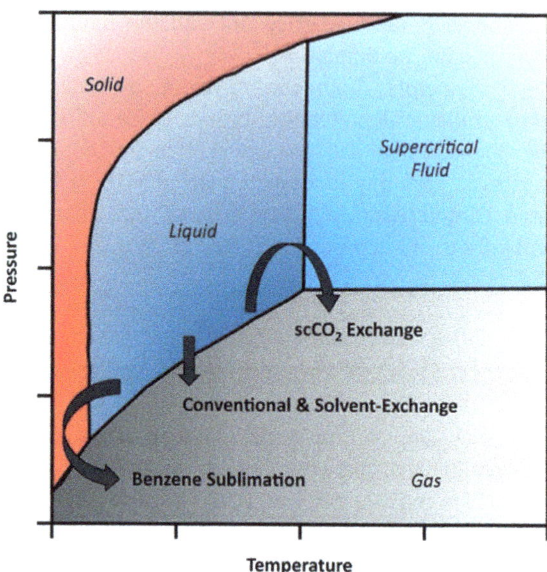

Fig. 16.21 Phase diagram of the physical phenomena encountered for conventional and solvent-exchange activation, $scCO_2$ exchange and benzene freeze-drying. [85] [Copyright: Royal Society of Chemistry]

was synthesized in DMF–chlorobenzene mixture. The framework could be activated by exchanging DMF–chlorobenzene with comparatively low boiling $CHCl_3$. The importance of the selection of suitable activation technique is reflected in the activation of IRMOFs as shown by Nelson et al. [88]. When conventional activation technique was employed, IRMOF-3 showed a BET surface area of 10 m^2g^{-1} while no N_2 uptake was observed for IRMOF-16. When DMF was exchange with $CHCl_3$, the surface areas were increased to 1800 m^2g^{-1} for IRMOF-3 and 470 m^2g^{-1} for IRMOF-16.

Activation by supercritical CO_2 ($scCO_2$) processing: Activation by supercritical CO_2 is the relatively new strategy for activation of MOFs. In this process, solvents which are miscible with liquid CO_2 are exchanged with liquid CO_2 at high pressure (>73 atm). Then the sample is brought above the supercritical temperature of CO_2 (i.e. 31 °C) so that the framework gets occupied with supercritical CO_2. Finally, the supercritical CO_2 apparatus is slowly vented while holding the temperature above the critical point. As a result of which the system is transforming directly from the supercritical phase to the gas phase without liquid-to-gas phase transition and hence avoiding the capillary forces.

Activation by freeze-drying: Another newly developed method for MOFs activation is the activation by benzene freeze-drying. Here first the solvent is exchanged with benzene and left in benzene. The MOF is then frozen to 0 °C and brought back to room temperature. The procedure is repeated several times. Upon the final freeze cycle, the MOF is placed under vacuum at a temperature and pressure below the solvent's triple point. Since in the final step the sample is warmed under reduced pressure, benzene directly sublimes which is a direct solid-to-gas phase transition and hence avoids the liquid-to-gas phase transition and hence associated capillary forces. Recent advances show that instead of benzene which is carcinogenic, cyclohexane also can be used.

16.3 Application of MOFs and COFs

MOFs and COFs possess some desirable properties like inherent porosity, tunable pore size, large surface area and ordered channel structure, low density, relatively high thermo-chemical stability, and designable functionality affirms the candidature of these materials for wide range of application (Fig. 16.22) like gas storage and separation, heterogeneous catalysis, optoelectronics, energy storage, chemical sensing and drug delivery. This section presents a detail discussion of some selected application of MOFs and COFs.

16.3.1 Gas Storage Application

Gases like H_2 and CH_4 are important from view point of energy production. Thus, medium for their effective storage is considered as an essential requirement. Various options are available for storing gases but these often require high pressure and multi-stage compressors. Moreover, these options are not viable from economic point of view. Thus, there is a need to substitute these materials by simple, easy to

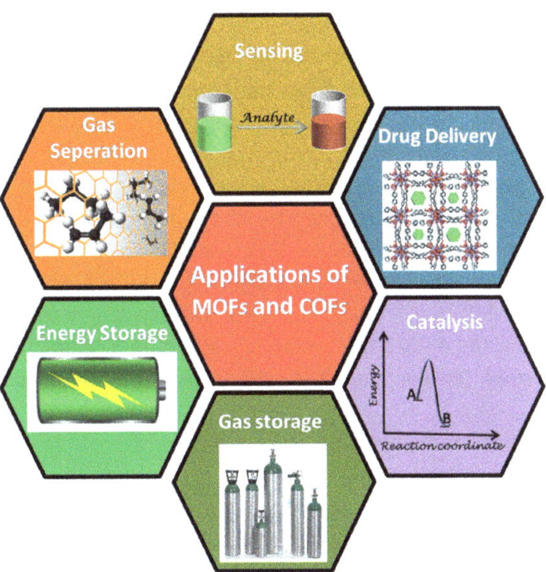

Fig. 16.22 Representation of some important application of MOFs and COFs

handle and cheaper options. In this regard, porous framework materials especially MOFs and COFs are preferable choice because of their easy synthetic procedures, high surface area, well-defined pores and channels with structural and compositional tunability. This section deals with the utility of MOFs and COFs for storage application of various important gases from energy and environment point of view with brief discussion over the factors responsible for effective storage.

16.3.1.1 Hydrogen Storage

To reduce the dependence upon the fossil fuels, there is an ongoing quest for clean energy sources with high energy density. Hydrogen gas (H_2) upon combustion produces water along with enormous amount of energy due to its high calorific value (120 MJ Kg^{-1}) and thus considered as clean and ideal source of energy. However, low volumetric storage capacity of H_2 due to its low density in gaseous state (0.08 kg m^{-3}) hinders its use as an efficient fuel. Thus, effective storage materials need to be investigated. The goal set by US Department of Energy (DOE) for hydrogen storage is 1.5 kg of H_2/Kg in gravimetric capacity and 1.0 kg of H_2/L of volumetric capacity at operating temperature of 233–333 K with a pressure of 100 atm by the year 2020 [89].

Several favourable characteristics of MOF like large surface areas, functionalized polar groups, open metal centres facilitate the storage of H_2 in it. Many MOFs have been tried so far for H_2 uptake capacity, and results showed that MOF possess the required characteristics to act as H_2 uptake material and one of the best options to meet the targets set by US department of Energy in near future. For example, MOF-177 derived from zinc acetate unit [$Zn_4O(CO_2)_6$] and the tri-topic linker 4,4′,4″- benzene-1,3,5-triyltribenzoate (BTB) to form a (6, 3) net exhibit remarkable H_2 uptake on a gravimetric basis (7.5 wt%). The performance of MOF-177 for H_2 uptake is attributed to its large surface area (5000 m^2g^{-1}) [90]. Similarly, IRMOF-20 derived from same metallic SBU and linear di-topic link thieno[3,2-b] thiophene-2,5-dicarboxylic acid (TTDC) showed substantial H_2 uptake on volumetric basis (34 g/L) [90]. The measurements were carried out at 70 bar and 77 K. The important thing to note from the performance of these MOFs that they dispels the common misconception that porous materials will inherently have poor volumetric storage capacity. There are several other MOFs which showed appreciable H_2 uptake capacity like MOF-74, HKUST-1, IRMOF-11, MIL-101, NOTT-102 and MOF-205 [91–94]. From studies, it is evident that MOFs with large surface area and open metal sites provide stronger interactions between metal nodes and H_2 which is the principle reason behind the success of MOFs for H_2 uptake. Theoretical studies suggest that doping of MOFs with certain specific metal ions can boost their H_2 uptake capacity.

On the other hand, COFs also pose as promising candidates for H_2 uptake due to their chemical robustness, high water stability and crystallinity along with high surface area and tunable pore size. Three-dimensional COFs showed better gas storage properties compared to two dimensional COFs. Various COFs have been explored for H_2 storage. For example, 3D COFs like COF-102 and COF-103 exhibit reasonably high H_2 uptake of 7.24 and 7.05 wt% at 77 K and 85 bar, which is at par with MOF-177 and IRMOF-20 [26]. COF-10 with BET surface area (1760 m^2g^{-1}) showed H_2 uptake of 3.9 wt% at 77 K and 85 bar which is highest among 2-D COFs [95].

From various studies, it is inferred that H_2 uptake capacity of COFs can be improved in two ways:

(i) Increasing the surface area of COFs while maintaining their microporous nature.
(ii) Metallation of the COFs with metal ions such as Li^+, Ca^{2+} and Pd^{2+} which can strongly interact with H_2.

16.3.1.2 CH_4 Storage

Methane gas is another most convenient, inexpensive and relatively clean source of energy. Natural gas also contains around 95% of methane. Gravimetric heat of combustion of methane (50 MJ Kg^{-1}) is much lower than that of hydrogen (120 MJ Kg^{-1}) but comparable to that of gasoline (44 MJ Kg^{-1}). Therefore, methane can replace gasoline and coal which are often responsible for air pollution. Cars powered by CH_4 produced relatively less carbon dioxide than gasoline but they require costly tanks and compressors to store CH_4. Therefore, for practical use of CH_4 as a fuel, an efficient storage material is required. US-DOE has set a target of 263 cm^3 of CH_4 per cm^3 of adsorbent [95].

MOFs and COFs have the ability to store methane at lower pressures because methane molecules can pack tightly in their pores. The basic criteria for high methane uptakes in porous materials are:

(i) High sorption capacity.
(ii) Good adsorption enthalpy.
(iii) An efficient charge–discharge rate.

First methane storage studies in MOFs were reported by Noro et al. [96]. They utilized [{$CuSiF_6$(4,4′-bipyridine)$_2$}$_n$] stabilized in 3D microporous network to store 0.21 g/ml of methane. Various other MOFs like [PCN-14 Cu_2(adip)] (16 wt% at 35 bar), HKUST-1 (15.7 wt% at 150 bar), MIL-101(14.2 wt% at 125 bar), IRMOF-1 [228 cm^3(STP) g^{-1} at 36 bar) showed substantial methane uptake capacity [97–99]. MOF-177 and Ni-MOF-74 exhibit 22 wt% at 100 bar and 190 cm^3 (STP)g^{-1} at 35 bar methane uptake capacity, respectively, which is the highest storage among the dry samples [100]. In case of MOFs, open metal sites and microporous nature along with high pore volume are the contributing factors responsible for high methane uptake.

Among COFs, COF-102 and COF-103 exhibit highest methane uptake of 187 mg g^{-1} and 175 mg g^{-}1,respectively, at 35 bar and 298 K which is comparable to best previously reported materials: Ni-MOF-74 [90, 101]. Theoretically, it was predicted that Li ion doping in these COFs can double their methane uptake capacity due to increased induced dipole interactions and London dispersion forces between methane molecules and doped lithium ions. From various studies, it is concluded that like H_2 uptake, CH_4 uptake in COFs is related to the surface area of the COFs and the concentration of doped ions which improves weak interactions.

16.3.1.3 CO_2 Storage

Every year, the uncontrolled combustion of fossil fuel for transportation and running industries is increasing the concentration of greenhouse gases such as carbon dioxide and carbon monoxide in the atmosphere. Sea level rising and dramatic change of the climatic conditions are some of the environmental sector which are under direct impact of air pollution due to increase amount of carbon dioxide in the atmosphere. According to one article published in the *OECD Environmental Outlook to 2050* released at the 2011 United Nations Climate Change Conference, discussed the need for negative emissions, stating "Achieving lower concentration targets up to 450 ppm for clean environment to sustain existence of life". Thus, the major quest in this regard is to develop a material for CO_2 storage and sequestration.

Various porous materials like activated carbon and zeolites have been explored for CO_2 adsorption. MOFs have also shown great potential for CO_2 storage due to their high internal surface area and the presence of polarity due to functional groups inside the pores. Variety of MOFs based on the functionality and surface area has been explored for CO_2 uptake. For example, MOF-177 and MIL-101 exhibit nearly 60 wt% of uptake capacity [101a, 103]. NU-100 (69.8 wt%, 40 bar), Mg-MOF-74 (68.9 wt%, 36 bar), MOF-5 (58 wt%, 10 bar) and HKUST-1 (19.8 wt%, 1 bar) are other well-known MOFs which show considerably high CO_2 uptake. MOF-210, due to its ultra-high surface area (10,450 m^2g^{-1}) exhibit very high CO_2 uptake (74.2 wt% at 50 bar) which is higher than that of any other porous material. Also, it has been established through theoretical as well as experimental studies that the presence of polar groups such as $-NH_2$, or free N containing heterocyclic residues facilitates the CO_2 uptake [102, 103].

Likewise, COFs have also shown great potential for CO_2 uptake. Yaghi and co-workers for the first time studied a series of COFs for CO_2 capture. Studies reveal that COF-102 exhibits highest uptake (27 mmol g^{-1} at 35 bar) which is at par with the uptake shown by MOF-5 [104]. Theoretical calculations have predicted that doping in COFs can improve their CO_2 uptake. For example, lithium doping in COF-102 and COF-105 improves their CO_2 storage capacity to 409 and 344 mg g^{-1}, respectively, which is manifolds higher than their pristine COFs [105]. Studies reveal that triazine-based, azine-based and imine-based COFs show high CO_2 uptake due to dipole-induced dipole interactions between polar groups and CO_2 molecules inside the framework pores [72]. It has also been observed that the presence of polar functional groups in the pores introduces the selectivity in adsorption of gases.

16.3.1.4 Ammonia Storage

Ammonia is widely used in chemical industries (for the production of nitrogeneous fertilizers) and pharmaceutical industry. It is also explored as an energy source for fuel cells [106]. However, its handling, storing and shipping are costly affair, and also, it requires special precautions due to its corrosive nature and toxicity. The US Occupational Safety and Health Administration (OSHA) sets a 15-min. exposure limit for 35 ppm gaseous ammonia [107]. Various porous materials like activated carbons have been explored for NH_3 uptake, but low ammonia affinity limits their wider utility [108]. MOFs, due to the presence of open metal sites and pores decorated with functional groups, show great potential for NH_3 uptake. For example, Dinca and co-workers [109] have explored three triazolate-based MOFs, among which Mn-MOF exhibit highest NH_3 uptake at 298 K and 1 bar. The study showed that uptake of NH_3 by MOF is better than activated carbons currently used commercially. The large NH_3 uptake is ascribed to high open metal sites in the frameworks while their excellent resistance towards ammonia is attributed to its triazolate-based ligands. Since NH_3 is a bronsted base, it has great affinity towards acidic group (such as sulphonic acid). The presence of such functional group besides open metal site can also act as strong ammonia capture site. Yaghi and co-workers modified UiO-66-NH_2 with anhydrous HCl which exhibits considerably high NH_3 uptake compared to pristine MOF at ambient temperature [110]. Fe-MIL-101 decorated with sulphonic acid group also exhibits great ammonia uptake due to the presence of sulphonic acid groups [107].

Similar to MOFs, various COFs have also been explored for NH_3 uptake. Yaghi and co-workers have synthesized boroxine and boronated ester-based COFs for storage of NH_3 gas. Boronate ester-based COFs (COF-10) having high density Lewis acid boron sites which can strongly interact with Lewis base (NH_3) exhibit an ammonia storage capacity of 15 mol kg^{-1} at 1 bar and 298 K the highest ammonia uptake showed by any porous materials [111]. The exceptional high uptake of ammonia by COF-10 is due to the formation of a classical ammonia borane coordination bond. Adding more to the advantage, the ammonia gas adsorbed by the COFs at room temperature can be easily recovered by application of heat which makes the process reversible in nature.

16.3.2 Heterogeneous Catalysis

MOFs and COFs are currently attracting considerable interest as heterogeneous catalysts at moderate temperatures. MOFs with exotic topologies, versatile chemical composition, organic and inorganic building units have shown outstanding catalytic performance in various organic transformation reactions such as oxidation, acetylation, hydroxylation, epoxidation, coupling, hydrogenation, condensation, alkylation and cyclization. In MOFs, both metal centres and organic ligands contribute to catalytic activities while pores provide space for small molecules. The metal centres act as Lewis acid site for catalytic activity are obtained by removing coordinated water/solvent while terminal ligands act as Lewis basic site. Various MOFs have been explored for variety of organic transformation reactions. For example, Cu-BDC has been used for the acetylation of alcohols with 80% efficiency, HKUST-1 for cyanosilylation of aldehydes, MOF-199 for oxidative C–C coupling, IRMOF-9-NH_2 for Knoevenagel condensation Allylic N-alkylation, NUGRH-1 for Friedel Craft reaction, etc. [112].

On the other hand, COFs have also emerged as an efficient catalyst owing to the presence of well-defined active sites such as N, O and S in the framework. Utilizing the pre- or post-synthetic modification approach, interior of the framework can be easily modified for catalysis. The metal ion can be introduced either in-situ or ex-situ approach in the framework for catalysis. Due to polymeric nature, COFs are insoluble in organic solvent and thus can be easily recovered after catalysis without significant loss in their catalytic performance. COFs can be further utilized to carry out the catalysis where organic molecules of large size can easily diffuse inside the pores. Wang et al. have incorporated Pd^{2+} ions in the imine-based COF (COF-LZU1) pore walls, bonding with the imine's nitrogen and present in between the two adjacent COFs layers. In Suzuki–Miyaura coupling reaction this Pd@COF-LZU-1 has shown high catalytic activity [113]. Likewise, cobalt loaded porphyrin-based COF (COF-366 Co) was utilized for the electrochemical reduction of CO_2 with decent activity and selectivity to a competing reaction (H_2 formation) [114]. Banerjee et al. utilized the ex-situ approach, to immobilize palladium (Pd) and gold (Au) nanoparticles inside the framework of TpPa-1 COF [115]. These metal loaded COFs are remarkable water stable even under acidic as well basic conditions. The resulting Pd loaded COF has showed an excellent activity towards C-H activation and C–C coupling reaction where Au nanoparticles loaded TpPa-1 was used for the reduction reaction of nitro compounds. In such metal loaded COFs, the COF skeleton imparted an extra chemical robustness to the active metal centres during catalysis [115].

16.3.3 Energy Storage

The porous framework materials containing redox-active centre can be easily converted into energy storage materials. MOFs have the natural advantage because the metal ion present in the framework can easily undergo different redox states on application of the potential. MOFs with high surface area and low density are promising electrode material for rechargeable batteries and super-capacitors of next generation. MOFs can be tailor made to suit the final application by choosing the specific metal site and tuning their pore sizes. The chemical interaction between the metal sites and functional linkers with polysulphides improves the cycling performance of Li–S batteries. Moreover, metal centres in MOFs are the active site for redox reactions while open framework structure supports reversible insertion and extraction of ions [116]. MIL-53(Fe) acts as a cathode material to reversibly insert

Li^+. Studies reveal that 0.6 Li^+ per Fe^{3+} could intercalate into MIL-53(Fe) at C/40 with no structural alteration [117]. The result reveals that MOFs are material of choice for lithium ion battery. Apart from being cathode material, MOFs can also perform the function of anode material. For example, Mn-BTC MOFs (Mn-1,3,5-benzenetricarboxylate) exhibit high specific capacity of 694 mAh g^{-1} and approximately 83% capacity retention over 100 cycles at 103 mA g^{-1}. The COO– groups in Mn-BTC MOFs play a significant role for the Li^+ insertion/ extraction [118].

On the other hand, owing to their ultra-high surface area COFs have the unique ability of integrating redox-active groups which pose them superior candidates as electrochemical capacitors. Different redox-active organic linkers such as quinone, naphthalene diimide (NDI) and pyridine were incorporated to serve the purpose. Dichtel et al. have synthesized an anthraquinone moiety containing Schiff-based COF (DAAQ-TFP COF) by the condensation reaction between 1,3,5-triformylphloroglucinol and 2,6-diaminoanthraquinone. The anthraquinone moiety itself is redox-active. The DAAQ-TFP COF displays reversible redox processes. Due to extra chemical stability from COFs framework, the DAAQ-TFP COF has displayed an excellent supercapacitance performance even after 5000 charge–discharge cycles. Taking the advantage of redox-activity of anthraquinone moiety, they have fabricated DAAQ-TFP COF as oriented thin films. The improved capacitance values during the capacitance measurement of thin films, in comparison with the randomly oriented COFs powder, have been observed [119]. Later on, Xu et al. came up with post-synthetic approach and introduced redox-active characteristics to NiP-COFs with organic radicals such as TEMPO which has displayed a high capacitance value of 167 Fg^{-1} [120]. In another interesting approach, COFs was fabricated over the amine-reduced graphene oxide. Adopting this approach, the stacking or aggregation between the graphene sheets was reduced. This has provided more exposed electrode surface area for energy storage application. Moreover, COFs can also act as host material for sulphur to extract positive effect on Li–S batteries. The electrode constructed by impregnation of sulphur into the pores of COFs exhibits stable cycling performance.

16.3.4 Drug Delivery

Design and development of bio-compatible drug delivery system with high drug uptake and administrable drug release are of prime importance to minimize the side effects and in turn enhance treatment efficacy [121]. Many different kinds of nano-carriers have been explored for this purpose such as mesoporous silica, metal nanoparticles, quantum dots, dendrimers and organic micelles. However, these nano-carriers suffer from either low loading capacity or unacceptable degradability and toxicity [122]. In next generation drug delivery system, MOFs and COFs have shown great potential to be used as effective drug carriers. The advantages of using MOFs as drug carriers are [122–124]:

(i) High surface area and porosity powered MOFs with high drug loading capacity.
(ii) MOFs can adopt diverse morphologies, composition and chemical properties owing to the versatility in their structures which favours them with multi-functionalities and stimuli responsive drug controlled release.
(iii) Any modification in MOFs structure will not alter their desirable physico-chemical properties.
(iv) Relatively weak coordinate bonds make MOFs biodegradable material.

Above favourable features of MOFs enable them as promising material for drug delivery. For example, MIL-100(Cr) and MIL-101(Cr) constructed from di- and tri-carboxylates show considerably high ibuprofen uptake of 0.35 g/g for dehydrated MIL-100(Cr) and 1.4 g/g for dehydrated MIL-101(Cr). The release kinetics of ibuprofen under physiological condition indicates a total release of ibuprofen from body in 3 days from MIL-100(Cr) and 6 days from MIL-101(Cr). Compared to MCM-41, MIL-100(Cr) exhibits similar ibuprofen dosage and kinetics while MIL-101(Cr) showed four times larger drug content and slower drug release kinetics. Although this study is based on chromium-based MOFs which are known for their toxicity, this work opens new avenue for MOFs-based drug delivery system [125]. Subsequently, MIL-53(Fe), a less toxic iron analogue, was developed which shows 20 wt% ibuprofen loading and total release took 21 days under physiological condition [126]. This work highlights the flexible nature of MOFs to optimize drug-matrix interaction. Likewise, MIL-100(Fe) has been used as drug delivery system for an antitumor drug doxorubicin (9 wt%, 2 weeks release time) [127]. Zn-based MOFs, Zn-TATAT (TATAT = 5,5′ 5″-(1,3,5-triazine-2,4,6-triyl) tris (azanediyl) triisophthalate) and Zn-CDDB (CDDB = 4,4′-(9-H carbazole-3,6-diyl) dibenzoic acid) were used as carrier for anticancer drug, 5-fluorouracil, which showed 33.3 wt% and 53.3 wt% drug loading capacity [128, 129]. It is found that hydrogen bonding interactions between the drug and MOFs are responsible for such high uptake of drug. Zirconium-based MOFs UiO-66 have attracted considerable interest as a drug carrier because of their two octahedral and tetrahedral cages, bio-compatibility and stability. It is synthesized from zirconium-oxo-clusters and terephthalate anions bearing different functional groups. Studies revealed that UiO-66 and its NH_2 functionalized analogue MOFs are effective carrier for caffeine and 5-fluorouracil. In another study, UiO MOF constructed from $ZrCl_4$ and aminotriphenyldicarboxylic acid (amino-TPDC) bridging ligands were used for the co-delivery of *cis*-platin and pooled small interfering RNAs (siRNAs) to enhance their therapeutic efficacy by overcoming drug resistance genes and resensitizing-resistant ovarian cancer cells to *cis*-platin treatment. The drugs *cis*-platin and siRNAs were sequentially loaded by encapsulating and coordinating to metal sites on the MOFs surface [122, 130]. Moreover, there are stimuli responsive MOFs which exhibited regulated delivery of loaded drugs upon in response to variety of stimuli such as pH, magnetic field, ions, temperature, light and pressure [122, 131–134].

Likewise, COFs also possess following excellent properties which enable this class of porous materials to be used in drug delivery application [135]:

(i) High surface area and porous volume with tunable pore structures for large drug uptake and controlled release.
(ii) π-π conjugated system which facilitates to load aromatic group-based drug through π-π stacking. Moreover, this π conjugation and laminated structure impart excellent photoelectric properties in COFs which enables them for bio-sensing and bio-imaging.
(iii) Dynamic covalent linkages for stimuli responsiveness: COFs are constructed from dynamic covalent linkages instead of weak interactions and coordinate bond, which makes them stable enough in normal conditions and degrade to release the drug by the application of stimuli such as change in pH [136].
(iv) Unique tailorable characteristics and outstanding modifiability: Reports show that surface modification in COFs can improve their bio-compatibility and targeting ability [137]. COFs can be attached to biological probe and drugs through post-synthetic modifications.

Due to such exciting properties of COFs, attempts were made to insert large drug molecule inside the framework for drug delivery applications. Yan et al. have utilized a 3D polyimide-based COF (PI-COF-4 and PI-COF-5) to load ibuprofen (IBU) drug molecule inside the pore. The drug loading was confirmed from the UV–Vis studies while a thermo-gravimetric study reveals the drug loading of 24 wt % in PI-COF-4 and 20 wt% in PI-COF-5 with respect to the COF [138]. In addition to IBU, PI-COFs were able to deliver captopril and caffeine too. Interestingly, this was the first example of applying COF as drug delivery system. This result set an example and demands a further development of COFs for pharmaceutical applications. Later on, Zhao et al. have demonstrated the loading of three different drug molecules, 5-fluorouracil, captopril and ibuprofen, inside the pore of the two-dimensional PI-2-COF and PI-3-COF [139]. Lotsch and co-workers have designed imine-based TTI-COF for the loading of Quercetin used to boost immunity. The drug was bound on the wall of pores through weak interactions [140]. In another report, a photo-responsive single layer COF was fabricated in which the azo-benzene group was introduced in the organic linker. Application of UV radiation leads to reversible photo-induced decomposition-recovering of COFs and exhibits controlled loading and release of copper phthalocyanine. Furthermore, doxorubicin loaded covalent triazine nanopolymer (CTNP) synthesized via the Friedel–crafts reaction acts as a potential nanocarrier for cancer therapy and imaging [141]. Banerjee and co-workers explored imine-based COFs for drug loading and release study. They also utilize COFs for targeted drug delivery. Post-synthetic modified covalent organic sheets with folic acid were used to load and deliver 5-fluorouracil [142]. Although the loading capacity of this modified COF was low, it possesses good anticancer activity compared to other reported COFs.

Apart from loading anticancer drugs, many functional COFs themselves possess anticancer activity. EDTFP-1 COF constructed from 2,4,6-triformylphloroglucinol and 4,4'-ethylenedianiline could accelerate ROS generation and caused the apoptosis of cancer cells [143].

Overall, it may be concluded that MOFs and COFs have shown potential to be used as next generation drug delivery system. However, challenges still exist regarding the targeted drug delivery and efficient clearing of framework once they have finished their job in vivo. Therefore, modification in design according to the need and detailed examination of their in vivo behaviours is of paramount importance to take this class of nano-carriers from bench to bed.

16.3.5 Separation

Separation is a process that splits the mixture into its components. It is opposite to the process of mixing, which is thermodynamically favoured process, and separation is generally not a spontaneous process. Separation of the components of a mixture is often based on selective adsorption. Adsorptive separation by a porous material is usually achieved in following ways [144]:

(i) Size and shape exclusion also known as molecular sieving effect: Based on size and shape, certain components are stopped from entering the pores of an adsorbent while others are allowed to enter where they are adsorbed. This is steric separation and is common in zeolites and molecular sieves.
(ii) Thermodynamic equilibrium effect: There is a different adsorbate packing interactions for different components over the surface of adsorbent which leads to selective adsorption.
(iii) Kinetic effect: Difference in the diffusing rates of the components also leads to selective adsorption.
(iv) Quantum sieving effect: Light molecules differ in their rate of diffusion in narrow micropores which assist in separating them.

MOFs are ideal candidate for gas separation applications. Structures and properties of MOFs can be well-designed and customized by the choice of metallic SBUs and organic linkers. This remarkable feature of MOFs is quite different from zeolites where customization of structure according to need is not possible. In addition to this, high porosity with tailorable pore size, diverse scope of functionalities, thermal and chemical stability is some of the attractive properties of MOFs which are elementary for separation applications. Owing to such excellent properties, various MOFs have been explored for the gas separation and purification purpose. For example, ytterbium-based PCN-17 MOF comprising of large cages linked by small aperture was able to separate H_2 and O_2 over N_2 and CO [145]. $Zn_2(cnc)_2(dpt)$ and MIL-96 were found to suitable for the separation of CO_2 and CH_4 [146, 147]. The separation is due to size/shape exclusion of MOFs leading to selective adsorption of CO_2 over CH_4. In another study, utility of Cu-based MOF [$Cu_2(pzdc)_2(pyz)$] was demonstrated for the separation of C_2H_2 and CO_2, and the task for which traditional porous materials like activated carbon and zeolites is of no use as these two molecules are very similar in size. The sorption isotherms of both gases show that MOFs binds preferably with C_2H_2 than CO_2 at ambient temperature and low pressure. The selectivity is due to strong binding interaction between surface O atoms and C_2H_2 through hydrogen bonding. Guo et al. have synthesized copper net supported HKUST-1 ($Cu_3(BTC)_2$) membranes and utilized them for the separation of H_2/CO_2, H_2/CH_4 and H_2/N_2 [148]. The separation of H_2 is due to its high permeation flux through MOF membrane compared to other gases. H_2 molecules being small can pass through HKUST-1 membrane more easily compared to CO_2, CH_4 and N_2, and this can be attributed to the structural and chemical feature of the MOF which favours stronger interaction with CO_2, CH_4 and N_2 than H_2. The selective separation/adsorption in some MOFs is due to the steric effects and the interactions between adsorbate molecules and surface of adsorbent. For example, ZIF-95 and ZIF-100 showed high CO_2 storage capacity compared to CH_4, CO and N_2 which was attributed to the cooperative effects of the pore apertures (similar to CO_2) and the strong quadrupolar interactions of $CO_2$with N atoms present of the pore walls [149].

From the above discussion, it can be concluded that MOFs showed great promise in gas separation because both their pore size/shape and their surface properties can be easily tuned by the choice of metal node and organic linkers. In addition to this, open metal site in MOFs also assists in the separation of polar and non-polar gas pairs such as CO_2/CH_4 [150].

However due to relatively low adhesion of MOFs-based membranes to a polymeric support and possible defects between crystals, limits the full potential of MOFs to be realized for separation applications. Therefore, COFs-based membrane was explored for the gas separation, heavy metal ion separation, nanofilteration and water treatment because of their exciting characteristics such as

(i) *Tunable pore size*: Size exclusion depending upon the pore size is an important criterion for porous membrane-based separation. The pore size of the COFs depends upon the geometry and connectivity of the linkers which can be modified either by changing the length and structure of organic linkers or by post-synthetic modification to introduce large side groups and functional groups [76, 151].
(ii) *Chemical stability*: COFs which can sustain their crystallinity and porosity in humid, organic solvents and strongly acidic conditions are of great utility in gas separation and water treatment. The stability of COFs can be improved by the rational selection of organic linkers for the COFs synthesis and by introducing intra-molecular and inter-molecular hydrogen bonding [115, 152].
(iii) *Hydrophilicity*: Hydrophilic COFs find extensive use in desalination, dye extraction and pervaporation [153–155].

(iv) *Surface charge*: Charge present over the surface of COFs plays a significant role in desalination and organic solvent nanofilteration.

Apart from above-mentioned characteristics of COFs, large surface area and adaptable functionality are also appreciated for their application in gas separation and water treatment. A computational study revealed that the monolayer CTF-0-based membrane can deliver a very high separating factor at room temperature for separation of H_2 from H_2/CO_2, H_2/N_2, H_2/CO and H_2/CH_4 mixture [156]. It is interesting to note that COFs membranes exhibit high H_2 permeability with highest being recorded for ultrathin continuous 2D-CTF-1 membrane. With increase in the thickness of COFs, selectivity of H_2/CO_2 improved while the permeance of gas is reduced [157]. Similarly, Gao et al. have synthesized 3D COF membrane (COF-320) on porous alumina ceramic support under solvothermal condition and utilized it for H_2/CH_4 and N_2/H_2 separation application. The COF-320 membrane exhibits high hydrogen permeation flux as compared to other gases leading to H_2 selectivity of COF membrane [158]. Banerjee et al. recently have prepared mixed matrix membranes (MMMs) of COF over PBI polymers support [159]. This membrane is flexible in nature and displays high degree of thermo-chemical stability which further enhances its potential in separation application. The observed moderate selectivity in CO_2/CH_4 and CO_2/N_2 separation is attributed to the existence of polymeric chains trapped inside during synthesis. The presence of these polymeric chains reduces the pore size and thus restricts it in achieving high selectivity for separation. In another report, self-standing thin films of COFs have been prepared via interfacial crystallization process. Depending upon the concentration of the reagents, the thickness of the films was tuned. In addition, it is easy to handle, flexible in nature and can be grown in different diameters. Although these self-standing thin films are not defect free but still, they have demonstrated remarkable solvent-performance and solute-rejection performance.

Not just limited to separation of gases, COFs can also be used for the separation and extraction of elements and nanofilteration of dyes, salts and other organics from wastewaters. For instance, benzimidizole-based 2D COF functionalized with carboxylic acid has been employed as a solid-state matrix for the separation and enrichment of uranium [160]. Wang et al. have demonstrated the utility of COF-based membranes for the removal of dyes from water. The membrane showed high pure water permeability ($50\ L\ m^2\ h^{-1}\ bar^{-1}$) and a high Congo rejection rate (99.5%) which was better than MOFs-based membranes [161]. With COFs (LZU1) membrane still higher water permeability ($75\ L\ m^2\ h^{-1}\ bar^{-1}$) and approximately similar rejection rate of methylene blue (99.2%) and Congo dye (98.6%) was achieved [154]. Another application of COFs membranes was in desalination. Tunablity of pore size, surface charge and hydrophilicity of COFs membranes improves selectivity and reduces membrane fouling. Simulation studies have revealed that water desalination through seven TpPa-X membranes with various functional groups showed over 95% NaCl rejection while the water permeance was three orders of magnitude higher than typical commercial seawater reverse osmosis (RO) [153]. COFs-based extraction systems have also been developed for enrichment and analysis of molecules such as sudan dyes which are present in chilli powder below detection limit.

Despite so many applications, challenges such as long-term stability of COF-based membrane in realistic separation, high-cost and time-consuming fabrications methods might hinder their industrial use.

16.3.6 Chemical Sensors

Highly sensitive and selective detection of gas, vapour phase analyte, heavy metal ions are of paramount importance for various applications such as chemical threat detection, medical diagnostic, occupational safety, environment and water monitoring. Highly porous, crystalline MOFs and COFs can detect gas and vapour phase analytes with high sensitivity as they can concentrate the analyte molecules at higher levels than are present in external atmosphere. This facilitates the direct detection of analytes in sorbent and eliminates the sample preparation step. Apart from the sorption capacity of MOFs/COFs, sensitivity of analyte detection depends upon the strength of analyte binding to MOFs/COFs, dynamics of analyte transport within the framework (slower transport will lead to long response time and poor signal) and pore size (it is observed that all else being equal, small pore will adsorb analytes more strongly compared to larger ones, leading to enhance sensitivity), while the selectivity of MOFs/COFs for specific analyte depends either on size/shape exclusion (molecular sieving) wherein analyte molecules smaller than pore aperture can be absorbed leaving the larger ones, or, on chemically specific interactions of analyte with pore surface [162]. MOFs/COFs surface or pore aperture can be tailored to host specific analyte by choice of building blocks or post-synthetic modification methods. This will increase the selectivity as well as the sensitivity of the detection of analytes. For instance, Eddaoudi and co-workers have constructed series of IRMOFs by varying the size and chemical functionality of linkers leading to the formation of *iso*-structural MOFs with pore aperture ranging from 30 to 3.8 Å [163].

MOFs/COFs-based sensors can be categorized on the basis of mode of signal transduction such as optical, electrical and electromechanical sensors. Based on different modes of signal transduction, many MOFs-based sensors have been explored. For example, luminescent Zn_3btc_2 exhibits size selective sensing of amines [164]. Smaller amines like ethylamine, dimethylamine and propylamine that can easily diffuse into the pores of MOFs showed decrease in the luminescence while aniline and butylamine showed no quenching, presumably due to size exclusion. Xi et al. [165] showed that MOFs decorated with phosphorescent iridium (III) complex exhibit emission quenching in the presence of O_2. The quenching occurs due to energy transfer to O_2 leading to the formation of triplet state. Same behaviour was found to be insensitive towards the presence of N_2. In addition to this, luminescent MOFs have used to detect nitro-aromatics (NAC), one of the major classes of secondary explosives. The detection of NACs is not easy because of their inferior vapour pressure and limited chemical reactivity. However, their electron deficient property is favourable for the formation of π-complexes with electron rich fluorophores which can be applied for their detection. The combination of porosity, luminescence and open metal site in MOFs makes them promising candidate for the sensing of NAC. For example, Li and co-workers have synthesized fluorescent $[Zn_2(bpdc)_2(bpee)]\cdot 2DMF$ (bpdc = 4,4'-biphenyldicarboxylate; bpee = 1,2-bipyridylethene; DMF = dimethyl formamide) for the detection of DNT vapours. The MOF shows rapid response to analyte, the quenching percentage reached 85% within 10 s [166]. Lanthanide-based MOFs find great utility for the detection of NAC. Tb^{3+}@NENU-522 displays high selectivity and recyclability in the detection of NAC explosives [167]. Interestingly, the sensing of TNT by the MOF can be easily distinguished by the naked eye.

Similarly to MOFs, COFs have also showed great potential in the field of sensing and detection. Recent studies have demonstrated the sensing applications of COFs which can be accredited to their diverse compositions and synergistic functionality which arises from the combination of well-defined porosity and semiconducting properties. Due to tunable porosity, COFs are capable of selectively hosting specific guest molecules, a prerequisite for sensing of ions or molecules. Various COFs-based sensors have been proposed for the detection of heavy metal ions [168], pH changes [169], organic explosives [170], etc. Jiang et al. [171] have demonstrated the synthesis of pyrene-based luminescent COF for sensing of nitro compounds. The resulted COF possess high sensitivity and selectivity for picric acid over other employed nitro compounds. Later on, Fang et al. [172] have reported the application of imine-based COFs as biosensor for bovine serum albumin and probe DNA immobilization owing to the strong electrostatic interactions. In another report, Wang et al. [168] have utilized thioether-based fluorescent COF (COF-LZU-8) for mercury, Hg^{2+} sensing and removal. In follow up, Liu et al. [173] utilized COF-JLU3 for selective sensing of Cu^{2+}. Banerjee et al. [169] have synthesized new imide-based COFs (TpBDH and TfpBDH) which have been transformed into thin-layered covalent organic nanosheets (CONs) by simple liquid-based exfoliation method. These 2D CONs have been employed for fast and highly selective detection of nitro-aromatic analytes through luminescence-based turn off/on sensing mechanism. Compared to CONs, sensing ability of COFs is limited due to aggregated π stacked layers which render poor electron mobility and ineffective interaction with analytes. On the other hand, π-π interactions are considerably weakened in CONs which is the reason behind the superior sensing ability of CONs compared to the bulk COFs. Detection of harmful volatile organic compounds in workplace or water content of gas and solvent streams in industrial process is another important area in the field of chemical sensing. In this direction, Bein et al. [174] developed tetrakis (4-aminophenyl)pyrene-based COFs which can act as solid-state supra-molecular solvatochromic sensors that exhibit a strong colour change when exposed to humidity or solvent vapours. It has also been observed that solvatochromic response of sensors which is dependent on vapour concentration and solvent polarity.

16.3.7 Optoelectronics

Designing the material at molecular scale allows a fine adjustment of the energy gap as well as HOMO/LUMO levels of the semiconductor. MOFs and COFs built from aromatic building blocks with periodic arrays offers exciting semi-conductive and photo-conductive behaviours, making them suitable for various photo-electronic applications [175]. However, certain issues related to the fabrication of competitive electronic devices need to be addressed. Three-dimensional orderings at atomic scale enable COFs to perform important role in organic electronics such as light emission, charge transfer and separation. Yaghi et al. [176] for the first time noticed charge carrier mobility in imine-based 2D-COFs. In another example, COF-366 is a p-type semiconductor that exhibits one-dimensional hole mobility of 8.1 $cm^2V^{-1}\ s^{-1}$ superior to inorganic amorphous silicon ($\sim 1\ cm^2V^{-1}\ s^{-1}$). In addition, the hole mobility value is significantly higher than those of common conjugated polymers and ordered crystalline organic semiconductors. The electrical conductivity of COF-366 across a gap of 2 mm between two Au electrodes was determined. The electric current of 0.75 nA for COF-366 measured at the end of electrodes affirms that the COF-366 is conductive in nature. Studies indicated that pore volume of COFs is directly related to the electron conduction [177]. Banerjee et al. for the first time mechanochemically synthesized bipyridine-based COFs which surpassed its conventional solvothermal counterparts by exhibiting a stable open-circuit voltage of 0.93 V at 50 °C and a proton conductivity of 0.014 S/cm [178]. In 2014, Cai et al. [179] synthesized the TTF-COF by reaction between tetrathiafulvalene tetra benzaldehyde and p-phenylenediamine under solvothermal conditions. The TTF-COF attains planar sheet conformation while a distorted confirmation was observed in case of TTF-Py-COF. This contrast in the structural behaviour of two COFs originates from the difference of linkers between TTF units in the COFs. The TTF-COF having compact layer structure is responsible for its high carrier mobility which is found to 10^{-5} to 1 $cm^2V^{-1}\ s^{-1}$. In 2013, Jiang et al. [180] published the results on the synthesis of conjugated organic framework having three-dimensionally ordered stable structure and delocalized π-clouds (CS-COFs) which is found to be hole conducting framework.

16.4 Conclusions

Great progress has been made in the synthesis and applications of metal organic and covalent organic frameworks (MOFs and COFs) in recent times. The chapter provides a clear in-depth understanding of structure–property relationships of MOFs and COFs. Various synthetic protocols summarized in this chapter could practically assist in design and synthesis of new MOFs and COFs. An important synthetic challenge in this field is the evolution of novel synthetic procedures that can be carried out at room temperature. The formative steps in this direction are the use simple yet important methods like mechanochemical grinding which leads to direct the formation of MOFs and COFs at room temperature. Second point that needs to be addressed is improving the stability of these porous materials. This can be done by the introduction of –OH functionality in the vicinity of Schiff base centres to establish intra-molecular hydrogen bonding or an irreversible enol to keto tautomerization following the Schiff base reaction leads to improved chemical stability of crystalline framework.

The aesthetic characteristics of MOFs and COFs, pore size tunability and chemical functionalization of the cavity are relevant from many perspectives, so much so that it enables MOFs and COFs as useful materials for virtually all aspects of storage, separation and catalysis. Based on various studies, it has been established beyond doubt that functionalized MOFs with open metal site and enhances H_2 and CO_2 uptake to meet desired standards.

In the culmination, MOFs and COFs provide infinite possibility to address the problem in energy, environment and health-related areas. The opportunity to efficiently utilize these architectures is limited only by imagination and our skills to prepare and characterize suitable well-defined structures.

References

1. Li Y, Xu Y, Yang W, Shen W, Xue H, Pang H (2018) MOF-derived metal oxide composites for advanced electrochemical energy storage. Small 14:1704435
2. Ramchandra P, Boucar D (2011) Green energy and technology. Springer, London Dordrecht Heidelberg, New York
3. Intergovernmental Panel on Climate Change (2007) Climate change 2007, Fourth Assessment Report. https://www.ipcc.ch/ipccreports/assessments-reports.html.
4. Hu B, DeBruler C, Rhodes Z, Liu TL (2017) Liu Tl long-cycling aqueous organic redox flow battery (aorfb) toward sustainable and safe energy storage. J Am Chem Soc 139:1207
5. Yuan S, Zhu Y-H, Li W, Wang S, Xu D, Li L, Zhang Y, Zhang X-B (2017) Surfactant-free aqueous synthesis of pure single-crystalline snse nanosheet clusters as anode for high energy- and power-density sodium-ion batteries. Adv Mater 29:1602469
6. Xu Y, Zheng S, Tang H, Guo X, Xue H, Pang H (2017) Prussian blue and its derivatives as electrode materials for electrochemical energy storage. Energy Storage Mater 9:11
7. Yang Z, Zhang J, Kintner-Meyer MCW, Lu X, Choi D, Lemmon JP, Liu J (2011) Electrochemical energy storage for green grid. Chem Rev 111:3577
8. Liang K, Li L, Yang Y (2017) Inorganic porous films for renewable energy storage. ACS Energy Lett 2:373
9. Furukawa H, Cordova KE, O'Keeffe M, Yaghi OM (2013) The chemistry and applications of metal-organic frameworks. Science 341:1230444
10. Eddaoudi M, Moler DB, Li H, Chen B, Reinekee TM, O'Keeffe M, Yaghi OM (2001) Modular chemistry: secondary building units as a basis for the design of highly porous and robust metal−organic carboxylate frameworks. Acc Chem Res 34:319
11. Chen B, Xiang S, Qian G (2010) Metal−organic frameworks with functional pores for recognition of small molecules. Acc Chem Res 43:1115
12. Historical production and use of carbon materials. https://www.caer.uky.edu/carbon/history/carbonhistory.shtml. Accessed 16 Nov 2013

13. (a) Lea FM (1956) The chemistry of cement and concrete. St. Martin's Press, New York (Chapter 1). (b) https://ciks.cbt.nist.gov/~garbocz/appendix1/node4.html
14. Zimmermann NER, Haranczyk M (2016) Metal–organic frameworks with functional pores for recognition of small molecules. Cryst Growth Des 16:3043
15. Davis ME, Lobo RF (1992) Zeolite and molecular sieve synthesis. Chem Mater 4:756
16. Flanigen EM, Broach RW, Wilson ST (2010) Zeolites in industrial separation and catalysis. Wiley-VCH Verlag, Weinheim, pp 1–26
17. Breck DW Crystalline Zeolite Y (1964) U.S. Patent 3130007
18. Argauer RJ, Landolt GR (1972) Crystalline zeolite zsm-5 and method of preparing the same. U.S. Patent 3702886 A
19. Peralta D, Chaplais G, Masseron AS, Barthelet K, Chizallet C, Quoineaud AA, Pirngruber GD (2012) Comparison of the behavior of metal–organic frameworks and zeolites for hydrocarbon separations. J Am Chem Soc 134:8115
20. Davis ME (2002) Ordered porous materials for emerging applications. Nature 417:813
21. Schuth F, Sing KSW, Weitkamp J (eds) (2002) Handbook of porous solids. Wiley-VCH Verlag GmbH
22. Li H, Eddaoudi M, O'Keeffe M, Yaghi OM (1999) Design and synthesis of an exceptionally stable and highly porous metal-organic framework. Nature 402:276
23. Yap MH, Fow KL, Chen GZ (2017) Synthesis and applications of MOF-derived porous nanostructure. Green Energy Environ 2:218
24. Wu DC, Xu F, Sun B, Fu RW, He HK, Matyjaszewski K (2012) Design and preparation of porous polymers. Chem Rev 112:3959
25. Cote AP, Benin AI, Ockwig NW, O'Keeffe M, Matzger AJ, Yaghi OM (2005) Porous, crystalline, covalent organic frameworks. Science 310:1166
26. El-Kaderi HM, Hunt JR, Mendoza-Cortés JL, Côté AP, Taylor RE, O'Keeffe M, Yaghi OM (2007) Designed synthesis of 3d covalent organic frameworks. Science 316:268
27. Hu X, An Q, Li G, Tao S, Liu J (2006) Imprinted photonic polymers for chiral recognition. Angew Chem Int Ed 45:8145
28. Rzayev J (2009) Synthesis of polystyrene–polylactide bottlebrush block copolymers and their melt self-assembly into large domain nanostructures. Macromolecules 42:2135
29. Weber J, Su Q, Antonietti M, Thomas A (2007) Exploring polymers of intrinsic microporosity–microporous, soluble polyamide and polyimide. Macromol Rapid Commun 28:1871
30. (a) https://en.wikipedia.org/wiki/Zeolite; (b) Morris RE, Čejka J (2015) Exploiting chemically selective weakness in solids as a route to new porous materials. Nat Chem 7:381; (c) Zimmermann NER, Haranczyk M (2016) History and utility of zeolite framework-type discovery from a data-science perspective. Cryst Growth Des 16:3043–3048
31. Mintova S, Jaber M, Valtchev V (2015) Nanosized microporous crystals: emerging applications. Chem Soc Rev 44:7207
32. Dey C, Kundu T, Biswal BP, Mallick A, Banerjee R (2014) Crystalline metal-organic frameworks (mofs): synthesis, structure and function. Acta Cryst B70:3
33. Cui Y, Xu H, Yue Y, Guo Z, Yu J, Chen Z, Gao J, Yang Y, Qian G, Chen B (2012) A luminescent mixed-lanthanide metal–organic framework thermometer. J Am Chem Soc 134:3979
34. Shultz AM, Farha OK, Hupp JT, Nguyen ST (2009) A catalytically active, permanently microporous MOF with metalloporphyrin struts. J Am Chem Soc 131:4204
35. Stavila V, Bhakta RK, Alam TM, Majzoub EH, Allendorf MD (2012) Reversible hydrogen storage by naalh4 confined within a titanium-functionalized MOF-74(mg) nanoreactor. ACS Nano 6:9807
36. O'Keeffe M, Peskov MA, Ramsden SJ, Yaghi OM (2008) The reticular chemistry structure resource (RCSR) database of, and symbols for, crystal nets. Acc Chem Res 41:1782
37. Blatov VA, Shevchenko AP, Serezhkin VN (2000) TOPOS3. 2: a new version of the program package for multipurpose crystal-chemical analysis. J Appl Cryst 33:1193

38. Delgado-Friedrichs O, O'Keeffe M (2003) Identification of and symmetry computation for crystal nets. Acta Cryst A59:351
39. Yaghi OM, O'Keeffe M, Ockwig NW, Chae HK, Eddaoudi M, Kim J (2003) Reticular synthesis and the design of new materials. Nature 423:705
40. Corey EJ (1988) Robert Robinson lecture. Retrosynthetic thinking—essentials and examples. Chem Soc Rev 17:111
41. Lehn JM (1996) Supramolecular chemistry and chemical synthesis. In: Chemical synthesis NATO ASI series, vol 320
42. Coudert FX (2015) Responsive metal–organic frameworks and framework materials: under pressure, taking the heat, in the spotlight, with friends. Chem Mater 27:1905
43. Liang Z, Marshall M, Chaffee AL (2009) CO2 adsorption-based separation by metal organic framework (cu-btc) versus zeolite (13x). Energy Fuels 23:2785
44. Burtch NC, Jasuja H, Walton KS (2014) Water stability and adsorption in metal–organic frameworks. Chem Rev 114:10575
45. Cheng G, Hasell T, Trewin A, Adams DJ, Cooper AI (2012) Soluble conjugated microporous polymers. Angew Chem Int Ed 51:12727
46. Jiang JX, Su FB, Trewin A, Wood CD, Campbell NL, Niu HJ, Dickinson C, Ganin AY, Rosseinsky MJ, Khimyak YZ, Cooper AI (2007) Conjugated microporous poly(aryleneethynylene) networks. Angew Chem Int Ed 46:8574
47. Chen Q, Wang JX, Yang F, Zhou D, Bian N, Zhang XJ, Yan CG, Han BH (2011) Tetraphenylethylene-based fluorescent porous organic polymers: preparation, gas sorption properties and photoluminescence properties. J Mater Chem 21:13554
48. Kou Y, Xu Y, Guo Z, Jiang D (2011) Supercapacitive energy storage and electric power supply using an aza-fused π-conjugated microporous framework. Angew Chem Int Ed 50:8753
49. Cooper AI (2009) Conjugated microporous polymers. Adv Mater 21:1291
50. Jiang J-X, Su F, Trewin A, Wood CD, Niu H, Jones JT, Khimyak YZ, Cooper AI (2008) Synthetic control of the pore dimension and surface area in conjugated microporous polymer and copolymer networks. J Am Chem Soc 130:7710
51. Diercks CS, Yaghi OM (2017) The atom, the molecule, and the covalent organic framework. Science 355:eaal1585
52. Feng X, Ding X, Jiang D (2012) Covalent organic frameworks, Chem Soc Rev 41:6010–6022
53. Waller PJ, Gandara F, Yaghi OM (2015) Chemistry of covalent organic frameworks. Acc Chem Res 48:3053
54. Spitler EL, Dichtel WR (2010) Lewis acid-catalysed formation of two-dimensional phthalocyanine covalent organic frameworks. Nat Chem 2:672
55. Kitagawa S, Kitaura R, Noro S–i (2004) Functional porous coordination polymers. Angew Chem Int Ed 43:2334
56. Lu W, Wei Z, Gu Z, Liu T, Park J, Park J, Tian J, Zhong M, Zhang Q, Rentle T III, Bosch M, Zhou H (2014) Tuning the structure and function of metal–organic frameworks via linker design. Chem Soc Rev 43:5561
57. Rowan SJ, Cantrill SJ, Cousins GRL, Sanders JKM, Stoddart JF (2002) Dynamic covalent chemistry. Angew Chem Int Ed 41:898
58. Spitler EL, Colson JW, Uribe-Romo FJ, Woll AR, Giorino MR, Saldirar A, Dichel WR (2012) Lattice expansion of highly oriented 2d phthalocyanine covalent organic framework films. Angew Chem Int Ed 51:2623
59. Koo BT, Heden RF, Clancy P (2017) Nucleation and growth of 2d covalent organic frameworks: polymerization and crystallization of COF monomers. Phys Chem Chem Phys 19:9745–9754
60. Wang L, Zeng C, Xu H, Yin P, Chen D, Deng J, Li M, Zheng N, Gu C, Ma Y (2019) A highly soluble, crystalline covalent organic framework compatible with device implementation. Chem Sci 10:1023–1028

61. (a) Ding XS, Guo J, Feng X, Honsho Y, Guo JD, Seki S, Maitarad P, Saeki A, Nagase S, Jiang D (2011) Synthesis of metallophthalocyanine covalent organic frameworks that exhibit high carrier mobility and photoconductivity. Angew Chem Int Ed 50:1289; (b) Tilford RW, Gemmill WR, zur Loye HC, Lavigne JJ (2006) Chem Mater 18:5296
62. Medina DD, Rotter JM, Hu Y, Dogru M, Werner V, Auras F, Markiewicz JT, Knochel P, Bein T (2015) Room temperature synthesis of covalent–organic framework films through vapor-assisted conversion. J Am Chem Soc 137:1016
63. Abrishamkar A, Rodríguez-San-Miguel D, Rodríguez Navarro JA, Rodriguez-Trujillo R, Amabilino DB, Mas-Ballesté R (2017) J Vis Exp 125:56020
64. Klinowski J, Almeida Paz FA, Silva P, Rocha J (2011) Microwave-assisted synthesis of metal–organic frameworks. Dalton Trans 40:321
65. (a) Campbell NL, Clowes R, Ritchie LK, Cooper AI (2009) Rapid microwave synthesis and purification of porous covalent organic frameworks. Chem Mater 21:204; (b) Dogru M, Sonnauer A, Gavryushin A, Knochel P, Bein T (2011) A covalent organic framework with 4 nm open pores. Chem Commun 47:1707
66. Biswal BP, Chandra S, Kandambeth S, Lukose B, Heine T, Banerjee R (2013) Mechanochemical synthesis of chemically stable isoreticular covalent organic frameworks (2013). J Am Chem Soc 135:5328
67. Pichon A, Lazuen-Garay A, James SL (2006) Solvent-free synthesis of a microporous metal–organic framework. Cryst Eng Comm 8:211
68. Bang JH, Suslick KS (2010) Applications of ultrasound to the synthesis of nanostructured materials. Adv Mater 22:1039
69. Mason TJ, Peters D (2003) In practical sonochemistry: power ultrasound uses and applications. Horwood Publishing, Chichester
70. Joaristi AM, Jaun-Alcaniz J, Serra-crespo P, Kapteijn F, Gascon J (2012) Electrochemical synthesis of some archetypical Zn^{2+}, Cu^{2+}, and Al^{3+} metal organic frameworks. Cryst Growth Des 12:3489
71. Cooper ER, Andrews CD, Wheatley PS, Webb PB, Wormald P, Morris RE (2004) Ionic liquids and eutectic mixtures as solvent and template in synthesis of zeolite analogues. Nature 430:1012
72. Kuhn P, Antonietti M, Thomas A (2008) Porous, covalent triazine-based frameworks prepared by ionothermal synthesis. Angew Chem Int Ed 18:3450
73. (a) Wan S, Guo J, Kim J, Ihee H, Jiang DL (2009) A photoconductive covalent organic framework: self-condensed arene cubes composed of eclipsed 2d polypyrene sheets for photocurrent generation. Angew Chem Int Ed 47:882
74. (a) Zwaneveld NAA, Pawlak R, Abel M, Catalin D, Gigmes D, Bertin D, Porte L (2008) Organized formation of 2D extended covalent organic frameworks at surfaces. J Am Chem Soc 130:6678; (b) Spitler EL, Colson JW, Uribe-Romo FJ, Woll AR, Giovino MR, Saldivar A, Dichtel WR (2012) Lattice expansion of highly oriented 2D phthalocyanine covalent organic framework films. Angew Chem Int Ed 51:2623; (c) Spitler EL, Koo BT, Novotney JL, Colson JW, Uribe-Romo FJ, Gutierrez GD, Clancy P, Dichtel WR (2011) A 2D covalent organic framework with 4.7-Nm pores and insight into its interlayer stacking. J Am Chem Soc 133:19416
75. Guan C-Z, Wang D, Wan L-J (2012) Construction and repair of highly ordered 2d covalent networks by chemical equilibrium regulation. Chem Commun 48:2943–2945
76. Clair S, Ourdjini O, Abel M, Porte L (2011) Tip- or electron beam-induced surface polymerization. Chem Commun 47:8028
77. Dey K, Pal M, Rout KC, Kunjattu S, Das HA, Mukherjee R, Kharul UK, Banerjee R (2017) Selective molecular separation by interfacially crystallized covalent organic framework thin films. J Am Chem Soc 139:13083
78. Guillerm V, Gross S, Serre C, Devic T, Bauer M, Féerey G (2010) A zirconium methacrylate oxocluster as precursor for the low-temperature synthesis of porous zirconium(iv) dicarboxylates. Chem Commun 46:767

79. Li JR, Timmons DJ, Zhou HC (2009) Interconversion between molecular polyhedra and metal−organic frameworks. J Am Chem Soc 131:6368
80. Wang Z, Cohen SM (2007) Postsynthetic covalent modification of a neutral metal−organic framework. J Am Chem Soc 129:12368
81. Cohen SM (2012) Postsynthetic methods for the functionalization of metal–organic frameworks. Chem Rev 112:970
82. Farha OK, Mulfort KL, Hupp JT (2008) An example of node-based postassembly elaboration of a hydrogen-sorbing, metal−organic framework material. Inorg Chem 47:10223
83. Chui SS-Y, Lo SM-F, Charmant JPH, Orpen AG, Williams ID (1999) A chemically functionalizable nanoporous material. Science 283:1148
84. Yamada T, Kitagawa H (2009) Protection and deprotection approach for the introduction of functional groups into metal−organic frameworks. J Am Chem Soc 131:6312
85. Mondloch JE, Karagiaridi O, Farha OK, Hupp JT (2013) Activation of metal–organic framework materials. Cryst Eng Comm 15:9258
86. Férey G, Mellot-Draznieks C, Serre C, Millange F, Dutour J, Surblé S, Margiolaki I (2005) A chromium terephthalate-based solid with unusually large pore volumes and surface area. Science 309:2040
87. Cavka JH, Jakobsen S, Olsbye U, Guillou N, Lamberti C, Bordiga S, Lillerud KP (2008) A new zirconium inorganic building brick forming metal organic frameworks with exceptional stability. J Am Chem Soc 130:13850
88. Nelson AP, Farha OK, Mulfort KL, Hupp JT (2009) Supercritical processing as a route to high internal surface areas and permanent microporosity in metal−organic framework materials. J Am Chem Soc 131:458
89. DOE technical targets for onboard hydrogen storage for light-duty vehicles. https://www.energy.gov/eere/fuelcells/doe-technical-targets-onboard-hydrogen-storage-light-duty-vehicles
90. Wong-Foy AG, Matzger AJ, Yaghi OM (2006) Exceptional H_2 saturation uptake in microporous metal−organic frameworks. J Am Chem Soc 128:3494
91. Rowsell JLC, Yaghi OM (2006) Effects of functionalization, catenation, and variation of the metal oxide and organic linking units on the low-pressure hydrogen adsorption properties of metal−organic frameworks. J Am Chem Soc 128:1304
92. Li JR, Kuppler RJ, Zhou HC (2009) Selective gas adsorption and separation in metal–organic frameworks. Chem Soc Rev 38:1477
93. Wang X-S, Ma S, Forster PM, Yuan D, Eckert J, Lo´ pez JJ, Murphy BJ, Parise JB, Zhou HC (2008b) Enhancing H_2 uptake by "close-packing" alignment of open copper sites in metal–organic frameworks. Angew Chem 120:7373
94. Farha OK, Yazaydin AO, Eryazici I, Malliakas CD, Hauser BG, Kanatzidis MG, Nguyen ST, Snurr RQ, Hupp JT (2010) De novo synthesis of a metal–organic framework material featuring ultrahigh surface area and gas storage capacities. Nat Chem 2:944
95. Metal-organic framework compound sets methane storage record, https://cen.acs.org/articles/95/web/2017/12/Metal-organic-framework-compound-sets.html
96. Noro S, Kitagawa S, Kondo M, Seki K (2000) A new, methane adsorbent, porous coordina-tion polymer [{cusif$_6$(4,4'-bipyridine)$_2$}n]. Angew Chem Int Ed 39:2082
97. Wang X, Ma S, Forster P, Yuan D, Eckert J, Lo´ pez J, Murphy B, Parise J, Zhou H (2008a) Enhancing H_2 uptake by "close-packing' alignment of open copper sites in metal-organic frameworks. Angew Chem Int Ed 47:7263
98. Perry JJ, Kravtsov V, McManus GJ, Zaworotko MJ (2007) Bottom up synthesis that does not start at the bottom: quadruple covalent cross-linking of nanoscale faceted polyhedra. J Am Chem Soc 129:10076
99. Peng Y, Krungleviciute V, Eryazici I, Hupp JT, Farha OK, Yildirim T (2013) Methane storage in metal–organic frameworks: current records, surprise findings, and challenges. J Am Chem Soc 135:11887
100. Wu H, Zhou W, Yildirim T (2009) J Am Chem Soc 131:4995

101. Mendoza-Cortes JL, Pascal TA, Goddard WA III (2011) Design of covalent organic frameworks for methane storag. J Phys Chem A 115:13852
102. An J, Geib SJ, Rosi NL (2010) High and selective CO_2 uptake in a cobalt adeninate metal −organic framework exhibiting pyrimidine- and amino-decorated pores. J Am Chem Soc 132:38
103. Panda T, Pachfule P, Chen Y, Jiang J, Banerjee R (2011) Amino functionalized zeolitic tetrazolate framework (ZTF) with high capacity for storage of carbon dioxide. Chem Commun 47:2011
104. Furukawa H, Yaghi OM (2009) Storage of hydrogen, methane, and carbon dioxide in highly porous covalent organic frameworks for clean energy applications. J Am Chem Soc 131:8875
105. Choi YJ, Choi JH, Choi KM, Kang JK (2011) Covalent organic frameworks for extremely high reversible CO_2 uptake capacity: a theoretical approach. J Mater Chem 21:1073
106. Schuth F et al (2012) Ammonia as a possible element in an energy infrastructure: catalysts for ammonia decomposition. Energy Environ Sci 5:6278
107. Van Humbeck JF et al (2014) Ammonia capture in porous organic polymers densely functionalized with brønsted acid groups. J Am Chem Soc 136:2432
108. Qajar A et al (2015) Enhanced ammonia adsorption on functionalized nanoporous carbons. Microporous Mesoporous Mater 218:15
109. Rieth AJ, Tulchinsky Y, Dinca M (2016) High and reversible ammonia uptake in mesoporous azolate metal–organic frameworks with open Mn, Co, and Ni sites. J Am Chem Soc 138:9401
110. Morris W, Doonan CJ, Yaghi OM (2011) Postsynthetic modification of a metal-organic framework for stabilization of a hemiaminal and ammonia uptake. Inorg Chem 50:6853
111. Doonan CJ, Tranchemontagne DJ, Glover TG, Hunt JR, Yaghi OM (2010) Exceptional ammonia uptake by a covalent organic framework. Nat Chem 2:235
112. Sabale S, Zheng J, Vemuri RS, Yu X-Y, McGrail BP, Motkuri RK (2016) Recent advances in metal-organic frameworks for heterogeneous catalyzed organic transformations. Synth Catal 1:1
113. Ding S-Y, Gao J, Wang Q, Zhang Y, Song W-G, Su C-Y, Wang W (2011) Construction of covalent organic framework for catalysis: Pd/COF-LZU1 in Suzuki–Miyaura coupling reaction. J Am Chem Soc 49:19816
114. Liu H, Chu J, Yin Z, Cai X, Zhuang L, Deng H (2018) Covalent organic frameworks linked by amine bonding for concerted electrochemical reduction of CO_2. Chem 4:1696
115. Kandambeth S, Mallick A, Lukose B, Mane MV, Heine T, Banerjee R (2012) Construction of crystalline 2d covalent organic frameworks with remarkable chemical (acid/base) stability via a combined reversible and irreversible route. J Am Chem Soc 48:19524
116. Goodenough JB, Kim Y (2010) Challenges for rechargeable li batteries. Chem Mater 22:587–603
117. de Combarieu G, Morcrette M, Millange F, Guillou N, Cabana J, Grey CP, Margiolaki I, Férey G, Tarascon J-M (2009) Influence of the benzoquinone sorption on the structure and electrochemical performance of the MIL-53 (Fe) hybrid porous material in a lithium-ion battery. Chem Mater 21:1602
118. Maiti S, Pramanik A, Manju U, Mahanty S, Appl ACS (2015) Reversible lithium storage in manganese 1,3,5-benzenetricarboxylate metal–organic framework with high capacity and rate performance. Mater Interfaces 7:16357
119. Mulzer CR, Shen L, Bisbey RP, McKone JR, Zhang N, Abruña HD, Dichtel WR (2016) Superior charge storage and power density of a conducting polymer-modified covalent organic framework. ACS Cent Sci 9:667–673
120. Xu F, Xu H, Chen X, Wu D, Wu Y, Liu H, Gu C, Fu R, Jiang D (2015) Radical covalent organic frameworks: a general strategy to immobilize open-accessible polyradicals for high-performance capacitive energy storage. Angew Chem Int Ed 54:6814
121. Petros RA, DeSimone JM (2010) Nat Rev Drug Discovery 9:615

122. Wu M-X, Yang Y-W (2017) Metal–organic framework (mof)-based drug/cargo delivery and cancer therapy. Adv Mater 29:1606134
123. Huxford RC, Della Rocca J, Lin W (2010) Metal-organic frameworks as potential drug carriers. Curr Opin Chem Biol 14:262
124. Farha OK, Hupp JT (2010) Rational design, synthesis, purification, and activation of metal −organic framework materials. Acc Chem Res 43:1166
125. Horcajada P, Serre C, Vallet-Regí M, Sebban M, Taulelle F, Férey G (2006) Metal–organic frameworks as efficient materials for drug delivery. Angew Chem 118:6120
126. Horcajada P, Serre C, Maurin G, Ramsahye NA, Balas F, Vallet-Regí M, Sebban M, Taulelle F, Férey G (2008) Flexible porous metal-organic frameworks for a controlled drug delivery. J Am Chem Soc 130:6774
127. Horcajada P, Chalati T, Serre C, Gillet B, Sebrie C, Baati T, Eubank JF, Heurtaux D, Clayette P, Kreuz C, Chang JS, Hwang YK, Marsaud V, Bories PN, Cynober L, Gil S, Ferey G, Couvreur P, Gref R (2010) Porous metal-organic-framework nanoscale carriers as a potential platform for drug delivery and imaging. Nat Mater 9:172
128. Sun CY, Qin C, Wang CG, Su ZM, Wang S, Wang XL, Yang GS, Shao KZ, Lan YQ, Wang EB (2011) Chiral nanoporous metal-organic frameworks with high porosity as materials for drug delivery. Adv Mater 23:5629
129. Bag PP, Wang D, Chen Z, Cao R (2016) Outstanding drug loading capacity by water stable microporous MOF: a potential drug carrier. Chem Commun 52:3669
130. He C, Lu K, Liu D, Lin W (2014) Nanoscale metal–organic frameworks for the co-delivery of cisplatin and pooled sirnas to enhance therapeutic efficacy in drug-resistant ovarian cancer cells. J Am Chem Soc 136:5181
131. Kundu T, Mitra S, Patra P, Goswami A, Diaz Diaz D, Banerjee R (2014) Mechanical downsizing of a gadolinium(iii)-based metal-organic framework for anticancer drug delivery. Chem Eur J 20:10514
132. Zhuang J, Kuo CH, Chou LY, Liu DY, Weerapana E, Tsung CK (2014) Optimized metal-organic- framework nanospheres for drug delivery: evaluation of small-molecule encapsulation. ACS Nano 8:2812
133. Zheng H, Zhang Y, Liu L, Wan W, Guo P, Nystrom AM, Zou X (2016) One-pot synthesis of metal– organic frameworks with encapsulated target molecules and their applications for controlled drug delivery. J Am Chem Soc 138:962
134. Nagata S, Kokado K, Sada K (2015) Metal–organic framework tethering pnipam for on–off controlled release in solution. Chem Commun 51:8614
135. Zhao F, Liu H, Mathe SDR, Dong A, Zhang J (2018) Covalent organic frameworks: from materials design to biomedical application. Nanomaterials 8:15
136. Mura S, Nicolas J (2013) Stimuli-responsive nanocarriers for drug delivery. Nat Mater 12:991–1003
137. Greenwald RB, Choe YH, McGuire J, Conover CD (2003) Effective drug delivery by pegylated drug conjugates. Adv Drug Deliv Rev 55:217
138. Fang Q, Wang J, Gu S, Kaspar RB, Zhuang Z, Zheng J, Guo H, Qiu S, Yan Y (2015) 3D porous crystalline polyimide covalent organic frameworks for drug delivery. J Am Chem Soc 137:8352–8355
139. Bai L, Phua SZ, Lim WQ, Jana A, Luo Z, Tham HP, Zhao L, Gao Q, Zhao Y (2016) Nanoscale covalent organic frameworks as smart carriers for drug delivery. Chem Commun 52:4128
140. Vyas VS, Vishwakarma M, Moudrakovski I, Haase F, Savasci G, Ochsenfeld C, Spatz JP, Lotsch BV (2016) Exploiting noncovalent interactions in an imine-based covalent organic framework for quercetin delivery. Adv Mater 28:8749
141. Rengaraj A, Puthiaraj P, Haldorai Y, Heo NS, Hwang SK, Han YK, Kwon S, Ahn WS, Huh YS (2016) Porous covalent triazine polymer as a potential nanocargo for cancer therapy and imaging. Appl ACS Mater Interfaces 8:8947

142. Mitra S, Sasmal HS, Kundu T, Kandambeth S, Illath K, Diaz Diaz D, Banerjee R (2017) Targeted drug delivery in covalent organic nanosheets (CONS) via sequential postsynthetic modification. J Am Chem Soc 139:4513
143. Zhang Y, Luo M, Zu YG, Fu YJ, Gu CB, Wang W, Yao LP, Efferth T (2012) Dryofragin, a phloroglucinol derivative, induces apoptosis in human breast cancer MCF-7 cells through ros-mediated mitochondrial pathway. Dryofragin Chem-Bio Interact 199:129-136
144. Rong J, Kuppler RJ, Zhou H-C (2009) Selective gas adsorption and separation in metal–organic frameworks. Chem Soc Rev 38:1477
145. Ma SQ, Wang XS, Yuan DQ, Zhou H-C (2008) A coordinatively linked Yb metal-organic framework demonstrates high thermal stability and uncommon gas-adsorption selectivity. Angew Chem Int Ed 47:4130
146. Xue M, Ma SQ, Jin Z, Schaffino RM, Zhu GS, Lobkovsky EB, Qiu SL, Chen BL (2008) Robust metal−organic framework enforced by triple-framework interpenetration exhibiting high H_2 storage density. Inorg Chem 47:6825
147. Loiseau T, Lecroq L, Volkringer C, Marrot J, Férey G, Haouas M, Taulelle F, Bourrelly S, Llewellyn PL, Latroche M (2006) MIL-96, a porous aluminum trimesate 3D structure constructed from a hexagonal network of 18-membered rings and μ3-oxo-centered trinuclear units. J Am Chem Soc 128:10223
148. Guo H, Zhu G, Hewitt IJ, Qiu S (2009) "Twin copper source" growth of metal-organic framework membrane: Cu(3)(BTC)(2) with high permeability and selectivity for recycling H(2). J Am Chem Soc 131:1646
149. Wang B, Côté AP, Furukawa H, ƠKeeffe M, Yaghi OM (2008) Colossal cages in zeolitic imidazolate frameworks as selective carbon dioxide reservoirs. Nature 453:207
150. Bae Y-S, Farha OK, Spokoyny AM, Mirkin CA, Hupp JT, Snurr RQ (2008) Carborane-based metal-organic frameworks as highly selective sorbents for CO_2 over methane. Chem Commun 21:4135
151. Nagai A, Guo Z, Feng X, Jin S, Chen X, Ding X, Jiang D (2011) Pore surface engineering in covalent organic frameworks. Nat Commun 2:536
152. Lohse MS, Bein T (2018) Covalent organic frameworks: structures, synthesis, and applications. Adv Funct Mater 28:1705553
153. Zhang K, He Z, Gupta KM, Jiang J (2017) Computational design of 2d functional covalent–organic framework membranes for water desalination. Environ Sci Water Res Technol 3:735
154. Fan H, Gu J, Meng H, Knebel A, Caro J (2018) High-flux membranes based on the covalent organic framework COF-LZU1 for selective dye separation by nanofiltration. Angew Chem Int Ed 57:4083
155. Fan H, Xie Y, Li J, Zhang L, Zheng Q, Zhang G (2018) Ultra-high selectivity COF-based membranes for biobutanol production. J Mater Chem A 6:17602
156. Wang Y, Li J, Yang Q, Zhong C (2016) Two-dimensional covalent triazine framework membrane for helium separation and hydrogen purification. ACS Appl Mater Interfaces 8:8694
157. Ying Y, Liu D, Ma J, Tong M, Zhang W, Huang H, Yang Q, Zhong C (2016) A go-assisted method for the preparation of ultrathin covalent organic framework membranes for gas separation. J Mater Chem A 4:13444
158. Lu H, Wang C, Chen J, Ge R, Leng W, Dong B, Huang J, Gao Y (2015) A novel 3D covalent organic framework membrane grown on a porous α-Al_2O_3 substrate under solvothermal conditions. Chem Commun 51:15562
159. Biswal BP, Chaudhari HD, Banerjee R, Kharul UK (2016) Chemically stable covalent organic framework (COF)-polybenzimidazole hybrid membranes: enhanced gas separation through pore modulation. Chem Eur J 22:4695
160. Chong SY-L (2017) Porous organic cages. Sci Comment 3:391
161. Wang R, Shi X, Xiao A, Zhou W, Wang Y (2018) Interfacial polymerization of covalent organic frameworks (COFS) on polymeric substrates for molecular separations. J Membr Sci 566:197–204

162. Kreno LE, Leong K, Farha OK, Allendorf M, Van Duyne RP, Hupp JT (2012) Metal-organic framework materials as chemical sensors. Chem Rev 112:1105
163. Eddaoudi M, Kim J, Rosi N, Vodak D, Wachter J, O'Keeffe M, Yaghi OM (2002) Systematic design of pore size and functionality in isoreticular MOFS and their application in methane storage. Science 295:469
164. Qiu L-G, Li Z-Q, Wu Y, Wang W, Xu T, Jiang X (2008) Facile synthesis of nanocrystals of a microporous metal–organic framework by an ultrasonic method and selective sensing of organoamines. Chem Commun 3642
165. Xie Z, Ma L, deKrafft KE, Jin A, Lin W (2010) Porous phosphorescent coordination polymers for oxygen sensing. J Am Chem Soc 132:922
166. Lan A, Li K,Wu H, Olson DH, Emge TJ, Ki W, Hong M, Li J (2009) A luminescent microporous metal organic framework for the fast and reversible detection of high explosives. Angew Chem Int Ed 48:2334
167. Zhang SR, Du DY, Qin JS, Bao SJ, Li SL, He WW, Lan YQ, Shen P, Su ZM (2014) A fluorescent sensor for highly selective detection of nitroaromatic explosives based on a 2D, extremely stable, metal organic framework. Chem Eur J 20:3589
168. Ding S-Y, Dong M, Wang Y-W, Chen Y-T, Wang H-Z, Su C-Y, Wang W (2016) Thioether-based fluorescent covalent organic framework for selective detection and facile removal of mercury (II). J Am Chem Soc 138:3031
169. Das G, Biswal BP, Kandambeth S, Venkatesh V, Kaur G, Addicoat M, Heine T, Verma S, Banerjee R (2015) Chemical sensing in two dimensional porous covalent organic nanosheets. Chem Sci 6:3931
170. Rao MR, Fang Y, De Feyter S, Perepichka DF (2017) Conjugated covalent organic frameworks via michael addition–elimination. J Am Chem Soc 139:2421
171. Dalapati S, Jin S, Gao J, Xu Y, Nagai A, Jiang D (2013) An azine-linked covalent organic framework. J Am Chem Soc 46:17310
172. Wang P, Kang M, Sun S, Liu Q, Zhang Z, Fang S (2014) Imine-linked covalent organic framework on surface for biosensor. C J C 32:838
173. Li Z, Zhang Y, Xia H, Mu Y, Liu X (2016) A robust and luminescent covalent organic framework as a highly sensitive and selective sensor for the detection of Cu^{2+} ions. Chem Commun 52:6613
174. Ascherl L, Evans EW, Hennemann M, Di Nuzzo D, Hufnagel AG, Beetz M, Friend RH, Clark T, Bein T, Auras F (2018) Solvatochromic covalent organic frameworks. Nat Commun 9:3802
175. Ding H, Li Y, Hu H (2014) A tetrathiafulvalene-based electroactive covalent organic framework. Chem Eur J 20:14614
176. Wan S, Gàndara F, Asano A, Furukawa H, Saeki A, Dey SK, Liao L, Ambrogio MW, Botros YY, Duan X, Seki S, Stoddart JF, Yaghi OM (2011) Covalent organic frameworks with high charge carrier mobility. Chem Mater 23:4094
177. Koo BT, Berard PG, Clancy P (2015) A kinetic monte carlo study of fullerene adsorption within a pcpbba covalent organic framework and implications for electron transport. J Chem Theory Comput 11:1172
178. Shinde DB, Aiyappa HB, Bhadra M, Biswal BP, Wadge P, Kandambeth S, Garai B, Kundu T, Kurungot S, Banerjee R (2016) A mechanochemically synthesized covalent organic framework as a proton conducting solid electrolyte. J Mater Chem A 4:2682
179. (a) Cai S-L, Zhang Y-B, Pun AB, Yang J, Toma FM, Sharp ID, Yaghi OM, Fan J, Zheng S-R, Zhang W-G, Liu Y (2014) Tunable electrical conductivity in oriented thin films of tetrathiafulvalene-based covalent organic framework. Chem Sci 5:4693; (b) Ding H, Li Y, Hu H, Sun Y, Wang J, Wang C, Wang C, Zhang G, Wang B, Xu W, Zhang D (2014) A tetrathiafulvalene-based electroactive covalent organic framework. Chem Eur J 20:14614
180. Guo J, Xu Y, Jin S, Chen L, Kaji T, Honsho Y, Addicoat MA, Kim J, Saeki A, Ihee H, Seki S, Irle S, Hiranoto M, Gao J, Jiang D (2013) Conjugated organic framework with three-dimensionally ordered stable structure and delocalized π clouds. Nat Commun 4:2736

Chapter 17
Green Chemistry Approach for Synthesis of Materials

Dibakar Goswami and Soumyaditya Mula

Abstract Green chemistry is the key to sustainability, not only for its basic concept to minimize the use and generation of hazardous materials but also due to its vast application towards one of the most efficient, problem-solving routes for the synthesis of advanced materials. The concept of green chemistry has been utilized almost in every sector of synthetic methods, starting from catalysis to more advanced stages of microwave-based and sonochemical syntheses. Lately, 'greener' approaches viz. use of renewable feedstocks, solvent engineering, etc. have also become an integral part of materials advancement. As a result, several bio-based 'green' materials, bio-fuel, materials for drug-delivery, bio-degradable fabrics, dyes, liquid crystals, etc. have emerged as high-end value-added materials for energy, health, and environmental benefits. This chapter describes, in a nutshell, various important applications of green chemistry in green manufacturing processes.

Keywords Green chemistry · Synthesis · Sustainability

Abbreviations

CAGR	Compound Annual Growth Rate
LD_{50}	Lethal Dose, 50%, meaning the amount of the substance required to kill 50% of the test population.
IPA	Isopropyl alcohol
HCl	Hydrochloric acid
ACS	American Chemical Society
API	Active Pharmaceutical Ingredient
PEG	Poly-ethylene glycol

D. Goswami (✉) · S. Mula
Bio-Organic Division, Bhabha Atomic Research Centre, Mumbai 400085, India
e-mail: dibakarg@barc.gov.in

S. Mula
e-mail: smula@barc.gov.in

D. Goswami · S. Mula
Homi Bhabha National Institute, Mumbai 400094, India

A. K. Tyagi and R. S. Ningthoujam (eds.), *Handbook on Synthesis Strategies for Advanced Materials*, Indian Institute of Metals Series,
https://doi.org/10.1007/978-981-16-1807-9_17

PAH	Poly-(allyl amine hydrochloride)
DMMA	Dimethylmaleic anhydride
CNT	Carbon nanotube
SWCNT	Single-wall carbon nanotube
NIR	Near Infra-red
BODIPY	4,4-Difluoro-4-bora-3*a*,4*a*-diaza-*s*-indacene
MWI	Microwave Irradiation
$ZnCl_2$	Zinc Chloride
DBU	1,8-Diazabicyclo[5.4.0]undec-7-ene
HBLC	Hydrogen bonded liquid crystal
BPy	4,4′-Bipyridine
TFE	Tetrafluoroethylene
$ScCO_2$	Supercritical carbon dioxide
FEP	Fluorinated ethylene propylene
PFA	Perfluoroalkoxy alkanes
ETFE	Ethylene tetrafluoroethylene

17.1 Introduction

The very urge of green chemistry lies in its nomenclature itself! Green relates us to the environment, chemistry relates us to study of materials. Over the decades, however, a general notion has been created that development via chemistry is not eco-friendly. The natural reason behind this is the generation of huge amounts of wastes, mainly because of doing chemical reactions for materials development. What if the waste is recycled for further chemical transformations which generate minimum waste? This very idea was transformed into a perspective by IUPAC named as "Green Chemistry", which was defined by IUPAC as "the invention, design, and application of chemical products and processes to reduce or to eliminate the use and generation of hazardous substances" [1]. In 1998, Paul Anastas and John Warner first coined the word "Green Chemistry" [2a], which became popular, and finally was framed using 12 principles which were basically targeted for promoting chemical reactions which are atom-economic, energy-efficient, and most importantly less waste-generating (Fig. 17.1). This led to a new beginning in chemical science. The importance of atom-economy was felt much more as never before. The idea of chemical recyclability changed the whole concept, and a new era of sustainability was born. Use of renewable and biodegradable materials for materials development was found unavoidable.

The very idea of less waste generation was meticulously quantified by Sheldon using the E (environmental) Factor (kgs waste/kg product) during the late 1980s. This E factor concept was pivotal to focus on a detrimental problem of waste generation in industries (Table 17.1). The E factor takes into account the chemical

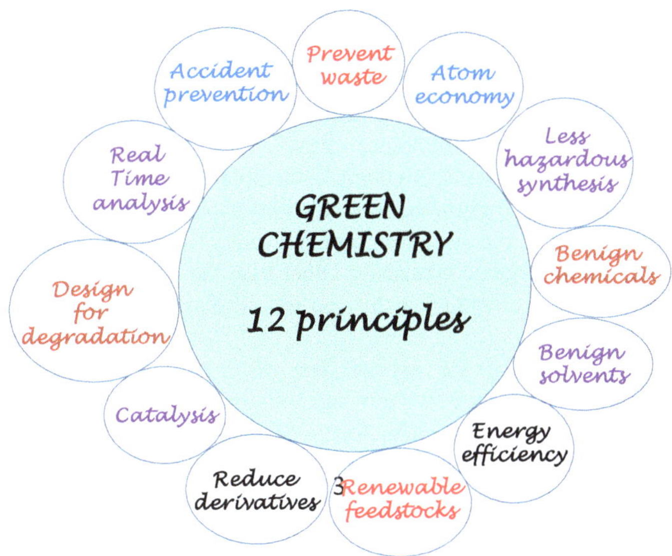

Fig. 17.1. 12 Principles of Green Chemistry [2b]

Table 17.1 The E-factor [Reproduced from Ref. [3b] with permission from the Royal Society of Chemistry]

Industry segment	Product tonnage	E- factor (Kg waste/ Kg product)
Oil refining	10^6–10^8	<0.1
Bulk chemicals	10^4–10^6	<1–5
Fine chemicals	10^2–10^4	5–>50
Pharmaceuticals	10–10^3	25–>100

yield, all reagents, and solvents (excluding water). Qualitatively, higher the E factor, higher is the amount of waste and, consequently, lower is the "greenness". In an ideal situation, E factor should be zero to achieve a perfectly green process [3].

Table 17.1 indicates that the E factor is more for industries dealing with fine chemicals and pharmaceuticals compared to those dealing with bulk materials. This is obvious since the synthesis of fine chemicals and pharmaceuticals involves multi-step syntheses, and the use of classical stoichiometric reagents for most of the steps. A paradigm shift was inevitable to focus on the green, less waste-generating processes, and also to eliminate the use of toxic/hazardous chemicals. It would not only reduce the amount of waste but also would improve the "quality" of waste.

What were the possible changes that a chemist could think of before jumping into sustainability? Chemical reactions involve hazardous chemicals, environment damaging organic solvents, and organic as well as organometallic wastes, which need to be disposed-off into environment. One after another, all these need replacements. To start with, the starting materials should originate from renewable, inexpensive, biodegradable feedstocks. In the last century, petrochemicals, biomass, and coal tar have been used a lot in chemical industries [4]. However, there is an increasing demand for evolving technologies with more and more use of materials with agricultural or biological origin.

In this regard, carbohydrates are being contemplated as an alternative feedstock. Polysaccharides are regularly used in textile and paper industries. Low molecular weight fragments like glucose, fructose, etc. are also being conceptualised as starting materials for organic syntheses. Other than these, agro-products like corn, starch, tarpenes, etc. are being transformed into important targets via a variety of processes.

Now let us discuss about the reagents. The less hazardous and safer reagents not only contribute towards development of safer and more efficient chemical reactions, but also enables green supply chain for consumables. The prime focus of green chemistry is not limited to eliminate the toxic reagents, but also to minimize the use of chemical reagents, such as the solvent or a chemical participating in the reactions.

Before further elaboration, let us take the example of synthesis of polycarbonates for better understanding. Polycarbonate is a crystalline thermoplastic polymer. It has immense application in medical industry, automobiles, electronics, etc., and its global market is expected to grow at a CAGR of 5.8% during 2017–2023 [5a]. It is traditionally synthesized by the reaction between bisphenol (BPA) and phosgene gas. However, the use of toxic phosgene gas and stoichiometric generation of HCl as side-product make this process hazardous. A recently commercialized process uses dimethylcarbonate (DMC) for synthesis of polycarbonates. DMC is considered to be a safe, green chemical owing to its low toxicity (LD_{50}, rat = 13,000 mg kg^{-1}) for human health. In this process, DMC is produced by the oxidadive carbonylation of methanol, followed by transesterification of DMC with bisphenol A in the melt-phase, and finally polymerized to give polycarbonates (Scheme 17.1). This is a classic example of substituting a toxic, hazardous reagent with a green, environment-friendly reagent [5b].

The target to minimize the use of reagents has led to the increased use of catalysts. This has a direct link to one of the most important objectives of green chemistry, i.e. to increase atom economy. Atom economy of a reaction is described as the efficiency to incorporate all the atoms in the reagents into the product, thereby reducing the amount of the waste that is generated in the process. Moreover, use of stoichiometric regents leads to less atom economy, compared to the same process where catalyst have been used. For example, the reduction of acetophenone, a much-used reaction in pharmaceutical industries, can be classically done either using excess sodium borohydride, or can be achieved using catalytic hydrogenation in IPA using (diphosphine)$RuCl_2$(diamine) precatalysts and KO*t*-Bu, as reported by Dowpharma (Scheme 17.2) [6]. Obviously, hydrogenation of the ketone by hydride reagent is associated with less atom economy, as well as production of boron-containing waste in stoichiometric amounts. However, the same reaction

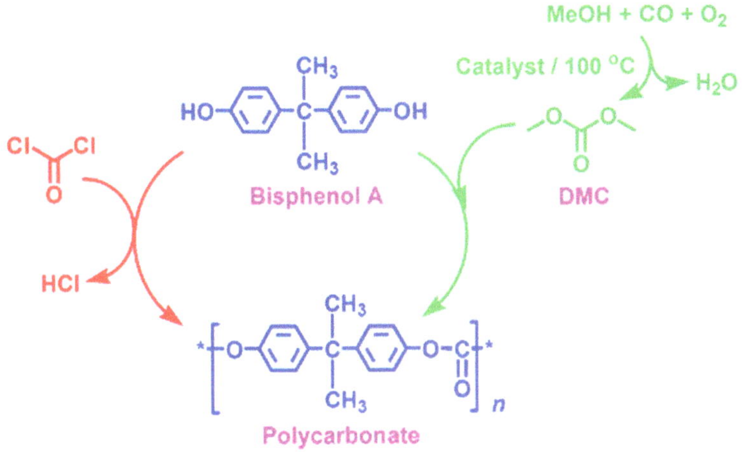

Scheme 17.1 Green synthesis of polycarbonate *vis-à-vis* the conventional route (red) [5b]

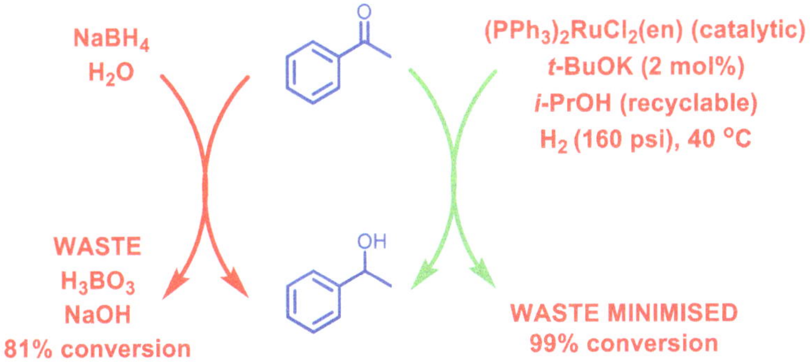

Scheme 17.2 Catalytic reduction (green) of acetophenone *vis-à-vis* the conventional route (red) [6]

carried out following a catalytic hydrogenation method is environment-friendly, has 99% conversion, and produces minimum waste.

The greenness of the solvent employed in the reaction also has a great role to play. The volatile, environment-damaging, carcinogenic organic solvents can be replaced by non-volatile, environment friendly, green solvents. Undoubtedly, water is the greenest solvent on earth. However, since many organic reagents are neither soluble nor stable in water, a substantial amount of research has been dedicated to the discovery and use of other green solvents. Room temperature ionic liquids (RTILs) have emerged as the leading candidate in this category. Several industrially important reactions, including olefin metathesis, Heck reaction, hydroformylation, Suzuki coupling, etc. have been achieved in RTILs. Several pharmaceutical processes utilize ionic liquids for manufacturing advanced intermediates.

A classic example is BASIL (Biphasic Acid Scavenging Utilizing Ionic Liquids) introduced by BASF for synthesis of alkoxyphenylphosphines used for preparing photoinitiators. The process yields HCl, which, when scavenged with conventional tertiary amines, produces a thick black slurry. This reduced the yield. Use of 1-methylimidazole as scavenger produced an ionic liquid, which appeared as a phase-separate clear liquid. This increased the yield, as well as shortened the reaction time. Thus, the productivity could be increased by a factor of 8×10^4 to 690,000 $kgm^{-3}\ h^{-1}$ [7].

To emphasize, these small applications in the classical areas have led the ionic liquids to become an important candidate for use in the generation of clean and efficient energy technologies, e.g., in batteries, solar cells, fuel cells, etc. Other green solvents e.g., supercritical fluid etc. are also being used in several industries. Supercritical fluid, e. g., supercritical CO_2 ($ScCO_2$), exists beyond critical point and represents a state of matter having properties of both gas and liquid. This is an environmentally friendly, non-flammable, and non-toxic fluid, and has been regarded as a "green" replacement for conventional organic solvents. $ScCO_2$ is being extensively used as a reaction medium for chemical synthesis, purification and crystallization of pharmaceuticals, polymerization and polymer processing as well as dyeing and cleaning of fibres and textiles [8a–c] (Fig. 17.2).

For example, perchloroethylene (PERC, $Cl_2C = CCl_2$), commonly used solvent for dry cleaning is known to be carcinogenic and contaminate groundwater. To avoid the use of PERC, a green technology, known as Micell Technology, was developed by using liquid CO_2 and a surfactant for dry cleaning of clothes [8d]. Dry cleaning machines are also available based on this technique. However, their uses are still in infancy, and will, hopefully, succeed in bringing a new era of green chemical engineering in the near future.

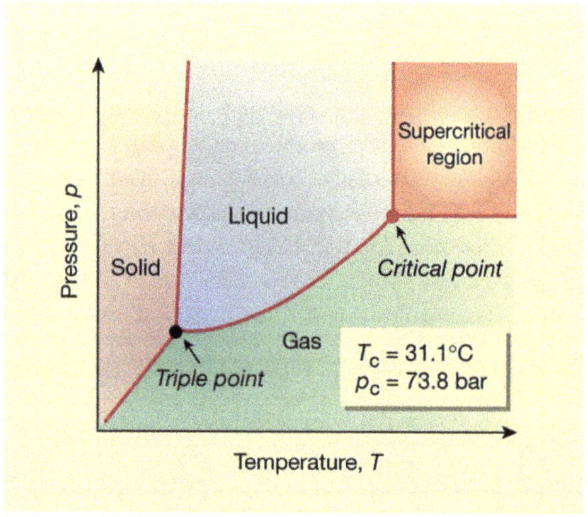

Fig. 17.2 Phase diagram of CO_2 (Reproduced from Ref. [8c] with permission from Springer Nature)

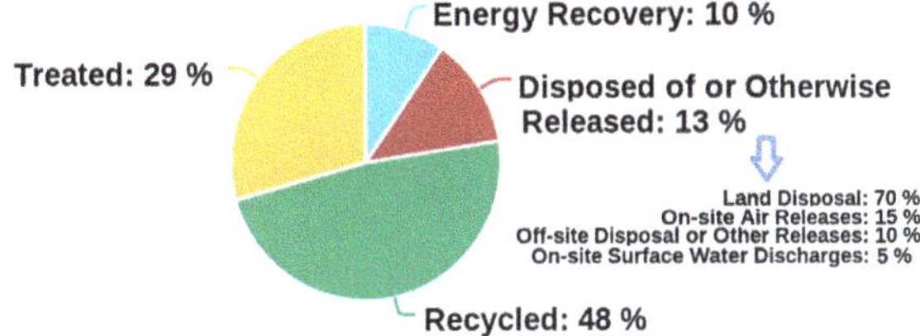

Fig. 17.3 Production-related waste management 2017. *Source* https://www.epa.gov/trinationalanalysis 21,456 facilities reported to TRI for 2017

The impact of introducing and promoting the green chemistry principles for use in the synthesis of advanced materials were visible from the fact that, in USA, the amount of chemical waste disposed to land, air, and water has decreased by 7% between 2007 and 2017, according to data collected by the Toxics Release Inventory (TRI, Fig. 17.3) of United States Environmental Protection Agency (USEPA).

In 2017, most releases were in the form of land disposal, primarily from metal mining operations. In addition, releases to air continued to decline in 2017. Since 2007, air releases reported to TRI have decreased by 57%. This was attributed to the good handling and process design, using green chemical pathways.

Thus, it was invariably established that green chemistry has the ability to improve the environmental profile of chemical processes and syntheses. Hence the industries came forward to apply green chemical principles in their process design. The next segment will discuss the applications of green chemistry in various industries, and how green chemistry is changing the world.

17.2 Application in Synthesis of Advanced Materials

17.2.1 Pharmaceuticals

In the USA, the ACS Green Chemistry Institute® Pharmaceutical Roundtable (GCIPR) was introduced in 2005 to promote green chemistry in pharmaceutical industries to adopt, resulting an enormous propagation of green chemistry research in areas like catalysis including bio-catalysis, solvent-free synthesis, C-H activation, flow chemistry, etc., which have impacted the innovation and sustainable development of many therapeutics including small molecules to biologics. As a result, the pharmaceutical industry has chosen to use green processes for preparation of advanced drugs, and have invariably reduced the use of hazardous chemicals, and eventually a major reduction in the generation of waste is evident (Fig. 17.4).

A classic example of adopting green chemistry in pharmaceutical industries is the preparation of Celecoxib by Pfizer (Celebrex®). It is a widely used as a cyclooxyginase-2 antiinflammatory agent. Shifting the traditional synthetic route to a green route has increased the yield from 63 to 84%, reduced waste, and has minimized the use of harmful hydrazine [9]. Additionally, the use of organic solvents has been reduced by more than 50%, and the use of solvents like methylene

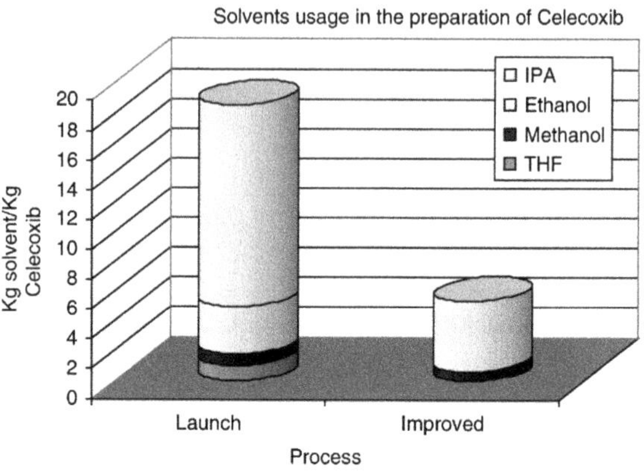

Fig. 17.4 Solvent usage in the preparation of celecoxib (launch method *vis-à-vis* improved method) [Reproduced from Ref. [9] with permission from Taylor & Francis Ltd (www.tandfonline.com)]

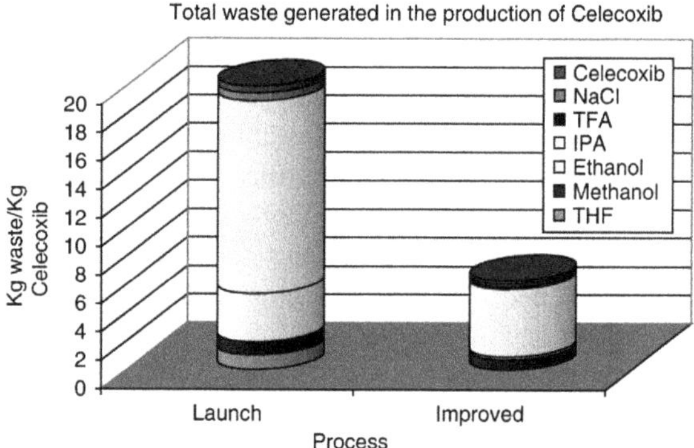

Fig. 17.5 Total waste generated in the preparation of celecoxib (launch method *vis-à-vis* improved method) [Reproduced from Ref. [9] with permission from Taylor & Francis Ltd (www.tandfonline.com)]

chloride and hexane was eliminated. Figure 17.5 and 17.6 demonstrate the essence of greenness in the improved manufacturing process of Celecoxib, compared to the conventional method.

The sustainable production of ibuprofen (Scheme 17.3), a widely used pain killer, by BASF (formerly known as BHC), is an innovative technology that has been found to be super-effective in bulk pharmaceutical manufacturing (approx. 2 billion tablets per year) [10a]. Instead of a traditional six-step process, this green protocol completes the synthesis in three steps. HF, used in first step, is recycled so that minimum waste is generated. Also, the subsequent steps use Raney nickel and palladium catalysts, which are recovered and reused, thus almost eliminating the waste generation, and also increases the atom economy to almost 77% compared with 40% in the traditional six-step synthesis. Most importantly, the process does not require the use of volatile organic solvents (VOS). For developing this, BHC was awarded with Presidential Green Chemistry Challenge: Greener Synthetic Pathways Award in 1997.

Fig. 17.6 The pilot-scale plant for supercritical carbon dioxide rope fabric dyeing machine and profile of the dyeing vessel (Reproduced from Ref. [19] published by IOP publishing)

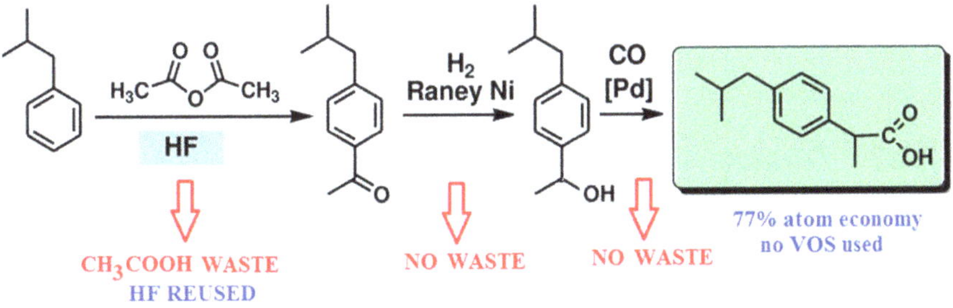

HF

H_2 Raney Ni

CO [Pd]

77% atom economy
no VOS used

CH_3COOH WASTE
HF REUSED

NO WASTE

NO WASTE

Scheme 17.3 Green synthesis of ibuprofen [10b]

Another success story revolves around the synthesis of sitagliptin, the API in Januvia™, an advanced treatment for Type-2 diabetes. Merck and Codexis developed a green synthesis of sitagliptin, using an *R*-selective transaminase enzyme that reduced waste by 19%, improved yield to 56%, and eliminated the need for a metal (Rhodium) catalyst, which was otherwise used in the traditional asymmetric synthesis of sitagliptin. This procedure was awarded with Presidential Green Chemistry Challenge: Greener Reaction Conditions Award in 2010 (Scheme 17.4) [11a].

An enzyme-based process was developed by Codexis for the production of the API atorvastatin for Pfizer's Lipitor®, the world's best-selling drug. This method increased yield, reduced the amount of organic solvents, and ultimately reduced the amount of waste. This has won its designer Presidential Green Chemistry Challenge Award from the US Environmental Protection Agency (USEPA) in 2006 [12].

Few of the other API's are also being produced via greener routes. Very recently, Veleva et al. [13] has tabulated the percent reduction in waste generation

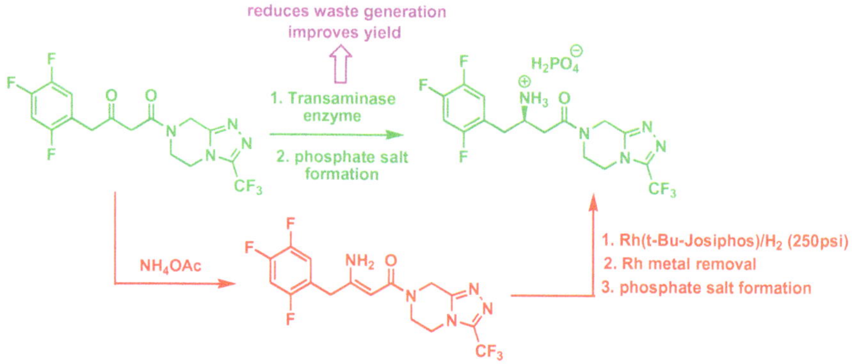

Scheme 17.4 Green synthesis of sitagliptin [11b]

by different industries through adopting green chemistry procedures during synthesis of the advanced drugs/active pharmaceutical intermediates (API's). This demonstrated impressive results and has established the efficacy of green chemistry to reduce environmental burdens.

17.2.2 Textiles

The importance of introducing green chemistry and engineering in textile industry can be presumed from the fact that textile sector represents 3% of all merchandize trade around the globe. As a major concern, nearly 10% of global carbon output, 20% of global water pollution, and 5% of landfill waste is related to textile industry [14]. The main reasons behind such a massive negative impact on environment are huge consumption of water during the processing of fabrics, excessive use of more than 8000 toxic chemicals, including non-biodegradable and heavy metal dyes, to make the fabric [15].

To minimize the use of toxic chemicals, and in turn to minimize the production of toxic wastes, many textile companies have adopted green engineering techniques to produce biodegradable polymers that can be converted into fibres, films, and rods for efficient use. For example, the Nature Works LLC. invented a process to produce polylactic acid (PLA) polymers from corn, thus eliminating the use of fossil fuel resources, and also the use of organic solvents and other hazardous materials. Apart from being natural and recyclable, clothing made from these fibres showed all desirable attributes essential in product performance [16a]. Additionally, the negative environmental impact of regenerating cellulose fibres (viz. rayons) in an organic conventional solvent containing carbon disulphide and cuprammonium hydroxide may be overcome by regenerating cellulose in a benign organic solvent, N-methyl morpholine-N-oxide (NMMO) hydrate, which is completely recycled. This process has given birth to lyocell fibre (popular tradenames: Tencel

(Courtaulds, USA), Lyo Cell (Lenzig, Austria), and New Cell (Akzo-Nobel, Germany)), which is more sustainable than synthetic (polyester, nylon, etc.) and natural fibres (cotton). These fibres have been shown to be superior than synthetic fibres (e.g. polyester, etc.) and natural fibres (e. g. cotton) [16b].

The use of enzymes in almost every stage of textile industry has emerged as the most important application of green chemistry in textile processes. Use of amylase bacteria for enzymatic desizing, lipase/cellulose enzymes for bio scouring, cellulose enzyme based softening and bio-stone washing and dye-decolourisation using laccase enzymes derived from fungi, etc. have revolutionized the textile industry [15]. In addition to this, a significant development towards greenness was achieved when Dyecat Limited introduced a process where the polyester is colored during its synthesis, instead of conventional dying the polyester fabric after it is made [17]. During the synthesis of the polyester, a chromophore of choice is bound in the molecular level. This invariably reduced the use of toxic dyes, and the post-treatment by a huge quantity of water.

In textile industry, supercritical carbon dioxide (ScCO2) is also used during textile dyeing processes [18]. Low viscosity coupled with high diffusibility results in shorter dyeing periods eliminating the need for water in the process, and essentially reduces the operation costs. This kind of anhydrous $ScCO_2$ based dyeing is widely applied in textile industry. Figure 17.6 represents the commercial plant based on $ScCO_2$ dyeing technologies [19].

17.2.3 Biofuel

Plant biomass such as corn, agricultural residues, and grasses, have recently been identified as one of the most important source for generating biofuels in solid, liquid, and gaseous forms [20]. These are important renewable sources that reduces the use of toxic chemicals and solvents, and thereby reducing the amount of waste generated. The production of biofuels from plant sources is a multistep process, which is accomplished using a high temperature (e.g. pyrolysis) or a low-temperature deconstruction using enzymes, followed by fermentation of the intermediates using microorganisms. However, generation of plant biomass derived biofuel requires cultivation of plants, and their use. As a spin-off, higher cultivation of crops and plants leads to better air quality, thereby reduces the environmental burden. Hence it is essential to promote a correct implementation of agricultural programmes to enable a potential development towards commercialization of biofuel.

However, the process of converting a complex biomass into valuable chemicals maintaining the atom economy is a very difficult and multistep process, and has only been partially realized. Horvath et al. proposed a multistep catalytic conversion to the valuable chemicals, which can be used as a feedstock for producing several biofuels, and biosolvents [21], a perfect example of zero-sum carbon cycling. This

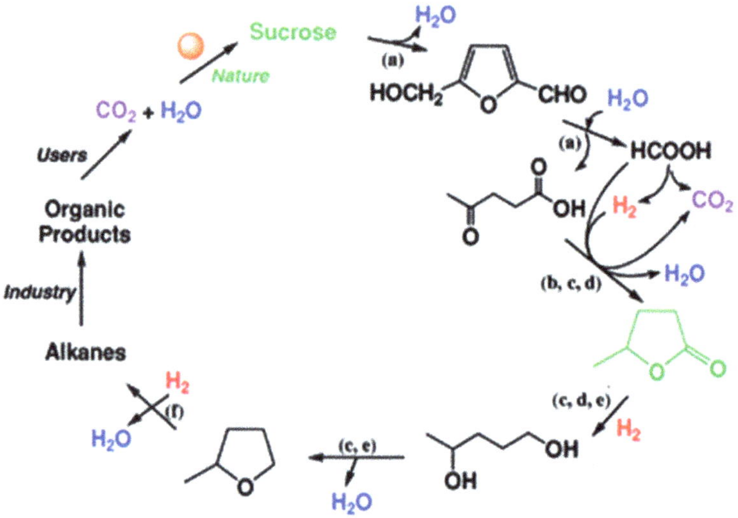

Fig. 17.7 Ways to convert feedstocks into valuables (Reproduced from Ref. [21] with permission from Springer Nature) **a** H_2SO_4/H_2O or Nafion-NR50/H_2O; **b** H_2/Ru(acac)$_3$/TPPTS/H_2O; **c** H_2/Ru(acac)$_3$/PBu$_3$/NH_4PF_6; **d** HCOONa/[(η^6-C_6Me_6)Ru(bpy)(H_2O)][SO_4]; **e** H_2/Ru(acac)$_3$/PBu$_3$; **f** H2/Pt(acac)$_2$/CF_3SO_3H

is a nice blend of all the necessary requirements of a proper green engineering concept (Fig. 17.7).

Bioethanol, produced from carbohydrates derived from plant sources viz. corn, sugarcane, etc. are being used as a gasoline additive (10% ethanol, 90% gasoline) to improve the quality of emissions from vehicles. Although pure ethanol is being considered to be used as a fuel, it is still not implemented. The French Futurol project targeting the development of an industrial-scale advanced bioethanol (2nd-generation) production technology, was completed at the end of 2018 [22]. In addition, biodiesel, a liquid fuel generated from renewable sources like vegetable oils, is considered as a replacement for petroleum-based fuels [23].

17.2.4 Nanomaterials

Nanomaterials, highly useful in healthcare and electronics, are generally manufactured using the top-down method or bottom-up approaches. Both the procedures are highly non-sustainable, particularly because of requirements of environmentally toxic reagents and solvents, requirement of specialized environments (high temperature and/or vacuum), and use of multistep purification procedures leading to generation of a large volume of toxic waste [24a]. Although the impacts of synthesized nanomaterials on environment and human health have been evaluated

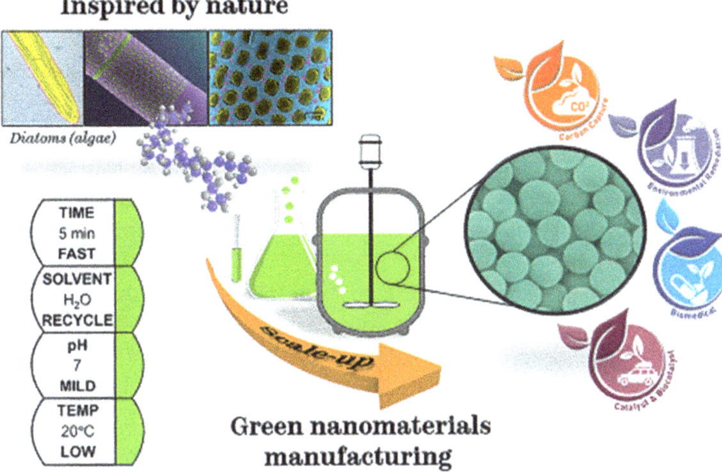

Fig. 17.8 Bioinspired nanomaterials synthesis (Reproduced from Ref. [24a] with permission from the authors)

using Life-cycle assessment (LCA) and risk assessment (RA) [24b–c], the environmental impacts of the synthetic procedures were evaluated using E factor. This revealed that for a production of 10^2–10^3 Kgs of nanomaterials, the E factor is around 100–10^6, which is almost 10^4 times higher than that in case of production of pharmaceuticals [24d]. Greener and sustainable routes for the synthesis of nanomaterials are being developed, although initially in a lab scale, with an aim to replace the traditional industrial methods. Although physical methods such as ball mill, ultrasound assisited, laser ablation strategy, UV-induced photochemical strategy, etc. have been used to circumvent the toxic chemical methods, they required specialized instruments, and in turn increases the cost of synthesis (Fig. 17.8).

In view of the above, bioinspired synthesis of nanomaterials, e.g., metals, oxides, sulphides, etc., using microorganisms, proteins, enzymes or highly specific biomolecules, have emerged as the most promising green synthetic pathways. Apart from being environment friendly, the ability of the biomolecules to guide the oriented growth of organic or inorganic substances have led to the belief that a large variety of inorganic structures, which are currently unattainable through any other methods, can be attained via this route [24e].

As an obvious example, synthesis of silver nanoparticles via the reduction of silver salt using plant extracts have been elaborately studied [24f]. Several plant systems have also been explored for the synthesis of silver nanoparticles [24g]. The secondary metabolites present in the plants reduce metal salts to form metal nano-particles without producing any toxic waste. Apart from plant extracts, several microorganisms e.g. bacteria, fungi and yeasts have been employed for the synthesis of silver nanoparticles. A comprehensive list of different methods has been tabulated by Rao et al. [24h].

Following a similar trend, gold nanoparticles have also been prepared. Extracellular (i.e., reduction-based mechanism of chloroauric ions in presence of bacteria, fungi, yeasts, etc.), intracellular, and plant-based methods (e.g., using Magnolia kobus leaves extract, biomass resulted from oats, leaf extracts from several medicinal plants, etc.) have been employed for this purpose. Rajeshwari et al. have tabulated different bioinspired methods for preparation of gold nanoparticles [24i].

Till today, several metal nanoparticles have been synthesized using different microorganisms, as well as different lower plants [24j]. Tables 17.2 and 17.3 summarizes different microorganisms/plants used for synthesis of nanoparticles through green routes. The vast potential of microorganisms and plant extracts has enabled researchers to produce metal nanoparticles in a green method, reducing the use of toxic chemicals and also minimizing the generation of wastes.

17.2.5 Drug Delivery

The concept of "green medicine" has emerged very recently which encompasses the area of green-technology driven drug delivery systems including nanometal formulations, polymers, and natural biomaterials especially originated from plant extracts [27]. Amongst these, the use of biocompatible, non-toxic magnetic nanoparticles (e.g. Fe_3O_4-nanoparticles) have emerged as a potential drug delivery agent due to their possible localization to the target using a magnetic field from outside of the body. The use of such magnetic nanoparticles has been so far extended for hyperthermic treatments, gene carriers and sensors for bioimaging, etc. Very recently, enhanced delivery of iron oxide nanoparticles to brain using lysophosphatidic acid (LPA) has also excited researchers towards a new way of treating the neuronal diseases [28a].

A well-defined green method using soybean phospholipids for drug delivery and subsequent tumor therapy was developed. Soybean phospholipids (SP) are coproducts of soybean oil processing, hence represents a natural feedstock of plant origin. A 2D molybdenum disulfide (MoS_2) nanosheet was developed and evaluated for their potential as a photothermal agent for tumor therapy. A composite of soybean phospholipid-MoS_2 nanosheet was prepared, and this showed excellent colloidal stability, minimal hemolysis, coagulation, and cytototoxicity, and good photothermal stability. These SP-MoS_2 nanosheets were found to show excellent anticancer properties both in vitro and in vivo under photothermal conditions [28b].

Till date, a variety of methods have been developed for nanoparticle drug delivery based on green principles [27a]. These include iron oxide nanoparticles for cancer radiation therapy [27c], magnetic nanoparticles for hyperthermia therapy [27d] etc. The use of graphene and graphene oxide has also been explored in drug delivery systems. Recently, a graphene–gold (G–Au) nanocomposite was prepared

Table 17.2 List of nanoparticles synthesized using microorganisms/plant extracts (Reproduced from Ref. [24j] with permission from Elsevier)

Microorganism	Extracellular/ Intracellular	Types of naoparticle	Shapes	Size (nm)	Applications	Ref
Bacteria						
Pseudomonas deceptionensis	Extracellular	Silver	Spherical	10–30	Antimicrobial and antibiofilm	[25a]
Weissella oryzae	Intracellular	Silver	Spherical	10–30	Antimicrobial and antibiofilm	[25b]
Bacillus methylotrophicus	Extracellular	Silver	Spherical	10–30	Antimicrobial	[25c]
Brevibacterium frigoritolerans	Extracellular	Silver	Spherical	10–30	Antimicrobial	[25d]
Bhargavaea indica	Extracellular	Silver and gold	Silver anisotropic; gold, flower	30–100	Antimicrobial	[25e, f]
Bacillus amyloliquefaciens	Extracellular	Cadmium sulfide	Cubic/hexagonal	3–4	–	[25g]
Bacillus pumilus, Bacillus persicus, and Bacillus licheniformis	Extracellular	Silver	Triangular, hexagonal, and spherical	77–92	Antiviral and Antibacterial	[25h]
Listeria monocytogenes, Bacillus subtilis, and Streptomyces anulatus	–	Silver	Anisotropic	Varied shape and sizes	Antimicrobial and mosquitocidal	[25i]
Fungus						
Neurospora crassa	Intra- and extracellular	Silver, gold, bimetallic silver and gold	Quasi-spherical	> 100	–	[25j]
Actinomycetes						
Streptomyces sp. LK3	–	Silver	Spherical	5	Acaricidal	[25k]

(continued)

Table 17.2 (continued)

Microorganism	Extracellular/ Intracellular	Types of naoparticle	Shapes	Size (nm)	Applications	Ref
Yeast						
Yarrowia lipolytica NCYC 789	Extracellular	Silver	Spherical	15	Antibiofilm	[25l]
Rhodosporidium diobovatum	Intracellular	Lead	–	2–5	–	[25m]
Extremophilic yeast	Extracellular	Silver and gold	Irregular	Silver, 20; gold, 30–100	–	[25n]
Candida utilis NCIM 3469	Extracellular	Silver	Spherical	20–80	Antibacterial	[25o]

Table 17.3 Synthesis and applications of biological nanoparticles from plants (Reproduced from Ref. [24j] with permission from Elsevier)

Plants	Plant tissues for extraction	Types of nanoparticle	Shapes	Size (nm)	Applications	Refs
Euphorbia prostrata	Leaves	Silver and titanium dioxide (TiO_2)	Spherical	Silver 10–15; TiO_2, 81.7–84.7	Leishmanicidal	[26a]
Sargassum algae	Alga	Palladium	Octahedral	5–10	Electrocatalytic activities towards hydrogen peroxide	[26b]
Ginkgo biloba	Leaves	Copper	Spherical	15–20	Catalytic	[26c]
Panax ginseng	Root	Silver and gold	Spherical	Silver, 10–30; gold, 10–40	Antibacterial	[26d]
Red ginseng	Root	Silver	Spherical	10–30	Antibacterial	[26e]
Cymbopogon citratus	Leaves	Gold	Spherical, triangular, hexagonal and rod	20–50	Mosquitocidal	[26f]
Azadirachta indica	Leaves	Silver	–	41–60	Biolarvicidal	[26g]
Nigella sativa	Leaves	Silver	Spherical	15	Cytotoxicity	[26h]
Cocos nucifera	Leaves	Lead	Spherical	47	Antibacterial and photocatalytic	[26i]
Catharanthus roseus	Leaves	Palladium	Spherical	40	Catalytic activity in dye degradation	[26j]
Pistacia atlantica	Seeds	Silver	Spherical	27	Antibacterial	[26k]
Banana	Peel	Cadmium sulfide	–	1.48	–	[26l]
Nyctanthes arbortristis	Flower	Silver	–	–	Antibacterial and cytotoxic	[26m]
Anogeissus latifolia	Gum powder	Silver	Spherical	5.5–5.9	Antibacterial	[26n]
Pinus densiflora	Cones	Silver	Oval in shape, few triangular shaped	30–80	Antimicrobial	[26o]
	Fruit	Zinc	Spherical	>20		[26p]

(continued)

Table 17.3 (continued)

Plants	Plant tissues for extraction	Types of nanoparticle	Shapes	Size (nm)	Applications	Refs
Artocarpus gomezianus					Luminescence, photocatalytic and antioxidant	
Citrus medica	Fruit	Copper	–	20	Antimicrobial	[26q]
Orange and pineapple	Fruits	Silver	Spherical	10–300	–	[26r]
Lawsonia inermis	Leaves	Iron	Hexagonal	21	Antibacterial	[26s]
Gardenia jasminoides	Leaves	Iron	Rock like appearance	32	Antibacterial	[26s]

using sonochemistry. This nanocomposite was used to prepare a new electrode material for detection of nitric oxide, a prominent biomarker of cancer [29a]. In another study, a green nanoparticle formulation with chitosan and graphene oxide in dilute acetic acid showed increased drug encapsulation and drug loading [29b].

Carbon dots (CDs) have also emerged as an efficient, green drug delivery vectors. For example, a cisplatin(IV)-loaded Carbon dots (CD-Pt[IV]-PEG-[PAH/DMMA]) was prepared and evaluated in vivo. These CDs exhibited high tumor localization and significant anti tumor properties [30a]. Apart from these, carbon nanotubes have also been utilized in cancer diagnosis and therapy. Huang et al. reported a reduction in cancer cell growth using single-walled CNTs (SWCNTs) and concomitant irradiation with a low power NIR laser for 10 min [30b]. In a similar line, SWCNTs conjugated with antitumor drug doxorubicin, and linked with folic acid, have been utilized for targeted breast cancer therapy, where the drug was released at only lower pH. Concomitant irradiation of the doxorubicin-conjugated SWCNTs using a NIR laser increased the antitumor potential of the conjugate [30c].

Apart from these, a plethora of materials viz. peptides, proteins, lipids, and polymers have been designed using plant extracts and biomaterials. These materials have low toxicities and high biocompatibilities. Few of such systems, have also been approved by FDA for clinical uses [31]. These have helped researchers to circumvent one of the biggest problems in drug delivery, i.e. toxicity. This has emerged as one of the most important advantages of applying green chemistry in targeted drug delivery.

17.2.6 Synthesis of Dyes

BODIPY dyes: BODIPY (dipyrromethene-BF_2, 4,4-difluoro-4-bora-3*a*,4*a*-diaza-*s*-indacene) dyes are very important due to their tunable fluorescence with high quantum yields and, high thermal and chemical stabilities [32]. These dyes are being used in diversified hi-tech applications such as sensors, lasers, artificial light-harvesting systems as well as for different biological applications [33].

Synthesis of BODIPY dyes from substituted/unsubstituted pyrroles under inert condition in dichloromethane (CH_2Cl_2) takes long time (several hours to several days) to furnish the dyes in poor to moderate (10–50%) yields. Two following procedures (Scheme 17.5) are generally being used: (a) acid (usually trifluoroacetic acid, TFA) catalyzed condensation of pyrroles with aromatic aldehyde, subsequent oxidation (using DDQ or p-chloranil) followed by reaction with Et_3N or Hunig's base and $BF_3.Et_2O$; (b) condensation of pyrroles with aromatic/aliphatic acid chloride followed by treatment with Et_3N or Hunig's base and $BF_3{\cdot}OEt_2$. Due to the availability of different aldehydes, their condensation with pyrroles (Procedure (a), Scheme 17.5) is generally used to synthesize variety of BODIPY dyes although the reaction yields of this procedure is less as compared to the Procedure (b) (Scheme 17.5) [33, 34].

S. V. Dzyuba et al. reported solvent-free mechano-chemical synthesis of BODIPY dyes by mixing the reagents with a pestle and mortar in open air. This procedure takes only about 5 min to synthesize the dyes with yields comparable to those obtained via traditional routes after several hours to days [34]. For example, 4-nitrobenzaldehyde was condensed with 2,4-dimethylpyrrole by grinding using a simple mortar and pestle followed by oxidation with p-chloranil. Subsequently, Et_3N and $BF_3{\cdot}OEt_2$ were treated with the reaction mixture by grinding to afford

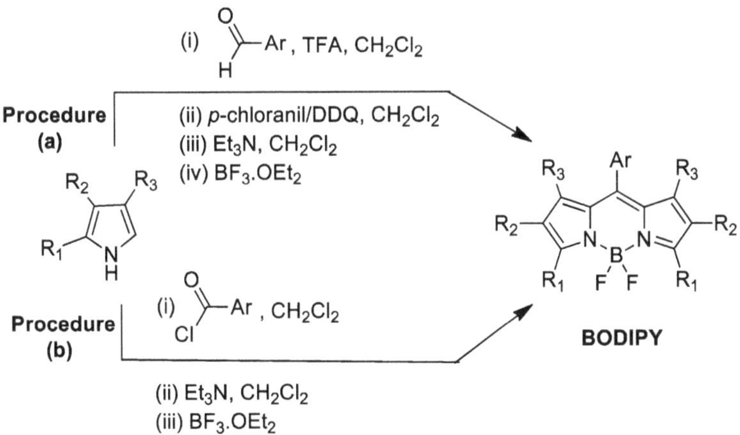

Scheme 17.5 General synthetic procedures of BODIPY dyes in CH_2Cl_2 (Reproduced from Ref. [34] with permission from Beilstein Journals)

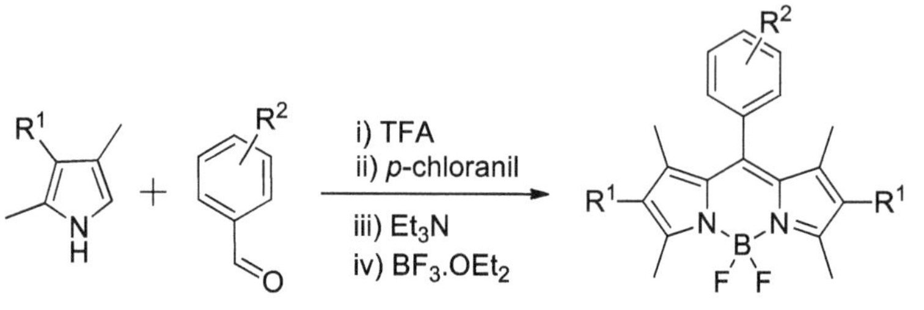

Scheme 17.6 Solvent free synthesis of BODIPY dyes (Reproduced from Ref. [34] with permission from Beilstein Journals)

BODIPY dye **1a** (Scheme 17.6). In all the steps, reagents were mixed together by grinding and each step was accompanied by a color change. The whole procedure takes only 5 min (as compared to 5 h to ca. 12 h (or overnight) reported for conventional methods) to furnish desired product **1a** with 29% overall yield, which was comparable to other conventional protocols, 24–40% (Table 17.4) [35]. Several other BODIPY dyes were also synthesized using the same protocol in 5 min which generally requires long time in conventional methods (Table 17.4).

The method is also very useful to synthesize BODIPY dyes starting from acetal. BODIPY dye **2** (Scheme 17.7) was synthesized by mechano-chemical process under solvent-free conditions using triethyl orthoformate as the component within 5 min in 29% yield. The conventional method for this BODIPY dye performed in solution required several hours [41]. Further, reactions of different acid chlorides with pyrroles (Procedure (b), Scheme 17.5) were also tried to synthesize BODIPYs.

Table 17.4 Synthesis of BODIPY dyes in mechano-chemical and conventional methods (Reproduced from Ref. [33] with permission from Beilstein Journals)

BODIPY	R^1	R^2	Mechano-chemical synthesis		Conventional synthesis	
			Reaction time	Yield, %	Reaction time	Yield, %
1a	H	4-NO_2	5 Min	29	5–12 h	24–40 [35]
1b	H	4-OCH_3	5 Min	15	5–19 h	28–38 [36]
1c	H	4-CN	5 Min	13	until aldehyde is consumed + 1 h	15 [37]
1d	H	4-ethynyl	5 Min	32	14–47 h	8–38 [38]
1e	Et	4-CF_3	5 Min	22	2 days	34 [39]
1f	Et	4-pyridyl	5 Min	15	1–4 days	9–40 [40]

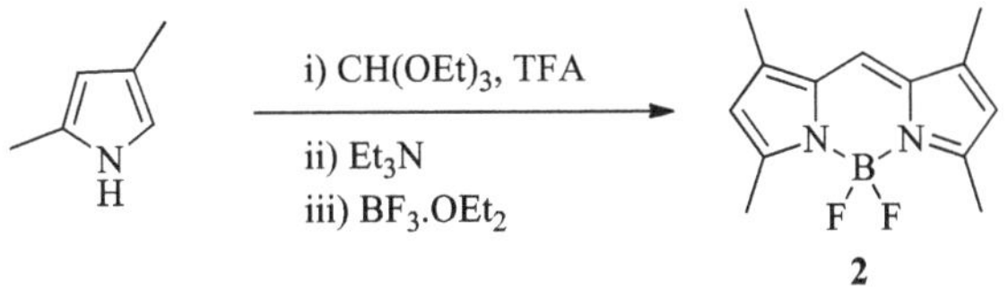

Scheme 17.7 Solvent-free synthesis of BODIPY dye **2** (Reproduced from Ref. [34] with permission from Beilstein Journals)

i) $CH(OEt)_3$, TFA
ii) Et_3N
iii) $BF_3.OEt_2$

But low yields of the dyes were obtained as the acid chlorides are moisture-sensitive in nature.

Hemicyanine dyes: Benzimidazole/pyridine based hemicyanine dyes are generally synthesized by the reaction of an aromatic aldehyde with reactive methyl group of a benzimidazole or pyridine quaternary salt in presence of a catalyst under refluxing condition in an organic solvent [42]. These require several hours of refluxing in organic solvents which are not environment friendly and also the isolation of the products is complex.

L.-Y. Wang et al. described an environmentally friendly pathway to synthesize hemicyanine dyes using solvent-free microwave-assisted reaction condition [43]. The reactions were reported to proceed well even with the solid starting reactants at reaction temperature below the melting points of both components. For example, piperidine catalyzed condensation of 1,2,3-trimethylbenzimidazolium iodide with respective aromatic aldehydes in solvent-free conditions under microwave irradiation at 252–324 W furnished benzimidazole based hemicyanine dyes, **3a-g** with 82–93% yield within 1–2 min. Using the similar reaction condition, pyridine based hemicyanine dyes, **4a-g** were also synthesized in good yield (75–99%) (Scheme 17.8) in few minutes. The pure dyes can be easily recrystallized from the diluted ethyl alcohol. Thus, in the new method is environment friendly as it avoided excess chemicals and reaction solvents used in classical synthesis of hemicyanine dyes.

Near-infrared fluorescent sphingosine derivatives: J. V. Frangioni et al. explored the use of microwave irradiation (MWI) in synthesis of near-infrared fluorescent sphingosine derivatives **6a-b** [44]. Reaction of NIR dyes,

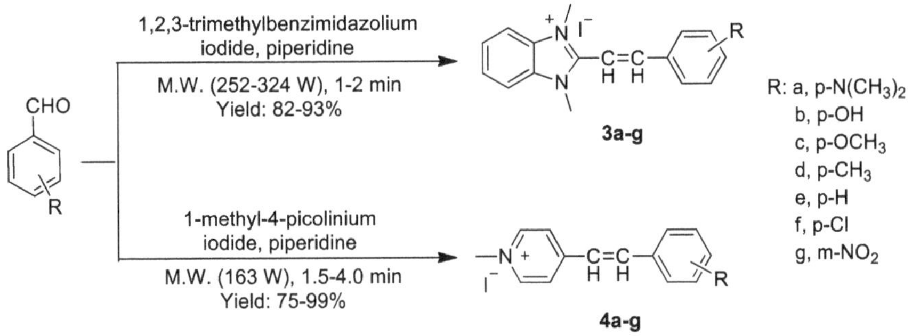

1,2,3-trimethylbenzimidazolium iodide, piperidine
M.W. (252-324 W), 1-2 min
Yield: 82-93%

1-methyl-4-picolinium iodide, piperidine
M.W. (163 W), 1.5-4.0 min
Yield: 75-99%

R: a, p-$N(CH_3)_2$; b, p-OH; c, p-OCH_3; d, p-CH_3; e, p-H; f, p-Cl; g, m-NO_2

Scheme 17.8 Synthesis of hemicyanine dyes in solvent-free microwave condition (Reproduced from Ref. [43] with permission from Elsevier)

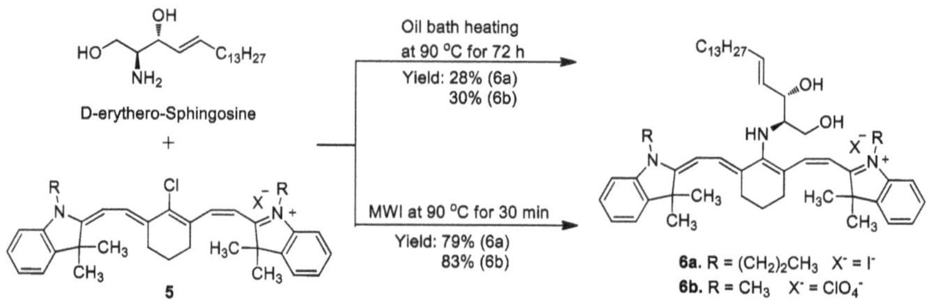

Scheme 17.9 Sphingosine derivatives syntheses using microwave (Reproduced from Ref. [44] with permission from the Royal Society of Chemistry)

heptamethineindocyanines IR-780 (**5a**) and IR-786 (**5b**), which differ in the aliphatic chains of their indole nitrogens (Scheme 17.9) and d-erythro-sphingosine as a nucleophile were carried out in dimethylformamide by both conventional heating and microwave irradiation. Importantly, under MWI, more than two-fold product yields (79% for **6a** and 83% for **6b**) were obtained in minutes as compared to conventional methods (28% for **6a** and 30% for **6b**) which take several days.

Azo dyes: Synthesis of Azo dyes are two steps process: diazotization of aromatic primary amines under strongly acidic conditions followed by the coupling of the diazonium salts with an activated aromatic compound (phenols or aromatic amines) at low temperature (0–5 °C) [45a]. This synthetic method has many limitations such as low-temperature reaction, use of acid–base catalyst, low stability of aryl diazonium salts at room temperature, long reaction times, moderate yields, etc. Also, these processes are environmentally incompatible due to the harmful effluents (acidic and basic) of the laboratory and industry [45a, b].

Javad Safari et al. [45c] developed an efficient solvent-free one-pot synthesis of azo dyes to overcome the drawbacks and limitations of the previous synthetic procedures. Aromatic amines, β-naphthol, and sulfonic acid functionalized magnetic Fe_3O_4 nanoparticles ($Fe_3O_4@SiO_2\text{-}SO_3H$) were grinded under solvent-free conditions at room temperature to furnish the azo dyes (Scheme 17.10).

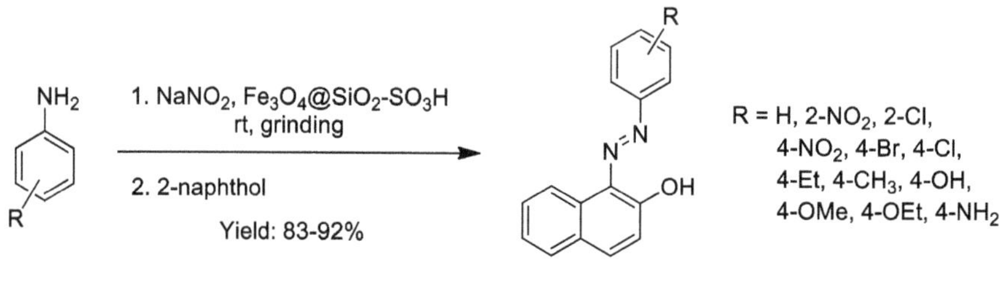

Scheme 17.10 Solvent-free diazotization and diazo coupling reactions (Reproduced from Ref. [45c] with permission from the Royal Society of Chemistry)

Using this method, several azo dyes were synthesized in high yields from different types of aromatic amine derivatives with electron-donating and electron-withdrawing groups (Scheme 17.10). The magnetite-supported catalysts can be separated by filtration as well as can easily be recovered from the reaction mixture using permanent magnet. Thus the time-consuming and laborious separation steps become easy. Thus, as compared to the traditional methods, this protocol is advantageous because of the rapid synthesis of Azo dyes with high yields under solvent-free condition at room temperature, high stability of azo dyes supported on $Fe_3O_4@SiO_2$-SO_3H catalysts and recyclability of the magnetic catalyst.

17.2.7 Synthesis of Liquid Crystals

Phthalocyanine-based liquid crystals: Since its first report in 1982, many phthalocyanine-based liquid crystals were synthesized [46a] by the conventional heating (in an oil bath) the reactants in *n*-hexanol (bp = 197.3 °C) or other high boiling alcohols as the reaction solvent. But it requires very long reaction time which is the main hurdle for the synthesis of this type of liquid crystals. For example, synthesis of C_{14}PcZn with moderate yield of 58% took 24 h by heating the mixture of 4,5-bis(3,4-ditetradecoxyphenoxy)-1,2-dicyanobenzene, $ZnCl_2$ and DBU in n-hexanol (Scheme 17.11). But using the microwave heating with or without phase transfer catalyst, the reaction yield was improved with drastic reduction of the reaction time [46b, c, d].

For example, **C_{10}PcZn** was synthesized in high yield (83%) in 30 min under microwave heating of 4,5-bis(3,4-didecoxyphenoxy)-1,2-dicyanobenzene at 180 ° C using ethylene glycol as the reaction solvent (Scheme 17.11). However, reaction yields gradually decreased for the longer alkyl chain containing liquid crystals (67% for **C_{12}PcZn** and 58% for **C_{14}PcZn)** due to the poor solubility of the long

Scheme 17.11 Structures of phthalocyanine derivatives [46d]

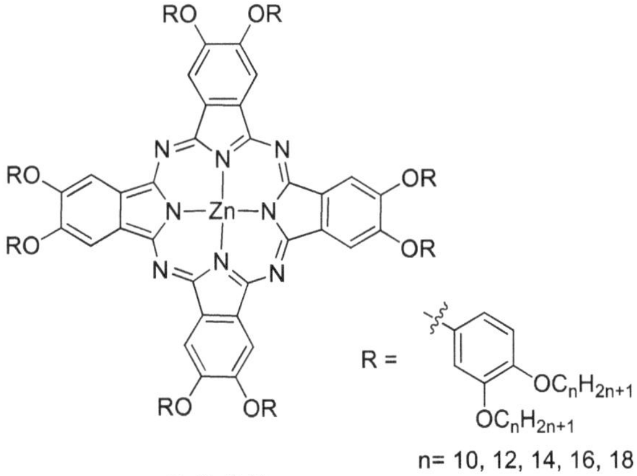

RO OR
N N N
RO OR
N—Zn—N
RO OR
N N N
RO OR
R = OC_nH_{2n+1} OC_nH_{2n+1}
n= 10, 12, 14, 16, 18
8: C_nPcZn

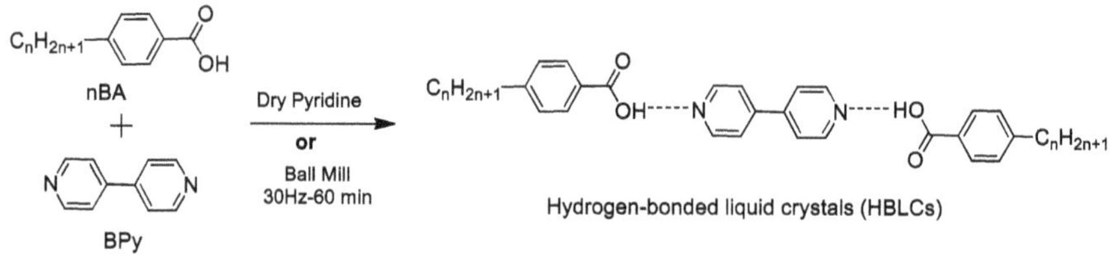

Scheme 17.12 Preparation of hydrogen-bonded complexes (Reproduced from Ref. [47c] Taylor & Francis Ltd (www.tandfonline.com))

alkyl-substituted starting materials (4,5-bis(3,4-dialkoxyphenoxy)-1,2-dicyanobenzene) in the polar ethylene glycol. Thus a phase transfer catalyst, Aliquat 336 (bp = 225 °C) was used to improve the solubility which furnished long chain-substituted phthalocyanine derivatives, $\mathbf{C_{16}PcZn}$ and $\mathbf{C_{18}PcZn}$, in good yields (62% and 59% respectively) under microwave heating condition at 160 °C in 30–60 min.

Solvent-free synthesis of self-assembled liquid crystals by mechano-chemistry: Kato and Fréchet showed that benzoic acid (BA) and stilbazole form 1:1 complex which leads to a well-defined mesogen structure by the intermolecular hydrogen bonding between the carboxylic acid and the pyridine ring [47a, b]. After that Micael D. Miranda et al. [47c] extended Kato's work synthesizing hetero-dimer hydrogen-bonded liquid crystals (HBLCs) between 4,4′-bipyridine (BPy) and some nBAs using a solvent-free synthetic technique (a green chemistry method), neat mechano-chemistry (Scheme 17.12).

Different nBAs (where n = 2, 5, 6, 7) were used as hydrogen bond donors and BPy was selected as hydrogen bond acceptor as its both pyridyl ends have the capability to recognize H-bond donor molecules. The hydrogen-bonded complexes were obtained by a mechano-synthesis technique by mixing of each s benzoic acid with BPy in 2:1 molar ratios in a Retsch MM400 ball mill system (Retsch Solution in Milling & Sieving, Haan, Germany) with a 10-mL stainless steel grinding jar and two 7-mm-diameter stainless steel balls per jar (Scheme 17.12).

Liquid-crystalline thermal properties of hydrogen-bonded mesogenic complexes synthesized through both the conventional and green method are similar indicating no change in the structure of the Liquid crystals. For example, both the $(5\ BA)_2$–BPy complexes are single mesogenic compounds with clear phase transitions with smectic A phases [47b, c].

17.2.8 Fluoropolymers Synthesis

Tetrafluoroethylene (TFE)-based copolymers are highly chemical and thermal resistance with melt processing capability. These are high performance materials which are being used in wide range of applications [48a]. This is prepared from small olefin monomer tetrafluoroethylene (TFE) which is difficult to handle. For example, air mixed TFE is flammable. Further, it has a high tendency to explode during its conversion from liquid phase to gas under pressure and in gas phase also, it is highly explosive at elevated temperatures. TFE undergoes auto polymerization in the presence of oxygen and this process is exothermic enough to ignite an explosion. Thus in conventional synthesis of TFE-based copolymers, special caution needs to be taken due to potential explosions which can be avoided by using $ScCO_2$ as reaction medium. Because TFE forms a "pseudo" azeotrope with CO_2, thus CO_2/TFE mixtures are far less susceptible to ignition *i.e.* handling and delivering of monomer becomes much safer [47b]. Using this method various TFE-based fluoropolymers were synthesized successfully in CO_2 such as PFA, TFE/vinyl acetate polymers, ETFE, FEP, Nafion®-type materials, and Teflon®AF-type materials. Importantly, the fluoropolymers synthesized using Sc-CO_2 contain significantly less amount of acid end groups responsible for synthesis of very high molecular weight materials as compared to that of synthesized by conventional organic and aqueous reaction systems [48c, d].

17.3 Conclusions

The major thrust areas of green chemistry can be summarized into three broad ideas, (i) to reduce the amount of waste, (ii) to reduce the use of toxic chemicals, and (iii) increased use of renewable sources. However, till date, the use of green chemistry in industries has been very limited, mainly because of the poor implementation of the government regulations. A sustainability versus cost parameters of the industries also has a subtle role to play. However, the time has come that the ideas confined in publications to be translated into production. In certain areas like pharmaceuticals, textiles, and biofuels, considerable developments towards sustainability have been achieved. However, they are still in their infancy. The industries should come up, without any further delay, with a determined goal to save the environment by adopting the green principles.

References

1. Tundo P, Anastas P, Black DS, Breen J, Collins T, Memoli S, Miyamoto J, Polyakoff M, Tumas W (2000) Synthetic pathways and processes in green chemistry. Introductory overview. Pure Appl Chem 72:1207–1228
2. (a) Anastas PT, Warner JC (1998) Green chemistry: theory and practice. Oxford University Press, New York; (b) https://www.acs.org/content/acs/en/greenchemistry/principles/12-principles-of-green-chemistry.html. Accessed on 20 June 2019
3. (a) Sheldon RA (1992) Organic synthesis. past, present and future. Chem Ind (London) 903–906; (b) Sheldon RA (2007) The E Factor: Fifteen years on. Green Chem 9:1273–1283

4. Lichtenthaler FW, Peters SCR (2004) Carbohydrates as green raw materials for the chemical industry.Carbohydrates as green raw materials for the chemical industry. Chimie 7:65–90
5. (a) https://www.globenewswire.com/news-release/2018/01/25/1304697/0/en/Polycarbonate-an-Impact-Resistant-and-Lightweight-Product-to-Witness-a-CAGR-of-5–8-during-2017–2023.html. Accessed on 09 Apr 2019; (b) Sun J, Kuckling D (2016) Synthesis of high-molecular-weight aliphatic polycarbonates by organo-catalysis. Polym Chem 7:1642–1649
6. de Koning PD, Jackson M, Lennon IC (2006) Use of achiral (Diphosphine) $RuCl_2$ (Diamine) precatalysts as a practicalaAlternative to sodium borohydride for ketone reduction. Org Process Res Dev 10:1054–1058
7. (a) https://www.pharmtech.com/ionic-liquids-chemical-synthesis-pharmaceuticals. Accessed on 09 Apr 2019; (b) Freemantle M (2003) Chemistry basf's smart ionic liquid: process scavenges acid on a large scale without producing solids. Chem Eng News 81:9; (c) Seddon KR (2003) A taste of the future. Nature Mater 2:363–365
8. (a) Zhang X, Heinonen S, Levänen E (2014) Applications of supercritical carbon dioxide in materials processing and synthesis. RSC Adv 4:61137–61152; (b) Laboureur L, Ollero M, Touboul D (2015) Lipidomics by supercritical fluid chromatography. Int J Mol Sci 16:13868–13884; (c) Leitner W (2000) Designed to dissolve. Nature 405:129–130; (d) Tullo A (2002) Pemex outlines strategy: bold expansion plans may put Mexican petrochemicals back on track. Chem Eng News 80:12
9. Cue BW, Zhang J (2009) Green process chemistry in the pharmaceutical industry.Green process chemistry in the pharmaceutical industry. Green Chem Lett Rev 2:193–211
10. (a) Sheldon RA (2012) Fundamentals of green chemistry: efficiency in reaction design. Chem Soc Rev 41:1437–1451; (b) Agee BM, Mullins G, Swartling DJ (2016) Progress towards a more sustainable synthetic pathway to ibuprofen through the use of solar heating. Sustain Chem Process 4:8–16
11. (a) https://www.epa.gov/greenchemistry/presidential-green-chemistry-challenge-2010-greener-reaction-conditions-award. Accessed on 9 Apr 2019; (b) Ringenberg MR, Ward TR (2011) Merging the best of two worlds: artificial metalloenzymes for enantioselective catalysis. Chem Commun 47:8470–8476
12. https://www.epa.gov/greenchemistry/presidential-green-chemistry-challenge-2006-greener-reaction-conditions-award. Accessed on 9 Apr 2019
13. Veleva VR, Cue BW Jr, Todorova S, Thakor H, Mehta NH, Padia KB (2018) Benchmarking green chemistry adoption by the Indian pharmaceutical supply chain. Green Chem Lett Rev 11:439–456
14. Scott A (2015) Cutting out textile pollution. Chem Eng News 93:18–19
15. Kumar PS, Gunasundari E (2018) Green chemistry in textiles. In: Muthu S (eds) Sustainable innovations in textile chemistry and dyes. Textile science and clothing technology. Springer, Singapore
16. (a) https://www.epa.gov/greenchemistry/presidential-green-chemistry-challenge-2002-greener-reaction-conditions-award. Accessed on 9 Apr 2019; (b) Choudhury AKR (2017) Green chemistry and textile industry. J Textile Eng Fashion Technol 2:351–361
17. https://www.leeds.ac.uk/news/article/1342/a_greener_way_to_your_little_black_dress. Accessed on 09 Apr 2019
18. Knittel D, Saus W, Schollmeyer E (1993) Application of supercritical carbon dioxide in finishing processes. J Text Inst 84:534–552
19. (a) Eren HA, Avinc O, Eren S (2017) Supercritical carbon dioxide for textile applications and recent developments. IOP Conf Ser Mater Sci Eng 254:082011. https://doi.org/10.1088/1757-899X/254/8/082011. Available under the terms of the Creative Commons Attribution License (CC BY)
20. Clarke CB, Tu WC, Levers O, Bröhl A, Hallett JP (2018) Green and sustainable solvents in chemical processes. Chem Rev 118:747–800
21. Mehdi H, Fábos V, Tuba R, Bodor A, Mika LT, Horváth IT (2008) Integration of homogeneous and heterogeneous catalytic processes for a multi-step conversion of biomass:

from sucrose to levulinic acid, γ-Valerolactone, 1,4-Pentanediol, 2-Methyl-tetrahydrofuran, and alkanes. Top Catal 48:49–54

22. https://www.ifpenergiesnouvelles.com/innovation-and-industry/our-expertise/renewable-energies/biofuels/our-solutions. Accessed on 9 Apr 2019
23. https://www.energy.gov/eere/bioenergy/biofuels-basics. Accessed on 9 Apr 2019
24. (a) Patwardhan SV, Manning JRH, Chiacchi M (2018) Bioinspired synthesis as a potential green method for the preparation of nanomaterials: Opportunities and challenges. Curr Opin Green Sustain Chem 12:110–116. Available under the terms of the Creative Commons Attribution License (CC BY); (b) Gallagher MJ, Allen C, Buchman JT, Qiu TA, Clement PL, Krause MOP, Gilbertson LM (2017) Research highlights: applications of life-cycle assessment as a tool for characterizing environmental impacts of engineered nanomaterials. Environ Sci Nano 4:276–281; (c) Tsang MP, Kikuchi-Uehara E, Sonnemann GW, Aymonier C, Hirao M (2017) Evaluating nanotechnology opportunities and risks through integration of life-cycle and risk assessment. Nature Nanotechnol 12:734–739; (d) Eckelman MJ, Zimmerman JB, Anastas PT (2008) Toward green nano. J Ind Ecol 12:316–328; (e) Gao S, Li Z, Zhang H (2010) Bioinspired green synthesis of nanomaterials and their applications. Curr Nanosci 6:452–468; (f) Roy A, Bulut O, Some S, Mandal AK, Yilmaz MD (2019) Green synthesis of silver nanoparticles: biomolecule-nanoparticle organizations targeting antimicrobial activity. RSC Adv 9:2673–2702, and the references cited therin; (g) Wei L, Lu J, Xu H, Patel A, Chen Z-S, Chen G (2015) Silver nanoparticles: synthesis, properties, and therapeutic applications. Drug Discov Today 20:595–601; (h) Siddiqi KS, Husen A, Rao RAK (2018) A review on biosynthesis of silver nanoparticles and their biocidal properties. J Nanobiotechnol 16:14–41; (i) Sengani M, Grumezescu AM, Rajeswari VD (2017) Recent trends and methodologies in gold nanoparticle synthesis – a prospective review on drug delivery aspect. OpenNano 2:37–46; (j) Singh P, Kim Y-J, Zhang D, Yang D-C (2016) Biological synthesis of nanoparticles from plants and microorganisms. Trends Biotechnol 34:588–599, and the references cited therein
25. (a) Jo JH, Singh P, Kim YJ, Wang C, Mathiyalagan R, Jin C-G, Yang DC (2016) Pseudomonas deceptionensis DC5- mediated synthesis of extracellular silver nanoparticles. Artif Cells Nanomed Biotechnol 44:1576–1581; (b) Singh P, Kim YJ, Wang C, Mathiyalagan R, Yang DC (2016) Weissellaoryzae DC6-facilitated green synthesis of silver nanoparticles and their antimicrobial potential. Artif Cells Nanomed Biotechnol 44:1569–1575; (c) Wang C, Kim YJ, Singh P, Mathiyalagan R, Jin Y, Yang DC (2016) Green synthesis of silver nanoparticles by Bacillus methylotrophicus, and their antimicrobial activity. Artif Cells Nanomed Biotechnol 44:1127–1132; (d) Singh P, Kim YJ, Singh H, Wang C, Hwang KH, Farh ME, Yang DC (2015) Biosynthesis, characterization, and antimicrobial applications of silver nanoparticles. Int J Nanomed 10:2567–2577; (e) Singh P, Kim YJ, Singh H, Mathiyalagan R, Wang C, Yang DC (2015) Biosynthesis of anisotropic silver nanoparticles by *bhargavaea indica* and theirs synergistic effect with antibiotics against pathogenic microorganisms. Nanomaterials 10. https://doi.org/10.1155/2015/234741; (f) Singh P, Kim YJ, Wang C, Mathiyalagan R, Yang DC (2016) Microbial synthesis of flower-shaped gold nanoparticles. Artif Cells Nanomed Biotechnol 44:1469–1474; (g) Singh BR, Dwivedi S, Al-Khedhairy AA, Musarrat J (2011) Synthesis of stable cadmium sulfide nanoparticles using surfactin produced by *Bacillus amyloliquifaciens* strain KSU-109. Colloids Surf B Biointerfaces 85:207–213; (h) Elbeshehy EKF, Elazzazy AM, Aggelis G (2015) Silver nanoparticles synthesis mediated by new isolates of Bacillus spp., nanoparticle characterization and their activity against bean yellow mosaic virus and human pathogens. Front Microbiol 6:453; (i) Soni N, Prakash S (2015) Antimicrobial and mosquitocidal activity of microbial synthesized silver nanoparticles. Parasitol Res 114:1023–1030; (j) Longoria EC, Vilchis-Nestor AR, Avalos-Borja M (2011) Biosynthesis of silver, gold and bimetallic nanoparticles using the filamentous fungus Neurospora crassa. Colloids Surf B Biointerfaces 83:42–48; (k) Karthik L, Kumar G, Kirthi AV, Rahuman AA, Rao KVB (2014) *Streptomyces sp.* LK3 mediated synthesis of silver nanoparticles and its biomedical application. Bioprocess Biosyst Eng 37:261–267; (l) Apte M, Sambre D, Gaikawad S, Joshi S, Bankar A, Kuma AR,

Zinjarde S (2013) Psychrotrophic yeast *yarrowialipolytica* NCYC 789 mediates the synthesis of antimicrobial silver nanoparticles via cell-associated melanin. AMB Express 3:32; (m) Seshadri S, Saranya K, Kowshik M (2011) Green synthesis of lead sulfide nanoparticles by the lead resistant marine yeast, *rhodosporidium diobovatum*. Biotechnol Prog 27:1464–1469; (n) Mourato A, Gadanho M, Lino AR, Tenreiro R (2011) Biosynthesis of crystalline silver and gold nanoparticles by extremophilic yeasts. Bioinorg Chem Appl 546074; (o) Waghmare SR, Mulla MN, Marathe SR, Sonawane KD (2015) Ecofriendly production of silver nanoparticles using Candida utilis and its mechanistic action against pathogenic microorganisms. Biotech 5:33–38

26. (a) Zahir AA, Chauhan IS, Bagavan A, Kamaraj C, Elango G, Shankar J, Arjaria N, Roopan SM, Rahuman AA, Singh N (2015) Green synthesis of silver and titanium dioxide nanoparticles using *euphorbia prostrata* extract shows shift from apoptosis to G0/G1 arrest followed by necrotic cell death in *leishmania donovani*. Antimicrob Agents Chemother 59:4782–4799; (b) Momeni S, Nabipour I (2015) A simple green synthesis of palladium nanoparticles with sargassum alga and their electrocatalytic activities towards hydrogen peroxide. Appl BiochemBiotechnol 176:1937–1949; (c) Nasrollahzadeh M, Sajadi SM (2015) Green synthesis of copper nanoparticles using Ginkgo biloba L. leaf extract and their catalytic activity for the Huisgen [3 + 2] cycloaddition of azides and alkynes at room temperature. J Colloid Interface Sci 457:141–147; (d) Singh P, Kim YJ, Yang DC (2016) A strategic approach for rapid synthesis of gold and silver nanoparticles by Panax *ginseng leaves*. Artif Cells NanomedBiotechnol 44:1949–1957; (e) Singh P, Kim YJ, Wang C, Mathiyalagan R, Farh MEA, Yang DC (2016) Biogenic silver and gold nanoparticles synthesized using red ginseng root extract, and their applications. Artif Cells Nanomed Biotechnol 44:811–816; (f) Murugan K, Benelli G, Panneerselvam C, Subramaniam J, Jeyalalitha T, Dinesh D, Nicoletti M, Hwang JS, Suresh U, Madhiyazhagan P (2015) Cymbopogon citratus-synthesized gold nanoparticles boost the predation efficiency of copepod Mesocyclopsaspericornis against malaria and dengue mosquitoes. Exp Parasitol 153:129–138; (g) Poopathi S, Britto LJD, Praba VL, Mani C, Praveen M (2015) Synthesis of silver nanoparticles from *Azadirachta indica*-a most effective method for mosquito control. Environ Sci Pollut Res Int 22:2956–2963; (h) Amooaghaie R, Saeri MR, Azizi M (2015) Synthesis, characterization and biocompatibility of silver nanoparticles synthesized from *Nigella sativa* leaf extract in comparison with chemical silver nanoparticles. Ecotoxicol Environ Saf 120:400–408; (i) Elango G, Roopan SM (2015) Green synthesis, spectroscopic investigation and photocatalytic activity of lead nanoparticles. Spectrochim Acta A Mol BiomolSpectrosc 139:367–373; (j) Kalaiselvi A, Roopan SM, Madhumitha G, Ramalingam C, Elango G (2015) Synthesis and characterization of palladium nanoparticles using *Catharanthus roseus* leaf extract and its application in the photo-catalytic degradation. Spectrochim Acta A Mol BiomolSpectrosc 135:116–119; (k) Sadeghi B, Rostami A, Momeni SS (2015) Facile green synthesis of silver nanoparticles using seed aqueous extract of Pistacia atlantica and its antibacterial activity. Spectrochim Acta A Mol BiomolSpectrosc 134:326–332; (l) Zhou GJ, Li SH, Zhang YC, Fu YZ (2014) Biosynthesis of CdS Nanoparticles in Banana Peel Extract J NanosciNanotechnol 14:4437–4442; (m) Gogoi N, Babu PJ, Mahanta C, Bora U (2015) Green synthesis and characterization of silver nanoparticles using alcoholic flower extract of *Nyctanthes arbortristis* and in vitro investigation of their antibacterial and cytotoxic activities. Mater Sci Eng C Mater Biol Appl 46:463–469; (n) Kora AJ, Beedu SR, Jayaraman A (2012) Size-controlled green synthesis of silver nanoparticles mediated by gum ghatti (*Anogeissus latifolia*) and its biological activity. Org Med Chem Lett 2:17; (o) Velmurugan P, Park JH, Lee SM, Jang JS, Lee KJ, Han SS, Lee SH, Cho M, Oh BT (2015) Synthesis and characterization of nanosilver with antibacterial properties using *Pinus densiflora* young cone extract. J PhotochemPhotobiol B 147:63–68; (p) Suresh D, Shobharani RM, Nethravathi PC, Kumar MAP, Nagabhushana H, Sharma SC (2015) *Artocarpus gomezianus* aided green synthesis of ZnO nanoparticles: Luminescence, photocatalytic and antioxidant properties. Spectrochim Acta A Mol Biomol Spectrosc 141:128–134; (q) Shende S, Ingle AP, Gade A, Rai M (2015) Green synthesis of copper nanoparticles by *Citrus medica Linn.* (Idilimbu) juice

and its antimicrobial activity. World J Microbiol Biotechnol 31:865–873; (r) Hyllested JÆ, Palanco ME, Hagen N, Mogensen KB, Kneipp K (2015) Green preparation and spectroscopic characterization of plasmonic silver nanoparticles using fruits as reducing agents. Beilstein J Nanotechnol 6:293–299; (s) Naseem T, Farrukh MA (2015) Antibacterial activity of green synthesis of iron nanoparticles using *lawsonia inermis* and *gardenia jasminoides* leaves extract J Chem 912342

27. (a) Jahangirian H, Lemraski EG, Webster TJ, Moghaddam RR, Abdollahi Y (2017) A review of drug delivery systems based on nanotechnology and green chemistry: green nanomedicine. Int J Nanomed 12:2957–2978, and the references cited therein; (b) White MA, Johnson JA, Koberstein JT (2006) Toward the syntheses of universal ligands for metal oxide surfaces: controlling surface functionality through click chemistry. J Am Chem Soc 128:11356–11357; (c) Soetaert F, Korangath P. Serantes, D, Fiering S, Ivkov R. (2020) Cancer therapy with iron oxide nanoparticles: agents of thermal and immune therapies. Adv Drug Del Rev 163-164: 65-83; (d) Kumar CSSR, Mohammad F (2011) Magnetic nanomaterials for hyperthermia-based therapy and controlled drug delivery. Adv Drug Deliv Rev 63:789–808; (e) Zhang Y, Sun C, Kohler N, Zhang M (2004) Self-assembled coatings on individual monodisperse magnetite nanoparticles for efficient intracellular uptake. Biomed Microdevices 6:33–40; (f) Liu J, Luo Z, Zhang J et al (2016) Hollow mesoporous silica nanoparticles facilitated drug delivery via cascade pH stimuli in tumor microenvironment for tumor therapy. Biomaterials 83:51–65; (g) Pozdnyakov AS, Emelyanov AI, Kuznetsova NP et al (2016) Nontoxic hydrophilic polymeric nanocomposites containing silver nanoparticles with strong antimicrobial activity. Int J Nanomed 11:1295–1304; (h) Li X, Gong Y, Zhou X et al (2016) Facile synthesis of soybean phospholipid-encapsulated MoS2 nanosheets for efficient in vitro and in vivo photothermal regression of breast tumor. Int J Nanomed 11:1819–1833; (i) Chen Z, Zhang T, Wu B, Zhang X (2016) Insights into the therapeutic potential of hypoxia-inducible factor-1α small interfering RNA in malignant melanoma delivered via folate-decorated cationic liposomes. Int J Nanomed 11:991–1002; (j) Liu Y, Deng L-Z, Sun H-P et al (2016) Sustained dual release of placental growth factor-2 and bone morphogenic protein-2 from heparin-based nanocomplexes for direct osteogenesis. Int J Nanomed 11:1147–1158; (k) Firouzamandi M, Moeini H, Hosseini SD et al (2016) Preparation, characterization, and in ovo vaccination of dextran-spermine nanoparticle DNA vaccine coexpressing the fusion and hemagglutinin genes against Newcastle disease. Int J Nanomed 11:259–267; (l) Erdal MS, Özhan G, Mat MC, Özsoy Y, Güngör S (2016) Colloidal nanocarriers for the enhanced cutaneous delivery of naftifine: characterization studies and in vitro and in vivo evaluations. Int J Nanomed 11:1027–1037; (m) Ilbasmis-Tamer S, Unsal H, Tugcu-Demiroz F, Kalaycioglu GD, Degim IT, Aydogan N (2016) Stimuli-responsive lipid nanotubes in gel formulations for the delivery of doxorubicin. Colloid Surf B 143:406–414; (n) Nguyen TTC, Nguyen CK, Nguyen TH, Tran NQ (2017) Highly lipophilic pluronics-conjugated polyamidoamine dendrimer nanocarriers as potential delivery system for hydrophobic drugs. Mater Sci Eng C 70:992–999; (o) Rauta PR, Das NM, Nayak D, Ashe S, Nayak N (2016) Enhanced efficacy of clindamycin hydrochloride encapsulated in PLA/PLGA based nanoparticle system for oral delivery. IET Nanobiotechnol 10:254–261; (p) Anna E, Czapar AE, Yao-Rong Zheng Y-R et al (2016) Tobacco mosaic virus delivery of phenanthriplatin for cancer therapy. ACS Nano 10:4119–4126; (q) Zhang D, Wang J, Xu D (2016) Cell-penetrating peptides as noninvasive transmembrane vectors for the development of novel multifunctional drug-delivery systems. J Control Release 229:130–139; (r) Xie X, Luo S, Mukerabigwi JF et al (2016) Targeted nanoparticles from xyloglucan–doxorubicin conjugate loaded with doxorubicin against drug resistance. RSC Adv 6:26137–26146; (s) Chen LC, Chen YC, Su CY, Hong CS, Ho HO, Sheu MT (2016) Development and characterization of self-assembling lecithin-based mixed polymeric micelles containing quercetin in cancer treatment and an in vivo pharmacokinetic study. Int J Nanomed 11:1557–1566; (t) Yang B, Dong X, Lei Q, Zhuo R, Feng JX (2015) Host–guest interaction-based self-engineering of nano-sized vesicles for co-delivery of genes and anticancer drugs. ACS Appl Mater Interfaces 7:22084–22094

28. (a) Sun Z, Worden M, Thliveris JA, Hombach-Klonisch S, Klonisch T, van Lierop J, Hegmann T, Miller DW (2016) Biodistribution of negatively charged iron oxide nanoparticles (IONPs) in mice and enhanced brain delivery using lysophosphatidic acid (LPA). Nanomedicine 12:1775–1784; (b) Li X, Gong Y, Zhou X, Jin H, Yan H, Wang S, Liu J (2016) Facile synthesis of soybean phospholipid-encapsulated MoS_2 nanosheets for efficient *in vitro* and *in vivo* photothermal regression of breast tumor. Int J Nanomed 11:1819–1833
29. (a) Renu GB, Kasturi M, Meifang Z, Muthupandian A, Nay MH, Sivakumar M (2017) Sonochemical and sustainable synthesis of graphene-gold (G-Au) nanocomposites for enzymeless and selective electrochemical detection of nitric oxide. Biosen Bioelectron 87:622–629; (b) Tu H, Lu Y, Wu Y, Tian J, Zhan Y, Zeng Z, Deng H, Jiang L (2015) Fabrication of rectorite-contained nanoparticles for drug delivery with a green and one-step synthesis method. Int J Pharm 493:426–433
30. (a) Feng T, Ai X, An G, Yang P, Zhao Y (2016) Charge-convertible carbon dots for imaging-guided drug delivery with enhanced *in vivo* cancer therapeutic efficiency. ACS Nano 10:4410–4420; (b) Huang N, Wang H, Zhao J, Lui H, Korbelik M, Zeng H (2010) Single-wall carbon nanotubes assisted photothermal cancer therapy: animal study with a murine model of squamous cell carcinoma. Laser Surg Med 42:798–808; (c) Jeyamohan P, Hasumura T, Nagaoka Y, Yoshida Y, Maekawa T, Kumar DS (2013) Accelerated killing of cancer cells using a multifunctional single-walled carbon nanotube-based system for targeted drug delivery in combination with photothermal therapy. Int J Nanomed 8:2653–2667
31. Lombardo D, Kiselev MA, Caccamo MT (2019) Smart nanoparticles for drug delivery application: development of versatile nanocarrier platforms in biotechnology and nanomedicine. J Nanomat 3702518:26. https://doi.org/10.1155/2019/3702518
32. (a) Jagtap KK, Shivran N, Mula S, Naik DB, Sarkar SK, Mukherjee T, Maity DK, Ray AK (2013) Change of boron substitution improves the lasing performance of bodipy dyes: amechanistic rationalisation. Chem Eur J 19:702–708; (b) Gupta M, Mula S, Tyagi M, Ghanty TK, Murudkar S, Ray AK, Chattopadhyay S (2013) Rational design of boradiazaindacene (bodipy)-based functional molecules. Chem Eur J 19:17766–17772
33. (a) Loudet A, Burgess K (2007) Bodipy dyes and their derivatives: syntheses and spectroscopic properties bodipy dyes and their derivatives: syntheses and spectroscopic properties. Chem Rev 107:4891–4932; (b) Boens N, Leen V, Dehaen W (2012) Fluorescent indicators based on BODIPY. Chem Soc Rev 41:1130–1172; (c) Kamkaew A, Lim SH, Lee HB, Kiew LV, Chung LY, Burgess K (2013) Bodipy dyes in photodynamic therapy. Chem Soc Rev 42:77–88; (d) Lu H, Mack J, Yang Y, Shen Z (2014) Structural modification strategies for the rational design of red/NIR region bodipys Structural modification strategies for the rational design of red/NIR region BODIPYs. Chem Soc Rev 43:4778–4823; (e) Bessette A, Hanan GS (2014) Design, synthesis and photophysical studies of dipyrromethene-based materials: insights into their applications in organic photovoltaic devices. Chem Soc Rev 43:3342–3405; (f) Mula S, Ray AK, Banerjee M, Chaudhuri T, Dasgupta K Chattopadhyay S (2008) Design and development of a new pyrromethene dye with improved photostability and lasing efficiency: theoretical rationalization of photophysical and photochemical properties. J Org Chem 73:2146–2154
34. Jameson LP, Dzyuba SV (2013) Expeditious, mechanochemical synthesis of bodipy dyes. Beilstein J Org Chem 9:786–790
35. (a) Lu H, Zhang S, Liu H, Wang Y, Shen Z, Liu C, You X (2009) Experimentation and theoretic calculation of a bodipy sensor based on photoinduced electron transfer for ions detection. J Phys Chem A 113:14081–14086; (b) Matsumoto T, Urano Y, Shoda T, Kojima H, Nagano T (2007) A Thiol-Reactive Fluorescence Probe Based on Donor-Excited Photoinduced Electron Transfer: Key Role of Ortho Substitution. Org Lett 9:3375–3377; (c) Ueno T, Urano Y, Kojima H, Nagano T (2006) Mechanism-based molecular design of highly selective fluorescence probes for nitrative stress. J Am Chem Soc 128:10640–10641
36. (a) Meng G, Velayudham S, Smith A, Luck R, Liu H (2009) Color Tuning of Polyfluorene Emission with BODIPY Monomers. Macromolecules 42:1995–2001; (b) Baruah M, Qin W, Flors C, Hofkens J, Vallée RAL, Beljonne D, Van der Auweraer M, De Borggraeve WM,

Boens N (2006) Solvent and pH dependent fluorescent properties of a dimethylaminostyryl borondipyrromethene dye in solution. J Phys Chem A 110:5998–6009; (c) Jiao L, Yu C, Li J, Wang Z, Wu M, Hao E (2009) β-formyl-bodipys from the vilsmeier−haack reaction. J Org Chem 74:7525–7528

37. Chen Y, Jiang J (2011) 4-(4,4-Difluoro-1,3,5,7-tetra¬methyl-3a-aza-4a-azonia-4-borata-s-indacen-8-yl) benzonitrile. Acta Cryst E 67:o908
38. (a) Kim KT, Kim BH (2013) A fluorescent probe for the 3'-overhang of telomeric DNA based on competition between two interstrand G-quadruplexes. Chem Commun 49:1717–1719; (b) Li Z, Bittman R (2007) Synthesis and spectral properties of cholesterol- and FTY720-containing boron dipyrromethene Dyes. J Org Chem 72:8376–8382; (c) Benstead M, Rosser GA, Beeby A, Mehl GH, Boyle RW (2011) Mesogenic bodipys: an investigation of the correlation between liquid crystalline behaviour and fluorescence intensity. Photochem Photobiol Sci 10:992–999; (d) Teki Y, Tamekuni H, Haruta K, Takeuchi J, Miura Y (2008) Design, synthesis, and uniquely electron-spin-polarized quartet photo-excited state of a π-conjugated spin system generated *via* the ion-pair state . J Mater Chem 18:381–391; (e) Kondo M, Furukawa S, Hirai K, Kitagawa S (2010) Coordinatively immobilized monolayers on porous coordination polymer crystals. Angew Chem Int Ed 49:5327–5330; (f) Hayek A, Bolze F, Bourgogne C, Baldeck PL, Didier P, Arntz Y, Mély Y, Nicoud J-F (2009) Boron containing two-photon absorbing chromophores. 2. Fine tuning of the one- and two-photon photophysical properties of pyrazabol based fluorescent bioprobes. Inorg Chem 48:9112–9119
39. (a) Crawford SM, Thompson A (2010) Conversion of 4,4-Difluoro-4-bora-3a,4a-diaza-s-indacenes (*F*-bodiyps) to dipyrrins with a microwave-promoted deprotection strategy. Org Lett 12:1424–1427; (b) Ali AA-S, Cipot-Wechsler J, Crawford SM, Selim O, Stoddard RL, Cameron TS, Thompson A (2010) The first series of alkali dipyrrinato complexes. Can J Chem 88:725–735
40. (a) Ulrich G, Ziessel R (2004) Convenient and efficient synthesis of functionalized oligopyridine ligands bearing accessory pyrromethene-BF_2 fluorophores. J Org Chem 69:2070–2083; (b) Ulrich G, Ziessel R (2004) Functional dyes: bipyridines and bipyrimidine based boradiazaindacene. Tetrahedron Lett 45:1949–1953; (c) Alamiry MAH, Harriman A, Mallon LJ, Ulrich G, Ziessel R (2008) Energy- and charge-transfer processes in a perylene–bodipy–pyridine tripartite array. Eur J Org Chem 2008:2774–2782
41. Sekiya M, Umezawa K, Sato A, Citterio D, Suzuki K (2009) A novel luciferin-based bright chemiluminescent probe for the detection of reactive oxygen species. Chem Commun 3047–3049
42. (a) Hamer FM (1964) The cyanine dyes and related compound. Interscience, New York; (b) Phillips AP (1949) Condensation of aromatic aldehydes with γ-picoline methiodide. J Org Chem 14:302–305; (c) Abdel MO, Khairy MH, Zarif HK, Vlademir DT (1976) Studies on cyanine dyes II. Preparation of ferrocene cyanines. J Appl Chem Biotechnol 26:71–78
43. Wang L-Y, Zhang X-G, Shi Y-P, Zhang Z-X (2004) Microwave-assisted solvent-free synthesis of some hemicyanine dyes. Dyes Pigm 62:21–25
44. Bhushan KR, Liu F, Misra P, Frangioni JV (2008) Microwave-assisted synthesis of near-infrared fluorescent sphingosine derivatives.. Chem Commun 4419–4421
45. (a) Rahimizadeh M, Eshghi H, Shiri A, Ghadamyari Z, Matin MM, Oroojalian F, Pordel P (2012) $Fe(HSO_4)_3$ as an Efficient Catalyst for Diazotization and Diazo Coupling Reactions. J Korean Chem Soc 56:716–719; (b) Zarei A, Hajipour AR, Khazdooz L, Mirjalili BF, Chermahini AN (2009) Rapid and efficient diazotization and diazo coupling reactions on silica sulfuric acid under solvent-free conditions. Dyes Pigm 81:240–244; (c) Safari J, Zarnegar Z (2015) An environmentally friendly approach to the green synthesis of azo dyes in the presence of magnetic solid acid catalysts. RSC Adv 5:17738–17745
46. (a) Ohta K, Nguyen-Tran H-D, Tauchi L, Kanai Y, Megumi T, Takagi Y (2011) Handbook of porphyrin science, volume 12: applications of porphyrins, phthalocyanines and related systems, Chap. 53: Liquid crystals of phthalocyanines, porphyrins and related compounds, World Scientific, Singapore, pp 1–120; (b) Ooi K, Maeda F, Ohta K, Takizawa T, Matsuse T

(2005) Rapid synthesis of phthalocyanine-based discotic liquid crystals by using a novel hand-made microwave-heating apparatus. J Porphyr Phthalocya 09:544–553; (c) Maeda F, Uno K, Ohta K, Sugibayashi M, Nakamura N, Matsuse T, Kimura M (2003) Unique molecular structure and properties of novel purple intermediates of phthalocyanine derivative. J Porphyr Phthalocya 07:58–69; (d) Akabane T, Ohta K, Takizawa T, Matsuse T, Kimura M (2017) Discotic liquid crystals of transition metal complexes, 54: Rapid microwave-assisted synthesis and homeotropic alignment of phthalocyanine-based liquid crystals. J Porphyr Phthalocya 21:476–492

47. (a) Kato T, Frechet JMJ (1989) A new approach to mesophase stabilization through hydrogen bonding molecular interactions in binary mixtures. J Am Chem Soc 111:8533–8534; (b) Kato T, Frechet JMJ, Wilson PG, Saito T, Uryu T, Fujishima A, Jin C, Kaneuchi F (1993) Hydrogen-bonded liquid crystals. Novel mesogens incorporating nonmesogenic bipyridyl compounds through complexation between hydrogen-bond donor and acceptor moieties. Chem Mater 5:1094–1100; (c) Miranda MD, Chávez FV, Maria TMR, Eusebio MES, Sebastião PJ, Silva MR (2014) Self-assembled liquid crystals by hydrogen bonding between bipyridyl and alkylbenzoic acids: solvent-free synthesis by mechanochemistry. Liq Cryst 41:1743–1751

48. (a) Fiering AE (1994) In: Banks RE, Smart BE, Tatlow JC (eds) Organofluorine chemistry: principles and commercial applications. Plenum Press, New York; (b) Van Bramer DJ, Schiflett MB, Yokozeki A (1994) Safe handling of tetrafluoroethylene. United States Patent 5:345, 013; (c) DeYoung JP, Romack TJ, DeSimone JM (1997) Fluoroolefin polymerization in carbon dioxide: Synthesis and characterization of TFE based fluoropolymers. Polym Prepr 38:424–425; (d) Romack TJ, DeSimone JM (1995) Synthesis of tetrafluoroethylene-based, nonaqueous fluoropolymers in supercritical carbon dioxide. Macromolecules 28:8429–8431

Chapter 18
Bio-inspired Synthesis of Nanomaterials

Mainak Roy and Poulomi Mukherjee

Abstract Over the years, human has emphatically developed the skill of synthesizing materials with fascinating properties that resulted in an unprecedented technological progress and industrial growth. Nature, on the other hand, makes exotic materials with unique properties in a seemingly simple (yet baffling at times) approach which is environmentally benign and debars intensive use of energy, high boiling solvents and corrosive chemicals. Such green synthesis recipe adopted by nature is in stark contrast to the industrial processing of materials that often sheds out toxic emissions and polluting discards to the environment. That is why it is sometimes compelling upon humankind to look up to the Mother Nature for efficient and non-polluting recipe for synthesis of materials. Biomimetic synthesis involves preparation of naturally occurring and functionally important materials using a set of chemical reactions that closely resemble nature's own method of synthesizing them in course of different biological processes. Biogenic synthesis, on the other hand, encompasses every possible approach that involves either molecules of biological origin or biological agents like plants and microbes for materials production. These reactions usually take place in aqueous phase at around room temperature and at biological pH and make use of catalysts and/or biocatalysts for moderating the reaction conditions and may produce non-toxic by-products and easily disposable wastes benign to the environment. Till date, a large number of different materials with exquisite morphology and diverse functionalities have been reported by this route. The present chapter aims at providing an overview of synthetic strategies developed on being inspired by natural processes.

M. Roy (✉)
Chemistry Division, Bhabha Atomic Research Centre, Mumbai 400085, India
e-mail: mainak73@barc.gov.in

P. Mukherjee (✉)
Nuclear and Biotechnology Division, Bhabha Atomic Research Centre, Mumbai 400085, India
e-mail: poulomi@barc.gov.in

M. Roy · P. Mukherjee
Homi Bhabha National Institute, Anushakti Nagar, Mumbai 400094, India

A. K. Tyagi and R. S. Ningthoujam (eds.), *Handbook on Synthesis Strategies for Advanced Materials*, Indian Institute of Metals Series,
https://doi.org/10.1007/978-981-16-1807-9_18

Keywords Nanomaterials · Biomimetic · Plants · Fungi · Virus · Bacteria · Actinomycete yeast

18.1 Introduction

Off late, nanoparticles are gaining importance in the field of materials research because of properties superior to bulk that has led to different novel applications and eventually technologies based on their size and morphology. Nanoparticles, by definition, should be less than 100 nm along one or more of its dimensions [1]. They have made an impact in the field of health care, drug delivery, cosmetics, energy storage, catalysis, electronics, nonlinear optics, space science and almost all the frontiers of science [2–4].

Broadly, there are two different classes of techniques for synthesizing nanoparticles [5, 6]. They are:

(i) *Top-down methods* wherein micro-crystallites of the bulk are crushed down to smaller dimension by means of ball milling, sputtering and chemical/thermal/ laser etching
(ii) *Bottom-up methods* wherein atoms and/molecules coalesce to form nuclei that grow in size to form nanocrystallites. Bottom-up methods include precipitation, vapour phase deposition, sol–gel methods, sonochemical technique, pyrolysis, etc. Methods involving chemical synthesis of nanoparticles may involve the use of polluting chemicals [2] and high boiling solvents, disposal of which and the worked-up residues are potentially hazardous to the environment. Physical methods, on the other hand, are energy intensive [2] and often make use of expensive setups involving high vacuum, high voltages, etc. These limitations of both physical and chemical synthesis of nanoparticles have called for yet another approach which is apparently free from the perils of polluting the environment. It is Mother Nature's own way of synthesizing materials and is proven to be safe being in vogue from time immemorial. Biogenic synthesis or its different adapted versions either directly make use of natural components such as plants and microbes or their functional elements and by-products. The green syntheses mostly take place under ambient conditions and are gradually becoming significant to nanotechnology. Most of these processes are time-consuming. However, on the brighter side they allow for controlled manipulation of material properties at a lower dimension and often resulting in exotic morphologies. Schematic representation of different synthetic approaches to nanoparticles is given in Fig. 18.1.

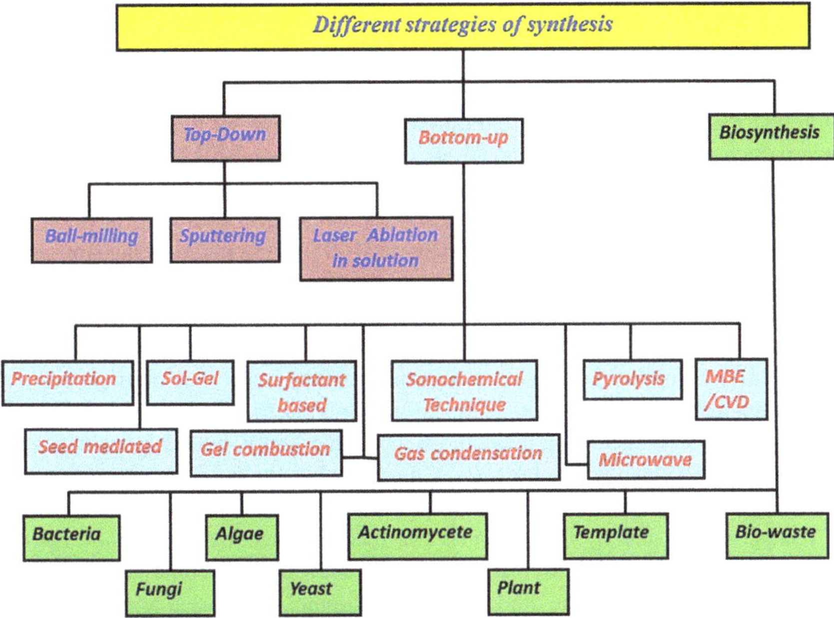

Fig. 18.1 Schematic representation of different synthetic approaches to nanoparticles

18.2 Overview of Biogenic Synthesis

Extracts and biomass obtained from different biological sources and even living organisms are being employed for the synthesis of nanoparticles in an eco-friendly approach. Sources include fruits, flowers, roots, leaves, seeds, barks of plants belonging to different families and species and almost the entire gamut of microbes including bacteria, fungi, algae, yeasts and even actinomycetes. Some of these micro-organisms may act as nano-sized templates with exotic architecture for nanoparticles with diverse shapes and sizes. Extracts from different biological wastes have also been employed in the synthesis of nanomaterials. Majority of the efforts have been made in producing metal and metal oxide nanoparticles. Reducing and capping properties of the biological agents present in the extracts are believed to have been instrumental in converting the metallic ions to respective nanoparticles. Till date, a large number of reviews have been devoted to the synthesis of nanoparticles using a variety of extracts/biomass and elucidation of the mechanisms of nanomaterial formation by this route [2, 7–11], derivative excerpts from which and other literature will be discussed in the subsequent chapters.

18.3 General Mechanism of Biogenic Synthesis of Nanoparticles

Different researchers have reported different mechanisms for biogenic synthesis of nanoparticles by different biological agents. Micro-organisms tend to accumulate metals within the cells. Whereas some metals are beneficial to them, some others are proven toxic. Accumulation of essential metals above certain physiological concentration also results in undue toxicity [12], which the micro-organisms tend to reduce as a means of their natural survival. Uptake of toxic metal ions essentially results from the inability of the transport mechanisms prevalent in the micro-organisms to distinguish between the toxic and beneficial metal ions. The process may be initiated either upon adsorption of the metal ions onto the cell wall or through diffusion across it, causing surface or intracellular reduction [13]. Metal nanoparticles may get accumulated on both the inner boundary (cytoplasmic membrane) and outside of the cell walls and also in the periplasmic space of micro-organisms [14] wherein metal ions have undergone chemical transformation to nano-sized crystals by the action of enzymes and/secondary metabolite such as alkaloids, and terpenoids. Sometimes, there is even preferential uptake of one particular metal ion over the other [15]. For example, the molecular mechanism of the bacterial formation of magnetite first involves invagination of the cytoplasmic membrane in a process similar to that of eukaryotic cells. It is followed by the accumulation of ferrous ions at a supersaturating concentration into the vesicles by trans-membrane transporters. Thereafter, the ferrous ion gets partially reduced to ferrihydrite which is subsequently dehydrated to magnetite. The process of bio-mineralization of iron is realized through the intervention of an oxido-reductase system and BacMP membrane associated proteins [16].

Similarly, extracellular synthesis of nanoparticles involves use of extract/culture filtrate from the biomass of micro-organisms for reduction and capping of the particles. Cadmium sulphide nanoparticle formation, for example, takes place through the cleavage of the disulphide linkages (S–S bonds) followed by the formation of cadmium–sulphur (Cd–S) bonds in the form of cadmium thiolate complex [17]. Different enzymes and bioactive molecules present within the cells or in the cell-free extract, and their specific roles in synthesizing different nanomaterials will be elaborated under the heading of the respective micro-organisms. In an analogous approach, plant extracts containing reducing sugars, aldehydes and ketones reduce metal ions and also stabilize the nanoparticles [18]. Different phytochemicals present in the secondary metabolites of plants therefore play an important role in synthesizing different nanoparticles.

18.4 Nanoparticle Synthesis Using Plant Extract

Plants, either in the form of extract or biomass, amongst all possible biological resources, perhaps have the maximum potential of getting scaled up to the industrial level because of the abundance in supply, faster production and ease of processing. In contrast to the microbial synthesis, this method is devoid of elaborate culture, cost of protocols for isolation and downstream processing, and most importantly, they are inherently least hazardous of the lot and non-pathogenic to human beings [19]. Plant extract contains active biomolecules such as proteins, sugars, phenols, alkaloids, steroids, carbohydrates, sapogenins, flavonoids, amino acids and enzymes [8]. Some of them are excellent reducing agents and reduce the metal ions to corresponding metal nanoparticles whereas others, in combination, take part in stabilizing the nanoparticles by capping. Extract from leaves, bark, fruits, seeds and practically every part of the plant has been made use of in synthesizing nanoparticles. Nanoparticle size depends on pH of the medium and also on the nature of bio-capping agents present in the extract. Silver nanoparticles with large variation in size were reported using plant extract [8]. Nanoparticles may be formed within the plant cell by way of accumulation. Minerals present in the soil, either in bare or chelated form, get solubilized and absorbed by the plants at very low levels. Sometimes, nanoparticles from different geological sources, viz. water, air, soil, etc., are absorbed by the plants through their cell membranes and cell walls of the root epidermis, which then get into the vascular bundle of xylem and thereafter transported to the leaves. Schematic representation of nanoparticles transport within the plants is shown in Fig. 18.2 (adapted with permission from the reference Tripathi et al. [20] with permission from Elsevier). The transport process is sometimes highly selective towards the nanoparticles, wherein cell membrane acting like a semi-permeable membrane allows the transport of the selected particles [20] through its pores. Gold nanoparticles were found to get accumulated in Alfalfa plants, in *Sesbania drummondii* seedlings [21] and also in *Chilopsis linearis* (desert willow) [22]. Alfalfa plants were probably one of the earliest to have been reported for phytosynthesis of nanoparticles and are known to have synthesized both gold and silver nanoparticles. Formation of nanoparticles of both silver and gold inside Alfalfa resulted via uptake of the precursor ions from an environment rich in these metal salts. *Sesbania drummondii* produced catalytically active spherical gold nanoparticles with narrow size dispersion [21]. Plants can even synthesize metal alloys in the presence of more than one metal ion. An intracellular alloy of gold–silver and copper (Au–Ag–Cu) was reportedly produced within the plant when *Brassica juncea* seeds were made to grow on a soil containing appropriate metal precursors [21]. Bioaccumulation of metal nanoparticles by plants (and transgenic plants) has often been utilized for recovering noble metals from lean ores in runoff mines and also cleaning up of contaminated sites [23, 24] in an eco-friendly and viable approach and are, respectively, termed as phytomining and phytoremediation [25]. However, nanoparticle uptake may cause severe toxicity in plants by way of producing reactive oxygen species (ROS) which in turn causes oxidative stress,

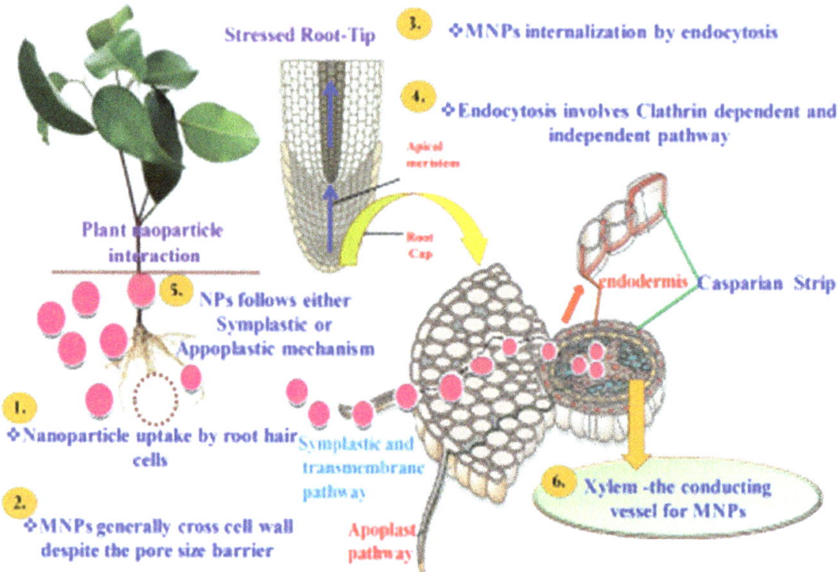

Fig. 18.2 Schematic representation of nanoparticles transport within plants (Adapted from Tripathi et al. [20], and reprinted *with permission from Elsevier*)

damaging DNA and proteins. Nanoparticle-induced phytotoxicity is manifested in the form of morphological and anatomical aberrations such as retardation of seed germination, reduction in root and shoot size and biomass [20].

Extracts of different plants have been employed for the extracellular synthesis of nanoparticles of gold, silver, copper, cobalt, nickel, platinum, palladium, titanium, selenium and zinc and also metal oxides [26]. Some of these nanoparticles also exhibit preventive and/or remedial measures against a large number of ailments including cancer, microbial and other infections. Amongst the different nanomaterials produced using plant extract, metal nanoparticles especially silver and gold nanoparticles have been the most extensively researched upon and reported because of their unique properties. Silver nanoparticles exhibit high electrical and thermal conductivities, excellent catalytic properties and high antimicrobial efficacy against drug-resistant microbes. They can also function as a surface enhanced resonance Raman spectroscopy (SERS) template [19]. Gold nanoparticles are also potential catalysts and antimicrobial agents. Excellent surface plasmon resonance (SPR) properties of gold nanoparticles may find application in plasmonic and photonic devices [19]. Several excellent articles and reviews highlighted on the different plant resources used for synthesizing silver and gold nanoparticles and their different parts such as fruits, leaves, roots, tubers, rhizome, bark, aerial soot, etc., in the synthetic protocol [19, 21, 25, 27, 28]. Silver nanoparticles have been synthesized using extract of *Ehretia laevis* that exhibited catalytic oxidation of aromatic alcohols [29]. Extract of fruits of *Crataegus douglasii* also produced silver nanoparticles [28]. Synthesis of silver nanoparticles has been carried out using extracts of *Acorus calamus* for prospective antibacterial and anticancer efficacies and using extract of *Boerhaavia diffusa* for efficacy against a variety of microbes [27]. Extract of different plants reported for the synthesis of silver nanoparticles included *Sesuvium portulacastrum* L., *Tribulus terrestris* L., *Abutilon indicum*, *Cocos nucifera* and tea [27]. A quick recipe of silver nanoparticles production was reported using extract of *Acalypha indica* [30]. Extract of the papaya fruit [31] and bark of *Cinnamomum zeylanicum* [32] also reportedly yielded silver nanoparticles. Phytosynthesis of metal nanoparticles depended largely on conditions used. It was observed for the leaf broth of *Magnolia kobus* and *Diospyros kaki* that the rates of chemical reactions involved in silver nanoparticle synthesis increased at increased temperatures, however, with concomitant decrease in particle size for *D. kaki* [21]. Concentration of the extract and pH of the medium also played a key role in synthesizing nanoparticles. Morphology of both gold and silver particles and also their optical properties could be manipulated by changing the concentration of plant extract [33, 34]. Gold nanoparticles were synthesized from the fresh leaves as well as different parts of the plant *Sphaeranthus indicus* [19]. Extract of *Moringa oleifera* [35], a plant containing large amount of bioactive molecules and antioxidants and also known for the traditional treatment of inflammation, bacterial/viral infection, hyperglycaemia and even cancer has been used for the synthesis of gold nanoparticles having improved therapeutic properties (anti-proliferative activities). Almost all parts of the plant, viz. flower, root, leaf, seedpod, have been used for the phytosynthesis of different kinds of nanoparticles. Neem (*Azadirachta indica*) [36, 37] and aloe vera [38] are versatile plant sources, and their extracts have been made use of by different groups of researchers for producing both gold and silver nanoparticles and even gold–silver core–shell in the case of the former. A long list of plant resources for the synthesis of gold is provided in the article by Teimuri-Mofrad et al. [28] which includes leaf extracts of banana, pomegranate, tamarind, olive and other citrus fruits and also flower extracts of *Lonicera japonica*, *Plumeria alba* and saffron. Extracts of the wheat biomass [39], fenugreek (*Trigonella foenum-graecum*) [40], lemon grass [41], dried clove buds [42], shells of the willow tree [43] and even glucan of mushroom [44] have been reported in the literature. Gold nanostructures ranging from few nanometres to few tens of nanometres in size were produced from the extracts of *Coriandrum sativum* (coriander) [45]. A rapid synthesis of gold nanoparticles with faceted morphology was achieved using extract of *Pelargonium graveolens* [46]. Room temperature synthesis of anisotropic gold and spherical or quasi-spherical silver nanoparticles were carried out using extract of phyllanthin [25]. Similar to silver nanoparticles, formation of gold nanoparticles is also influenced by the temperature of the reaction bath. Although most of the green synthesis using plant extract had been reported at ambient temperatures, some of the experiments were conducted at elevated temperatures and it was found in the case of *Cymbopogon flexuosus* that morphologies of gold nanoparticles changed if temperature was changed [47]. Effect of pH was also studied for extracellular synthesis [48] of gold nanoparticles. Even nanoparticle size synthesized within living plants or using plant extracts was found to depend on the pH of the medium [25]. Apart from gold and silver, copper oxide nanoparticles have been reported using leaf extract of *Abutilon indicum* [49]. Iron oxide nanoparticles have been synthesized from the leaf extract of *Moringa oleifera* [50]. Iron oxide nanoparticles have also been synthesized from different plant extracts, and they exhibit different functionalities [51]. Iron oxide nanoparticles have also been made using soya bean sprouts [52]. Further, different kinds of citrus fruits produced iron oxide particles. These nanoparticles reportedly have diverse biomedical applications and could be potentially manipulated using magnetic field for targeted drug delivery [51]. Similarly, copper oxide nanoparticles have been synthesized using extracts of variety of plants. Some of these precursors are also known for their therapeutic applications whereas the copper oxide nanoparticles so produced have the potential for photocatalytic applications [19 and references therein]. Nickel and zinc oxide were synthesized using extracts of *Moringa oleifera* and *Agathosma betulina* [53–56], whereas cobalt oxide was synthesized using extract of *Aspalathus linearis* [57].

As already discussed briefly, the mechanisms adopted by the plants for nanoparticles synthesis are very much different from those perceived for conventional methods. Different mechanisms adopted during phytosynthesis of nanoparticles have been discussed at length in several of the reviews [18, 26]. Nanocrystals are formed due to the reduction of the corresponding metal ions. Once formed, these particles tend to get agglomerated spontaneously into larger particles in a bid to reduce surface energy. A very common ploy to arrest such spontaneous aggregation of nanocrystallites is to make use of a suitable capping agent. Plants are known to produce large variety chemicals such as alkaloids, terpenoids, sugars, flavonoids and other secondary metabolites which possibly reduce and also stabilize nanoparticles during their phytosynthesis [18, 26]. Gold nano-triangles are reportedly formed upon reduction of tetrachloroaurate ions with lemon grass extract [41], whereas particle growth is promoted by aldehydes and/or ketones in the extract. Different types of proteins, lipids and polyphenols are produced as by-products of plants extract. One such important class of polyphenols is flavonoids which have been extensively explored for the phytosynthesis of different nanomaterials [26]. Amino acids, polysaccharides, vitamins and citrates [25] also present in the extract of plants have been reported for the synthesis of nanoparticles with desired shapes and sizes. Shapes and sizes of nanoparticles may be manipulated by varying the concentration of plant extracts and hence relative amounts of these bioactive chemicals. Proteins, terpenoids and other biomolecules present in the extracts of *Magnolia kobus* contain functional groups such as amines, alcohols, aldehydes and carboxylic acids [25] which are reportedly instrumental in the synthesis of nanoparticles. Wheat biomass, which is an agricultural by-product, was found to reduce trivalent gold ions to zero valent metallic gold by virtue of the functional groups on its cell walls after getting bound to Au^{3+} that resulted in faceted nanoparticles [39]. Mechanism of nanoparticle formation within the living plants has been studied with reference to the intracellularly formed silver and gold nanoparticles in *Brassica juncea*, wherein the process is believed to have been mediated through the reducing sugar molecules present within the plant's chloroplast [25]. Purification is an important step for nanoparticle synthesis, and the workup is generally done through repeated washing followed by centrifugation, chromatographic and electrophoretic separation of the particles from rest of the biomass and/or dissolved or colloidal components remaining in the system. Sometimes, segregation of particles according to their sizes is also achieved by ultracentrifugation [25].

18.5 Nanomaterials Using Biowastes

Nanomaterials may be synthesized employing different types of wastes generated in industries, especially food and agro-industries, in municipalities and even in our household kitchens. They are low-cost raw materials, available in plenty, and otherwise, their disposal is sometimes a major issue to the society. Biowaste is a general term given to all possible organic wastes that are generated in and around us after processing of the plant and animal products and majority of them include peels of fruits and vegetables and animal remains.

Fruit peels constitute a form of plant waste often discarded after consuming the fruit pulp. Peels of fruits like pomegranate, lemon, oranges, mango, grapes, banana, papaya and plum were successfully used for the production of nanoparticles. Some fruit peels are known to contain natural antioxidants such as flavonoids, carotenoids, ascorbic acid and sugars [58] and were possibly responsible for reduction of the metal ions forming nanoparticles. Because of their high reduction potentials, noble metals are prone to get reduced easily by diverse reducing agents be it of inorganic or organic origin. Different metals, viz. silver, gold and palladium nanoparticles, have been synthesized by reducing the respective metal salts with extracts from appropriate plant wastes. For example, extract from the peels of Satsuma Mandarin, a source of traditional medicine, has been used in synthesizing silver nanoparticles without the aid of any additional capping agents [59]. An extract of the fruit peel was used along with silver nitrate for the preparation of nanoparticles. A similar approach to producing silver nanoparticles was adopted by Kaviya et al. [60], wherein *Citrus sinensis* was used in the form of peel extract for reducing the silver salt at room temperature and also at an elevated temperature. The silver nanoparticles exhibited antibacterial activity against both Gram negative and Gram positive bacteria. A narrow distribution of starch supported silver nanoparticles was prepared using aqueous extract of the peel of *Citrus sinensis* under ambient conditions that exhibited antimicrobial, cytocompatibility and free radical scavenging properties [58]. Silver nanoparticles with high antioxidant property were produced at room temperature from *Citrus sinensis* peel extract by another research group [61]. Grape skin, stalk and seeds were used in room temperature synthesis of stable nanoparticles of gold [62]. Nanoparticle growth was achieved in a short time and in a single step. Polyphenolic antioxidants such as catechin and proanthocyanidins present in grape skin, stalks and seeds might have resulted in transforming gold(III) ions from $HAuCl_4$ to nanoparticles and subsequently their stabilization [62]. Studies on nanoparticle formation due to the effect of a reference compound catechin were also reported. Mango is again a tropical juicy edible fruit belonging to the genus *Mangifera*, grown in abundance in Asia. Au and Ag nanoparticles were produced by mango peel extract [63, 64]. The silver nanoparticles loaded onto non-woven fabrics were explored for their antibacterial properties. The authors also investigated into the mechanism of silver nanoparticles formation using FTIR that revealed the possible participation of hydroxyl, ketone, aldehydes and carboxylic groups of cellulose, flavonoid, pectin, lignin present in the mango peel in the process. A plausible mechanistic pathway to reducing Ag^+ by mango peel extract was reported [63]. Gold nanoparticles produced from mango peel extract did not show biological cytotoxicity on some specific cells up to a concentration of 160 μg/ml [64]. Banana peel extract [65] was used in synthesizing silver nanoparticles with moderate stability. Variation in pH on nanoparticle synthesis and role of different functional groups involved in the bio-reduction of Ag^+ was studied. The silver nanoparticles showed antibacterial and free radical scavenging properties. Apart from mango, banana, orange and grapes, peel extracts of pineapple [66], papaya [67] and bilberry wastes and spent coffee grounds [68] were used to synthesize silver nanoparticles. However, in some cases, extract was obtained from biowastes using solvents such as ethanol. Gold nanoparticles were synthesized from the peel extract of *Garcinia mangostana* [69]. Rice bran extract was used for the bio-mineralization of gold [70]. The possible bioactive components of rice bran instrumental in reducing Au^{3+} were investigated by high-performance liquid chromatography (HPLC) and liquid chromatography–mass spectrometry (LC–MS) that showed ferulic acid was the primary reducing agent. Grass, an important fodder for cattle, is often left to dry in fields as a plant waste. An interesting conversion of the waste dry grass into a useful resource for producing silver nanoparticles was reported [71]. The nanoparticles were investigated for their anticancer, antifungal and antibacterial effects. Another potential form of agricultural waste originates from the vegetable markets. Vegetable skin that is usually discarded after peeling off vegetables along with the discarded vegetables and thrown away leaves constitute the vegetable waste. Everyday huge amount of such vegetable wastes are generated at the domestic as well as commercials fronts. In a report, vegetable wastes were made use of in producing silver nanoparticles [72]. Antibacterial properties of the silver nanoparticles so produced against both Gram positive and Gram negative bacteria were reported. Silver nanorods and nanoparticles were also synthesized from industrial milk wastes [73]. Whereas nanorods were obtained at room temperature and in alkaline pH, nanoparticles were produced at higher temperatures and reduced pH. These nanostructures were also found to increase the stability of the milk (keeping quality). Spherical nanoparticles of palladium were synthesized [74] using watermelon rind and were used as catalyst for Suzuki coupling reaction. Nanoparticles of zero valent iron were synthesized using extracts from peel and other wastes of fruits like orange, lemon, mandarin [75]. Apart from metal nanoparticles, researchers have produced nanocrystalline oxides using biowastes. Peel extract of drum sticks (*Moringa oleifera)* was used to synthesize zinc oxide nanoparticles [76], which were employed for photocatalytic dye degradation and studying their antifungal, antibacterial and hemolytic properties. An innovative method of producing gold nanoparticles on eggshell membrane for biosensing applications was also reported [77].

18.6 Microbial Synthesis of Nanoparticles

Apart from terrestrial plants and animals, multi-cellular organisms which include most of the fungi, many algae were used for nanomaterials production. Not only multi-cellular organisms, unicellular organisms such as bacteria, unicellular algae, unicellular fungi, which constitute one of the oldest habitats of this world also took part in synthesizing nanomaterials. Together, they come under the category of microbial synthesis. *Microbial synthesis of nanoparticles* is therefore an eco-friendly and inexpensive protocol that employs different biological organisms in synthesizing nanoparticles.

Micro-organisms such as bacteria [78, 79] and fungal species [12] have long been used for studying their adaptation to and sometimes remediation of toxic metals, either by complexing with metal ions or by reducing them, which at times might have produced metal particles by chance. But deliberate search to finding biological agents with intent of nanoparticle synthesis is rather a recent venture. Reduction of toxic metals within the cells forms a part of the defence mechanism of the microbes [80] for survival in diverse and extreme environment. It is not fully understood as to how the biological agents synthesize nanoparticles, since different microbes adopt different mechanisms for nanoparticles synthesis. Sometimes it is different for intracellular and extracellular methods. However, extensive studies have been reported in the literature [3] that throws some light on the plausible mechanisms of microbial synthesis. In brief, some bacteria promote nucleation of metal clusters first which interacts further with the bacterial cells to form nano-sized particles and then move through the cell walls. Nanoparticles formation by fungal cells occurs via their electrostatic interaction with the metal ions followed by their enzymatic reduction on cell wall. In case of actinomycetes, metal gets reduced on the mycelial surface and also on cytoplasmic membrane [3]. Amongst these microbes, which include bacteria, actinomycetes (both are prokaryotes; i.e. organisms made up of cells without distinct nucleus) and algae, yeasts and fungi (eukaryotes, i.e. organism having cells with well-defined nucleus enclosed by cell membranes), bacteria emerged as one of the most popular choice for microbial synthesis of nanoparticles.

18.6.1 Bacteria Mediated Synthesis of Nanoparticles

Bacteria are prokaryotic micro-organisms which can survive in almost all kind of habitats including the human gut and some living in radioactive environments too [81]. Bacteria may be of different shapes and sizes and hence form an impending bio-factory of exotic nanomaterials. There are innumerable instances where bacteria naturally synthesize nanoparticles. Some notable examples include magnetotactic bacteria synthesizing magnetite nanoparticles [82], diatoms synthesizing nano-structured siliceous materials [83] and bacterial S-layer synthesizing fine grain gypsum and calcite [84]. Magnetotactic bacteria can synthesize magnetic nanoparticles either in the form of oxides (Fe_3O_4), or sulphides (Fe_3S_4) or both depending on its nature [85, 86]. Synthesizing such bio-inorganic nanomaterials with exceedingly complex functionalities and hierarchical ordering at different length scale is still a far cry for most of the laboratory based synthetic protocols. Some of the earliest reports on bacterial synthesis of nanoparticles probably dealt with the removal from mines and leachates of valuable metals, wherein the nanoparticles got accumulated intracellularly within the bacterial cell. Nanoparticles of silver were formed when metal-resistant bacterium *Pseudomonas stutzeri* AG259 was made to grow in the presence of silver. A hard coating of carbon–silver composite was formed when silver nanoparticles embedded within the bacterial biomass were given heat treatment [87, 88]. Moreover, optical properties of these composite coatings could be tuned depending on the amount of silver loading in them and nature of heat treatment. Gold accumulation was reported for *Pedomicrobium* [89] and on the cell wall of *Bacillus subtilis* 168 [90]. Gold nanoparticles formation from gold thiosulphate complex by sulphate reducing bacteria was also reported [91]. Gold nanoparticles were produced within the periplasm of *Geobacter ferrireducens* [92]. A filamentous cyanobacterium *Plectonema boryanum* UTEX 485 was reported for producing gold nanoparticles with controlled morphology [86 and references therein]. $Au(S_2O_3)_2^{3-}$ and $AuCl_4^-$ separately upon reaction with the bacterium precipitated, respectively, cubic nanoparticles and octahedral platelets depending on the conditions of the reaction [93]. Bioaccumulation of gold from solution of gold chloride took place via formation of amorphous gold (I)-sulphide on the cell wall followed by the precipitation of octahedral gold (III) platelets close to the surface and also in solution [94]. Spherical gold particles were reported for *E. coli* DH5α [95]. Trivalent ions of gold reportedly got bio-adsorbed onto *Rhodobacter capsulatus* and reduced to the respective nanoparticles [96]. Both gold and silver ions got reduced to their respective metal nanoparticles when separately exposed to *Lactobacillus* strains present in buttermilk. In the presence of suitable precursors, alloys of the two metals were also formed [97]. Accumulation of silver nanoparticles in the periplasmic space of the cells was observed for atmospheric isolate of *Bacillus* sp. [98].

Extracellular synthesis takes place when reductive enzymes and/or biomolecules secreted outside the cells by the microbes and/or adhered onto the outer surface of the cell wall are involved in producing the nanoparticles and are often a more viable method as compared to the intracellular approach. Room temperature synthesis of nano-gold with diverse shapes and particle sizes was reported for *Rhodopseudomonas capsulata* at different pH of the solution [99].

Extract of *P. aeruginosa* (ATCC 90271, strain 1 and strain 2) reportedly produced gold nanoparticles. Surface plasmon resonance (SPR) properties of these nanoparticles tend to get modified along with particle size producing a systematic change in their colour [100]. Uniform and well-dispersed nanoparticles of silver were produced when hydroxyl ions were present in the reaction milieu along with dried cells of *Aeromonas sp.* SH10 [101]. Synthesis of silver nanoparticles was also reported using supernatant of *Klebsiella pneumonia, E. coli* and *Enterobacter cloacae cultures* [102]. In an analogous approach, *B. licheniformis* and *Morganella* sp. produced nano-silver [103, 104]. Divalent palladium reportedly got reduced to zero valent palladium by *Desulfovibrio desulfuricans* NCIMB 8307 with the help of an external electron donating agent [105]. Extracellular production of gold nanoparticles via reduction of gold ions with hydrogen as electron donor and in the presence of Au(III) reductase was reported for bacteria like *Pyrobaculum islandicum, Pyrococcus furiosus* and *Thermotoga maritima* [85, 92]. Off late, a strategy has been adopted for the synthesis of exotic nanomaterials especially those of metallic nanoparticles with well-defined morphologies using bacterial hollows as template [106]. These are naturally occurring templates often in the form of bacterial S-layers with predefined dimension and structure that allowed formation of materials within them with unique size and crystallographic modifications. At this stage, it might be difficult to further manipulate and control the morphology of nanoparticles produced from a preset combination of proteins acting as templates. Nevertheless, the possibility of selectively producing proteins and other biologically active molecules with desired morphology and cavity size is being explored through genetic engineering. Template-based synthetic protocols will be discussed under a separate section.

As briefly mentioned earlier, intracellular nanoparticle synthesis by microbes is a part of their defence mechanism in response to the metal toxicity. Predominant mechanisms include biosorption, bioaccumulation, extracellular complexation or precipitation of metals, lack of specific metal transport system, efflux system, toxicity due to reduction and oxidation and alteration of solubility [3, 107 and reference therein]. Cell wall plays a keen role in the synthesis of nanoparticles. Silver uptake by *P. stutzeri* led to metal efflux and metal binding and formation of silver nanoparticles in the vacuole-like granules in the periplasm [3]. NADH-dependent reductase enzyme acted as the electron source for reducing silver ion, and it got oxidized to NAD^+ [2, 3]. *Rhodopseudomonas capsulata* produced gold nanoparticles extracellularly through the intervention of enzymes depending on NADH and also cofactor NADH secreted by microbes. The source of electron had been NADH-dependent reductase [99]. Silver nanoparticle formation was also reported via involvement of NADH-dependent reductase. Again NADH served as the reducing agent. Alternately nitrate-dependent reductase took part in reducing silver ions [3]. Charged carboxylic acid groups of the peptide chains and residual amino acids got bound to silver amine complex on the cell wall of treated *Corynebacterium* sp. along with other reducing agents that followed reduction of silver ions forming nanoparticles. Mms6 protein associated with magnetotactic bacteria *Magnetospirillum magneticum* AMB-1 is reportedly known to have mediated the formation of magnetite particles with cubo-octahedral geometry [108].

18.6.2 Actinomycetes Mediated Synthesis of Nanoparticles

The term "*Actinomycetes*" originates from the Greek terms "*aktis*" and "*mykes*" which stands for ray and fungus, respectively [109]. Earlier *Actinomycetes* were considered as ray fungus [110]. Actinomycetes belong to the group of Gram positive, aerobic bacteria [111] with fungus-like branched filamentous growth [112, 113]. Actinomycetes are similar to bacteria in terms of cell wall composition [109], but resemble both fungi and bacteria [3]. They are present both on the soil and also in marine environment. They actively take part in nitrogen fixation and also act as growth promoters and bio-control agents to plant pathogens [113]. Some actinomycetes are benign to higher plants and animal, whereas some others may be pathogenic to them. Again, some may be commercially important being the sources of many industrially and medically relevant compounds such as antibiotics (antibacterial and antifungal) and immune-suppressant [114]. Actinomycetes are known to produce novel secondary metabolites with anticancer properties that may control cholesterol and exhibit immunosuppressive properties [113]. Actinomycetes reportedly synthesized different nanomaterials having good stability, polydispersity [115] and also exhibit antifungal, antibacterial and anti-parasitic activities amongst a host of other properties [116].

Actinomycetes were reported for synthesizing nanoparticles intracellularly as well as extracellularly [117]. An early report on Actinomycetes induced extracellular synthesis of nanoparticles was due to *Thermomonospora* sp. which produced narrowly dispersed gold nanoparticles with rather small particle size [118]. Streptomyces with more than 500 species forms the largest genus of Actinobacteria. Silver nanoparticles were synthesized using *Streptomyces* sp, and a variety of land acetomycetes was found to be effective against different multi-drug-resistant human pathogens [114]. Silver nanoparticles were synthesized using extracts of different species of the Streptomyces [119, 120] that inhibited several clinically potent bacterial species and also exhibited anticancer effect [121]. Gold nanoparticles were reportedly synthesized using *Streptomyces griseoruber*, actinomycetes [122]. The nanoparticles exhibited catalytic degradation of dye. *Rhodococcus* sp. accumulated intracellularly well-dispersed gold particles on its cell wall and also on cytoplasmic membrane [118]. The nanoparticles had no toxic effect on the cells since their growth continued despite their occurrence. Actinomycetes reportedly synthesized nanoparticles of copper and copper oxide [123, 124].

Actinomycetes produce significant quantities of proteins, enzymes and other secondary metabolites that are believed to have been responsible for the improved reduction of metal ions [120], making actinomycetes prolific biosynthetic agents for nanoparticles. The possible mechanism of nanoparticles formation and stabilization by actinomycetes was reported in different review articles [3, 117 and references therein]. Four specific protein molecules in the secretion of actinomycetes were identified by means of electrophoretic analysis. These protein molecules possibly originating from the enzymes contained free amine groups and/or cysteine residues that bound to the gold and silver nanoparticles thereby stabilizing them. FTIR investigation revealed that stabilization of gold nanoparticles synthesized by *Thermomonospora* sp. was possibly due to protein molecules on the surface of the particles [115]. Extracellular synthesis of silver nanoparticles by *Streptomyces* sp. was reported [125], which showed from FTIR studies the possible participation of OH in reducing silver ions. Role of nitrate reductase was also discussed. It was proposed in another report [112] that positively charged silver ions got trapped by the negatively charged carboxylate ions of the mycelia cell wall of the actinomycetes and got enzymatically reduced to silver nuclei and eventually to bigger particles. Further, according to mechanisms for intracellular nanoparticle formation, metal ions got reduced on the mycelia surface and also on the cytoplasmic membrane of the actinomycetes on being acted by enzymes there [118].

18.6.3 Algae Mediated Synthesis of Nanoparticles

Algae are both unicellular and multi-cellular eukaryotic (also prokaryotic at times) organism belonging to the kingdom Protista and are predominantly aquatic in nature [126] but also found in the soil [127]. Depending on their size, algae are of two types, viz. microalgae and macro-algae, and have been extensively used in pigment, agricultural, medical, bio-fuel, cosmetic and pharmaceutical industries [126, 127]. They are photosynthetic in nature just like plants and were also employed for the production of nanomaterials. There are excellent reviews on eco-friendly algal synthesis of nanoparticles [127, 128]. Possibility of bulk and inexpensive production of algae projects it as a cost-effective and green source of nanoparticles. Often algae develop specific charges on cell surface, thereby accelerating the formation of particles via nucleation and growth [4, 127]. A variety of nanoparticles were synthesized using algae. However, majority of the reports in the literature are on metallic nanoparticles. Algae tend to synthesize metals nanoparticles by reducing them both inside and outside the cell. Because of this property, algae have become an upcoming contender for nanoparticle synthesis. Metabolites from algal culture are known to contain polysaccharides, proteins and other reducing agents which are believed to have reduced metal ions, whereas proteins containing amino groups and sulphated polysaccharides help in capping them [127]. Different classes of algae including diatoms and euglenoids were used for the synthesis of gold–silver, palladium nanoparticles [127]. Diatoms are known to synthesize gold as well as silica gold biocomposite [129]. Gold nanoparticles were accumulated in suspended *C. vulgaris* [130]. Gold got attached to the algal cells and subsequently got reduced to nanoparticles. Extract of a brown seaweed *Sargassum* sp. was employed for the synthesis of gold nanoplates at room temperature. Its size was found to vary depending on the reaction conditions [131]. Initial studies showed that protein molecules in the extract were possibly the bioactive species. However, one major disadvantage with algal synthesis and perhaps common to most of the intracellular or cell associated microbial synthesis protocols is the difficulty in separating the nanoparticles from the remaining components. Brown algae *Fucus vesiculosus* reduced gold (III) ions by virtue of the hydroxyl groups of polysaccharides [132]. Such algal reduction of metals might provide a benign methodology of gold recovery from leachates.

A large number of marine microalgae were effective in synthesizing silver nanoparticles. Algal synthesis of silver nanoparticles carried out with the excreted metabolites was reported for their antibacterial efficacies and tuneable optical properties [4, 133]. Seaweed (*Chaetomorpha linum)* was reported for the reduction of silver nitrate. Algal metabolites such as flavonoids and terpenoids reduced Ag^+ and stabilized the nanoparticles [134]. Polysaccharides also helped in manipulating dimension and morphology of silver particles [4]. A number of toxic metals including mercury, lead and cadmium were reportedly removed from their aqueous solution by *different* algal species [135].

18.6.4 Fungi Mediated Synthesis of Nanoparticles

Fungi are one of the oldest members of the eukaryotic family of micro-organism and belong to the kingdom "fungi". In contrast to algae, fungi are primarily heterotrophs and obtain food from both living and dead plants and animals. They are known to digest the food outside their cell and assimilate nutrients through the cell wall. Depending on the mode of their acquiring food, fungi may be classified into biotrophs, saprotrophs or necrotrophs and can be either unicellular or multi-cellular.

Fungi were used for the synthesis of both metals and oxides nanoparticles. They were relatively more tolerant towards metals, produced large amount of biomasses and involved less complicated protocols [136–140]. Most importantly, the size and morphology of fungal synthesized nanoparticles could be manipulated easily by tweaking the culture condition, viz. time, pH, temperature and the amount of biomass. Intracellular formation provided better control over particle size since their growth was limited by the availability of space inside the organism, whereas extracellular synthesis is far more hassle-free mode because of easier down-processing of nanoparticles. One of the earliest reports on gold nanoparticles synthesis was due to *Verticillium* sp. [14]. Formation of gold nanoparticles occurred both intracellularly and on the surface of the cells. It was hypothesized that the negatively charged gold (III) chloride ions were electrostatically attached to the positively charged groups in the mycelial enzyme, got reduced to metal aggregates and then transported to the cytoplasmic membrane. The enzymes in the cytoplasmic membrane might have also contributed in reducing the metal ions [14]. Formation of nanoparticles of gold both within the cells and outside, with diverse morphology ranging from spheres, rods and triangles was observed for *Trichothecium* sp.

The fungal material produces specific proteins including enzymes being subjected to different experimental conditions [141]. Accumulation of gold nanoparticles in *Verticillium luteoalbum* biomass grown for different duration was studied. It showed that with ageing of the cells, the number of nanoparticles decreased while particle shape remained practically unaltered, implying higher concentration of enzyme formation at earlier stages of growth [142]. Spherical gold nanoparticles were synthesized at an acidic pH 3, whereas a combination of spherical, rod-shaped, triangular and hexagonal particles was formed at slightly higher pH 5. Increasing the pH further to 7 and 9 resulted in predominantly irregular shaped particles [143]. Incubation temperature of *V. luteoalbum* biomass reportedly increased the rate of nanoparticles formation and particle size increased with incubation time. Selective accumulation of silver was reported for *Phoma* sp. [15]. It was observed that silver got selectively accumulated in *Phoma* species, whereas metals like nickel, copper, lead and cadmium got released partially into the medium.

Verticillium sp. biomass accumulated silver nanoparticles primarily below the cell surface [13, 144], whereas *Aspergillus flavus* produced silver particles over surface [145]. Both extracellular and intracellular syntheses of platinum nanoparticles of varying sizes and shapes were reported, respectively, in the associated medium and on the cell membrane or cell wall of *Fusarium oxysporum* fungal strain [146].

Fungi are known to secrete large amount biomolecules responsible for the production of nanoparticles [137]. Endophytic fungus *Colletotrichum* sp. rapidly synthesized spherical gold nanoparticles [46]. A rapid synthesis of silver nanoparticles was reported for *A. fumigatus* [147]. Silver nanoparticles with pyramidal morphology were synthesized on the mycelial surface of white rot fungus (*Phanerochaete chrysosporium*) [148]. Highly stable silver nanoparticles with spherical morphology were produced upon reaction of the culture filtrate of *Fusarium semitectum* with silver ions [149]. Silver nanoparticles from *Aspergillus niger* were spherical in shape and ca. 20 nm in diameter. Again involvement of proteins in capping and nitrate reductase along with quinone in extracellular electron transfer was established [150]. Silver nanoparticles were produced extracellularly by *Cladosporium cladosporioides* [151]. Similarly, proteins of *Coriolus versicolor* were reported for both intracellular and extracellular syntheses of silver nanoparticles [152]. Cell-free extract of *Trichoderma asperellum* was found to synthesize gold nano-triangles as well as spherical silver nanoparticles with antibacterial efficacy [153, 154]. Other fungal species such as *Fusarium solani* [155], *Phoma glomerata* [156], *Penicillium fellutanum* [157], *Trichoderma viride* [158], *Penicillium brevicompactum* WA 2315 [159] were also reported for extracellular production of silver nanoparticles. Extract of *Volvariella volvacea* produced nano-silver, gold and silver–gold particles [160]. *F. oxysporum* also synthesized all the three types of nanoparticles, namely gold, silver and alloys of both [144, 161, 162]. Iron oxide nanoparticles were formed extracellularly when fungus *Fusarium oxysporum* and *Verticillium* sp. [163] were made to react with both Fe(II) and Fe (III) salts. Formation of iron oxide nanoparticles by the fungus was believed to have taken place via hydrolysis of magnetite precursors (anionic iron complexes) by cationic proteins [163]. *Fusarium oxysporum* mediated synthesis of zirconia [136], silica, titania [164], barium titanate [165], strontium carbonate [166] and CdSe quantum dots [167] were also reported, making it a versatile organism for nano-synthesis.

Different proteins, bio-membranes and carbohydrates were reported for their active role in the synthesis of nanoparticles by fungus, and different mechanisms were proposed. According to one such popular mechanism, silver hydrosols were produced by the action of α–NADPH-dependent enzyme nitrate reductase, often in the presence of quinone derivatives like naphthoquinones and anthraquinones [168, 169]. The quinone derivatives acted as electron shuttling agents and peptides served as stabilizers. In an analogous approach, synthesis of gold nanoparticles from gold (III) chloride was mediated by phytochelatin and sulphite reductase enzyme that depended on α–NADPH [170]. Here also capping of gold particles was done by peptide molecules. Formation of nano-gold outside its cell was reported using *F. oxysporum* [161], which testified participation of amide linkages and protein molecules ranging between molecular mass of 66 and 10 kDa.

18.6.5 Yeast Mediated Synthesis of Nanoparticles

Yeasts are also members of the kingdom fungi and are single-celled eukaryotic micro-organisms which need food, moisture and warmth for their survival. They are not able to harness sun light for their growth, hence produce energy from organic matter and are therefore called *chemoorganotrophs*. They are probably the most useful microbes being capable of fermenting sugar and starch to produce carbon dioxide and alcohol. That is why they are indispensable to the bakery and brewery industries. Just like fungi, yeasts grow very fast on simple nutrients and their growth may be controlled easily under laboratory conditions [171, 172]. Yeasts secrete numerous reducing enzymes which have resulted in the mass production of nanoparticles via both intracellular and extracellular mechanisms [173]. Large quantities of highly toxic metals get accumulated in different genera of yeasts. Such intracellular accumulation of metals results in severe toxicity to the cells, and the effect is overcome by means of different processes such as sorption, chelation onto extracellular peptides or polysaccharides, enzymatic oxidation and reduction and controlled cell membrane transport [171]. Yeasts also form complex with metal ions and stabilize them. There are excellent reviews on nanoparticle synthesis by yeasts [3, 86, 173]. Initially, yeasts were known for the producing cadmium sulphide and lead sulphide often called "semiconductor crystals" or "quantum semiconductor crystals" [174]. The first report on biosynthesis of nanocrystalline CdS was due to *Candida glabrata* and *Schizosaccharomyces pombe* [86]. Quantum crystallites were produced by the yeasts through the binding of the metal to γ-glutamyl peptide forming a metal peptide complex [174]. Yield of nanoparticles was found to depend on the growth phase of the yeast. Maximum yield was achieved upon incubating the yeast cells in their mid-log phase of growth in 1 mM cadmium [86]. Lead sulphide was produced intracellularly by yeast *Torulopsis* [175]. Baker's yeast (*Saccharomyces cerevisiae*) was employed in the bioremediation of cadmium and lead and also in the synthesis oxide nanoparticles of antimony (Sb_2O_3) [176]. Gold nanoparticles with varying sizes were synthesized using

Pichia jadinii [143], wherein enzymes and proteins in cytoplasm and on the cell wall reduced gold ions to nanoparticles. Capping of particles was done by the hydrophobic sheath of peptide layer protecting them from getting agglomerated [3]. Nano-gold particles were also synthesized in the peptidoglycan layer of the cell wall of the biomass of Baker's yeast [177] and also by the yeast *Yarrowia lipolytica* [178]. In the presence of soluble silver, Ag nanoparticles with controlled size were synthesized extracellularly by yeast strains MKY3, upon binding of Ag+ to the protein molecules secreted under silver stress [179].

Yeasts give rise to glutathione (GSH) and ligands that bind to metals such as metallothioneins and phytochelatins (PC) which are primarily responsible for the nanoparticle formation by yeast [3]. Mechanism of stress-induced intracellular formation of CdS nanocrystals was discussed in the literature [86]. Multiple biochemical reactions are initiated in the yeast cells following metal toxicity induced stress in them. Phytochelatin synthase produced phytochelatins that in turn would complex with cytoplasmic metal and get transported across the vacuolar membrane. *Y. lipolytica* reportedly produced metallothioneins in the presence of toxic metals such as nickel and cadmium [180], which is why it finds application in heavy metal remediation. On the other hand, aldehydes groups present in the reducing sugars were possibly effective in reducing gold ions in the case of Baker's yeasts [177].

18.6.6 Virus Based Synthesis of Nanoparticles

Virus are linear or circular sub-microscopic infectious agent consisting of either a single-stranded or a double-stranded nucleic acid housed within a protein shell (capsid) and can multiply rapidly within living cells of an organism. They are found almost everywhere in ecosystem and can be pathogenic to plants, animals and even microbes.

Virus interact with metal ions through the thick outer coating of capsid proteins [181]. These protein may act as a template [182] for symmetric, polyvalent and mono-disperse viral nanoparticles with high-aspect ratio, surface area and exotic architecture for drug delivery, energy, sensing, imaging, catalytic and therapeutic applications [183]. Virus may also be used for the production of metal nanoconjugates and nanocomposites. Gahlawat et al. [2] reviewed the role of virus in synthesizing different kinds of nanoparticles. M13 bacteriophage was explored for the synthesis of semiconducting CdS and ZnS nanocrystals [184]. Different combination of virus were also used as additives to plant extracts for nanoparticle production that resulted in increased number and reduced size of the nanoparticles [2].

18.7 Template Bound Biomimetic Approach

There is an increasing demand for nanocrystalline materials having exotic morphologies for advanced device applications. Templates offer a convenient and easy route to the fabrication of such exotic nanostructures. Since time immemorial, nature has made use of a wide variety of templates for synthesizing different materials. Biomimetics encompasses a host of such different synthetic strategies for materials production adopted in pursuit of mimicking biological systems. Synthesis using biomimetic approach to nanomaterials with complex morphologies in many a cases made use of biological structures as templates. Various templates in the form of DNA [185], proteins [186], peptides [187], viruses [188] and even pollen grains [189] were used for the synthesis of nanostructures. This section summarizes different types of templates of biological origin and their application in synthesizing nanomaterials.

A variety of inorganic oxides including calcium carbonates [190], hydroxyapatite [191], zinc oxide [192], iron oxide [193] and silica [83] were synthesized using frameworks of biological templates. Silica is abundantly available on the earth's crust. It is found in major agricultural wastes and also in trace amount forms an ingredient for collagen in human body, which is why it finds application in diverse therapeutic procedures. Because of its many important properties and compatibility to the biological systems, SiO_2 nanoparticles (NPs) find applications as nanocarriers in drug/gene target delivery and imaging diagnosis [194]. Silica nanoparticles with tailored compositions, shapes and sizes were also synthesized by using bio-templates based on proteins and other macromolecules. Kröger et al. [83] isolated polycationic peptides (named silaffins) from cell walls of diatoms (a major group of microalgae generally found in water bodies and known for consuming silicon from the same) and synthesized silica nanospheres by reacting silaffins with silicic acid. The reaction took place rapidly. It is believed that modification in the covalently bonded lysine–lysine units present in the peptide molecule was responsible for the paradigm of nano-structured silica. In an analogous approach, formation of ordered silica structures was demonstrated by Cha et al. [195] using synthetic cysteine–lysine block copolypeptides. Silicatein formed structured aggregates due to self-assembly and hydrolysed tetraethoxysilane forming silica structures. Cysteine–lysine block copolypeptides also exhibited similar silica forming properties. It was observed that reduced form of the sulphydryl groups of cysteine produced hard silica spheres, whereas its oxidized form resulted in columns of amorphous silica. Formation of silica structures from aqueous solution of its salts in the presence of proteins and thereby the role of individual amino acids and small peptide oligomers on their formation was investigated by Belton et al. [196]. A strong correlation between the sizes of silica particles synthesized and isoelectric point (pI) and hydrophobicity of the amino acids used was reported. Whereas the presence of hydroxyl and other hydrophobic groups produced relatively smaller particles, nitrogen containing molecules resulted in larger sizes. Moreover, it was found that the number of lysine units present in the peptides influenced the surface area and porosity of silica thus produced. However, glycine units in the additive did not show up significant effects on the morphology of silica particles so produced. Apart from silica, nanoparticles of titania was also synthesized in presence of silicatein. It was reported [197] that the TiO_2 nanoparticles obtained from nanoscopically structure-directing proteins were very much different from those synthesized chemically. Amorphous forms of calcite are precipitated in nature in fascinating morphologies and quickly transform to its crystalline polymorphs. Aizenberg and co-workers [190] explored the possible function of protein rich macromolecules in realizing such biogenically formed stable amorphous and crystalline calcite phases. Ferritin is an iron containing blood cell protein molecule primarily known for its biological functions of iron storage and heme production. It consists of a self-assembly of multiple subunits forming a cage like structure. The cavity is made up of iron atoms in the form of a ferrihydrite (iron (III) oxyhydroxide) core [198, 199]. Such nanoscopic enclosure of ferritin was extensively employed for inorganic nanomaterials synthesis. Researchers synthesized mixed valence iron oxide (Fe_3O_4) [198, 200] using ferritin. They also prepared nanoparticles of iron sulphide upon reaction of the native protein core with H_2S or Na_2S. Similarly, manganese chloride when reacted with a buffered solution of apoferritin (demetalized form of ferritin obtained by reductive dissolution of the iron oxide core), amorphous manganese oxyhydroxide (MnOOH) was formed [201], whereas reaction of $UO_2(O_2CCH_3)_2$ with apoferritin resulted in the binding of the UO_2^{2+} species with the protein molecule forming polymerized uranyl oxyhydroxide [199]. The researchers also demonstrated the formation of cadmium sulphide ferritin nanocomposite upon reaction of appropriate reagents within these nano-sized reaction vessels of ferritin [202]. A number of peptide molecules having selective affinity towards different inorganic moieties were identified and designed for the synthesis of inorganic nanomaterials with engineered composition, crystallographic orientation and/or morphology [203, 204]. Gold [205] and silver [203] nanoparticles were synthesized using such selectively binding peptides identified through combinatorial phage display library. Synthesis of iron oxide nanowires and nanoparticles was carried out with suitably designed and isolated peptide molecules by bacterial peptide display method [206]. S-layers (surface layers) of bacteria consist of crystalline layers of proteins or glycoprotein subunits enveloping the surface of prokaryotic cells via self-assembly. S-layers can be few nanometres thick depending on the nature of the species and can have pores with diameters ranging from 2 to 8 nm [207, 208]. Gold nanoparticles having square superlattice and a uniform distribution were synthesized using S-layer of *Bacillus sphaericus* CCM2177 as a template [106]. Intact pollen grains were used as templates for silica, calcium carbonate and calcium phosphate with complex morphologies. Complex morphologies were obtained by soaking freeze-dried pollens in different precursor solutions and then removing the template by heat treatment, thereby creating inorganic pollen replicas of the materials [189]. Roy and his co-workers [209] used cysteine molecule to synthesize at room temperature silver nanoparticles with antibacterial efficacy.

Deoxyribonucleic acid (DNA) consists of a double helix structure of phosphate sugars and nitrogen containing bases like cytosine, guanine, adenine and thiamine covalently bonded to each other. It is an essential element of life for majority of known organisms and viruses and responsible for their genetic transmission, growth and development. DNA has dimension of a few nanometres and often exhibits high degree of specificity towards inorganic materials [210] that make it an effective template for nanomaterials. James J. Storhoff and Chad A. Mirkin wrote a wonderful review on the use of DNA for synthesis of inorganic and organic building blocks [211]. Platinum necklaces with extremely small thickness and extremely high regularity were fabricated using DNA template [212]. It was reported that DNA bound Pt(II) complexes upon reduction formed strong Pt–Pt bonds and served as the nucleation site without disrupting the metal cluster chain. DNA templated synthesis of conducting silver nanowires was also reported [211].

Apart from metallic wires, nanowires of semiconducting materials such as zinc oxide [192], lead sulphide [213] and cadmium sulphide [214] were produced using DNA-based designer templates. Virus-based templates were also successfully used for the synthesis of nanomaterials. A typical example being tobacco mosaic virus (TMV), which were used for the synthesis nanotube composites of silica by sol–gel condensation, iron oxide by oxidative hydrolysis and PbS and CdS by co-precipitation method [188]. Nucleation of materials under appropriate pH took place on the exposed surface of dispersed TMV, thereby resulting in heavily mineralized outer crust and internal cores and hence producing protein-confined exotic inorganic nanowires and composites with high-aspect ratio. TMV particle having an internal channel at high concentration has the natural tendency of getting assembled into nematic liquid crystals which were exploited for the synthesis of mesoporous silica having periodic structure [215]. Subsequently, nanometre-sized channels are made free from the protein materials by heat treatment. Similarly, different nanomaterials were synthesized using cowpea chlorotic mottle virus (CCMV)-based templates [216]. An interesting property of CCMV is that it swells under the influence of pH of the medium and thereby offers manipulation of the pore size and hence size of the entrapped nanoparticles.

18.8 Hurdles in Biogenic Synthesis

The merits of biogenic synthesis of nanoparticles largely overshadow its limitations, albeit there are few. The predominant shortcoming is the inherent slow kinetics of the synthetic methods. Whereas nanoparticles may be produced in minutes or hours in a conventional setup, microbial synthesis may take few days to weeks' time for completion of the process. Products are formed by the microbes in very small amount. Hence, for the sake of upscaling of production, they are required to be grown in large quantities that involve elaborate culturing. Further, isolation of nanoparticles from microbes often requires extensive protocols and downstream processing, especially when they are formed within the organism and therefore increases the cost of nanoparticle production. Some of the biological agents employed for nanoparticle synthesis are pathogenic to plants and human beings and therefore might increase the risk while handling them and also during their disposal.

18.9 Summary

Synthesis of nanoparticles by different biological agents has been discussed briefly in this chapter. Plants hold the advantage of being renewable and a perennial source of innumerable biomolecules for production of different nanomaterials at an affordable rate and price, thereby making it a benign and competitive alternative to the prevalent chemical synthesis routes. Microbial synthesis, on the other hand, encompasses wide spectrum of protocols for synthesis of nanoparticles both within and outside the cells of different microbes. Mechanisms of such nanoparticles synthesis by the respective biological agents have also been discussed.

18.10 Future Scope

Plants and microbes are found to synthesize nanoparticles naturally with stringent dimensional restrictions and exquisite morphology in a highly reproducible manner, albeit in a slow pace and relatively minuscule amount. Nevertheless, it is said that *Boon* often comes in the guise of a *Bane*. The slow kinetics of nanoparticles formation involves intricate mechanisms of biomolecular reduction aided by catalysts acting as electron transferring agents and their subsequent stabilization by different biomolecules, allowing controlled manipulation of the nanoparticles size and morphology. Sometimes, the biomolecules themselves act as templates allowing the formation and growth of nanomaterials inside them having unusual crystallographic modifications. Mimicking nature's own recipe and scaling up the processes would perhaps allow affordable production of fascinatingly designed nanoparticles in an environmentally benign fashion. Since shape and size of nanoparticles are known to impact significantly their optical, catalytic, magnetic, electronic and biological properties, biogenic synthesis would certainly open up the feasibility of manipulating materials at the nanoscale and thereby novel applications hitherto unknown to humankind. One such possibility could be the realization of inorganic–organic hybrids having unique properties through bio-templating approach. Efforts are being made in deciphering the mechanism of nanoparticle formation by biogenic route and identifying the active molecules involved in the process. These aspects have been discussed at length in different sections of the present chapter. Based on the research inputs, technology might seek to exploit genetic engineering for the production of active metabolites, enzymes and other interesting biomolecules which would enable facile manoeuvring of synthesis of nanoparticles with tailored exotic properties.

References

1. Thakkar KN, Mhatre SS, Parikh RY (2010) Biological synthesis of metallic nanoparticles, nanomedicine: nanotechnology. Biol Med 6:257
2. Gahlawat G, Choudhury AR (2019) A review on the biosynthesis of metal and metal salt nanoparticles by microbes. RSC Adv 9:12944
3. Hulkoti N, Taranath T (2014) Biosynthesis of nanoparticles using microbes—a review, colloids and surfaces. B, Biointerfaces 121:474
4. Sharma D, Kanchi S, Bisetty K (2019) biogenic synthesis of nanoparticles: a review. Arabian J Chem 12:3576
5. Rao CNR, Müller A, Ramachandra CN, Cheetham AK (2004) The chemistry of nanomaterials: synthesis, properties and applications in 2 volumes. Wiley-VCH
6. Singh J, Dutta T, Kim K-H, Rawat M, Samddar P, Kumar P (2018) Green synthesis of metals and their oxide nanoparticles: applications for environmental remediation. J Nanobiotechnol 16:84
7. Agarwal H, Venkat Kumar S, Rajeshkumar S (2017) A review on green synthesis of zinc oxide nanoparticles—an eco-friendly approach. Resour-Effic Technol 3:406
8. Ahmed S, Ahmad M, Swami BL, Ikram S (2016) A review on plants extract mediated synthesis of silver nanoparticles for antimicrobial applications: a green expertise. J Adv Res 7:17
9. Devi T, Thangamathi P, Ananth S, Soundari G, Lavanya M (2017) A review on nanoparticles synthesis using entomopathogenic fungi. Int J Curr Innov Res 3:887
10. Rauwel P, Küünal S, Ferdov S, Rauwel E (2015) A review on the green synthesis of silver nanoparticles and their morphologies studied via TEM. Adv Mater Sci Eng 2015:1
11. Siddiqi KS, Husen A, Rao RAK (2018) A review on biosynthesis of silver nanoparticles and their biocidal properties. J Nanobiotechnol 16:14
12. Mehra RK, Winge DR (1991) Metal ion resistance in fungi: molecular mechanisms and their regulated expression. J Cell Biochem 45:30
13. Mukherjee P, Ahmad A, Mandal D, Senapati S, Sainkar SR, Khan MI, Parishcha R, Ajaykumar P, Alam M, Kumar R (2001) Fungus-mediated synthesis of silver nanoparticles and their immobilization in the mycelial matrix: a novel biological approach to nanoparticle synthesis. Nano Lett 1:515
14. Mukherjee P, Ahmad A, Mandal D, Senapati S, Sainkar SR, Khan MI, Ramani R, Parischa R, Ajayakumar PV, Alam M, Sastry M, Kumar R (2001) Bioreduction of $AuCl_4$— ions by the fungus, Verticillium sp. and surface trapping of the gold nanoparticles formed. Angewandte Chemie Int Edn 40:3585
15. Pighi L, Pümpel T, Schinner F (1989) Selective accumulation of silver by fungi. Biotech Lett 11:275
16. Arakaki A, Nakazawa H, Nemoto M, Mori T, Matsunaga T (2008) Formation of magnetite by bacteria and its application. J R Soc Interface 5:977
17. Sanghi R, Verma P (2009) A facile green extracellular biosynthesis of CdS nanoparticles by immobilized fungus. Chem Eng J 155:886
18. Mie R, Samsudin MW, Din LB (2013) A review on biosynthesis of nanoparticles using plant extract: an emerging green nanotechnology, in nanosynthesis and nanodevice, vol 667, pp 251–254. Trans Tech Publications Ltd
19. Yadi M, Mostafavi E, Saleh B, Davaran S, Aliyeva I, Khalilov R, Nikzamir M, Nikzamir N, Akbarzadeh A, Panahi Y, Milani M (2018) Current developments in green synthesis of metallic nanoparticles using plant extracts: a review. Artif Cells Nanomed Biotechnol 46: S336
20. Tripathi DK, Singh Shweta S, Singh S, Pandey R, Singh VP, Sharma NC, Prasad SM, Dubey NK, Chauhan DK (2017) An overview on manufactured nanoparticles in plants: uptake, translocation, accumulation and phytotoxicity. Plant Physiol Biochem 110:2
21. Iravani S (2011) Green synthesis of metal nanoparticles using plants. Green Chem 13:2638

22. Rodriguez E, Parsons JG, Peralta-Videa JR, Cruz-Jimenez G, Romero-Gonzalez J, Sanchez-Salcido BE, Saupe GB, Duarte-Gardea M, Gardea-Torresdey JL (2007) Potential of chilopsis linearis for gold phytomining: using XAS to determine gold reduction and nanoparticle formation within plant tissues. Int J Phytorem 9:133
23. Bennett LE, Burkhead JL, Hale KL, Terry N, Pilon M, Pilon-Smits EA (2003) Analysis of transgenic indian mustard plants for phytoremediation of metal-contaminated mine tailings. J Environ Qual 32:432
24. Terry N, Banuelos GS (1999) Phytoremediation of contaminated soil and water. CRC Press
25. Baker S, Rakshith D, Kavitha KS, Santosh P, Kavitha HU, Rao Y, Satish S (2013) Plants: emerging as nanofactories towards facile route in synthesis of nanoparticles. Bioimpacts 3:111
26. Latif MS, Abbas S, Kormin F, Mustafa MK (2019) Green synthesis of plant-mediated metal nanoparticles: the role of polyphenols. Asian J Pharm Clin Res 12:75
27. Subhani MA, Wahab N, Ibrahim M, Rehman HU, Asad M, Ullah W, Kamil M, Ikramullah M, Arif M, Zubair M, Shah W, Nawaz M (2019) Shuja, Synthesis of silver nanoparticles from plant extracts and their antimicrobial application. Int J Biosci 14:243
28. Teimuri-Mofrad R, Hadi R, Tahmasebi B, Farhoudian S, Mehravar M, Ramin N (2017) Green synthesis of gold nanoparticles using plant extract: mini-review. Nanochem Res 2:8
29. Warghane UK, Dhankar RP (2019) Novel biosynthesis of silver nanoparticles for catalytic oxidation of alcohols containing aromatic ring. Mater Today: Proc 15:526
30. Krishnaraj C, Jagan E, Rajasekar S, Selvakumar P, Kalaichelvan P, Mohan N (2010) Synthesis of silver nanoparticles using acalypha indica leaf extracts and its antibacterial activity against water borne pathogens. Colloids Surf B 76:50
31. Jain D, Daima HK, Kachhwaha S, Kothari S (2009) Synthesis of plant-mediated silver nanoparticles using papaya fruit extract and evaluation of their anti microbial activities. Digest J Nanomater Biostructures 4:557
32. Sathishkumar M, Sneha K, Won SW, Cho C-W, Kim S, Yun Y-S (2009) Cinnamon zeylanicum bark extract and powder mediated green synthesis of nano-crystalline silver particles and its bactericidal activity. Colloids Surf B 73:332
33. Kasthuri J, Kathiravan K, Rajendiran N (2008) Phyllanthin-assisted biosynthesis of silver and gold nanoparticles: a novel biological approach. J Nanopart Res 11:1075
34. Kasthuri J, Veerapandian S, Rajendiran N (2009) Biological synthesis of silver and gold nanoparticles using apiin as reducing agent. Colloids Surf B 68:55
35. Tiloke C, Anand K, Gengan RM, Chuturgoon AA (2018) Moringa oleifera and their phytonanoparticles: potential antiproliferative agents against cancer. Biomed Pharmacother 108:457
36. Anuradha J, Abbasi T, Abbasi S (2010) Green'synthesis of gold nanoparticles with aqueous extracts of neem (Azadirachta Indica). Res J Biotechnol 5:75
37. Shankar SS, Rai A, Ahmad A, Sastry M (2004) Rapid synthesis of Au, Ag, and bimetallic Au Core–Ag shell nanoparticles using neem (Azadirachta Indica) leaf broth. J Colloid Interface Sci 275:496
38. Chandran SP, Chaudhary M, Pasricha R, Ahmad A, Sastry M (2006) Synthesis of gold nanotriangles and silver nanoparticles using aloevera plant extract. Biotechnol Prog 22:577
39. Armendariz V, Jose-Yacaman M, Moller AD, Peralta-Videa J, Troiani H, Herrera I, Gardea-Torresdey J (2004) HRTEM characterization of gold nanoparticles produced by wheat biomass. Revista Mexicana de Física 50:7
40. Aromal SA, Philip D (2012) Green synthesis of gold nanoparticles using trigonella foenum-graecum and its size-dependent catalytic activity. Spectrochim Acta Part A Mol Biomol Spectrosc 97:1
41. Shankar SS, Rai A, Ahmad A, Sastry M (2005) Controlling the optical properties of lemongrass extract synthesized gold nanotriangles and potential application in infrared-absorbing optical coatings. Chem Mater 17:566

42. Raghunandan D, Bedre MD, Basavaraja S, Sawle B, Manjunath S, Venkataraman A (2010) Rapid biosynthesis of irregular shaped gold nanoparticles from macerated aqueous extracellular dried clove buds (Syzygium Aromaticum) solution. Colloids Surf B 79:235
43. Bahram M, Mohammadzadeh E (2014) Green synthesis of gold nanoparticles with willow tree bark extract: a sensitive colourimetric sensor for cysteine detection. Anal Methods 6:6916
44. Sen IK, Maity K, Islam SS (2013) Green synthesis of gold nanoparticles using a glucan of an edible mushroom and study of catalytic activity. Carbohyd Polym 91:518
45. Narayanan KB, Sakthivel N (2008) Coriander leaf mediated biosynthesis of gold nanoparticles. Mater Lett 62:4588
46. Shankar SS, Ahmad A, Pasrichaa R, Sastry M (2003) Bioreduction of chloroaurate ions by geranium leaves and its endophytic fungus yields gold nanoparticles of different shapes. J Mater Chem 13:1822
47. Rai A, Singh A, Ahmad A, Sastry M (2006) Role of halide ions and temperature on the morphology of biologically synthesized gold nanotriangles. Langmuir 22:736
48. Armendariz V, Herrera I, Peralta-Videa JR, Jose-Yacaman M, Troiani H, Santiago P, Gardea-Torresdey JL (2004) Size controlled gold nanoparticle formation by avena sativa biomass: use of plants in nanobiotechnology. J Nanopart Res 6:377
49. Ijaz F, Shahid S, Khan SA, Ahmad W, Zaman S (2017) Green synthesis of copper oxide nanoparticles using abutilon indicum leaf extract: antimicrobial, antioxidant and photocatalytic dye degradation activitie. Trop J Pharm Res 16:743
50. Silveira C, Shimabuku QL, Fernandes Silva M, Bergamasco R (2018) Iron-oxide nanoparticles by the green synthesis method using moringa oleifera leaf extract for fluoride removal. Environ Technol 39:2926
51. Yew YP, Shameli K, Miyake M, Bt Ahmad Khairudin NB, Bt Mohamad SE, Naiki T, Xin K Lee (2020), Green biosynthesis of superparamagnetic magnetite Fe_3O_4 nanoparticles and biomedical applications in targeted anticancer drug delivery system: a review. Arabian J Chem 13:2287
52. Cai J, Ruffieux P, Jaafar R, Bieri M, Braun T, Blankenburg S, Muoth M, Seitsonen AP, Saleh M, Feng X (2010) Atomically precise bottom-up fabrication of graphene nanoribbons. Nature 466:470
53. Ezhilarasi AA, Vijaya JJ, Kaviyarasu K, Maaza M, Ayeshamariam A, Kennedy LJ (2016) Green synthesis of NiO nanoparticles using moringa oleifera extract and their biomedical applications: cytotoxicity effect of nanoparticles against HT-29 cancer cells. J Photochem Photobiol, B 164:352
54. Matinise N, Fuku XG, Kaviyarasu K, Mayedwa N, Maaza M (2017) ZnO nanoparticles via moringa oleifera green synthesis: physical properties and mechanism of formation. Appl Surf Sci 406:339
55. Thema FT, Manikandan E, Dhlamini MS, Maaza M (2015) Green synthesis of ZnO nanoparticles via agathosma betulina natural extract. Mater Lett 161:124
56. Thema FT, Manikandan E, Gurib-Fakim A, Maaza M (2016) Single phase bunsenite NiO nanoparticles green synthesis by Agathosma betulina natural extract. J Alloy Compd 657:655
57. Diallo A, Beye AC, Doyle TB, Park E, Maaza M (2015) Green synthesis of Co_3O_4 nanoparticles via aspalathus linearis: physical properties. Null 8:30
58. Konwarh R, Gogoi B, Philip R, Laskar M, Karak N (2011) Biomimetic preparation of polymer-supported free radical scavenging, cytocompatible and antimicrobial "green" silver nanoparticles using aqueous extract of citrus sinensis peel. Colloids Surf B 84:338
59. Basavegowda N, Lee YR (2013) Synthesis of silver nanoparticles using satsuma mandarin (citrus unshiu) peel extract: a novel approach towards waste utilization. Mater Lett 109:31
60. Kaviya S, Santhanalakshmi J, Viswanathan B, Muthumary J, Srinivasan K (2011) Biosynthesis of silver nanoparticles using citrus sinensis peel extract and its antibacterial activity. Spectrochim Acta Part A Mol Biomol Spectrosc 79:594

61. Kokila T, Ramesh P, Geetha D (2015) A biogenic approach for green synthesis of silver nanoparticles using peel extract of citrus sinensis and its application. Int J Chem Tech Res 7:804
62. Krishnaswamy K, Vali H, Orsat V (2014) Value-Adding to grape waste: green synthesis of gold nanoparticles. J Food Eng 142:210
63. Yang N, Li W-H (2013) Mango peel extract mediated novel route for synthesis of silver nanoparticles and antibacterial application of silver nanoparticles loaded onto non-woven fabrics. Ind Crops Prod 48:81
64. Yang N, WeiHong L, Hao L (2014) Biosynthesis of Au nanoparticles using agricultural waste mango peel extract and its in vitro cytotoxic effect on two normal cells. Mater Lett 134:67
65. Kokila T, Ramesh P, Geetha D (2015) Biosynthesis of silver nanoparticles from cavendish banana peel extract and its antibacterial and free radical scavenging assay: a novel biological approach. Appli Nanosci 5:911
66. Poadang S, Yongvanich N, Phongtongpasuk S (2017) Synthesis, characterization, and antibacterial properties of silver nanoparticles prepared from aqueous peel extract of pineapple, ananas comosus. CMU J Nat Sci 16:123
67. Kokila T, Ramesh P, Geetha D (2016) Biosynthesis of AgNPs using carica papaya peel extract and evaluation of its antioxidant and antimicrobial activities. Ecotoxicol Environ Saf 134:467
68. Baiocco D, Lavecchia R, Natali S, Zuorro A (2016) Production of metal nanoparticles by agro-industrial wastes: a green opportunity for nanotechnology. Chem Eng Trans 47:67
69. Xin Lee K, Shameli K, Miyake M, Kuwano N, Khairudin BA, Bahiyah N, Mohamad B, Eva S, Yew YP (2016) Green synthesis of gold nanoparticles using aqueous extract of garcinia mangostana fruit peels. J Nanomater 2016:8489094
70. Malhotra A, Sharma N, Kumar N, Dolma K, Sharma D, Nandanwar HS, Choudhury AR (2014) Multi-analytical approach to understand biomineralization of gold using rice bran: a novel and economical route. RSC Advances 4:39484
71. Khatami M, Sharifi I, Nobre MA, Zafarnia N, Aflatoonian MR (2018) Waste-grass-mediated green synthesis of silver nanoparticles and evaluation of their anticancer, antifungal antibacterial activity. Green Chem Lett Rev 11:125
72. Mythili R, Selvankumar T, Kamala-Kannan S, Sudhakar C, Ameen F, Al-Sabri A, Selvam K, Govarthanan M, Kim H (2018) Utilization of market vegetable waste for silver nanoparticle synthesis and its antibacterial activity. Mater Lett 225:101
73. Sivakumar P, Sivakumar P, Anbarasu K, Pandian K, Renganathan S (2013) Synthesis of silver nanorods from food industrial waste and their application in improving the keeping quality of milk. Ind Eng Chem Res 52:17676
74. Lakshmipathy R, Reddy BP, Sarada N, Chidambaram K, Pasha SK (2015) Watermelon rind-mediated green synthesis of noble palladium nanoparticles: catalytic application. Appl Nanosci 5:223
75. Machado S, Grosso J, Nouws H, Albergaria JT, Delerue-Matos C (2014) Utilization of food industry wastes for the production of zero-valent iron nanoparticles. Sci Total Environ 496:233
76. Surendra T, Roopan SM, Al-Dhabi NA, Arasu MV, Sarkar G, Suthindhiran K (2016) Vegetable peel waste for the production of ZnO nanoparticles and its toxicological efficiency, antifungal, hemolytic, and antibacterial activities. Nanoscale Res Lett 11:546
77. Zheng B, Qian L, Yuan H, Xiao D, Yang X, Paau MC, Choi MMF (2010) Preparation of gold nanoparticles on eggshell membrane and their biosensing application. Talanta 82:177
78. Aiking H, Kok K, van Heerikhuizen H, van't Riet J (1982) Adaptation to cadmium by klebsiella aerogenes growing in continuous culture proceeds mainly via formation of cadmium sulfide. Appl Environ Microbiol 44:938
79. Stephen JR, Macnaughtont SJ (1999) Developments in terrestrial bacterial remediation of metals. Curr Opin Biotechnol 10:230

80. Iravani S (2014) Bacteria in nanoparticle synthesis: current status and future prospects. Int Sch Res Not 2014:359316
81. Fredrickson JK, Zachara JM, Balkwill DL, Kennedy D, Li SW, Kostandarithes HM, Daly MJ, Romine MF, Brockman FJ (2004) Geomicrobiology of high-level nuclear waste-contaminated vadose sediments at the hanford site, washington state. Appl Environ Microbiol 70:4230
82. Philipse AP, Maas D (2002) Magnetic colloids from magnetotactic bacteria: Chain formation and colloidal stability. Langmuir 18:9977
83. Kröger N, Deutzmann R, Sumper M (1999) Polycationic peptides from diatom biosilica that direct silica nanosphere formation. Science 286:1129
84. Schultze-Lam S, Harauz G, Beveridge T (1992) Participation of a cyanobacterial S layer in fine-grain mineral formation. J Bacteriol 174:7971
85. Khandel P, Shahi SK (2018) Mycogenic nanoparticles and their bio-prospective applications: current status and future challenges. J Nanostruct Chem 8:369
86. Mohanpuria P, Rana NK, Yadav SK (2008) Biosynthesis of nanoparticles: technological concepts and future applications. J Nanopart Res 10:507
87. Joerger R, Klaus T, Granqvist CG (2000) Biologically produced silver-carbon composite materials for optically functional thin-film coatings. Adv Mater 12:407
88. Klaus T, Joerger R, Olsson E, Granqvist C-G (1999) Silver-Based crystalline nanoparticles, microbially fabricated. Proc Natl Acad Sci 96:13611
89. Mann S (1992) Bacteria and the midas touch. Nature 357:358
90. Beveridge T, Murray R (1980) Sites of metal deposition in the cell wall of bacillus subtilis. J Bacteriol 141:876
91. Lengke M, Southam G (2006) Bioaccumulation of gold by sulfate-reducing bacteria cultured in the presence of gold (I)-thiosulfate complex. Geochim Cosmochim Acta 70:3646
92. Kashefi K, Tor JM, Nevin KP, Lovley DR (2001) Reductive precipitation of gold by dissimilatory Fe (III)-reducing bacteria andarchaea. Appl Environ Microbiol 67:3275
93. Lengke MF, Fleet ME, Southam G (2006) Morphology of gold nanoparticles synthesized by filamentous cyanobacteria from gold (I)—thiosulfate and Gold (III)—chloride Complexes. Langmuir 22:2780
94. Lengke MF, Ravel B, Fleet ME, Wanger G, Gordon RA, Southam G (2006) Mechanisms of gold bioaccumulation by filamentous cyanobacteria from gold (III)—chloride complex. Environ Sci Technol 40:6304
95. Du L, Jiang H, Liu X, Wang E (2007) Biosynthesis of gold nanoparticles assisted by escherichia coli DH5α and its application on direct electrochemistry of hemoglobin. Electrochem Commun 9:1165
96. Feng Y, Yu Y, Wang Y, Lin X (2007) Biosorption and bioreduction of trivalent aurum by photosynthetic bacteria Rhodobacter capsulatus. Curr Microbiol 55:402
97. Nair B, Pradeep T (2002) Coalescence of nanoclusters and formation of submicron crystallites assisted by lactobacillus strains. Cryst Growth Des 2:293
98. Pugazhenthiran N, Anandan S, Kathiravan G, Prakash NKU, Crawford S, Ashokkumar M (2009) Microbial synthesis of silver nanoparticles by Bacillus sp. J Nanopart Res 11:1811
99. He S, Guo Z, Zhang Y, Zhang S, Wang J, Gu N (2007) Biosynthesis of gold nanoparticles using the bacteria Rhodopseudomonas capsulata. Mater Lett 61:3984
100. Husseiny M, El-Aziz MA, Badr Y, Mahmoud M (2007) Biosynthesis of gold nanoparticles using Pseudomonas aeruginosa. Spectrochim Acta Part A Mol Biomol Spectrosc 67:1003
101. Mouxing F, Qingbiao L, Daohua S, Yinghua L, Ning H, Xu D, Huixuan W, Huang J (2006) Rapid preparation process of silver nanoparticles by bioreduction and their characterizations. Chin J Chem Eng 14:114
102. Shahverdi AR, Minaeian S, Shahverdi HR, Jamalifar H, Nohi A-A (2007) Rapid synthesis of silver nanoparticles using culture supernatants of enterobacteria: a novel biological approach. Process Biochem 42:919

103. Kalishwaralal K, Deepak V, Ramkumarpandian S, Nellaiah H, Sangiliyandi G (2008) Extracellular biosynthesis of silver nanoparticles by the culture supernatant of Bacillus licheniformis. Mater Lett 62:4411
104. Parikh RY, Singh S, Prasad B, Patole MS, Sastry M, Shouche YS (2008) Extracellular synthesis of crystalline silver nanoparticles and molecular evidence of silver resistance from Morganella sp.: towards understanding biochemical synthesis mechanism. Chem Bio Chem 9:1415
105. Yong P, Rowson NA, Farr JPG, Harris IR, Macaskie LE (2002) Bioreduction and biocrystallization of palladium by Desulfovibrio desulfuricans NCIMB 8307. Biotechnol Bioeng 80:369
106. Dieluweit S, Pum D, Sleytr UB (1998) Formation of a gold superlattice on an S-layer with square lattice symmetry. Supramol Sci 5:15
107. Ankit C, Singh A, Sharma M (2014) Biological synthesis of nanoparticles using bacteria and their applications. Am J Pharmatech Res 4:38
108. Arakaki A, Masuda F, Amemiya Y, Tanaka T, Matsunaga T (2010) Control of the morphology and size of magnetite particles with peptides mimicking the Mms6 protein from magnetotactic bacteria. J Colloid Interface Sci 343:65
109. Das S, Lyla P, Khan SA (2008) Distribution and generic composition of culturable marine actinomycetes from the sediments of Indian continental slope of Bay of Bengal. Chin J Oceanol Limnol 26:166
110. Srinivasan M, Laxman R, Deshpande M (1991) Physiology and nutritional aspects of actinomycetes: an overview. World J Microbiol Biotechnol 7:171
111. Gupta A, Singh D, Singh SK, Singh VK, Singh AV, Kumar A (2019) 10—Role of actinomycetes in bioactive and nanoparticle synthesis, in role of plant growth promoting microorganisms in sustainable agriculture and nanotechnology. In: Kumar A, Singh AK, Choudhary KK (eds) Woodhead Publishing, pp 163–182
112. Abdeen S, Geo S, Sukanya S, Praseetha PK (2014) Biosynthesis of silver nanoparticles from actinomycetes for therapeutic applications. Int J Nano Dimens 5:155
113. Bhatti AA, Haq S, Bhat RA (2017) Actinomycetes benefaction role in soil and plant health. Microb Pathog 111:458
114. Saminathan K (2015) Biosynthesis of silver nanoparticles using soil actinomycetes Streptomyces sp. Int J Curr Microbiol App Sci 4:1073
115. Ahmad A, Senapati S, Khan MI, Kumar R, Sastry M (2003) Extracellular biosynthesis of monodisperse gold nanoparticles by a novel extremophilic actinomycete, Thermomonospora sp. Langmuir 19:3550
116. Manivasagan P, Venkatesan J, Sivakumar K, Kim S-K (2016) Actinobacteria mediated synthesis of nanoparticles and their biological properties: a review. Crit Rev Microbiol 42:209
117. Golinska P, Wypij M, Ingle AP, Gupta I, Dahm H, Rai M (2014) Biogenic synthesis of metal nanoparticles from actinomycetes: biomedical applications and cytotoxicity. Appl Microbiol Biotechnol 98:8083
118. Ahmad A, Senapati S, Khan MI, Kumar R, Ramani R, Srinivas V, Sastry M (2003) Intracellular synthesis of gold nanoparticles by a novel alkalotolerant actinomycete, Rhodococcus species. Nanotechnology 14:824
119. Al-Dhabi NA, Mohammed Ghilan A-K, Arasu MV (2018) Characterization of silver nanomaterials derived from marine Streptomyces sp. Al-Dhabi-87 and its in vitro application against multidrug resistant and extended-spectrum beta-lactamase clinical pathogens. Nanomaterials 8:279
120. Vijayabharathi R, Sathya A, Gopalakrishnan S (2018) Extracellular biosynthesis of silver nanoparticles using Streptomyces griseoplanus SAI-25 and its antifungal activity against Macrophomina phaseolina, the charcoal rot pathogen of sorghum. Biocatal Agric Biotechnol 14:166

121. Abd-Elnaby HM, Abo-Elala GM, Abdel-Raouf UM, Hamed MM (2016) Antibacterial and anticancer activity of extracellular synthesized silver nanoparticles from marine Streptomyces rochei MHM13. Egypt J Aquat Res 42:301
122. Ranjitha VR, Rai VR (2017) Actinomycetes mediated synthesis of gold nanoparticles from the culture supernatant of streptomyces griseoruber with special reference to catalytic activity. 3 Biotech 7:299
123. Hassan SE-D, Salem SS, Fouda A, Awad MA, El-Gamal MS, Abdo AM (2018) New approach for antimicrobial activity and bio-control of various pathogens by biosynthesized copper nanoparticles using endophytic actinomycetes. J Radiat Res Appl Sci 11:262
124. Nabila MI, Kannabiran K (2018) Biosynthesis, characterization and antibacterial activity of copper oxide nanoparticles (CuO NPs) from actinomycetes. Biocatal Agric Biotechnol 15:56
125. Karthik L, Kumar G, Kirthi AV, Rahuman A, Rao KB (2014) Streptomyces sp. LK3 mediated synthesis of silver nanoparticles and its biomedical application. Bioprocess Biosyst Eng 37:261
126. Poonam S, Nivedita S (2017) Industrial and biotechnological applications of algae: a review. J Adv Plant Biol 1:01
127. Sharma A, Sharma S, Sharma K, Chetri S, Vashishtha A, Singh P, Kumar R, Rathi B, Agrawal V (2016) Algae as crucial organisms in advancing nanotechnology: a systematic review. J Appl Phycol 28:1759
128. Vincy W, Mahathalana TJ, Sukumaran S, Jeeva S (2017) Algae as a source for synthesis of nanoparticles–a review. Int J Latest Trends Eng Technol 5
129. Schröfel A, Kratošová G, Bohunická M, Dobročka E, Vávra I (2011) Biosynthesis of gold nanoparticles using diatoms—silica-gold and EPS-gold bionanocomposite formation. J Nanopart Res 13:8
130. Hosea M, Greene B, Mcpherson R, Henzl M, Alexander MD, Darnall DW (1986) Accumulation of elemental gold on the alga chlorella vulgaris. Inorg Chim Acta 123:161
131. Liu B, Xie J, Lee J, Ting Y, Chen JP (2005) Optimization of high-yield biological synthesis of single-crystalline gold nanoplates. J Phys Chem B 109:15256
132. Mata Y, Torres E, Blazquez M, Ballester A, González F, Munoz J (2009) Gold (III) biosorption and bioreduction with the brown alga fucus vesiculosus. J Hazard Mater 166:612
133. Merin DD, Prakash S, Bhimba BV (2010) Antibacterial screening of silver nanoparticles synthesized by marine micro algae. Asian Pacific J Trop Med 3:797
134. Kannan RRR, Arumugam R, Ramya D, Manivannan K, Anantharaman P (2013) Green synthesis of silver nanoparticles using marine macroalga Chaetomorpha linum. Appl Nanosci 3:229
135. Shanab S, Shalaby E, Essa A (2012) Bioremediation of heavy metals by microalgal species
136. Bansal V, Rautaray D, Ahmad A, Sastry M (2004) Biosynthesis of zirconia nanoparticles using the fungus Fusarium oxysporum. J Mater Chem 14:3303
137. Guilger-Casagrande M, de Lima R (2019) Synthesis of silver nanoparticles mediated by fungi: a review. Front Bioeng Biotechnol 7:287
138. Kitching M, Ramani M, Marsili E (2015) Fungal biosynthesis of gold nanoparticles: mechanism and scale up. Microb Biotechnol 8:904
139. Zhao X, Zhou L, Riaz Rajoka MS, Yan L, Jiang C, Shao D, Zhu J, Shi J, Huang Q, Yang H, Jin M (2018) Fungal silver nanoparticles: synthesis, application and challenges. Crit Rev Biotechnol 38:817
140. Zielonka A, Klimek-Ochab M (2017) Fungal synthesis of size-defined nanoparticles. Adv Nat Sci: Nanosci Nanotechnol 8:9
141. Ahmad A, Senapati S, Khan MI, Kumar R, Sastry M (2005) Extra-/intracellular biosynthesis of gold nanoparticles by an Alkalotolerant Fungus, Trichothecium Sp. J Biomed Nanotechnol 1:47
142. Gericke M, Pinches A (2006) Biological synthesis of metal nanoparticles. Hydrometallurgy 83:132

143. Gericke M, Pinches A (2006) Microbial production of gold nanoparticles. Gold Bulletin 39:22
144. Senapati S, Mandal D, Ahmad A, Khan MI, Sastry M, Kumar R (2004) Fungus mediated synthesis of silver nanoparticles: a novel biological approach. Indian J Phys 78:101
145. Vigneshwaran N, Ashtaputre N, Varadarajan P, Nachane R, Paralikar K, Balasubramanya R (2007) Biological synthesis of silver nanoparticles using the fungus Aspergillus flavus. Mater Lett 61:1413
146. Riddin T, Gericke M, Whiteley C (2006) Analysis of the inter-and extracellular formation of platinum nanoparticles by Fusarium oxysporum f. sp. lycopersici using response surface methodology. Nanotechnology 17:3482
147. Bhainsa KC, D'souza S (2006) Extracellular biosynthesis of silver nanoparticles using the fungus Aspergillus fumigatus. Colloids Surf B: Biointerfaces 47:160
148. Vigneshwaran N, Kathe AA, Varadarajan P, Nachane RP, Balasubramanya R (2006) Biomimetics of silver nanoparticles by white rot fungus, Phaenerochaete chrysosporium. Colloids Surf B 53:55
149. Basavaraja S, Balaji S, Lagashetty A, Rajasab A, Venkataraman A (2008) Extracellular biosynthesis of silver nanoparticles using the fungus Fusarium semitectum. Mater Res Bull 43:1164
150. Gade A, Bonde P, Ingle A, Marcato P, Duran N, Rai M (2008) Exploitation of Aspergillus niger for synthesis of silver nanoparticles. J Biobased Mater Bioenergy 2:243
151. Balaji D, Basavaraja S, Deshpande R, Mahesh DB, Prabhakar B, Venkataraman A (2009) Extracellular biosynthesis of functionalized silver nanoparticles by strains of Cladosporium cladosporioides fungus. Colloids Surf B 68:88
152. Sanghi R, Verma P (2009) Biomimetic synthesis and characterisation of protein capped silver nanoparticles. Biores Technol 100:501
153. Mukherjee P, Roy M, Mandal B, Dey G, Mukherjee P, Ghatak J, Tyagi A, Kale S (2008) Green synthesis of highly stabilized nanocrystalline silver particles by a non-pathogenic and agriculturally important fungus T. Asperellum. Nanotechnology 19:075103
154. Mukherjee P, Roy M, Mandal B, Choudhury S, Tewari R, Tyagi A, Kale S (2012) Synthesis of uniform gold nanoparticles using non-pathogenic bio-control agent: evolution of morphology from nano-spheres to triangular nanoprisms. J Colloid Interface Sci 367:148
155. Ingle A, Rai M, Gade A, Bawaskar M (2009) Fusarium solani: a novel biological agent for the extracellular synthesis of silver nanoparticles. J Nanopart Res 11:2079
156. Birla SS, Tiwari VV, Gade AK, Ingle AP, Yadav AP, Rai MK (2009) Fabrication of silver nanoparticles by Phoma glomerata and its combined effect against Escherichia coli, Pseudomonas aeruginosa and Staphylococcus aureus. Lett Appl Microbiol 48:173
157. Kathiresan K, Manivannan S, Nabeel M, Dhivya B (2009) Studies on silver nanoparticles synthesized by a marine fungus, Penicillium fellutanum isolated from coastal mangrove sediment. Colloids Surf B 71:133
158. Fayaz AM, Balaji K, Girilal M, Yadav R, Kalaichelvan PT, Venketesan R (2010) Biogenic synthesis of silver nanoparticles and their synergistic effect with antibiotics: a study against gram-positive and gram-negative bacteria, nanomedicine: nanotechnology. Biol Med 6:103
159. Shaligram NS, Bule M, Bhambure R, Singhal RS, Singh SK, Szakacs G, Pandey A (2009) Biosynthesis of silver nanoparticles using aqueous extract from the compactin producing fungal strain. Process Biochem 44:939
160. Philip D (2009) Biosynthesis of Au, Ag and Au–Ag nanoparticles using edible mushroom extract. Spectrochim Acta Part A Mol Biomol Spectrosc 73:374
161. Mukherjee P, Senapati S, Mandal D, Ahmad A, Khan MI, Kumar R, Sastry M (2002) Extracellular synthesis of gold nanoparticles by the fungus Fusarium oxysporum. ChemBioChem 3:461
162. Senapati S, Ahmad A, Khan MI, Sastry M, Kumar R (2005) Extracellular biosynthesis of bimetallic Au–Ag alloy nanoparticles. Small 1:517
163. Bharde A, Rautaray D, Bansal V, Ahmad A, Sarkar I, Yusuf SM, Sanyal M, Sastry M (2006) Extracellular biosynthesis of magnetite using fungi. Small 2:135

164. Bansal V, Rautaray D, Bharde A, Ahire K, Sanyal A, Ahmad A, Sastry M (2005) Fungus-mediated biosynthesis of silica and titania particles. J Mater Chem 15:2583
165. Bansal V, Poddar P, Ahmad A, Sastry M (2006) Room-temperature biosynthesis of ferroelectric barium titanate nanoparticles. J Am Chem Soc 128:11958
166. Rautaray D, Sanyal A, Adyanthaya SD, Ahmad A, Sastry M (2004) Biological synthesis of strontium carbonate crystals using the fungus Fusarium oxysporum. Langmuir 20:6827
167. Kumar SA, Ansary AA, Ahmad A, Khan M (2007) Extracellular biosynthesis of CdSe quantum dots by the fungus, Fusarium oxysporum. J sBiomed Nanotechnol 3:190
168. Durán N, Marcato PD, Alves OL, De Souza GI, Esposito E (2005) Mechanistic aspects of biosynthesis of silver nanoparticles by several Fusarium oxysporum strains. J Nanobiotechnol 3:8
169. Kumar SA, Abyaneh MK, Gosavi S, Kulkarni SK, Pasricha R, Ahmad A, Khan M (2007) Nitrate reductase-mediated synthesis of silver nanoparticles from $AgNO_3$. Biotech Lett 29:439
170. Kumar SA, Abyaneh MK, Gosavi SW, Kulkarni SK, Ahmad A, Khan MI (2007) Sulfite reductase-mediated synthesis of gold nanoparticles capped with phytochelatin. Biotechnol Appl Biochem 47:191
171. Morata A, Loira I (2017) Yeast—industrial applications
172. Jha AK, Prasad K, Kulkarni AR (2008) Yeast mediated synthesis of silver nanoparticles. Int J Nanosci Nanotechnol 4:17
173. Boroumand Moghaddam A, Namvar F, Moniri M, Tahir PM, Azizi S, Mohamad R (2015) Nanoparticles biosynthesized by fungi and yeast: a review of their preparation, properties, and medical applications. Molecules, Basel, Switzerland 20:16540
174. Dameron C, Reese R, Mehra R, Kortan A, Carroll P, Steigerwald M, Brus L, Winge D (1989) Biosynthesis of cadmium sulphide quantum semiconductor crystallites. Nature 338:596
175. Kowshik M, Vogel W, Urban J, Kulkarni SK, Paknikar KM (2002) Microbial synthesis of semiconductor PbS nanocrystallites. Adv Mater 14:815
176. Jha AK, Prasad K, Prasad K (2009) A green low-cost biosynthesis of Sb_2O_3 nanoparticles. Biochem Eng J 43:303
177. Lin Z, Wu J, Xue R, Yang Y (2005) Spectroscopic characterization of Au_3+ biosorption by waste biomass of saccharomyces cerevisiae. Spectrochim Acta Part A Mol Biomol Spectrosc 61:761
178. Agnihotri M, Joshi S, Kumar AR, Zinjarde S, Kulkarni S (2009) Biosynthesis of gold nanoparticles by the tropical marine yeast yarrowia lipolytica NCIM 3589. Mater Lett 63:1231
179. Kowshik M, Ashtaputre S, Kharrazi S, Vogel W, Urban J, Kulkarni S, Paknikar K (2003) Extracellular synthesis of silver nanoparticles by a silver-tolerant yeast strain MKY3. Nanotechnology 14:95
180. Strouhal M, Kizek R, Vacek J, Trnková L, Němec M (2003) Electrochemical study of heavy metals and metallothionein in yeast Yarrowia lipolytica. Bioelectrochemistry 60:29
181. Kobayashi M, Tomita S, Sawada K, Shiba K, Yanagi H, Yamashita I, Uraoka Y (2012) Chiral meta-molecules consisting of gold nanoparticles and genetically engineered tobacco mosaic virus. Opt Express 20:24856
182. Narayanan KB, Han SS (2017) Helical plant viral nanoparticles—bioinspired synthesis of nanomaterials and nanostructures. Bioinspir Biomim 12:031001
183. Zeng Q, Wen H, Wen Q, Chen X, Wang Y, Xuan W, Liang J, Wan S (2013) Cucumber mosaic virus as drug delivery vehicle for doxorubicin. Biomaterials 34:4632
184. Mao C, Flynn CE, Hayhurst A, Sweeney R, Qi J, Georgiou G, Iverson B, Belcher AM (2003) Viral assembly of oriented quantum dot nanowires. Proc Natl Acad Sci 100:6946
185. Ongaro A, Griffin F, Beecher P, Nagle L, Iacopino D, Quinn A, Redmond G, Fitzmaurice D (2005) DNA-templated assembly of conducting gold nanowires between gold electrodes on a silicon oxide substrate. Chem Mater 17:1959

186. Hall SR, Shenton W, Engelhardt H, Mann S (2001) Site-specific organization of gold nanoparticles by biomolecular templating. ChemPhysChem 2:184
187. Naik RR, Jones SE, Murray CJ, McAuliffe JC, Vaia RA, Stone MO (2004) Peptide templates for nanoparticle synthesis derived from polymerase chain reaction-driven phage display. Adv Func Mater 14:25
188. Shenton W, Douglas T, Young M, Stubbs G, Mann S (1999) Inorganic-organic nanotube composites from template mineralization of tobacco mosaic virus. Adv Mater 11:253
189. Hall SR, Bolger H, Mann S (2003) Morphosynthesis of complex inorganic forms using pollen grain templates. Chem Commun 22:2784
190. Aizenberg J, Addadi L, Weiner S, Lambert G (1996) Stabilization of amorphous calcium carbonate by specialized macromolecules in biological and synthetic precipitates. Adv Mater 8:222
191. Liu J, Wu Q, Ding Y (2005) Self-assembly and fluorescent modification of hydroxyapatite nanoribbon spherulites. Eur J Inorg Chem 2005:4145
192. Lazareck AD, Cloutier SG, Kuo T-F, Taft BJ, Kelley SO, Xu JM (2006) DNA-directed synthesis of zinc oxide nanowires on carbon nanotube tips. Nanotechnology 17:2661
193. Shenton W, Mann S, Cölfen H, Bacher A, Fischer M (2001) Synthesis of nanophase iron oxide in lumazine synthase capsids. Angew Chem Int Ed 40:442
194. Bharti C, Nagaich U, Pal AK, Gulati N (2015) Mesoporous silica nanoparticles in target drug delivery system: a review. Int J Pharm Investig 5:124
195. Cha JN, Stucky GD, Morse DE, Deming TJ (2000) Biomimetic synthesis of ordered silica structures mediated by block copolypeptides. Nature 403:289
196. Belton D, Paine G, Patwardhan SV, Perry CC (2004) Towards an understanding of (bio) silicification: the role of amino acids and lysine oligomers in silicification. J Mater Chem 14:2231
197. Sumerel JL, Yang W, Kisailus D, Weaver JC, Choi JH, Morse DE (2003) Biocatalytically templated synthesis of titanium dioxide. Chem Mater 15:4804
198. Meldrum F, Heywood B, Mann S (1992) Magnetoferritin, in vitro synthesis of a novel magnetic protein. Science 257:522
199. Meldrum FC, Wade VJ, Nimmo DL, Heywood BR, Mann S (1991) Synthesis of inorganic nanophase materials in supramolecular protein cages. Nature 349:684
200. Wong KKW, Douglas T, Gider S, Awschalom DD, Mann S (1998) Biomimetic synthesis and characterization of magnetic proteins (Magnetoferritin). Chem Mater 10:279
201. Meldrum FC, Douglas T, Levi S, Arosio P, Mann S (1995) Reconstitution of manganese oxide cores in horse spleen and recombinant ferritins. J Inorg Biochem 58:59
202. Wong KK, Mann S (1996) Biomimetic synthesis of cadmium sulfide-ferritin nanocomposites. Adv Mater 8:928
203. Naik RR, Brott LL, Clarson SJ, Stone MO (2002) Silica-precipitating peptides isolated from a combinatorial phage display peptide library. J Nanosci Nanotechnol 2:95
204. Whaley SR, English D, Hu EL, Barbara PF, Belcher AM (2000) Selection of peptides with semiconductor binding specificity for directed nanocrystal assembly. Nature 405:665
205. Slocik JM, Stone MO, Naik RR (2005) Synthesis of gold nanoparticles using multifunctional peptides. Small 1:1048
206. Mao C, Solis DJ, Reiss BD, Kottmann ST, Sweeney RY, Hayhurst A, Georgiou G, Iverson B, Belcher AM (2004) Virus-based toolkit for the directed synthesis of magnetic and semiconducting nanowires. Science 303:213
207. Albers S-V, Meyer BH (2011) The archaeal cell envelope. Nat Rev Microbiol 9:414
208. Sleytr UB, Schuster B, Egelseer E-M, Pum D (2014) S-layers: principles and applications. FEMS Microbiol Rev 38:823
209. Roy M, Mukherjee P, Mandal BP, Sharma RK, Tyagi AK, Kale SP (2012) Biomimetic synthesis of nanocrystalline silver sol using cysteine: stability aspects and antibacterial activities. RSC Adv 2:6496
210. Mertig M, Pompe W, Niemeyer C, Mirkin C (2004) Nanobiotechnology—concepts, applications and perspectives

211. Storhoff JJ, Mirkin CA (1999) Programmed materials synthesis with DNA. Chem Rev 99:1849
212. Mertig M, Colombi Ciacchi L, Seidel R, Pompe W, De Vita A (2002) DNA as a selective metallization template. Nano Lett 2:841
213. Levina L, Sukhovatkin V, Musikhin S, Cauchi S, Nisman R, Bazett-Jones DP, Sargent EH (2005) Efficient infrared-emitting PbS quantum dots grown on DNA and stable in aqueous solution and blood plasma. Adv Mater 17:1854
214. Bigham SR, Coffer JL (2000) Thermochemical passivation of DNA-stabilized Q-cadmium sulfide nanoparticles. J Cluster Sci 11:359
215. Fowler CE, Shenton W, Stubbs G, Mann S (2001) Tobacco mosaic virus liquid crystals as templates for the interior design of silica mesophases and nanoparticles. Adv Mater 13:1266
216. Douglas T, Young M (1998) Host-guest encapsulation of materials by assembled virus protein cages. Nature 393:152

Chapter 19
Photo- and Radiation-Induced Synthesis of Nanomaterials

Madhab Chandra Rath

Abstract Nanomaterials of noble metals and semiconductors are of immense use in optoelectronic devices, sensors, biological applications and many more. Their synthesis always remains an important topic of research, because of a strong correlation between their optical properties and shapes/sizes. Out of several synthetic routes, chemical route is a preferred one for its easiness and simplicity. Among chemical routes, photochemical and radiation chemical methods are very efficient and powerful. In these two processes, the synthesis of nanomaterials proceeds through the reactions of the precursor ions with free radicals, which are generated upon photo and/or high energy radiations such as gamma and electron beam irradiation. Various nanomaterials have been synthesized by this process and their shapes/sizes could be easily tuned by controlling the experimental parameters like precursor concentration, types of radiation, absorbed dose, dose rate, etc. The chapter provides an account of the synthesis of a wide variety of semiconductor nanomaterials of different shapes and sizes by these methods. These nanomaterials were found to possess very unusual optical properties as compared to those synthesized by normal chemical routes. Various other nanomaterials have also been synthesized through this method and the processes have been optimized. The mechanism of formation of such nanomaterials has been elucidated by time-resolved absorption measurements.

Keywords Photochemical · Radiation chemical · Nanomaterials · Synthesis · Free radicals

---

M. C. Rath (✉)
Radiation and Photochemistry Division, Bhabha Atomic Research Centre, Homi Bhabha National Institute, Mumbai 400085, India
e-mail: madhab@barc.gov.in

A. K. Tyagi and R. S. Ningthoujam (eds.), *Handbook on Synthesis Strategies for Advanced Materials*, Indian Institute of Metals Series,
https://doi.org/10.1007/978-981-16-1807-9_19

19.1 Introduction

Nanomaterials have continued to play an immense role in our day-to-day life starting from the vedic era. Prior to the development of various sophisticated imaging techniques like transmission electron microscope (TEM), scanning electron microscope (SEM), atomic force microscope (AFM), scanning tunnelling microscope (STM), etc. which directly provide the information about size and shape of the nanomaterials; the term 'nano' was missing from the literature. Researchers would rather use to mention 'ultra-small particle', 'very fine particle', etc. in the discussions. Several Ayurvedic medicines like 'bhasma' and other 'jadi-buti' were known to contain these so-called very very fine particles of 'swarna' (Gold), 'roupya' (Silver) and many other metals. Nowadays, these 'very fine particles' of metals are termed as 'noble metal nanoparticles' since their sizes are in the nanometre scale. Nanoparticles had been in use in ancient India. In the ancient Greek and Roman civilizations, people used nanosized gold and silver for colouring of different objects like glass, walls, etc.

The mother nature also knows the usefulness of such, nanosized particles since ages. For example, the leaves of Lotus and Lilly plants float on the surface of water and their top surface still does not get wet. Even if a drop of waterfalls on its surface, it retains its shape as a drop only and never gets spread over it. With the help of the above-mentioned sophisticated imaging techniques, it was discovered that the top surface of lotus leaves is composed of arrays of nanoshaped structures of leaf pigments [1, 2]. Therefore, when water is put on these nanostructure arrays, it does not get spread over such surface and hence remains as a drop only, so the surface of the leaf behaves like a hydrophobic material, this phenomenon is termed as 'lotus effect'. Similarly, the colours of the feathers of peacock as well as the wings of butterfly are mainly due to the structural colouration. This originates mainly due to the optical interference arising from the light reflection from the nanoscale photonic lattices present in the arrays of barbules which are the major constituents of feathers in peacock.

However, in the modern days after the invention of microscopic techniques, like optical microscopy as well as electron microscopy such as scanning electron microscopy (SEM) and transmission electron microscopy (TEM), the size of these ultra-small particles could be determined and thereafter named as nanosized particles.

Several researchers have tried to synthesize various types of nanomaterials of different sizes and shapes by chemical, solvothermal, hydrothermal, sonochemical, photochemical, radiation chemical and organometallic routes. In this context, it may be noted that photochemical and radiation chemical routes are considered as green chemistry routes since there is no use of any (i) hazardous chemicals, (ii) high temperature, (iii) inert atmosphere and (iv) no threat to environment. Radiation-induced synthesis of noble metals like silver and gold nanoparticles has been extensively studied by several researchers [3–7]. In the recent years, radiation-induced synthesis of various transition metal nanoparticles like nickel and copper, etc. has been reported in the literature [8–12]. Similarly, radiation-induced synthesis of compound semiconductor nanoparticles like CdS, CdSe, ZnS, etc. has been extensively studied in the recent past by several researchers across the globe [13–22]. Radiation-induced synthesis of metal oxide nanoparticles has also been reported by several groups [23–28]. Photochemical synthesis of noble metals as well as other transition metal nanoparticles has been reported by various researchers [29–34]. Similarly, photochemical synthesis of compound semiconductor nanoparticles like CdSe, CdS has also been studied by various groups [34–39]. Photochemical synthesis of metal oxides has been investigated by various researchers [40, 41]. Ichimura and his group have reported the photochemical deposition of semiconductor and metal oxide thin films over a suitable substrate immersed in the reaction solution through photo-irradiation with UV light [35–37]. The radiation and photochemical synthesis of nanomaterials or nanofilms is an ever-expanding field of research and the technique is being applied for synthesizing a wide range of nanomaterials of interest.

19.1.1 Synthesis Methods

Several methods have been employed for the synthesis of nanomaterials, which are classified under two main groups (i) physical methods and (ii) chemical methods. Physical methods are based on ‘top-down’ principle that is breaking of bulk materials into atoms and then subsequently putting them together to form nanomaterials in the form of thin films, quantum dots or quantum well, etc. This is an energy intensive process which requires high cost and sophisticated machines as well as skilled manpower. On the contrary, chemical methods are based on ‘bottom-up’ principle that is nanomaterials are build up by atoms or molecules those are made available from the ions present in the solution. There are various chemical methods used for the synthesis/growth of nanomaterials, such as (i) chemical vapour deposition (CVD), (ii) chemical bath deposition (CBD), (iii) high temperature chemical synthesis, i.e. solvothermal or hydrothermal route, (iv) organometallic, (v) sonochemical, (vi) electrochemical, (vi) catalytic, (vii) photochemical and (viii) radiation chemical routes. Among these, photochemical and radiation chemical routes are considered as green chemistry methods.

The photochemical and radiation chemical reactions leading to the formation of various types of nanomaterials involve the use of different types of radiations. UV and visible lights (energy within a few eV only, e.g. 1–4 eV) are used for the photochemical reactions, whereas high energy photons like x-ray, gamma ray, etc. are used for the radiation chemical reactions. All these radiations are the part of electromagnetic radiations (Scheme 19.1). Different UV and visible light sources apart from direct sunlight are often used as the light sources for the photochemical reactions. Cobalt-60 gamma chamber is used as the source for the gamma radiation (1.25 MeV). However, apart from these electromagnetic radiations, radiation chemical reactions can also be carried out with the high energy (keV to MeV)

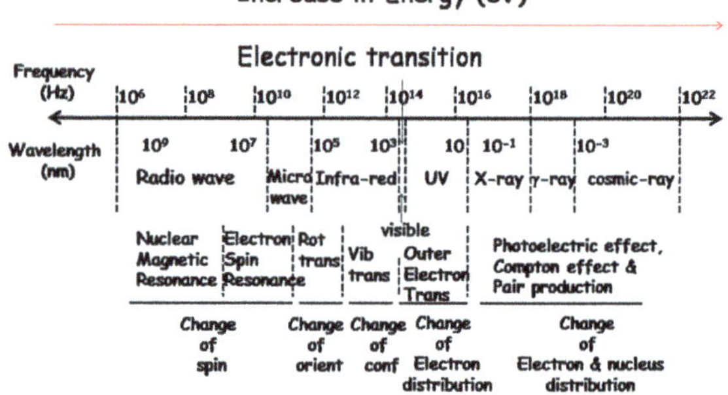

Scheme 19.1 Various types of processes occur in the entire spectrum of electromagnetic radiation

ionizing radiations like electron beam, proton beam and other higher mass particle beams. These are usually produced in various accelerators, for example liner electron accelerator (LINAC) is used for generating high energy electron beam.

When the radiation (both low and high energy radiations) interacts with the precursor solution, there happen various processes at room temperature, which eventually leads to the formation of desired nanomaterials in the form of nanoparticles, nanorods, nanoflowers, nanoribbons, nanosheets, quantum dots, core-shell nanoparticles, etc. in the final colloidal solution (Scheme 19.2). However, in both the situations, certain transient intermediate species are formed which react with the precursors leading to the formation of final products in the form of various kinds of nanomaterials depending on the nature of nanomaterials as well as the surface passivating agents. In the case of photochemical processes, the light photon is absorbed by specific solute molecules to get photoexcited to higher energy states like singlet or triplets states. Then the photochemical reactions take place in these photoexcited states, which produce either intermediate species like 'free radicals' or stable products (Scheme 19.3). In the case of radiation chemical processes, the high energy radiation, e.g. electron beam or gamma ray deposits energy in the solvent matrix and thereafter super-excited states as well as ions and radicals of the solvent molecules are generated instantaneously which subsequently leads to the formation of various free radicals (Scheme 19.4) [42]. In the case of water as the solvent, this phenomenon leads to the formation of free radicals like hydrated electrons (e_{aq}^-), hydrogen atom (H) and hydroxyl radicals ($OH^•$) along with

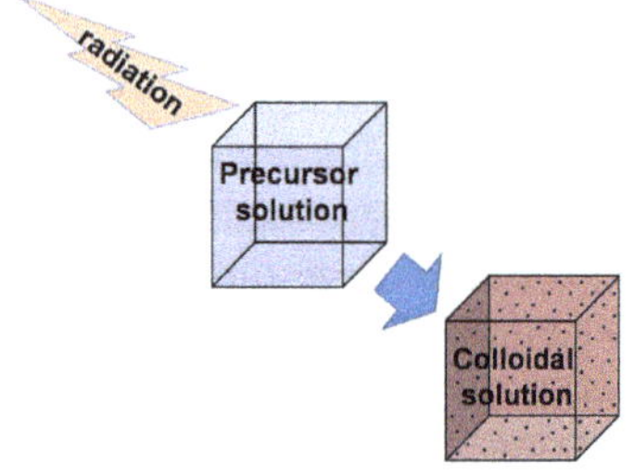

Scheme 19.2 Schematic representation of the radiation (light photons as well as high energy radiations) induced synthesis of nanomaterials

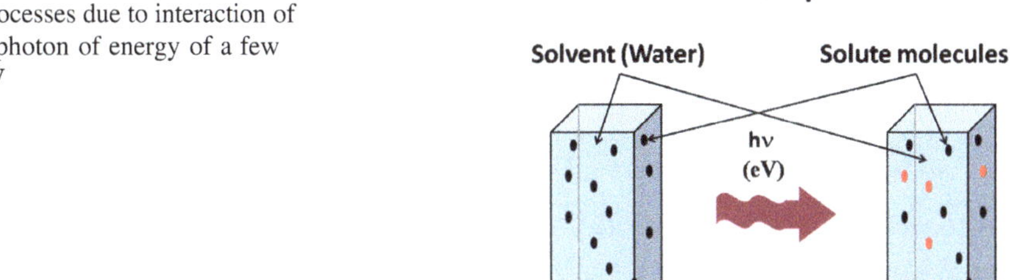

Scheme 19.3 Photochemical processes due to interaction of a photon of energy of a few eV

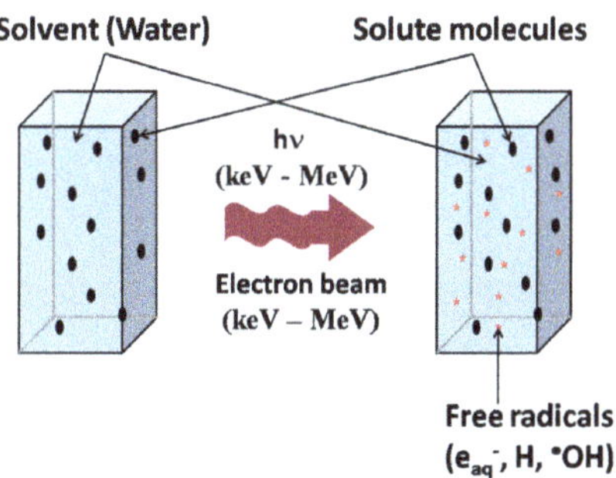

Scheme 19.4 Radiation chemical process, occurring due to the interaction of photon, of energy keV to MeV or electron, and beam of energy keV to MeV

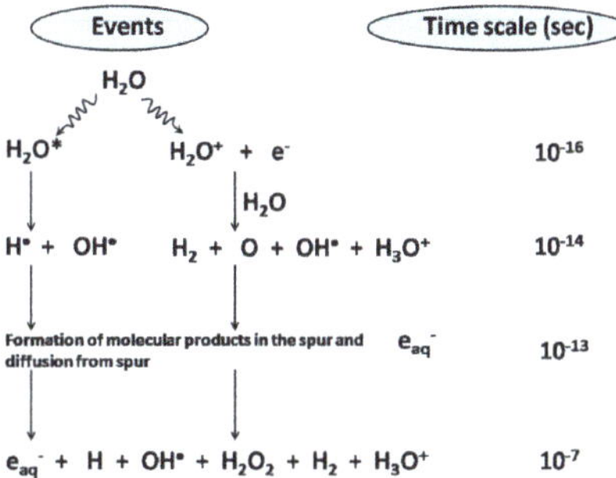

Scheme 19.5 Radiation chemical process occurring due to the interaction of photon of energy keV to MeV or electron beam of energy keV to MeV

molecular species like H_2, H_3O^+ and H_2O_2 at a later time of about a few microseconds. The radiolysis of water is summarized in Scheme 19.5.

Therefore, it is now understood that in both the situations (photochemical and radiation chemical reactions), there is formation of free radicals (Scheme 19.6) in the solution and which can be effectively utilized for the synthesis of various kinds of nanomaterials.

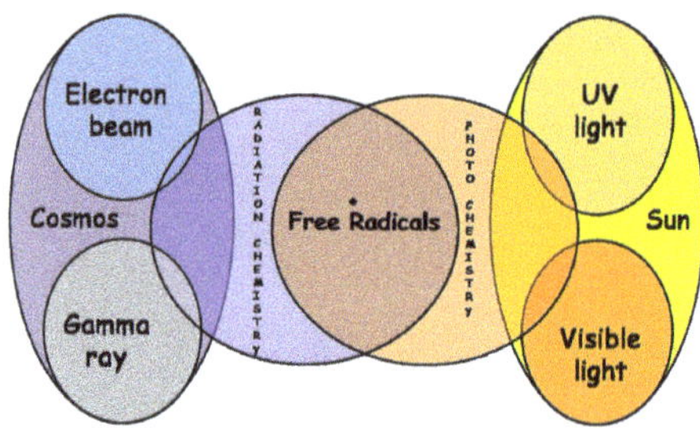

Scheme 19.6 Free radicals produced via both photochemical and radiation chemical routes

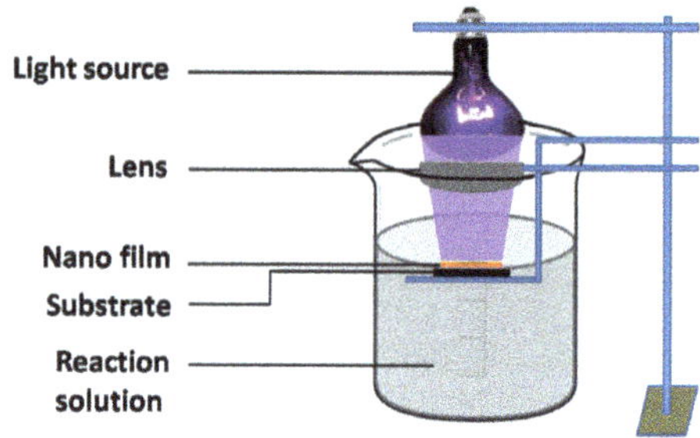

Scheme 19.7 Experimental setup showing the photochemical deposition of nanofilms

Various types of nanofilms of metallic as well as semiconductors can be grown over suitable substrates by both photochemical and radiation chemical routes. In the case of photochemical route, the substrate is immersed in the reaction solution, which is photo-irradiated from the top as shown in Scheme 19.7. Whereas, in the case of irradiation chemical route, the substrate is immersed inside the reaction solution and the whole solution is irradiated. In this way, the nanomaterials get deposited over the substrate in both the cases and we get a desired nanofilm.

The photo- and radiation-induced synthesis of nanomaterials takes place through several steps. First, the free radicals reduce the precursor ions to form either metastable and reduced ions or atoms or molecules as the case may be. Then these species undergo further reaction to form stable molecules or clusters of atoms and/or ions. This process is called as nucleation. Further addition of atoms/molecules takes place onto these nucleation centres to form nanoparticles, the process is called as growth. Finally, these nanoparticles undergo ageing, through the Ostwald ripening process which enhances their stability. The overall processes are summarized in Scheme 19.8.

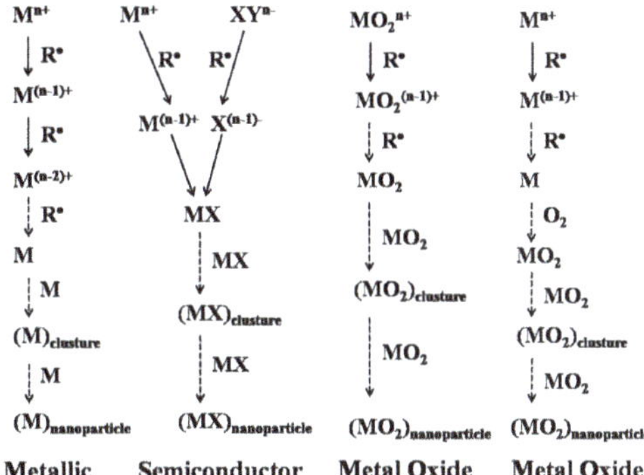

Scheme 19.8 Schematic representation showing the synthesis steps of formation of nanoparticles

19.1.2 Nanomaterials

Nanomaterials are considered to have size less than about 100 nm at least in along one dimension. The formation of these nanomaterials is strongly dependent on both thermodynamic as well as kinetic parameters. Moreover, the shape and size of nanomaterials are mainly achieved by the controlled synthesis that is dependent on (i) precursor concentrations, (ii) solvent properties, (iii) temperature, (iv) capping agents, (v) dose and dose rate (in the case of radiation-induced synthesis) and (vi) light intensity as well as the concentrations of reagents required for the generation of free radicals (in the case of photochemical synthesis).

The reduction in the size of the materials leads to the evolution of various types of nanomaterials and this is called confinement in particularly semiconductors. Confinement in one direction leads to the formation of thin films, and when a lower band gap material is sandwiched between two thin films of higher band gap, this is called a quantum well structure. The confinement in two directions leads to the formation of quantum rods and that in all the three directions leads to the formation of quantum dots (Scheme 19.9). When the size of nanoparticles becomes less than or equal to the Bohr radius of exciton or de-Broglie wavelength of the charge carriers then these are called quantum dots. In all these quantum confinements, the dimension of the nanomaterials should be very less (i.e. Bohr radius of exciton). In this situation, the exciton is unable to move freely in the crystal lattice and therefore there is a complete confinement in all the three dimensions (Scheme 19.10).

Confinement in materials

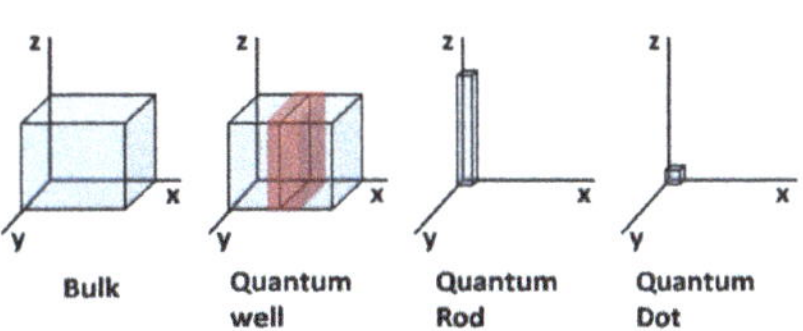

Scheme 19.9 Confinement in materials with respect to their size

Scheme 19.10 An artistic imagination of quantum dots in a colloidal solution

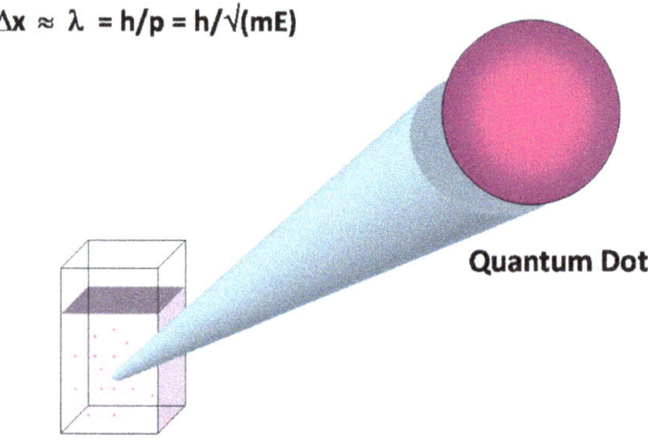

Nanomaterials of various nature like metallic, semiconductor and metal oxides could be easily synthesized by photochemical and radiation chemical methods using suitable radiation sources. In this way, one can synthesize nanoparticles, nanofilms and different nanosize materials.

19.2 Photochemical Synthesis of Nanomaterials

In the case of photochemical synthesis of nanomaterials, the UV or visible light gets absorbed by a particular species which produces certain free radicals, which have reducing property. In the case of UV absorption, usually acetone along with 2-propanol is used, where acetone absorbs light and gets photoexcited. In the photoexcited state, it is oxidizing in nature, and therefore, it abstracts one H atom from the nearby 2-propanol molecule, and in this process, there is a formation of two 2-hydroxy-2-propyl radicals, $(CH_3)_2C^{\bullet}OH$ in the solution (Scheme 19.11). However, in the case of photo-irradiation with visible light, one has to use

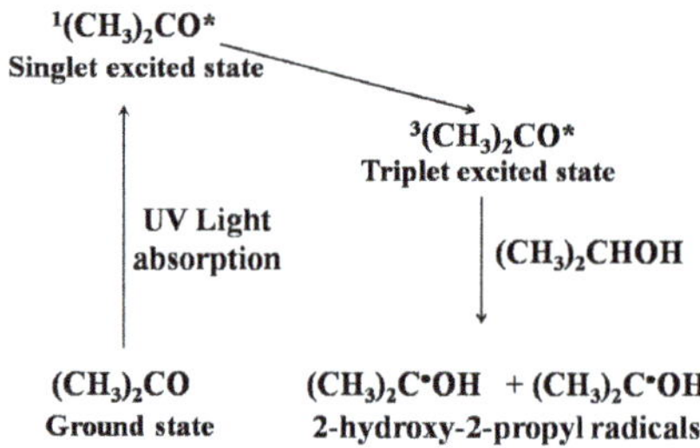

Scheme 19.11 Photochemical processes happening in the case of photoexcitation of acetone

acetophenone instead of acetone as the scintillator. Similarly, acetophenone absorbs the visible light and gets photoexcited, where it abstracts one H atom from 2-propanol to form two different radicals. In both the cases, the radicals formed are very much reducing in nature. Apart from these two types of generation of free radicals, there is also another method that happens in the case of uranyl ion photoexcitation. In this case, the uranyl ion gets photoexcited by both UV and visible light and in the photoexcited state it is highly oxidizing in nature, therefore it abstracts one H atom from the nearby molecules including even water. In all three cases, one can use any aliphatic alcohols having an α C atom attached to at least one H atom. Therefore, tert-butanol is not a suitable compound for this purpose.

In other case, for example the photochemical synthesis of CdS, CuS or ZnS nanoparticles, the precursor for S is usually used as $Na_2S_2O_3$. In this case, $S_2O_3^{2-}$ ions absorb the UV light and release S atom as well as electrons. These react with corresponding metal ions to form the metal sulphides in either thin films or nanoparticles [35–37]. Separate examples of photochemical synthesis of metal oxide, II–VI semiconductor and metallic nanoparticles in aqueous solution are being discussed in the following sections.

19.2.1 Photochemical Synthesis of UO_2 Nanoparticles in Aqueous Solutions

In this section [40], formation of UO_2 nanoparticles via UV photo-irradiation has been discussed. Usually UO_2 is being produced through the pyrolytic reduction of ammonium diuranate (ADU) using hydrogen [43]. Nenoff and co-workers [25], Roth et al. [24] and Rath and co-workers [26, 27] have investigated the radiolytic synthesis of UO_2 nanoparticles. Formation of UO_2 nanoparticles takes place via the reduction of UO_2^{2+} by 2-hydroxy-2-propyl radicals, $(CH_3)_2C^{\bullet}OH$.

Photochemical properties of UO_2^{2+} are complex in nature [44]. Excited state of UO_2^{2+} is known to be oxidizing in nature and abstracts H atom from the nearby alcohol molecule like 2-propanol in the reaction medium [45]. During this process, a $(CH_3)_2C^{\bullet}OH$ radical is produced, which could easily reduce further UO_2^{2+} to UO_2^{+} or UO_2^{+} to UO_2. Reduction of UO_2^{2+} to U (IV) through photochemical route has been reported [46–49]. Pavelkova and co-workers have investigated the synthesis of UO_2, ThO_2 and UO_2–ThO_2 nanoparticles through photochemical route [50]. In their study, nanoparticles formation takes place in two steps, i.e. formation of a solid amorphous precursor followed by heat treatment of this precursor. Recently, Rath and co-workers have investigated a direct, one step synthesis of UO_2 nanoparticles from the aqueous solution of uranyl nitrate through UV irradiation [40]. The formation of these nanoparticles takes place through the reduction of UO_2^{2+} by $(CH_3)_2C^{\bullet}OH$ radicals.

2-hydroxy-2-propyl radical, $(CH_3)_2C^{\bullet}OH$ is generated upon UV light irradiation in acetone in the presence of 2-propanol, which is a strong reducing agent

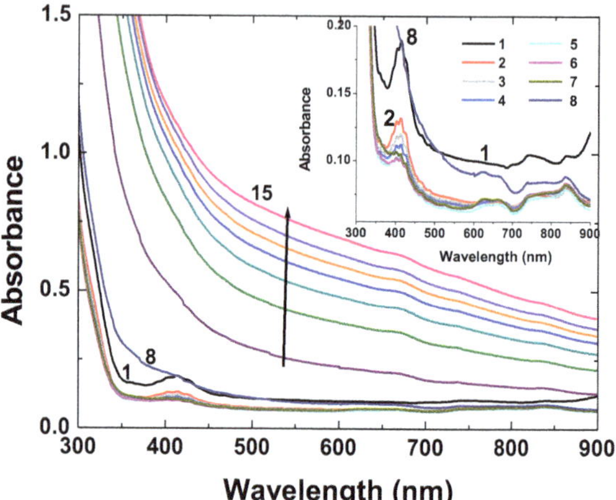

Fig. 19.1 Absorption spectra 10 mM $UO_2(NO_3)_2$, 1% (v/v) 2-propanol and 1% (v/v) acetone, at different time intervals of photo-irradiation under 350 nm UV lamp. Up to 70 min. Inset: absorption spectra in an expanded scale (Reproduced with permission from American Scientific publishers [40])

(E^1 = −1.6 V at pH 3), which can reduce uranyl ions to form UO_2. The absorption spectra of the reaction mixture (10 mM uranyl nitrate, 1% (v/v) acetone and 1% (v/v) 2-propanol) recorded at several time intervals of photo-irradiations inside a photoreactor confirm the formation of new product as there was a systematic increase in the absorption spectra. Absorption spectra of the reaction mixture inside a capped quartz cell were recorded at different photo-irradiation time intervals (Fig. 19.1).

The absorbance values monitored at 670 nm (corresponds to the absorption due to UO_2) are plotted against photo-irradiated time for different solvent mixtures are shown in Fig. 19.2. It is evident from this figure that the formation of UO_2 nanoparticles is definitely associated with an induction time up to about 25 min, where the increase in the absorbance value at 670 nm is nil. But after this time,

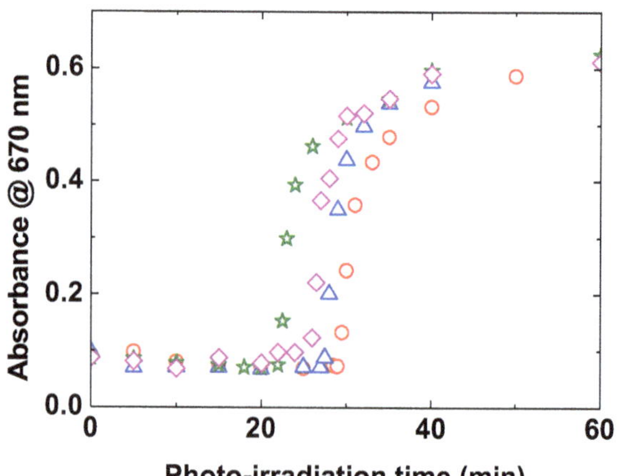

Fig. 19.2 Absorbance at 670 nm versus photo-irradiation time plot obtained in 10 mM $UO_2(NO_3)_2$ solution containing (o) 1% 2-propanol, (Δ) 1% 2-propanol and 1% acetone, (◇) 2% 2-propanol and 1% acetone and (⋆) 3% 2-propanol and 1% acetone at pH 3.4 irradiated with 350 nm UV lamp (Reproduced with permission from American Scientific publishers [40])

there is a fast increase in the absorbance value which gets saturated after about 50 min of photo-irradiation.

Concentration of 1% (v/v) 2-propanol in the solution is 130 mM and which is sufficient for the complete reduction of 10 mM UO_2^{2+} to UO_2. Apart from the photoexcitation of acetone, UO_2^{2+} can also get photoexcited with this 350 nm UV light and can undergo reduction by H atom abstraction from 2-propanol to form UO_2^{+} and $(CH_3)_2C^{\bullet}OH$ radical which further reacts to produce UO_2 [45]. The probable reactions those must be occurring in the reaction mixture for the formation of UO_2 nanoparticles are shown below:

$$(CH_3)_2CO \xrightarrow{hv} *(CH_3)_2CO \tag{19.1}$$

$$*(CH_3)_2CO + (CH_3)_2CHOH \rightarrow 2(CH_3)_2C^{\bullet}OH \tag{19.2}$$

$$(CH_3)_2C^{\bullet}OH + UO_2^{2+} \rightarrow UO_2^{+} + (CH_3)_2CO + H^{+} \tag{19.3}$$

$$UO_2^{+} + UO_2^{+} \rightarrow UO_2 + UO_2^{2+} \tag{19.4}$$

$$(CH_3)_2C^{\bullet}OH + UO_2^{+} \rightarrow UO_2 + (CH_3)_2CO + H^{+} \tag{19.5}$$

$$UO_2^{2+} \xrightarrow{hv} *UO_2^{2+} \tag{19.6}$$

$$*UO_2^{2+} + (CH_3)_2CHOH \rightarrow [UO_2^{+} + (CH_3)_2C^{\bullet}OH] + H^{+} \tag{19.7}$$

$$[UO_2^{+} + (CH_3)_2C^{\bullet}OH] \rightarrow UO_2 + (CH_3)_2CO + H^{+} \tag{19.8}$$

$$UO_2 + nUO_2 \rightarrow (UO_2)\ \text{nanoparticles} \tag{19.9}$$

UO_2 is formed through the Reactions 19.4, 19.5 and 19.8 above. It was observed that there was a decrease in the induction time with an increase in the 2-propanol content. It was noticed that there was no formation of UO_2 nanoparticles in the reaction mixture containing 10 mM uranyl nitrate in the absence of 2-propanol or acetone even upon a very long time exposure to UV lamps. This confirms definite role of 2-propanol in the formation of UO_2 nanoparticles.

The formation of UO_2 nanoparticles was also found to occur both in the de-aerated as well as in the aerated solutions. So the dissolved oxygen seems to have no negative role in the formation of UO_2 nanoparticles. The probable reactions those might be taking place in the presence of oxygen are shown below:

$$(CH_3)_2C^{\bullet}OH + O_2 \rightarrow (CH_3)_2CO + HO_2^{\bullet} \tag{19.10}$$

$$HO_2^{\bullet} + HO_2^{\bullet} \rightarrow H_2O_2 + O_2 \tag{19.11}$$

$$(CH_3)_2C^{\bullet}OH + H_2O_2 \rightarrow (CH_3)_2CO + OH^{\bullet} + H_2O \quad (19.12)$$

$$OH^{\bullet} + (CH_3)_2CHOH \rightarrow (CH_3)_2C^{\bullet}OH + H_2O \quad (19.13)$$

However, the UO_2 nanoparticles were found to decompose when the colloidal sol is exposed to air/oxygen. The decomposed colloidal sol again produced UO_2 nanoparticles upon exposure to UV light and this cycle could be repeated for several times. These UO_2 nanoparticles were found to be very stable under de-aerated condition for several months. The colloidal stability of these nanoparticles against agglomeration could be due to the coulombic repulsion among them, as the H^+ ions adsorbed on the surface of UO_2 nanoparticles makes the surface positively charged (zeta potential = +21 mV).

The products have been characterized by XRD, dynamic light scattering (DLS) and SEM. The XRD pattern of the products confirms the formation of UO_2 nanoparticles with cubic crystalline phase. The particle size as obtained from the DLS measurements was found to be 23 ± 5 nm. The SEM image of these UO_2 nanoparticles confirms that the particles have spherical shape (size about 1 μm). SEM image of the UO_2 nanoparticles confirms the formation of spherical-shaped particles with size less than 100 nm in the reaction mixtures.

It is being noticed here that there is a slow transformation from the pale yellow to a pale black colour appearance in the solution kept inside the photoreactor. In this situation, $(CH_3)_2C^{\bullet}OH$ radicals are being continuously generated in the solution during UV irradiation time. So it is obvious that UO_2^+ and UO_2 are also being continuously produced in the solution. These nascent species can come closure to form smaller clusters, where there is no increase in the absorption spectra, as these small clusters do not have any absorption in the monitoring wavelength range.

This time period can be named as the induction time period. Then these small clusters come closure to grow further and attaining a particular size of about a few nanometres. These small nanoparticles would have an absorption along with a scattering in the UV-visible range. This process is very fast and therefore there is fast increase in the absorbance at 670 nm which finally saturates indicating no further growth. Photochemical formation of UO_2 nanoparticles is schematically shown in Scheme 19.12. It was also observed that the UV irradiation for a few

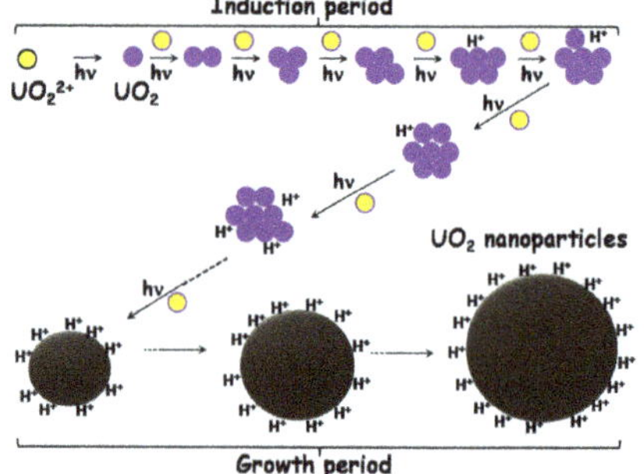

Scheme 19.12 Schematic representation of formation and growth of UO_2 nanoparticles through photochemical route (Reproduced with permission from American Scientific publishers [40])

minutes, which is equivalent to just up to the end of the induction period which is about 20–25 min, could form UO_2 nanoparticles in the capped cell containing the reaction mixture, without further photo-irradiation. It was established that the photochemical synthesis of UO_2 nanoparticles in aqueous solution is always associated with an induction time period.

19.2.2 Photochemical Synthesis of Starch Capped CdSe Quantum Dots in Aqueous Solution

Cadmium selenide (CdSe) quantum dots (QDs) have become one of the advanced semiconducting materials, as their optical properties could be easily tuned by controlling their size [51]. Various methods like electrochemical, sonochemical, radiation and photochemical routes have been employed for their bottom-up approach synthesis, where the later one is very important because there is no use of additional reducing agents, no requirement of stringent laboratory conditions like inert atmosphere and high temperature [52]. Various biomolecules like proteins, amino acids, carbohydrates provide a good capping of these QDs and reduce their cytotoxicity effects [53]. The aqueous phase synthesis is very important as their size and shape can be perfectly tuned as well as they get application in aqueous solutions [54]. Murray et al. [55] synthesized CdSe QDs in organic solvents and subsequently various researchers synthesized it in aqueous solution [56, 57].

The reaction mixture (0.5 mM of $CdSO_4$, 0.5 mM of Na_2SeSO_3, 0.5 mg/mL of starch solution, 2% v/v of acetone and 2% v/v of 2-propanol) was irradiated with 300 nm UV light inside a UV photoreactor [58]. Acetone in the presence of 2-propanol produce two 2-hydroxy-2-propyl radicals, $(CH_3)_2C^{\bullet}OH$ upon UV irradiation, which are strong reducing in nature ($E^0 = -1.5$ V vs. NHE) [59]. These radicals can reduce Cd^{2+} to Cd^{+} and release Se^{-} from $SeSO_3{}^{2-}$ through the following reactions and lead to the formation of CdSe QDs. The probable reaction mechanism is given below:

$$(CH_3)_2C^{\bullet}OH + \left[Cd(NH_3)_4\right]^{2+} \rightarrow Cd^{\bullet+} + (CH_3)_2CO + H^{+} \quad (19.14)$$

$$SeSO_3^{2-} + (CH_3)_2C^{\bullet}OH \rightarrow Se^{\bullet-} + SO_3^{2-} + (CH_3)_2CO + H^{+} \quad (19.15)$$

$$Cd^{\bullet+} + Se^{\bullet-} \overset{hv}{\rightarrow} CdSe \rightarrow (CdSe)_n \quad (19.16)$$

$$(CdSe)_n \overset{starch}{\rightarrow} starch - (CdSe)_{quantum\ dot} \quad (19.17)$$

The growth of CdSe QDs was investigated by monitoring the absorption spectra at different photo-irradiation time (Fig. 19.3a). It is seen from this figure that the absorbance value at 480 nm did not increase further after about 10 min of

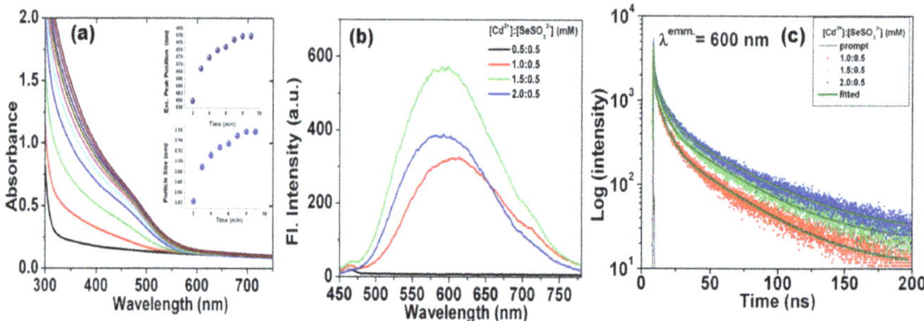

Fig. 19.3 **a** Absorption spectra of the sol containing CdSe QDs obtained with Cd:Se ratio 1.5:0.5 mM, insets: excitonic peak position and particle size with time of photo-irradiation, **b** fluorescence (Fl) spectra and **c** fluorescence decay curves of QDs synthesized at different Cd:Se ratios (Reproduced with permission of Elsevier [58])

photo-irradiation, indicating a completion of the formation of CdSe QDs in the reaction mixture. This is confirmed from the plots of excitonic peak position and the particle size versus photo-irradiation time (insets of Fig. 19.3a). The colour of the reaction mixture does not change in the absence of acetone indicating the reaction is initiated by the excited state of acetone only.

The room temperature fluorescence spectra of the QDs were measured for different QDs (Fig. 19.3b) synthesized in different reaction mixtures. The fluorescence intensity was found to increase with an increase in the Cd:Se ratio, and a higher intensity was obtained in the case of reaction mixture with a composition of Cd:Se as 1.5:0.5 mM. Whereas those synthesized with Cd:Se ratio 0.5:0.5 mM and 0.5:1.0 mM do not exhibit any fluorescent as seen from the above figure. The fluorescence lifetimes of these QDs (Fig. 19.3c) were found to consist of two components (τ_1 and τ_2), which could be originated from the band gap (τ_1) and trap states (τ_2) [60]. As the e–h pair recombine very fast, the shorter component, i.e. τ_1 which is less than 10 ns has been assigned to this process. On the other hand, the larger component, i.e. τ_2 which is more than 30 ns has been assigned to the trap state e–h pair recombination, as it is a slower process. Nevertheless, it was observed that the average lifetime $<\tau>$ value of these QDs was found to increase with an increase in the Cd content in the reaction mixture (see Table 19.1). It could be explained as the QDs with a higher Cd:Se ratio are better capped by the starch molecules and thereby exhibit a better fluorescence, whereas the QDs synthesized

Table 19.1 Fluorescence lifetime values of CdSe QDs prepared with various compositions of precursors (Reproduced with permission of Elsevier [58])

$[Cd^{2+}]:[Se^{2-}]$ (mM)	Lifetime (ns)				
	τ_1	A_1	τ_2	A_2	$<\tau>$
1.0:0.5	6.16	0.38	38.3	0.62	26.1
1.5:0.5	5.93	0.31	39.6	0.69	29.2
2.0:0.5	6.95	0.30	43.4	0.70	32.5

with lower Cd:Se ratio such as 0.5:0.5 and 0.5:1.0 mM are poorly capped by starch molecules and thereby exhibit either very poor or no fluorescence.

These QDs were extracted by freezing the colloidal sol at 0 °C and subsequently allowed to come back to liquid form at room temperature. In this process, it is known that amylose and amylopectin parts of starch get settled down upon freezing [61]. CdSe QDs along with starch also get separated from the sol in the bottom of the beaker in this process, which could be easily recovered from the colloidal sol.

The XRD patterns of these QDs confirm the formation of cubic zinc blend nanocrystal of CdSe [55]. The crystallite size was calculated using Scherrer formula and was found to be about 2–3 nm in all these cases. Raman spectra of these QDs exhibit two peaks at 204 and 409 cm^{-1} that confirm the formation of CdSe nanoparticles and correspond to 1st and 2nd order longitudinal modes of CdSe QDs [62]. From the FTIR spectra, it is clear that the OH stretching peak (3650–3000 cm^{-1}) present in starch is very broad which indicates the presence of many OH groups and inter and intramolecular H bonding there. The other peaks are at 2900 cm^{-1}(asymmetric C–H stretching), 1640 cm^{-1} (O–H bending of water in starch), 1340 cm^{-1} (angular deformation of C–H bond), 1150 cm^{-1} (C–O and C–C stretching), 1075 cm^{-1} (C–O–H bending), 998 cm^{-1} (skeletal vibration of α 1–4 glycosidic linkage (C–O–C), 926 cm^{-1} (C–C stretching) and 850 cm^{-1} (C_2 deformation) [63]. The starch capped CdSe samples give distinct peaks which indicate the existence of starch on CdSe surface.

The TEM images confirmed that the nanoparticles are well dispersed and spherical in shape with sizes about 3 nm. The HRTEM image clearly showed (111) and (200) planes with interplannar distance of 3.5 Å and 2.9 Å, respectively, which matches very well with the cubic zinc blende structure of CdSe nanoparticles. The SEM images also further confirm the spherical nature of the particles, but of relatively bigger size, indicating an agglomeration of these QDs. Starch capped CdSe QDs have been successfully synthesized in aqueous solution through photochemical route using 300 nm UV light, which is very simple, efficient and powerful. Their optical properties were found to strongly depend upon Cd to Se ratio.

19.2.3 Photochemical Synthesis of Metal Nanoparticles

Photochemical synthesis of various metallic nanoparticles has been reported in the literature till date. Among these, different noble metals like silver (Ag) and gold (Au), other metals like nickel (Ni) and copper (Cu) have been reported by various authors [29–34]. The free radicals generated in solution reduce Ag^{+} ions to Ag, Au^{3+} ions to Au, Ni^{2+} ions to Ni and Cu^{2+} ions to Cu. Then these reduced ions called atoms get agglomerated to form nucleation centres and then growth of nanoparticles takes place upon addition of successive atoms to the nucleation centres. In the absence of any capping agents, these metastable particles undergo an uncontrolled growth to get precipitated out. However, because of their surface charge either negative or positive, the nanoparticles can be stable even without any surface capping agents. This is due to Coulombic repulsive forces among the nanoparticles. In certain cases, these nanoparticles are stabilized by the use of suitable polymers, block copolymers and surfactants [33, 34, 64–67]. In these studies, UV light from high pressure mercury lamp is being used. UV light induces the formation of free radicals which are reducing in nature, so that the reduction of metal ions can take place under UV illumination.

19.3 Radiation Chemical Synthesis of Nanomaterials

The radiolysis of water generates various free radicals along with different molecular products. Out of which hydrated electrons, e_{aq}^- and H atoms are reducing radicals and $OH^•$ is an oxidizing radical [68, 69]. Their radiation chemical yields (G = μmol/J) are mentioned in Table 19.2. However, in order to carry out any specific reaction the radiolytic solution has to be either reducing or oxidizing in nature. In this context, the synthesis of nanomaterials (metallic, semiconductor or metal oxides) requires a reducing environment. Such a condition can only be achieved by scavenging the $OH^•$ radicals and which is often carried out by adding tert-butanol in water. Tert-butanol reacts with $OH^•$ radicals to form an intermediate species, which are very less reactive (Scheme 19.13).

In this situation, the remaining e_{aq}^- and H atoms make the system completely reducing in nature. In certain cases, other aliphatic alcohol like 2-propanol, 1-propanol, 1-butanol, ethanol or methanol is added to water for scavenging $OH^•$ radicals. In this case, the reaction product between the $OH^•$ radical and these alcohols is also a reducing radical which increases the reducing environment. In

Table 19.2 Radiation chemical yields, G values (μmol/J) in irradiated water

Type of radiation	e_{aq}^-	H	$OH^•$	H_2	H_2O_2	$HO_2^•$
Gamma and high energy electron beam	0.28	0.062	0.28	0.047	0.073	0.0027

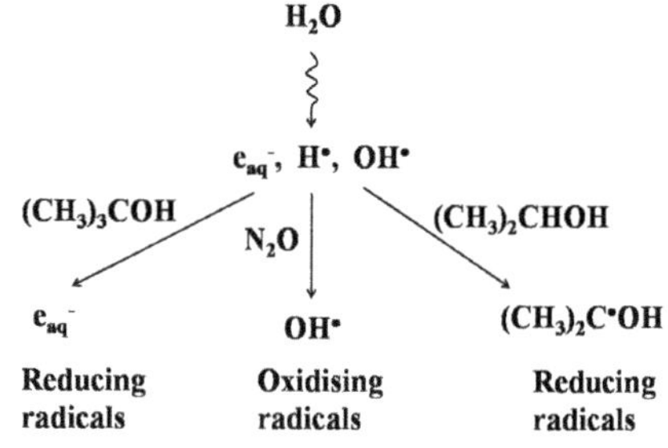

Scheme 19.13 Radiation chemical process occurring in water for getting an oxiding or reducing condition

Scheme 19.14 Radiation chemical processes occurring in water for getting an oxidizing condition

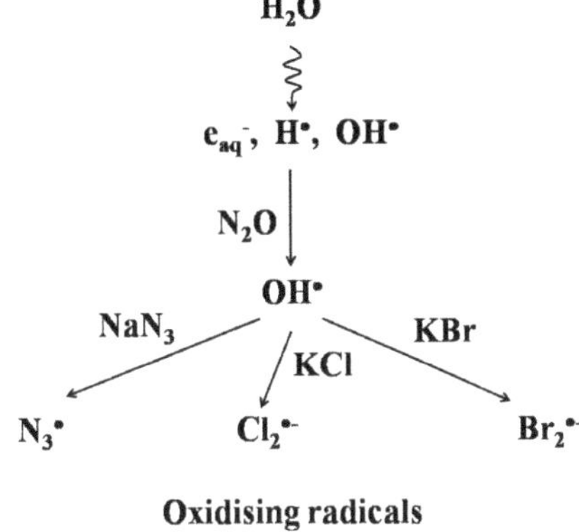

another condition, acetone is also added along with 2-propanol in order to generate predominantly the $(CH_3)_2C^{\bullet}OH$ radicals, which are highly reducing in nature. In this case, acetone reacts with e_{aq}^- and 2-propanol reacts with both H and $OH^{\bullet}$ radicals and both the reactions lead to the formation of $(CH_3)_2C^{\bullet}OH$ radicals with a higher radiation chemical yield. Similarly, one can create an oxidizing condition in the system by scavenging hydrated electrons, e_{aq}^-. This is predominantly achieved by saturating N_2O gas in water prior to irradiation with either electron beam or cobalt-60 gamma radiation. In this case, N_2O reacts with e_{aq}^- and generates a $OH^{\bullet}$ radical, which increase radical yield of $OH^{\bullet}$ radicals. In some other conditions, different secondary oxidizing radicals are created in order to study the oxidation reactions. These include $N_3^{\bullet}$, $Br_2^{\bullet -}$, $Cl_2^{\bullet -}$ radicals which are obtained by adding different salts NaN_3, KBr, KCl, respectively, after saturating the solution with N_2O (Scheme 19.14) [69]. In this situation, the $OH^{\bullet}$ radical reacts with the anion of these salts to generate corresponding anion radical, which dimerizes with the anion itself to form a relatively stable dimer radical as mentioned above. However, in the case of synthesis of nanomaterials, a reducing environment is desired.

Separate examples of radiation chemical synthesis of metal oxide, II–VI semiconductor and metallic nanoparticles in aqueous solution are being discussed in the following sections.

19.3.1 Radiolytic Synthesis of UO_2 Nanoparticles in Aqueous Solutions

Radiolytic synthesis of UO_2 nanoparticles has been reported by various researchers in the literature [24–28]. A mechanism for the radiolytic formation of these nanoparticles has been discussed here [26]. Atinault et al. have investigated the radiolytic oxidation of U (IV) to $(UO_2)^{2+}$ [15]. In this study, it was observed that there was a time lapse between the formation of UO_2 nanoparticle and the electron beam (eb) irradiation in the reaction mixture. These studies have been thoroughly investigated under different experimental conditions. The reactions of 2-hydroxy-2-propyl radicals, $(CH_3)_2C^{\bullet}OH$ with UO_2^{2+} lead to the formation of these nanoparticles in aqueous solution. At the natural pH 3.4 in the uranyl solution, e_{aq}^-, can react with H^+, UO_2^{2+} as well as NO_3^- ions. Reactions occurring in the radiolysis of water and subsequent reactions leading to the formation UO_2 nanoparticles are given below.

$$H_2O \rightsquigarrow e_{aq}^-, H^{\bullet}, OH^{\bullet}, HO_2^{\bullet}, H_2O_2, H_3O^+ \quad (19.18)$$

$$e_{aq}^- + H^+ \rightarrow H^{\bullet} \quad (19.19)$$

$$H^{\bullet}/OH^{\bullet} + (CH_3)_2CHOH \rightarrow (CH_3)_2C^{\bullet}OH + H_2/H_2O \quad (19.20)$$

$$e_{aq}^- + UO_2^{2+} \rightarrow UO_2^+ \quad (19.21)$$

$$e_{aq}^- + NO_3^- \rightarrow NO_3^{2-} \quad (19.22)$$

$$UO_2^{2+} + (CH_3)_2C^{\bullet}OH \rightarrow UO_2^+ + (CH_3)_2CO + H^+ \quad (19.23)$$

$$UO_2^+ + (CH_3)_2C^{\bullet}OH \rightarrow UO_2 + (CH_3)_2CO + H^+ \quad (19.24)$$

$$UO_2 + n\ UO_2 \rightarrow (UO_2)_{nanoparticle} \quad (19.25)$$

It is obvious from these reactions that about 20 mM radicals will be needed for the conversion of whole of 10 mM UO_2^{2+} to UO_2. Assuming the G value of the primary radicals in this solvent mixture 0.6 μmol J^{-1}, the overall radical concentrations could be about 30 mM in the case of an absorbed dose of 50 kGy. This higher absorbed dose has been given to the reaction mixture in order to take care of the reaction between e_{aq}^- and NO_3^-, (see the Reaction 19.22). Reaction 19.22 does not have any role in the formation of UO_2 nanoparticles as understood from the experiments in the reaction mixture containing higher amount of NO_3^- ions. However, NO_3^- ions do scavenge e_{aq}^- to which indirectly affects the formation of UO_2 nanoparticles, as the formation of these nanoparticles happens through a reduction pathway. Reactions 19.23, 19.24 and 19.25 as shown above are the simplified presentations for the formation of UO_2 nanoparticles. Uranyl ions exist in different speciation forms at different pH and in the presence of different counter anions. However, at the natural pH 3.4, it exists predominantly as UO_2^{2+}. These species (UO_2 and/or other U(IV) species if any) slowly give rise to the formation of UO_2 nanoparticles (Scheme 19.15). It is now understood from this scheme that there could be an induction period during the formation of UO_2 nanoparticles, which was investigated in detail in the following sections.

It is observed from the above reactions that during the reduction of UO_2^{2+} to UO_2, two H^+ ions are generated. Hence, 20 mM H^+ ions are expected to be formed during the reduction of 10 mM UO_2^{2+} in the solution. In such situation, the pH of

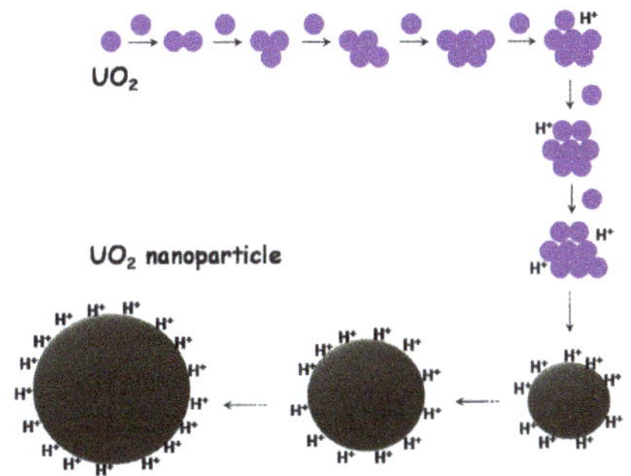

Scheme 19.15 A simplified presentation showing the nucleation and growth of UO_2 nanoparticles (Reproduced with permission from Elsevier [27])

the final radiolysed solution should be 1.7. On the contrary, the pH of the colloidal sol was found to be only 3.1. The as-grown UO_2 nanoparticles in the colloidal solution are stable under a de-aerated condition for weeks and get decomposed only when exposed to air or oxygen. Hence, it is obvious that the excess H^+ ions generated during the formation of UO_2 nanoparticles must have been utilized by these nanoparticles in increasing their colloidal stability. The H^+ ions might have been adsorbed on the surface of these nanoparticles (Scheme 19.15) and provide a coulombic repulsion between them, thereby hindering their aggregation to form any precipitate. This was confirmed by measuring the zeta potential of the irradiated colloidal sol ($\zeta = +21$ mV). However, a minor adsorption of H^+ ions even on the bigger clusters could have occurred in the solution, as these H^+ ions are being formed from the beginning of the formation of UO_2 molecular species as shown in Scheme 19.15.

Radiolytic synthesis of UO_2 nanoparticles has been performed in the pH range 2–5 in aqueous solutions containing 10 mM uranyl nitrate and 10% 2-PrOH [28]. The formation of UO_2 nanoparticles was found to occur in the pH range 2.5–3.5. The speciation of uranyl ions at different pHs is the main cause for such observation, and at this pH range it exits as UO_2^{2+}. In the case of high acidic pH, i.e. <pH 2, UO_2 nanoparticles does not occur due to these reactions [15]:

$$2(UO_2)^+ + 4H^+ \rightarrow (UO_2)^{2+} + U^{4+} + 2H_2O \tag{19.26}$$

$$2(UO_2)^+ \rightarrow \left((UO_2)^+\right)_2 \rightarrow 2(UO_2)^{2+} \tag{19.27}$$

So, pH of solutions was just the natural pH 3.4. Therefore, the excess H^+ produced during the formation of UO_2 nanoparticles must have been utilized in their colloidal stability, by making the surface positively charged which provide a coulombic repulsion among them.

Reaction mixtures of different sets have been made by 10 mM uranyl nitrate and 2-propanol in the range 1%–20% (v/v), where the concentration of 2-PrOH was 130 mM in the case of 1% and 2.6 M in the case of 20%, respectively. 2-PrOH reacts with primary radicals ($H^{\bullet}$ and $OH^{\bullet}$) to form 2-hydroxy-2-propyl, $(CH_3)_2C^{\bullet}OH$ radicals. The appearance of black coloured colloidal sol was found to

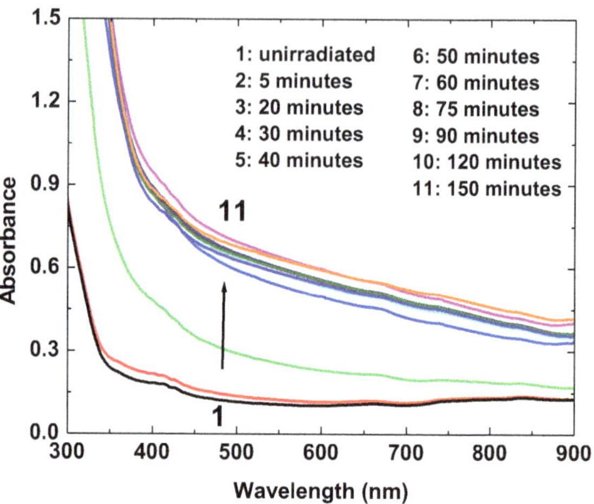

Fig. 19.4 Absorption spectra of the irradiated sol containing 10 mM uranyl nitrate and 5% 2-PrOH at various time lapse (Reproduced with permission from Elsevier [27])

take place after electron beam irradiation was over, and the time taken in the case of 1% 2-PrOH was higher as compared to that in the case of 10% 2-PrOH. The irradiation time was only a few seconds, whereas the time delay between the irradiation and the appearance of blackish sol was of several minutes. Such delay time occurs only during the post-irradiation process which is named as induction time. The black coloured sol containing UO_2 nanoparticles has absorption at around 670 nm along with scattering as well as absorption in the whole UV–visible range. The absorption spectrum of the sol was recorded after irradiation at various time lapse and the absorbance value at 670 nm was measured at different time intervals in order to monitor the growth of these nanoparticles (Fig. 19.4). It was noticed that during the post-irradiation induction time, there was almost no change in the absorption spectra from that of the unirradiated solution. However, there was increase in the absorbance value at each wavelength and an appearance of a well-defined spectrum, once the appearance of black colour happens in the sol. The absorbance value at 670 nm plotted against post-irradiation time lapse for solutions with different % (v/v) of 2-PrOH are shown in Fig. 19.5.

It is certainly evidenced from these observations that the growth of UO_2 nanoparticles starts only after a time delay from the irradiation. The growth of these nanoparticles is found to be favoured in the case of a higher % (v/v) of 2-PrOH. From Fig. 19.5, it is clearly seen that there is a remarkable increase in the post-irradiation time lapse from 10 to 1% (v/v), whereas there is no appearance of the absorption spectral pattern. But, once the appearance of black colour starts in the sol, it takes a very small time to reach saturation in all the cases. Therefore, it is certain that 2-PrOH molecules play an important role in the nucleation as well as the growth of these nanoparticles in the present study. The post-irradiation time lapse in the formation of UO_2 nanoparticles is being named as induction time, which is nothing but nucleation time. The time taken from the beginning of increase in the absorbance value at 670 nm up to the saturation value, is being named as

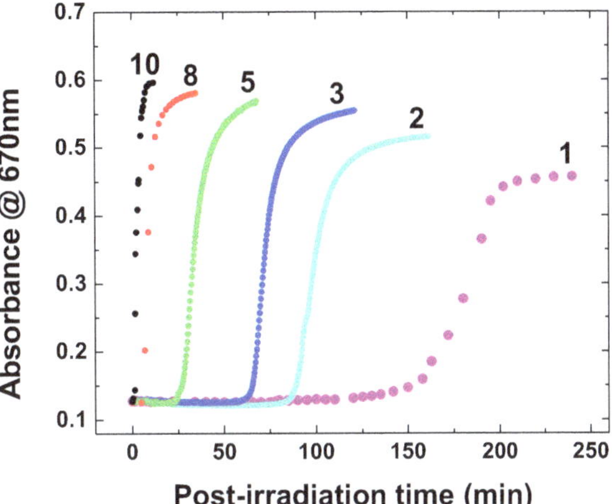

Fig. 19.5 Plots of absorbance at 670 nm versus post-irradiation time lapse in sol containing different % (v/v) 2-PrOH (Reproduced with permission from Elsevier [27])

growth time (Scheme 19.15). The induction time and formation time associated with the formation of UO_2 nanoparticles are determined from Fig. 19.5.

The formation time (= induction time + growth time) was found to decrease with the increase in the concentration of 2-PrOH in the range from 130 mM to 2.6 M and follows a non-linear pattern. In this situation, 2-PrOH molecules could be used in the solvation of UO_2 that favours the nucleation and growth of the nanoparticles. However, in the later stage, H^+ ions could play a crucial role in providing their stability during the growth process. The morphology of these nanoparticles was studied by scanning electron microscopy (SEM). Rod (length 2 ± 1 μm and width 300 ± 200 nm) as well as spherical-shaped particles were formed in all these cases (Fig. 19.6). Rod-shapes were almost identical in all the cases of 2-PrOH, however, the spherical nanoparticles were decreased with increase in the % (v/v) 2-PrOH.

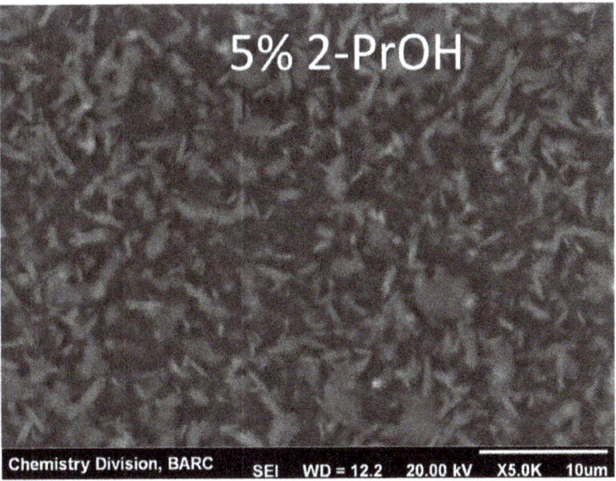

Fig. 19.6 SEM image of UO_2 nanoparticles formed in the sol containing 10 mM uranyl nitrate solutions and 5% (v/v) 2-PrOH via electron beam irradiation (Reproduced with permission from Elsevier [27])

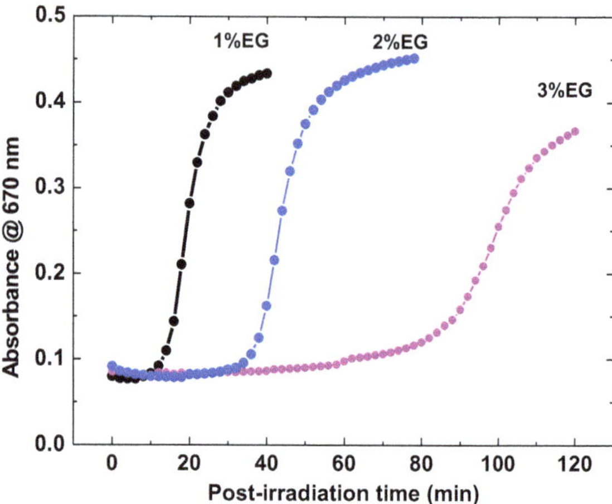

Fig. 19.7 Plots of absorbance at 670 nm versus post-irradiation time lapse for sols containing 10% (v/v) 2-PrOH and different % (v/v) of EG (Reproduced with permission from Elsevier [27])

From these observations it is understood that the nascent UO_2 coalesced in the presence of 2-PrOH in the aqueous solution in the above-mentioned pH range, to form UO_2 nanoparticles. Therefore, it is expected that such process should also be solvent viscosity dependent. Three reaction mixtures were used for this study: (i) 10 mM uranyl nitrate, 10% (v/v) 2-PrOH and 1% (v/v) ethylene glycol, (ii) 10 mM uranyl nitrate, 10% (v/v) 2-PrOH and 2% (v/v) ethylene glycol, (iii) 10 mM uranyl nitrate, 10% (v/v) 2-PrOH and 3% (v/v) ethylene glycol. Separate experiments were also carried out in 10% (v/v) EG as solvent. The total absorbed dose was also kept 50 kGy in all these experiments as in the previous cases. In this case, there was no formation of black colloidal sol after the electron beam irradiation, which confirms that EG alone does not favour the radiolytic reduction of UO_2^{2+} ions. It only provides a higher viscosity of the solution. The absorbance value at 670 nm was monitored with post-irradiation time lapse for all these three sets (shown in Fig. 19.7). There was an increase in the induction time with an increase in % (v/v) of EG. This clearly indicates that the nascent UO_2 gets difficulty to come closure to form UO_2 nanoparticles in case of a higher solvent viscosity. The formation times of these nanoparticles are found to increase with solvent viscosity up to 1.4 cP and follow a non-linear pattern.

These above-mentioned experiments have been carried out at room temperature. Experiments were also carried out at different temperatures, 25, 32.5 and 40 °C. The plots of absorbance at 670 nm versus post-irradiation time lapse at three different temperatures are shown in Fig. 19.8 for reaction mixtures (i) 10 mM uranyl nitrate and 3% 2-PrOH and (ii) 10 mM uranyl nitrate and 5% 2-PrOH. The total absorbed dose was also kept at 50 kGy in all these experiments as in the previous cases. There was a clear decrease in the induction time for the formation of UO_2 nanoparticles with an increase in the temperature from room temperature to 40 °C.

An increase in the temperature could lead to an increase in the kinetic energy of UO_2 species in the sol. This will certainly increase interactions among UO_2 and

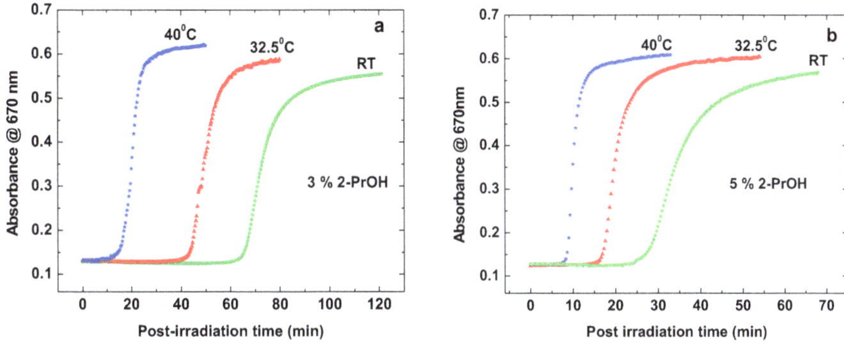

Fig. 19.8 Plots of absorbance at 670 nm versus post-irradiation time lapse at room temperature (RT), 32.5 and 40 °C maintained during irradiation as well as absorbance measurement period for the reaction mixture containing **a** 10 mM uranyl nitrate and 3% 2-PrOH and **b** 10 mM uranyl nitrate and 5% 2-PrOH (Reproduced with permission from Elsevier [27])

thereby a higher chance to form agglomerate and which finally give rise to the formation of these nanoparticles in the sol. After observing the information on the post-irradiation induction time from the above studies, experiments were also carried out with different uranyl nitrate concentrations in order to investigate the effect of precursor concentrations in the formation of these nanoparticles. The absorbed doses were kept different for the different concentrations of uranyl nitrate, for example 25, 50, 75 and 100 kGy for 5, 10, 15 and 20 mM uranyl nitrate, respectively. The % (v/v) of 2-PrOH was fixed at 3% in all these cases. The absorbance value at 670 nm was monitored at different time lapses at room temperature after the irradiation was over (Fig. 19.9).

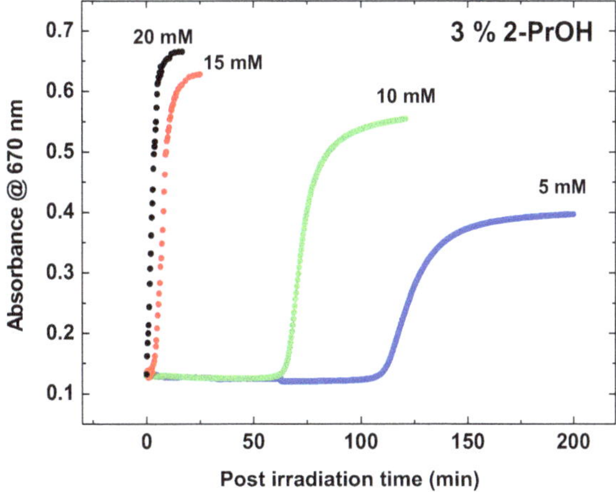

Fig. 19.9 Plots of absorbance values at 670 nm versu post-irradiation time lapse for the reaction mixtures containing 5, 10, 15 and 20 mM uranyl nitrate with fixed 3% (v/v) of 2-PrOH. The total absorbed doses in these cases were, 25, 50, 75 and 100 kGy, respectively (Reproduced with permission from Elsevier [27])

It is noticed from this figure that there is an increase in the induction time whereas a decrease in the saturation absorbance value with decrease in the precursor concentration from 20 to 5 mM. This indicates that the aggregation of nascent UO_2 species to form clusters and nanoparticles is higher in the case of a higher precursor concentration (e.g. 20 mM), than that in the case of a lower precursor concentration (5 mM). Because of a lower precursor concentration, the concentration of UO_2 species would be less and therefore the concentration of UO_2 nanoparticles would also be less and therefore the saturation absorbance value was lower in the case of 5 mM uranyl nitrate than that the case of 20 mM uranyl nitrate. From all these studies it is clearly understood that the radiolytic formation of UO_2 nanoparticles in aqueous solution always associated with a post-irradiation induction time, which could be judiciously tuned by various parameters like, solvent composition, viscosity, temperature and precursor concentrations.

19.3.2 Radiolytic Synthesis of CdSe Nanoparticles in Aqueous Solutions

Radiolytic synthesis of CdS nanoparticles [13, 14, 16, 70] has encouraged the researchers to synthesize similar related nanoparticles, like CdSe through the same process. CdSe nanoparticles are always favoured for their tremendous usefulness in various device applications [71–73]. Various synthetic methods have been reported for getting CdSe nanostructures in the recent past [55, 74, 75]. Different capping agents like TOP, TOPO and others are being used in these studies. In such methods, stringent laboratory conditions like high temperature, glove box and moisture-free atmosphere are required. Synthesis of CdSe nanoparticles has also been carried out in the polymeric host matrices under ambient laboratory conditions in order to avoid those stringent requirements [76, 77]. Radiolytic synthesis of such nanoparticles using cobalt-60 gamma radiation has been reported in the literature [78–81]. The radiolytic synthesis of CdSe quantum dots in PVA matrix using ammoniated cadmium sulphate ($CdSO_4$) and sodium selenosulphate (Na_2SeSO_3) as the starting precursor materials has also reported [82]. The polymer film, PVA containing CdSe quantum dots were very stable and exhibit room temperature excitonic absorption patterns. The reversibility effect on the formation of CdSe nanoparticles in aqueous solutions by high energy electron beam irradiation has been reported by Rath and co-workers [19] using the above-mentioned precursors. The as-grown CdSe nanoparticles in aqueous solutions were found to decompose upon exposure to air/oxygen and the decomposed solution again produced CdSe nanoparticles upon irradiation with electron beams and this cycle could be repeated several times. However, these nanoparticles were quite stable when they are dispersed in organic solvents.

In the radiation-induced synthesis of nanoparticles, usually a reducing condition is maintained in the reaction mixture. This is achieved by adding tert-butanol

($CH_3(CH_3)_2COH$) to the reaction mixture, which scavenge $OH^•/H^•$ radicals, and hence, e^-_{aq} remains behind in the solution. In this way, a reducing condition is maintained in the reaction mixture, where the reduction takes place through e^-_{aq}. In the case, the solution is saturated with N_2O gas, which is an electron scavenger, these hydrated electrons will undergo competing reactions between reagent ions and N_2O. Such a situation is being regarded as a partially reducing condition. Therefore, the yield of formation of the expected product will depend on the reactivity between these two reactions. Radiolysis of water has been mentioned in the Reaction 19.18 in the previous sub-section. Following reactions are important reactions pertinent to the above conditions [83].

$$OH^\bullet(H^\bullet) + CH_3(CH_3)_2COH \rightarrow {}^\bullet CH_2(CH_3)_2COH + H_2O\ (H_2) \tag{19.28}$$

$$e^-_{aq} + N_2O \overset{H_2O}{\rightarrow} N_2 + OH^- + OH^\bullet \left(k = 9.1 \times 10^9\ M^{-1}s^{-1}\right) \tag{19.29}$$

Cobalt-60 gamma radiation was used for irradiating in the reaction mixture (de-aerated aqueous solution containing freshly prepared 10 mM ammoniated $CdSO_4$, 10 mM Na_2SeSO_3 and 1 M $CH_3(CH_3)_2COH$). The reaction mixture was continuously given an absorbed dose of about 100 kGy, and finally, after the irradiation, an orange coloured colloidal sol was formed. The product thus obtained was used for further characterizations.

Blank experiment was performed by keeping the above-mentioned reaction mixture in a capped glass test tube for a long time without any irradiation. Absorption spectra of the reaction mixture were recorded (i) before irradiation, (ii) blank run, (iii) supernatant of the radiolyzed solution and (iv) the recovered product dispersed in methanol (Fig. 19.10).

No reaction was found to take place in the case of blank experiment in the above-mentioned reaction mixture containing low precursor concentrations. Because the formation of these nanoparticles takes place only in the case of

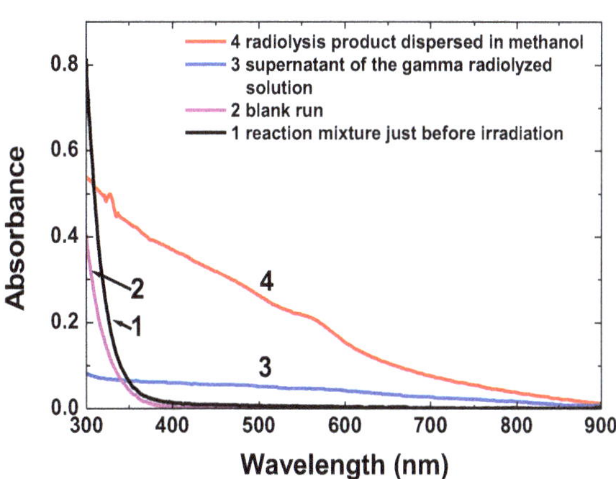

Fig. 19.10 Absorption spectra of the unirradiated and gamma irradiated solutions containing 10 mM ammoniated $CdSO_4$, 10 mM Na_2SeSO_3 and 1 M tert-butanol. Absorbed dose = 100 kGy (Reproduced with permission of Elsevier [18])

irradiation, so the radiation-induced synthesis route can be considered as a very powerful and efficient method. This method of synthesis has an advantage over other chemical routes as it does not require any additional reducing reagents and any stringent conditions. The supernatant of the radiolyzed solution was clear and transparent which does not show any absorption in the entire visible range indicating that the reaction is completed with the above-mentioned absorbed dose. Therefore, it is certainly clear that the formation of CdSe nanoparticles happens only via reactions initiated through the hydrated electrons upon radiolysis. The CdSe nanoparticles dispersed in methanol exhibit a broad excitonic absorption peak at about 580 nm. The band edge absorption beyond this peak position suggests that nanoparticles must have a broad size distribution as well as various trap states. This might be due to no capping agents present in the present reaction mixture. The absorption spectrum of CdSe nanoparticles obtained here also does not match with those reported in the literature through chemical routes using TOPO and TOP as capping agents. The particle size of the CdSe nanoparticles in this case, was found to be 3–5 nm as obtained from the absorption spectra [84, 85].

$$E_g = E_g(0) + a/d^2 \tag{19.30}$$

where, $\alpha = 3.7$ eV nm^2, $E_g(0) = 1.7$ eV, d = particle size (nm) and E_g = band gap value in eV. The molar absorption coefficient of CdSe nanoparticles at 560 nm in methanol was estimated to be 1–3 $\times$ 10^4 M^{-1} cm^{-1} in the present study.

The product obtained was characterized by XRD measurements. The broad XRD peaks are the signature of CdSe nanoparticles. The effective lattice strain and size of these nanoparticles have been determined by using the following equation:

$$\frac{\beta \cos\theta}{\lambda} = \frac{1}{\varepsilon} + \frac{\eta \sin\theta}{\lambda} \tag{19.31}$$

where β is the measured FWHM in radians, θ is the Bragg angle of the peak, λ is the X-ray wavelength (0.154 nm), ε is the particle size, η is the effective strain in the nanostructures [86]. The particle size calculated from the intercept of the linear fit of the data points from the plot of (β cosθ)/λ versus (sinθ)/λ was found to be 3–5 nm. The negative slope obtained here was due to a compressive strain in their lattice structures. Similar observations were also obtained in the case of radiation-induced ZnO nanoparticles [23]. This kind of negative lattice strains is also reported in the case of PbS nanocrystallites synthesized by chemical route [86].

TEM and selected area electron diffraction (SAED) patterns of CdSe nanoparticles are shown in Fig. 19.11a. The formation of flower-like structures of dimensions about 100 nm composed of primary nanoparticles of size about 3–5 nm is clearly evidenced from this TEM image. These results are in agreement with the particle size as obtained from XRD studies and also absorption studies. SAED patterns of CdSe nanoparticles confirm the formation of cubic CdSe nanoparticles in the present case.

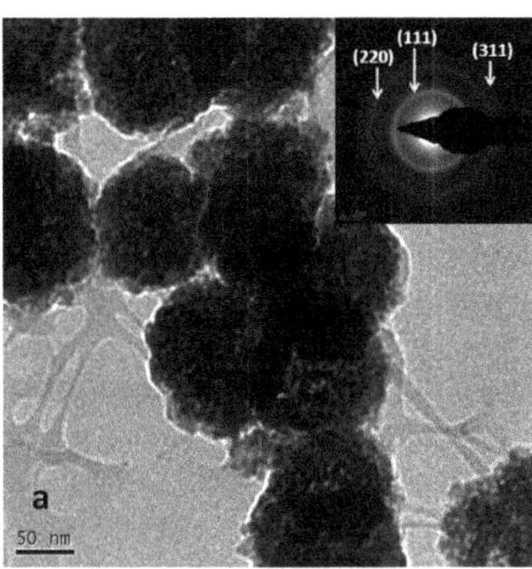

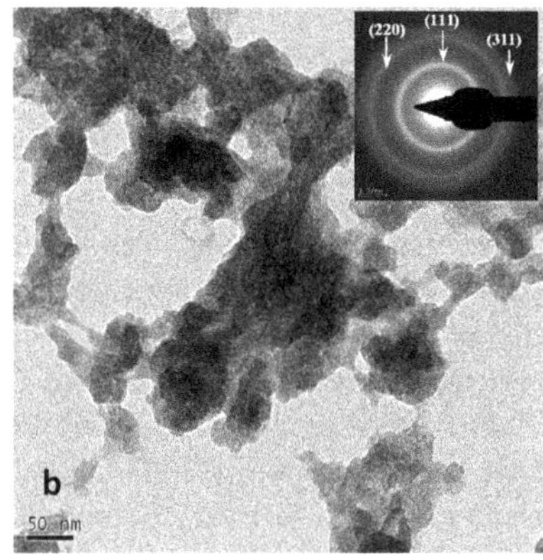

Fig. 19.11 TEM image of CdSe nanoparticles, obtained from **a** gamma irradiation and **b** electron beam irradiation of de-aerated aqueous solution containing 10 mM ammoniated $CdSO_4$, 10 mM Na_2SeSO_3 and 1 M tert-butanol. Insert shows SAED pattern of the nanoparticles (Reproduced with permission of Elsevier [18])

Above mentioned reaction mixture saturated with N_2O gas was also irradiated with gamma radiation with the same absorbed dose 100 kGy as in the previous studies. In this situation, the hydrated electrons react with N_2O to form oxidizing radicals $OH^{\bullet}$ (Reaction 19.29) and also with the individual precursors such as $[Cd(NH_3)_4]^{2+}$ and $[SeSO_3]^{2-}$ as follows [87].

$$e_{aq}^{-} + \left[Cd(NH_3)_4\right]^{2+} \rightarrow \left[Cd(NH_3)_4\right]^{+} \quad (k = 3.1 \times 10^{10} M^{-1} s^{-1}) \tag{19.32}$$

The reaction rate constant ($k = 2.27 \pm 0.06 \times 10^9\ M^{-1}\ s^{-1}$) between the e_{aq}^{-} and $[SeSO_3]^{2-}$ ions (Reaction 19.33) has been determined by monitoring the decay rate of e_{aq}^{-} in the present study. ${SeSO_3}^{2-}$ ions are analogous of ${S_2O_3}^{2-}$ ions. It is reported that S^- ions are released from ${S_2O_3}^{2-}$ ions upon radiolysis ($e_{aq}^{-} + [S_2O_3]^{2-} \rightarrow [SO_3]^{2-} + S^-$, $k = 1.5 \times 10^8\ M^{-1}\ s^{-1}$) [87]. Therefore, it is expected that a similar reaction could also occur in the case of ${SeSO_3}^{2-}$ ions [88].

$$e_{aq}^{-} + [SeSO_3]^{2-} \rightarrow [SO_3]^{2-} + Se^{-} \quad (k = 2.27 \times 10^{9} M^{-1} s^{-1}) \tag{19.33}$$

Cd^+ or $[Cd(NH_3)_4]^+$ ions can react with $[SeSO_3]^{2-}$ and Se^- ions to form nascent CdSe molecular species written as $(CdSe)_{intermediate}$ which finally undergo nucleation and growth to form CdSe nanoparticles (Reaction 19.34). $(CdSe)_{intermediate}$ mentioned in the Reaction 19.34 is a representation of various possible species, which finally lead to the formation of stable CdSe nanoparticles. Nevertheless, the possibility of various other reactions like radical-radical, radical-ionic species involving Cd^{2+}, Cd^+, $[Cd(NH_3)_4]^+$, ${SeSO_3}^{2-}$, ${SeSO_3}^{-}$, and other ionic or molecular species cannot be ruled out here. It was observed that the formation of CdSe nanoparticles takes place only under a strong reducing condition, and when both the precursor ions are present in the reaction mixture. So the Reaction 19.34 could be mentioned for the formation of CdSe nanoparticles in the present study.

$$Cd^{+} + [SeSO_3]^{2-}/Se^{-} \rightarrow (CdSe)_{intermediate} \rightarrow (CdSe)_{nanoparticle} \quad (19.34)$$

This reaction mechanism was further supported by carrying out experiments in the above-mentioned reaction mixture saturated with N_2O gas. The yield of formation of CdSe was found to be reduced in this situation as compared to the above fully reducing condition, which is too expected as per the Reaction 19.29. Based on the reactivity values for Reactions 19.29, 19.32 and 19.33, by considering the saturated concentration of N_2O in aqueous solution to be 25 mM, and the reagent concentrations 5 mM each, it could be found out that the yield of CdSe nanoparticles under N_2O-saturated condition would be about 60% of that under a fully reducing environment with the same absorbed dose. There was a formation of CdSe nanoparticles under a strong reducing condition even in the case of 0.5 mM precursor concentrations. Due to very low precursor concentrations, there was no formation of CdSe nanoparticles under a N_2O saturated condition.

From these above gamma radiolysis studies, it is now certain that the formation of CdSe nanoparticles takes place only through the reactions of e_{aq}^{-} with the precursor ions. In another set of experiments, CdSe nanoparticles were also synthesized using the above-mentioned reaction mixture (10 mM ammoniated $CdSO_4$, 10 mM Na_2SeSO_3 and 1 M tert-butanol) via electron beam irradiation in a 7 MeV LINAC under similar reducing conditions. The as-obtained product was recovered from the sol and used for characterizations by XRD and TEM measurements. These results confirmed the formation of CdSe nanoparticles in this case with the primary size of about 2–3 nm. Relatively, smaller particle size obtained in this case as compared to that in the case of gamma irradiation could be due to a higher dose rate is the case of electron beam irradiation [9, 13, 16]. In the case of a gamma irradiation, a slower reduction rate of the precursor ions, slower interaction of nascent Cd^{+} and Se^{-} ions with excess precursors probably produces larger complexes leading to the growth of larger particles. On the contrary, in the case of electron beam irradiation, the reduction of precursor ions is faster than their interaction and association leading to the growth smaller particles. Further, the agglomerates obtained in the case of electron beam irradiation are smaller (<100 nm) as compared to those ($\sim$100 nm) in the case of gamma irradiation. Due to a higher dose rate in the electron beam irradiation, the agglomerates cannot be compared to those obtained in the case of gamma irradiation. So the agglomerates in the case of gamma irradiation look to be uniform as seen in the TEM images, whereas those obtained in the case of electron beam irradiation are non-uniform in nature as seen from the TEM images.

The *dc* magnetization (M) of these CdSe nanoparticles as a function of the magnetic field (H) was carried out at room temperature (Fig. 19.12). It is seen from this result that these nanoparticles exhibit symmetric hysteresis loops which is a signature of ferromagnetic materials. The saturation magnetization value M*s* was

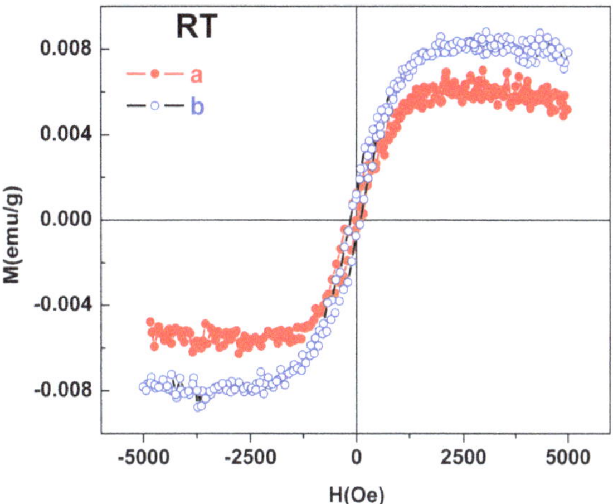

Fig. 19.12 Plot of magnetization (M) versu magnetic field (H) for CdSe nanoparticles obtained from **a** gamma irradiation and **b** electron beam irradiation measured at room temperature (Reproduced with permission of Elsevier [18])

found to be 0.006 emu/g for CdSe nanoparticles obtained via gamma irradiation whereas 0.0085 emu/g for those obtained in the case of electron beam irradiation. The room temperature ferromagnetism (RTFM) behaviour of these nanoparticles was further confirmed from the experiments on the magnetization versus temperature. The difference in the saturation magnetization values and increased magnetic moment could be due to the decreased particle size in the case of CdSe nanoparticles obtained by electron beam irradiation as compared to the nanoparticles obtained by gamma irradiation [89].

Such kind of magnetic properties of undoped semiconductor nanoparticles might come from the unsaturation of the surface atoms and the defects [89, 90]. The surface layer of the CdSe nanoparticles in this study might contain Cd^{2+} and SeO_2 which leads to the existence of surface defect states and surface unsaturation in these nanoparticles [91, 92]. The RTFM behaviour for the doped and capped CdSe quantum dots was explained on the basis of the electronic effects at the crystal surfaces due to the presence of the capping agents in the literature [93, 94]. About 70% Cd and 30% Se lie on the surface of TOPO capped CdSe nanocrystals prepared by the high temperature pyrolysis method. The presence of the excess Cd has been assigned to a superior capping property of TOPO towards the surface Cd [95]. The metal to ligand charge transfer in the surface of CdSe nanoparticles creates a vacancy in the d-orbital of Cd, which is the origin of the RTFM in these nanoparticles [93].

The RTFM behaviour in these nanoparticles is attributed to a comparable value with that of the reported M*s* values for different doped and capped CdSe nanoparticles (cf. Table 19.3). From this study it is confirmed that the RTFM behaviour of the CdSe nanoparticles obtained through radiation chemical routes is stronger than the thiol and TOPO capped CdSe nanoparticles. However, it is comparable with copper (1.8%) doped CdSe nanoparticles. The higher value of the RTFM behaviour in the case of capped and uncapped particles arises due to the

Table 19.3 The saturation magnetization, M*s* values in different CdSe nanoparticles obtained at room temperature from the *dc* magnetization (*M*) measurements as a function of the magnetic field (*H*) (Reproduced with permission of Elsevier [18])

CdSe nanoparticle	M*s* (emu/g)	Particle size (nm)	Remarks
1.8% Cu-doped	+0.0095 and −0.0095	2–7	[93, 94]
Thiol capped	+0.0035 and −0.0035	2–7	[93, 94]
TOPO capped	+0.005 and −0.005	2–7	[93]
Bare and undoped (gamma irradiation)	+0.006 and −0.006	3–5	[18]
Bare and undoped (electron beam irradiation)	+0.008 and −0.008	2–3	[18]

smaller the size of the nanoparticles; whereas the dopant leads to an increase in the RTFM behaviour in the case of doped nanoparticles. It was found that the CdSe nanoparticles obtained via the electron beam irradiation exhibited about 30% higher RTFM behaviour as compared to those obtained by gamma irradiation. Such observation could be explained on the basis of their smaller size and higher disordered structures as compared to those obtained in the case of gamma irradiation.

19.3.3 Radiation Chemical Synthesis of Metal Nanoparticles

Radiation chemical synthesis of various metallic nanoparticles has been reported in the literature. Different noble metals like silver (Ag) and gold (Au), other metals like nickel (Ni) and copper (Cu) have been reported by various authors [3–12]. In these studies, the reducing radicals such as e_{aq}^-, different alcohol radicals (e.g. $(CH_3)_2C^{\bullet}OH$)), and $CO_2^{\bullet-}$ are being used for the reduction of metal ions to the corresponding metal atoms. Subsequently, these metal ions get agglomerated to form nanoparticles of different shapes and sizes depending on the nature of solvents and capping agents, etc. Belloni et al. have reported a review of articles on the dose and dose rate effect on the radiolytic synthesis of metal nanoparticles in aqueous solution [9, 16].

Photochemical and radiation chemical synthesis of various nanoparticles, metallic, semiconductor and metal oxides are complimentary to each other. Both the routes are considered as green chemistry methods. In both cases, similar products can be obtained with ease, which makes these methods so important. Many more unexplored syntheses could be explored in coming years with the use of proper experimental conditions.

19.4 Limitations of Photochemical and Radiation Chemical Synthesis

The synthesis of various types of nanomaterials including noble metals, metal oxides and chalcogenides could be carried out through the photochemical as well as radiation chemical methods as discussed in the sections. However, it is very difficult to synthesize nanoparticles of metals either whose reduction potentials are too low (e.g. alkali and alkaline earth metals, lanthanides and actinides) or highly unstable in the zero oxidation state (e.g. transition metals excluding noble metals) in aqueous solution. This is because the reducing agents often used in such processes are either hydrated electrons, e_{aq}^{-} (E^1 = −2.8 V versus NHE) or 2-hydroxy-2-propyl radicals, $(CH_3)_2C^{\bullet}OH$ (E^1 = −2.1 V versus NHE) which are generated in aqueous solutions only. So a radiation-induced synthesis using such type of free radicals is not possible in non-aqueous solution, where the metals might be stable in their zero-valent state. Similarly, formation of an alloy of two or more metals would also be difficult to expect. Except for the metal chalcogenides and metal oxides of certain metals, there is hardly any literature available on the synthesis of similar types of compounds of other metal ions with different anions. This could also be due to either difficult to reduce those metal ions or the instability of the reduced metal ions in the aqueous solution. It is to be noted here that in the case of these two types of synthesis, the reduction reaction occurs in the step of one-electron reduction reaction only. Therefore, unless the stability as well as the reduction potentials of each oxidation state of metal ions is suitable, it would be extremely difficult rather not possible to synthesize the whole lot of metal nanoparticles as well as their compounds.

The photochemical and radiation chemical synthesis requires the desired radiation sources like UV or visible light sources and high energy radiation sources like cobalt-60 gamma chamber or electron accelerator, respectively. So all the setups need to be installed in specific laboratory. It is relatively easier to instal the light sources but extremely difficult to instal either cobalt-60 gamma chamber or electron beam accelerator. Such types of experiments could be carried out only where these facilities are installed and maintained properly.

19.5 Conclusions and Future Scope

Photo- and radiation-induced synthesis of different nanomaterials of various shapes and sizes are certainly very efficient and eco-friendly as compared to the conventional chemical routes. However, suitable light or radiation sources are required in order to carry out such synthesis. The size and shapes of the nanoparticles could be controlled by suitably adjusting (i) light intensity, (ii) irradiation time, (iii) concentration, (iv) absorbed dose, (v) solution pH, (vi) solvent composition, etc. One can even control the induction time associated with the formation of the nanoparticles. However, further research is necessary to understand the exact kinetics and dynamics of the formation of the nanomaterials formed through the reactions of free radicals with the precursors. Synthesis of metallic, intermetallic as well as metal oxide nanoparticles through such processes need to be investigated, which would be very much useful for the scientific community. Thin film deposition of various nanomaterials through such methods would be of high demand as far as device fabrication is concerned. Noble metal extraction from the solution phase using such methods will be another important field and requires to be explored in detail. It is expected that the synthesis of various nanoparticles could be possible by utilizing the direct Sunlight.

References

1. Guo Z, Zhou F, Hao J, Liu W (2005) J Am Chem Soc 127:15670–15671
2. Ensikat HJ, Kuru PD, Neinhuis C, Barthlott W (2011) Beilstein J Nanotechnol 2:152–161
3. Henglein A (1989) Chem Rev 89:1861–1873
4. Hayes D, Micic OI, Nenadovic MT, Swayambunathan V, Meisel D (1989) J Phys Chem 93:4603–4615
5. Swayambunathan V, Hayes D, Schmidt H, Liao YX, Meisel D (1990) J Am Chem Soc 112:3831–3837
6. Marignier JL, Belloni J, Delcourt MO, Chevalier JP (1985) Nature 317:344–345
7. Abedini A, Daud AR, Hamid MAA, Othman NK, Saion E (2013) Nanoscale Res Lett 8:474–483
8. Jushi SS, Pat SF, Iyer V, Mahumuni S (1998) Nano Stru Mater 10:1135–1144
9. Belloni J (2006) Catal Today 113:141–156
10. Joshi R, Mukherjee T (2007) Radiat Phys Chem 76:811–817
11. Kapoor S, Joshi R, Mukherjee T (2003) J Coll Int Sci 267:74–77
12. Kapoor S, Joshi R, Mukherjee T, Mittal JP (2001) Res Chem Intermed 27:747–754
13. Mostafavi M, Liu YP, Pernot P, Belloni J (2000) Radiat Phys Chem 59:49–59
14. Souici AH, Keghouche N, Delaire JA, Remita H, Mostafavi M (2006) Chem Phys Lett 422:25–29
15. Atinault E, De Waele V, Belloni J, Le Naour C, Fattahi M, Mostafavi M (2010) J Phys Chem A 114:2080–2085
16. Belloni J, Mostafavi M (2001) Studies in physical and theoretical chemistry 87, radiation chemistry: present status and future trends. In: Jonah CD, Rao BSM (eds) Elsevier, p 411
17. Singh S, Guleria A, Rath MC, Singh AK, Adhikari S, Sarkar SK (1815) Mater Lett 65:2011
18. Singh S, Rath MC, Singh AK, Mukherjee T, Jayakumar OD, Tyagi AK, Sarkar SK (2011) Radiat Phys Chem 80:736
19. Singh S, Rath MC, Singh AK, Sarkar SK, Mukherjee T (2010) Mat Chem Phys 124:6
20. Singh S, Guleria A, Rath MC, Singh AK, Adhikari S, Sarkar SK (2013) J Nanosci Nanotech 13:5365
21. Singh S, Guleria A, Singh AK, Rath MC, Adhikari S, Sarkar SK (2013) J Coll Inter Sci 398:112
22. Singh S, Rath MC, Sarkar SK (2011) J Phys Chem A 115:13251
23. Rath MC, Sunitha Y, Ghosh HN, Sarkar SK, Mukherjee T (2008) Radiat Phys Chem 78:77–80
24. Roth O, Hasselberg H, Jonsson M (2009) J Nucl Mater 383:231
25. Nenoff TM, Jacobs BW, Robinson DB, Provencio PP, Huang J, Ferreira S, Hanson DJ (2011) Chem Mater 23:5185

26. Rath MC, Naik DB, Sarkar SK (2013) J Nucl Mater 438:26
27. Rath MC, Naik DB (2014) J Nucl Mater 454:54–59
28. Rath MC, Keny SJ, Naik DB (2016) Radiat Phys Chem 126:85–89
29. Scaiano JC, Billone P, Gonzalez CM, Maretti L, Luisa Marin M, McGilvray KL, Yuan N (2009) Pure Appl Chem 81:635–647
30. Marin ML, McGilvray KL, Juan CS (2008) J Am Chem Soc 130:16572–16584
31. Gu H, Yang Y, Tian J, Shi G (2013) ACS Appl Mater Interfaces 5:6762–6768
32. Dong S-A, Zhou S-P (2007) Mat Sci Eng B 140:153–159
33. Kapoor S (1998) Langmuir 14:1021–1025
34. Kapoor S, Palit DK, Mukherjee T (2002) Chem Phys Lett 355:383–387
35. Kobayashi R, Sato N, Ichimura M, Arai E (2003) J Optoele Adv Mat 5:893–898
36. Ichimura M, Kobayashi R, Miyawaki T (2004) Jpn J Appl Phys 43:L1196
37. Podder J, Kobayashi R, Ichimura M (2005) Thin Solid Films 472:71–75
38. Kumar VN, Suriakarthick R, Shyju TS, Gopalakrishnan R (2013) AIP Conf Proc 1536:347–348
39. Singh A, Kunwar A, Rath MC (2018) J Nanosci Nanotechnol 18:3419–3426
40. Rath MC, Keny S, Naik DB (2016) J Nanosci Nanotech 16:9575–9582
41. Su C-Y, Lan W-J, Chu C-Y, Liu X-J, Kao W-Y, Chen C-H (2016) Electrochim Acta 190:588–595
42. Dispenza C, Grimaldi N, Sabatino MA, Soroka IL, Jonsson M (2015) J Nanosci Nanotechnol 15:3445–3467
43. Janov J, Alfredson PG, Vilkaitis VK (1972) J Nucl Mater 44:161
44. Rabinowitch E, Belford RL (1964) Spectroscopy and photochemistry of uranyl compounds. The Macmillan Company, New York
45. Tsushima S (2009) Inorg Chem 48:4856
46. Jingxin H, Xianye Z, Yunfu D, Zhihong Z, Honggui X (1986) J Less-Comm Metals 122:287
47. DePoorter GL, Rofer-DePoorter CK (1895) J Inorg Nucl Chem 40:1978
48. Park YY, Harada M, Tomiyasu H, Ikeda Y (1991) J Nucl Sci Technol 28:418
49. Salomone VN, Meichtry JM, Schinelli G, Leyvac AG, Litter MI (2014) J Photochem Photobiol A Chem 277:19
50. Pavelkova T, Cuba V, Šebesta F (2013) J Nucl Mater 442:29
51. Bera D, Qian L, Tseng T, Holloway PH (2010) Materials 3:2266–2345
52. Sakamoto M, Fujistuka M, Majima T (2009) J Photochem Photobiol C Photochem Rev 10:33–56
53. Sapsford KE, Algar WR, Berti L, Gemmill KB, Casey BJ, Oh E, Stewart MH, Medintz IL (2013) Chem Rev 113:1904–2074
54. Jing L, Kershaw SV, Li Y, Huang X, Li Y, Rogach AL, Gao M (2016) Chem Rev 116:1063–1073
55. Murray CB, Norris DJ, Bawendi MG (1993) J Am Chem Soc 115:8706–8715
56. Rogach AL, Kornowski A, Gao M, Eychmuller A, Weller H (1999) J Phys Chem B 103:3065–3069
57. Kalasad MN, Rabinal MK, Mulimani BG (2009) Langmuir 25:1272–1273
58. Singh A, Guleria A, Neogy S, Rath MC (article in press) Arab J Chem. https://doi.org/10.1016/j.arabjc.2018.09.006
59. Wardman P (1989) J Phys Chem Ref Data 18:1637–1755
60. Javier A, Magana D, Jennings T, Strouse GF (2003) Appl Phys Lett 83:1423–1425
61. Spoehr HA, Milner HW (1936) J Biol Chem 116:493–502
62. Dzhagan VM, Valakh MY, Raevskaya AE, Stroyuk AL, Kuchmiy SY, Zahn DRT (2008) Nanotechnology 19:305707–305712
63. Kizil R, Irudayaraj J, Seetharaman K (2002) J Agric Food Chem 50:3912–3918
64. Kamat PV (2002) J Phys Chem B 106:7729–7744
65. Kapoor S, Mukherjee T (2003) Chem Phys Lett 370:83–87
66. Zewde B, Ambave A, Stubbs J III, Raghavan D (2016) J S M Nanotechnol Nanomed 4:1043
67. Kim F, Song JH, Yang P (2002) J Am Chem Soc 124:14316–14317

68. Le Caër S (2011) Water 3:235–253
69. Spinks JWT, Woods RJ (1976) An introduction radiation chemistry, 3rd edn. John Wiley & Sons, New York
70. Kharisov BI, Kharissova OV, Mendez UO (eds) (2013) Radiation synthesis of materials and compounds. CRC Press, Boca Raton
71. Ashkenasy G, Cahen D, Cohen R, Shanzer A, Vilan A (2002) Acc Chem Res 35:121–128
72. Oertel DC, Bawendi MG, Arango AC, Bulovic V (2005) Appl Phys Lett 87:213505–213507
73. Kamat PV (2008) J Phys Chem C 112:18737–18753
74. Peng ZA, Peng X (2001) J Am Chem Soc 123:1389–1395
75. Qu L, Peng ZA, Peng X (2001) Nano Lett 1:333–337
76. Khanna PK, Singh N, Charan S, Lonkar SP, Reddy AS, Patil Y, Viswanath AK (2006) Mat Chem Phys 97:288–294
77. Dayal S, Kopidakis N, Olson DC, Ginley DS, Rumbles G (2009) J Am Chem Soc 131:17726-1–17726-7
78. Xie Y, Qiao Z, Chen M, Liu X, Qian Y (1999) Adv Mater 11:1512–1515
79. Hu Y, Chen W, Chen J, Zhang S (2003) Mat Lett 57:3137–3139
80. Ge X, Ni Y, Liu H, Ye Q, Zhang Z (2001) Mat Res Bull 36:1609–1613
81. Qiao Z, Xie Y, Huang J, Zhu Y, Qian Y (2000) Radiat Phys Chem 58:287–292
82. Biswal J, Singh S, Rath MC, Ramnani SP, Sarkar SK, Sabharwal S (2010) Int J Nanotechnol 7:1013–1026
83. Janata E, Kelm M, Ershov BG (2002) Radiat Phys Chem 63:157–160
84. Simurda M, Nemec P, Preclikova J, Trojanek F, Miyoshi T, Kasatani K, Maly P (2006) Thin Solid Films 503:64–68
85. Lippens PE, Lannoo M (1989) Phys Rev B 39:10935–10942
86. Qadri SB, Yang JP, Skelton EF, Ratna BR (1997) Appl Phys Lett 70:1020–1021
87. Buxton GV, Greenstock CL, Helman WP, Ross AB (1988) J Phys Chem Ref Data 17:513–886
88. Guha SN, Moorthy PN, Kishore K, Naik DB, Rao KN (1987) Proc Ind Acad Sci (Chem Sci) 99:261–268
89. Arbuzova TI, Naumov SV, Samokhvalov AA, Gizhevskii BA, Arbuzov VI, Shal'nov KV (2001) Phys Solid State 43:878–883
90. Sundaresan A, Bhargavi R, Rangarajan N, Siddesh U, Rao CNR (2006) Phys Rev B.74:1613061–1613064
91. Bowen KJE, Colvin VL, Alivisatos AP (1994) J Phys Chem 98:4109–4117
92. Henglein A (1988) Top Curr Chem 143:113–180
93. Seehra MS, Dutta P, Neeleshwar S, Chen YY, Chen CL, Chou SW, Chen CC, Dong CL, Chang CL (2008) Adv Mater 20:1656–1660
94. Singh SB, Limaye MV, Date SK, Kulkarni SK (2008) Chem Phys Lett 464:208–210
95. Taylor J, Kippeny T, Rosenthal SJ (2001) J Cluster Sci 12:571–582

Chapter 20
Mechanochemistry: Synthesis that Uses Force

Dipa Dutta Pathak and V. Grover

Abstract Grinding is a basic physical process, and the grinding tools "mortar and pestle" have been in use since times immemorial. It has been practiced in almost all spheres of human life from kitchen to laboratories as well as in large industrial processes. Chemical synthesis by applying force or the "mechanochemistry" has been employed as a synthetic procedure for a long time but now the need to adopt "greener", cost-effective and less harmful methods of synthesis has brought back the mechanochemistry to forefront in last decade. It has emerged as the one of the most efficient, advantageous and environmentally benign alternatives to traditional synthesis routes for the preparation of nanomaterials for advanced applications. The features such as ease of operation, simplicity of equipment, high reproducibility, relatively mild reaction conditions and the solvent-free condition (in case of dry milling) have made it the synthesis technique of choice for the synthetic chemist. It is used for synthesizing a wide variety of both single-phasic and composite materials varying from inorganic solids (oxides and non oxides), organic compounds, polymers, metal complexes, metal–organic frameworks. Materials with applications in varied areas such as hydrogen storage materials, energy applications, pharmaceuticals, as well as advanced nanocatalysts have been synthesized using this method. In recent times, the dry grinding or milling has been further modified by addition of a small amount of solvent or polymer, also called liquid-assisted grinding or polymer-assisted grinding that yields different products, speeds up the reaction and also ensures better usage of reactants. The fact that mechanical force or shear is the driving force for the reaction, and it also presents a novel way to obtain hitherto unknown (and interesting) products. The chapter discusses the basics of mechanochemical synthesis along with the above-mentioned points in the details.

D. D. Pathak · V. Grover (✉)
Chemistry Division, Bhabha Atomic Research Centre, Mumbai 400085, India
e-mail: vinita@barc.gov.in

V. Grover
Homi Bhabha National Institute, Mumbai 400094, India

A. K. Tyagi and R. S. Ningthoujam (eds.), *Handbook on Synthesis Strategies for Advanced Materials*, Indian Institute of Metals Series,
https://doi.org/10.1007/978-981-16-1807-9_20

Keywords Ball milling · Grinding · Green synthesis · Solvent-free · Sustainability · Liquid-assisted grinding · Ball-to-powder ratio · Oxides · Chalcogenides · Porous materials

20.1 Introduction

In today's world when "sustainability" is the buzzword in each and every sphere related to the mankind, how could science remain behind? All the technological and scientific advances are directly or indirectly related to the materials synthesis. The urge to make the synthetic procedures more environment-friendly, less wasteful and hence more acceptable and sustainable, synthetic chemists have been on the lookout for better synthesis techniques. This has led to a lot of emphasis on synthesis routes that conform to the principles of green chemistry [1–3]. The "mechanochemistry" has been one of the forefront runners among the green routes of synthesis. In fact, mechanochemistry has been recently acknowledged by IUPAC as one of the top ten emerging technologies in chemistry that have the potential to answer the increasing demand for clean and sustainable processes.

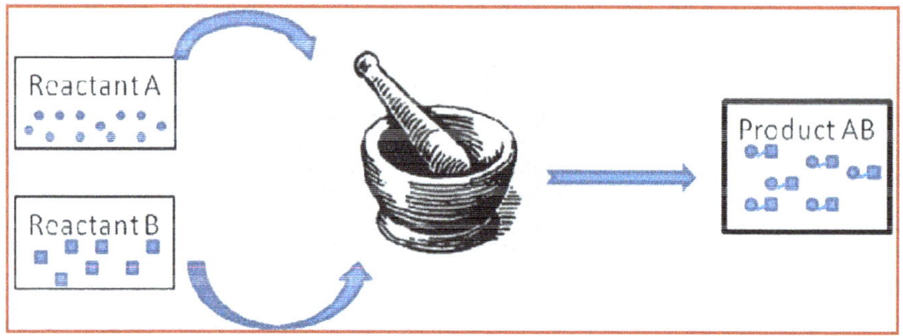

Schematic 1: Schematic of a mechanochemical reaction employing mortar and pestle

20.1.1 *What Is Mechanochemistry?*

As the name indicates, mechanochemistry means the chemistry or the chemical changes brought about by mechanical force that could be compression, shear or friction. In this route of synthesis, application of mechanical energy, by means of grinding or high-energy ball milling is exploited to prepare useful materials and also to improve the efficiency of complex synthesis processes. It is known that mechanical force can cause wear and tear of the material; similarly at molecular level, it causes weakening of chemical bonds, creates defects, destabilizes them and makes them more reactive [4]. The reaction caused by heat, light or electric potential and the corresponding fields are termed as photochemistry, thermochemistry and electrochemistry. In the similar way, the reactions brought about by mechanical stress are classified under the name mechanochemistry. The mechanical actions yield smaller particle sizes, creates active sites, generate fresh active surfaces, increase the surface area, and all these aid in promoting the chemical synthesis.

The advantages of this unique class of synthesis technique are manifold.

1. The first one is the fact that it is a solvent-free (or involves very little solvent) preparative route. In any chemical reaction, the solvent or the dissolving media makes up most of the mass of the reaction system, so the removal of solvent from the synthesis protocol, which is many times hazardous, makes it automatically "greener". In addition, in the cases where the reactions are additive types, the reactants combine to give the final product and hence the work-up of the reaction to remove the by-products can be avoided. Also, some of the reactants and/or products are sensitive to the solvents, and hence, synthesis by this route provides an excellent pathway to access them.
2. Generally, the synthetic chemical reactions are carried out in the homogeneous medium and it requires that at least one of the reactants must be soluble in the solvent (or dissolving media). However, in mechanochemical synthesis, the solubility of the reactants is not a necessary criterion, and hence, this has opened up a world of materials that can now be used as the starting materials for reactions such as minerals, rocks, metals and cellulose. For example, Hammere et al. [5] carried out the solvent-free enzymatic cleavage of cellulose (which is a difficult step in generation of bioethanol from biomass) by employing mechanochemistry and could demonstrate glucose concentrations much higher than that observed by conventional methods.
3. It is an energy-saving and a low-temperature route. Since the synthesis route does not involve high temperature, there is a better probability of obtaining metastable products or the products which are generally not envisaged by the synthesis routes employing high temperatures. The possibility of different reaction pathways in a reaction when carried out mechanochemically, as compared to when it is carried out by other stimuli such as heat, temperature and electricity, gives access to the products that are hitherto unachievable by other routes.
4. Improved selectivity and quantitative yields are other advantages of this technique.

20.2 Equipments Used: Tools of the Trade

Mechanochemical synthesis essentially requires a grinding device that put shear stress and friction on the reactants to bring about the desired chemical/physical change. Traditionally, mortar and pestle have been used as the grinding equipment

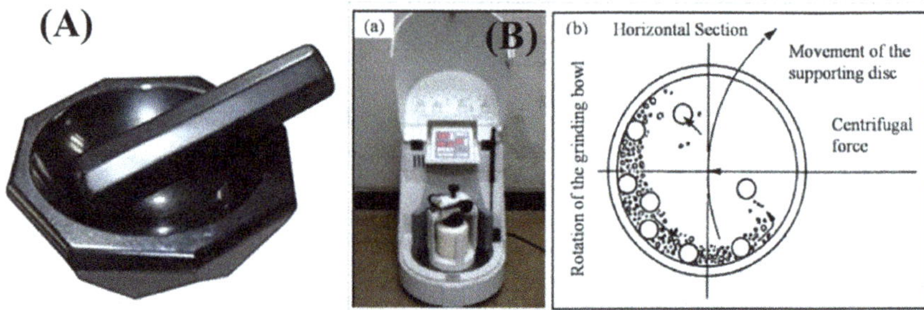

Fig. 20.1 Typical (**A**) mortar–pestle and (**B**) planetary ball mill (*a*) the schematic of motion of the balls inside ball mill (*b*). (**B**) is reproduced from "Coatings 2015, 5(3), 425-464; https://doi.org/10.3390/coatings5030425" (Ref. 8b)

to carry out simple mechanochemical reactions. Since the grinding by mortar and pestle involves manual intervention, it introduces many uncertainties and variable factors such as the experimenter, i.e. the person who is carrying out the synthesis, the force applied by the person and other factors prevailing during the course of synthesis. Hence, to overcome these drawbacks, the automated grinding/milling instruments are being employed, also known as ball mills. Figure 20.1 shows a typical mortar–pestle (agate) and a ball mill used for grinding/milling. These ball mills employ close jars or vessels with ball bearings that act as grinding media. In these ball mills, the reactants can be ground together for longer durations of time and without minimum interference from surrounding environment [6]. The ball mills are available in many different sizes and designs. In fact, they have been used since long for many non-synthetic purposes as well such as mixing, blending and mechanical alloying. Similarly, attrition mills are also available that can mechanically reduce the solid particle size. Ball mills are primarily of two types: (1) shakers [7] and (2) planetary ball mills [8]. In shakers, the reactant powders are put in the closed vessels, loaded with ball bearings and shaken at a pre-determined or desired frequency to grind the reactants together, whereas in planetary ball mills, the closed jars spin at the desired frequency in the counter-direction to that of the spinning disk mounted on them. Effectively the term planetary comes from the rotation of the jars around that of the spinning wheel, just like the motion of planets around the Sun. In addition, *twin–screw extruders* have also been employed for carrying out the mechanochemical reactions. This device consists of two long and wide screws that rotate in opposite directions placed inside an extrusion barrel. The samples are transported through the extrusion barrel, and during the passage, they are vigorously ground or kneaded between the screws and the reaction is carried out [8a]. Extruders also put a compressive and shear stress on the molecules during the process. While ball mills can be utilized for the batch synthesis, extruders can be employed for continuous synthesis. Many organic reactions such as Knoevenagal synthesis can be carried out by employing extruders. Crawford et al. [9] have carried out several organic condensation reactions using twin extruders such as

Knoevenagel condensation, imine formation, aldol reaction and Michael addition with an added advantage that there was no need to purify the end products, not to mention that the synthesis procedure is continuous and solvent-free. It must be mentioned that like ball mills, even the twin extruders are being widely used in industries for non-synthetic and in many cases for the non-scientific purposes. Some examples are making cereals, crisps and pizza dough and also for processing plastics. Twin extruders are also used in the pharmaceutical industry for consistently mixing ingredients for multiple formulation [2].

20.3 History of Mechanochemical Synthesis

The basic tools of mechanochemistry, i.e. mortar and pestle, are known to exist since stone age, and hence, it can be regarded as one of the earliest engineering technologies [10]. The fact that grinding action of mortar and pestle would also be associated with some chemical change as well gives indication that mechanochemistry indeed has its origins in pre-historic times. Takacs [11] has written a very lucid review on historical relevance and evolution of mechanochemical synthesis which is strongly recommended to the reader. Earliest documented reference of mechanochemistry is from 315 BC wherein the use of copper mortar and copper pestle is mentioned to obtain mercury from cinnabar. Though the term mechanochemistry is not mentioned, it is definitely an example of chemical change brought about by the mechanical force [12]. Also, even though there is no explicit documentation of this synthesis technique, the mortar and pestle were in regular use in the laboratory of early chemists and alchemists. Then in 1820, a paper by Faraday documents the reduction of silver chloride performed by grinding with tin, iron, copper, zinc [13]. Takacs [11] mentions that Faraday's choice of words along with others such Johnston and Adam in [14, 15] hint towards the existing common knowledge of chemical changes brought about by grinding even though there are no references probably because of lack of documentation. Technically W. Spring (Belgium) and M. Carey Lea (Pennsylvania) are known as the pioneers in the branch of chemistry that deals with chemical changes brought about by mechanical force. Spring had reported the combination reactions of several metals and sulphur/arsenic in 1883 [16]. This was unarguably the first reported large-scale systematic investigation of the chemical changes/processes brought about by mechanical action [11]. On the other hand, Lea [17] could demonstrate that the mechanical action can induce chemical change and also those chemical changes could be different from the changes brought about by effect of heat. Mechanochemistry, however, is believed to have become a separate accepted branch of chemistry when Ostwald included mechanochemistry in his chemical systematic in 1919, along with thermochemistry, electrochemistry and photochemistry [18]. An early work on solvent-free mechanochemical reaction was reported by Ling and Baker in [19]. Also, the research on mechanochemical reactions of organic polymers, such as cellulose, was done in 1920s [20]. However,

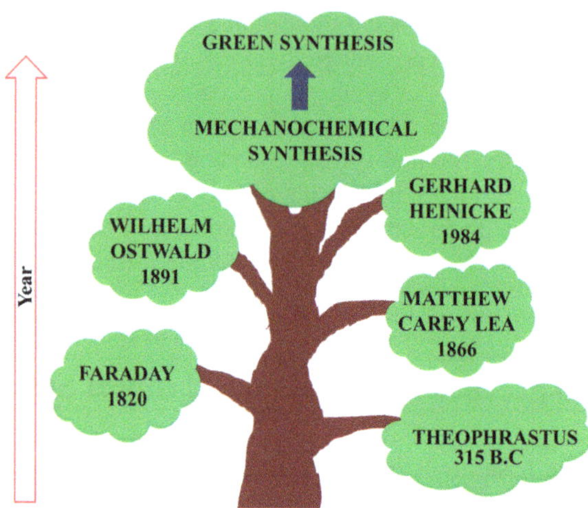

Fig. 20.2 Timeline of the evolution of mechanochemistry

it must be noted that mechanosynthesis had traditionally been employed largely for insoluble inorganic materials (alloys and metal oxides) and in the cases when there no solvent-based alternative was available [21]. In 1984, almost a century later, the presently accepted definition of mechanochemistry was proposed by Gerhard Heinicke as "branch of chemistry concerned with chemical and physical changes of solids induced by the action of mechanical influence" [22]. Molecular mechanochemistry was promoted in eighties [23–26], which could show that mechanochemistry is not just an alternative synthetic route; but in some cases, it can yield products unobtainable by other chemical routes. The organic, organometallic and supramolecular synthesis, their methodologies and products have gained popularity in last decade or so [27]. Figure 20.2 depicts the timeline of evolution of mechanochemistry as the synthesis procedure.

20.4 Effects of Mechanical Force on Materials?

The mechanical energy imparted on the material during the grinding process through ball milling can be significant and can bring about various changes in the reactants. The central occurrence in mechanical milling is the collisions among balls and powder. During milling, the powder particles are trapped between the balls which are undergoing constant collisions and are consequently subjected to deformation and fracture processes which define the structure of the powder. Figure 20.3 shows the effect of mechanochemical force on reactants and their conversion into products. To start with, high-energy ball milling is capable of providing sufficient mechanical energy to break the order of the crystalline and produce nanometre-scaled particles. It is in fact a very lucrative way to produce nanomaterials [28, 29]. These nano-sized seeds, produced by mechanical shear, can further assemble into larger clusters and

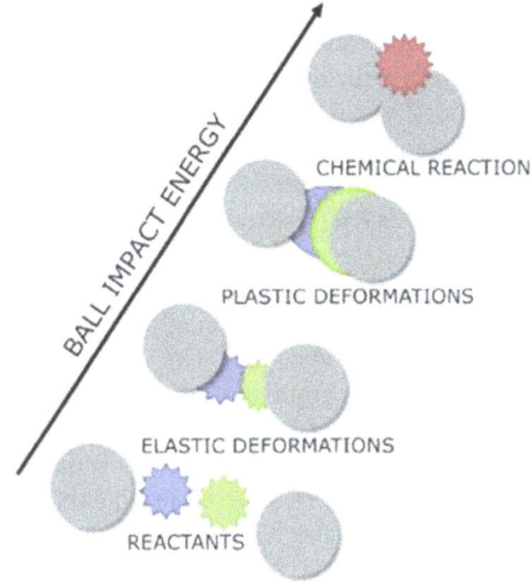

Fig. 20.3 Mechanochemical reactions: From reactants to products (Reproduced from Cova, C.M., Luque, R. Advances in mechanochemical processes for biomass valorization. BMC Chem Eng 1, 16 (2019). https://doi.org/10.1186/s42480-019-0015-7)

provide an effective channel for nanostructures growth [30]. The two major effects that caused by ball milling are fracture and welding [31]. As the reactants are continuously under the strong mechanical impact generated by collisions between the balls and balls and walls of container, it creates a large amount of both structural and microstructural defects (Fig. 20.3). The progressive accumulation of these defects leads to particle size reduction or also to chemical reaction. It must be noted that creation of this large pool of defects in the milled material creates abundant possibilities of obtaining various metastable states with controlled defects and functionalities either in the milled solid or by additional processing treatments. This may also lead to other effects such as disordering of crystalline lattice, amorphisation and polymorphic phase transitions. Whereas the normal solid-state reactions proceed by diffusion which is actually the rate determining step during the solid-state synthesis, in case of mechanochemical synthesis the rate depends on various parameters such as balls to powder weight ratio, milling speed and time, and atmosphere employed during milling [32]. In addition, the nature of the obtained product also depends upon other factors such as chemical composition of the reactant powder mixtures and chemical nature of the grinding tools.

20.4.1 Phase Transformations Caused by Mechanochemical Force

An interesting case study is the conversion on Y_2O_3 into various possible polymorphs depending on the nature of the grinding tool. The stable polymorph of Y_2O_3 is C-type (bixbyite). It was found to convert into fluorite-type cubic structure on being milled with zirconia tools. However, it was found to be not due to Zr incorporation from grinding media. On using steel grinding media, conversion to monoclinic structure was observed [33–35]. Another technologically interesting material, ZrO_2, is known to exist in monoclinic modification at room temperature and is known to transform to tetragonal and cubic modifications at ~1200 °C and ~2400 °C, respectively. It is observed that grinding monoclinic zirconia yields a nanostructured material which is locally similar to monoclinic zirconia but exhibits cubic-type ordering at nano-metre scale [36]. The stabilization of the cubic structure has been ascribed to significant structural distortions and defects induced by the high-energy ball-milling process [36] or due to incorporation of aliovalent cations, e.g. Fe^{2+}/Fe^{3+} and Cr^{3+} into the sample (due to the wear and tear and oxidation of the grinding media) [37, 38]. Phase transformations are in general a combined effect of plastic and shear effect and the local temperature generated during the milling process. Most of the time the phases observed after milling is the high T/high P phases obtained by equilibrium conditions. It must be noted that this does not imply that high T/high P conditions are generated during the milling process [34, 38–41]. Many a times, it is the rearrangement of the basic structural unit. For example, the bixbyite structure of Y_2O_3 is the ordered arrangement of three oxygen deficient F-type unit cells. Upon milling, these units are randomised yielding F-type phase which is the observed high-temperature phase for Y_2O_3 (obtained by subjecting Y_2O_3 to high T) [42, 43].

Many examples are reported in the literature in context of the altering various physico-chemical properties of solids by ball milling. Li et al. [44] synthesized carbon nanoparticles by milling carbon nanotubes wherein they could observe onion—like carbon nanoparticles by milling for 15 min which converted to amorphous carbon after a milling time of 60 min. Giri et al. synthesized ZnO nanoparticles [29] in the size range of 7–35 nm by ball-milling technique. Similarly, various one-dimensional nanomaterials have been synthesized using high-energy ball-milling and annealing process [45–48]. It is interesting to note that many a times, the structure of the oxide nanopowders obtained after ball milling is different from that synthesized by other chemical methods [39, 49–54]. The oxide nanopowders obtained mechanochemically contain amorphous/disordered regions with nanocrystalline grains embedded in amorphous grain boundaries. Similarly, the defect creation by ball milling has also been exploited by various researchers to understand the defect controlled physico-chemical properties. For example, the mechanochemically treated vanadium phosphate catalysts were investigated for their catalytic behaviour as a function of the nature of milling media. The milling process affected the reactivity of the oxygen species linked to various oxidation states of vanadium (V^{5+} and V^{4+}), and consequently, an increase in the selectivity was observed [55].

20.5 Modifications of Mechanochemistry

In some cases, the mechanochemical synthesis is also aided by various additives which have led to development of modified mechanochemical techniques that help in augmenting reactivities. Some of them are briefly discussed as follows:

Liquid-assisted grinding (LAG): This modification involves use of small amount of liquid/solvent along with the solid reactants in the grinding media. It is known that very small amounts of added liquid can significantly accelerate and sometimes enable the mechanochemical reactions between solids. Usually, the molar equivalents of solvent are added, and in this case, the mechanochemical approach is termed as to be with "minimal solvent" rather than "solvent-free". The original term used to describe them was "solvent drop grinding" but now it is known as "liquid-assisted grinding". Liquid-assisted grinding is equivalent to the term "kneading" [56, 57]. The amount of liquid taken is defined by the parameter η, which is the ratio of the volume of the liquid to the weight of the reactant. The value of η can be used to characterize the liquid-assisted grinding reactions. Various parameters such as polarity of the liquid, choice of anion and the parameter η can be optimized to control liquid-assisted grinding reactions.

Another modification of LAG is **ion- and liquid-assisted grinding**. In this case, small amounts of salt (5 mol% or less) is added in addition to liquid to grinding jar containing solid reactants to activate systems that do not react or react only partially by LAG. This method has been used to quantitatively obtain pillared MOFs [58], zeolitic imidazolate frameworks (ZIFs) [59], API bismuth subsalicylate, etc., directly from metal oxides [60]. The salt additives enable selective templating of MOF polymorphs.

Polymer-assisted grinding (POLAG): Another approach to the neat (or dry) mechanochemical synthesis is utilization of polymer additives, i.e. polymer-assisted grinding (POLAG) [61]. As the name suggests, this particular technique employs polymers to carry out, facilitate or accelerate the desired mechanochemical transformations [2, 62]. The liquid-assisted grinding uses solvents in limited quantities, but the advantage with polymeric materials is that they are capable of exhibiting different structures on a wide range of length scales, which aids in obtaining materials of higher degree of complexity. The factors that may affect solid-state reactions are the heat released or absorbed, humidity (especially in reactions such as those involving dehydration). The polymeric molecules have varying capabilities to absorb and dissipate heat and humidity, and hence, they act differently as compared to smaller molecules. This has repercussions on the synthesis reaction and may control the nature of product formation. Scaramuzza et al. [63] investigated the dehydration of carbamazepine dihydrate which is an anticonvulsant drug and is known to possess five polymorphs, under the simultaneous effects of milling and polymeric excipients. It was shown that though milling alone did not cause any dehydration, but the presence of specific polymers could lead to partial or complete dehydration [63]. Interestingly, the polymer chain length was observed to be a major factor in controlling the kinetics of the solid-state reaction. A suitable combination of the amount of the polymer and the milling time could be tuned to isolate different polymorphic forms of dehydrated carbamazepine solid. Various in situ techniques have been utilized for monitoring the grinding-based synthetic reactions to have an insight into the mechanism [64]. The readers may refer to a very comprehensive article on how the potential of mechanochemistry is augmented by specific additives such as enzymes (mechano-enzymatic reactions) and light (photo-mechanochemistry) that are used to enhance, direct or enable the reactivity [65].

20.6 Examples Of Different Classes of Functional Compounds Synthesized by Mechanochemical Synthesis

As it is mentioned earlier that in mechanochemical synthesis, energy for the chemical reactions is provided by mechanical forces like shearing, compression, grinding, frictional and rotational. Often, this process allows for fast solid-phase reactions without heating at high temperature. In addition, transportation of energy occurs under solvent-free conditions (or with minimum amount of solvents), and hence, it significantly reduces the solvent consumption. Mechanochemical synthesis is undoubtedly an environmentally benign chemical process with minimum waste generated [65–69]. Some typical examples of various classes of compounds synthesized by mechanochemical synthesis are listed below.

20.6.1 Synthesis of Oxides

For oxides, high-energy ball-milling (HEBM) technique was first used to reduce grain size down to the nanometre range. Further, along with grain size refinement phase transformation was observed during dry ball milling in various oxides [70]. The phase transformations induced by HEBM in oxides can be polymorphic transformations, amorphization and disordering of oxides including effects such as surface amorphization and grain boundary creation, annihilation and disordering [70]. For example, in metastable polymorphs of alumina (viz γ-Al_2O_3,) grinding leads to the formation of stable α-Al_2O_3 [71]. Typically, such transformations are brought about by high pressure or high temperature. Another example of phase transformations induced by grinding is the α-PbO_2-type structure (often named TiO_2II) of titania [72]. Besides three crystalline forms anatase, brookite and rutile, this fourth high-pressure polymorph was prepared from crystalline phases by static high-temperature heating or subjecting to high pressure and shock wave. In addition, the nature of the grinding tool also dictates conversion of Y_2O_3 into various possible polymorphs, and details of such conversions are described in previous section [42].

Further, different types of phase pure nanocrystalline oxides have been synthesized by heterogeneous mechanochemical method. Wise selection of starting materials and milling conditions can be used to prepare a wide range of nanocrystalline oxide by mechanochemical reaction [73–75]. Ao et al. [74] used mixture of $ZnCl_2$ and Na_2CO_3 as reagents and NaCl as diluents to fabricate pure ZnO (21 nm). In another work, Yang and co-worker [75] prepared 13 nm cobalt oxide (Co_3O_4) nanoparticles via mechanochemical reaction of cobalt salt ($Co(NO_3)_2 \cdot 6H_2O$) with NH_4HCO_3 followed by thermal treatment. Typically for pure phase oxides synthesis, the mixture of the precursor was ball milled and subsequently subjected to calcinations. For mixed phase oxides, the process is similar, but mixture of different oxides is used as starting material. For example, powder mixture of BaO and TiO_2 was used for preparation of $BaTiO_3$ perovskite phase [76].

20.6.1.1 Oxide-Based Composites

Hu et al. [77] reported ternary SnO_2–graphite composites with transition metal (M) (M = Fe, Mn, Co) prepared by a simple ball-milling method and explored their potential towards anode materials for Li-ion batteries. To ensure the close contact between nanostructured SnO_2 and metal, first a mixture of SnO_2 and metal was milled for 15 h. They demonstrated the positive effects of interfacial intermetallic phase Sn_xM_y for suppressing long diffusion of Sn during cycling. To synthesize the ternary composite, as-prepared SnO_2–M was further milled with the graphite for 5 h.

20.6.1.2 Porous Oxide

To overcome the drawback of the template-assisted method for porous oxide preparation, which commonly involves wet conditions, soluble metal oxide precursors and long time for drying, a facile mechanochemical nanocasting method was demonstrated by Xiao et al. [78]. The ball-milling process could yield, a series of highly porous metal oxides (ZrO_2, Fe_2O_3, CeO, CuO_x–CeO_y, CuO_x–CoO_y–CeO_z) in much smaller time duration. For single-phase oxide synthesis, mixture of same amount of metal salt with commercial SiO_2 was ball milled for 60 min at a vibrational frequency of 30 Hz followed by etching with NaOH at room temperature. For multi-component oxide composite synthesis, initially the metal salts were milled for some time to homogenize them before milling with silica. Figure 20.4 depicts the schematics of the mechanochemical nanocasting route to obtain porous products.

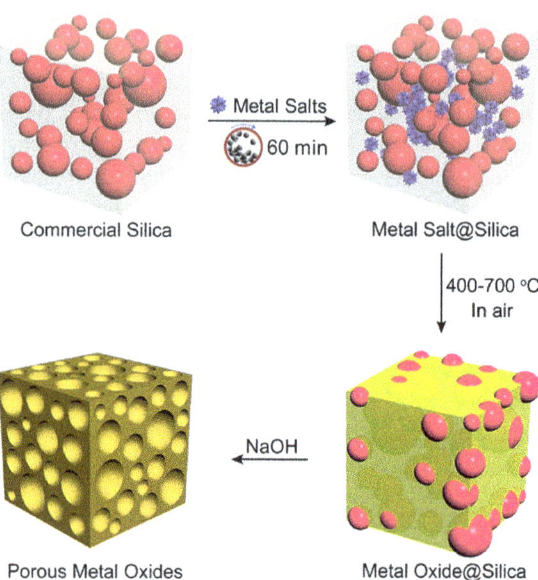

Fig. 20.4 Mechanochemical nanocasting route for the synthesis of porous metal oxides [78]. Reprinted with permission from "W. Xiao, S. Yang, P. Zhang, P. Li, P. Wu, M. Li, N. Chen, K. Jie, C. Huang, N. Zhang, S. Dai, Facile Synthesis of Highly Porous Metal Oxides by Mechanochemical Nanocasting, Chemistry of Materials, 30 (2018) 2924–2929" Copyright (2018) American Chemical Society

20.6.2 *Synthesis of Chalcogenides*

Chalcogenides, as advanced perspective materials, exhibit a great variety of applications like catalyst, Li-ion batteries, hydrogen evolution and storage, fluorescent labels in biomedical application and solar energy conversion [79–81]. Several metal sulphides can be directly synthesized by mechanochemical processing of elemental metal and sulphur [79, 80]. This type of dry mode synthesis requires nitrogen or argon atmosphere to prevent post-synthesis solid–gas reactions. Otherwise, the surface of the synthesized particles gets covered with different species like sulphates, hydroxysulphates, oxysulphates, thiosulphates and sulphites. Wet mode of mechanochemical synthesis process is advantageous for chalcogenide nanocrystals fabrication and is known as "acetate route". This mode of synthesis is favourable for smaller and monodispersed particles [79].

20.6.2.1 Pristine/Phase Pure Chalcogenides

Balaz and co-worker synthesized well-crystallized monodispersed nanoparticles of ZnS, CdS and PbS by the mechanochemical route using the corresponding acetates and Na_2S as starting material [80]. The by-products of the acetate route synthesis are water soluble so the as-prepared products needed extra washing. These ZnS nanocrystal have been explored as bio-markers [82]. Around the same time, $CuInS_2$ nanoparticles were prepared from elemental copper, indium and sulphur powders by high-energy milling in an inert atmosphere of argon [83].

20.6.2.2 Composites of Chalcogenides

In recent time, metal chalcogenide nanocomposites have attracted considerable research interest due to superior outcome of their numerous applications. [81, 84–86] Composite chalcogenides (MX/NS) (M = Cd; X = S, As, Se; N = Zn) were synthesized by two-step solid-state mechanochemical methods with two different approaches. (1) First a stoichiometric mixture of M-acetate and Na_2S was milled to prepare metal sulphides (MS). This was followed by addition of a stoichiometric mixture of N-acetate and Na_2S into the same milling pot containing MS, and the mixture was milled further to obtain NS, thus yielding MS/NS nanocomposites. (2) Instead of using acetate precursor, elemental M and S is used in the first step of this approach.

Using the first approach described in earlier section, Balaz and co-worker [84] fabricated CdS/ZnS composite chalcogenides and exploited the composite for its photocatalytic activity towards the degradation of methyl orange (MO). The same group also fabricated CdS/ZnS and CdSe@ZnS nanocomposites following the above-mentioned approaches and explored the materials as fluorescent labels in biomedical engineering [85, 86]. The fluorescent properties of chalcogenides chitosan-coated InAs/ZnS-based bio-marker composite have been explored for the labelling of Caco-2, HCT116, HeLa and MCF7 cancer cell lines by Bujňáková el al. [87]. Instead of acetate root, the author preferred the dry mode for their composite chalcogenides preparation. Another important work on ternary chalcogenide semiconductor was reported by Dutková et al. [81]. They studied the fabrication of $CuInS_2$/ZnS nanocrystals by a two-step mechanochemical synthesis. Copper, indium and sulphur precursors were used for fabrication of tetragonal $CuInS_2$ followed by further co-milling with the precursors for cubic ZnS. The same group also fabricated $CuInSe_2$/ZnS following the two-step synthesis method and used their fluorescent activity in bio-labelling (Fig. 20.5) [88].

20.6.3 Mechanochemistry for Organic Synthesis

Traditionally, the organic synthesis has been carried out in solvent, whereas the solventless solid-state route has been used for synthesizing inorganic compounds. However, Toda and co-workers, during the 1980s, did pioneering work that proved that many organic reactions can be prepared by solid state too [89, 90]. These solid-state syntheses employed ball milling as the updated version of traditional grinding chemistry [91]. The simpler and facile synthesis of C–C bonds is always desired by synthetic organic chemists. The mechano milling has been found to be a convenient route to carry out C–C bond forming reactions with increases efficiency and reproducibility [92]. Many useful reactions such as aldol reaction [93], Michael condensation [94], Wittig reaction [95], Suzuki coupling [96] as well as many addition reactions have been carried out by mechanochemical method. Some of the additional advantages of carrying out the mechanochemical synthesis of organic

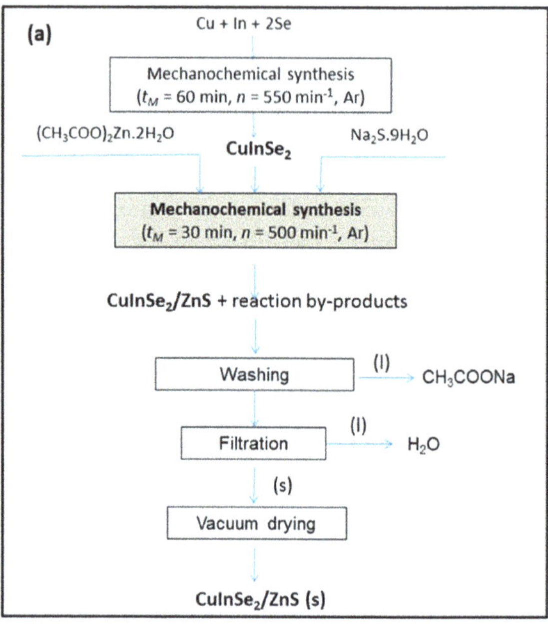

Fig. 20.5 Flow-sheet to prepare CuInSe$_2$/ZnS nanocrystals [88]. Reprinted from "E. A.-O. Dutková, Z.L. Bujňáková, O.A.-O. Sphotyuk, J. Jakubíková, D. Cholujová, V. Šišková, N.A.-O. Daneu, M.A.-O. Baláž, J. Kováč, J. Kováč, Jr., J. Briančin, P. Demchenko, SDS-Stabilized CuInSe(2)/ ZnS Multinanocomposites Prepared by Mechanochemical Synthesis for Advanced Biomedical Application"

materials are ease of purification procedures, quantitative conversion and minimum by-products. For example, 2-iodoxybenzoic acid (IBX) yields various important products but the drawback is its insolubility in common organic solvents (except DMSO) and its explosive nature at higher T [97]. These limitations were overcome by Mal and co-workers by using IBX under solvent-free mechano-milling conditions [98]. Mechanochemistry also provides a promising way to isolate the intermediates which are non-isolable by solution routes and hence helps in mechanistic investigations. The bis(benzotriazolyl)methanethione-assisted thiocarbamoylation of anilines is known to proceed through aryl-N-thiocarbamoylbenzotriazole, which is an unisolable reactive intermediate and is known to rapidly decompose to the corresponding isothiocyanate in organic solvent [99]. Štrukil et al. [100] could successfully show the formation of aryl-N-thiocarbamoylbenzotriazole under liquid-assisted grinding synthesis wherein it could be easily separated and investigated by spectroscopic techniques. The functionalization of inert C–H bonds in organic molecules provides a huge scope for synthesis of a wide range of compounds but it requires harsh reaction conditions like high T, longer reaction durations and also involves handling large amount of toxic organic solvents and sensitive metal catalyst. Mechanochemical method has been a boon in this aspect as was shown by Bolm and co-workers [101]. The author could carry out rhodium (III)-catalyzed C–H bond functionalization under mechanochemical conditions under solvent-free conditions and at room temperature [101].

20.6.4 Polymer Synthesis

After the synthesis of poly(phenylene)vinylenes (PPVs) by mechanochemical Gilch reaction by Ravnsbæk et al. [102], the activity related to polymer synthesis by ball-milling gained researchers' attraction. Ravnsbæk et al. [102] reported a rapid, solvent-free methodology for solid-state Gilch polymerizations of PPVs by ball milling. The poly(2-methoxy-5-2′-ethylhexyloxy phenylene vinylene) (MEH-PPV) monomer and three equivalents of potassium tert-butoxide were used as starting materials. The starting materials were subjected to vibrational ball milling at 30 Hz for 30 min. In another work, the L. Borchardt and his group [103] demonstrated the polycondensation between a diamine and a dialdehyde by a mechanochemical ball-milling technique. The study analysed the effect of various ball millings parameters on the polymerizations. However, the issue of chain degradation during mechanochemical ball-milling technique remains unexplored. Kim et al. [104] also investigated the effect of the ball-milling synthesis parameters such as milling time, vibrational frequency, mass of ball media on the degree of lactide ring-opening polymerization. Especially in the case of liquid-assisted grinding (LAG) mode, these are the key factors for achieving a high degree polymerization.

Chitosan is used for fabrication of many functional bio-materials for everyday use. The process of extracting of this biopolymer from the deacetylation of chitin is generally very difficult and accompanied by depolymerisation. This leads to low molecular weight chitosan. Nardo et al. [105] reported a novel method to prepare high molecular weight chitosan by the combination of mechanochemistry and ageing. This method can be applicable for all chitin sources.

As the synthesis cost and stringent reaction conditions of porous polymer materials limit their true commercialization, many research groups tried to find out different approaches for their alternative preparation route. Lee and co-worker [106] demonstrated a green, low-cost mechanochemical synthesis route for both Friedel–Crafts "knitting" and Scholl-coupling (SC) reactions of hyper-cross-linked polymers (HCPs) without using any toxic solvent. The microporous HCPs, obtained in only 5 min, have surface areas up to 782 m^2g^{-1}. LAG of the SC model mechanochemical synthesis resulted in much higher porosity than the traditional solution-based synthesis techniques.

20.6.5 Synthesis of Porous Materials

High specific surface area with large pore volume and uniform pore size distribution of highly porous materials makes them promising materials for energy storage, adsorbent for pollutants removal, fuel cells and supercapacitors [67, 107, 108]. Mechanochemical-assisted green methods have been exploited for the fabrication of highly porous materials like porous carbon, metal–organic frameworks (MOF), covalent organic frameworks (COF).

20.6.5.1 Porous Carbon

Among the different porous structures, porous activated carbon is the most exploited and the desired material. Lin et al. [109] used activation-free green synthesis method for synthesis of highly porous carbons from coconut shells. The coconut shells were smashed into tiny microparticles with abundant surface defects. After carbonization of as-prepared microparticles at 900 °C for 9 h, carbon possessing high specific surface is (SSA) of 1770 $m^2\ g^{-1}$ and large pore volume (Fig. 20.6). In another work, Casco and co-worker [110] described carbonization/activation-free room temperature mechanochemical synthesis of N-doped porous carbons. The ultra-fast (5 min) mechanochemical method resulted a high SSA up to 1080 $m^2 \cdot g^{-1}$ and exhibited high N-content (16 wt%).

To improve the porosity of the material, activating agent/pore-creating agents like KOH, H_3PO_4, $ZnCl_2$ and K_2CO_3 have also been ball milled with carbon precursor during thermal activation process. Tiruye et al. [107] and Schneidermann et al. [111] used NaCl/$ZnCl_2$ and K_2CO_3 as pore-creating agent, and the resulted carbon exhibited SSA of 1570 and 3040 m^2g^{-1}, respectively. Recently, a mechanochemical activation process is used for fabrication of highly porous activated carbons. This process involves compaction, an additional compression step, between the grinding of carbon precursor with activating agent subsequent thermal activation under inert atmosphere [108, 112]. Of late, fabrication of ordered mesoporous carbons (OMCs) by mechanochemistry has gathered researchers' attention due to previous costly, energy- and time-consuming hard templating synthesis of OMCs. These hard templating approaches also use hazardous chemicals such as HF. On the other hand, soft-templating strategy includes surfactants or co-polymers for uniform pores in the structure after decomposing during heat treatment. Ball-milling procedure provides sufficient energy to ensure uniform

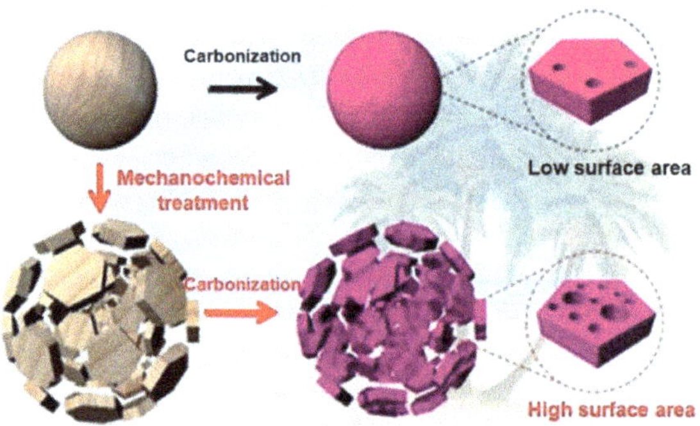

Fig. 20.6 Schematic for synthesis of highly porous carbon from coconut shells [109]. Reprinted with permission from "X. Lin, Y. Liang, Z. Lu, H. Lou, X. Zhang, S. Liu, B. Zheng, R. Liu, R. Fu, D. Wu, Mechanochemistry: A Green, Activation-Free and Top-Down Strategy to High-Surface-Area Carbon Materials, ACS Sustainable Chemistry & Engineering, 5 (2017) 8535–8540." Copyright (2017), American Chemical Society

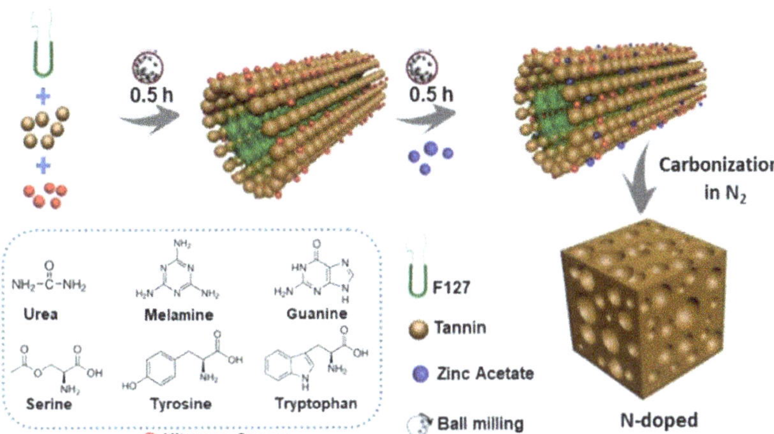

Fig. 20.7 Schematic illustration of N-doped mesoporous carbon synthesis from various precursors [114]. Reprinted with permission from "Chemical Engineering Journal, 381, J. Zhao, W. Shan, P. Zhang, S. Dai, Solvent-free and mechanochemical synthesis of N-doped mesoporous carbon from tannin and related gas sorption property, 122579 (2020)" with permission from Elsevier

dispersion of the polymer into the template voids and initializing the coordination reactions. Zhang and co-worker [113] demonstrated synthesis of OMC using bipyridine and copper chloride as precursors and silica as template. The mechanochemical reaction followed by carbonization at 500 °C fabricated OMC after SiO_2 and Cu removal, resulted in SSA of 1030 $m^2 \cdot g^{-1}$. Recently, Zhao et al. [114] fabricated a series of N-doped OMC using tannin as a carbon precursor, zinc acetate as a linker and Pluronic F127 as pore-creating agent (Fig. 20.7). The group used urea, melamine, tyrosine and tryptophan as nitrogen source. To remove the template and evaporate metallic Zn after solvent-free ball milling, the as-prepared coordination polymers were carbonized at high temperature (800 °C).

20.6.6 Synthesis of Metal–Organic Frameworks

Metal–organic frameworks (MOF) are another class of widely researched porous architecture with variety of applications. In MOF structures, coordination polymers, consisting of metal ions with organic ligands, create 3D crystalline structures. MOFs have gained significant importance due to their tailorable shape, pore sizes and surface functionalization along with possibility of obtaining high porosity.

In 2006, Pichon and co-workers [115] first reported the synthesis of MOF [Cu $(INA)_2$] by mechanochemical reaction. To prepare Cu-containing MOF, the group milled copper acetate with isonicotinic acid [NC_5H_4–4–CO_2H (INAH), INA] for

10 min without heat. The grinding of metal salts together with bridging organic ligands demonstrated a suitable and efficient fabrication method for a robust 3D framework with high yield. The group also investigated the possibility of synthesizing other Cu-based MOFs under similar solvent-free conditions. They prepared Cu-based MOFs $Cu_3(BTC)_2$ (BTC 5 1,3,5-benzene-tricarboxylate) by grinding Cu (OAc)·H_2O with benzene-1,3,5-tricarboxylic acid (H_3BTC) for 10 min under same conditions [116]. It is worth mentioning that this is a highly cost-effective, scalable and environment-friendly approach to fabricate MOFs. In recent years, zinc-containing MOFs like MOF-5, ZIF-8 and MOF-74 have been successfully synthesized using mechanochemical reactions [117–119].

20.6.7 *Synthesis of Catalysts*

Over the last few years, the synthesis of catalysts with enhanced activity and selectivity has seen a great push by employing mechanochemical routes. It provides a simpler methodology to synthesize catalytic materials for various applications which is tremendously advantageous over conventional synthesis routes which generally involve many steps [120]. The grinding or the milling accumulates excess potential energy in the material which along with shear and friction forces creates defects in the material that leads to enhanced reactivity. Solventless, hassle-free synthesis of Ag nanoparticles decorated MWNT could be achieved by simply mixing and grinding CH_3COO–Ag with MWNT and then heating in inert atmosphere [121]. Two kinds of morphologies were observed, and strong interaction of Ag NPs with MWCNTs could be established by Raman spectroscopy. This facile and scalable method could also be applied to other metal acetates such as Pd, Ni, Ag and Co. Li et al. reported an efficient and an easy method of mechanochemical synthesis of metal oxide-doped graphene [122]. SnO_2@graphene nanocomposites could be synthesized by a wet mechanochemical procedure by employing graphene oxide (GO) and $SnCl_2$ as precursors. SnO_2 nanoparticles could be in situ anchored on graphene sheets yielding uniform and resilient nanocomposites [122]. This provides a promising synthesis procedure for GO-based nanocomposite synthesis for multifold applications in fields such as energy storage and catalysis.

20.6.8 *Graphene-Based Materials*

Graphene layers have a strong tendency to restack which significantly affects the porosity and physical properties. Mechanochemistry has been widely used to synthesize various graphene-containing composites to retain the unique properties and minimized the re-stacking. Wang et al. [123] synthesized composite of graphitic carbon nitride (g–C_3N_4) with N-doped graphene (N–G) by ball milling using 100 g stainless steel balls (diameter-5-mm) at 500 rpm for 24 h. The superior oxygen reduction reaction (ORR) performance by the composite can be attributed to the close contact of nitrogen-rich g–C_3N_4 and N–G leading to a good conductivity. The mechanochemical synthesis process ensures scalable production of high-performance, low-cost, metal-free catalysts. Another composite of graphene and oxide was reported by Mashkouri et al. [124] Graphene sheets were synthesized using stainless steel double-walled ball mill beaker with two 30 mm stainless steel balls. Typically, 0.3 g graphite with 1.5 g potassium permanganate and 2.1 g citric acid in 5 ml of water were used as starting materials. The milling experiment was performed at 15 Hz frequency for 2 h at room temperature. For composite preparation, wet mixture of few layers of graphene sheet from ball milling was refluxed for 12 h with potassium permanganate followed by the addition of H_2O_2.

In addition to these various other important classes of complex and advanced materials such as fullerenes and related compounds, co-crystals, ring and cage structures can be synthesized using this very versatile and facile method of synthesis. Numerous examples of such syntheses can be easily found in the literature.

20.7 Limitations of Mechanochemical Route of Synthesis

In recent years, the attention of synthetic chemists has turned towards the application of mechanical energy to induce reactions which is one of the "greener" synthesis approaches. Despite a number of advantages, mechanochemical synthesis has its associated limitations.

1. The first and foremost disadvantage is that for all the materials that are prepared by milling or the grinding method, the contamination of surface and interface is the major concern. For example, the contamination by milling tool such as Fe or WC, and also, by the atmospheric gases is an issue with high-energy ball milling. In this regard optimization of milling time and milling speed helps in overcoming this drawback up to an extent. Also, when dealing with the ductile materials, sometimes a thin coating layer is formed on milling media which also prevents the contamination.
2. In case of mechanochemical synthesis by high-energy ball milling, it is difficult to precisely control powder properties such as particle morphology, agglomerates and the residual strain in observed crystallized phase. Also, it is almost impossible to obtain single crystals from mechanochemical synthesis.
3. An important concern regarding this particular route is that the monitoring of reactions/transformation is not the easiest job. The comprehensive understanding that makes a synthesis strategy predictive is not available in the case of mechanochemical synthesis and most of it is still on trial-and-error basis [125]. Hence, it is difficult to come up with standard protocols for large-scale synthesis.

20.8 Conclusion and Outlook

Mechanochemistry is indeed an interdisciplinary field that encompasses a broad group of disciplines. Various practicing researchers from different branches of science have varied approaches to mechanochemistry. The objectives the desired outcomes and exchange of ideas would beneficial for development of this branch as the whole. Mechanochemistry has historically been a sidewalk approach to synthesis, but it is progressing by leaps and bounds and making its big splash into the mainstream synthetic chemistry because not only it is practical and advantageous, but also due to the opportunities, it provides in developing more sustainable methods. It offers a large band width of synthesis right from downsizing materials to prepare nanoparticles to synthesizing ceramics and pharmaceuticals. Now this technique has been extended to include organic and organometallic synthesis as well. It indeed has many advantages in terms of energy saving, less time consumption, no involved post-reaction work-up as well as access to exotic products.

In terms of future scope, there is an ardent need to develop more suitable and customized tools required for this branch of synthesis as was done for solution-based syntheses techniques. For example, in situ monitoring of mechanochemical reactions is highly desirable to understand the reaction pathways and mechanisms so as to set up standard procedures for making it viable for scaling up for industrial level. Such kind of monitoring is, however, difficult due to the fact that there is a continuous violent motion, and also, most of the real-time monitoring techniques such as UV-visible, IR and Raman need to shine light (and need transparent medium), whereas the tools used in mechanosynthesis are mostly made up of opaque materials. However, there has been some significant progress in this field where real-time monitoring of reaction products could be done using transparent grinding media and synchrotron X-ray beam [126]. The better equipments that can enable "live" monitoring of reaction and better understanding of the mechanism will definitely help a great in increasing the scope of this synthesis technique.

The idea of bringing about the selective reaction by force can be taken beyond ball mills as has been done by researchers at Stanford University, USA. They have come up with a novel idea wherein the molecules are designed with the rigid part and a softer core and are akin to molecular anvils which on applying pressure causes the rigid parts to move and compress the softer parts, thus bringing about the desired electron transfer reactions.

As mentioned earlier, synthesis by applying force (or grinding) has been into existence since historic times; however, with new modifications this technique is taking a re-birth as mechanochemical synthesis. However, when it comes to tools, mechanochemistry is still in infancy. There is no doubting the fact that with the development of specialized and customized tools to carry out the synthesis and understand the mechanisms, the mechanochemistry will step into the bigger synthesis arena with huge commercial impact. The future of mechanochemistry is indeed optimistic!

References

1. Lancaster M (2002) Green chemistry: an introductory text. The Royal Society of Chemistry, Cambridge UK
2. Anastas PT, Kirchhoff MM (2002) Origins, current status, and future challenges of green chemistry. Acc Chem Res 35:686–694
3. Sheldon RA (2014) Green and sustainable manufacture of chemicals from biomass: state of the art. Green Chem 16:950–963
4. Gilman JJ (1996) Mechanochemistry. Science 274:65
5. Hammerer F, Loots LA-O, Do JL, Therien JPD, Nickels CW, Friščić TA-O, Auclair KA-O (2018) Solvent-free enzyme activity: quick, high-yielding mechanoenzymatic hydrolysis of cellulose into glucose. Angew Chem 130:2651–2654
6. Mack J, Fulmer D, Stofel S, Santos N (2007) The first solvent-free method for the reduction of esters. Green Chem 9:1041–1043
7. Burmeister CF, Kwade A (2013) Process engineering with planetary ball mills. Chem Soc Rev 42:7660–7667
8. (a) Jonoobi M, Harun J, Mathew AP, Oksman K (2015) Mechanical properties of cellulose nanofiber (CNF) reinforced polylactic acid (PLA) prepared by twin screw extrusion. Compos Sci Technol 70:1742–1747. (b) Lu Y, Guan S, Hao L, Yoshida H (2015) Review on the photocatalyst coatings of TiO_2: fabrication by mechanical coating technique and its application. Coatings 5(3):425–464
9. Crawford DE, Miskimmin CKG, Albadarin AB, Walker G, James SL (2017) Organic synthesis by Twin Screw Extrusion (TSE): continuous, scalable and solvent-free. Green Chem 19:1507–1518
10. Lynch AJ, Rowland CA (2005) The history of grinding, society for mining, metallurgy, and exploration 209
11. Takacs L (2013) The historical development of mechanochemistry. Chem Soc Rev 42:7649–7659
12. Takacs L (2000) Quicksilver from cinnabar: the first documented mechanochemical reaction? JOM 52:12–13
13. Faraday M (1820) Q Jl Sci Lit Arts 8:374–376
14. Takacs L (2007) The mechanochemical reduction of AgCl with metals. J Therm Anal Calorim 90:81–84
15. Johnston J, Adams LH (1913) Effect of high pressure on the physical and chemical behavior of solids. Am J Sci 35:205–253
16. Spring W (1883) Formation de Quelques Sulfures par l'Actionde la Pression. Considérations qui en Découlent Touchantles Propriétés des États Allotropiques du Phosphore et du Carbone. Bull Soc Chim Fr 40:641–647
17. Lea MC (1893) On endothermic reactions effected by mechanical force. Am J Sci 46:241–244
18. Ostwald W (1919) Handbook of general chemistry, chemical literature and the organization of science. Akademische Verlagsgesellschaft, Leipzig
19. Ling AR, Baker JL (1893) XCVI—halogen derivatives of quinone. Part III. Derivatives of quinhydrone. J Chem Soc Trans 63:1314–1327
20. Baláž P (2008) Mechanochemistry in minerals engineering. In: Mechanochemistry in nanoscience and minerals engineering. Springer, Berlin, Heidelberg
21. James SL, Adams CJ, Bolm C, Braga D, Collier P, Friščić T, Grepioni F, Harris KDM, Hyett G, Jones W, Krebs A, Mack J, Maini L, Orpen AG, Parkin IP, Shearouse WC, Steed JW, Waddell DC (2012) Mechanochemistry: opportunities for new and cleaner synthesis. Chem Soc Rev 41:413–447
22. Petruschke M, Tribochemistry von G. HEINICKE. Berlin: Akademie-Verlag 1984. Bestellnummer: 7631993(6746). 495 S., 329 Bilder, 106 Tabellen, 98,– M, Acta Polym., 36 (1985) 400–401

23. Patil AO, Curtin DY, Paul IC (1984) Solid-state formation of quinhydrones from their components. Use of solid-solid reactions to prepare compounds not accessible from solution. J Am Chem Soc 106:348–353
24. Toda F, Tanaka K, Sekikawa A (1987) Host–guest complex formation by a solid–solid reaction. J Chem Soc Chem Commun 279–280
25. Etter MC, Urbanczyk-Lipkowska Z, Zia-Ebrahimi M, Panunto TW (1990) Hydrogen bond-directed cocrystallization and molecular recognition properties of diarylureas. J Am Chem Soc 112:8415–8426
26. Pedireddi VR, Jones W, Chorlton AP, Docherty R (1996) Creation of crystalline supramolecular arrays: a comparison of co-crystal formation from solution and by solid-state grinding. Chem Commun 987–988
27. Garay AL, Pichon A, James SL (2007) Solvent-free synthesis of metal complexes. Chem Soc Rev 36:846–855
28. Valiev R (2004) Nanostructuring of metals by severe plastic deformation for advanced properties. Nat Mater 3:511–516
29. Giri PK, Bhattacharyya S, Singh DK, Kesavamoorthy R, Panigrahi BK, Nair KGM (2007) Correlation between microstructure and optical properties of ZnO nanoparticles synthesized by ball milling. J Appl Phys 102:
30. Buyanov RA, Molchanov VV, Boldyrev VV (2009) Mechanochemical activation as a tool of increasing catalytic activity. Catal Today 144:212–218
31. Mateti S, Mathesh M, Liu Z, Tao T, Ramireddy T, Glushenkov AM, Yang W, Chen YI (2021) Mechanochemistry: a force in disguise and conditional effects towards chemical reactions. Chem Commun 57:1080–1092
32. Volkov VA, El'kin IA, Zagainov AV, Protasov AV, Elsukov EP (2014) Dynamic equilibria of phases in the processes of the mechanosynthesis of an alloy with composition $Fe_{72.6}C_{24.5}O_{1.1}N_{1.8}$. Phys Met Metallogr 115:557–565
33. Michel D, Faudot F, Gaffet E, Mazerolles L (1993) Oxydes céramiques élaborés par voie mécanochimique. Rev Met Paris 90:219–226
34. Begin-Colin S, Le Caër G, Zandona M, Bouzy E, Malaman B (1995) Influence of the nature of milling media on phase transformations induced by grinding in some oxides. J. Alloys Compd 227:157–166
35. Gajović A, Tomašić N, Djerdj I, Su DS, Furić K (2008) Influence of mechanochemical processing to luminescence properties in Y_2O_3 powder. J. Alloys Compd. 456:313–319
36. Gateshki M, Petkov V, Williams G, Pradhan SK, Ren Y (2005) Atomic-scale structure of nanocrystalline ZrO_2 prepared by high-energy ball milling. Phys. Rev. B 71:
37. Štefanić G, Musić S, Gajović A (2006) Structural and microstructural changes in monoclinic ZrO_2 during the ball-milling with stainless steel assembly. Mater Res Bull 41:764–777
38. Gajović A, Furić K, Štefanić G, Musić S (2005) In situ high temperature study of ZrO_2 ball-milled to nanometer sizes. J Mol Struct 744–747:127–133
39. Michel D, Gaffet E, Berthet P (1995) Structure of nanosized refractory oxde powders. Nanostruct Mater 6:667–670
40. Michel D, Mazerolles L, Berthet P, Gaffet E (1995) Nanocrystalline and amorphous oxide powders prepared by high-energy ball milling. Eur J Solid State Inorg Chem 32:673–682
41. Michel D, Mazerolles L, Gaffet E (1993) Nanocrystallline oxide powders prepared by ball-milling. Third Euro-Ceramics, Iberica 1:255–260
42. Katagiri S, Ishizawa N, Marumo F (2013) A new high temperature modification of face-centered cubic Y_2O_3. Powder Diffr 8:60
43. Lacroix B, Paumier F, Gaboriaud RJ (2011) Crystal defects and related stress in Y_2O_3 thin films: origin, modeling, and consequence on the stability of the C-type structure. Phys Rev B 84:
44. Li YB, Wei BQ, Liang J, Yu Q, Wu DH (1999) Transformation of carbon nanotubes to nanoparticles by ball milling process. Carbon 37:493–497
45. Chen Y, Fitz Gerald J, Williams JS, Bulcock S (1999) Synthesis of boron nitride nanotubes at low temperatures using reactive ball milling. Chem Phys Lett 299:260–264

46. Fitz Gerald JD, Chen Y, Conway MJ (2003) Nanotube growth during annealing of mechanically milled Boron. Appl Phys A 76:107–110
47. Chen Y, Li CP, Chen H, Chen Y (2006) One-dimensional nanomaterials synthesized using high-energy ball milling and annealing process. Sci Technol Adv Mater 7:839–846
48. Chen Y, Fitz Gerald J, Chadderton L, Chaffron L (1999) Investigation of nanoporous carbon powders produced by high energy ball milling and formation of carbon nanotubes during subsequent annealing. Mater Sci Forum 312:375–380
49. Chadwick AV, Pooley MJ, Rammutla KE, Savin SL, Rougier A (2003) A comparison of the extended x-ray absorption fine structure of nanocrystalline ZrO_2 prepared by high-energy ball milling and other methods. J Phys Condens Matter 15:431
50. Scholz G, Stösser R, Klein J, Silly G, Buzaré J, Laligant Y, Ziemer B (2002) Local structural orders in nanostructured Al_2O_3 prepared by high-energy ball milling. J Phys Condens Matter 14:2101
51. Indris S, Bork D, Heitjans P (2000) Nanocrystalline oxide ceramics prepared by high-energy ball milling. J Mater Synth Process 8:245–250
52. Tkáčová K, Šepelák V, Števulová N, Boldyrev VV (1996) Structure-reactivity study of mechanically activated zinc ferrite. J Solid State Chem 123:100–108
53. Shen TD, Koch CC, McCormick TL, Nemanich RJ, Huang JY, Huang JG (2011) The structure and property characteristics of amorphous/nanocrystalline silicon produced by ball milling. J Mater Res 10:139–148
54. Arbain R, Othman M, Palaniandy S (2011) Preparation of iron oxide nanoparticles by mechanical milling. Miner Eng 24:1–9
55. Taufiq-Yap YH, Goh CK, Hutchings GJ, Dummer N, Bartley JK (2011) Influence of milling media on the physicochemicals and catalytic properties of mechanochemical treated vanadium phosphate catalysts. Catal Lett 141:400–407
56. Friščić T, Jones W (2009) Recent advances in understanding the mechanism of cocrystal formation via grinding. Cryst Growth Des 9:1621–1637
57. IUPAC (1997) Compendium of chemical terminology, 2nd ed (the "Gold Book"). Blackwell Scientific Publications, Oxford
58. Friscić T, Reid Dg Fau – Halasz I, Halasz I Fau – Stein RS, Stein Rs Fau – Dinnebier RE, Dinnebier Re Fau – Duer MJ, Duer MJ (2010) Ion- and liquid-assisted grinding: improved mechanochemical synthesis of metal-organic frameworks reveals salt inclusion and anion templating. Angew Chem Int Ed 49:712–715
59. Beldon PJ, Fábián L, Stein RS, Thirumurugan A, Cheetham AK, Friščić T (2010) Rapid room-temperature synthesis of zeolitic imidazolate frameworks by using mechanochemistry. Angew Chem Int Ed 49:9640–9643
60. André V, Hardeman A Fau – Halasz I, Halasz I Fau – Stein RS, Stein Rs Fau – Jackson GJ, Jackson Gj Fau – Reid DG, Reid Dg Fau – Duer MJ, Duer Mj Fau – Curfs C, Curfs C Fau – Duarte MT, Duarte Mt Fau – Friščić T, Friščić T (2011) Mechanosynthesis of the metallodrug bismuth subsalicylate from Bi_2O_3 and structure of bismuth salicylate without auxiliary organic ligands. Angew Chem Int Ed 50:7858–7861
61. Hasa D, Schneider Rauber G, Voinovich D, Jones W (2015) Cocrystal formation through mechanochemistry: from neat and liquid-assisted grinding to polymer-assisted grinding. Angew Chem Int Ed 54:7371–7375
62. Konnert L, Dimassi M, Gonnet L, Lamaty F, Martinez J, Colacino E (2016) Poly(ethylene) glycols and mechanochemistry for the preparation of bioactive 3,5-disubstituted hydantoins. RSC Adv 6:36978–36986
63. Scaramuzza D, Schneider Rauber G, Voinovich D, Hasa D (2018) Dehydration without heating: use of polymer-assisted grinding for understanding the stability of hydrates in the presence of polymeric excipients. Cryst Growth Des 18:5245–5253
64. Germann LS, Emmerling ST, Wilke M, Dinnebier RE, Moneghini M, Hasa D (2020) Monitoring polymer-assisted mechanochemical cocrystallisation through in situ X-ray powder diffraction. Chem Commun 56:8743–8746

65. Friščić T, Mottillo C, Titi HM (2020) Mechanochemistry for synthesis. Angew Chem Int Ed 59:1018–1029
66. Jones W, Eddleston MD (2014) Introductory lecture: mechanochemistry, a versatile synthesis strategy for new materials. Faraday Discuss 170:9–34
67. Szczęśniak B, Borysiuk S, Choma J, Jaroniec M (2020) Mechanochemical synthesis of highly porous materials. Mater Horiz 7:1457–1473
68. Tan D, García F (2019) Main group mechanochemistry: from curiosity to established protocols. Chem Soc Rev 48:2274–2292
69. Do J-L, Friščić T (2017) Mechanochemistry: a force of synthesis. ACS Cent Sci 3:13–19
70. Šepelák V, Bégin-Colin S, Le Caër G (2012) Transformations in oxides induced by high-energy ball-milling. Dalton Trans 41:11927–11948
71. Bodaghi M, Mirhabibi AR, Zolfonun H, Tahriri M, Karimi M (2008) Investigation of phase transition of γ-alumina to α-alumina via mechanical milling method. Phase Transitions 81:571–580
72. Sekiya T, Ohta S, Kamei S, Hanakawa M, Kurita S (2001) Raman spectroscopy and phase transition of anatase TiO_2 under high pressure. J Phys Chem Solids 62:717–721
73. Tsuzuki T, McCormick PG (2004) Mechanochemical synthesis of nanoparticles. J Mater Sci 39:5143–5146
74. Ao W, Li J, Yang H, Zeng X, Ma X (2006) Mechanochemical synthesis of zinc oxide nanocrystalline. Powder Technol 168:148–151
75. Yang H, Hu Y, Zhang X, Qiu G (2004) Mechanochemical synthesis of cobalt oxide nanoparticles. Mater Lett 58:387–389
76. Stojanovic BD, Simoes AZ, Paiva-Santos CO, Jovalekic C, Mitic VV, Varela JA (2005) Mechanochemical synthesis of barium titanate. J Eur Ceram Soc 25:1985–1989
77. Hu R, Ouyang Y, Liang T, Wang H, Liu J, Chen J, Yang C, Yang L, Zhu M (2017) Stabilizing the nanostructure of SnO_2 anodes by transition metals: a route to achieve high initial coulombic efficiency and stable capacities for lithium storage. Adv Mater 29:1605006
78. Xiao W, Yang S, Zhang P, Li P, Wu P, Li M, Chen N, Jie K, Huang C, Zhang N, Dai S (2018) Facile synthesis of highly porous metal oxides by mechanochemical nanocasting. Chem Mater 30:2924–2929
79. Baláž P, Baláž M, Achimovičová M, Bujňáková Z, Dutková E (2017) Chalcogenide mechanochemistry in materials science: insight into synthesis and applications (a review). J Mater Sci 52:11851–11890
80. Baláž P, Boldižárová E, Godočı X, Ková E, Briančin J (2003) Mechanochemical route for sulphide nanoparticles preparation. Mater Lett 57:1585–1589
81. Dutková E, Daneu N, Lukáčová Bujňáková Z, Baláž M, Kováč J, Kováč J Jr, Baláž P (2019) Mechanochemical synthesis and characterization of $CuInS_2/ZnS$ nanocrystals. Molecules 24:1031
82. Bujňáková Z, Dutková E, Kello M, Mojžiš J, Baláž M, Baláž P, Shpotyuk O (2017) Mechanochemistry of chitosan-coated zinc sulfide (ZnS) nanocrystals for bio-imaging applications. Nanoscale Res Lett 12:328
83. Dutková E, Sayagués MJ, Briančin J, Zorkovská A, Bujňáková Z, Kováč J, Kováč J, Baláž P, Ficeriová J (2016) Synthesis and characterization of $CuInS_2$ nanocrystalline semiconductor prepared by high-energy milling. J Mater Sci 51:1978–1984
84. Baláž P, Baláž M, Dutková E, Zorkovská A, Kováč J, Hronec P, Kováč J, Čaplovičová M, Mojžiš J, Mojžišová G, Eliyas A, Kostova NG (2016) CdS/ZnS nanocomposites: from mechanochemical synthesis to cytotoxicity issues. Mater Sci Eng C 58:1016–1023
85. Baláž P, Sayagués MJ, Baláž M, Zorkovská A, Hronec P, Kováč J, Kováč J, Dutková E, Mojžišová G, Mojžiš J (2014) CdSe@ZnS nanocomposites prepared by a mechanochemical route: no release of Cd^{2+} ions and negligible in vitro cytotoxicity. Mater Res Bull 49:302–309
86. Bujňáková Z, Baláž M, Dutková E, Baláž P, Kello M, Mojžišová G, Mojžiš J, Vilková M, Imrich J, Psotka M (2017) Mechanochemical approach for the capping of mixed core CdS/ZnS nanocrystals: elimination of cadmium toxicity. J Colloid Interface Sci 486:97–111

87. Bujňáková Z, Dutková E, Zorkovská A, Baláž M, Kováč J, Kello M, Mojžiš J, Briančin J, Baláž P (2017) Mechanochemical synthesis and in vitro studies of chitosan-coated InAs/ZnS mixed nanocrystals. J Mater Sci 52:721–735
88. Dutková EA-O, Bujňáková ZL, Sphotyuk OA-O, Jakubíková J, Cholujová D, Šišková V, Daneu NA-O, Baláž MA-O, Kováč J, Kováč J Jr, Briančin J, Demchenko P (2021) SDS-stabilized $CuInSe_2$/ZnS multinanocomposites prepared by mechanochemical synthesis for advanced biomedical application. Nanomaterials 11:69
89. Toda F (1995) Solid state organic chemistry: efficient reactions, remarkable yields, and stereoselectivity. Acc Chem Res 28:480–486
90. Tanaka K, Toda F (2000) Solvent-free organic synthesis. Chem Rev 100:1025–1074
91. Toda F, Tanaka K, Iwata S (1989) Oxidative coupling reactions of phenols with iron(III) chloride in the solid state. J Org Chem 54:3007–3009
92. Margetić D, Štrukil V (2016) Mechanochemical organic synthesis. Elsevier
93. Machuca E, Juaristi E (2015) Organocatalytic activity of α,α-dipeptide derivatives of (S)-proline in the asymmetric aldol reaction in absence of solvent. Evidence for non-covalent π–π interactions in the transition state. Tetrahedron Lett 56:1144–1148
94. Zhang Z, Dong Y-W, Wang G-W, Komatsu K (2004) Mechanochemical michael reactions of chalcones and azachalcones with ethyl acetoacetate catalyzed by K_2CO_3 under solvent-free conditions. Chem Lett 33:168–169
95. Balema VP, Wiench JW, Pruski M, Pecharsky VK (2002) Mechanically induced solid-state generation of phosphorus ylides and the solvent-free wittig reaction. J Am Chem Soc 124:6244–6245
96. Nielsen SF, Peters D, Axelsson O (2000) The suzuki reaction under solvent-free conditions. Synth Commun 30:3501–3509
97. Ladziata U, Zhdankin VV (2006) Hypervalent iodine(V) reagents in organic synthesis, ARKIVOC ix:26–58
98. Achar TK, Maiti S, Mal P (2014) IBX works efficiently under solvent free conditions in ball milling. RSC Adv 4:12834–12839
99. Katritzky AR, Khashab NM, Bobrov S, Yoshioka M (2006) Synthesis of mono- and symmetrical Di-N-hydroxy- and N-Aminoguanidines. J Org Chem 71:6753–6758
100. Štrukil V, Gracin D, Magdysyuk OV, Dinnebier RE, Friščić T (2015) Trapping reactive intermediates by mechanochemistry: elusive aryl N-thiocarbamoylbenzotriazoles as bench-stable reagents. Angew Chem Int Ed 54:8440–8443
101. Hermann GN, Becker P, Bolm C (2015) Mechanochemical rhodium(III)-catalyzed C–H bond functionalization of acetanilides under solventless conditions in a ball mill. Angew Chem Int Ed 54:7414–7417
102. Ravnsbæk JB, Swager TM (2014) Mechanochemical synthesis of poly(phenylene vinylenes). ACS Macro Lett. 3:305–309
103. Grätz S, Borchardt L (2016) Mechanochemical polymerization—controlling a polycondensation reaction between a diamine and a dialdehyde in a ball mill. RSC Adv 6:64799–64802
104. Ohn N, Shin J, Kim SS, Kim JG (2017) Mechanochemical ring-opening polymerization of lactide: liquid-assisted grinding for the green synthesis of poly(lactic acid) with high molecular weight. Chemsuschem 10:3529–3533
105. Di Nardo T, Hadad C, Nguyen Van Nhien A, Moores A (2019) Synthesis of high molecular weight chitosan from chitin by mechanochemistry and aging. Green Chem 21:3276–3285
106. Lee J-SM, Kurihara T, Horike S (2020) Five-Minute mechanosynthesis of hypercrosslinked microporous polymers. Chem Mater 32:7694–7702
107. Tiruye GA, Muñoz-Torrero D, Berthold T, Palma J, Antonietti M, Fechler N, Marcilla R (2017) Functional porous carbon nanospheres from sustainable precursors for high performance supercapacitors. J Mater Chem A 5:16263–16272
108. Rajendiran R, Nallal M, Park KH, Li OL, Kim H-J, Prabakar K (2019) Mechanochemical assisted synthesis of heteroatoms inherited highly porous carbon from biomass for electrochemical capacitor and oxygen reduction reaction electrocatalysis. Electrochim Acta 317:1–9

109. Lin X, Liang Y, Lu Z, Lou H, Zhang X, Liu S, Zheng B, Liu R, Fu R, Wu D (2017) Mechanochemistry: a green, activation-free and top-down strategy to high-surface-area carbon materials. ACS Sustainable Chem Eng 5:8535–8540
110. Casco ME, Kirchhoff S, Leistenschneider D, Rauche M, Brunner E, Borchardt L (2019) Mechanochemical synthesis of N-doped porous carbon at room temperature. Nanoscale 11:4712–4718
111. Schneidermann C, Jäckel N, Oswald S, Giebeler L, Presser V, Borchardt L (2017) Solvent-free mechanochemical synthesis of nitrogen-doped nanoporous carbon for electrochemical energy storage. Chemsuschem 10:2416–2424
112. Balahmar N, Mitchell AC, Mokaya R (2015) Generalized mechanochemical synthesis of biomass-derived sustainable carbons for high performance CO_2 storage. Adv Energy Mater 5:1500867
113. Zhang E, Hao G-P, Casco ME, Bon V, Grätz S, Borchardt L (2018) Nanocasting in ball mills—combining ultra-hydrophilicity and ordered mesoporosity in carbon materials. J Mater Chem A 6:859–865
114. Zhao J, Shan W, Zhang P, Dai S (2020) Solvent-free and mechanochemical synthesis of N-doped mesoporous carbon from tannin and related gas sorption property. Chem Eng J 381:
115. Pichon A, Lazuen-Garay A, James SL (2006) Solvent-free synthesis of a microporous metal–organic framework. CrystEngComm 8:211–214
116. Yuan W, Garay AL, Pichon A, Clowes R, Wood CD, Cooper AI, James SL (2010) Study of the mechanochemical formation and resulting properties of an archetypal MOF: $Cu_3(BTC)_2$ (BTC = 1,3,5-benzenetricarboxylate). CrystEngComm 12:4063–4065
117. Prochowicz D, Sokołowski K, Justyniak I, Kornowicz A, Fairen-Jimenez D, Friščić T, Lewiński J (2015) A mechanochemical strategy for IRMOF assembly based on pre-designed oxo-zinc precursors. Chem Commun 51:4032–4035
118. Tanaka S, Kida K, Nagaoka T, Ota T, Miyake Y (2013) Mechanochemical dry conversion of zinc oxide to zeolitic imidazolate framework. Chem Commun 49:7884–7886
119. Julien PA, Užarević K, Katsenis AD, Kimber SAJ, Wang T, Farha OK, Zhang Y, Casaban J, Germann LS, Etter M, Dinnebier RE, James SL, Halasz I, Friščić T (2016) In situ monitoring and mechanism of the mechanochemical formation of a microporous MOF-74 framework. J Am Chem Soc 138:2929–2932
120. Xu C, De S, Balu AM, Ojeda M, Luque R (2015) Mechanochemical synthesis of advanced nanomaterials for catalytic applications. Chem Commun 51:6698–6713
121. Lin Y, Watson KA, Fallbach MJ, Ghose S, Smith JG, Delozier DM, Cao W, Crooks RE, Connell JW (2009) Rapid, solventless, bulk preparation of metal nanoparticle-decorated carbon nanotubes. ACS Nano 3:871–884
122. Li S, Wang Y, Lai C, Qiu J, Ling M, Martens W, Zhao H, Zhang S (2014) Directional synthesis of tin oxide@graphene nanocomposites via a one-step up-scalable wet-mechanochemical route for lithium ion batteries. J Mater Chem A 2:10211–10217
123. Wang M, Wu Z, Dai L (2015) Graphitic carbon nitrides supported by nitrogen-doped graphene as efficient metal-free electrocatalysts for oxygen reduction. J Electroanal Chem 753:16–20
124. Mashkouri S, Arsalani N, Mostafavi H (2017) Wet mechanochemical approach assistance to the green synthesis of graphene sheet at room temperature and in situ anchored with MnO_2 in a green method. J Alloys Compd 715:486–493
125. Bowmaker GA (2013) Solvent-assisted mechanochemistry. Chem Commun 49:334–348
126. Friščić T, Halasz I, Beldon PJ, Belenguer AM, Adams F, Kimber SAJ, Honkimäki V, Dinnebier RE (2013) Real-time and in situ monitoring of mechanochemical milling reactions. Nat Chem 5:66–73

GPSR Compliance
The European Union's (EU) General Product Safety Regulation (GPSR) is a set of rules that requires consumer products to be safe and our obligations to ensure this.

If you have any concerns about our products, you can contact us on

ProductSafety@springernature.com

In case Publisher is established outside the EU, the EU authorized representative is:

Springer Nature Customer Service Center GmbH
Europaplatz 3
69115 Heidelberg, Germany

www.ingramcontent.com/pod-product-compliance
Ingram Content Group UK Ltd.
Pitfield, Milton Keynes, MK11 3LW, UK
UKHW021832270726
14058UKWH00001B/103

* 9 7 8 9 8 1 1 6 1 8 0 9 3 *